Dynamics, Geometry, Number Theory

Dynamics, Geometry, Number Theory

The Impact of Margulis on Modern Mathematics

Edited by David Fisher, Dmitry Kleinbock, and Gregory Soifer

THE UNIVERSITY OF CHICAGO PRESS • CHICAGO AND LONDON

The University of Chicago Press, Chicago 60637
The University of Chicago Press, Ltd., London

Published 2022
Printed in the United States of America

31 30 29 28 27 26 25 24 23 22   1 2 3 4 5

ISBN-13: 978-0-226-80402-6 (cloth)
ISBN-13: 978-0-226-80416-3 (e-book)

DOI: https://doi.org/10.7208/chicago/9780226804163.001.0001

Library of Congress Cataloging-in-Publication Data

Names: Fisher, David (David Michael), editor. | Kleinbock, Dmitry, editor. Soifer, Gregory, editor.
Title: Dynamics, geometry, number theory : the impact of Margulis on modern mathematics / David Fisher, Dmitry Kleinbock, and Gregory Soifer.
Description: Chicago : University of Chicago Press, 2022.
Identifiers: LCCN 2021029509 | ISBN 9780226804026 (cloth) | ISBN 9780226804163 (ebook)
Subjects: LCSH: Margulis, G. A. (Gregori Aleksandrovitsch), 1946– | Dynamics. | Number theory.
Classification: LCC QA29.M355 D96 2022 | DDC 515/.39—dc23
LC record available at https://lccn.loc.gov/2021029509

♾ This paper meets the requirements of ANSI/NISO Z39.48-1992 (Permanence of Paper).

Contents

INTRODUCTION

The impact of the work and ideas of Gregory Margulis on modern mathematics is broad, deep, and profound. Margulis's work developed and explored key connections between ergodic theory, Lie theory, geometry, and number theory that have had a tremendous impact on mathematics. The goal of this volume is to provide the reader with an overview of many of the areas in which Margulis made contributions. His contributions range from deep insights into the structure of discrete subgroups of Lie groups that play a key role in many geometric topics to compelling contributions to homogeneneous dynamics that played a key role in making it an essential tool for number theory. Instead of emphasizing the applications to many fields of mathematics that can be found in the individual contributions, the aim of this introduction will be to point to the unity of the area defined by Margulis's contributions. This field still lacks a good name, but the best one we know of was proposed by Francois Ledrappier who termed the area *ergodic geometry*.

We now provide an overview of the parts and chapters of the book. This lays down for the reader a rough map of the larger area of research. At the end of the introduction we will point to several developments both recent and older that tie together the disparate parts of this volume and that illustrate some essential unities in Margulis's work.

The book is organized into four main parts. The first concerns arithmeticity, superrigidity, and normal subgroups for lattices in Lie groups. The first chapter in that part, by Fisher, is a survey of developments stemming from Margulis's seminal work in the 1970s with some emphasis on open questions and problems. The other two contributions to this part are modern reimaginings of Margulis's proofs of two major results. The first, by Bader and

David Fisher. Department of Mathematics. Indiana University Bloomington, IN 47405
fisherdm@indiana.edu

Furman, gives a new proof of Margulis's superrigidity theorem in terms of a new language of *algebraic representations* or *gates* discovered by those authors. The second, by Brown, Rodriguez Hertz, and Wang, gives a new proof of the normal subgroup theorem, or at least the half of that result that depends on a theorem about factors of actions [Mar5]. The original proof of Margulis was in terms of invariant sub σ-algebras; the new one is in terms of invariant measures. All of these ideas provide additional deep connections to the field of homogeneous dynamics discussed below. In particular homogeneous dynamics and Margulis's work in that area play a key role in the solution of Zimmer's conjecture by Brown, Fisher, and Hurtado and in results connecting arithmeticity to totally geodesic surfaces by both Margulis-Mohammadi and Bader, Fisher, Miller, and Stover [BFH1, BFH2, MM, BFMS1, BFMS2].

The second part of the book contains two additional chapters on discrete subgroups. The first, by Danciger, Drumm, Goldman, and Smilga, concerns subgroups of affine transformations acting properly on affine space. Much work in this area was motivated by either Auslander's conjecture that compact complete affine manifolds are solvmanifolds or by Margulis's construction of examples that show that the word *compact* is necessary in that conjecture. The next chapter, by Gelander, Glasner, and Soifer, concerns maximal subgroups. Again this area was pioneered by the construction by Margulis and Soifer of infinite index maximal subgroups in certain higher rank lattices, answering a question of Platonov's. The two chapters in this part are unified because both evolve from work of Margulis and Margulis-Soifer in which a use of ping pong to produce discrete subgroups was developed [MS1, MS2, Mar6].

The third part contains a relatively diverse set of three chapters, all concerned in one way or another with representation theory and spectral theory. An initial chapter by Benoist and Kobayashi determines exactly what homogeneous spaces for simple Lie groups give rise to tempered representations. As pointed out in their introduction, tempered representations play a key role in Margulis's work on a wide range of topics, ranging from the construction of expanders to homogeneous dynamics to geometry of homogeneous spaces. The second contribution to this part is a survey on recent progress on expanders by Breuillard and Lubotzky. Margulis's construction of explicit families of expanders was a breakthrough that eventually led to dramatic new connections with additive combinatorics and other areas of geometry and analysis [Mar3]. The last chapter in this part is an essay by Karlsson exploring a novel sense of metric spectral theory. While the motivations for this theory are manifold, some motivations come from Karlsson's early work with Margulis on multiplicative ergodic theorems, work that was itself motivated by Margulis's work on superrigidity [Mar4, KM1].

The final, and largest, part of this book is devoted to homogeneous dynamics. Margulis's work on this topic spans many individual contributions. Probably foremost among them is his solution to the Oppenheim conjecture [Mar7]. This part opens with a survey by Beresnevich and Kleinbock, surveying the topic of Diophantine approximation on manifolds with emphasis on approaches from homogeneous dynamics that were first pioneered by Kleinbock and Margulis [KM2]. The second contribution, by Eskin and Mozes, describes the role of Margulis functions in homogeneous dynamics and beyond. The first construction of these functions was given in the paper of Eskin, Mozes, and Margulis [EMM] and they have been applied in many areas of dynamics as described in this contribution. The next chapter, by Lindenstrauss, gives an update on the important topic of measure rigidity for diagonal actions. A major focus in this area has been the conjectures of Furstenberg, Katok-Spatzier, and Margulis on classification of invariant measures. Despite the main conjectures remaining very much open, known special cases provide numerous applications to important questions in number theory. This essay is followed by two on the currently very hot topic of effective results in homogeneous dynamics. This area was pioneered by Margulis in joint work with Einsiedler and Venkatesh and also Mohammadi [EMV, EMMV]. The first, by Einsiedler and Mohammadi, gives a survey of recent results in this area. The second, by Einsiedler and Wirth, illustrates proof techniques in any interesting special case. This part and the book close with a survey by Hee Oh of another interesting and new topic, the development of homogeneous dynamics in infinite covolume with a particular focus on discrete subgroups of $SL(2, \mathbb{C})$.

One key idea of Margulis's that indicates the unity of large parts of the work presented here is the nondivergence of unipotent orbits [Mar1]. Margulis originally developed this idea to prove that nonuniform lattices in higher rank semisimple Lie groups were arithmetic [Mar2]. Later he used it in his work proving the Oppenheim conjecture in number theory using techniques from homogeneous dynamics [Mar7]. This key result played a central role in his theorems, which an outsider might think of as belonging to two different areas. This idea is central to the theory, with versions, variants, and strengthening playing a key role in many of the chapters on homogeneous dynamics but also in several rigidity results in the first part of the book.

The connections between the different parts and chapters are too numerous to list here, as are the connections to other areas of mathematics, but we will point to a few. Both the first chapter (by Fisher) and the last chapter (by Oh) have some focus on recent results about totally geodesic submanifolds in hyperbolic manifolds. Our understanding of these submanifolds is informed by combinations of ideas from rigidity theory and homogeneous dynamics. In addition, the contribution of Bader and Furman, which is used in some of this work, brings into focus a fact that was hidden in previous proofs of superrigidity—namely, that a key step is that objects invariant under some group T are also invariant under the normalizer $N(T)$ of T. A similar idea in a very different context plays a prominent role in Margulis's solution of the Oppenheim conjecture, work that features here in the contributions of Eskin and Mozes, Einsiedler and Mohammadi, and Einsiedler and Wirth. This idea also plays a key role in many other results in homogeneous dynamics, notably including Ratner's proof of her measure classification for groups generated by unipotent elements.

An additional connection between rigidity theory and homogeneous dynamics is the focus of the chapter by Brown, Rodriguez Hertz, and Wang. They give an alternate proof of Margulis's normal subgroups theorem, where a key step is written in terms of classifying invariant measures. This brings this result into closer contact with the ideas that permeate the part on homogeneous dynamics. Most particularly there is a central connection to the rigidity of abelian actions that is the focus of Lindenstrauss's contribution. Similar ideas appear in the resolution of Zimmer's conjecture that is discussed in Fisher's contribution.

Other deep connections visible here are older. For example, tempered representations as discussed by Benoist and Kobayashi play an important role in proving exponential decay of matrix coefficients. This is then central in many results in homogeneous dynamics.

The casual reader of this volume will find a sequence of introductions to various individual subfields. But the careful reader of this book will find many other interesting juxtapositions and connections between the different contributions. We hope that this will lead to both a greater understanding of existing work and to new insights into the areas pioneered by Margulis.

References

[BFMS1] U. Bader, D. Fisher, N. Miller, and M. Stover. Arithmeticity, superrigidity, and totally geodesic submanifolds. *Preprint*, 2019.

[BFMS2] U. Bader, D. Fisher, N. Miller, and M. Stover. Arithmeticity, superrigidity and totally geodesic submanifolds II: $SU(n, 1)$. *Preprint*, 2020.

[BFH1] A. Brown, D. Fisher, and S. Hurtado. Zimmer's conjecture: Subexponential growth, measure rigidity, and strong property (T). *Invent. Math.*, 221(3):1001–1060, 2020. *arXiv:1608.04995.*

[BFH2] A. Brown, D. Fisher, and S. Hurtado. Zimmer's conjecture for actions of $\mathrm{SL}(m, \mathbb{Z})$. *Preprint*, 2017.

[EMMV] M. Einsiedler, G. Margulis, A. Mohammadi, and A. Venkatesh. Effective equidistribution and property (τ). *J. Amer. Math. Soc.*, 33(1):223–289, 2020.

[EMV] M. Einsiedler, G. Margulis, and A. Venkatesh. Effective equidistribution for closed orbits of semisimple groups on homogeneous spaces. *Invent. Math.*, 177(1):137–212, 2009.

[EMM] A. Eskin, G. Margulis, and S. Mozes. Upper bounds and asymptotics in a quantitative version of the Oppenheim conjecture. *Ann. of Math. (2)*, 147(1):93–141, 1998.

[KM1] A. Karlsson and G. A. Margulis. A multiplicative ergodic theorem and nonpositively curved spaces. *Comm. Math. Phys.*, 208(1):107–123, 1999.

[KM2] D. Y. Kleinbock and G. A. Margulis. Flows on homogeneous spaces and Diophantine approximation on manifolds. *Ann. of Math. (2)*, 148(1):339–360, 1998.

[Mar1] G. A. Margulis. The action of unipotent groups in a lattice space. *Mat. Sb. (N.S.)*, 86(128):552–556, 1971.

[Mar2] G. A. Margulis. Arithmeticity of nonuniform lattices. *Funkcional. Anal. i Priložen.*, 7(3):88–89, 1973.

[Mar3] G. A. Margulis. Explicit constructions of expanders. *Problemy Peredači Informacii*, 9(4):71–80, 1973.

[Mar4] G. A. Margulis. Discrete groups of motions of manifolds of nonpositive curvature. In *Proceedings of the International Congress of Mathematicians (Vancouver, BC, 1974), Vol. 2*, pages 21–34. Canad. Math. Congress, Montreal, QC, 1975.

[Mar5] G. A. Margulis. Factor groups of discrete subgroups and measure theory. *Funktsional. Anal. i Prilozhen.*, 12(4):64–76, 1978.

[Mar6] G. A. Margulis. Complete affine locally flat manifolds with a free fundamental group. Volume 134, pages 190–205. Zap. Nauchn. Sem. Leningrad. Otdel. Mat. Inst. Steklov. (LOMI) 1984. *Automorphic functions and number theory, II.*

[Mar7] G. A. Margulis. Discrete subgroups and ergodic theory. In *Number theory, trace formulas and discrete groups (Oslo, 1987)*, pages 377–398. Academic Press, Boston, MA, 1989.

[MM] G. Margulis and A. Mohammadi. Arithmeticity of hyperbolic 3-manifolds containing infinitely many totally geodesic surfaces. *Preprint* 2019.

[MS1] G. A. Margulis and G. A. Soifer. Nonfree maximal subgroups of infinite index of the group $SL_n(\mathbb{Z})$. *Uspekhi Mat. Nauk*, 34(4(208)):203–204, 1979.

[MS2] G. A. Margulis and G. A. Soifer. Maximal subgroups of infinite index in finitely generated linear groups. *J. Algebra*, 69(1):1–23, 1981.

PART I

Arithmeticity, superrigidity, normal subgroups

1

DAVID FISHER

SUPERRIGIDITY, ARITHMETICITY, NORMAL SUBGROUPS: RESULTS, RAMIFICATIONS, AND DIRECTIONS

To Grisha Margulis for revealing so many vistas

Abstract. This essay points to many of the interesting ramifications of Margulis's arithmeticity theorem, superrigidity theorem, and normal subgroup theorem. We provide some history and background, but the main goal is to point to interesting open questions that stem directly or indirectly from Margulis's work and its antecedents.

1 Introduction

We begin with an informal overview of the events that inspire this essay and the work it describes. For formal definitions and theorems, the reader will need to look into later sections, particularly section 2.

During a few years in the early 1970s, Gregory Margulis transformed the study of lattices in semisimple Lie groups. In this section and the next, G is a semisimple Lie group of real rank at least 2 with finite center, and Γ is an irreducible lattice in G. For brevity we will refer to these lattices as *higher rank lattices*. The reader new to the subject can always assume G is $SL(n,\mathbb{R})$ with $n > 2$. We recall that a lattice is a discrete group where the volume of G/Γ is finite and that Γ is called uniform if G/Γ is compact and nonuniform otherwise. In 1971, Margulis proved that nonuniform higher rank lattices are arithmetic—that is, that they are commensurable to the integer points in some realization of G as a matrix group [Mar2]. The proof used a result Margulis had proven slightly earlier on the nondivergence of unipotent orbits in the space G/Γ [Mar1]. This result on nondivergence of unipotent orbits has since played a fundamental role in homogeneous dynamics and its applications to number theory, a topic treated in many other essays in this volume. Margulis's arithmeticity theorem had been conjectured by Selberg and Piatetski-Shapiro. Piatetski-Shapiro had also conjectured the result on

David Fisher. Department of Mathematics. Indiana University Bloomington, IN 47405
fisherdm@indiana.edu

nondivergence of unipotent orbits [Sel]. Both Selberg and Piatetski-Shapiro had also conjectured the arithmeticity result for uniform lattices, but it was clear that that case requires a different proof, since the space G/Γ is compact and questions of divergence of orbits do not make sense.

In 1974, Margulis resolved the arithmeticity question in truly surprising manner. He proved his superrigidity theorem, which classified the linear representations of a higher rank lattice Γ over any local field of characteristic zero and used this understanding of linear representations to prove arithmeticity [Mar3]. Connections between arithmetic properties of lattices and the rigidity of their representations had been observed earlier by Selberg [Sel]. Important rigidity results had been proven in the local setting by Selberg, Weil, Calabi-Vesentini, and others and in a more global setting by Mostow [Sel, Wei2, Wei1, Cal, CV, Mos1]. Despite this, the proof of the superrigidity theorem and this avenue to proving arithmeticity were quite surprising at the time. The proof of the superrigidity theorem, though inspired by Mostow's study of boundary maps in his rigidity theorem, was also quite novel in the combination of ideas from ergodic theory and the study of algebraic groups.

Four years after proving his superrigidity and arithmeticity theorems, Margulis proved another remarkable theorem about higher rank lattices, the normal subgroup theorem. Margulis's proofs of both superrigidity and the normal subgroup theorem were essentially dynamical and cemented ergodic theory as a central tool for studying discrete subgroups of Lie groups.

The main goal of this chapter is to give some narrative of the repercussions and echoes of Margulis's arithmeticity, superrigidity, and normal subgroup theorems and the related results they have inspired in various areas of mathematics with some focus on open problems. To keep true to the spirit of Margulis's work, some emphasis will be placed on connections to arithmeticity questions, but we will also feature some applications to settings where there is no well-defined notion of arithmeticity. For a history of the ideas that led up to the superrigidity theorem, we point the reader to a survey written by Mostow at the time [Mos4] and to a discussion of history in another survey by this author [Fis2, section 3].

In the next section of this essay we give precise statements of Margulis's results. Afterward we discuss various later developments with an emphasis on open questions. We do not attempt to give a totally comprehensive history. In some cases, we mention results without giving full definitions and statements, simply in order to indicate the full breadth and impact of Margulis's results without ending up with an essay several times the length of the current one. We mostly refrain from discussing proofs or only discuss them in outline. For a modern proof of superrigidity theorems, we refer the reader to the paper of Bader and Furman in this volume [BF3]. The proof is certainly along the lines of Margulis's original proof, but the presentation is particularly elegant and streamlined.

2 Arithmeticity and superrigidity: Margulis's results

For the purposes of this essay, we will always consider semisimple Lie groups with finite center and use the fact that these groups can be realized as algebraic groups. We will also have occasion to mention algebraic groups over other local fields, but we will keep the main focus on the case of Lie groups for simplicity. Given a semisimple Lie group G, the real rank of G is the dimension of the largest subgroup of G diagonalizable over $\mathbb{R}$.

Given an algebraic group G defined over $\mathbb{Q}$, one can consider the integer points of the group, which we will denote by $G(\mathbb{Z})$. Arithmetic groups are a (slight) generalization of this construction. We say two subgroups A_1 and A_2 of G are commensurable if their intersection is finite index in each of them—that is, $[A_1 \cap A_2 : A_i] < \infty$ for $i = 1, 2$.

A lattice $\Gamma < G$ is *arithmetic* if the following holds: there is another semisimple algebraic Lie group G' defined over $\mathbb{Q}$ with a homomorphism $\pi : G' \to G$ with $\ker(\pi) = K$, a compact group such that Γ is commensurable to $\pi(G(\mathbb{Z}))$.

A lattice Γ in a product of groups $G_1 \times G_2$ is *irreducible* if the projection to each factor is indiscrete. In most contexts this is equivalent to Γ not being commensurable to a product of a lattice Γ_1 in G_1 and a lattice Γ_2 in G_2. Irreducibility for a lattice in a product with more than two factors is defined similarly. We can now state the Margulis arithmeticity theorem formally.

THEOREM 2.1 (Margulis arithmeticity).
Let G be a semisimple Lie group of real rank at least 2 and $\Gamma < G$ an irreducible lattice; then Γ is arithmetic.

We will now state the superrigidity theorems and then briefly sketch the reduction of arithmeticity to superrigidity. This requires considering representations over fields other than $\mathbb{R}$ or $\mathbb{C}$—namely, representations of finite extensions of the p-adic fields $\mathbb{Q}_p$. Together, these are all the local fields of characteristic zero. Superrigidity and arithmeticitiy are also known for groups over local fields of positive characteristic as both source and target by combined works of Margulis and Venkataramana and for targets groups over valued fields that are not necessarily local by the work of Bader and Furman in this volume [Mar5, Ven, BF3].

To state the superrigidity theorem cleanly, we recall a definition. Given a lattice $\Gamma < G$ and topological group H, we say a homomorphism $\rho : \Gamma \to H$ *almost extends to a homomorphism of* G if there are representations $\rho_G : G \to H$ and $\rho' : \Gamma \to H$ such that ρ_G is continuous, $\rho'(\Gamma)$ is precompact and commutes with $\rho_G(G)$, and $\rho(\gamma) = \rho_G(\gamma)\rho'(\gamma)$ for all γ in Γ. We can now state the strongest form of Margulis's superrigidity theorem that holds in our context:

THEOREM 2.2 (Margulis superrigidity).
Let G be a semisimple Lie group of real rank at least 2, let $\Gamma < G$ be an irreducible lattice, and let k be a local field of characteristic zero. Then any homomorphism $\rho : \Gamma \to \mathrm{GL}(n, k)$ almost extends to a homomorphism of G.

In many contexts this theorem is stated differently, with assumptions on the image of ρ. Assumptions often are chosen to allow ρ to extend to G rather than almost extend or to extend on a subgroup of finite index. These assumptions are typically that $\rho(\Gamma)$ has simple Zariski closure and is not precompact, which guarantees extension on a finite index subgroup, and that the Zariski closure is center-free to guarantees an extension on all of Γ. In many contexts where Margulis's theorem is generalized beyond linear representations to homomorphisms to more general groups, only this type of special case generalizes. The version we state here is essentially contained in [Mar5], at least when G has finite center. The case of infinite center is clarified in [FM]. We will raise some related open questions later.

We sketch a proof of arithmeticity from superrigidity; for more details see, for example, [Zim4, chapter 6.1] or [Mar5, chapter IX]. First, notice that since Γ is finitely generated, the matrix entries of Γ lie in a finitely generated field k that is an extension of $\mathbb{Q}$. Assume G is simple and center-free. Then Theorem 2.2 implies that every representation of Γ either extends to G or has bounded image. Note that $\mathrm{Aut}(\mathbb{C})$ acts transitively on the set transcendental numbers. So if we assume k contains transcendentals, we can take the defining representation of Γ and compose it with a sequence of automorphisms of $\mathbb{C}$ that send the trace of the image of some particular γ to infinity. It is obvious that this can't happen in a representation with bounded image; it is also not hard to check that it cannot happen in one that extends to G. This means that k is a number field, so $\Gamma \subset G(k)$ and we want to show that $\Gamma \subset G(\mathcal{O}_k)$. To see that $\Gamma \subset G(\mathcal{O}_k)$, assume not. Then there is a prime $\mathfrak{p}$ of k such that the image of Γ in $G(k_\mathfrak{p})$ is unbounded where $k_\mathfrak{p}$ is the completion of k for its $\mathfrak{p}$-adic valuation. But this contradicts Theorem 2.2 since this unbounded representation should almost extend to G with ρ_G nontrivial and continuous and such ρ_G

cannot exist since $G(k_{\mathfrak{p}})$ is totally disconnected. To complete the proof, we want to show that Γ is commensurable to $G(\mathcal{O}_k)$. Assuming that k is of minimal possible degree over $\mathbb{Q}$, we establish this by showing that $G(\mathcal{O}_k)$ is already a lattice in G. We do this by showing that for any Galois automorphism σ of k other than the identity, the map $\Gamma \to G(\sigma(\mathcal{O}_k))$ obtained by composing the identity with σ has bounded image. This follows from the superrigidity theorem again simply because Galois conjugation does not extend to a continuous automorphism of the real points of G.

We mention next one additional application of the superrigidity theorem. Let V be a vector space, and assume a higher rank lattice Γ acts on V linearly. A natural object of study with many applications is the cohomology of Γ with coefficients in V. The first cohomology is particularly useful for applications.

THEOREM 2.3 (Margulis first cohomology).
Let Γ be a higher rank lattice and V a vector space on which Γ acts linearly; then $H^1(\Gamma, V) = 0$.

Let H be the Zariski closure of Γ in $\mathrm{GL}(V)$. The proof results from realizing that cocycles valued in V correspond to representation into $H \ltimes V$, applying superrigidity to see that all these representations must be conjugate into H, and realizing that this implies the cocycle is trivial. An important part of this argument is that we can apply superrigidity to the group $H \ltimes V$, which is neither semisimple nor reductive, since V is contained in the unipotent radical. We remark that if the image of Γ in $\mathrm{GL}(V)$ is precompact and all simple factors of G have higher rank, the result follows from property (T) for Γ. There are other ways of computing $H^1(\Gamma, V)$ using techniques from geometry and representation theory, but as far as this author knows, none of these quite recover the full statement of Theorem 2.3 in the case of nonuniform lattices; see, for example, [BW]. These geometric and representation theoretic methods can also be used to show vanishing theorems concerning higher degree cohomology that are not accessible by Margulis's methods.

Margulis also proved a variant of superrigidity and arithmeticity for lattices with dense commensurators. For a subgroup $\Gamma < G$ we define

$$\mathrm{Comm}_G(\Gamma) = \{g \in G \mid g\Gamma g^{-1} \text{ and } \Gamma \text{ are commensurable}\}.$$

The next theorem was proved by Margulis at essentially the same time as the superrigidity theorem for higher rank lattices [Mar3]. The proof works independently of the rank of the ambient noncompact simple group G, but given Theorem 2.2, it is most interesting when the rank of G is 1.

THEOREM 2.4 (Margulis commensurator superrigidity).
Let G be a semisimple Lie group without compact factors, let $\Gamma < G$ be an irreducible lattice, and let $\Lambda < \mathrm{Comm}_G(\Gamma)$ be dense in G and k a local field of characteristic zero. Then any homomorphism $\rho : \Lambda \to \mathrm{GL}(n, k)$ almost extends to a homomorphism of G.

As before Margulis obtained a corollary concerning arithmeticity, which again is most interesting when the rank of G is 1.

COROLLARY 2.5 (Margulis commensurator arithmeticity). *Let G be a semisimple Lie group, let $\Gamma < G$ be an irreducible lattice, and let $\Lambda < \mathrm{Comm}_G(\Gamma)$ be dense in G; then Γ is arithmetic.*

The argument that Theorem 2.4 implies Corollary 2.5 is essentially the same as the argument that Theorem 2.2 implies Theorem 2.1. The converse to Corollary 2.5, that the commensurator of an arithmetic lattice is dense, was already known at the time of Margulis's work and is due to Borel [Bor].

An important related theorem of Margulis is the normal subgroup theorem. We state here the version for lattices in Lie groups [Mar4].

THEOREM 2.6.
Let G be a semisimple real Lie group of real rank at least 2 and $\Gamma < G$ an irreducible lattice. Then any normal subgroup $N \lhd \Gamma$ is either finite or finite index.

One can view this statement as being about some kind of superrigidity of homomorphisms of Γ to discrete groups: either the representation is almost faithful or the image is bounded. Knowing Theorem 2.1, Theorem 2.6 can also be viewed as an arithmeticity theorem saying that any infinite normal subgroup of a higher rank arithmetic lattice is still an arithmetic lattice. The proof of Theorem 2.6 is quite different from the proof of Theorem 2.2, but there is a long-standing desire to unify these phenomena in the context of higher rank lattices.

3 Superrigidity and arithmeticity in rank 1

The purpose of this section is to discuss lattices in rank 1 simple Lie groups. We discuss both known rigidity results and known constructions and raise some, mostly long-standing, questions. The rank 1 Lie groups are the isometry groups of various hyperbolic spaces:

(1) The group $SO(n,1)$ is locally isomorphic to the isometry group of the n-dimensional hyperbolic space $\mathbb{H}^n$.

(2) The group $SU(n,1)$ is locally isomorphic to the isometry group of the n-(complex)-dimensional complex hyperbolic space $\mathbb{CH}^n$.

(3) The group $Sp(n,1)$ is locally isomorphic to the isometry group of the n-(quaternionic)-dimensional quaternionic hyperbolic space $\mathbb{HH}^n$.

(4) The group F_4^{-20} is the isometry group of the two-dimensional Cayley hyperbolic plane $\mathbb{OH}^2$.

Exceptional isogenies between Lie groups yield isometries between some low-dimensional hyperbolic spaces—namely, that $\mathbb{H}^2 = \mathbb{CH}^1$, that $\mathbb{HH}^1 = \mathbb{H}^4$, and that $\mathbb{OH}^1 = \mathbb{H}^8$.

The strongest superrigidity and arithmeticity results for rank 1 groups generalize Margulis's results completely to lattices in $Sp(n,1)$ and F_4^{-20}. There are also numerous interesting partial results for lattices in the other two families of rank 1 Lie groups $SO(n,1)$ and $SU(n,1)$.

At the time of Margulis's proof of arithmeticity, nonarithmetic lattices were only known to exist in $SO(n,1)$ when $2 \leq k \leq 5$. No nonarithmetic lattices were known in the other rank 1 simple groups. Margulis asked about the other cases in [Mar3]. In this section we will also discuss known results, including other criteria for arithmeticity of lattices in rank 1 groups and known examples of nonarithmetic lattices.

3.1 QUATERNIONIC AND CAYLEY HYPERBOLIC SPACES.

In this subsection we describe the developments that proved that all lattices in $Sp(n,1)$ for $n>1$ and F_4^{-20} are arithmetic. The first major result in this direction, concerning rigidity of quaternionic and Cayley hyperbolic lattices, was proved by Corlette [Cor2].

THEOREM 3.1 (Corlette).

Let $G = Sp(n,1)$ for $n>1$ or $G = F_4^{-20}$ and $\Gamma < G$ be a lattice. Let H be a real simple Lie group with finite center and $\rho : \Gamma \to H$ a homomorphism with unbounded Zariski dense image. Then ρ almost extends to G.

REMARK 3.2.

(1) When $n=1$, the group $Sp(1,1)$ is isomorphic to $SO(4,1)$.

(2) In this setting, one can replace that "ρ almost extends" with the statement that "ρ extends on a subgroup of finite index."

The proof of Corlette's theorem has two main steps. The first is the existence of a Γ-equivariant harmonic map from $\mathbb{HH}^n$ or $\mathbb{OH}^2$ to H/K, the symmetric space associated to H. This step is contained in earlier work of Corlette or Donaldson; see also Labourie [Cor1, Don, Lab1]. Corlette then proves a Bochner formula that allows him to conclude the harmonic map is totally geodesic, from which the result follows relatively easily. This work is inspired by earlier work of Siu that proved generalizations of Mostow rigidity using harmonic map techniques [Siu]. The idea of using harmonic maps to prove superrigidity theorems was well-known at the time of Corlette's work and is often attributed to Calabi.

Following Corlette's work, Gromov and Schoen developed the existence and regularity theory of harmonic maps to buildings in order to prove the following [GS]:

THEOREM 3.3 (Gromov-Schoen).
Let $G = \mathrm{Sp}(n,1)$ for $n > 1$ or $G = F_4^{-20}$ and $\Gamma < G$ be a lattice. Let H be a simple algebraic group over a non-Archimedean local field with finite center and let $\rho : \Gamma \to H$ be a homomorphism with Zariski dense image. Then ρ has bounded image.

The main novelty in the work of Gromov and Schoen is to prove existence of a harmonic map into certain singular spaces with enough regularity of the harmonic map to apply Corlette's Bochner inequality argument. The harmonic map is to the Euclidean building associated to H by Bruhat and Tits [BT], and it is easy to see that there are no totally geodesic maps from hyperbolic spaces to Euclidean buildings.

Combining these two results with arguments of Margulis's deduction of arithmeticity from superrigidity, we can deduce the following:

THEOREM 3.4.
Let $G = \mathrm{Sp}(n,1)$ for $n > 1$ or $G = F_4^{-20}$ and $\Gamma < G$ be a lattice; then Γ is arithmetic.

We mention here a related result of Bass-Lubotzky that answered a question of Platonov [BL, Lub]. Namely, Platonov asked if any linear group that satisfied the conclusion of the superrigidity theorem was necessarily an arithmetic lattice. Bass and Lubotzky produce counter-examples as subgroups $\Delta < \Gamma \times \Gamma$ such that $\mathrm{diag}(\Gamma) < \Delta$, where $\Gamma < G$ is a lattice and G is either F_4^{-20} or $\mathrm{Sp}(n,1)$ for $n > 1$. The proofs involve a number of new ideas but depend pivotally on the work of Corlette and Gromov-Schoen to prove the required superrigidity results. In the examples produced by Bass and Lubotzky, the proof that Δ is superrigid is always deduced from the known superrigidity of diag(Γ). The fact that Γ is a hyperbolic group in the sense of Gromov plays a key role in constructing Δ.

QUESTION 3.5. *Are there other superrigid nonlattices? Can one find a superrigid nonlattice that is Zariski dense in higher rank simple Lie groups? Can one find a superrigid nonlattice that does not contain a superrigid lattice? Can one find a superrigid nonlattice that is a discrete subgroup of a simple noncompact Lie group?*

3.2 RESULTS IN REAL AND COMPLEX HYPERBOLIC GEOMETRY.

3.2.1 Nonarithmetic lattices: Constructions and questions

To begin this subsection I will discuss the known construction of nonarithmetic lattices in $SO(n,1)$ and $SU(n,1)$. To begin slightly out of order, we emphasize one of the most important open problems in the area, borrowing wording from Margulis in [Mar6].

QUESTION 3.6. *For what values of n does there exist a nonarithmetic lattice in* $SU(n,1)$?

The answer is known to include 2 and 3. The first examples were constructed by Mostow in [Mos2] using reflection group techniques. The list was slightly expanded by Mostow and Deligne using monodromy of hypergeometric functions [DM, Mos3]. The exact same list of examples was rediscovered/reinterpreted by Thurston in terms of conical flat structures on the 2 sphere [Thu]; see also [Sch]. There is an additional approach via algebraic geometry suggested by Hirzebruch and developed by him in collaboration with Barthels and Höfer [BHH]. More examples have been discovered recently by Couwenberg, Heckman, and Looijenga using the Hirzebruch-style techniques and by Deraux, Parker, and Paupert using complex reflection group techniques [CHL, DPP1, DPP2, Der]. But as of this writing there are only 22 commensurability classes of nonarithmetic lattices known in $SU(2,1)$ and only 2 known in $SU(3,1)$. An obvious refinement of Question 3.6 is as follows:

QUESTION 3.7. *For what values of n do there exist infinitely many commensurability classes of nonarithmetic lattice in* $SU(n,1)$?

We remark here that the approach via conical flat structures was extended by Veech and studied further by Ghazouani and Pirio [Vee, GP2]. Regrettably, this approach does not yield more non-arithmetic examples. It seems that the reach of this approach is roughly equivalent to the reach of the approach via monodromy of hypergeometric functions; see [GP1]. There appears to be some consensus among experts that the answer to both Questions 3.6 and 3.7 should be "for all n"; see, for example, [Kap1, conjecture 10.8]. Margulis's own wording as used in Question 3.5 is more guarded.

At the time of Margulis's work the only known nonarithmetic lattices in $SO(n,1)$ for $n > 2$ were constructed by Makarov and Vinberg by reflection group methods [Mak, Vin1]. It is known by work of Vinberg that these methods will only produce nonarithmetic lattices in dimension less than 30 [Vin2]. The largest known nonarithmetic lattice produced by these methods is in dimension 18 by Vinberg, and the full limits of reflection group constructions is not well understood [Vin3]. We refer the reader to [Bel] for a detailed survey. The following question seems natural:

QUESTION 3.8. *In what dimensions do there exist lattices in* $SO(n,1)$ *or* $SU(n,1)$ *that are commensurable to nonarithmetic reflection groups? In what dimensions do there exist lattices in* $SO(n,1)$ *or* $SU(n,1)$ *that are commensurable to arithmetic reflection groups?*

For the real hyperbolic setting, there are known upper bounds of 30 for nonarithmetic lattices and 997 for any lattices. The upper bound of 30 also applies for arithmetic uniform hyperbolic lattices [Vin2, Bel]. In the complex hyperbolic setting, there seem to be no known upper bounds, but a similar question recently appeared in, for example, [Kap1, question 10.10]. For a much more detailed survey of reflection groups in hyperbolic spaces, see [Bel].

A dramatic result of Gromov and Piatetski-Shapiro vastly increased our stock of nonarithmetic lattices in $SO(n,1)$ by an entirely new technique [GPS]:

THEOREM 3.9 (Gromov, Piatetski-Shapiro).
For each n there exist infinitely many commensurability classes of nonarithmetic uniform and nonuniform lattices in $SO(n,1)$.

The construction in [GPS] involves building hybrids of two arithmetic manifolds by cutting and pasting along totally geodesic codimension 1 submanifolds. The key observation is that noncommensurable arithmetic manifolds can contain isometric totally geodesic codimension 1 submanifolds.

This method has been extended and explored by many authors for a variety of purposes; see, for example, [Ago1], [BT], [ABB+2] and [GL]. It has also been proposed that one might build nonarithmetic complex hyperbolic lattices using a variant of this method, though that proposal has largely been stymied by the lack of codimension 1 totally geodesic 1 submanifolds. The absence of codimension 1 submanifolds makes it difficult to show that attempted "hybrid" constructions yield discrete groups. For more information see, for example, [Pau], [PW], [Wel], and [Kap1, conjecture 10.9]. We point out here that the results of Esnault and Groechenig discussed below as Theorem 3.19 imply that the "inbreeding" variant of Agol and Belolipetsky-Thomson [Ago1, BT] cannot produce nonarithmetic manifolds in the complex hyperbolic setting even if the original method of Gromov and Piatetski-Shapiro does.

In [GPS], Gromov and Piatetski-Shapiro ask the following intriguing question:

QUESTION 3.10. *Is it true that, in high enough dimensions, all lattices in* $SO(n, 1)$ *are built from sub-arithmetic pieces?*

The question is somewhat vague, and *sub-arithmetic* is not defined in [GPS], so a more precise starting point is as follows:

QUESTION 3.11. *For* $n > 3$, *is it true that any nonarithmetic lattice in* $\Gamma <$ $SO(n, 1)$ *intersects some conjugate of* $SO(n-1, 1)$ *in a lattice?*

This is equivalent to asking whether every finite volume nonarithmetic hyperbolic manifold contains a closed codimension 1 totally geodesic submanifold. Both reflection group constructions and hybrid constructions contain such submanifolds. It seems the consensus in the field is that the answer to this question should be no, but we know of no solid evidence for that belief. It is also not known to what extent the hybrid constructions and reflection group constructions build distinct examples. Some first results, indicating that the classes are different, are contained in [FLMS, theorem 1.7] and in [Mil, theorem 1.5].

It is worth mentioning that our understanding of lattices in $SO(2, 1)$ and $SO(3, 1)$ is both more developed and very different. Lattices in $SO(2, 1)$ are completely classified, but there are many of them, with the typical isomorphism class of lattices having many nonconjugate realizations as lattices, parameterized by moduli space. In $SO(3, 1)$, Mostow rigidity means there are no moduli spaces. But Thurston-Jorgensen hyperbolic Dehn surgery still allows one to construct many "more" examples of lattices, including ones that yield a negative answer to Question 3.11. There remains an interesting sense in which the answer to Question 3.10 could still be yes, even for dimension 3.

QUESTION 3.12. *Can every finite volume hyperbolic 3-manifold be obtained as Dehn surgery on an arithmetic manifold?*

To clarify the question, it is known that every finite volume hyperbolic 3-manifold is obtained as a topological manifold by Dehn surgery on some cover of the figure 8 knot complement, which is known to be the only arithmetic knot complement [HLM, Rei1]. What is not known is whether one can obtain the geometric structure on the resulting 3-manifold as geometric deformation of the complete geometric structure on the arithmetic manifold on which one performs Dehn surgery.

3.2.2 Arithmeticity, superrigidity, and totally geodesic submanifolds

This section concerns recent results by Bader, this author, Miller, and Stover, motivated by questions of McMullen and Reid in the case of real hyperbolic manifolds. Throughout this section *a geodesic submanifold* will mean a closed, immersed, totally geodesic submanifold. (In fact all results can be stated also for orbifolds but we ignore this technicality here.) A geodesic submanifold is *maximal* if it is not contained in a proper geodesic submanifold of smaller codimension.

For arithmetic manifolds, the presence of one maximal geodesic submanifold can be seen to imply the existence of infinitely many. The argument involves lifting the submanifold S to a finite cover $\tilde{M}$ where an element λ of the commensurator acts as an isometry. It is easy to check that $\lambda(S)$ can be pushed back down to a geodesic submanifold of M that is distinct from S. This was perhaps first made precise in dimension 3 by Maclachlan-Reid and Reid [MR, Rei2], who also exhibited the first hyperbolic 3-manifolds with no totally geodesic surfaces.

In the real hyperbolic setting the main result from [BFMS1] is as follows:

THEOREM 3.13 (Bader, Fisher, Miller, Stover).
Let Γ be a lattice in $\mathrm{SO}_0(n,1)$. If the associated locally symmetric space contains infinitely many maximal geodesic submanifolds of dimension at least 2, then Γ is arithmetic.

REMARK 3.14.

(1) The proof of this result involves proving a superrigidity theorem for *certain* representations of the lattice in $SO(n, 1)$. As the required conditions become a bit technical, we refer the interested reader to [BFMS1]. The superrigidity is proven in the language introduced in [BF3].

(2) At about the same time, Margulis and Mohammadi gave a different proof for the case $n = 3$ and Γ cocompact [MM]. They also proved a superrigidity theorem, but both the statement and the proof are quite different from [BFMS1].

(3) A special case of this result was obtained a year earlier by this author, Lafont, Miller, and Stover [FLMS]. There we prove that a large class of nonarithmetic manifolds have only finitely many maximal totally geodesic submanifolds. This includes all the manifolds constructed by Gromov and Piatetski-Shapiro but not the examples constructed by Agol and Belolipetsky-Thomson.

In the context of Margulis's work it is certainly worth mentioning that Theorem 3.13 has a reformulation entirely in terms of homogeneous dynamics and that homogenenous dynamics play a key role in the proof. It is also interesting that a key role is also played by dynamics that are not quite homogeneous but that take place on a projective bundle over the homogeneous space G/Γ.

Even more recently the same authors have extended this result to cover the case of complex hyperbolic manifolds.

THEOREM 3.15 (Bader, Fisher, Miller, Stover).
Let $n \geq 2$ *and* $\Gamma < SU(n, 1)$ *be a lattice and* $M = \mathbb{CH}^n / \Gamma$. *Suppose that* M *contains infinitely many maximal totally geodesic submanifolds of dimension at least* 2. *Then* Γ *is arithmetic.*

As before, this is proven using homogeneous dynamics, dynamics on a projective bundle over G/Γ, and a superrigidity theorem. Here the superrigidity theorem is even more complicated than before and depends also on results of Simpson and Pozzetti [Sim, Poz].

The results in this section provide new evidence that totally geodesic manifolds play a very special role in nonarithmetic lattices and perhaps provide some evidence that the conventional wisdom on Questions 3.11 and 3.6 should be reconsidered.

3.2.3 *Other superrigidity and arithmeticity results for lattices in* SO$(n, 1)$ *and* SU$(n, 1)$

The combination of the results in the previous section and Margulis's commensurator superrigidity theorem, as well as questions in 3.2.1, raise the following:

QUESTION 3.16. *Let* $\Gamma < G$ *be a lattice where* $G = \mathrm{SO}(n, 1)$ *or* $\mathrm{SU}(n, 1)$. *What conditions on a representation* $\rho : \Gamma \to \mathrm{GL}(m, k)$ *imply that* ρ *extends or almost extends? What conditions on* Γ *imply that* Γ *is arithmetic?*

For $\mathrm{SU}(n, 1)$ Margulis asks a similar, but more restricted, question [Mar6]. He asks whether there might be particular lattices in $\mathrm{SU}(n, 1)$ where superrigidity holds without restrictions on ρ as in the higher rank, quaternionic hyperbolic and Cayley hyperbolic cases.

A very first remark is that for many Γ as above it is known that there are surjections of Γ on both abelian and nonabelian free groups. This suggests that one might want to study faithful representations or ones with finite kernel, though surprisingly very few known superrigidity results explicitly assume faithfulness of the representation. The main counterexample to this is the following theorem of Shalom [Sha2]. We recall that for a discrete group Δ of a rank 1 simple Lie group, $\delta(\Delta)$ is the Hausdorff dimension of the limit set of Δ. The limit set admits many equivalent definitions; see, for example, [Sha2] for discussion.

THEOREM 3.17 (Shalom).
Let $\Gamma < G$ *be a lattice where* $G = \mathrm{SO}(n, 1)$ *or* $\mathrm{SU}(n, 1)$. *Let* $\rho : \Gamma \to H$ *be a discrete, faithful representation where* H *is either* $\mathrm{SO}(m, 1)$ *or* $\mathrm{SU}(m, 1)$. *Then* $\delta(\Gamma) \leq \delta(\rho(\Gamma))$.

Shalom actually proves a result for nonfaithful discrete representations as well, relating the dimension of the limit set of the image and the kernel to the dimension of the limit set of the lattice. Shortly after Shalom proved the above theorem, Besson, Courtois, and Gallot proved that equality only occurs in the case where the representation almost extends [BCG]. The methods of Besson, Courtois, and Gallot, the so-called barycenter mapping, have been used in many contexts. The key ingredient in Shalom's proofs, understanding precise decay rates of matrix coefficients, has not been exploited nearly as thoroughly for applications to rigidity. For either Shalom's techniques or the barycenter map technique, the utility of the methods are currently limited by the requirement that the representation have discrete image.

Relatively few other superrigidity or arithmeticity-type results are known for real hyperbolic manifolds, but a plethora of other interesting phenomena have been discovered in the complex hyperbolic setting. We begin with some of the most recent, which involve a bit of a detour in a surprising direction.

Simpson's work on Higgs bundles and local systems focuses broadly on the representation theory of $\pi_1(M)$, where M is a complex projetive variety, or more generally a complex quasi-projective variety [Sim]. This is related to our concerns because when $G = SU(n, 1)$, then $M = K\backslash G/\Gamma$ is a projective variety when Γ is compact and quasi-projective when it is not. We say a representation $\rho : \Gamma \to H$ is *rigid* or *infinitesimally rigid* if the first cohomology $H^1(\Gamma, \mathfrak{h})$ vanishes where $\mathfrak{h}$ is the Lie algebra of H. For $G = SU(n, 1)$, $H = G$, and ρ the defining representation $\rho : \Gamma \to H$, the vanishing of this cohomology group is a result of Calabi-Vesentini [CV]. We state Simpson's main conjecture only in the projective case to avoid technicalities [Sim]:

CONJECTURE 3.18. *Let M be a projective variety and $\rho : \pi_1(M) \to SL(n, \mathbb{C})$ an infinitesimally rigid representation. Then $\rho(\Gamma)$ is integral—that is, there is a number field k such that $\rho(\pi_1(M))$ is contained in the integer points* $SL(n, \mathcal{O}_k)$.

We state the conjecture for SL targets rather than GL targets to avoid a technical finite determinant condition. Higher rank irreducible Kähler locally symmetric spaces of finite volume provide examples where Simpson's conjecture follows from Margulis's arithmeticity theorem. Recent work of Esnault and Groechenig prove this result in many cases [EG2, EG1]. In particular their results have the following as a (very) special case:

THEOREM 3.19 (Esnault-Groechenig).
Let $\Gamma < SU(n, 1)$ be a lattice with $n > 1$; then Γ is integral—that is, there is a number field k and k structure on $SU(n, 1)$ *such that* $\Gamma < SU(n, 1)(\mathcal{O}_k)$.

The theorem is immediate from the results in [EG1] for the case of cocompact lattices. For an explanation of how it also follows in the noncocompact case see [BFMS2]. A construction of Agol as extended by Belilopetsky-Thomson shows that the analogous result fails in $SO(n, 1)$ [Agol, BT]—that is, there are nonintegral lattices, both cocompact and noncocompact, in $SO(n, 1)$ for every n.

We note here that the proof by Esnault and Groechenig does not pass through a superrigidity theorem. In the context of this paper, one might expect this, but the methods of [EG1] depend on algebraic geometry and deep results of Lafforgue on the Langlands program [Laf]. However, in this context one might also ask the following:

QUESTION 3.20. *Let* $\Gamma < \mathrm{SU}(n,1)$ *be a lattice and* $n > 1$*. Assume* k *is a totally disconnected local field,* H *is a simple algebraic group over* k*, and* $\rho : \Gamma \to H$ *is a Zariski dense, faithful representation. Is* $\rho(\Gamma)$ *compact?*

We mention here a question from our paper with Larsen, Stover, and Spatzier [FLSS] that aims at understanding the degree to which a lattice in $\mathrm{SO}(n,1)$ can fail to be integral by studying the p-adic representation theory of these groups.

QUESTION 3.21. *Let* Σ_g *be a surface group of genus* $g \geq 2$*. Is there a discrete and faithful representation of* Σ_g *into* $\mathrm{Aut}(Y)$ *for* Y*, a locally compact Euclidean building? Can we take* Y *to be a finite product of bounded valence trees?*

More generally one can ask the same questions with Σ_g replaced by a lattice Γ in $G = \mathrm{SO}(n,1)$. Once $n > 2$, it is known that Γ is contained in the k points of G for some number field k. To understand the extent to which Γ fails to be integral, it suffices to consider the case where Y is the building associated to some p-adic group $G(k_p)$.

There is one other context in which enough superrigidity results are known to imply arithmeticity—namely, Klingler's work on fake projective planes [Kli1].

DEFINITION 3.22. A fake projective plane is a complex projective surface with the same Betti numbers as $P(\mathbb{C}^2)$ that is not biholomorphic to $P(\mathbb{C}^2)$.

Results of Yau on the Calabi conjecture show that any fake projective plane is of the form $\mathbb{CH}^2/\Gamma$ with Γ a cocompact lattice [Yau]. Let $G = \mathrm{SU}(2,1)$; we can further assume that $K\backslash G/\Gamma = M$ satisfies the condition that $c_1^2 = 3c_2 = 9$, where c_1 and c_2 are the first and second Chern numbers of M. Yau's work implies complex ball quotients satisfying these conditions are exactly the fake projective planes. Klingler then shows that Γ is arithmetic.

THEOREM 3.23 (Klingler).
If M *is a fake projective plane, then* $\Gamma = \pi_1(M)$ *is arithmetic.*

This result is striking since the condition for arithmeticity is purely topological. The proof uses superrigidity theorems proven using harmonic map techniques as in subsection 3.1. Following Klingler's work, the fake projective planes were classified and further studied by Prasad-Yeung and Cartwright-Steger [PY, CS]. There turn out to be exactly 50 examples. We note that this is precisely 50 and not 50 up to commensurability and that some of these examples are commensurable. The topological condition of being a fake projective plane is not invariant under passage to finite covers.

Two more recent results of Klingler and collaborators are also intriguing in this context. In the first of these papers he shows that for certain lattices Γ in $SU(n,1)$, the representation theory of Γ is very restricted as long as one considers representations in dimension below $n-1$ [Kli2]. The results there are proven by showing that holomorphic symmetric differentials control the linear representation theory of fundamental groups of compact Kähler manifolds. In a later paper by Brunebarbe, Klingler, and Totaro, the authors extend this to investigate the case of compact Kähler manifolds without holomorphic symmetric differentials [BKT].

A different direction for the study of representations of complex hyperbolic lattices was introduced by Burger and Iozzi in [BI]. They introduce a notion of a maximal representation of a lattice Γ in $G=SU(n,1)$ generalizing a definition of Toledo in the case of $SU(1,1)\cong SL(2,\mathbb{R})$ [Tol]. Burger-Iozzi show that maximal representation of Γ into $SU(m,1)$ extends to G. The proof uses a result on incidence geometry generalizing an earlier result of Cartan to the measurable setting [Car]. The definitions and results of this paper were further extended to the case of $SU(p,q)$ targets when $p\neq q$ by Pozzetti in her thesis [Poz]. The proof of Theorem 3.15 uses Pozzetti's version of the Cartan theorem. More recently Koziarz and Maubon extended the result to include the case where $p=q$ and reproved all earlier results using techniques of harmonic maps and Higgs bundles [KM3].

4 Orbit equivalence rigidity

This section mostly serves to point to a broad area of research that we will not attempt to summarize or survey in any depth.

DEFINITION 4.1. Let (S,μ) be a finite measure space with an ergodic G action and (S',μ) a finite measure space with an ergodic G' action. We say the actions are *orbit equivalent* if there are conull Borel sets $S_0\subset S$ and $S'_0\subset S'$ and a measure class preserving isomorphism $\phi: S_0\to S'_0$ such that s and t are in the same G orbit if and only if $\phi(s)$ and $\phi(t)$ are in the same G' orbit.

In a remarkable result in [Zim1], Zimmer further developed the ideas in Margulis's proof of superrigidity to prove the following:

THEOREM 4.2.
Let G_1 and G_2 be center-free connected simple Lie groups and assume $\mathbb{R}$-rank$(G_1) > 1$. Let (S_i, μ_i) be probability measure spaces with free ergodic actions of G_i for $i=1,2$. If the actions are orbit equivalent, then they are conjugate.

The key ingredient in the proof of Theorem 4.2 is Zimmer's cocycle superrigidity theorem. We do not state this here but point the reader to [Zim4] and [FM] for detailed discussions.

An important further development in the theory comes in work of Furman, who extends Zimmer's results on orbit equivalence to lattices [Fur1, Fur2]. We do not give a comprehensive discussion but state one result.

THEOREM 4.3.
Let G be a center-free connected simple Lie group and assume $\mathbb{R}$-rank$(G) > 1$. Let $\Gamma_1 < G$ be a lattice and let Γ_2 be any finitely generated group. Let (S_i, μ_i) be probability measure spaces with free ergodic actions of Γ_i for $i=1,2$. If the actions of Γ_i on (S, μ_i) are orbit equivalent, then Γ_2 is virtually a lattice in G.

Here *virtually* means there is a finite index subgroup of Γ_2 whose quotient by a finite normal subgroup is a lattice in G. Furman also shows that there is a unique obstruction to conjugacy of the actions.

Following these results, the study of orbit equivalence rigidity became a rich topic in which many rigidity results are known, many of which depend on cocycle superrigidity theorems. We do not attempt a survey but point to one written earlier by Furman [Fur3].

5 The Zimmer program

In 1983, Zimmer proposed a number of conjectures about actions of higher rank simple Lie groups and their lattices on compact manifolds [Zim3, Zim5]. These conjectures were motivated by a number of Zimmer's own theorems, including the cocycle superrigidity theorem mentioned in the previous section. But perhaps the clearest motivation is as a *nonlinear* analogue of Margulis's superrigidity theorem. These conjectures have led to a tremendous amount of activity; see this author's earlier survey and recent update [Fis2, Fis3] for more information. Here we focus only on two aspects: the recent breakthrough made by Brown, this author, and Hurtado, and a statement of a general conjectural superrigidity theorem for $\mathrm{Diff}(M)$ targets.

The clearest conjecture made by Zimmer predicted that any action of a higher rank lattice on a compact manifold of sufficiently small dimension should preserve a Riemannian metric. Since the isometry group of a compact manifold is a compact Lie group, this, together with Margulis's superrigidity theorem, often implies the action factors through a finite quotient of the lattice. The recent work of Brown, this author, and Hurtado makes dramatic progress on this conjecture and completely resolves it in several key cases [BFH1, BFH2, BFH3]. For example we have the following:

THEOREM 5.1 (Brown, Fisher, Hurtado).
Let Γ be a lattice in $\mathrm{SL}(n,\mathbb{R})$, let M be a compact manifold, and let $\rho:\Gamma\to\mathrm{Diff}(M)$ be a homomorphism. Then

(1) if $\dim(M)<n-1$, the image of ρ is finite;
(2) if $\dim(M)<n$ and $\rho(\Gamma)$ preserves a volume form on M, then the image of ρ is finite.

This result is sharp, since $\mathrm{SL}(n,\mathbb{R})$ acts on the projective space $P(\mathbb{R}^n)$ and $\mathrm{SL}(n,\mathbb{Z})$ acts on the torus $\mathbb{T}^n$. The papers with Brown and Hurtado prove results about all lattices in all simple Lie groups G of higher real rank, but are only sharp for certain choices of G. In particular, results about volume preserving actions are only sharp for $\mathrm{SL}(n,\mathbb{R})$ and $\mathrm{Sp}(2n,\mathbb{R})$, while results about actions not assumed to preserve volume are sharp for all split simple groups. See [Can] and [BFH3] for more discussion.

The most naive version of the Zimmer program is perhaps the following:

QUESTION 5.2. *Let G be a simple Lie group of higher real rank, $\Gamma<G$ a lattice, and M a compact manifold. Can one understand all homomorphisms $\rho:\Gamma\to\mathrm{Diff}(M)$? If ω is a volume form on M, can one classify all homomorphisms $\rho:\Gamma\to\mathrm{Diff}(M,\omega)$?*

The careful reader will notice a slight variation in wording in the two questions. This is due to the fact that non–volume preserving actions are known to be nonclassifiable. In particular the parabolic induction described by Stuck in [Stu2] shows that even homomorphisms $\rho:G\to\mathrm{Diff}(M)$ cannot be classified. In particular Stuck shows that given two vector fields X and Y on a compact manifold M and a parabolic subgroup Q in G, one can construct two homomorphisms $\rho_X,\rho_Y:G\to\mathrm{Diff}((G\times M)/Q)$ such that ρ_X and ρ_Y

are conjugate if and only if the flows generated by X and Y on M are conjugate.

We briefly describe Stuck's construction. Any parabolic subgroup $Q < G$ admits a homomorphism $\phi : Q \to \mathbb{R}$. Any vector field X on M defines an $\mathbb{R}$ action, which we denote by $\bar{\rho}_X : \mathbb{R} \times M \to M$. We define a Q action on $G \times M$ by $(g, m)q = (gq^{-1}, \bar{\rho}(\phi(q)))$. As this commutes with the left G action on the first variable, we obtain an action ρ_X of G on $(G \times M)/Q$. The space $(G \times M)/Q$ is a manifold and in fact an M fiber bundle over G/Q. It is transparent that applying the construction to two vector fields X and Y on manifolds M and M', the G actions are conjugate if and only if $\bar{\rho}_X$ and $\bar{\rho}_Y$ are. The following seems accessible:

PROBLEM 5.3. *If ρ_X and ρ_Y are conjugate as Γ actions, are $\bar{\rho}_X$ and $\bar{\rho}_Y$ conjugate as $\mathbb{R}$ actions?*

The main goal of this section is to describe a conjectural picture of all Γ-actions on compact manifolds M in terms of G actions. This is very much in the spirit of Margulis's superrigidity theorem, and we begin by slightly restating that theorem. Given a higher rank lattice Γ in Lie group G there is a natural compact extension $G \times K$ of G in which Γ sits diagonally as a lattice. Here K is a product of two groups $K = K_1 \times K_2$, where K_1 is totally disconnected and K_2 is a Lie group. The group K_1 is the profinite completion of Γ. The group K_2 is the compact Lie group such that Γ is commensurable to the integral points in $G' = G \times K_2$. The definition of *arithmeticity* in section 2 ensures that K_2 exists. We note here that K_2 is semisimple and any simple factor of K_2 has the same complexification as some simple factor of G. We refer to $G \times K$ as the *canonical envelope* of Γ. Given $G \times K$, Margulis's superrigidity theorem can be restated as follows:

THEOREM 5.4 (Margulis superrigidity variant).
Let G be a semisimple Lie group of real rank at least 2; let $\Gamma < G$ be an irreducible lattice and k a local field of characteristic zero. Then any homomorphism $\rho : \Gamma \to GL(n, k)$ extends to a homomorphism of $G \times K$, the canonical envelope of Γ.

We now describe a conjecture analogous to Theorem 5.4 but with $\mathrm{Diff}(M)$ targets. We begin by defining a *local action* of a group in a way that is similar to standard definitions of pseudo-groups or groupoids but is adapted to our purposes. The definition is complicated because we need to be able to restrict local actions of topological groups to local or global actions of their discrete subgroups. So it does not suffice for our purposes to have the germ of the group action near the identity in the acting group, but rather we need it near every element in the acting group.

DEFINITION 5.5. Let D be a group and M a manifold. We say D has *local action* on M if

(1) for every point $x \in M$ and every element $d \in D$ there is an open neighborhood $V_{x,d}$ of $d \in D$ and an open neighborhood $U_{x,d}$ of x in M and a *local action* map $\rho_{x,d} : V_{x,d} \times U_{x,d} \to M$, and,

(2) given d, d' in D, whenever $z \in U_{x,d}$ and $g, hg \in V_{x,d}$ and for every y such that $\rho_{x,d}(g, x) \in U_{g,d'}$ and $h \in V_{g,d'}$, we have $\rho_{x,d}(hg, z) = \rho_{g,d'}(h, \rho_{x,d}(g, z))$.

It may be possible to offer a simpler or more transparent variant of the definition. Point (1) gives local diffeomorphisms at every point corresponding to elements of D, and point (2) requires that the collection of such local diffeomorphisms remembers the group multiplication on D whenever possible. Even when M is compact, one cannot restrict attention to a finite collection of local action maps unless D is also compact. The paradigmatic example to keep in mind is that $SL(n, \mathbb{R})$ acts locally on $\mathbb{T}^n$. Only $SL(n, \mathbb{Z})$ has a globally defined action, but the lift to $\mathbb{R}^n$ one immediately has a global $SL(n, \mathbb{R})$ action, which can easily be seen to give a local action on $\mathbb{T}^n$. If one carries out the construction of "blowing up" the origin in $\mathbb{T}^n$ as in [KL], one obtains a manifold M with a local $SL(n, \mathbb{Z})$ action, which does not extend to $SL(n, \mathbb{R})$ on any cover.

It is clear that one way to have a local action is to have a global action. We say a local action *restricts* from a global action $\rho : D \times M \to M$ if we have that

$$\rho_{x,d} = \rho|_{V_{x,d} \times U_{x,d}}$$

for every x in M and $d \in D$. Given a subgroup $C < D$ and a global C action $\rho_C : C \times M \to$ we say that ρ_C restricts from local D action if

$$\rho_{x,d}|_{(C \cap V_{x,d}) \times U_{x,d}} = \rho|_{(V_{x,d} \cap C) \times U_{x,d}}$$

for all $x \in X$ and $d \in D$.

With these definitions in hand, we can state a general superrigidity conjecture, which we believe is at the heart of the phenomena observed so far in the Zimmer program.

CONJECTURE 5.6. *Let G be a simple Lie group of real rank at least 2 and $\Gamma < G$ a lattice. Let $G \times K$ be the canonical envelope of Γ described above. Then for any* *compact manifold M and any homomorphism* $\rho:\Gamma \to \mathrm{Diff}(M)$ *there is a local action of* $G \times K$ *on M that restricts to the* Γ *action.*

REMARK 5.7.

(1) For a fixed action, one expects that the local action is trivial on a finite index subgroup of K_1—that is, that the local action is one of $G \times K_1' \times K_2$, where K_1' is a finite quotient of K. This is true for K_1 actions by the smooth version of the Hilbert-Smith conjecture, which has been known for some time.

(2) The example of $\mathrm{SL}(n,\mathbb{Z})$ acting on $\mathbb{T}^n$ shows that one needs some notion of local action to state the conjecture. The existence of isometric actions that extend to K_2 justify the need for the compact extension of G. The existence of actions through finite quotients of Γ justify the need for K_1.

At the moment, Conjecture 5.6 incorporates all known ideas for building "exotic" actions of lattices Γ in higher rank simple Lie groups. In addition to the parabolic induction examples discussed above, there is the blow-up construction introduced by Katok and Lewis, which by now has several variants [KL, Fis1, BF, FW, KRH]. The conjecture is also of a similar flavor to a conjecture stated in various forms by Labourie, Margulis, and Zimmer that a manifold admitting a higher rank lattice action should, under some circumstances, be homogeneous on an open dense set [Lab2, Mar6]

In complete generality, Conjecture 5.6 seems very far out of reach. It does seem most accessible for the case where $G = \mathrm{SL}(n,\mathbb{R})$ and $\dim(M) = n$ and perhaps when the action is analytic. By the work of Brown, this author, and Hurtado and that of Brown, Rodriguez Hertz, and Wang, the conjecture is known for lattices in $\mathrm{SL}(n,\mathbb{R})$ for $\dim(M) < n$. A very interesting and overlooked paper by Uchida from 1979 classifies all analytic actions of $\mathrm{SL}(n,\mathbb{R})$ on the sphere S^n [Uch]. This suggests starting with the following:

PROBLEM 5.8. *Classify analytic* $\mathrm{SL}(n,\mathbb{R})$ *actions on manifolds of dimension n. Classify analytic local actions of* $\mathrm{SL}(n,\mathbb{R})$ *on manifolds of dimension n.*

The second part of the problem is clearly harder than the first. For both parts, it should be useful to look at [CG] and [Stu1].

Other contexts in which Conjecture 5.6 might be more accessible is when one assumes additional geometric or dynamical properties of the action. Key contexts include Anosov actions [BRHW2, Fis3] and actions preserving rigid geometric structures [Gro2, Zim6]. Both hyperbolicity of the dynamics and existence of geometric structures can be used to produce additional Lie groups acting on a manifold or at least on certain foliations, so these hypotheses should be helpful to find some kind of local action given a Γ action.

6 Other sources, other targets

Another topic that we can touch on only briefly here is the generalization of Margulis's superrigidity theorem to other sources and targets. The set of targets considered is quite often spaces of nonpositive curvature, frequently without any assumption that the dimension is finite. The set of sources is often broadened to more general locally compact groups. Since there is no good analogue of rank without linear structure, the most common assumption is that one has a locally compact, compactly generated group G that is a product $G = G_1 \times \cdots \times G_k$ and that one has an irreducible lattice $\Gamma < G$. To give an indication, we state one particularly nice result due to Monod. To do so we need to define a term. We assume for the definition that X is a geodesic metric space.

DEFINITION 6.1. A subgroup $L < \mathrm{Isom}(X)$ is *reduced* if there is no unbounded closed convex subset $Y \subsetneq X$ such that gY is finite (Hausdorff) distance from Y for all $g \in L$.

Reduced is one possible geometric substitute for considering subgroups whose Zariski closure is simple or semisimple. We note that Monod proves other results in [Mon] that require weaker variants of this hypotheses, but these are more difficult to state.

THEOREM 6.2.
Let Γ be an irreducible uniform lattice in a product $G = G_1 \times \cdots \times G_n$ of non-compact locally compact σ-compact groups with $n > 1$. Let $H < \mathrm{Isom}(X)$ be a closed subgroup, where X is any complete CAT(0) *space not isometric to a finite-dimensional Euclidean space. Let $\tau : \Gamma \to H$ be a homomorphism with reduced unbounded image. Then τ extends to a continuous homomorphism $\tilde{\tau} : G \to H$.*

In the theorem, X is not assumed to be locally compact. The theorem holds for nonuniform lattices with a mild assumption of *square integrability*. For a survey of earlier results, we point the reader to Burger's ICM address [Bur]. In this context, we also mention that Gelander, Karlsson, and Margulis have extended Monod's results to a broader class of nonpositively curved spaces [GKM]. A key context for application of these kinds of results are lattices in isometry groups of products of trees [BMZ] and to Kac-Moody groups [CR], which provide many examples of lattices in products of locally compact, compactly generated groups.

A major difference between existing geometric superrigidity theorems like Monod's and Margulis's Theorem 2.2 is that Margulis does not need any assumption like *reduced*. It is a major open problem in the area to prove some analogue of this fact. We state here a version of this question. Since X is not locally compact, we need to modify the notion of a representation *almost extending* slightly. If $\Gamma < G$ is a lattice and X is a nonpositively curved space, we say $\rho : \Gamma \to \mathrm{Isom}(X)$ *almost extends* if there exists $\rho_1 : G \to \mathrm{Isom}(X)$ and $\rho_2 : \Gamma \to \mathrm{Isom}(X)$, where ρ_2 has bounded image, $\rho_1(G)$ commutes with $\rho_2(\Gamma)$, and $\rho(\gamma) = \rho_1(\gamma)\rho_2(\gamma)$ for all $\gamma \in \Gamma$.

QUESTION 6.3. *Let $G = G_1 \times \cdots \times G_n$ for $n > 2$, where each G_i is a locally compact group with Kazhdan's property (T), or let G be a simple Lie group of higher real rank. Assume $\Gamma \in G$ is a cocompact lattice and that X is a* CAT(0) *space. Given $\rho : \Gamma \to \mathrm{Isom}(X)$, is there a Γ-invariant subspace $Z \subset X$ such that $\rho : \Gamma \to \mathrm{Isom}(Z)$ almost extends to G?*

One can easily see that the passage from X to Z is necessary by taking the G action on G/K, the symmetric space for G, restricting to the Γ action, and adding a discrete Γ periodic family of rays to G/K. One might assume something weaker than property (T) for each G_i and should not really require the lattice to be cocompact, but solving the question as formulated above would be a good first step.

Recently Bader and Furman have deeply rethought the proof of Margulis's superrigidity theorem [BF3, BF2, BF1]. This work is used in the proof of Theorems 3.13 and 3.15. It is also used quite strikingly in a proof of superrigidity theorems for groups that are not lattices in any locally compact group. The groups in question are isometry groups of the so-called exotic $\tilde{A}_2$ buildings. The isometry groups of these buildings are known to be, in many cases, discrete and cocompact. Bader, Caprace, and Lécureux prove a superrigidity theorem for large enough groups of isometries of buildings of type A_2 and use this to show that a lattice in the isometry groups of a building of type A_2 has an infinite image linear representation if and only if the building is classical and thus the isometry group is a linear group over a totally disconnected local field [BCL].

7 The normal subgroup theorem, commensurators, attempts at unification

In this section we describe some results and questions related to Theorems 2.4 and 2.6. We also describe some attempts to unify the phenomena behind Theorems 2.2 and 2.6.

As described in [BRHW3] in this volume, Margulis's proof of the normal subgroups theorem follows a remarkable strategy. He proves that given a higher rank lattice Γ and a normal subgroup N, the quotient group Γ/N has property (T) and is amenable. From this, one trivially deduces that Γ/N is a finite subgroup. For more discussion see chapter 3, by Brown, Rodriguez Hertz and Wang, in this volume.

We begin by mentioning that the normal subgroup theorem has also been generalized to contexts of products of fairly arbitrary locally compact, compactly generated groups. This was first done by Burger and Mozes in the special case where $G = G_1 \times \cdots \times G_k$, where each G_i is a large enough subgroup of $\mathrm{Aut}(T_i)$, where T_i is a regular tree. Burger and Mozes used this result in order to show that certain irreducible lattices they construct in such G are infinite simple groups [BM2, BM1]. These new simple lattices are (a) finitely presented, (b) torsion-free, (c) fundamental groups of finite, locally CAT(0) complexes, (c) of cohomological dimension 2, (d) biautomatic, and (e) the free product of two isomorphic free groups F_1 and F_2 over a common finite index subgroup. The existence of such simple groups is quite surprising. In later work, Bader and Shalom proved a much more general result about normal subgroups of lattices in fairly arbitrary products of locally compact, second countable, compactly generated groups [BS]. To be clear, the Bader-Shalom paper gives the "amenability half" of the proof—that is, that Γ/N is amenable. Shalom had proven earlier that Γ/N has property (T) in [Sha1]. These results were used by Caprace and Remy to show that certain Kac-Moody groups are also simple groups [CR].

We note that uniform lattices in rank 1 simple Lie groups are hyperbolic. This means, in particular, that they have infinitely many, infinite index normal subgroups by Gromov's geometric variants on small cancellation theory. In particular, the normal closure N of any large enough element $\gamma \in \Gamma$ has the property that Γ/N is an infinite hyperbolic group [Gro1]. The following interesting question is open.

QUESTION 7.1. *Let G be a rank 1 simple Lie group and $\Gamma < G$ a lattice. Assume N is a finitely generated infinite normal subgroup of Γ. If G is* $\mathrm{Sp}(n, 1)$ *of* F_4^{-20}, *is*

N necessarily finite index? If G is $SU(n,1)$ *or* $SU(n,1)$ *is* Γ/N *necessarily a-(T)-menable?*

A-(T)-menability is a strong negation of property (T) introduced by Gromov. One way to prove the question would be to prove that for all G and all Γ and N, the group Γ/N is a-(T)-menable. This would resemble Margulis's proof of the normal subgroup theorem where a key step is proving the quotient group is amenable. For $SO(n,1)$ with $n>2$ and for all n, many lattices are known to have finitely generated normal subgroups N where Γ/N is abelian; see, for example, [Ago2] and [Kie]. For $SU(n,1)$, both abelian groups and surface groups are known to occur [Kap2, Sto]. For $\Gamma < SO(2,1)$ it is relatively elementary that there are no infinite index finitely generated normal subgroups. This author first learned a variant of this question around 2006 from Farb.

An older question related to Theorem 2.6 was raised in conversation between Zimmer and Margulis in the late 1970s. Given a lattice $\Gamma < G$, we say a subgroup $C < \Gamma$ is *commensurated* if $\Gamma < \mathrm{Comm}_G(C)$.

CONJECTURE 7.2. *Let G be a simple Lie group of real rank at least 2 and* $\Gamma < G$ *a lattice. Let N be a commensurated subgroup of* Γ. *Then N is either finite or finite index in* Γ.

For a fairly large set of nonuniform lattices the conjecture is known by work of Venkataramana and Shalom-Willis [SW, Ven]. Shalom and Willis also formulate a natural generalization for irreducible lattices in products including S arithmetic lattices. We do not include it here in the interest of brevity.

It has been known since the conversation between Margulis and Zimmer that Conjecture 7.2 can be formulated as a question about homomorphisms from Γ to a certain locally compact group that is a kind of completion of Γ with respect to N. Shalom and Willis prove their results on Conjecture 7.2 by proving a superrigidity theorem for homomorphisms of a special class of lattices to general locally compact groups. They also formulate an intriguing superrigidity conjecture for homomorphisms from any higher rank lattice Γ to locally compact groups. They demonstrate that their conjecture implies not only Conjecture 7.2 but also Theorems 2.2 and 2.6 and also the congruence subgroup conjecture; see [SW, conjecture 7.7] and the surrounding discussion.

In the context of that work, Shalom raised a question about an interesting analogue of Corollary 2.5.

QUESTION 7.3. *Let G be a simple Lie group and* Γ *a Zariski dense discrete subgroup. Assuming* $\mathrm{Comm}_G(\Gamma)$ *is not discrete, is* Γ *an arithmetic lattice?*

Since $\Gamma < \mathrm{Comm}_G(\Gamma)$, it is relatively easy to see that the simplicity of G implies this is equivalent to assuming $\mathrm{Comm}_G(\Gamma)$ is dense in G. For finitely generated subgroups of $SO(3, 1)$ the question was answered by Mj, building on work of Leininger, Long, and Reid [Mj, LLR]. For finitely generated subgroups in $SO(2, 1)$ the conjecture is easily resolved by noting the limit set is a proper closed subset of S^1. Mj also shows that in general it suffices to consider the case where the limit set is full. Recent work of Koberda and Mj studies the case where there is an arithmetic lattice Γ_0 such that $\Gamma \lhd \Gamma_0$ and resolves this case in many settings, including when Γ_0/Γ is abelian [KM1, KM2].

Another very interesting variant of the normal subgroup theorem was raised by Margulis in response to the proof of results of Abert et al. in [ABB$^+$1].

CONJECTURE 7.4. *Let G be a simple Lie group of real rank at least 2 and $\Gamma < G$ a discrete subgroup. Further assume that the injectivity radius is bounded on $K\backslash G/\Gamma$. Then Γ is a lattice in G.*

It is easy to see that this conjecture implies the normal subgroup theorem. The results in [ABB$^+$1] are proven using a theorem of Stuck and Zimmer, which itself is proven by using elements of Margulis's proof of the normal subgroup theorem; see [SZ] and [Zim2].[1]

The work of Stuck and Zimmer was the precursor of a long sequence of works by Nevo and Zimmer concerning actions of higher rank simple groups. Their most striking result is the following:

THEOREM 7.5 (Nevo-Zimmer).
Let G be a simple Lie group of real rank at least 2. Let μ be a measure on G whose support generates G and which is absolutely continuous with respect to Haar measure. Assume G acts on compact metric space X, and let ν be a μ-stationary measure on X. Then either ν is G invariant or there exists a ν-measurable G-equivariant map $X \to G/Q$ for $Q < G$, a proper parabolic subgroup.

One can view the Nevo-Zimmer theorem as providing complete obstructions to the existence of G-invariant measures in terms of *projective factors* G/Q. They prove a similar but slightly more technical result for actions of lattices $\Gamma < G$. One can view this as a tool for studying actions of G or Γ on compact manifolds M. A central element of the proof of Theorem 5.1 is finding enough Γ-invariant measures on M to control growth of derivatives of the group action. One difficulty for using Theorem 7.5 is that it is hard

[1] Conjecture 7.4 was resolved by Fraczyk and Gelander while this paper was in press.

to determine in practice when a measurable projective factor exists. Another difficulty is that to control growth of derivatives, one needs to control a wider class of measures than the μ-stationary ones.

In the proof of Theorem 5.1 on Zimmer's conjecture we use a different method of detecting invariant measures and projective factors for Γ actions on compact manifolds M that is more effective for applications. This is developed by Brown, Rodriguez Hertz, and Wang in [BRHW1]. Where Nevo and Zimmer follow Margulis and study G-invariant σ-algebras of measurable sets on X to find the projective factor, Brown, Rodriguez Hertz, and Wang instead study invariant measures. See the paper of Brown, Rodriguez Hertz, and Wang in this volume for an account of how to prove Theorem 2.6 by their methods [BRHW3].

Their approach is particularly intriguing since they earlier used a variant of the same method in place of Zimmer's cocycle superrigidity theorem [BRHW2]. The philosophy behind the approach is generally referred to as *Nonresonance implies invariance*. We close this section with a brief description of this philosophy and one implementation of it.

To apply this philosophy to actions of a lattice Γ, one always need to pass to the induced G action on $(G \times M)/\Gamma$. This allows one to use the structure of G—namely, the root data associated to a choice of Cartan subalgebra. To explain this philosophy better, we recall some basic facts. The Cartan subgroup A of G is the largest subgroup diagonalizable over $\mathbb{R}$; the Cartan subalgebra $\mathfrak{a}$ is its Lie algebra. It has been known since the work of Élie Cartan that a finite-dimensional linear representation ρ of G is completely determined by linear functionals on $\mathfrak{a}$ that arise as generalized eigenvalues of the restriction of ρ to A. Here we use that there is always a simultaneous eigenspace decomposition for groups of commuting symmetric matrices and that this makes the eigenvalues into linear functionals. These linear functionals are referred to as the *weights* of the representation. For the adjoint representation of G on its own Lie algebra, the weights are given the special name of *roots*. Corresponding to each root β there is a unipotent subgroup $G_\beta < G$ called a *root subgroup*; and it is well-known that large enough collections of root subgroups generate G. Two linear functionals are called *resonant* if one is a positive multiple of the other. Abstractly, given a G-action and an A-invariant object O, one may try to associate to O a class of linear functionals Ω. *Nonresonance implies invariance* is the observation that, given any root β of G that is not resonant to an element of Ω, the object O will automatically be invariant under the unipotent root group G^β. If one can find enough such nonresonant roots, the object O is automatically G-invariant. We will illustrate this philosphy by sketching the proof of the following theorem from [BRHW1].

THEOREM 7.6.

Let G be a simple Lie group of real rank at least 2, and let $\Gamma < G$ *be a lattice. Let Q be a maximal parabolic in G of minimal codimension. Assume M is a compact manifold and* $\rho : \Gamma \to \mathrm{Diff}(M)$ *and* $\dim(M) < \dim(G/Q)$. *Then* Γ *preserves a measure on M.*

We begin as above by inducing the action to a G action on $(G \times M)/\Gamma$ and noting that Γ-invariant measures on M correspond exactly to G-invariant measures on $(G \times M)/\Gamma$. Taking the minimal parabolic $P < G$ and using that P is amenable, one finds a P-invariant measure μ. The goal is to prove that μ is G-invariant. Once μ is G-invariant, disintegerating μ over the map $(G \times M)/\Gamma \to G/\Gamma$ yields a Γ-invariant measure on M. Since the measure μ is P-invariant and $A < P$, μ is also clearly invariant under the Cartan subgroup A, so one can try to apply the philosophy that nonresonance implies invariance by associating some linear functionals to the pair (A, μ). The linear functionals we consider are the Lyapunov exponents for the A-action.

More precisely we consider the Lyapunov exponents for the restriction of the derivative of A action to the subbundle F of $T((G \times M)/\Gamma)$ defined by directions tangent to the M fibers in that bundle over G/Γ. We refer to this collection of linear functionals as *fiberwise Lyapunov exponents*. In this context [BRHW1, proposition 5.3] shows that, given an A-invariant measure on X that projects to Haar measure on G/Γ, if a root β of G is not resonant with any fiberwise Lyapunov exponent, then the measure is invariant by the root subgroup G_β. The rest of the proof is quite simple. The stabilizer of μ contains P, which implies that the projection of μ to G/Γ is Haar measure, so the proposition just described applies. The stabilizer G_μ of μ in G is a closed subgroup containing P. We also know that G_μ contains the group generated by the G_β for all roots β not resonant with any fiberwise Lyapunov exponent. We also know that the number of distinct fiberwise Lyapunov exponents is bounded by the dimension of M. Since any closed subgroup of G containing P is parabolic, G_μ is parabolic. So either $G_\mu = G$ or the number of resonant roots needs to be at least the dimension of G/Q for Q, a maximal proper parabolic. This is because given any single root β with $G_\beta \not< Q$, the group generated by G_β and Q is G. Our assumption on the dimension of M immediately implies there are not enough fiberwise Lyapunov exponents to produce $\dim(G/Q)$ resonant roots, so μ is G-invariant.

We say a few words here on why this philosophy also works to prove superrigidity-type results. One view of the proof of superrigidity, introduced by Margulis in [Mar5, chapter VII] is that one starts with an A-invariant section of some vector bundle over G/Γ and then proceeds to produce a finite-dimensional space of sections that is G invariant. While the proof does not rely on the nonresonance condition, it should be clear that the objects considered in that proof might be amenable to an analysis like the one above.

8 Other criteria for a subgroup to be a lattice

Many of the results and conjectures discussed so far concern criteria for when a discrete subgroup Γ in G is actually a lattice or even an arithmetic lattice. We end this chapter by pointing to some more theorems and questions giving criteria for Zariski dense discrete subgroups to be lattices. One is the recent resolution by Benoist and Miquel of a conjecture of Margulis, building on the earlier work of Oh. Another is a question of Prasad and Spatzier that can be seen as similar to the Benoist-Miquel theorem. Finally, we mention a question of Nori, that is a variant of both of these phenomena and point to some other results on Nori's question by Chatterji and Venkataramana.

To state the theorem, we recall some definitions. Let G be a simple Lie group. It is possible to state a version of the theorem for G semisimple as well, but we avoid this for simplicity. A subgroup U is *horospherical* if it is the stable group of an element g in G—that is, $U := \{u \in G \mid \lim_{n\to\infty} g^n u g^{-n} = e\}$. Horospherical subgroups are always nilpotent, so a lattice $\Delta < U$ is always a discrete cocompact subgroup.

THEOREM 8.1 (Oh, Benoist-Miquel).
Let G be a simple Lie group of real rank at least 2 and $\Gamma < G$ be a discrete Zariski dense subgroup. Assume Γ contains a lattice Δ in some horospherical subgroup U of G. Then Γ is an arithmetic lattice in G.

This result was conjectured by Margulis, inspired by some elements of his original proof of arithmeticity for nonuniform lattices. Uniform lattices do not intersect unipotent subgroups, so the Γ appearing in the theorem is necessarily a nonuniform lattice. For many semisimple groups G, including all split groups but $SL(3,\mathbb{R})$, the result was proved by Hee Oh in her thesis [Oh1]. In subsequent work, including joint work with Benoist, Oh covered many additional cases [Oh2, BO1, BO2]. Recently Benoist and Miquel have presented a proof that works in full generality [BM].

We present a conjecture of Ralf Spatzier that has also been stated elsewhere as a question by Gopal Prasad. The conjecture is formally somewhat similar Theorem 8.1, but was more inspired by work on rank rigidity in differential geometry [Bal, BS]. To make the conjecture we require a definition. Given a countable group Γ, let A_i be the subset of Γ consisting of elements whose centralizer contains a free abelian subgroup of rank at most i as a finite index subgroup. The rank of Γ, sometimes called the Prasad-Raghnathan rank, is the minimal number i such that $\Gamma = \gamma_1 A_i \cup \cdots \cup \gamma_m A_i$ for some finite set $\gamma_1, \ldots, \gamma_m \in \Gamma$. Note that any torsion-free Γ has rank at least 1.

CONJECTURE 8.2. *Let G be a simple Lie group with real rank at least 2. Let Γ be a Zariski dense discrete subgroup of G whose rank is the real rank of G. Then Γ is a lattice in G.*

Spatzier also asked whether it was enough for $\Gamma < G$ to have rank at least 2 to be a lattice. It seems to be generally believed that the statement analogous to the one in the Beniost-Miquel theorem is false—that is, that there is an infinite covolume group $\Gamma < G$ such that some maximal diagonalizable subgroup $A < G$ such that $A \cap \Gamma$ is a lattice in A. One can attempt to do this by taking the group generated by some lattice $\Gamma_A < A$ and some other hyperbolic element $\gamma' \in G$, where γ' and Γ_A play ping pong in an appropriate sense. Verifying that this works seems somewhat tricky, and we do not know a complete argument.

As pointed out by Chatterji and Venkataramana [VC], there is a more general question of Nori related to Theorem 8.1.

QUESTION 8.3 (Nori, 1983). *If H is a real algebraic subgroup of a real semisimple algebraic group G, can one find sufficient conditions on H and G such that any Zariski dense discrete subgroup Γ of G that intersects H in a lattice in H is itself a lattice in G?*

Chatterji and Venkataramana conjecture that the answer to this question is yes in the case where G is simple and noncompact and $H < G$ is a proper simple noncompact subgroup. They also prove this conjecture in many cases when the real rank of H is at least 2. Key ingredients in their proof are borrowed from Margulis's proofs of superrigidity and arithmeticity. A case they leave open that is of particular interest is when $H = SL(2, \mathbb{R})$ and $G = \mathrm{SL}(n, \mathbb{R})$. Even the case $n = 3$ embedding $SL(2, \mathbb{R})$ in $SL(3, \mathbb{R})$.

References

[ABB+1] M. Abert, N. Bergeron, I. Biringer, T. Gelander, N. Nikolov, J. Raimbault, and I. Samet. On the growth of L^2-invariants for sequences of lattices in Lie groups. *Ann. of Math. (2)*, 185(3):711–790, 2017.

[ABB+2] M. Abert, N. Bergeron, I. Biringer, T. Gelander, N. Nikolov, J. Raimbault, and I. Samet. On the growth of L^2-invariants of locally symmetric spaces, II: exotic invariant random subgroups in rank one. *Int. Math. Res. Not. IMRN* (9): 2588–2625, 2020.

[Ago1] I. Agol. Systoles of hyperbolic 4-manifolds. *Preprint* 2006.

[Ago2] I. Agol. The virtual Haken conjecture. *Doc. Math.*, 18:1045–1087, 2013. *With an appendix by Agol, Daniel Groves, and Jason Manning.*

[BCL] U. Bader, P.-E. Caprace, and J. Lécureux. On the linearity of lattices in affine buildings and ergodicity of the singular Cartan flow. *J. Amer. Math. Soc.*, 32(2):491–562, 2019.

[BFMS1] U. Bader, D. Fisher, N. Miller, and M. Stover. Arithmeticity, superrigidity, and totally geodesic submanifolds. *Preprint* 2019.

[BFMS2] U. Bader, D. Fisher, N. Miller, and M. Stover. Arithmeticity, superrigidity and totally geodesic submanifolds II: $SU(n,1)$. *Preprint*, 2020.

[BF1] U. Bader and A. Furman. Superrigidity for non-lattice. *Preprint*, 2018.

[BF2] U. Bader and A. Furman. Super-rigidity and non-linearity for lattices in products. *Compos. Math.*, 156(1):158–178, 2020.

[BF3] U. Bader and A. Furman. An extension of Margulis's superrigidity theorem. In *Dynamics, geometry, number theory: The impact of Margulis on modern mathematics*, edited by David Fisher, Dmitry Kleinbock, and Gregory Soifer, pages 47–65. University of Chicago Press, Chicago, 2021.

[BS] U. Bader and Y. Shalom. Factor and normal subgroup theorems for lattices in products of groups. *Invent. Math.*, 163(2):415–454, 2006.

[Bal] W. Ballmann. Nonpositively curved manifolds of higher rank. *Ann. of Math. (2)*, 122(3):597–609, 1985.

[BHH] G. Barthel, F. Hirzebruch, and T. Höfer. *Geradenkonfigurationen und Algebraische Flächen.* Aspects of Mathematics, D4. Friedr. Vieweg & Sohn, Braunschweig, 1987.

[BL] H. Bass and A. Lubotzky. Nonarithmetic superrigid groups: counterexamples to Platonov's conjecture. *Ann. of Math. (2)*, 151(3):1151–1173, 2000.

[Bel] M. Belolipetsky. Arithmetic hyperbolic reflection groups. *Bull. Amer. Math. Soc. (N.S.)*, 53(3):437–475, 2016.

[BT] M. V. Belolipetsky and S. A. Thomson. Systoles of hyperbolic manifolds. *Algebr. Geom. Topol.*, 11(3):1455–1469, 2011.

[BM] Y. Benoist and S. Miquel. Arithmeticity of discrete subgroups containing horospherical lattices. *Duke Math. J.* 169(8):1485–1539, 2020.

[BO1] Y. Benoist and H. Oh. Discrete subgroups of $SL_3(\mathbb{R})$ generated by triangular matrices. *Int. Math. Res. Not. IMRN*, (4):619–632, 2010.

[BO2] Y. Benoist and H. Oh. Discreteness criterion for subgroups of products of SL(2). *Transform. Groups*, 15:503–515, 2010.

[BF] E. J. Benveniste and D. Fisher. Nonexistence of invariant rigid structures and invariant almost rigid structures. *Comm. Anal. Geom.*, 13(1):89–111, 2005.

[BCG] G. Besson, G. Courtois, and S. Gallot. Lemme de Schwarz réel et applications géométriques. *Acta Math.*, 183(2):145–169, 1999.

[Bor] A. Borel. Density and maximality of arithmetic subgroups. *J. Reine Angew. Math.*, 224:78–89, 1966.

[BW] A. Borel and N. Wallach. *Continuous cohomology, discrete subgroups, and representations of reductive groups*, volume 67 of *Mathematical surveys and monographs.* American Mathematical Society, Providence, RI, second edition, 2000.

[BFH1] A. Brown, D. Fisher, and S. Hurtado. Zimmer's conjecture: Subexponential growth, measure rigidity, and strong property (T). *Preprint*, 2016. *arXiv:1608.04995*.

[BFH2] A. Brown, D. Fisher, and S. Hurtado. Zimmer's conjecture for actions of $SL(m, \mathbb{Z})$. *Invent. Math.* 221(3):1001–1060, 2020.

[BFH3] A. Brown, D. Fisher, and S. Hurtado. Zimmer's conjecture for non-uniform lattices and escape of mass. *Preprint*, 2019.

[BRHW1] A. Brown, F. Rodriguez Hertz, and Z. Wang. Invariant measures and measurable projective factors for actions of higher-rank lattices on manifolds. *Preprint*, 2016. arXiv:1609.05565.

[BRHW2] A. Brown, F. Rodriguez Hertz, and Z. Wang. Global smooth and topological rigidity of hyperbolic lattice actions. *Ann. of Math. (2)*, 186(3):913–972, 2017.

[BRHW3] A. Brown, F. Rodriguez Hertz, and Z. Wang. The normal subgroup theorem through measure rigidity. *Preprint*, 2019.

[BT] F. Bruhat and J. Tits. Schémas en groupes et immeubles des groupes classiques sur un corps local. *Bull. Soc. Math. France*, 112(2):259–301, 1984.

[BKT] Y. Brunebarbe, B. Klingler, and B. Totaro. Symmetric differentials and the fundamental group. *Duke Math. J.*, 162(14):2797–2813, 2013.

[Bur] M. Burger. Rigidity properties of group actions on CAT(0)-spaces. In *Proceedings of the International Congress of Mathematicians, Vol. 1, 2 (Zürich, 1994)*, pages 761–769. Birkhäuser, Basel, 1995.

[BI] M. Burger and A. Iozzi. A measurable Cartan theorem and applications to deformation rigidity in complex hyperbolic geometry. *Pure Appl. Math. Q.*, 4(1, Special Issue: In honor of Grigory Margulis. Part 2):181–202, 2008.

[BM1] M. Burger and S. Mozes. Groups acting on trees: from local to global structure. *Inst. Hautes Études Sci. Publ. Math.*, (92):113–150, 2000.

[BM2] M. Burger and S. Mozes. Lattices in product of trees. *Inst. Hautes Études Sci. Publ. Math.*, (92):151–194, 2000.

[BMZ] M. Burger, S. Mozes, and R. J. Zimmer. Linear representations and arithmeticity of lattices in products of trees. In *Essays in geometric group theory*, volume 9 of *Ramanujan Mathematical Society lecture notes series*, pages 1–25. Ramanujan Mathematical Society Mysore, 2009.

[BS] K. Burns and R. Spatzier. Manifolds of nonpositive curvature and their buildings. *Inst. Hautes Études Sci. Publ. Math.*, (65):35–59, 1987.

[CG] G. Cairns and E. Ghys. The local linearization problem for smooth SL(n)-actions. *Enseign. Math. (2)*, 43(1–2):133–171, 1997.

[Cal] E. Calabi. On compact, Riemannian manifolds with constant curvature. I. In *Proc. Sympos. Pure Math., Vol. III*, pages 155–180. American Mathematical Society, Providence, RI, 1961.

[CV] E. Calabi and E. Vesentini. On compact, locally symmetric Kähler manifolds. *Ann. of Math. (2)*, 71:472–507, 1960.

[Can] S. Cantat. Progrès récents concernant le programme de Zimmer [d'après A. Brown, D. Fisher, et S. Hurtado]. *Preprint*, 2017.

[CR] P.-E. Caprace and B. Rémy. Simplicity and superrigidity of twin building lattices. *Invent. Math.*, 176(1):169–221, 2009.

[Car] E. Cartan. Sur le groupe de la géométrie hypersphérique. *Comment. Math. Helv.*, 4(1):158–171, 1932.

[CS] D. I. Cartwright and T. Steger. Enumeration of the 50 fake projective planes. *C. R. Math. Acad. Sci. Paris*, 348(1–2):11–13, 2010.

[Cor1] K. Corlette. Flat *G*-bundles with canonical metrics. *J. Differential Geom.*, 28(3):361–382, 1988.

[Cor2] K. Corlette. Archimedean superrigidity and hyperbolic geometry. *Ann. of Math. (2)*, 135(1):165–182, 1992.

[CHL] W. Couwenberg, G. Heckman, and E. Looijenga. Geometric structures on the complement of a projective arrangement. *Publ. Math. Inst. Hautes Études Sci.*, (101):69–161, 2005.

[DM] P. Deligne and G. D. Mostow. Monodromy of hypergeometric functions and nonlattice integral monodromy. *Inst. Hautes Études Sci. Publ. Math.*, (63):5–89, 1986.

[Der] M. Deraux. A new non-arithmetic lattice in $\mathrm{PU}(3,1)$. *Algebr. Geom. Topol.*, 20(2):925–963, 2020.

[DPP1] M. Deraux, J. R. Parker, and J. Paupert. New non-arithmetic complex hyperbolic lattices II. *Preprint* 2016.

[DPP2] M. Deraux, J. R. Parker, and J. Paupert. New non-arithmetic complex hyperbolic lattices. *Invent. Math.*, 203(3):681–771, 2016.

[Don] S. K. Donaldson. Twisted harmonic maps and the self-duality equations. *Proc. London Math. Soc. (3)*, 55(1):127–131, 1987.

[EG1] H. Esnault and M. Groechenig. Cohomologically rigid local systems and integrality. *Selecta Math. (N.S.)*, 24(5):4279–4292, 2018.

[EG2] H. Esnault and M. Groechenig. Rigid connections and F-isocrystals. *Acta Math.* 225(1):103–158, 2020.

[Fis1] D. Fisher. Deformations of group actions. *Trans. Amer. Math. Soc.*, 360(1):491–505, 2008.

[Fis2] D. Fisher. Groups acting on manifolds: around the Zimmer program. In *Geometry, rigidity, and group actions*, Chicago Lectures in Mathematics, pages 72–157. University of Chicago Press, Chicago, IL, 2011.

[Fis3] D. Fisher. Recent progress in the Zimmer program. In *Group actions in ergodic theory, geometry, and topology*. University of Chicago Press, Chicago, 2019.

[FLMS] D. Fisher, J.-F. Lafont, N. Miller, and M. Stover. Finiteness of maximal geodesic submanifolds in hyperbolic hybrids. *Forthcoming, J. Eur. Math. Soc.*

[FLSS] D. Fisher, M. Larsen, R. Spatzier, and M. Stover. Character varieties and actions on products of trees. *Israel J. Math.*, 225(2):889–907, 2018.

[FM] D. Fisher and G. A. Margulis. Local rigidity for cocycles. In *Surveys in differential geometry, Vol. VIII (Boston, MA, 2002)*, pages 191–234. International Press, Somerville, MA, 2003.

[FW] D. Fisher and K. Whyte. Continuous quotients for lattice actions on compact spaces. *Geom. Dedicata*, 87(1–3):181–189, 2001.

[Fur1] A. Furman. Gromov's measure equivalence and rigidity of higher rank lattices. *Ann. of Math. (2)*, 150(3):1059–1081, 1999.

[Fur2] A. Furman. Orbit equivalence rigidity. *Ann. of Math. (2)*, 150(3):1083–1108, 1999.

[Fur3] A. Furman. A survey of measured group theory. In *Geometry, rigidity, and group actions*, Chicago Lectures in Math., pages 296–374. University of Chicago Press, Chicago, IL, 2011.

[GKM] T. Gelander, A. Karlsson, and G. A. Margulis. Superrigidity, generalized harmonic maps and uniformly convex spaces. *Geom. Funct. Anal.*, 17(5):1524–1550, 2008.

[GL] T. Gelander and A. Levit. Counting commensurability classes of hyperbolic manifolds. *Geom. Funct. Anal.*, 24(5):1431–1447, 2014.

[GP1] S. Ghazouani and L. Pirio. Moduli spaces of flat tori and elliptic hypergeometric functions. *Preprint*, 2016.

[GP2] S. Ghazouani and L. Pirio. Moduli spaces of flat tori with prescribed holonomy. *Geom. Funct. Anal.*, 27(6):1289–1366, 2017.

[Gro1] M. Gromov. Hyperbolic groups. In *Essays in group theory*, volume 8 of *Math. Sci. Res. Inst. Publ.*, pages 75–263. Springer, New York, 1987.

[Gro2] M. Gromov. Rigid transformations groups. In *Géométrie différentielle (Paris, 1986)*, volume 33 of *Travaux en Cours*, pages 65–139. Hermann, Paris, 1988.

[GPS] M. Gromov and I. Piatetski-Shapiro. Nonarithmetic groups in Lobachevsky spaces. *Inst. Hautes Études Sci. Publ. Math.*, (66):93–103, 1988.

[GS] M. Gromov and R. Schoen. Harmonic maps into singular spaces and *p*-adic superrigidity for lattices in groups of rank one. *Inst. Hautes Études Sci. Publ. Math.*, (76):165–246, 1992.

[HLM] H. M. Hilden, M. T. Lozano, and J. M. Montesinos. On knots that are universal. *Topology*, 24(4):499–504, 1985.

[Kap1] M. Kapovich. Lectures on complex hyperbolic Kleinian groups. *Preprint* 2019.

[Kap2] M. Kapovich. On normal subgroups in the fundamental groups of complex surfaces. *Preprint* https://arxiv.org/abs/math/9808085, 1998.

[KL] A. Katok and J. Lewis. Global rigidity results for lattice actions on tori and new examples of volume-preserving actions. *Israel J. Math.*, 93:253–280, 1996.

[KRH] A. Katok and F. Rodriguez Hertz. Arithmeticity and topology of smooth actions of higher rank abelian groups. *J. Mod. Dyn.*, 10:135–172, 2016.

[Kie] D. Kielak. Residually finite rationally solvable groups and virtual fibring. *Preprint*, 2018.

[Kli1] B. Klingler. Sur la rigidité de certains groupes fondamentaux, l'arithméticité des réseaux hyperboliques complexes, et les "faux plans projectifs." *Invent. Math.*, 153(1):105–143, 2003.

[Kli2] B. Klingler. Symmetric differentials, Kähler groups and ball quotients. *Invent. Math.*, 192(2):257–286, 2013.

[KM1] T. Koberda and M. Mj. Commutators, commensurators, and $PSL_2(\mathbb{Z})$. *Preprint*, 2018.

[KM2] T. Koberda and M. Mj. Commensurators of thin normal subgroups. *Preprint*, 2019.

[KM3] V. Koziarz and J. Maubon. Maximal representations of uniform complex hyperbolic lattices. *Ann. of Math. (2)*, 185(2):493–540, 2017.

[Lab1] F. Labourie. Existence d'applications harmoniques tordues à valeurs dans les variétés à courbure négative. *Proc. Amer. Math. Soc.*, 111(3):877–882, 1991.

[Lab2] F. Labourie. Large groups actions on manifolds. In *Proceedings of the International Congress of Mathematicians, Vol. II (Berlin, 1998)*, *Doc. Math.* Extra Vol. II, 371–380, 1998.

[Laf] L. Lafforgue. Chtoucas de Drinfeld et correspondance de Langlands. *Invent. Math.*, 147(1):1–241, 2002.

[LLR] C. Leininger, D. D. Long, and A. W. Reid. Commensurators of finitely generated nonfree Kleinian groups. *Algebr. Geom. Topol.*, 11(1):605–624, 2011.

[Lub] A. Lubotzky. Some more non-arithmetic rigid groups. In *Geometry, spectral theory, groups, and dynamics*, volume 387 of *Contemp. Math.*, pages 237–244. Amer. Math. Soc., Providence, RI, 2005.

[MR] C. Maclachlan and A. W. Reid. Commensurability classes of arithmetic Kleinian groups and their Fuchsian subgroups. *Math. Proc. Cambridge Philos. Soc.*, 102(2):251–257, 1987.

[Mak] V. S. Makarov. On a certain class of discrete groups of Lobačevskiĭspace having an infinite fundamental region of finite measure. *Dokl. Akad. Nauk SSSR*, 167:30–33, 1966.

[Mar1] G. A. Margulis. The action of unipotent groups in a lattice space. *Mat. Sb. (N.S.)*, 86(128):552–556, 1971.

[Mar2] G. A. Margulis. Arithmeticity of nonuniform lattices in weakly noncompact groups. *Funkcional. Anal. i Priložen.*, 9(1):35–44, 1975.

[Mar3] G. A. Margulis. Discrete groups of motions of manifolds of nonpositive curvature. In *Proceedings of the International Congress of Mathematicians (Vancouver, B.C., 1974), Vol. 2*, pages 21–34, 1975.

[Mar4] G. A. Margulis. Factor groups of discrete subgroups and measure theory. *Funktsional. Anal. i Prilozhen.*, 12(4):64–76, 1978.

[Mar5] G. A. Margulis. *Discrete subgroups of semisimple Lie groups*, volume 17 of *Ergebnisse der Mathematik und ihrer Grenzgebiete (3) [Results in mathematics and related areas (3)]*. Springer-Verlag, Berlin, 1991.

[Mar6] G. Margulis. Problems and conjectures in rigidity theory. In *Mathematics: frontiers and perspectives*, pages 161–174. American Mathematical Society, Providence, RI, 2000.

[MM] G. Margulis and A. Mohammadi. Arithmeticity of hyperbolic 3-manifolds containing infinitely many totally geodesic surfaces. *Preprint* 2019.

[Mil] O. Mila. The trace field of hyperbolic gluings.

[Mj] M. Mj. On discreteness of commensurators. *Geom. Topol.*, 15(1):331–350, 2011.

[Mon] N. Monod. Superrigidity for irreducible lattices and geometric splitting. *J. Amer. Math. Soc.*, 19(4):781–814, 2006.

[Mos1] G. D. Mostow. *Strong rigidity of locally symmetric spaces. Annals of mathematics studies, No. 78*. Princeton University Press, Princeton, NJ, 1973.

[Mos2] G. D. Mostow. On a remarkable class of polyhedra in complex hyperbolic space. *Pacific J. Math.*, 86(1):171–276, 1980.

[Mos3] G. D. Mostow. Generalized Picard lattices arising from half-integral conditions. *Inst. Hautes Études Sci. Publ. Math.*, (63):91–106, 1986.

[Mos4] G. D. Mostow. Selberg's work on the arithmeticity of lattices and its ramifications. In *Number theory, trace formulas and discrete groups (Oslo, 1987)*, pages 169–183. Academic Press, Boston, 1989.

[Oh1] H. Oh. Discrete subgroups generated by lattices in opposite horospherical subgroups. *J. Algebra*, 203(2):621–676, 1998.

[Oh2] H. Oh. On discrete subgroups containing a lattice in a horospherical subgroup. *Israel J. Math.*, 110:333–340, 1999.

[Pau] J. Paupert. Non-discrete hybrids in SU(2, 1). *Geom. Dedicata*, 157:259–268, 2012.

[PW] J. Paupert and J. Wells. Hybrid lattices and thin subgroups of Picard modular groups. *Topology Appl.*, 269:106918, 2020.

[Poz] M. B. Pozzetti. Maximal representations of complex hyperbolic lattices into $SU(M, N)$. *Geom. Funct. Anal.*, 25(4):1290–1332, 2015.

[PY] G. Prasad and S.-K. Yeung. Fake projective planes. *Invent. Math.*, 168(2):321–370, 2007.

[Rei1] A. W. Reid. Arithmeticity of knot complements. *J. London Math. Soc. (2)*, 43(1):171–184, 1991.

[Rei2] A. W. Reid. Totally geodesic surfaces in hyperbolic 3-manifolds. *Proc. Edinburgh Math. Soc. (2)*, 34(1):77–88, 1991.

[Sch] R. Schwartz. Notes on shapes of polyhedra. *Preprint*, 2015.

[Sel] A. Selberg. On discontinuous groups in higher-dimensional symmetric spaces. In *Contributions to function theory (Internat. Colloq. Function Theory, Bombay, 1960)*, pages 147–164. Tata Institute of Fundamental Research, Bombay, 1960.

[Sha1] Y. Shalom. Rigidity of commensurators and irreducible lattices. *Invent. Math.*, 141(1):1–54, 2000.

[Sha2] Y. Shalom. Rigidity, unitary representations of semisimple groups, and fundamental groups of manifolds with rank one transformation group. *Ann. of Math. (2)*, 152(1):113–182, 2000.

[SW] Y. Shalom and G. A. Willis. Commensurated subgroups of arithmetic groups, totally disconnected groups and adelic rigidity. *Geom. Funct. Anal.*, 23(5):1631–1683, 2013.

[Sim] C. T. Simpson. Higgs bundles and local systems. *Inst. Hautes Études Sci. Publ. Math.*, (75):5–95, 1992.

[Siu] Y. T. Siu. The complex-analyticity of harmonic maps and the strong rigidity of compact Kähler manifolds. *Ann. of Math. (2)*, 112(1):73–111, 1980.

[Sto] M. Stover. Cusp and b1 growth for ball quotients and maps onto Z with finitely generated kernel. *Preprint*, 2018.

[Stu1] G. Stuck. Low-dimensional actions of semisimple groups. *Israel J. Math.*, 76(1-2):27–71, 1991.

[Stu2] G. Stuck. Minimal actions of semisimple groups. *Ergodic Theory Dynam. Systems*, 16(4):821–831, 1996.

[SZ] G. Stuck and R. J. Zimmer. Stabilizers for ergodic actions of higher rank semisimple groups. *Ann. of Math. (2)*, 139(3):723–747, 1994.

[Thu] W. P. Thurston. Shapes of polyhedra and triangulations of the sphere. In *The Epstein birthday schrift*, volume 1 of *Geometry and Topology Monographs*, pages 511–549. Geometry & Topology Publications, Coventry, 1998.

[Tol] D. Toledo. Representations of surface groups in complex hyperbolic space. *J. Differential Geom.*, 29(1):125–133, 1989.

[Uch] F. Uchida. Classification of real analytic $SL(n, \mathbf{R})$ actions on n-sphere. *Osaka Math. J.*, 16(3):561–579, 1979.

[Vee] W. A. Veech. Flat surfaces. *Amer. J. Math.*, 115(3):589–689, 1993.

[Ven] T. N. Venkataramana. On superrigidity and arithmeticity of lattices in semisimple groups over local fields of arbitrary characteristic. *Invent. Math.*, 92(2):255–306, 1988.

[VC] T. N. Venkataramana and I. Chatterji. Discrete linear groups containing arithmetic groups. *Preprint*, 2009.

[Vin1] E. B. Vinberg. Discrete groups generated by reflections in Lobačevskiĭ spaces. *Mat. Sb. (N.S.)*, 72(114):471–488, 1966; correction, ibid. 73(115):303, 1967.

[Vin2] E. B. Vinberg. The nonexistence of crystallographic reflection groups in Lobachevskiĭ spaces of large dimension. *Funktsional. Anal. i Prilozhen.*, 15(2):67–68, 1981.

[Vin3] E. B. Vinberg. Non-arithmetic hyperbolic reflection groups in higher dimensions. *Mosc. Math. J.*, 15(3):593–602, 606, 2015.

[Wei1] A. Weil. On discrete subgroups of Lie groups. *Ann. of Math. (2)*, 72:369–384, 1960.

[Wei2] A. Weil. On discrete subgroups of Lie groups. II. *Ann. of Math. (2)*, 75:578–602, 1962.

[Wel] J. Wells. Non-arithmetic hybrid lattices in PU(2, 1). *Preprint*, 2019.

[Yau] S. T. Yau. Calabi's conjecture and some new results in algebraic geometry. *Proc. Nat. Acad. Sci. U.S.A.*, 74(5):1798–1799, 1977.

[Zim1] R. J. Zimmer. Strong rigidity for ergodic actions of semisimple Lie groups. *Ann. of Math. (2)*, 112(3):511–529, 1980.

[Zim2] R. J. Zimmer. Ergodic theory, semisimple Lie groups, and foliations by manifolds of negative curvature. *Inst. Hautes Études Sci. Publ. Math.*, (55):37–62, 1982.

[Zim3] R. J. Zimmer. Arithmetic groups acting on compact manifolds. *Bull. Amer. Math. Soc. (N.S.)*, 8(1):90–92, 1983.

[Zim4] R. J. Zimmer. *Ergodic theory and semisimple groups*, volume 81 of *Monographs in mathematics*. Birkhäuser Verlag, Basel, 1984.

[Zim5] R. J. Zimmer. Actions of semisimple groups and discrete subgroups. In *Proceedings of the International Congress of Mathematicians, Vol. 1, 2 (Berkeley, Calif., 1986)*, pages 1247–1258. American Mathematical Society, Providence, RI, 1987.

[Zim6] R. J. Zimmer. Lattices in semisimple groups and invariant geometric structures on compact manifolds. In *Discrete groups in geometry and analysis (New Haven, Conn., 1984)*, volume 67 of *Progr. Math.*, pages 152–210. Boston: Birkhäuser, Boston, Boston, MA, 1987.

2 AN EXTENSION OF MARGULIS'S SUPERRIGIDITY THEOREM

URI BADER AND ALEX FURMAN

To Gregory Margulis with gratitude and admiration

Abstract. We give an extension of Margulis's superrigidity for higher rank lattices. In our approach the target group could be defined over any complete valued field. Our proof is based on the notion of Algebraic Representation of Ergodic Actions.

1 Introduction

In this essay we present a proof of Margulis's Superrigidity theorem [10, Theorem 7.5.6] with algebraic target groups defined over valued fields that are not necessarily local.

THEOREM 1.1 (Margulis's superrigidity for arbitrary target fields). *Let ℓ be a local field, $H=\mathbf{H}(\ell)$ be the locally compact group formed by the ℓ-points of a connected, semisimple, algebraic group defined over ℓ. Assume that the ℓ-rank of $\mathbf{H}$ is at least two. Let $\Gamma < H$ be a lattice, and assume that the projection of Γ in H/N is nondiscrete whenever $N \lhd H$ is the ℓ-points of a proper normal ℓ-isotropic subgroup.*

Let k be a field with an absolute value, so that as a metric space k is complete. Let $G=\mathbf{G}(k)$ be the k-points of a connected, adjoint, k-simple, algebraic group $\mathbf{G}$ defined over k. Let $\rho:\Gamma\to G$ be a homomorphism, and assume that $\rho(\Gamma)$ is Zariski dense and unbounded in G. Then there exists a unique continuous homomorphism $\hat{\rho}:H\to G$ such that $\rho=\hat{\rho}|_\Gamma$.

Uri Bader. Weizmann Institute, Rehovot, Israel
bader@weizmann.ac.il

Alex Furman. University of Illinois at Chicago, Chicago, USA
furman@uic.edu

Bader and Furman were supported in part by the BSF grant 2008267.
Bader was supported in part by the ISF grant 704/08.
Furman was supported in part by the NSF grant DMS 1611765.

Note that the homomorphism $\hat{\rho}$ appearing in Theorem 1.1 is necessarily given by some algebraic data—this will be properly explained in Corollary 8.1. Theorem 1.1 has a generalization to the so-called S-arithmetic case; in that case the group H is assumed to be a product of semisimple algebraic groups over different local fields. However, this generalization already follows from our result regarding superrigidity for irreducible lattices in products of general locally compact groups [3, theorem 1.2] (alternatively, see [11] or [8]), so we do not discuss it here. Likewise, in proving Theorem 1.1 the case where H has more than one noncompact factor follows essentially from [3, theorem 1.2]. Thus, our main concern here is the case where H has a unique noncompact factor. This case will follow from Theorem 1.3 below. Before stating this theorem we present two properties of topological groups.

- **(A)** We say that a topological group S satisfies condition (A) if every continuous, isometric S-action without global fixed points on a metric space is topologically proper.
- **(B)** We say that a topological group S satisfies condition (B) if S is topologically generated by closed noncompact subgroups $T_0, \dots, T_n$ such that, in a cyclic order, for every i, T_{i+1} normalizes T_i and at least one of the T_i's is amenable.

EXAMPLE 1.2. Let ℓ be a local field. Let $H = \mathbf{H}(\ell)$ be the ℓ-points of a connected almost simple algebraic group $\mathbf{H}$ defined over ℓ. If $\mathbf{H}$ is simply connected, then, by [4, theorem 6.1], H satisfies condition (A). Furthermore, if the ℓ-rank of $\mathbf{H}$ is at least two, then H satisfies condition (B). In fact, the sequence of subgroups T_i could be chosen from the root groups, properly ordered. For example, for $H = \mathrm{SL}_3(\ell)$, we can use the sequence of subgroups

$$\begin{bmatrix} 1 & * & \\ & 1 & \\ & & 1 \end{bmatrix}, \begin{bmatrix} 1 & & * \\ & 1 & \\ & & 1 \end{bmatrix}, \begin{bmatrix} 1 & & \\ & 1 & * \\ & & 1 \end{bmatrix}, \begin{bmatrix} 1 & & \\ * & 1 & \\ & & 1 \end{bmatrix}, \begin{bmatrix} 1 & & \\ & 1 & \\ * & & 1 \end{bmatrix}, \begin{bmatrix} 1 & & \\ & 1 & \\ & * & 1 \end{bmatrix}.$$

THEOREM 1.3.

Let S be a second countable locally compact topological group, and let $\Gamma < S$ be a lattice. Assume S satisfies conditions (A) and (B).

Let k be a field with an absolute value, so that as a metric space k is complete. Let $G = \mathbf{G}(k)$ be the k-points of a connected, adjoint, k-simple algebraic group $\mathbf{G}$ defined over k. Let $\rho : \Gamma \to G$ be a homomorphism, and assume that $\rho(\Gamma)$ is Zariski dense and unbounded in G. Then there exists a unique continuous homomorphism $\hat{\rho} : S \to G$ such that $\rho = \hat{\rho}|_\Gamma$.

The proof of Theorem 1.3 will be given in section 3, and the detailed reduction of Theorem 1.1 to Theorem 1.3 will be carried out in section 7. Currently, we do not know examples of locally compact groups satisfying conditions (A) and (B) that are not, essentially, higher rank semi-simple groups over local fields. Nevertheless we find the formulation of Theorem 1.3 useful not only for its potential applications but also for psychological reasons, as it clarifies the different role played by the topological group S and the algebraic group G.

1.1 ACKNOWLEDGMENTS AND DISCLAIMERS. The content of this essay is essentially contained in our manuscript [2], which we do not intend to publish as, in retrospect, we find it hard to read. In our presentation here we do rely on [3], which contains other parts of the content of [2]. The manuscript contains further results regarding general cocycle superrigidity à la Zimmer, on which we intend to elaborate in a forthcoming paper. We also rely here on the foundational work done in [1]. In our discussion in section 6 and in the reduction of Theorem 1.1 to Theorem 1.3 given in section 7, we rely heavily on the work of Borel and Tits, which we refer to via [10].

It is our pleasure to thank Bruno Duchesne and Jean Lécureux for their contribution to this project. We are grateful to Michael Puschnigg for spotting an inaccuracy in the definition of a morphism of T-algebraic representations in an early draft of [2]. We would also like to thank Tsachik Gelander for numerous discussions. Above all, we owe a huge mathematical debt to Gregory Margulis, whose incredible insight is reflected everywhere in this work.

2 Ergodic theoretical preliminaries

In this section we set our ergodic theoretical framework and notations. Recall that a *Polish space* is a topological space that is homeomorphic to a complete separable metric space. By a *measurable space* we mean a set endowed with a σ-algebra. A *standard Borel space* is a measurable space that admits a measurable bijection to a Polish topological space, equipped with the σ-algebra generated by its topology. A *Lebesgue space* is a standard Borel space endowed with the measure class of a probability measure and the completion of the Borel σ-algebra obtained by adding all subsets of null sets. For a Lebesgue space X, we denote by $L^1(X)$, $L^2(X)$, and $L^\infty(X)$ the Banach spaces of equivalence classes of integrable, square-integrable, and bounded measurable functions $X \to \mathbb{C}$, respectively; where two functions are equivalent if they agree, *a.e.* We will also consider the space $L^0(X)$, consisting of equivalence classes of all measurable functions $X \to \mathbb{C}$. Endowing this space with the topology of convergence in measure, this is a Polish topological space.

Every coset space of a locally compact second countable group is a Lebesgue space when endowed with its Haar measure class. Unless otherwise stated, we will always regard the Haar measure class when considering locally compact second countable groups or their coset spaces as Lebesgue spaces. Given a locally compact second countable group S, a *Lebesgue S-space* is a Lebesgue space X endowed with a measurable and measure class preserving action of S.

Let S be a locally compact second countable group and X be a Lebesgue S-space. Then S acts on $L^\infty(X)$ via $sf(x) = f(s^{-1}x)$. This S-action is isometric, but in general it is not continuous. However, it is continuous when $L^\infty(X)$ is taken with the weak-topology induced by $L^1(X)$. The action of S on X is said to be *ergodic* if the only S-invariant function classes in $L^\infty(X)$ are the constant ones.

If the S-action preserves a finite measure in the given measure class on X, we say that the S-action is *finite measure preserving*. In such a case, the S-isometric action on $L^\infty(X)$ extends to an S-isometric action on $L^2(X)$, which is norm continuous, hence unitary. For finite measure preserving actions, the S-action on X is ergodic if and only if the only invariant function classes in $L^2(X)$ are the constant ones.

The action of S on X is said to be *metrically ergodic* if for every separable metric space (U, d) on which S acts continuously by isometries, any a.e. defined S-equivariant map $\phi : X \to U$ is essentially constant.

EXAMPLE 2.1. Let S be a locally compact second countable group satisfying condition (A) and let $T < S$ be a noncompact closed subgroup. Then the action of S on S/T is metrically ergodic.

Recall that a finite measure preserving action of S on X is *weakly mixing* if the diagonal S-action on $X \times X$ is ergodic. Weak mixing is equivalent to the condition that the only S-invariant finite-dimensional subspace of $L^2(X)$ is the constant functions. For finite measure preserving actions, metric ergodicity is equivalent to weak mixing (cf. [9, theorem 2.1]).

LEMMA 2.2. *Let S be a locally compact second countable group satisfying condition (A) and let X be a finite measure preserving S-ergodic Lebesgue space. Then for every noncompact closed subgroup $T < S$, the restricted T-action on X is weakly mixing.*

Proof. The isometric action of S on the unit sphere U of the orthogonal complement of the constant functions in $L^2(X)$ is continuous and has no fixed points, by the ergodicity assumption. Therefore the S-action on U is proper. It follows that there is no T-invariant compact subset in U. Therefore T has no finite-dimensional subrepresentations in $L^2(X)$ except for the constant functions. □

The action of S on X is said to be *amenable* if for every S-Borel space V and an essentially surjective S-equivariant Borel map $\pi : V \to X$ with compact convex fibers, such that the S-action restricted to the fibers is by continuous affine maps, one has an a.e. defined S-invariant measurable section (see [12, definition 4.3.1]).

EXAMPLE 2.3 ([12, proposition 4.3.2]). Let $T < S$ be an amenable closed subgroup. Then the action of S on S/T is amenable.

3 Algebraic representation of ergodic actions

In this section we fix a field k with a nontrivial absolute value that is separable and complete (as a metric space) and a k-algebraic group $\mathbf{G}$. We note that $\mathbf{G}(k)$, when endowed with the k-analytic topology, is a Polish topological group; see [1, proposition 2.2]. We also fix a locally compact second countable group T and a homomorphism $\tau : T \to \mathbf{G}(k)$, which is continuous, considering $\mathbf{G}(k)$ with its analytic topology. Let us also fix a Lebesgue T-space X.

DEFINITION 3.1. Given all the data above, an *algebraic representation* of X consists of a k-$\mathbf{G}$-algebraic variety $\mathbf{V}$ and an a.e. defined measurable map $\phi : X \to \mathbf{V}(k)$ such that for every $t \in T$ and for a.e. $x \in X$,

$$\phi(tx) = \tau(t)\phi(x).$$

We shall say that $\mathbf{V}$ *is an algebraic representation of* X and denote ϕ by $\phi_{\mathbf{V}}$ for clarity. A *morphism* from the algebraic representation $\mathbf{U}$ to the algebraic representation $\mathbf{V}$ consists of a $\mathbf{G}$-equivariant k-morphism $\pi : \mathbf{U} \to \mathbf{V}$ such that $\phi_{\mathbf{V}}$ agrees almost everywhere with $\pi \circ \phi_{\mathbf{U}}$. An algebraic representation $\mathbf{V}$ of X is said to be a *coset algebraic representation* if in addition $\mathbf{V}$ is isomorphic as an algebraic representation to a coset variety $\mathbf{G}/\mathbf{H}$ for some k-algebraic subgroup $\mathbf{H} < \mathbf{G}$.

Ergodic properties of X are reflected in its category of algebraic representations.

PROPOSITION 3.2 ([3, proposition 4.2]). *Assume X is T-ergodic. Then for every algebraic representation $\phi_{\mathbf{V}}: X \to \mathbf{V}(k)$ there exists a coset representation $\phi_{\mathbf{G}/\mathbf{H}}: X \to \mathbf{G}/\mathbf{H}(k)$ and a morphism of algebraic representations $\pi: \mathbf{G}/\mathbf{H} \to \mathbf{V}$—that is, a $\mathbf{G}$-equivariant k-morphism π such that for a.e. $x \in X$, $\phi_{\mathbf{V}}(x) = \pi \circ \phi_{\mathbf{G}/\mathbf{H}}(x)$.*

In case the T-action on X is weakly mixing (and in particular, finite measure preserving), the category of representation of X is essentially trivial.

PROPOSITION 3.3. *Assume X is T-weakly mixing. Then for every algebraic representation $\phi: X \to \mathbf{V}(k)$, ϕ is essentially constant. Further, if $\tau(T)$ is Zariski dense in $\mathbf{G}$, then the essential image of ϕ is $\mathbf{G}$-invariant.*

Proof. Letting $\mu \in \mathrm{Prob}(\mathbf{V}(k))$ be the push forward by ϕ of the measure on X and L be the closure of $\tau(T)$ in $\mathbf{G}(k)$, it follows from [1, corollary 1.13] that ϕ is essentially constant and its essential image is L-fixed. □

The following theorem guarantees nontriviality of the category of representations of X (here, the T-action on X is not assumed to preserve a finite measure).

THEOREM 3.4 ([3, theorem 4.5], [1, theorem 1.17]).
Assume the T-Lebesgue space X is both amenable and metrically ergodic. Assume the k-algebraic group $\mathbf{G}$ is connected, k-simple, and adjoint, and assume that $\tau(T)$ is Zariski dense and unbounded in $\mathbf{G}(k)$. Then there exists a coset representation $\phi: X \to \mathbf{G}/\mathbf{H}(k)$ for some proper k-subgroup $\mathbf{H} \lneq \mathbf{G}$.

4 *T*-algebraic representations of *S*

Throughout this section we fix a locally compact second countable group S and a lattice $\Gamma < S$. We endow S with its Haar measure and regard it as a Lebesgue space. We also fix a field k endowed with a nontrivial absolute value that is separable and complete (as a metric space) and a k-algebraic group $\mathbf{G}$. We denote by G the Polish group $\mathbf{G}(k)$. Finally, we fix a homomorphism $\rho: \Gamma \to G$.

DEFINITION 4.1. Given all the data above, for a closed subgroup $T < S$, a *T-algebraic representation of* S consists of the following data:

- A k-algebraic group $\mathbf{L}$

- A k-($\mathbf{G} \times \mathbf{L}$)-algebraic variety $\mathbf{V}$, regarded as a left $\mathbf{G}$, right $\mathbf{L}$ space, on which the $\mathbf{L}$-action is faithful
- A homomorphism $\tau : T \to \mathbf{L}(k)$ with a Zariski dense image
- An associated algebraic representation of the $\Gamma \times T$-space S on $\mathbf{V}$, where Γ acts on the left and T acts on the right of the Lebesgue space S—that is, a Haar a.e. defined measurable map $\phi : S \to \mathbf{V}(k)$ such that for almost every $s \in S$, every $\gamma \in \Gamma$, and every $t \in T$,

$$\phi(\gamma st) = \rho(\gamma)\phi(s)\tau(t).$$

We abbreviate the notation by saying that $\mathbf{V}$ is a T-algebraic representation of S, denoting the extra data by $\mathbf{L_V}$, $\tau_\mathbf{V}$, and $\phi_\mathbf{V}$. Given another T-algebraic representation $\mathbf{U}$, we let $\mathbf{L_{U,V}} < \mathbf{L_U} \times \mathbf{L_V}$ be the Zariski closure of the image of $\tau_\mathbf{U} \times \tau_\mathbf{V} : T \to \mathbf{L_U} \times \mathbf{L_V}$. Note that $\mathbf{L_{U,V}}$ acts on $\mathbf{U}$ and $\mathbf{V}$ via its projections to $\mathbf{L_U}$ and $\mathbf{L_V}$ correspondingly. A *morphism* of T-algebraic representations of S from the T-algebraic representation $\mathbf{U}$ to the T-algebraic representation $\mathbf{V}$ is a $\mathbf{G} \times \mathbf{L_{U,V}}$-equivariant k-morphism $\pi : \mathbf{U} \to \mathbf{V}$ such that $\phi_\mathbf{V}$ agrees a.e. with $\pi \circ \phi_\mathbf{U}$.

Fix a k-subgroup $\mathbf{H} < \mathbf{G}$ and denote $\mathbf{N} = N_\mathbf{G}(\mathbf{H})$. This is again a k-subgroup. Any element $n \in \mathbf{N}$ gives a $\mathbf{G}$-automorphism of $\mathbf{G}/\mathbf{H}$ by $g\mathbf{H} \mapsto gn^{-1}\mathbf{H}$. It is easy to see that the homomorphism $\mathbf{N} \to \mathrm{Aut}_\mathbf{G}(\mathbf{G}/\mathbf{H})$ thus obtained is surjective and its kernel is $\mathbf{H}$. Under the obtained identification $\mathbf{N}/\mathbf{H} \cong \mathrm{Aut}_\mathbf{G}(\mathbf{G}/\mathbf{H})$, the k-points of the k-group $\mathbf{N}/\mathbf{H}$ are identified with the k-$\mathbf{G}$-automorphisms of $\mathbf{G}/\mathbf{H}$.

DEFINITION 4.2. A T-algebraic representation of S is said to be a *coset T-algebraic representation* if it is isomorphic as a T-algebraic representation to $\mathbf{G}/\mathbf{H}$ for some k-algebraic subgroup $\mathbf{H} < \mathbf{G}$ and $\mathbf{L}$ corresponds to a k-subgroup of $N_\mathbf{G}(\mathbf{H})/\mathbf{H}$ that acts on $\mathbf{G}/\mathbf{H}$ as described above.

It is clear that the collection of T-algebraic representations of S and their morphisms form a category.

THEOREM 4.3.
Assume the T-action on S/Γ is weakly mixing. Then the category of T-algebraic representations of S has an initial object and this initial object is a coset T-algebraic representation.

We will first prove the following lemma.

LEMMA 4.4. *Assume the T-action on S/Γ is weakly mixing. Let* $\mathbf{V}$ *be a T-algebraic representation of S. Then there exists a coset T-algebraic representation of S for some k-algebraic subgroup* $\mathbf{H}<\mathbf{G}$ *and a morphism of T-algebraic representations* $\pi:\mathbf{G}/\mathbf{H}\to\mathbf{V}$.

Proof. The T-action on S/Γ is weakly mixing, and in particular ergodic; thus the $\Gamma\times T$-action on S is ergodic. Applying Proposition 3.2 we get that there exists a coset representation $(\mathbf{G}\times\mathbf{L})/\mathbf{M}$ for some k-algebraic subgroup $\mathbf{M}<\mathbf{G}\times\mathbf{L}$ and a morphism of algebraic representations $\pi:(\mathbf{G}\times\mathbf{L})/\mathbf{M}\to\mathbf{V}$. We are thus reduced to the case $\mathbf{V}=(\mathbf{G}\times\mathbf{L})/\mathbf{M}$. Denote the obvious projection from $\mathbf{G}\times\mathbf{L}$ to $\mathbf{G}$ and $\mathbf{L}$ correspondingly by pr_1 and pr_2. The composition of the map $\phi:S\to(\mathbf{G}\times\mathbf{L})/\mathbf{M}(k)$ with the $\mathbf{G}(k)$-invariant map

$$(\mathbf{G}\times\mathbf{L})/\mathbf{M}(k)\to\mathbf{L}/\,\mathrm{pr}_2(\mathbf{M})(k)$$

clearly factors through S/Γ and thus gives a coset representation of the T-Lebesgue space $X=S/\Gamma$ on $\mathbf{L}/\,\mathrm{pr}_2(\mathbf{M})$. Applying Proposition 3.3, we conclude that $\mathbf{L}/\,\mathrm{pr}_2(\mathbf{M})$ contains a $\mathbf{G}$-invariant point; thus $\mathrm{pr}_2(\mathbf{M})=\mathbf{L}$. It follows that as $\mathbf{G}$-varieties, $(\mathbf{G}\times\mathbf{L})/\mathbf{M}\cong\mathbf{G}/\mathbf{H}$ for $\mathbf{H}=\mathrm{pr}_1(\mathbf{M}\cap(\mathbf{G}\times\{e\}))<\mathbf{G}$, and the lemma follows. □

Proof of Theorem 4.3. We consider the collection

$\{\mathbf{H}<\mathbf{G}\mid\mathbf{H}$ is defined over k and there exists a coset T-representation

This is a nonempty collection as it contains $\mathbf{G}$. By the Noetherian property, this collection contains a minimal element. We choose such a minimal element $\mathbf{H}_0$ and fix corresponding algebraic k-subgroup $\mathbf{L}_0<N_{\mathbf{G}}(\mathbf{H}_0)/\mathbf{H}_0$, homomorphism $\tau_0:T\to\mathbf{L}_0(k)$, and a representation $\phi_0:S\to(\mathbf{G}/\mathbf{H}_0)(k)$. We argue to show that this coset T-representation is the required initial object.

Fix any T-algebraic representation of S, $\mathbf{V}$. It is clear that, if it exists, a morphism of T-algebraic representations from $\mathbf{G}/\mathbf{H}_0$ to $\mathbf{V}$ is unique, as two different $\mathbf{G}$-maps $\mathbf{G}/\mathbf{H}_0\to\mathbf{V}$ agree nowhere. We are left to show existence. To this end we consider the product T-algebraic representation $\mathbf{V}\times\mathbf{G}/\mathbf{H}_0$ given by the data $\phi=\phi_{\mathbf{V}}\times\phi_0$, $\tau=\tau_{\mathbf{V}}\times\tau_0$, and $\mathbf{L}$ being the Zariski closure of $\tau(T)$ in $\mathbf{L}_{\mathbf{V}}\times\mathbf{L}_0$. Applying Lemma 4.4 to this product T-algebraic representation, we obtain the following commutative diagram:

$$\begin{array}{ccccc} & & \overset{\phi_0}{\frown\!\!\!\longrightarrow} & & \\ S & \cdots\cdots\rightarrow & \mathbf{G}/\mathbf{H}(k) & & \\ \downarrow \phi_{\mathbf{V}} & \overset{\phi}{\searrow} & \vdots\ \pi & & \\ \mathbf{V}(k) & \xleftarrow{\ \mathrm{pr}_1\ } & \mathbf{V}\times\mathbf{G}/\mathbf{H}_0(k) & \xrightarrow{\ \mathrm{pr}_2\ } & \mathbf{G}/\mathbf{H}_0(k) \end{array}$$

By the minimality of $\mathbf{H}_0$, the $\mathbf{G}$-morphism $\mathrm{pr}_2 \circ \pi : \mathbf{G}/\mathbf{H} \to \mathbf{G}/\mathbf{H}_0$ must be a k-isomorphism and hence an isomorphism of T-algebraic representation. We thus obtain the morphism of T-algebraic representations

$$\mathrm{pr}_1 \circ \pi \circ (\mathrm{pr}_2 \circ \pi)^{-1} : \mathbf{G}/\mathbf{H}_0(k) \to \mathbf{V}(k).$$

This completes the proof of Theorem 4.3. □

REMARK 4.5. Let the data $\mathbf{G}/\mathbf{H}$, $\tau : T \to \mathbf{L}(k) < N_{\mathbf{G}}(\mathbf{H})/\mathbf{H}(k)$ and $\phi : S \to (\mathbf{G}/\mathbf{H})(k)$ form an initial object in the category of T-algebraic representation of S. For $g \in \mathbf{G}(k)$ we get a $\mathbf{G}$-equivariant k-isomorphism $\pi_g : \mathbf{G}/\mathbf{H} \to \mathbf{G}/\mathbf{H}^g$ given by $x\mathbf{H} \mapsto xg^{-1}\mathbf{H}^g$. Denoting by $\mathrm{inn}(g) : N_{\mathbf{G}}(\mathbf{H})/\mathbf{H} \to N_{\mathbf{G}}(\mathbf{H}^g)/\mathbf{H}^g$ the k-isomorphism $n\mathbf{H} \mapsto n^g\mathbf{H}^g$ and by $\mathbf{L}^g$ the image of $\mathbf{L}$ under inn(g) we get that the data

$$\mathbf{G}/\mathbf{H}^g, \quad \mathrm{inn}(g) \circ \tau : T \to \mathbf{L}^g(k) < N_{\mathbf{G}}(\mathbf{H}^g)/\mathbf{H}^g(k), \quad \pi_g \circ \phi : S \to \mathbf{G}/\mathbf{H}(k)$$

form another T-algebraic coset representation of S, isomorphic to the one given above, thus again an initial object in the category of T-algebraic representations of S. Furthermore, it is easy to verify that any actual coset presentation of the initial object in the category of T-algebraic representations of S is of the above form, for some $g \in \mathbf{G}(k)$.

It turns out that an initial object in the category of T-algebraic representations of S extends naturally to an N-algebraic representation of S, where N denotes the normalizer of T in S.

THEOREM 4.6.
Assume the action of T on S/Γ is weakly mixing and let $\mathbf{G}/\mathbf{H}$, $\tau : T \to \mathbf{L}(k) < N_{\mathbf{G}}(\mathbf{H})/\mathbf{H}(k)$ and $\phi : S \to \mathbf{G}/\mathbf{H}(k)$ be an initial object in the category of T-algebraic representations of S, as guaranteed by Theorem 4.3. Then the map

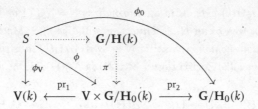

$\tau : T \to N_{\mathbf{G}}(\mathbf{H})/\mathbf{H}(k)$ extends to the normalizer $N = N_S(T)$ of T in S, and the map ϕ could be seen as an N-algebraic representation of S. More precisely, there exists a continuous homomorphism $\bar{\tau} : N \to N_{\mathbf{G}}(\mathbf{H})/\mathbf{H}(k)$ satisfying $\bar{\tau}|_T = \tau$ such that, denoting by $\bar{\mathbf{L}}$ the Zariski closure of $\bar{\tau}(N)$ in $N_{\mathbf{G}}(\mathbf{H})/\mathbf{H}$, the data

$$\mathbf{G}/\mathbf{H}, \quad \bar{\tau} : N \to \bar{\mathbf{L}}(k) < N_{\mathbf{G}}(\mathbf{H})/\mathbf{H}(k), \quad \phi : S \to \mathbf{G}/\mathbf{H}(k)$$

form an N-algebraic coset representation. Moreover, this N-algebraic coset representation is an initial object in the category of N-algebraic representations.

Proof. Fix $n \in N$. Set $\tau' = \tau \circ \mathrm{inn}(n) : T \to \mathbf{L}(k)$, where $\mathrm{inn}(n) : T \to T$ denotes the inner automorphism $t \mapsto t^n = ntn^{-1}$, and $\phi' = \phi \circ R_n : S \to \mathbf{G}/\mathbf{H}(k)$, where $R_n : S \to S$ denotes the right regular action $s \mapsto sn^{-1}$. We claim that the data $\mathbf{L}$, $\mathbf{G}/\mathbf{H}$, τ', and ϕ' form a new T-representation of S. Indeed, for almost every $s \in S$, every $\gamma \in \Gamma$, and every $t \in T$,

$$\phi'(\gamma st) = \phi(\gamma stn^{-1}) = \phi(\gamma sn^{-1}t^n) = \rho(\gamma)\phi'(s)\tau'(t).$$

Since the T-algebraic representation of S given by $\mathbf{L}$, $\mathbf{G}/\mathbf{H}$, τ, and ϕ forms an initial object, we get the dashed vertical arrow, which we denote $\bar{\tau}(n)$, in the following diagram:

$$\begin{array}{ccc} S & \xrightarrow{\ \phi\ } & \mathbf{G}/\mathbf{H}(k) \\ & \underset{\phi'}{\searrow} & \Big\downarrow{\scriptstyle \bar{\tau}(n)} \\ & & \mathbf{G}/\mathbf{H}(k) \end{array}$$

It follows from the uniqueness of the dashed arrow that the map $n \mapsto \bar{\tau}(n)$ is a homomorphism from N to the group of k-$\mathbf{G}$-automorphism of $\mathbf{G}/\mathbf{H}$, which we identify with $N_{\mathbf{G}}(\mathbf{H})/\mathbf{H}(k)$. For $n \in T$ the map $\tau(n) : \mathbf{G}/\mathbf{H} \to \mathbf{G}/\mathbf{H}$ could also be taken to be the dashed arrow; thus $\bar{\tau}|_T = \tau$, by uniqueness. The fact that the homomorphism $\bar{\tau} : N \to N_{\mathbf{G}}(\mathbf{H})/\mathbf{H}(k)$ is necessarily continuous is explained in the proof of [3, theorem 4.7]. We define $\bar{\mathbf{L}}$ to be the Zariski closure of $\bar{\tau}(N)$ in $N_{\mathbf{G}}(\mathbf{H})/\mathbf{H}$. We thus indeed obtain an N-representation of S, given by the algebraic group $\bar{\mathbf{L}}$, the variety $\mathbf{G}/\mathbf{H}$, the homomorphism $\bar{\tau} : N \to \bar{\mathbf{L}}(k)$, and the (same old) map $\phi : S \to \mathbf{G}/\mathbf{H}(k)$.

The final part, showing that the above data forms an initial object in the category of N-algebraic representations of S, is left to the reader, as we will not use this fact in the sequel. $\square$

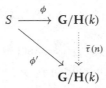

COROLLARY 4.7. *Let $T_1, T_2 < S$ be closed subgroups and assume for each $i \in \{1,2\}$ the action of T_i on S/Γ is weakly mixing, and let*

$$\mathbf{G}/\mathbf{H}_i, \quad \tau_i : T_i \to \mathbf{L}_i(k) < N_{\mathbf{G}}(\mathbf{H}_i)/\mathbf{H}_i(k), \quad \phi_i : S \to (\mathbf{G}/\mathbf{H}_i)(k)$$

be the corresponding initial objects in the categories of T_i-algebraic representations of S. Assume that T_2 normalizes T_1. Then a conjugate of $\mathbf{H}_2$ is contained in $\mathbf{H}_1$, and if $\mathbf{H}_2 = \mathbf{H}_1$, then also $\phi_2 = \phi_1$.

5 Proof of Theorem 1.3

In the proof below we shall need the following general Lemma that allows us to assemble a continuous homomorphism $\tau : S \to G$ from continuous homomorphisms of subgroups $\tau_i : T_i \to G$.

LEMMA 5.1. *Let S be a locally compact second countable group and let $\{T_i < S\}_{i \in I}$ be a countable family of closed subgroups that together topologically generate S. Let G be a Polish topological group, and for each $i \in I$ let $\tau_i : T_i \to G$ be a continuous homomorphism.*

Let X be a Lebesgue S-space, and assume that there exists a single measurable map $\phi : X \to G$ so that for every $i \in I$ and every $t \in T_i$ for a.e. $x \in X$,

$$\phi(tx) = \phi(x)\tau_i(t)^{-1}.$$

Then there exists a continuous homomorphism $\tau : S \to G$ so that $\tau|_{T_i} = \tau_i$ and

$$\phi(sx) = \phi(x)\tau(s)^{-1}$$

for every $s \in S$ and a.e. $x \in X$.

Proof. Let us fix a probability measure m in the given measure class on X and consider the space $L^0(X, G)$ of (equivalence classes of) measurable functions $\psi : X \to G$ taken with the topology of convergence in measure: given an open neighborhood U of $1 \in G$ and $\epsilon > 0$, a (U, ϵ) neighborhood of ψ consists of classes of those measurable functions $\psi' : X \to G$ for which $m\{x \in X \mid \psi'(x) \in \psi(x)U\} > 1 - \epsilon$. This topology is Polish.

The right translation action of G on $L^0(X, G)$ given by $g : \psi(x) \mapsto \psi(x)g^{-1}$ is clearly free and continuous. In fact, every G-orbit is homeomorphic to G and is closed in $L^0(X, G)$. To see this we assume $g_i\psi \to \psi'$ in measure and argue to show that there exists $g \in G$ such that $\psi'(x) = \psi(x)g^{-1}$. The given converges in measure imply that there exists a subsequence g_{i_j} for which $\psi(x)g_{i_j}^{-1} \to \psi'(x)$ for a.e. $x \in X$. Therefore $g_{i_j} \to \psi'(x)^{-1}\psi(x)$ for a.e. $x \in X$; thus indeed there is a limit $g = \lim_{j\to\infty} g_{i_j}$ in G such that $\psi'(x) = \psi(x)g^{-1}$.

The group S acts on $L^0(X, G)$ by precomposition $s: \psi(x) \mapsto \psi(s^{-1}x)$. This action is also continuous, because any measurable measure class preserving action $S \times X \to X$ of a locally compact group has the property that given $\epsilon > 0$, there exists $\delta > 0$ and a neighborhood V of $1 \in S$, so that for any measurable $E \subset X$ with $m(E) < \delta$, one has $m(sE) < \epsilon$ for every $s \in V$.

Now consider $\phi \in L^0(X, G)$ as in the Lemma. By the assumption, for each $i \in I$ the T_i-orbit of ϕ lies in the G-orbit of ϕ. Since S is topologically generated by $\cup_I T_i$, the S-action is continuous, and the G-orbit of ϕ is closed, it follows that the S-orbit of ϕ is contained in the G-orbit of ϕ. Hence, for every $s \in S$ there is $\tau(s) \in G$ so that a.e. on X,

$$\phi(xs^{-1}) = \phi(x)\tau(s)^{-1}.$$

This defines a homomorphism $\tau : S \to G$ that extends all $\tau_i : T_i \to G$. Continuity of the homomorphism follows from the fact that the S-action is continuous and G is homeomorphic to the G-orbit of ϕ. □

We now return to the proof of Theorem 1.3. We let $\mathbf{G}$ be a connected, adjoint, k-simple algebraic group defined over k as in Theorem 1.3. By the fact that $\rho(\Gamma)$ is unbounded in $\mathbf{G}(k)$ we get that the given absolute value on k is nontrivial. Further, by the countability of Γ, we may replace k with a complete and separable (in the topological sense) subfield k' such that $\rho(\Gamma) \subset \mathbf{G}(k')$. We will therefore assume below that the given absolute value on the field k is nontrivial and that k is complete and separable as a metric space. Accordingly, we will regard $G = \mathbf{G}(k)$ as a Polish group. We let $T_0, \ldots, T_n < S$ be subgroups as guaranteed by condition (B) and assume, as we may, that T_0 is amenable.

We fix $i \in \{0, \ldots, n\}$. By Lemma 2.2 the action of T_i on S/Γ is weakly mixing, thus by Theorem 4.3 the category of T_i-algebraic representations of S has an initial object that is a coset T_i-algebraic representation. We denote by

$$\mathbf{G}/\mathbf{H}_i, \quad \tau_i : T_i \to \mathbf{L}_i(k) < N_{\mathbf{G}}(\mathbf{H}_i)/\mathbf{H}_i(k), \quad \phi_i : S \to \mathbf{G}/\mathbf{H}_i(k)$$

the data forming this initial object.

By Examples 2.1 and 2.3 the S-action on S/T_0 is metrically ergodic and amenable. It follows from [3, lemma 3.5] that the Γ-action on S/T_0 is also metrically ergodic and amenable. By Proposition 3.4 we get that there exists a coset representation $\phi' : S/T_0 \to \mathbf{G}/\mathbf{H}'(k)$ for some proper k-subgroup $\mathbf{H}' \lneq \mathbf{G}$. Setting $\mathbf{L}' = \{e\} < N_{\mathbf{G}}(\mathbf{H}')/\mathbf{H}'$ and letting $\tau' : T_0 \to \mathbf{L}'(k)$ be the trivial homomorphism we view ϕ' as a T_0-equivariant map from S to $\mathbf{G}/\mathbf{H}'(k)$, thus getting a nontrivial T_0-algebraic coset representation of S. It follows that $\mathbf{H}_0$ is a proper subgroup of $\mathbf{G}$.

By Corollary 4.7 for each $i \in \{0, \ldots, n\}$ a conjugate of $\mathbf{H}_{i-1}$ is contained in $\mathbf{H}_i$, where i is taken in a cyclic order. Going a full cycle we get that the groups $\mathbf{H}_i$ are all conjugated. Using Remark 4.5 we assume that they all coincide and we denote this common group by $\mathbf{H}$. In particular, we have $\mathbf{H} = \mathbf{H}_0 \lneq \mathbf{G}$. Using again Corollary 4.7, we obtain that the maps $\phi_i : S \to \mathbf{G}/\mathbf{H}(k)$ all coincide and we denote this common map by

$$\phi : S \to \mathbf{G}/\mathbf{H}(k).$$

Let $\mathbf{N} = N_{\mathbf{G}}(\mathbf{H})$ and $\mathbf{L}$ be the algebraic subgroup generated by $\mathbf{L}_0, \ldots, \mathbf{L}_n$ in $\mathbf{N}/\mathbf{H}$. By [5, 2.1(b)], $\mathbf{L} < \mathbf{N}/\mathbf{H}$ is a k-algebraic subgroup. Denote by $\hat{\mathbf{L}}$ the preimage of $\mathbf{L}$ under the quotient map $\mathbf{N} \to \mathbf{N}/\mathbf{H}$ and note that $\hat{\mathbf{L}} < \mathbf{N}$ is a k-algebraic subgroup. We conclude that the k-$\mathbf{G}$-morphism $\pi : \mathbf{G}/\mathbf{H} \to \mathbf{G}/\hat{\mathbf{L}}$ is $\mathbf{L}_i$-invariant for every $i \in \{0, \ldots, n\}$. It follows that $\pi \circ \phi : S \to \mathbf{G}/\hat{\mathbf{L}}(k)$ is T_i-invariant for every i. Since S is topologically generated by the groups T_i, we conclude that $\pi \circ \phi : S \to \mathbf{G}/\hat{\mathbf{L}}(k)$ is S-invariant. Thus ϕ is a constant map and its essential image is a $\rho(\Gamma)$-invariant point in $\mathbf{G}/\hat{\mathbf{L}}$. Since $\rho(\Gamma)$ is Zariski dense in $\mathbf{G}$, this point is $\mathbf{G}$-invariant. We conclude that $\hat{\mathbf{L}} = \mathbf{G}$. Since $\hat{\mathbf{L}} < \mathbf{N} < \mathbf{G} = \hat{\mathbf{L}}$ we get that $\hat{\mathbf{L}} = \mathbf{N} = \mathbf{G}$. Thus, $\mathbf{G}$ normalizes $\mathbf{H}$. Since $\mathbf{G}$ is k-simple and $\mathbf{H} \lneq \mathbf{G}$ we conclude that $\mathbf{H}$ is trivial. In particular $\mathbf{L} = \hat{\mathbf{L}} = \mathbf{G}$ and it acts on $\mathbf{G}/\mathbf{H} = \mathbf{G}$ by right multiplication.

To summarize: We have an a.e. defined measurable map $\phi : S \to \mathbf{G}(k)$ and for every $i \in \{1, \ldots, n\}$ a continuous homomorphism $\tau_i : T_i \to \mathbf{G}(k)$ such that for every $\gamma \in \Gamma$, $t \in T_i$ for a.e. $x \in S$:

$$\phi(\gamma x t^{-1}) = \rho(\gamma)\phi(x)\tau_i(t)^{-1} \tag{5.1}$$

We also have that the algebraic group generated by $\tau_1(T_1), \ldots, \tau_n(T_n)$ is $\mathbf{G}$.

Taking $\gamma = 1$ in Equation (5.1) and applying Lemma 5.1 for $X = S$ endowed with the S-action $s : x \mapsto xs^{-1}$, we get a continuous homomorphism $\tau : S \to \mathbf{G}(k)$ with $\tau|_{T_i} = \tau_i$ so that $\phi(xs^{-1}) = \phi(x)\tau(s)^{-1}$ for every $s \in S$ and for a.e. $x \in S$. It follows now from equation (5.1) that for every $\gamma \in \Gamma$, every $s \in S$, and for a.e. $x \in S$,

$$\phi(\gamma x s^{-1}) = \rho(\gamma)\phi(x)\tau(s)^{-1}. \tag{5.2}$$

The a.e. defined measurable map $\Phi : S \to G$ given by $\Phi(s) = \phi(s)\tau(s)^{-1}$ is S-invariant, hence essentially constant. Denoting its essential image by $g \in \mathbf{G}(k)$ we get that for a.e. $s \in S$, $\phi(s) = g\tau(s)$. Equation (5.2) gives that for every $\gamma \in \Gamma$, $s \in S$, and for a.e. $x \in S$,

$$g\tau(\gamma x s) = \rho(\gamma) g\tau(x)\tau(s).$$

As the above is an a.e. satisfied equation of continuous functions in the parameter s, it is satisfied everywhere. Taking $x = s = e$, we get that for every $\gamma \in \Gamma$,

$$g\tau(\gamma) = \rho(\gamma) g.$$

Setting $\hat{\rho}(s) = g\tau(s)g^{-1}$ we get that $\hat{\rho} : S \to G$ is a continuous homomorphism such that $\hat{\rho}|_\Gamma = \rho$. The uniqueness of $\hat{\rho}$ follows from [3, lemma 6.3], and the proof of Theorem 1.3 is completed.

6 Continuous homomorphisms of algebraic groups

LEMMA 6.1. *Let k be a complete valued field and k' a finite field extension, endowed with the extended absolute value. Let $\mathbf{V}$ be an affine k'-variety and denote by $\mathbf{U}$ its restriction of scalar to k. Endow the spaces $\mathbf{V}(k')$ and $\mathbf{U}(k)$ with the corresponding analytic topologies and consider the natural maps $\mathbf{U}(k) \to \mathbf{U}(k')$ and $\mathbf{U}(k') \to \mathbf{V}(k')$. Then the composed map $\mathbf{U}(k) \to \mathbf{V}(k')$ is a homeomorphism.*

Proof. By the functoriality of the restriction of scalars and its compatibility with products, it is enough to prove the lemma for $\mathbf{V} = \mathbb{A}^1$, the one-dimensional affine space. In this case we get $\mathbf{V}(k') \simeq k'$ and $\mathbf{U}(k) \simeq k^{[k':k]}$, and the composed map $k^{[k':k]} \to k'$, which is a k-vector spaces isomorphism, is indeed a homeomorphism by [1, theorem 4.6]. □

PROPOSITION 6.2. *Let k be a field and let $\mathbf{G}$ be a connected, adjoint, k-isotropic, k-simple k-algebraic group. Then there exists a finite field extension k' of k, unique up to equivalence of extensions, and there exists a connected, adjoint, k'-isotropic, absolutely simple k'-algebraic group $\mathbf{G}'$, unique up to k'-isomorphism, such that $\mathbf{G}$ is k-isomorphic to the restriction of scalars of $\mathbf{G}'$ from k' to k. In particular, the natural k'-morphism $\mathbf{G} \to \mathbf{G}'$ gives rise to a group isomorphism $\mathbf{G}(k) \simeq \mathbf{G}'(k')$.*

If k is endowed with a complete absolute value, endowing k' with the extended absolute value, the isomorphism $\mathbf{G}(k) \simeq \mathbf{G}'(k')$ is also a homeomorphism with respect to the corresponding analytic topologies.

Note that the above proposition is trivial if $\mathbf{G}$ is absolutely simple to begin with, in which case we have that $k' = k$ and $\mathbf{G}' = \mathbf{G}$.

Proof. The existence of the field k' and the group $\mathbf{G}'$ follows from the discussion in [10, I.1.7]. Their uniqueness follows from the uniqueness statement in [10, Theorem I.1.8]. The last part follows from Lemma 6.1. □

Given a filed ℓ and an ℓ-algebraic group $\mathbf{H}$, we denote by $\mathbf{H}(\ell)^+$ the subgroup of $\mathbf{H}(\ell)$ generated by all the groups of ℓ-points of the unipotent radicals of all parabolic ℓ-subgroups of $\mathbf{H}$; see [10, I.1.5.2].

PROPOSITION 6.3. *Let ℓ and k be fields endowed with complete absolute values. Let $\mathbf{H}$ be a connected, semisimple algebraic group defined over ℓ. Assume that $\mathbf{H}$ has no ℓ-anisotropic factors. Let H be an intermediate closed subgroup $\mathbf{H}(\ell)^+ < H < \mathbf{H}(\ell)$. Let $\mathbf{G}$ be a connected, adjoint, absolutely simple algebraic group defined over k. Let $\theta : H \to \mathbf{G}(k)$ be a group homomorphism that is continuous with respect to the corresponding analytic topologies, and let its image be Zariski dense in $\mathbf{G}$. Then there exists a unique field embedding $i : \ell \to k$, which is continuous, and a corresponding unique k-algebraic groups morphism $\mathbf{H} \to \mathbf{G}$ such that θ coincides with the precomposition of the corresponding map $\mathbf{H}(k) \to \mathbf{G}(k)$ with the injection $H < \mathbf{H}(\ell) \to \mathbf{H}(k)$.*

Furthermore, if ℓ is a local field, then the assumption that $\mathbf{H}$ has no ℓ-anisotropic factors could be replaced by the assumption that the image of θ is unbounded in $\mathbf{G}(k)$.

Proof. The first paragraph is proven in [10, Proposition VII.5.3(a)]. Note that in this proof the fields are assumed to be local fields, but this assumption is used in the proof only via [10, Remark I.1.8.2(IIIa)], which applies equally well for complete valued fields.

Assume now that ℓ is a local field and $\theta(H)$ is unbounded in $\mathbf{G}(k)$. We first remark that the field embedding $i : \ell \to k$, if exists, is unique as it must coincide with the corresponding unique field embedding we get by replacing $\mathbf{H}$ by the group $\mathbf{H}_i$ given by the almost direct product of all ℓ-isotropic simple factors of $\mathbf{H}$, replacing H by the intermediate closed group $\mathbf{H}_i(\ell)^+ < H \cap \mathbf{H}_i(\ell) < \mathbf{H}_i(\ell)$ and replacing θ by its restriction to $H \cap \mathbf{H}_i(\ell)$, noting that indeed $H \cap \mathbf{H}_i(\ell)$ contains $\mathbf{H}_i(\ell)^+$ by [10, I.1.5.4(iv)]. Second, we remark that the corresponding k-algebraic groups morphism $\mathbf{H} \to \mathbf{G}$, if exists, is unique. To see this, we will assume i is given and $\alpha, \beta : \mathbf{H} \to \mathbf{G}$ are two such corresponding k-morphisms and argue to show that $\alpha = \beta$. By definition, $\alpha|_H = \beta|_H$. By [10, Proposition I.2.3.1(b)], H^+ is a closed cocompact normal subgroup in $\mathbf{H}(\ell)$. It follows that H is a cofinite volume subgroup of $\mathbf{H}(\ell)$. Applying [3, theorem 6.3] we get that $\alpha|_{\mathbf{H}(\ell)} = \beta|_{\mathbf{H}(\ell)}$ (note that in this reference we regard lattices, but the proof applies equally well to closed subgroups of cofinite volume). By [5, corollary 18.3] we have that $\mathbf{H}(\ell)$ is Zariski dense in $\mathbf{H}$; thus indeed $\alpha = \beta$. In the sequel we will argue to show the existence of the field embedding $i : \ell \to k$ and the corresponding k-algebraic groups morphism $\mathbf{H} \to \mathbf{G}$.

We let $\mathbf{H}_a$ be the almost direct products of all ℓ-anisotropic factors of $\mathbf{H}$ and denote $H_a = \mathbf{H}_a(\ell) \cap H$. We also consider the group $H^+ = \mathbf{H}(\ell)^+$. Note that H_a and H^+ are commuting normal subgroups of H; thus the Zariski closures of the images of these groups under θ are commuting algebraic normal subgroups of the simple group $\mathbf{G}$. As H^+ is cocompact in H, its image under θ is unbounded and in particular nontrivial. It follows that $\theta(H^+)$ is Zariski dense in $\mathbf{G}$; thus $\theta(H_a)$ is central and hence trivial. We set $\mathbf{H}' = \mathbf{H}/\mathbf{H}_a$ and let H' be the image of H in $\mathbf{H}'$ under the quotient map $\mathbf{H}(\ell) \to \mathbf{H}/\mathbf{H}_a(\ell)$. We get that θ factors via $\theta' : H' \to \mathbf{G}(k)$. Note that $\mathbf{H}'$ has no ℓ-anisotroic factors and that H' contains the group $\mathbf{H}'(\ell)^+$, by [10, I.1.5.4(iv) and I.1.5.5]. We conclude having a continuous $i : \ell \to k$ and a corresponding k-algebraic group morphism $\mathbf{H}' \to \mathbf{G}$ such that θ' coincides with the precomposition of the corresponding map $\mathbf{H}'(k) \to \mathbf{G}(k)$ with the injection $H' < \mathbf{H}'(\ell) \to \mathbf{H}'(k)$. We are done by considering the composed k-morphism $\mathbf{H} \to \mathbf{H}' \to \mathbf{G}$, noting that indeed θ coincides with the precomposition of the corresponding map $\mathbf{H}(k) \to \mathbf{G}(k)$ with the injection $H < \mathbf{H}(\ell) \to \mathbf{H}(k)$. □

Combining Proposition 6.3 with Proposition 6.2 we readily get the following result, in which the target group $\mathbf{G}$ is assumed to be k-simple rather than absolutely simple.

COROLLARY 6.4. *Let ℓ and k be fields endowed with complete absolute values. Let $\mathbf{H}$ be a connected, semisimple algebraic group defined over ℓ. Assume that $\mathbf{H}$ has no ℓ-anisotropic factors. Let H be an intermediate closed subgroup $\mathbf{H}(\ell)^+ < H < \mathbf{H}(\ell)$. Let $\mathbf{G}$ be a connected, adjoint, k-simple algebraic group defined over k. Let $\theta : H \to \mathbf{G}(k)$ be a group homomorphism that is continuous with respect to the corresponding analytic topologies, and its image is Zariski dense in $\mathbf{G}$. Then there exists a group homomorphism $\hat{\theta} : \mathbf{H}(\ell) \to \mathbf{G}(k)$ that is continuous with respect to the corresponding analytic topologies such that $\theta = \hat{\theta}|_H$.*

Furthermore, if ℓ is a local field, then the assumption that $\mathbf{H}$ has no ℓ-anisotropic factors could be replaced by the assumption that the image of θ is unbounded in $\mathbf{G}(k)$.

7 Deducing Theorem 1.1 from Theorem 1.3

We let k be a complete valued field and let $\mathbf{G}$ be a connected, adjoint, k-simple k-algebraic group. Assuming having $\Gamma < H$ and $\rho : \Gamma \to \mathbf{G}(k)$ as in the theorem, by [3, lemma 6.3] we have that a continuous homomorphism $\hat{\rho} : H \to G$ such that $\hat{\rho}|_\Gamma = \rho$, if exists, is uniquely given. Below we will prove its existence.

We fix a local field ℓ. We let $\mathcal{H}$ be the collection of all connected semisimple ℓ-algebraic groups $\mathbf{H}$ that satisfy the theorem—that is, for every lattice Γ in $H = \mathbf{H}(\ell)$, whose projection modulo $\mathbf{N}(\ell)$ for each proper normal ℓ-isotropic subgroup $\mathbf{N} \lhd \mathbf{H}$ is nondiscrete, and every homomorphism $\rho : \Gamma \to G$ with unbounded and Zariski dense image $\rho(\Gamma)$ in $\mathbf{G}$, there exists a continuous homomorphism $\hat{\rho} : H \to G$ satisfying $\rho = \hat{\rho}|_\Gamma$. We argue to show that $\mathcal{H}$ contains all connected semisimple ℓ-algebraic groups of ℓ-rank at least two.

Given a connected semisimple ℓ-algebraic group $\mathbf{H}$, let us denote by $\bar{\mathbf{H}}$ the associated adjoint group and by $p : \mathbf{H} \to \bar{\mathbf{H}}$ the corresponding ℓ-isogeny. The group $\bar{\mathbf{H}}$ is ℓ-isomorphic to the product of its factors. Let us denote by $\bar{\mathbf{H}}_0$ the product of all ℓ-isotropic factors of $\bar{\mathbf{H}}$ and let $q : \bar{\mathbf{H}} \to \bar{\mathbf{H}}_0$ be the corresponding projection. We claim that if $\bar{\mathbf{H}}_0$ is in $\mathcal{H}$, then so is $\mathbf{H}$. First note that both maps $p : \mathbf{H}(\ell) \to \bar{\mathbf{H}}(\ell)$ and $q : \bar{\mathbf{H}}(\ell) \to \bar{\mathbf{H}}_0(\ell)$ have compact kernels and closed cocompact images. This is clear for the projection q, and for p it follows from [10, Proposition I.2.3.3(i)]. In particular, we get that $q \circ p : \mathbf{H}(\ell) \to \bar{\mathbf{H}}_0(\ell)$ has a compact kernel and a closed cocompact image. Thus, if $\Gamma < \mathbf{H}(\ell)$ is a lattice, then $q \circ p(\Gamma) < \bar{\mathbf{H}}_0(\ell)$ is a lattice, and $\Lambda = \ker q \circ p|_\Gamma$ is a finite normal subgroup of Γ. Note that for each proper normal ℓ-isotropic subgroup $\mathbf{N} \lhd \bar{\mathbf{H}}_0$, the projection of $q \circ p(\Gamma)$ modulo $\mathbf{N}(\ell)$ is nondiscrete if the corresponding assumption applies to $\Gamma < \mathbf{H}(\ell)$. Note also that Λ is in the kernel of any Zariski dense homomorphism $\rho : \Gamma \to G$, by [10, Theorem I.1.5.6(i)], as $\rho(\Lambda)$ is a finite normal subgroup in the adjoint group $\mathbf{G}$. It follows that ρ factors through $q \circ p(\Gamma)$, and this proves our claim. Noting that the groups $\mathbf{H}$ and $\bar{\mathbf{H}}_0$ have equal ℓ-ranks, we are left to show that $\mathcal{H}$ contains all connected semisimple ℓ-algebraic groups of ℓ-rank at least two that are adjoint and have no ℓ-anisotropic factors. We proceed to do so.

Assume first that $\mathbf{H}$ is adjoint and has at least two k-isotropic factors. Up to replacing $\mathbf{H}$ by an ℓ-isomorphic copy, we assume that $\mathbf{H} = \mathbf{H}_1 \times \mathbf{H}_2$, where the groups $\mathbf{H}_1$ and $\mathbf{H}_2$ are connected, semi-simple, isotropic, adjoint ℓ-algebraic groups. We fix a lattice Γ in $H = \mathbf{H}(\ell)$, for which projection modulo $\mathbf{N}(\ell)$ for each proper normal ℓ-isotropic subgroup $\mathbf{N} \lhd \bar{\mathbf{H}}_0$ is nondiscrete, and a homomorphism $\rho : \Gamma \to G$ such that $\rho(\Gamma)$ is unbounded and Zariski dense in $\mathbf{G}$, and we argue to show that there exists a continuous homomorphism $\hat{\rho} : H \to$

G satisfying $\rho = \hat{\rho}|_\Gamma$. We let $\mathrm{pr}_i : \mathbf{H} \to \mathbf{H}_i$ be the corresponding projections, denote by H_i' the closure of $\mathrm{pr}_i(\Gamma)$ in $\mathbf{H}_i(\ell)$, and let $H' = H_1' \times H_2'$. Therefore $\Gamma < H'$ is a lattice with dense projections in the sense of [3, definition 1.1]; thus by [3, theorem 1.2] there exists a continuous homomorphism $\hat{\rho}' : H' \to G$ satisfying $\rho = \hat{\rho}'|_\Gamma$. By [10, Theorem II.6.7(a)] we have that $\mathbf{H}(\ell)^+ < H' < \mathbf{H}(\ell)$ and by Corollary 6.4, there exists a continuous homomorphism $\hat{\rho} : H \to G$ satisfying $\hat{\rho}' = \hat{\rho}|_{H'}$. We conclude that $\rho = \hat{\rho}|_\Gamma$, which finishes the proof in this case.

We are left with the case that $\mathbf{H}$ is ℓ-simple and of ℓ-rank at least two. We fix a lattice Γ in $H = \mathbf{H}(\ell)$ and note that by [10, Theorem III.5.7(b)] Γ has a finite abelianization. We set $H^+ = \mathbf{H}(\ell)^+$, $\Gamma^+ = \Gamma \cap H^+$ and conclude by [10, Theorem I.2.3.1(c)] that $\Gamma^+ < \Gamma$ is of finite index. In particular, $\Gamma^+ < H$ is a lattice; thus Γ^+ is also a lattice in the closed subgroup H^+ in which it is contained. We now set $S = H^+$ and note that it satisfies conditions (A) and (B): condition (B) follows from the assumption of higher rank, and condition (A) follows from [4, theorem 6.1]. By Theorem 1.3 there exists a continuous homomorphism $\hat{\rho}^+ : H^+ \to G$ such that $\rho|_{\Gamma^+} = \hat{\rho}^+|_{\Gamma^+}$. By Corollary 6.4 we get a continuous homomorphism $\hat{\rho} : H \to G$ such that $\hat{\rho}|_{H^+} = \hat{\rho}^+$. Considering Γ^+ as a lattice in Γ and noting that $\hat{\rho}|_\Gamma, \rho : \Gamma \to G$ are two homomorphisms that coincide on Γ^+, we deduce by [3, lemma 6.3] that $\hat{\rho}|_\Gamma = \rho$. This finishes the proof.

8 A fine version of Theorem 1.1

Taking $\theta = \hat{\rho}$ in Proposition 6.3 in case the target group $\mathbf{G}$ is absolutely simple and using Proposition 6.2 otherwise, we get the following corollary of Theorem 1.1, which could be viewed as a finer version of this theorem.

COROLLARY 8.1. *In the setting of Theorem 1.1, if* $\mathbf{G}$ *is absolutely simple, then there exist a unique field embedding* $i : \ell \to k$*, which is continuous, and a corresponding unique k-algebraic groups morphism* $\mathbf{H} \to \mathbf{G}$ *such that* ρ *coincides with the precomposition of the corresponding map* $\mathbf{H}(k) \to \mathbf{G}(k)$ *with the injection* $\Gamma < \mathbf{H}(\ell) \to \mathbf{H}(k)$.

In the general case, where $\mathbf{G}$ *is merely k-simple, considering the finite field extension* k' *of* k *and the* k'*-algebraic group* $\mathbf{G}'$ *given in Proposition 6.2, there exists a unique field embedding* $i : \ell \to k'$*, which is continuous, and a corresponding unique k-algebraic group morphism* $\mathbf{H} \to \mathbf{G}'$ *such that* ρ *coincides with the composition of the corresponding maps* $\Gamma < \mathbf{H}(\ell)$, $\mathbf{H}(\ell) \to \mathbf{H}(k')$, $\mathbf{H}(k') \to \mathbf{G}'(k')$, *and* $\mathbf{G}'(k') \simeq \mathbf{G}(k)$.

References

[1] Uri Bader, Bruno Duchesne, and Jean Lecureux, *Almost algebraic actions of algebraic groups and applications to algebraic representations*, Groups Geom. Dyn. **11** (2017), no. 2, 705–738, DOI 10.4171/GGD/413.

[2] U. Bader and A. Furman, *Algebraic representations of ergodic actions and super-rigidity*, ArXiv e-prints (2013), available at 1311.3696.

[3] ———, *Super-rigidity and non-linearity for lattices in products*, Compos. Math. **156** (2020), no. 1, 158–178, DOI 10.1112/s0010437x19007607, available at 1802.09931. MR4036451.

[4] U. Bader and T. Gelander, *Equicontinuous actions of semisimple groups*, Groups Geom. Dyn. **11** (2017), no. 3, 1003–1039, DOI 10.4171/GGD/420.

[5] A. Borel, *Linear algebraic groups*, 2nd. enl. ed., Graduate Texts in Mathematics, vol. 126, Springer-Verlag, New York, 1991.

[6] S. Bosch, U. Güntzer, and R. Remmert, *Non-Archimedean analysis* **261** (1984), xii+436, DOI 10.1007/978-3-642-52229-1. A systematic approach to rigid analytic geometry.

[7] E. G. Effros, *Transformation groups and C^*-algebras*, Ann. of Math. (2) **81** (1965), 38–55.

[8] T. Gelander, A. Karlsson, and G. A. Margulis, *Superrigidity, generalized harmonic maps and uniformly convex spaces*, Geom. Funct. Anal. **17** (2008), no. 5, 1524–1550.

[9] E. Glasner and B. Weiss, *Weak mixing properties for non-singular actions*, Ergodic Theory Dynam. Systems **36** (2016), no. 7, 2203–2217.

[10] G. A. Margulis, *Discrete subgroups of semisimple Lie groups*, Ergebnisse der Mathematik und ihrer Grenzgebiete (3) [Results in mathematics and related areas (3)], vol. 17, Springer-Verlag, Berlin, 1991.

[11] N. Monod, *Superrigidity for irreducible lattices and geometric splitting*, J. Amer. Math. Soc. **19** (2006), no. 4, 781–814.

[12] R. J. Zimmer, *Ergodic theory and semisimple groups*, Monographs in Mathematics, vol. 81, Birkhäuser Verlag, Basel, 1984. MR776417 (86j:22014).

3

AARON BROWN, FEDERICO RODRIGUEZ HERTZ, AND ZHIREN WANG

THE NORMAL SUBGROUP THEOREM THROUGH MEASURE RIGIDITY

Abstract. We present an expository proof of Margulis's normal subgroup theorem and measurable factor theorem using tools of measure rigidity for actions of higher-rank abelian groups in homogeneous dynamics.

1 Introduction and main results

1.1 INTRODUCTION. We present an expository proof of Margulis's normal subgroup theorem, Theorem 1.1 below, which appeared in [15] and [17] as translated in [16] and [18]. For certain discrete subgroups Γ (namely, for irreducible lattices in higher-rank semisimple Lie groups and for some more general groups), the normal subgroup theorem asserts that any normal subgroup N of Γ is either of finite index in Γ or is contained in the center of Γ.

The proof of the normal subgroup theorem follows in two steps: First, one establishes that Γ/N has Kazhdan's property (T). When all simple factors of G have higher (real) rank this fact is well known. When Γ is irreducible and G has rank-1 factors, additional arguments are needed to show noncentral normal subgroups of Γ have property (T); these appear as [17, theorem 1.3.2, theorem 1.4] combined with Margulis's arithmeticity theorem. The second step in the proof is to show that Γ/N is amenable whenever N is noncentral. This follows from Margulis's measurable factor theorem, Theorem 1.2 below,

Aaron Brown. Northwestern University, Evanston, IL 60208, USA
awb@northwestern.edu

Federico Rodriguez Hertz. Pennsylvania State University, State College, PA 16802, USA
hertz@math.psu.edu

Zhiren Wang. Pennsylvania State University, State College, PA 16802, USA
zhirenw@psu.edu

which appears as [15, theorem 1.14.2]. See also [19, chapter IV] for more general statements and complete proofs of the normal subgroup and measurable factor theorems.

Our proof of the normal subgroup theorem follows Margulis's proof. We present an alternative proof of the measurable factor theorem. Margulis's proof of the measurable factor theorem in [15] and [19] may be viewed as a result on the rigidity of certain σ-algebras. The proof we give is based on the rigidity of invariant measures for actions of higher-rank abelian groups in homogeneous dynamics. This approach is inspired by arguments from our paper [2]. However, none of the arguments presented here are original to us. In particular, the proof we present below is highly derivative of [5]–[8], [11], and [12].

1.2 DEFINITIONS. We begin with some definitions. Let $\mathfrak{g}$ be a real Lie algebra. Recall that $\mathfrak{g}$ is *semisimple* if $[\mathfrak{g}, \mathfrak{g}] = \mathfrak{g}$. The (real) *rank* of $\mathfrak{g}$ is, roughly, the dimension of the maximal ad-semisimple, abelian subalgebra $\mathfrak{a} \subset \mathfrak{g}$. (See section 8.1 below for a more precise definition.) A Lie group G is *semisimple* if its Lie algebra $\mathfrak{g}$ is semisimple and the *rank* of G is the rank of $\mathfrak{g}$. We will always assume G is connected. A semisimple Lie group admits a bi-invariant, locally finite volume form called the *Haar measure*. A *lattice* in G is a discrete subgroup such that the quotient G/Γ has finite volume.

A semisimple Lie group G has an almost direct product structure into normal subgroups $G = \Pi_{i=1}^{k} G_i$ of positive dimension. When no normal factor G_i is compact, we say a lattice $\Gamma \subset G$ is *irreducible* if, for every proper subset $C \subset \{1, \ldots, k\}$, the image of Γ under the natural projection $G \to \Pi_{i \in C} G_i$ is dense; this implies for any normal subgroup $H \subset G$ of positive dimension that $H\Gamma$ is dense in G.

Let G be semisimple and let $G = KAN$ be a choice of Iwasawa decomposition of G. (See section 8.1 below for details.) In particular, if G has finite center then K is a maximal compact subgroup, A is a maximal connected abelian subgroup whose image under the adjoint representation is $\mathbb{R}$-diagonalizable, and N is a connected subgroup normalized by A whose image under the adjoint representation is unipotent. The subgroup A has Lie algebra $\mathfrak{a}$ and dimension the rank of G. Let $M = C_K(A)$ be the centralizer of A in K and let $P = MAN$. Then P is a *minimal parabolic subgroup*. A *parabolic subgroup* of G is a closed subgroup Q containing a minimal parabolic subgroup for some choice of Iwasawa decomposition.

Given a closed subgroup $Q \subset G$, let λ_Q denote the (left) Haar measure on Q. Given a parabolic subgroup $Q \subset G$, we also denote by $\lambda_{Q\backslash G}$ the (right)

K-invariant volume form on $Q\backslash G$; this measure is always in the Lebesgue class. Similarly, given a lattice $\Gamma \subset G$, we write $\lambda_{G/\Gamma}$ for the normalized Haar measure on G/Γ.

1.3 MAIN RESULTS. We denote by $Z(G)$ the center of the group G. In [15] and [17], Margulis established the following rigidity of normal subgroups of irreducible lattices in higher-rank Lie groups.

THEOREM 1.1 (normal subgroup theorem; [19, theorem IV.4.10]). *Let G be a connected semisimple Lie group with rank at least 2 and no nontrivial compact factors. Let Γ be an irreducible lattice subgroup. If $N \lhd \Gamma$ is a normal subgroup of Γ, then either $N \subset Z(G)$ or N has finite index in Γ.*

In many situations, such as when G is linear, the center $Z(G)$ is finite; in this case, Theorem 1.1 asserts that every normal subgroup of Γ is either finite or cofinite.

The normal subgroup theorem, Theorem 1.1, follows from the following theorem characterizing measurable factors of the right action of Γ on $P\backslash G$. This action does not preserve any Borel probability measure on $P\backslash G$; however, it preserves the Lebesgue measure class $\lambda = \lambda_{P\backslash G}$. Given a standard measure space (X, μ), we say a (left) Borel action of Γ on (X, μ) is *nonsingular* if $\gamma_*\mu$ is equivalent to μ for every $\gamma \in \Gamma$. Roughly, Margulis's measurable factor theorem states that if a nonsingular left action of Γ on a measure space (X, μ) is a measurable factor of the right Γ-action on $(P\backslash G, \lambda_{P\backslash G})$, then (X, μ) is measurably isomorphic to $(Q\backslash G, \lambda_{Q\backslash G})$ for some parabolic subgroup $Q \supset P$; moreover, this isomorphism intertwines the left Γ-action on (X, μ) with the right Γ-action on $(Q\backslash G, \lambda_{Q\backslash G})$.

THEOREM 1.2 (measurable factor theorem; [19, corollary IV.2.13]). *Let G be a connected semisimple Lie group with rank at least 2 and no nontrivial compact factors. Let Γ be an irreducible lattice subgroup.*

Let Γ act on a Borel space X and let $p\colon P\backslash G \to X$ be a Borel map defined $\lambda_{P\backslash G}$-a.e. Assume that p is Γ-equivariant: for $\lambda_{P\backslash G}$-a.e. g and every $\gamma \in \Gamma$, we have

$$p(Pg\gamma) = \gamma^{-1} \cdot p(Pg).$$

Let $\mu = p_*\lambda_{P\backslash G}$ be the image of $\lambda_{P\backslash G}$ under p. Then there is a parabolic subgroup $Q \supset P$ such that (X, μ) is Γ-equivariantly isomorphic to $(Q\backslash G, \lambda_{Q\backslash G})$: there is a Γ-equivariant isomorphism of measure spaces $H\colon (X, \mu) \to (Q\backslash G, \lambda_{Q\backslash G})$ such that *if* $\pi : P\backslash G \to Q\backslash G$ *is the natural map, then the following diagram commutes:*

$$\begin{array}{ccc} (P\backslash G, \lambda_{P\backslash G}) & \xrightarrow{\ p\ } & (X, \mu) \\ & \overset{\pi}{\searrow} & \downarrow{\scriptstyle H} \\ & & (Q\backslash G, \lambda_{Q\backslash G}) \end{array}$$

In the above theorem, the Γ-equivariance of p implies the measure $\mu = p_* \lambda_{P\backslash G}$ is nonsingular for the Γ-action on X. Since the action of Γ on $P\backslash G$ is on the right, the Γ-equivariance of the isomorphism $H\colon (X, \mu) \to (Q\backslash G, \lambda_{Q\backslash G})$ asserts that $H(\gamma \cdot x) = H(x) \cdot \gamma^{-1}$ for μ-a.e. x and every $\gamma \in \Gamma$. The isomorphism H in Theorem 1.2 need not be defined everywhere but only on a set of full μ-measure.

A natural setting in which a measurable factor of the Γ-action on $P\backslash G$ appears is stated in Lemma 3.1 below. This forms a key step in the proof of Theorem 1.1 through Lemma 4.1.

One may ask if analogous results hold when the map p in Theorem 1.2 is assumed to be continuous or smooth. In [3], Dani proves a result analogous to Theorem 1.2 for continuous factors—that is, assuming the map p is a continuous surjection. More recently, Gorodnik and Spatzier studied in [10] smooth factors and (under an additional mild hypothesis) establish a smooth analogue of Theorem 1.2.

2 Representations, property (T), and amenability

To establish Theorem 1.1 we introduce the concepts of amenability and of property (T) groups. We begin with the following definition.

DEFINITION 2.1 (almost-invariant vectors). Let H be a locally compact topological group and let π be a unitary representation of H. We say that π admits *almost-invariant vectors* if, for every $\epsilon > 0$ and every compact subset $C \subset H$, there is a unit vector v with

$$\sup_{h \in C} \| \pi(h)v - v \| < \epsilon.$$

2.1 PROPERTY (T) GROUPS. We have the following definition.

DEFINITION 2.2 (property (T) groups). A locally compact topological group H has Kazhdan's *property (T)* if every unitary representation admitting almost-invariant vectors has a nontrivial invariant vector.

REMARK 2.3 (facts on property (T) groups). We collect several well-known facts about property (T) groups. See [1, chapter 1] or [23, chapter 13] for detailed exposition on property (T).

(1) Compact groups have property (T).
(2) A product group $G_1 \times G_2$ has property (T) if and only if both the factors G_1 and G_2 have property (T).
(3) If G is a connected simple Lie group with real rank at least 2, then G has property (T); more generally, if every almost simple factor of G has real rank at least 2, then G has property (T). (See [19, corollary III.5.4].)
(4) A Lie group G has property (T) if and only if every lattice subgroup Γ of G has property (T). (See [19, theorem III.2.12].)
(5) If a Lie group G has property (T) and if H is a closed normal subgroup of G, then the quotient group G/H has property (T).
(6) Suppose G is a semisimple Lie group with no compact factors and at least one almost-simple factor of G has real rank at least 2. Let Γ be an irreducible lattice. Then for any noncentral normal subgroup $N \subset \Gamma$, the quotient Γ/N has property (T). See [19, theorem III.5.9(B)]. See also [18, theorem 1.4].
(7) More generally, suppose that a semisimple Lie group G has real rank at least 2 and no compact factors and that Γ is an irreducible lattice. Then for any noncentral normal subgroup $N \subset \Gamma$, the quotient Γ/N has property (T). This follows from [19, theorem IV.3.9] and Margulis's arithmeticity theorem [19, theorem IV.3.9]. See also [18, theorem 1.3.2].

The group $\Gamma = \mathrm{SL}(2, \mathbb{Z}[\sqrt{2}])$ is an irreducible lattice in $\mathrm{SL}(2,\mathbb{R}) \times \mathrm{SL}(2,\mathbb{R})$. It is well known that $\mathrm{SL}(2,\mathbb{R})$ fails to have property (T). It follows that $\mathrm{SL}(2,\mathbb{R}) \times \mathrm{SL}(2,\mathbb{R})$ and hence Γ fail to have property (T). However, from Remark 2.3(7), the quotient Γ/N has property (T) for any noncentral normal subgroup $N \subset \Gamma$.

2.2 AMENABILITY. Let λ_H be a left-invariant Haar measure on a locally compact topological group H. Then $L^2(H, \lambda_H)$ is a Hilbert space and the action of H on itself by left translation induces a unitary representation of H on $L^2(H, \lambda_H)$ called the left-regular representation.

DEFINITION 2.4 (amenability). A locally compact topological group H is *amenable* if the left-regular representation of H in $L^2(H, \lambda_H)$ admits almost-invariant vectors.

Examples of amenable groups include all compact groups and all abelian, nilpotent, or solvable Lie groups.

We recall an equivalent notion of amenability that we use in the sequel. See, for instance, [19, (5.5.1)], [1, appendix G], or [23, section 12.3] for other characterizations of amenability.

LEMMA 2.5 (equivalent characterization of amenability). *H is amenable if and only if any continuous action of H on any compact metric space admits an invariant Borel probability measure.*

2.3 PROPERTY (T) AND AMENABILITY. We have the following well-known fact.

LEMMA 2.6. *If a locally compact topological group H has property (T) and is amenable, then H is compact. In particular, if a countable discrete group has property (T) and is amenable, then it is finite.*

Proof. Suppose the left-regular representation of H in $L^2(H, \lambda_H)$ admits almost-invariant vectors. If H has property (T), then this representation admits a nontrivial invariant vector. Such a vector coincides with a nonzero constant function $\varphi\colon H \to \mathbb{C}$. However, if a nonzero constant function φ is an element of $L^2(H, \lambda_H)$, then λ_H must be finite, whence H is compact. □

3 Suspension space, induced G-action, and Furstenberg's lemma

We present a key construction on which our proof of Theorem 1.2 depends. We also recall a classical result of Furstenberg and give a proof based on this construction.

3.1 SUSPENSION SPACE AND INDUCED G-ACTION. As above, let G be a semisimple Lie group and let Γ be a lattice subgroup. Consider a continuous action of Γ on a compact metric space X. We recall a standard construction from which we induce a continuous G-action on an auxiliary space.

On $G \times X$ consider the right action of Γ given by

$$(g, x) \cdot \gamma = (g\gamma, \gamma^{-1} \cdot x)$$

and a left action of G given by

$$g' \cdot (g, x) = (g'g, x).$$

These actions commute. Let X^Γ denote the quotient

$$X^\Gamma = (G \times X)/\Gamma$$

by the Γ action. The G-action on $G \times X$ descends to a G-action on X^Γ. Writing $[g, x]$ for the Γ-equivalence class of $(g, x) \in G \times X$ in X^Γ, we have $g' \cdot [g, x] = [g'g, x]$.

Note that X^Γ has a fiber-bundle structure over G/Γ with fibers homeomorphic to X:

$$\begin{array}{ccc} X & \xrightarrow{\iota} & X^\Gamma \\ & & \downarrow{\scriptstyle \pi} \\ & & G/\Gamma \end{array}$$

The G-action on X^Γ fibers over the G-action on G/Γ. If Γ is cocompact in G, then X^Γ is compact. If Γ is nonuniform, then we may equip X^Γ with a metric such that X^Γ is a complete, second countable, locally compact metric space and such that the projection $X^\Gamma \to G/\Gamma$ is distance nonincreasing with respect to some fixed choice of right-invariant distance on G.

3.2 FURSTENBERG'S LEMMA. We have the following lemma due to Furstenberg; see [9, theorem 15.1]. Let Γ act continuously on a compact metric space X. Let $\mathcal{P}(X)$ denote the space of Borel probability measures on X equipped with the weak-$*$ topology. The Γ-action on X naturally induces a continuous action of Γ on $\mathcal{P}(X)$: given $\mu \in \mathcal{P}(X)$, $\gamma_* \mu \in \mathcal{P}(X)$ is the Borel measure

$$\gamma_* \mu(B) = \mu(\gamma^{-1} \cdot B).$$

LEMMA 3.1 (Furstenberg's lemma). *There exists a Borel measurable function*

$$h \colon G \to \mathcal{P}(X)$$

such that

(1) *$h(g\gamma^{-1}) = \gamma_* h(g)$ for λ_G-a.e. g and every $\gamma \in \Gamma$, and*
(2) *$h(pg) = h(g)$ for every $p \in P$ and λ_G-a.e. g.*

In particular, h descends to a Γ-equivariant map $h\colon P\backslash G \to \mathcal{P}(X)$ defined $\lambda_{P\backslash G}$-a.e.

Proof. Let X^Γ be the suspension space associated to the Γ-action on X. Let $P = MAN$ be a minimal parabolic subgroup of G. As P is a compact extension of a solvable group, P is amenable. Although X^Γ need not be compact, the set of Borel probability measures on X^Γ projecting to the normalized Haar measure $\lambda_{G/\Gamma}$ on G/Γ is a compact, P-invariant subset of the space of Borel probability measures on X^Γ. Since P is amenable, by a slight extension of the characterization of amenability in Lemma 2.5, there exists a P-invariant Borel probability measure ν on X^Γ projecting to the normalized Haar measure $\lambda_{G/\Gamma}$ on G/Γ.

Fix such a P-invariant Borel probability measure ν on X^Γ. There exists a unique lift $\widetilde{\nu}$ of ν to a locally finite Borel measure on $G \times X$. The measure $\widetilde{\nu}$ is a Γ-invariant, P-invariant Borel measure that is finite on compact sets. The partition of $G \times X$ into elements of the form $\{g\} \times X$ is measurable and hence admits a family of conditional measures $\widetilde{\nu}_g$ (see Definition 6.1 and Lemma 6.2 below) parameterized by $g \in G$. Identifying each $\{g\} \times X$ with X, we view each $\widetilde{\nu}_g$ as a Borel probability measure on X and obtain a measurable map $h\colon G \to \mathcal{P}(X)$ given by $h\colon g \mapsto \widetilde{\nu}_g$.

By the P-invariance of $\widetilde{\nu}$, we have $\widetilde{\nu}_g = \widetilde{\nu}_{pg}$ for $p \in P$ and hence h descends to a well-defined function $h\colon P\backslash G \to \mathcal{P}(X)$ and (2) follows. Moreover, since the lifted measure $\widetilde{\nu}$ on $G \times X$ is Γ-invariant, we obtain Γ-equivariance of the measures $\{\widetilde{\nu}_g\}$,

$$\nu_{g\gamma} = \gamma_*^{-1}\widetilde{\nu}_g,$$

and (1) follows. □

4 The measurable factor theorem implies the normal subgroup theorem

The proof of Theorem 1.1 follows immediately from the following lemma, which we derive from Theorem 1.2.

LEMMA 4.1 (see [15, theorem 2.7]). *Let G be a connected semisimple Lie group with real rank at least 2 and no nontrivial compact factors. Let Γ be an irreducible lattice subgroup and let N be a normal subgroup of Γ. Then either $N \subset Z(G)$ or Γ/N is amenable.*

Proof of the normal subgroup theorem, Theorem 1.1. Let N be a non-central normal subgroup of Γ. As discussed in Remark 2.3, the quotient $H = \Gamma/N$ has property (T). By Lemma 4.1, if N is noncentral, then $H=\Gamma/N$ is amenable and hence finite by Lemma 2.6. Thus N is of finite index in Γ whenever N is noncentral. □

To establish Lemma 4.1, we use the characterization of amenability in Lemma 2.5. Fix a compact metric space X. Consider an action of $H=\Gamma/N$ by homeomorphisms of X. Note that the action of H induces an action of Γ for which every element in N acts as the identity transformation. Assuming that N is noncentral, we will show there exists an invariant Borel probability measure for this action; as X was arbitrary, it follows from Lemma 2.5 that H is amenable.

Proof of Lemma 4.1. Let $Y=\mathcal{P}(X)$ denote the set of Borel probability measures on X equipped with the weak-$*$ topology. The continuous action of Γ on X induces a continuous action of Γ on Y. By Lemma 3.1, we obtain a Γ-equivariant Borel measurable map $h\colon P\backslash G\to Y$ defined $\lambda_{P\backslash G}$-a.e. Let $\mu=h_*\lambda_{P\backslash G}$. Then μ is a nonsingular measure for the action of Γ on Y and (Y,μ) is a Γ-equivariant factor of the Γ-action on $(P\backslash G,\lambda_{P\backslash G})$. By Theorem 1.2, there is a parabolic subgroup Q and a Γ-equivariant, measurable isomorphism $H\colon (Q\backslash G,\lambda_{Q\backslash G})\to(Y,\mu)$ such that $h=H\circ\pi$ where $\pi: P\backslash G\to Q\backslash G$ is the natural projection.

We claim that $Q=G$ whenever N is noncentral. In this case, the quotient $Q\backslash G$ is a singleton, whence the measure μ on $Y=\mathcal{P}(X)$ is a point-mass, $\mu=\delta_{\mu_0}$, for some $\mu_0\in Y=\mathcal{P}(X)$. It follows that μ_0 is a fixed point for the Γ-action on $\mathcal{P}(X)$, whence μ_0 is a Γ-invariant Borel probability measure on X.

To complete the proof, it suffices to show that $Q\neq G$ implies $N\subset Z(G)$. Recall that $N\subset\Gamma$ acts trivially on X and hence also acts trivially on $Y=\mathcal{P}(X)$. By the measurable identification of (Y,μ) with $(Q\backslash G,\lambda_{Q\backslash G})$, we have that N acts trivially as a group of measurable transformations of $(Q\backslash G,\lambda_{Q\backslash G})$; as N acts continuously on $Q\backslash G$ and as $\lambda_{Q\backslash G}$ has full support, we have that N acts on $Q\backslash G$ by the identity homeomorphism.

Let $L\subset G$ denote the kernel of the right action of G on $Q\backslash G$; that is,

$$L=\{h\in G: Qgh=Qg \text{ for all } g\in G\}.$$

We have that L is a closed normal subgroup of G. If $L\neq G$, then G may be written as an almost direct product $G=H\cdot L$ for some normal subgroup $H\subset G$ of positive dimension. Since $N\subset L$ and H commutes with L, we have that H is contained in $N_G(N)$, the normalizer of N in G. Moreover, as N is normal in Γ we have $\Gamma \subset N_G(N)$. We thus have

$$\overline{H \cdot \Gamma} \subset N_G(N).$$

However, since Γ is irreducible, we have $\overline{H \cdot \Gamma} = G$. It follows that N is a discrete normal subgroup of G. This implies $N \subset Z(G)$ by a standard fact we recall in Lemma 4.2 below. □

To complete the proof of Lemma 4.1, we recall the following well-known fact and its proof.

LEMMA 4.2. *If N is a discrete normal subgroup of G, then N is central.*

Proof. Fix $n \in N$. Since N is discrete and normal, there is a compact neighborhood of the identity, $C \subset G$, such that $gng^{-1} = n$ for all $g \in C$. Since G is connected, C generates G and it follows that $gng^{-1} = n$ for all $g \in G$. □

5 Preliminaries and reformulation of Theorem 1.2

Let X be a standard Borel space. Let Γ be as in Theorem 1.2 and consider a Borel action of Γ on X. In the setting of the proof of Theorem 1.1, the natural action used in the proof of Lemma 4.1 was a continuous action. We have the following, which, in the abstract setting of Theorem 1.2, allows us to assume that the action is continuous.

LEMMA 5.1 (See [22, theorem 3.2]). *There exists a compact metric space Z, a continuous Γ-action on Z, and an injective, Γ-equivariant Borel map $\iota\colon X \to Z$.*

Pushing forward the measure on X to a measure on Z, we obtain an almost surjective, Γ-equivariant function $p\colon P\backslash G \to Z$. Replacing X with Z, we may thus assume for the remainder that the Γ-action on X is continuous.

We follow the notation of Theorem 1.2. Lift the Γ-equivariant measurable map $p\colon P\backslash G \to X$ to a Γ-equivariant map $\hat{p}\colon G \to X$,

$$\hat{p}(g) = p(Pg).$$

Let

$$Q := \{g \in G : \hat{p}(gx) = \hat{p}(x) \text{ for } \lambda_G\text{-a.e. } x \in G\}.$$

We have $P \subset Q$. Moreover, the definition implies that Q is a subgroup of G. Indeed if $g_1, g_2 \in Q$, then there are full measure subsets $R_1, R_2 \subset G$ such that

$\hat{p}(g_i x) = \hat{p}(x)$ for all $x \in R_i$; then $R = (g_2^{-1} \cdot R_1) \cap R_2$ has full measure in G and for $x \in R$ we have

$$\hat{p}(g_1 g_2 x) = \hat{p}(g_2 x) = \hat{p}(x),$$

whence $g_1 g_2 \in Q$. Similarly, the set $R' = g_1 \cdot R_1$ has full measure in G and for $x \in R'$,

$$\hat{p}(x) = \hat{p}(g_1 g_1^{-1} x) = \hat{p}(g_1^{-1} x),$$

whence $g_1^{-1} \in Q$.

Although not clear from the above definition, it will follow from observations below that Q is a closed subgroup. In particular, Q is a parabolic subgroup of G.

By definition of Q, the function $p\colon P\backslash G \to X$ in Theorem 1.2 descends to a well-defined, Γ-equivariant function $Q\backslash G \to X$. To establish Theorem 1.2, it remains to show that induced function $Q\backslash G \to X$ is $\lambda_{Q\backslash G}$-a.s. injective. That is, we show for a full-measure subset of G that the preimages of $\hat{p}\colon G \to X$ are Q-orbits. In particular, the proof of Theorem 1.2 follows from the following.

LEMMA 5.2. *There exists a full λ_G-measure subset $\hat{R} \subset G$ such that if $g \in \hat{R}$, then*

$$\hat{p}^{-1}(\hat{p}(g)) \cap \hat{R} \subset Qg.$$

To begin the proof of Lemma 5.2, we construct a Borel probability measure ν on X^Γ. To construct this measure ν, let $\hat{P}\colon G \to G \times X$ denote the inclusion of G into the graph of $\hat{p}$; that is,

$$\hat{P}(g) = (g, \hat{p}(g)).$$

Let $\widetilde{\nu} = \hat{P}_* \lambda_G$ denote the image of the Haar measure on G under $\hat{P}$. Then $\widetilde{\nu}$ is a locally finite Borel measure on $G \times X$. By the Γ-equivariance of $\hat{p}$, the measure $\widetilde{\nu}$ is right Γ-invariant and hence descends to a finite Borel measure ν on X^Γ, which we normalize to be a probability measure.

By the definition of Q we have that $Q \subset G$ coincides with the stabilizers of $\widetilde{\nu}$ and ν. In particular, this shows that Q is a closed subgroup of G.

Lemma 5.2 may be reformulated in terms of the *leaf-wise measures* of ν along orbits of certain subgroups of G acting on X^Γ. See section 6.3 for details on leaf-wise measures. Here we simply describe the properties that we use. Let N^- denote the subgroup opposite to N; that is, if N is generated by positive root spaces, then N^- is generated by negative root spaces (see section 8.1). Then PN^- is a dense open subset of G.

Associated to each subgroup $H \subset G$ (for which ν-almost every H-orbit on X^Γ is free) we construct in section 6.3 a measurable family of locally finite (hence Radon) Borel measures $\{\nu_x^H : x \in X\}$ on H called the *leaf-wise* measures of ν associated to the subgroup H.

Given $x \in X^\Gamma$ with free H-orbit, let

$$\Phi_x \colon H \to X, \quad \Phi_x \colon h \mapsto h \cdot x$$

be the canonical parametrization of the orbit $H \cdot x$. We may then push forward each measure ν_x^H to a Borel (in the intrinsic orbit topology on $H \cdot x$) measure $(\Phi_x)_* \nu_x^H$ on the orbit $H \cdot x$. Recall that two locally finite Borel measures μ_1, μ_2 on H are *proportional*, written $\mu_1 \propto \mu_2$, if there is $c > 0$ such that

$$\mu_1(B) = c\mu_2(B)$$

for all Borel sets B. The family of leaf-wise measures $\{\nu_x^H : x \in X^\Gamma\}$ on H have the following properties:

(1) If $E \subset X^\Gamma$ is a Borel set, then $\nu(E) = 0$ if and only if for ν-a.e. x,

$$(\Phi_x)_* \nu_x^H(E) = 0.$$

(2) There is a full measure subset $E \subset X^\Gamma$ such that for $x \in E$ and $h \in H$ such that $h \cdot x \in E$,

$$(\Phi_{hx}^{-1})_* (\Phi_x)_* \nu_x^H \propto \nu_{h \cdot x}^H.$$

Note that if $y = h \cdot x \in H \cdot x$, then $\Phi_y^{-1} \circ \Phi_x \colon H \to H$ corresponds to right translation by h^{-1}; indeed,

$$\Phi_y^{-1} \circ \Phi_x(h') = \Phi_y^{-1}(h'h^{-1}hx) = h'h^{-1}.$$

In particular, property (2) above implies

$$(r_{h^{-1}})_* \nu_x^H \propto \nu_{h \cdot x}^H,$$

where $r_h \colon H \to H$ denotes right translation on H by h.

Write $Q^- := N^- \cap Q$. Lemma 5.2 is equivalent to the following proposition, whose proof occupies sections 6–9.

PROPOSITION 5.3. *There exists a set $R' \subset X^\Gamma$ of full ν-measure such that for $x \in R'$, the measure $\nu_x^{N^-}$ is supported on Q^-. In particular, for $x \in R'$,*

$$\nu_x^{N^-} = \nu_x^{Q^-}.$$

We note that since ν is assumed to be Q^--invariant, we have that $\nu_x^{Q^-}$ is the Haar measure λ_{Q^-} on Q^- for ν-almost every $x \in R'$. (See Claim 6.5 below.)

We show Proposition 5.3 implies Lemma 5.2.

Proof of Lemma 5.2. From Proposition 5.3, the properties of leaf-wise measures discussed above, and the Q-invariance of ν, we may find a subset $R_0 \subset X^\Gamma$ with $\nu(R_0) = 1$ on which the following properties hold:

(1) For $x \in R_0$, we have $\nu_x^{N^-} = \lambda_{Q^-}$.

(2) For $x \in R_0$ and $h \in N^-$ such that $h \cdot x \in R_0$, we have

$$(\Phi_{hx}^{-1})_*(\Phi_x)_*\nu_x^{N^-} \propto \nu_{h\cdot x}^{N^-};$$

in particular, for all such x and h, we have $h^{-1} \in Q^-$ and hence $h \in Q^-$.

(3) For $x \in R_0$ and λ_Q-almost every $q \in Q$, we have $qx \in R_0$.

Viewing $G \times X$ as a covering space of X^Γ, we lift $R_0 \subset X^\Gamma$ to a Γ-invariant conull subset $\widetilde{R}_0 \subset G \times X$. Let $\hat{R}$ denote the image of $\widetilde{R}_0$ under the projection $G \times X \to G$. We claim Lemma 5.2 holds with this $\hat{R}$.

Take $g \in \hat{R}$ and write $y = \hat{p}(g) \in X$. Then $(g, y) \in \widetilde{R}_0$ and (following the notation from section 3.1) we have $x = [g, y] \in R_0$. Consider $g' \in \hat{R}$ such that $\hat{p}(g') = y$. Write $h = g'g^{-1}$. Then $h \cdot x = [g', y] \in R_0$.

To complete the proof, we claim $h \in Q$. Every element $h \in G$ can be written in the form $h = q_2^{-1} n q_1$, where $q_i \in Q$ and $n \in N^-$; moreover, for each fixed $h \in G$ there is an open set of $q_1 \in Q$ for which such q_2 and n exist and depend rationally on q_1. In particular, there are $q_1, q_2 \in Q$ such that $q_2 h$ is contained in the N^--orbit of q_1; since $x \in R_0$ and $h \cdot x \in R_0$ and since rational maps are locally Lipschitz (and hence preserve λ_Q-null sets), we may moreover assume q_1 and q_2 are chosen so that $q_1 \cdot x \in R_0$ and $q_2 h \cdot x \in R_0$.

Since $q_2 h \cdot x \in R_0$ and is contained in the N^--orbit of $q_1 \cdot x$, we have that $q_2 h q_1^{-1}$ is in the support of $\nu_{q_1\cdot x}^{N^-}$. Since $q_1 \cdot x \in R_0$, it follows that $q_2 h q_1^{-1} \in Q^-$, whence $h \in Q$. □

The proof of Proposition 5.3 is carried out in the next four sections using tools from smooth ergodic theory and measure rigidity for homogeneous dynamics. The main tools we use are derivative of [4], [5], and [11].

6 Conditional and leaf-wise measures

6.1 MEASURE THEORY. Let $(X, \mathcal{A}, \mu)$ be a complete probability space. We say $(X, \mathcal{A}, \mu)$ is *standard* if X may be equipped with a topology of a Polish space such that $\mathcal{A}$ is the μ-completion of the σ-algebra of Borel set $\mathcal{B}$ in this topology. If $(X, \mathcal{A}, \mu)$ is standard, it is measurably isomorphic to the union of an interval $[0, a)$, $0 \leq a \leq 1$ equipped with the Lebesgue measure and countably many point-masses.

Let $(X, \mathcal{A}, \mu)$ be a standard probability space. Let $\mathcal{P}$ and $\mathcal{Q}$ be partitions of X by μ-measurable sets. We say that $\mathcal{P}$ is finer than $\mathcal{Q}$ (or that $\mathcal{Q}$ is coarser than $\mathcal{P}$), written $\mathcal{Q} \prec \mathcal{P}$, if there is a full measure subset $Y \subset X$ such that

$$\mathcal{P}(x) \cap Y \subset \mathcal{Q}(x) \cap Y$$

for all $x \in Y$. Given a partition $\mathcal{P}$, we say a measurable subset $A \in \mathcal{A}$ is $\mathcal{P}$-saturated if for all $x \in A$, $\mathcal{P}(x) \subset A$.

DEFINITION 6.1 (measurable partitions and the measurable hull). A partition $\mathcal{P}$ of $(X, \mathcal{A}, \mu)$ of a standard probability space is *measurable* if there is a countable collection $\{A_i\}$ of $\mathcal{P}$-saturated sets such that for every $x \in X$ and every $y \notin \mathcal{P}(x)$ there is A_j such that either $\mathcal{P}(x) \subset A_j$ and $\mathcal{P}(y) \subset X \smallsetminus A_j$ or $\mathcal{P}(x) \subset X \smallsetminus A_j$ and $\mathcal{P}(y) \subset A_j$.

Given an arbitrary partition $\mathcal{P}$, the *measurable hull* of a partition $\mathcal{P}$ is the finest measurable partition $\mathcal{Q}$ with $\mathcal{Q} \prec \mathcal{P}$.

6.2 CONDITIONAL MEASURES. We now fix X to be a second countable, locally compact, complete metric space. Let $\mathcal{M}$ denote the set of all Borel probability measures on X. Fix $\mu \in \mathcal{M}$. Then (X, μ) is a standard probability space when equipped with the μ-completion of the Borel σ-algebra.

We have the following standard construction.

LEMMA 6.2 (conditional measures; see [21]). *Given a measurable partition $\mathcal{P}$ of (X, μ), there is a measurable function $X \mapsto \mathcal{M}$, written $x \mapsto \mu_x^{\mathcal{P}}$, with the following properties:*

(1) $\mu_x^{\mathcal{P}}$ is a Borel probability measure on X with $\mu_x^{\mathcal{P}}(\mathcal{P}(x)) = 1$;
(2) for a.e. x and every $y \in \mathcal{P}(x)$, we have $\mu_x^{\mathcal{P}} = \mu_y^{\mathcal{P}}$;
(3) for every bounded Borel function $\varphi \colon X \to \mathbb{R}$,

$$\int \varphi \, d\mu = \int \int \varphi(z) \, d\mu_x^{\mathcal{P}}(z) \, d\mu(x).$$

Moreover, up to a null set, the family $x \to \mu_x^{\mathcal{P}}$ *is uniquely determined by the above properties.*

6.3 LEAF-WISE MEASURES. Consider a connected, locally compact topological group H equipped with a right-invariant metric. Suppose H acts continuously (on the left) on the complete, second countable, locally compact metric space X. We will moreover assume the action is *locally free*: for every $x \in X$ there is an open neighborhood $U \subset H$ of the identity on which $h \mapsto h \cdot x$ is injective.

Let μ be a Borel probability measure on X. There is a natural partition of X into the orbits of H, which we denote by $\mathcal{H}$. In general (and in most situations of interest here) the partition $\mathcal{H}$ of (X, μ) is not a measurable partition. We describe a procedure that associates to each orbit $H \cdot x \in \mathcal{H}$ a locally finite (in the intrinsic topology on the orbit $H \cdot x$ inherited from H) Borel measure that has similar properties to conditional measures associated to measurable partitions.

We begin with the following definition.

DEFINITION 6.3 (partitions subordinate to orbits). A measurable partition $\mathcal{P}$ is *subordinate* to the partition $\mathcal{H}$ into H-orbits if, for μ-a.e. $x \in X$, the following hold:

(1) $\mathcal{P}(x) \subset H \cdot x$,
(2) $\mathcal{P}(x)$ contains an open (in the orbit topology) neighborhood of x in $H \cdot x$, and
(3) $\mathcal{P}(x)$ is precompact (in the orbit topology) in $H \cdot x$.

For simplicity, in what follows we will moreover assume that for μ-almost every $x \in H$, the orbit $H \cdot x$ is free; that is, for μ-a.e. x, we assume the map $h \mapsto h \cdot x$ is injective. This holds for all groups H considered in the setting of the proof of Proposition 5.3. For such x, we have a canonical parametrization of the orbit $H \cdot x$ given by $\Phi_x \colon H \to X$, $\Phi_x(h) = h \cdot x$. Recall (as discussed in section 5) that if $y = h' \cdot x \in H \cdot x$, then

$$\Phi_y^{-1} \circ \Phi_x \colon H \to H$$

corresponds to right translation by h'^{-1}.

Let $\mathcal{R}(H)$ denote the space of locally finite Borel (hence Radon) measures on H equipped with the standard topology (dual to compactly supported functions). Given $r>0$, let $B_H(r)\subset H$ denote the ball of radius r in H centered at the identity with respect to the fixed right-invariant metric on H.

PROPOSITION 6.4. *There exists a μ-measurable function $X\to\mathcal{R}(H)$, denoted*

$$x\mapsto \nu_x^H$$

such that the following properties hold:

(1) ν_x^H is normalized so that $\nu_x^H(B_H(1))=1$ for μ-a.e. x.
(2) For any Borel set $E\subset X$, $\mu(E)=0$ if and only if for μ-almost every $x\in X$,

$$\nu_x^H(\Phi_x^{-1}(E))=0.$$

(3) There exists a subset $E_0\subset X$ of full μ measure such that for $x\in E_0$ and $h\in H$ with $h\cdot x\in E_0$,

$$(\Phi_x)_*\nu_x^H \propto (\Phi_{h\cdot x})_*\nu_{h\cdot x}^H.$$

(4) For any measurable partition $\mathcal{P}$ subordinate to the partition into H-orbits, μ-almost every x, and $A\subset\mathcal{P}(x)$, we have

$$\mu_x^{\mathcal{P}}(A)=\frac{(\Phi_x)_*\nu_x^H(A)}{(\Phi_x)_*\nu_x^H(\mathcal{P}(x))}.$$

Moreover, the above properties uniquely determine the family $\{\nu_x^H\}$ modulo null sets.

For a detailed proof, we refer to [8, theorem 6.3]. Below, we outline a construction of the leaf-wise measures $\{\nu_x^H\}$.

Proof outline. Fix any $R>1$. Fix $x\in X$ for which the H-orbit of x is free. There are open neighborhoods $x\in W\subset U$ of x such that

(a) for $y\in U$, the connected component of $H\cdot y\cap U$ containing y is a topologically embedded $\dim(H)$-dimensional disk D_y;
(b) the disks $y\mapsto D_y$ depend continuously on y for $y\in U$;
(c) the partition of U into disks $\{D_y : y\in U\}$ is measurable; and
(d) $B_H(R)\cdot y\subset D_y$ for every $y\in W$; in particular, $B_H(R)\cdot x\subset D_x$.

Given $y\in U$, let μ_y^U be the conditional measure for the normalized restriction of μ to U relative to the measurable partition $\{D_y : y\in U\}$ of U. Given any Borel subset $E\subset B_H(R)$ and y satisfying property (d), define

$$\nu_y^H(E):=\frac{\mu_y^U(E\cdot y)}{\mu_y^U(B_H(1)\cdot y)}.$$

We may check that $\nu_y^H(E)$ is defined (modulo μ) independently of the choice of R or U. By a countable exhaustion of the space by partitions of the form $\{D_y : y \in U\}$ as $R \to \infty$ as above, for μ-a.e. y the quantity $\nu_y^H(E)$ is defined for any compact subset $E \subset H$. We then obtain a family of measures $\{\nu_x^H\}$ with the desired properties. □

We may write $\mu_x^H := (\Phi_x)_* \nu_x^H$ for the locally finite Borel (with respect to the orbit topology) measure on the orbit $H \cdot x$. The family of measures $\{\mu_x^H\}$ may be more natural to consider geometrically. However, it is more convenient in what follows to consider the family $\{\nu_x^H\}$ as each ν_x^H is supported on H and hence we can compare ν_x^H and ν_y^H for $x \neq y$. For the family $\{\mu_x^H\}$, it only makes sense to compare μ_x^H and μ_y^H when $y = h \cdot x$ for some $h \in H$ (in which case we have $\mu_x^H \propto \mu_y^H$.)

As there is no canonical normalization of each ν_x^H, we write

$$[\nu_x^H] := \{c\nu_x^H : c > 0\}$$

for the projective class of measures on H that are positively proportional to ν_x^H. We have the following straightforward claim whose proof is a standard exercise.

CLAIM 6.5. *A Borel probability measure μ on X is H-invariant if and only if for almost every $x \in X$, the projective class of the leaf-wise measure $[\nu_x^H]$ coincides with the projective class of left Haar measures on H.*

7 Measure rigidity

Let H be a connected Lie group. Equip the Lie algebra $\mathfrak{h}$ of H with an inner product and equip H with an induced right-invariant metric. Write $\mathrm{Isom}(H)$ for the group of isometries of H. Write $\mathrm{Isom}^H(H) \subset \mathrm{Isom}(H)$ for the subgroup of isometries given by right translations. We canonically identify $\mathrm{Isom}^H(H)$ with H. Given a locally finite Borel measure ν on H, recall that $[\nu]$ is the equivalence class of measures positively proportional to ν. Write

$$\mathrm{Isom}^H(H;[\nu]) := \{g \in \mathrm{Isom}^H(H) : [g_*\nu] = [\nu]\} = \{g \in \mathrm{Isom}^H(H) : g_*\nu \propto \nu\}$$

and let

$$\mathrm{Isom}^H(H;\nu) := \{g \in \mathrm{Isom}^H(H) : g_*\nu = \nu\}.$$

We have the following elementary fact.

CLAIM 7.1. *The groups* $\mathrm{Isom}^H(H;[\nu])$ *and* $\mathrm{Isom}^H(H;\nu)$ *are closed subgroups of* H.

Let H act continuously on a complete, second countable, locally compact metric space X. Let μ be a Borel probability measure on X. We assume that μ-a.e. H-orbit is free. Write $\mathcal{H}$ for the partition of (X,μ) into H-orbits.

Write $\mathrm{Aut}(X,\mu)$ for the set of bijective, measurable, μ-preserving maps of X. Since (X,μ) is standard, we have $f^{-1}\in\mathrm{Aut}(X,\mu)$ for every $f\in\mathrm{Aut}(X,\mu)$ (see, e.g., [13, corollary 15.2]). Given $f\in\mathrm{Aut}(X,\mu)$, the measure μ need not be f-ergodic. Write $\mathcal{E}(f)$ for the *ergodic decomposition* of μ with respect to f; precisely, $\mathcal{E}(f)$ is the measurable hull of the partition of (X,μ) into f-orbits.

PROPOSITION 7.2. *Suppose there exists* $f\in\mathrm{Aut}(X,\mu)$ *with the following properties:*

(1) *f intertwines almost every H-orbit and commutes with the H-action: for a.e. $x\in X$ and every $h\in H$,*

$$f(h\cdot x)=h\cdot f(x);\ \textit{and}$$

(2) $\mathcal{E}(f)\prec\mathcal{H}$.

Then, for μ-a.e. $x\in X$, the group $\mathrm{Isom}^H(H;[\nu_x^H])$ acts transitively on the support of ν_x^H.

Proof. By f-invariance of μ and the assumption that f preserves the canonical parametrizations of H-orbits, we have that $\nu_{f(x)}^H=\nu_x^H$ for almost every x; in particular, the measurable map $x\mapsto\nu_x^H$ is constant on f-ergodic components of μ.

The assumption that $\mathcal{E}(f)\prec\mathcal{H}$ implies for μ-a.e. x and ν_x^H-a.e. $h\in H$ that x and $h\cdot x$ are in the same f-ergodic component of μ; in particular for μ-a.e. x and ν_x^H-a.e. $h\in H$, we have

$$\nu_{h\cdot x}^H=\nu_x^H.$$

Since

$$(\Phi_x)_*\nu_x^H\propto(\Phi_{h\cdot x})_*\nu_{h\cdot x}^H$$

and since $\Phi_{h\cdot x}^{-1}\circ\Phi_x$ corresponds to right translation by h^{-1}, we have that h^{-1}, and hence h, are elements of the group $\mathrm{Isom}^H(H;[\nu_x^H])$. Since

$\mathrm{Isom}^H(H; [\nu_x^H])$ is a closed subgroup of H, its orbit in H is closed and the conclusion follows. □

Given an automorphism φ of H, write $d\varphi$ for the corresponding automorphism of the Lie algebra $\mathfrak{h}$. Assuming there exists a transformation $g \in \mathrm{Aut}(X, \mu)$ that intertwines H-orbits and acts on H-orbits by a contracting automorphism φ, we obtain stronger properties of the groups $\mathrm{Isom}^H(H; [\nu_x^H])$.

PROPOSITION 7.3. *Suppose there exists $g \in \mathrm{Aut}(X, \mu)$ and an automorphism $\varphi \in \mathrm{Aut}(H)$ with $\| d\varphi \| < 1$ such that for a.e. $x \in X$ and every $h \in H$,*

$$g(h \cdot x) = \varphi(h) \cdot g(x).$$

Moreover, suppose for μ-a.e. $x \in X$ that $\mathrm{Isom}^H(H; [\nu_x^H])$ acts transitively on the support of ν_x^H.

Then for μ-a.e. $x \in X$,

(1) $\mathrm{Isom}^H(H; [\nu_x^H]) = \mathrm{Isom}^H(H; \nu_x^H)$, *and*
(2) ν_x^H *coincides (up to a choice of normalization) with the left Haar measure on a connected Lie subgroup $L_x \subset H$.*

Proof. For (1), given $x \in X$ and $h \in \mathrm{Isom}^H(H; [\nu_x^H])$, set

$$c_x(h) = \frac{\nu_x^H(B^H(1) \cdot h)}{\nu_x^H(B^H(1))}.$$

For any $E \subset H$ with $\nu_x^H(E) > 0$, we then have

$$\nu_x^H(E \cdot h) = c_x(h)\nu_x^H(E).$$

We check the following hold for almost every x:

(a) For $h_1, h_2 \in \mathrm{Isom}^H(H; [\nu_x^H])$, we have

$$c_x(h_1 h_2) = c_x(h_2) c_x(h_1)$$

and hence obtain a homomorphism $c_x \colon \mathrm{Isom}^H(H; [\nu_x^H]) \to (\mathbb{R}^+, \times)$.
(b) $c_x(h) = c_{g(x)}(\varphi(h))$.
(c) $c_x \colon \mathrm{Isom}^H(H; [\nu_x^H]) \to \mathbb{R}^+$ is continuous.

Indeed (a) and (b) follow from properties of leaf-wise measures and g-invariance of μ. For (c), we have that $c_x \colon H \to \mathbb{R}^+$,

$$c_x \colon h \mapsto \frac{\nu_x^H(E \cdot h)}{\nu_x^H(E)}$$

is both lower- and upper-semicontinuous by considering $E \subset H$, respectively, open or closed.

Given $\delta > 0$ and $x \in X$, let ϵ_x be such that for all $h \in B^H(\epsilon_x)$,

$$|1 - c_x(h)| < \delta.$$

We have that $\epsilon_x > 0$ for almost every $x \in X$. Given $R > 0$ and any $\epsilon > 0$, we have that $\varphi^n(B^H(R)) \subset B^H(\epsilon)$ for all sufficiently large n. By Poincaré recurrence of orbits of g to sets on which $x \mapsto \epsilon_x$ is bounded from below and applying (b), we see for almost every $x \in X$ that

$$|1 - c_x(h)| < \delta$$

for all $h \in \operatorname{Isom}^H(H; [\nu_x^H])$. Taking $\delta \to 0$, conclusion (1) then follows.

For (2), given $x \in X$, set $L_x := \operatorname{Isom}^H(H; \nu_x^H)$. We have that L_x is a closed subgroup of H and hence a Lie group. We claim L_x is connected for a.e. x. Indeed, let L_x' denote the connected component of L_x through the identity. Let r_x denote the minimal distance from the identity to any L_x'-orbit not containing the identity. If $L_x' \neq L_x$, then $r_x > 0$. On the other hand, from (b) above and the assumptions on φ, we have $r_{g(x)} < \kappa r_x$ for some $0 < \kappa < 1$. If $L_x' \neq L_x$ for a positive measure set of x, we would thus obtain a contradiction with Poincaré recurrence to sets on which $x \mapsto r_x$ is bounded from below.

We thus have that ν_x^H coincides with a right Haar measure on $L_x \subset H$ for almost every $x \in X$. The assumption on φ ensures that H is nilpotent. It thus follows that L_x is nilpotent. As nilpotent groups are unimodular, ν_x^H also coincides with a left Haar measure on $L_x \subset H$ a.s. □

8 Structure theory of G and stable classes of roots

8.1 CARTAN AND IWASAWA DECOMPOSITIONS. Let $\mathfrak{g}$ denote the Lie algebra of G. Fix a Cartan involution θ of $\mathfrak{g}$ and write $\mathfrak{k}$ and $\mathfrak{p}$, respectively, for the $+1$ and -1 eigenspaces of θ. Let $\mathfrak{a}$ be a maximal abelian subalgebra of $\mathfrak{p}$. Let $\mathfrak{m}$ be the centralizer of $\mathfrak{a}$ in $\mathfrak{k}$. Recall that $\dim_{\mathbb{R}}(\mathfrak{a})$ is the $\mathbb{R}$-*rank* of G.

We let Σ denote the set of restricted roots of $\mathfrak{g}$ with respect to $\mathfrak{a}$; elements of Σ are real linear functionals on $\mathfrak{a}$. A *base* (or a collection of *simple roots*) for Σ is a subset $\Pi \subset \Sigma$ that is a basis for the vector space $\mathfrak{a}^*$ and such that every nonzero root $\beta \in \Sigma$ is either a positive or a negative integer combination of elements of Π. For a choice of Π, elements $\alpha \in \Pi$ are called *simple* (positive) roots. Relative to a choice of base Π, let $\Sigma_+ \subset \Sigma$ be the collection of positive roots and let Σ_- be the corresponding set of negative roots.

For $\beta \in \Sigma$ write $\mathfrak{g}^\beta$ for the associated root space. Then $\mathfrak{n} = \bigoplus_{\beta \in \Sigma_+} \mathfrak{g}^\beta$ is a nilpotent subalgebra. We have that $\theta(\mathfrak{n}) = \bigoplus_{\beta \in \Sigma_-} \mathfrak{g}^\beta$. Write $\mathfrak{n}^- = \theta(\mathfrak{n})$.

Let A, N, and K be the analytic subgroups of G corresponding to $\mathfrak{a}, \mathfrak{n}$, and $\mathfrak{k}$. We also write $N^- = \theta(N)$ for the analytic subgroup corresponding to $\mathfrak{n}^-$. Then $G = KAN$ is the corresponding *Iwasawa decomposition* of G. When G has finite center, K is compact. Note that the Lie exponential $\exp\colon \mathfrak{g} \to G$ restricts to diffeomorphisms between $\mathfrak{a}$ and A and $\mathfrak{n}$ and N. Write $M = C_K(\mathfrak{a})$ for the centralizer of $\mathfrak{a}$ in K. Then $P = MAN$ is the *standard minimal parabolic subgroup*. We have that P is amenable as it is a compact extension of a solvable Lie group.

8.2 STABLE COLLECTIONS OF ROOTS. We have the following definition.

DEFINITION 8.1 (stable collection of roots). A collection of roots $C \subset \Sigma$ is called *stable* if there exist $s_1, \ldots, s_k \in \mathfrak{a}$ such that

$$C = \{\beta \in \Sigma : \beta(s_i) < 0 \text{ for all } 1 \le i \le k\}.$$

If $C \subset \Sigma$ is a stable collection of roots, then the vector subspace

$$\mathfrak{u}^C := \bigoplus_{\beta \in C} \mathfrak{g}^\beta$$

is a nilpotent Lie subalgebra of $\mathfrak{g}$. Write U^C for the corresponding analytic subgroup of G. A maximal stable collection of roots $C \subset \Sigma$ corresponds to a choice of ordering of Σ and corresponding collection of negative roots. A minimal stable collection of roots corresponds to a positive proportionality class of roots in Σ, which we refer to as a *coarse root* and typically denote by $\chi \subset \Sigma$. By the structure of abstract root systems, we have that every coarse root $\chi \subset \Sigma$ is either a singleton $\chi = \{\beta\}$ or is of the form $\chi = \{\beta, 2\beta\}$.

As a primary example, we consider the following construction.

EXAMPLE 8.2 (stable collection of roots transverse to a parabolic subgroup). Let $G = KAN$ be an Iwasawa decomposition, $P = MAN$ a minimal parabolic subgroup, and $Q \supset P$ a parabolic subgroup. Let $\mathfrak{m}$, $\mathfrak{a}$, $\mathfrak{n}$, and $\mathfrak{q}$ denote, respectively, the Lie algebras of M, A, N, and Q. Let Π be a base of simple positive roots so that $\mathfrak{n}$ is spanned by roots spaces corresponding to positive roots. Let $\mathfrak{n}^- = \theta(\mathfrak{n})$ be the Lie subalgebra spanned by roots spaces corresponding to negative roots relative to this ordering. Let $\mathfrak{q}^- = \mathfrak{q} \cap \mathfrak{n}^-$.

We claim there exists a stable collection of roots $C \subset \Sigma$ such that $\mathfrak{q}^-$ and $\mathfrak{u}^C$ are transverse and of complementary dimension in $\mathfrak{n}^-$ so that $\mathfrak{n}^- = \mathfrak{q}^- \oplus \mathfrak{u}^C$. Indeed by the structure of parabolic subalgebras (see, for instance, [14]) we have that

$$\mathfrak{q} := \mathfrak{m} \oplus \mathfrak{a} \oplus \bigoplus_{\beta \in \Sigma_+} \mathfrak{g}^\beta \oplus \bigoplus_{\beta \in \mathbb{Z}_{\leq 0}\text{-span}(\Delta)} \mathfrak{g}^\beta$$

for some subset $\Delta \subset \Pi$. The last direct sum is taken over all roots β that are nonpositive integer combinations of elements of Δ.

Take C to be the collection of all negative roots

$$\beta = \sum_{\alpha \in \Pi} c_\alpha \alpha$$

(so c_α is a nonpositive integer for every $\alpha \in \Pi$) such that $c_\alpha \neq 0$ for some $\alpha \in \Pi \setminus \Delta$. We may find $s \in \mathfrak{a}$ such that

(1) $\alpha(s) = 0$ for $\alpha \in \Delta$ and
(2) $\alpha(s) > 0$ for $\alpha \in \Pi \setminus \Delta$.

It follows that $\beta(s) < 0$ for all $\beta \in C$ and that $\beta(s) \geq 0$ for all $\beta \in \Sigma \setminus C$; in particular, C is a stable collection of roots.

REMARK 8.3. Fix a norm on $\mathfrak{g}$. Let C be a stable collection of roots. Fix $s \in \mathfrak{a}$ with $\beta(s) < 0$ for all $\beta \in C$. Let $\mathfrak{u}^C$ be the unipotent subalgebra associated with C and let $U := U^C = \exp(\mathfrak{u}^C)$ be the corresponding subgroup of G. Then $\| \mathrm{Ad}(\exp(s))|_{\mathfrak{u}^C} \| < \kappa$ for some $0 < \kappa < 1$.

Let G act continuously on a metric space X. Let $f : X \to X$ be $f(x) = \exp(s) \cdot x$. Given $x \in X$ and $W \in \mathfrak{u}^C$ we have

$$f(\exp(W) \cdot x) = \exp(\mathrm{Ad}(\exp(s))(W)) \cdot f(x).$$

In particular, we have

$$f^n(\exp(W) \cdot x) = \exp(W_n) \cdot (f^n(x)),$$

where $W_n = \exp(\mathrm{Ad}(\exp(ns))(W))$ so $\| W_n \| \leq \kappa^n \| W \|$ for all $n \geq 0$. Assuming the metric on U^C-orbits induced by $\| \cdot \|$ is compatible with the ambient metric on X, it follows that the U^C-orbit of x is contained in the stable set of x for the action of f.

8.3 STRUCTURE OF LEAF-WISE MEASURES FOR STABLE COLLECTIONS OF ROOTS. Let X be a complete, second countable, locally compact metric space and let G act continuously on X. Fix an Iwasawa decomposition $G = KAN$ of G and let μ be an ergodic, A-invariant Borel probability measure on X.

Given a stable collection of roots $C \subset \Sigma$, let

$$x \mapsto \nu_x^C := \nu_x^{U^C}$$

denote the family of leaf-wise measure on U^C associated to μ. We have the following proposition, which is a simplification of the "product structure" of leaf-wise measures established by Einsiedler and Katok (see [4, proposition 5.1, corollary 5.2] and [5, theorem 8.4]). Roughly, we have that if ν_x^C has nontrivial support, then ν_x^χ has nontrivial support for some coarse root $\chi \subset C$.

PROPOSITION 8.4. *Suppose there is a stable collection of roots $C \subset \Sigma$ such that for almost every $x \in X$, the leaf-wise measure ν_x^C on U^C associated to μ is not a point-mass supported at the identity.*

Then there exists a coarse root $\chi \subset C$ such that for almost every $x \in X$, the leaf-wise measure ν_x^χ on U^χ associated to μ is not a point-mass supported at the identity.

Proof. Given any stable collection of roots $C \subset \Sigma$ we may find a coarse root $\chi \subset C$ and $s \in \mathfrak{a}$ such that $\beta(s) = 0$ for all $\beta \in \chi$ and $\beta'(s) < 0$ for all $\beta' \in C \smallsetminus \chi$.

Let $C' = C \smallsetminus \chi$. Then C' is a stable collection of roots. If $\nu_x^C = \nu_x^{C'}$ for almost every x, we may replace C with C' and proceed by induction on the cardinality of C. Thus, without loss of generality, we may assume there is $E \subset X$ with $\mu(E) > 0$ such that ν_x^C is not supported on $U^{C'}$ for $x \in E$. Since we assume μ is A-invariant and ergodic and since A normalizes U^C, $U^{C'}$, and U^χ, it follows that ν_x^C is not supported on $U^{C'}$ for almost every $x \in X$.

Let $f \colon X \to X$ be the map $f(x) = \exp(s) \cdot x$. Since $\chi(s) = 0$, $\exp(s)$ commutes with elements of U^χ. Since μ is f-invariant, we then have $\nu_x^\chi = \nu_{f(x)}^\chi$ for μ-a.e. x.

Since the assignment $x \mapsto \nu_x^\chi$ is measurable, by Lusin's theorem, given $\epsilon > 0$ there is a compact $K_\epsilon \subset X$ with $\mu(K_\epsilon) > 1 - \epsilon$ on which the map $x \mapsto \nu_x^\chi$

is continuous. For almost every $x \in X$ and ν_x^C-a.e. $u \in U^C \smallsetminus U^{C'}$ we have the following:

(1) The measure ν_x^C is not supported on $U^{C'}$.
(2) The identity element of U^X is contained in the support of ν_x^X and $\nu_{u \cdot x}^X$.
(3) There exist arbitrarily small values $\epsilon > 0$, such that $x \in K_\epsilon$ and $u \cdot x \in K_\epsilon$.
(4) For every $n \in \mathbb{N}$, we have $\nu_x^X = \nu_{f^n(x)}^X$ and $\nu_{u \cdot x}^X = \nu_{f^n(u \cdot x)}^X$.
(5) Taking $\epsilon < \frac{1}{2}$ sufficiently small, there exists $n_j \to \infty$ such that $f^{n_j}(x) \in K_\epsilon$ and $f^{n_j}(u \cdot x) \in K_\epsilon$.
(6) $u = hv$ where $h \in U^{C'}$ and $v \in U^X$; moreover for every $n \geq 0$, $f^n(u \cdot x) = u_n f^n(x)$ where $u_n = h_n v$ and $h_n \to 0$ as $n \to \infty$.

Taking $n_j \to \infty$ and passing to further subsequences, we may assume $f^{n_j}(x)$ and $f^{n_j}(u \cdot x)$ converge, respectively, to some $x_\infty \in K_\epsilon$ and $y_\infty \in K_\epsilon$. We have $y_\infty = v \cdot x_\infty$. We have that $\nu_{x_\infty}^X = \nu_x^X$ and $\nu_{y_\infty}^X = \nu_{u \cdot x}^X$; in particular, the identity element in U^X is contained in the supports of $\nu_{x_\infty}^X$ and $\nu_{y_\infty}^X$. However, since $y_\infty = v \cdot x_\infty$, it follows that $\nu_{x_\infty}^X$, in particular, ν_x^X is supported at $v \in U^X$; in particular, ν_x^X is not an atom supported at the identity. □

9 Proof of Proposition 5.3

We use the results and constructions from sections 6–8 to prove Proposition 5.3. From the discussion in section 5, this completes the proof of Theorem 1.2.

Proof of Proposition 5.3. Let ν be the Q-invariant measure on X^Γ constructed in section 5. By construction, the projection $(X^\Gamma, \nu) \to (G/\Gamma, \lambda_{G/\Gamma})$ is essentially injective; in particular, the projection $(X^\Gamma, \nu) \to (G/\Gamma, \lambda_{G/\Gamma})$ is a measurable isomorphism. It is well-known that $\lambda_{G/\Gamma}$ is ergodic under the action of any noncompact subgroup $H \subset G$. In particular, it follows for any noncompact subgroup $H \subset Q$ that the action of H on (X^Γ, ν) is ergodic.

Suppose for the sake of contradiction that there exists a positive ν-measure subset of $x \in X^\Gamma$ for which the leaf-wise measure $\nu_x^{N^-}$ is not supported on Q^-. Since A preserves ν and acts ergodically on (X^Γ, ν) and since Q^- is normalized by A, it follows that $\nu_x^{N^-}$ is not supported on Q^- for ν-almost every $x \in X^\Gamma$.

Let C be the stable collection of roots as in Example 8.2 with $\mathfrak{u}^C$ transverse to $\mathfrak{q}^-$ in $\mathfrak{n}^-$. Then the subgroup $U^C \subset N^-$ is transverse to Q^--orbits in N^-. Since $\nu_x^{N^-}$ is Q^--invariant for ν-almost every x, it follows that $\nu_x^{U^C}$ is not supported at the identity for almost every x.

By Proposition 8.4, we may find a coarse root $\chi \in C$ such that ν_x^χ is not supported at the identity for a positive measure subset of $x \in X^\Gamma$; by ergodicity, this then holds for ν-almost every $x \in X^\Gamma$.

Fix $s \in \mathfrak{a} \smallsetminus \{0\}$ with $\chi(s) = 0$ and $s' \in \mathfrak{a} \smallsetminus \{0\}$ with $\chi(s') < 0$. Set $f, g \in \mathrm{Aut}(X^\Gamma, \nu)$ to be

$$f(x) = \exp(s) \cdot x, \qquad g(x) = \exp(s') \cdot x.$$

We have that f is an ergodic transformation of (X^Γ, ν). By Propositions 7.2 and 7.3 (with the above f and g), for ν-a.e. $x \in X^\Gamma$ the measure ν_x^χ is the left Haar measure on a connected Lie subgroup $L_x \subset U^\chi$. Since ν_x^χ is not supported at the identity, L_x has positive dimension. Moreover, since f intertwines the parametrizations of H-orbits and preserves the measure μ, the Lie algebras of L_x and $L_{f(x)}$ coincide whence the map $x \mapsto L_x$ is f-invariant. By f-ergodicity of ν, there is a subgroup $L \subset U^\chi$ such that $L_x = L$ for ν-almost every x. It follows from Claim 6.5 that μ is L-invariant. But L is not a subgroup of Q, contradicting the choice of Q as the (maximal) stabilizer of ν. □

References

[1] B. Bekka, P. de la Harpe, and A. Valette, *Kazhdan's property (T)*, New Mathematical Monographs, vol. 11, Cambridge University Press, Cambridge, 2008.

[2] A. Brown, F. Rodriguez Hertz, and Z. Wang, *Invariant measures and measurable projective factors for actions of higher-rank lattices on manifolds*, 2016. Preprint, arXiv:1609.05565.

[3] S. G. Dani, *Continuous equivariant images of lattice-actions on boundaries*, Ann. of Math. (2) **119** (1984), no. 1, 111–119.

[4] M. Einsiedler and A. Katok, *Invariant measures on G/Γ for split simple Lie groups G*, Comm. Pure Appl. Math. **56** (2003), no. 8, 1184–1221.

[5] M. Einsiedler and A. Katok, *Rigidity of measures—the high entropy case and non-commuting foliations*, Israel J. Math. **148** (2005), 169–238.

[6] M. Einsiedler, A. Katok, and E. Lindenstrauss, *Invariant measures and the set of exceptions to Littlewood's conjecture*, Ann. of Math. (2) **164** (2006), no. 2, 513–560.

[7] M. Einsiedler and E. Lindenstrauss, *Rigidity properties of $\mathbb{Z}^d$-actions on tori and solenoids*, Electron. Res. Announc. Amer. Math. Soc. **9** (2003), 99–110.

[8] M. Einsiedler and E. Lindenstrauss, *Diagonal actions on locally homogeneous spaces*, Clay Math Proceedings, vol. 10, American Mathematical Society, Providence, RI, 2010.

[9] H. Furstenberg, *Boundary theory and stochastic processes on homogeneous spaces*. In *Harmonic analysis on homogeneous spaces (Proc. Sympos. Pure Math., Vol. XXVI, Williams Coll., Williamstown, Mass., 1972)*, Amer. Math. Soc., Providence, RI, 1973, 193–229.

[10] A. Gorodnik and R. Spatzier, *Smooth factors of projective actions of higherrank lattices and rigidity*, Geom. Topol. **22** (2018), no. 2, 1227–1266.

[11] A. Katok and R. J. Spatzier, *Invariant measures for higher-rank hyperbolic abelian actions*, Ergodic Theory Dynam. Systems **16** (1996), no. 4, 751–778.

[12] A. Katok and R. J. Spatzier, Corrections to: *"Invariant measures for higherrank hyperbolic abelian actions" [Ergodic Theory Dynam. Systems **16** (1996), no. 4, 751–778]*, Ergodic Theory Dynam. Systems **18** (1998), no. 2, 503–507.

[13] A. S. Kechris, *Classical descriptive set theory*, Graduate Texts in Mathematics, vol. 156, Springer-Verlag, New York, 1995.

[14] A. W. Knapp, *Lie groups beyond an introduction*, 2nd ed., Progress in Mathematics, vol. 140, Birkhäuser, Boston, 2002.

[15] G. A. Margulis, *Factor groups of discrete subgroups and measure theory*, Funktsional. Anal. i Prilozhen. **12** (1978), no. 4, 64–76.

[16] G. A. Margulis, *Quotient groups of discrete subgroups and measure theory*, Functional Anal. Appl. **12** (1978), no. 4, 295–305.

[17] G. A. Margulis, *Finiteness of quotient groups of discrete subgroups*, Funktsional. Anal. i Prilozhen. **13** (1979), no. 3, 28–39.

[18] G. A. Margulis, *Finiteness of quotient groups of discrete subgroups*, Functional Anal. Appl. **13** (1979), no. 3, 178–187.

[19] G. A. Margulis, *Discrete subgroups of semisimple Lie groups*, Ergebnisse der Mathematik und ihrer Grenzgebiete (3) [Results in mathematics and related areas (3)], vol. 17, Springer-Verlag, Berlin, 1991.

[20] G. A. Margulis and G. M. Tomanov, *Invariant measures for actions of unipotent groups over local fields on homogeneous spaces*, Invent. Math. **116** (1994), no. 1–3, 347–392.

[21] V. A. Rohlin, *On the fundamental ideas of measure theory*, Amer. Math. Soc. Translation **10** (1952), 1–52.

[22] V. S. Varadarajan, *Groups of automorphisms of Borel spaces*, Trans. Amer. Math. Soc. **109** (1963), 191–220.

[23] D. Witte Morris, *Introduction to arithmetic groups*, Deductive Press, 2015.

PART II

Discrete subgroups

4

JEFFREY DANCIGER, TODD A. DRUMM,
WILLIAM M. GOLDMAN, AND ILIA SMILGA

PROPER ACTIONS OF DISCRETE GROUPS OF AFFINE TRANSFORMATIONS

Dedicated to Grisha Margulis on the occasion of his 70th birthday

Abstract. In the early 1980s Margulis startled the world by showing the existence of proper affine actions of free groups on 3-space, answering a provocative and suggestive question Milnor posed in 1977. In this paper we discuss the historical background motivating this question, recent progress on this subject, and future directions inspired by this discovery.

1 Introduction

The theory of flat Riemannian manifolds, also known as Euclidean manifolds, is well understood. Starting with its nineteenth-century origins in theoretical crystallography, Euclidean crystallographic groups and complete flat Riemannian manifolds have a satisfying and cohesive structure theory. In particular, the Bieberbach theorems imply that every closed flat Riemannian manifold is finitely covered by a torus or, equivalently, any Euclidean crystallographic

Jeffrey Danciger. Department of Mathematics. University of Texas, Austin, Austin, TX 78712 USA
jdanciger@math.utexas.edu

Todd A. Drumm. Department of Mathematics. Howard University, Washington, DC 20059 USA
tdrumm@howard.edu

William M. Goldman. Department of Mathematics. University of Maryland, College Park, MD 20742 USA
wmg@math.umd.edu

Ilia Smilga. Institut de Mathématiques de Marseille (UMR 7373), Aix-Marseille Université & CNRS, 39 rue F. Joliot Curie, 13453 MARSEILLE Cedex, France
ilia.smilga@normalesup.org

Danciger was partially supported by NSF grants DMS 151025, DMS 1812216, DMS 1945493 and by an Alfred P. Sloan Foundation fellowship. Goldman gratefully acknowledges research support from NSF Grants DMS1065965, DMS1406281, DMS1709791. Smilga acknowledges support from NSF grant DMS1709952 and from the European Research Council (ERC) under the European Union Horizon 2020 research and innovation program, grant 647133 (ICHAOS). The authors gratefully acknowledge support from the Research Network in the Mathematical Sciences DMS 1107452, DMS 1107263, DMS 1107367 (GEAR).

group is virtually a lattice in $\mathbb{R}^n$, in particular is virtually free abelian. This survey concerns *complete affine manifolds,* a natural generalization of complete Euclidean manifolds whose structure theory, by contrast, remains tantalizingly mysterious and poorly understood. The famous *Auslander conjecture*—that every compact complete affine manifold has virtually polycyclic fundamental group—has been a focal point for research in this field. Now known to be true in dimensions < 7, it remains open in general.

The last 40 years have seen major advances in the theory of complete affine manifolds. A significant breakthrough was Margulis's discovery in 1983 of proper affine actions of nonabelian free groups in dimension 3 and the subsequent classification of complete affine 3-manifolds. As any proper affine action by a free group in dimension 3 preserves a Lorentzian structure, the corresponding complete affine 3-manifolds are, more specifically, complete flat Lorentzian 3-manifolds. They are known today as *Margulis spacetimes.*

Associated to a Margulis spacetime M^3 is a (necessarily noncompact) complete hyperbolic surface Σ homotopy equivalent to M^3, and we call M^3 an *affine deformation* of Σ. Hence, the deformation space of Margulis spacetimes whose associated hyperbolic surface Σ has a fixed topological type S naturally projects down to the Fricke-Teichmüller space $\mathfrak{F}(S)$ of S. The fiber of this projection consists of equivalence classes of *proper affine deformations* of Σ. A clear picture has emerged of the fiber of this projection as an open convex cone in the space of infinitesimal deformations of the hyperbolic structure on Σ. Much of this essay describes this point of view. Crucial is the properness criterion for affine deformations developed by Goldman, Labourie, and Margulis [73]. Along the way, we will also collect the known results on the topology and geometry of Margulis spacetimes and give an overview of the current state of the art in higher dimensional affine geometry.

The fundamental problem in Euclidean crystallography was, in modern parlance, the classification of fundamental polyhedra for Euclidean crystallographic groups. However, in the setting of Margulis spacetimes, standard constructions for fundamental polyhedra do not work. The introduction by Drumm of *crooked polyhedra* around 1990—about a decade after Margulis's discovery—provided tools for building fundamental domains and led in particular to the discovery that there exist Margulis spacetimes that are affine deformations of *any* noncompact complete hyperbolic surface of finite type. This kindled momentum for the subject and marked the beginning of a classification program for Margulis spacetimes, which was completed only recently.

The outline of this essay is as follows. Section 3 summarizes the early history of the subject, beginning with Bieberbach's "classification" of Euclidean manifolds, and its subsequent generalizations. These generalizations—due to Zassenhaus, Wang, Auslander, Mostow, and others—set the stage for the classification of complete affine manifolds with virtually solvable fundamental group. In 1977, Milnor asked whether every complete affine manifold has virtually solvable fundamental group, or equivalently, if proper affine actions of the two-generator free group F_2 *do not exist*. Shortly thereafter, Margulis surprised everyone by showing the existence of complete affine manifolds with fundamental group F_2.

Section 4 begins the construction and classification of Margulis spacetimes, modeled on the geometric construction and classification of hyperbolic surfaces. Crooked geometry is developed, including the disjointness criteria for crooked planes, which is fundamental in setting up the geometric conditions necessary for building Schottky groups. We briefly describe a compactification of M^3 as a flat $\mathbb{RP}^3$-manifold due to Suhyoung Choi, which implies that M^3 is homeomorphic to an open solid handlebody.

Section 5 introduces the *marked Lorentzian signed length spectrum*, or *Margulis invariant*, denoted α. The Margulis invariant is an $\mathbb{R}$-valued class function on $\pi_1(M) \cong \Gamma$, and ever since Margulis introduced this quantity, it has played an important role in the geometry of Margulis spacetimes. The simplest type of Margulis spacetime occurs when the associated hyperbolic surface has compact convex core (or equivalently, $\pi_1\Sigma = \Gamma_0 < \mathsf{Isom}(\mathsf{H}^2)$ is *convex cocompact*). In this case, every holonomy transformation is hyperbolic, and every essential loop is freely homotopic to a closed geodesic. A classical result in hyperbolic geometry asserts that such hyperbolic structures are determined up to isometry by their *marked length spectrum*, the $\mathbb{R}_+$-valued class function on Γ associating to $\gamma \in \pi_1(\Sigma)$ the *hyperbolic length* $\ell(\gamma)$ of the unique closed geodesic in Σ that is homotopic to γ. The *magnitude* of Margulis's invariant $\alpha(\gamma)$

$$\pi_1(\Sigma) \xrightarrow{|\alpha|} \mathbb{R}_+$$

$$\gamma \longmapsto |\alpha(\gamma)|$$

corresponds to the *Lorentzian length* of a closed geodesic homotopic to γ. In particular, the isometry type of a Margulis spacetime M^3 is determined by the marked length spectrum ℓ of Σ and the absolute value $|\alpha|$ of the Margulis invariant.

In fact, only $|\alpha|$ is needed to determine the isometry type of M^3. We discuss extensions of the definition of the Margulis invariant and of these results to the setting where Σ has cusps.

Section 6 develops a properness criterion for actions of free groups in three-dimensional affine geometry. This turns out to be closely related to the *direction* or *sign* of α. As originally noted by Margulis (the *Opposite Sign Lemma*), for any Margulis spacetime M, the sign of

$$\Gamma = \pi_1 M \xrightarrow{\alpha} \mathbb{R}$$

is constant, either positive or negative. A simple proof, given by Goldman–Labourie–Margulis, involves the continuous extension of the *normalized Margulis invariant* α/ℓ to the connected convex set of *geodesic currents* on Σ. This leads to the description of the deformation space of Margulis spacetimes with fixed hyperbolic surface Σ as an open convex cone in the vector space of affine deformations of Γ_0, naturally the cohomology group $H^1(\Gamma_0, \mathbb{R}^{2,1})$, where

$$\Gamma_0 = \pi_1 \Sigma < \mathsf{Isom}(\mathsf{H}^2) = \mathsf{SO}(2,1)$$

is the holonomy group of Σ.

Section 7 develops the connection between affine actions in three-dimensional flat Lorentzian geometry and infinitesimal deformations of hyperbolic surfaces. Due to the low-dimensional coincidence that the standard action of $\mathsf{SO}(2,1)$ on $\mathbb{R}^3$ is isomorphic to the adjoint action on the lie algebra $\mathfrak{so}(2,1)$, the space $H^1(\Gamma_0, \mathbb{R}^{2,1})$ of affine deformations of the surface group $\Gamma_0 < \mathsf{SO}(2,1)$ is in natural bijection with the space of infinitesimal deformations of the representation $\Gamma_0 \hookrightarrow \mathsf{SO}(2,1)$, which in turn identifies with the space of infinitesimal deformations of the hyperbolic surface Σ. This interpretation leads to the fundamental result of Mess that the hyperbolic surface Σ associated to a Margulis spacetime cannot be closed. In particular, if Γ is a nonsolvable discrete group acting affinely on 3-space, then Γ must be virtually free.

The infinitesimal deformations of hyperbolic structures on Σ that arise from proper affine actions may be represented by what Danciger–Guéritaud–Kassel call contracting *lipschitz* vector fields, which are the infinitesimal analogs of contracting Lipschitz maps on the hyperbolic plane.

Section 7 develops the theory of lipschitz vector fields and a structure theorem for Margulis spacetimes: *M^3 is a bundle of timelike lines over the hyperbolic surface Σ*. This gave an independent proof of the topological characterization of Margulis spacetimes referenced above. A discretized version of contracting lipschitz vector fields, known as infinitesimal strip deformations, was used by Danciger-Guéritaud-Kassel to parameterize the deformation space of Margulis spacetimes associated to Σ in terms of the *arc complex* of Σ. We describe concretely the consequences of this general theory in the case that Σ has Euler characteristic -1. The qualitative behavior depends on the topology of Σ, which is one of four possibilities: the one-holed torus, the three-holed sphere, the two-holed projective plane (or cross-surface), or the one-holed Klein bottle. Section 7 ends with a discussion of the construction, due to Danciger-Guéritaud-Kassel, of proper affine actions of right-angled Coxeter groups in higher dimensions. Similar in spirit to the case of Margulis spacetimes, these proper actions come from certain contracting deformations of hyperbolic and pseudo-hyperbolic reflection orbifolds.

Section 8 discusses other directions in higher dimensional affine geometry. As in dimension 3, a general approach to Auslander's conjecture involves classifying which groups can arise as Zariski closures of the linear holonomy group. Say that a connected subgroup $G < \mathsf{GL}(n)$ is *Milnor* if no proper affine action of F_2 with Zariski closure of the linear part equal to G exists. Margulis's original result can be restated by saying that $\mathsf{SO}(2,1)^0$ is *not* Milnor, and it is the groups G that are not Milnor which must be examined in order to study the Auslander conjecture. Smilga gives a general sufficient condition for a linear representation of a semisimple Lie group to be non-Milnor. For example, the adjoint representation of a noncompact semisimple Lie group is not Milnor. Some other known results in higher dimensions are discussed in Section 8, concluding with a summary of the current state of Auslander's conjecture and a brief discussion of the proof of Abels-Margulis-Soifer for dimension < 7.

1.1 ACKNOWLEDGMENTS. The authors thank François Guéritaud, and the two anonymous referees for their extensive and excellent comments that helped to improve this manuscript. Goldman also expresses thanks to the Clay Institute for Mathematical Sciences, Institute for Computational and Experimental Research in Mathematics (ICERM), and the Mathematical Sciences Research Institute (MSRI) where this manuscript was completed.

With great sadness, we must acknowledge that Todd Drumm, the second named author, passed away during the final stages of publication of this manuscript. His profound contribution to this subject is documented in this essay, but that will hardly compensate for our great personal loss.

2 Notations and terminology

We always work over the field $\mathbb{R}$ of real numbers, unless otherwise noted. Finitely generated free groups of rank $n \geq 1$ are denoted F_n. Discrete groups will be assumed to be finitely generated, unless otherwise indicated. Denote the group of isometries of a space X by $\mathsf{Isom}(X)$. If $G < \mathsf{GL}(N)$ is a matrix group, denote its Zariski closure (algebraic hull) by $\overline{G}^{\mathsf{Zar}} < \mathsf{GL}(N)$.

If G is a group and $S \subset G$, denote by $\langle S \rangle$ the subgroup of G generated by S. Similarly, denote the cyclic group generated by an element $A \in G$ by $\langle A \rangle$.

Denote the group of inner automorphisms of a group G by $\mathsf{Inn}(G)$. Denote the cohomological dimension of a group Γ by $\mathsf{cd}(\Gamma)$. Denote the identity component of a topological group G by G^0.

VECTOR SPACES AND AFFINE SPACES. Let A be an affine space. The (simply transitive) group of translations of A is a vector space V called the *vector space underlying* A. We denote the group of linear automorphisms of V by $\mathsf{GL(V)}$ and the group of affine transformations of A by $\mathsf{Aff(A)}$. Let $o \in \mathsf{A}$ be a choice of basepoint. Then each element $g \in \mathsf{Aff(A)}$ is given by a pair $(A, \mathbf{b})$, where $A \in \mathsf{GL(V)}$ is the *linear part* and $\mathbf{b} \in \mathsf{V}$ is the *translational part:*

$$g(x) = o + A(x - o) + \mathbf{b}.$$

We will henceforth suppress the basepoint $o \in \mathsf{A}$ and identify V with A via the map $v \in \mathsf{V} \mapsto o + v \in \mathsf{A}$. When $\dim \mathsf{A} = \dim \mathsf{V} = n$, we often further identify V with $\mathbb{R}^n$ and write $\mathsf{A} = \mathsf{A}^n$ to denote the affine space of $\mathsf{V} = \mathbb{R}^n$. Then, $\mathsf{GL(V)}$ identifies with invertible $n \times n$ (real) matrices. Writing $A = \mathscr{L}(g)$ and $\mathbf{b} = \mathscr{U}(g)$, the affine transformation g is the composition of an $n \times n$ matrix $\mathscr{L}(g)$ and a translation $\mathscr{U}(g)$ acting on A^n:

$$g(x) = \mathscr{L}(g)(x) + \mathscr{U}(g)$$

The linear part of g identifies with the differential

$$\mathbb{R}^n \cong \mathsf{T}_x\mathsf{A}^n \xrightarrow{\mathsf{D}_x g} \mathsf{T}_{g(x)}\mathsf{A}^n \cong \mathbb{R}^n$$

of g, for every $x \in \mathsf{A}^n$.

Composing $g, h \in \mathsf{Aff(A)}$, we find that

- $\mathscr{L}$ is a homomorphism of groups: $\mathscr{L}(g \circ h) = \mathscr{L}(g)\mathscr{L}(h)$; and
- $\mathscr{U}$ is a V-valued 1-cocycle, where V denotes the $\mathsf{Aff(A)}$-module defined by $\mathscr{L}$:

$$\mathscr{U}(g \circ h) = \mathscr{U}(g) + \mathscr{L}(g)\mathscr{U}(h).$$

If $\Gamma \xrightarrow{\rho} \mathsf{Aff(A)}$ defines an affine action of Γ on an affine space A, then the linear part $\mathsf{L} := \mathscr{L} \circ \rho$ defines a linear representation $\Gamma \xrightarrow{\mathsf{L}} \mathsf{GL(V)}$ where V is the vector space underlying A. Fixing $\mathsf{L} := \mathscr{L} \circ \rho$, we say that ρ is an *affine deformation* of L. Evidently an affine deformation of L is determined by the translational part $\mathsf{u} := \mathscr{U} \circ \rho$:

$$\Gamma \xrightarrow{\mathsf{u}} \mathsf{V}$$

so that, for $x \in \mathsf{A}$:

$$\rho(\gamma)(x) = \mathsf{L}(\gamma)x + \mathsf{u}(\gamma). \tag{4.1}$$

This map u satisfies the *cocycle identity*:

$$\mathsf{u}(\gamma\eta) = \mathsf{u}(\gamma) + \mathsf{L}(\gamma)\big(\mathsf{u}(\eta)\big) \tag{4.2}$$

for $\gamma, \eta \in \Gamma$. Denote the vector space of such cocycles $\Gamma \xrightarrow{\mathsf{u}} \mathsf{V}$ by $\mathsf{Z}^1(\Gamma, \mathsf{V})$.

If $\mathbf{v} \in \mathsf{V}$, define its *coboundary* $\delta(\mathbf{v}) \in \mathsf{Z}^1(\Gamma, \mathsf{V})$ as:

$$\gamma \overset{\delta\mathbf{v}}{\longmapsto} \mathbf{v} - \mathsf{L}(\gamma)\mathbf{v}.$$

Denote the image $\delta(\mathsf{V}) < \mathsf{Z}^1(\Gamma, \mathsf{V})$ by $\mathsf{B}^1(\Gamma, \mathsf{V})$. Two cocycles are *cohomologous* if their difference is a coboundary.

Conjugating an affine representation by translation by **v** preserves the linear part but changes the translational part by adding $\delta\mathbf{v}$. Thus translational conjugacy classes of affine deformations with fixed linear part L are cohomology classes of cocycles, comprising the *cohomology*

$$\mathsf{H}^1(\Gamma, \mathsf{V}) := \mathsf{Z}^1(\Gamma, \mathsf{V})/\mathsf{B}^1(\Gamma, \mathsf{V}).$$

3 History and motivation

We briefly review the efforts of nineteenth-century crystallographers leading to Bieberbach's work on Euclidean manifolds and lattices in $\mathsf{Isom}(\mathsf{E}^n)$, where we denote by E^n the Euclidean *n*-space (i.e., the affine space A^n endowed with a flat Euclidean metric). Then we discuss extensions of these ideas to affine crystallographic groups and the question of Milnor on virtual polycyclicity of discrete groups of affine transformations acting properly. The section ends describing Margulis's unexpected discovery of proper affine actions of nonabelian free groups.

3.1 EUCLIDEAN CRYSTALLOGRAPHY. In the nineteenth century crystallographers asked which groups of isometries of Euclidean 3-space E^3 can preserve a periodic tiling by polyhedra. The symmetries of such a tiling form a group Γ of isometries of E^3 such that the quotient space, or orbit space, $\Gamma\backslash\mathsf{E}^3$ is compact.

This led to a classification of *crystallographic space groups,* independently, by Schönflies and Fedorov in 1891; compare Milnor [106] for a historical discussion. Since the interiors of the tiles are disjoint, the elements of Γ cannot accumulate and the group must be discrete (with respect to the induced topology). Henceforth, we assume Γ is discrete.

Define a *Euclidean space group* to be a discrete subgroup $\Gamma < \mathsf{Isom}(\mathsf{E}^n)$ satisfying any of the following equivalent properties:

- The quotient
$$M = \Gamma\backslash\mathsf{E}^n$$
is compact.
- There exists a compact *fundamental polyhedron* $\Delta \subset \mathsf{E}^n$ for the action of the group Γ:
  - The interiors of the images $\gamma(\Delta)$, for $\gamma \in \Gamma$, are disjoint; and
  - $\mathsf{E}^n = \bigcup_{\gamma\in\Gamma} \gamma(\Delta)$.

Since the subgroup Γ is discrete and acts isometrically on E^n, its action is *proper.* In particular the quotient $\Gamma\backslash\mathsf{E}^n$ is Hausdorff. When Γ is not assumed to be a group of isometries of a metric space, criteria for a discrete group to act properly become a central issue.

In 1911–1912 Bieberbach found a general group-theoretic criterion for such groups in arbitrary dimension. In modern parlance, the discrete cocompact group Γ is called a *lattice* in $\mathsf{Isom}(\mathsf{E}^n)$. Furthermore, $\mathsf{Isom}(\mathsf{E}^n)$ decomposes as a semidirect product $\mathbb{R}^n \rtimes \mathsf{O}(n)$, where $\mathbb{R}^n$ is the vector space of *translations.* Indeed, an affine automorphism is a Euclidean isometry if and only if its linear part lies in the orthogonal group $\mathsf{O}(n)$.

Bieberbach showed the following:

- $\Gamma \cap \mathbb{R}^n$ is a lattice $\Lambda < \mathbb{R}^n$.
- The quotient Γ/Λ is a finite group, mapped isomorphically into $\mathsf{O}(n)$ by $\mathscr{L}$.
- Any isomorphism $\Gamma_1 \longrightarrow \Gamma_2$ between Euclidean crystallographic groups $\Gamma_1, \Gamma_2 < \mathsf{Isom}(\mathsf{E}^n)$ is induced by an *affine automorphism* $\mathsf{E}^n \longrightarrow \mathsf{E}^n$.
- There are only finitely many isomorphism classes of crystallographic subgroups of $\mathsf{Isom}(\mathsf{E}^n)$.

A *Euclidean manifold* is a flat Riemannian manifold—that is, a Riemannian manifold of zero curvature. A Euclidean manifold is *complete* if the underlying metric space is complete. By the Hopf-Rinow theorem, completeness is equivalent to geodesic completeness.

A torsion-free Euclidean crystallographic group $\Gamma < \mathsf{Isom}(\mathsf{E}^n)$ acts freely on E^n, and the quotient $\Gamma\backslash\mathsf{E}^n$ is a compact complete Euclidean manifold. Conversely, every compact complete Euclidean manifold is a quotient of E^n by a torsion-free crystallographic group. Bieberbach's theorems have the following geometric interpretation:

- Every compact complete Euclidean manifold is a quotient of a flat torus $\Lambda\backslash\mathsf{E}^n$, where $\Lambda < \mathbb{R}^n$ is a lattice of translations, by a finite group of isometries acting freely on $\Lambda\backslash\mathsf{E}^n$.
- Any homotopy equivalence $M_1 \longrightarrow M_2$ of compact complete Euclidean manifolds is homotopic to an affine diffeomorphism.
- There are only finitely many affine isomorphism classes of compact complete Euclidean manifolds in each dimension n.

3.2 CRYSTALLOGRAPHIC HULLS. Bieberbach's theorems provide a satisfactory qualitative picture of compact Euclidean manifolds, or (essentially) equivalently, cocompact Euclidean crystallographic groups. Does a similar picture hold for *affine crystallographic groups*—that is, discrete subgroups $\Gamma < \mathsf{Aff}(\mathsf{A}^n)$ that act properly and cocompactly on A^n?

Auslander and Markus [9] constructed examples of *flat Lorentzian crystallographic groups* Γ in dimension 3 for which all three Bieberbach theorems directly fail. In their examples, the quotients $M^3 = \Gamma\backslash\mathsf{A}^3$ are flat Lorentzian manifolds. Topologically, these 3-manifolds are all 2-torus bundles over S^1; conversely, every torus bundle over the circle admits such a structure. Their fundamental groups are semidirect products $\mathbb{Z}^2 \rtimes \mathbb{Z}$ and are therefore *polycyclic*—that is, iterated extensions of cyclic groups.

More generally, a group is *virtually polycyclic* if it contains a polycyclic subgroup of finite index. A *discrete* virtually solvable group of real matrices is virtually polycyclic.

These examples arise from a more general construction: namely, Γ embeds as a lattice in a closed Lie subgroup $G < \mathsf{Aff}(\mathsf{A})$ with finitely many connected components and whose identity component G^0 acts *simply transitively* on A.

Since $\Gamma^0 := \Gamma \cap G^0$ has finite index in Γ, the flat Lorentz manifold M^3 is finitely covered by the homogeneous space $\Gamma^0\backslash G^0$. Necessarily, G^0 is simply connected and solvable. The group G^0 plays the role of the translation group $\mathbb{R}^n$ acting by translations on A^n. The group G is called the *crystallographic hull* in Fried-Goldman [61].

3.2.1 Syndetic hulls

A weaker version of this construction was known to H. Zassenhaus, H. C. Wang, and L. Auslander (compare Raghunathan [113]), defined in [61], and improved in Grunewald–Segal [79].

If $\Gamma < \mathsf{GL}(n)$ is a solvable group, then a *syndetic hull* for Γ is a subgroup G such that

- $\Gamma < G < \overline{\Gamma}^{\mathsf{Zar}}$, where we recall that $\overline{\Gamma}^{\mathsf{Zar}} < \mathsf{GL}(n)$ is the Zariski closure (algebraic hull) of Γ in $\mathsf{GL}(n)$,
- G is a closed subgroup having finitely many connected components, and
- $\Gamma \backslash G$ is compact (although not necessarily Hausdorff).

The last condition is sometimes called *syndetic,* since "cocompact" is usually reserved for subgroups whose coset space is compact *and Hausdorff.* (This terminology follows Gottschalk–Hedlund [77].) Equivalently, $\Gamma < G$ is syndetic if and only if there exists $K \subset G$ that is compact and meets every left coset $g\Gamma$, for $g \in G$.

In general, syndetic hulls fail to be unique.

3.2.2 Solvable examples and polynomial structures

The theory of affine group actions is dramatically different for solvable and nonsolvable groups. Milnor [107] proved that every virtually polycyclic group admits a proper affine action. Later Benoist [13] found examples of virtually polycyclic groups for which no crystallographic affine action exists. Dekimpe and his collaborators [50, 45, 46, 47, 51, 48, 15, 49] replace complete affine structures by *polynomial structures*— that is, quotients of A^n by proper actions of discrete subgroups of the group of polynomial diffeomorphisms $\mathsf{A}^n \longrightarrow \mathsf{A}^n$. They show that every virtually polycyclic group admits a polynomial crystallographic action.

Polynomial structures satisfy a suggestive *uniqueness property* for affine crystallographic groups similar to the role complete affine structures play for Euclidean crystallographic groups. Fried-Goldman [61] prove that two isomorphic affine crystallographic groups are *polynomially equivalent.*

A simple example, seen in Figure 4.1, occurs in dimension 2, where a polynomial diffeomorphism of degree 2,

$$\mathsf{A}^2 \xrightarrow{f} \mathsf{A}^2$$

$$(x, y) \longmapsto (x + y^2, y),$$

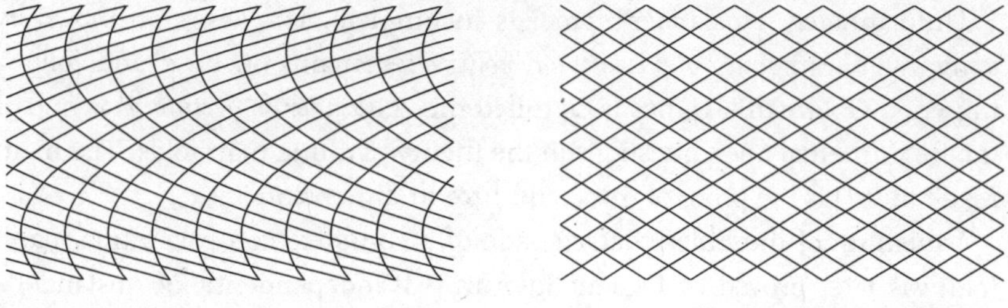

Figure 4.1. Tilings corresponding to some complete affine structures on the 2-torus

conjugates the affine crystallographic actions of $\mathbb{Z}^2$. Namely, f conjugates translation τ by $(u,v) \in \mathbb{R}^2$ to the affine transformation

$$f \circ \tau \circ f^{-1} : p \mapsto \begin{bmatrix} 1 & 2v \\ 0 & 1 \end{bmatrix} p + \begin{bmatrix} u + v^2 \\ v \end{bmatrix}. \tag{4.3}$$

The conjugate $f\mathsf{V}f^{-1}$ is a simply transitive vector group of affine transformations, where $\mathsf{V} \cong \mathbb{R}^2$ is the group of translations.

For different choices of lattices $\Lambda < \mathsf{V}$, the group $f\Lambda f^{-1}$ achieves all affine crystallographic actions of $\mathbb{Z}^2$ other than lattices of translations. Baues [11] showed that the deformation space of *marked complete affine structures* is homeomorphic to $\mathbb{R}^2$. The effect of changing the marking is the usual linear action of $\mathsf{GL}(2,\mathbb{Z})$, the mapping class group of the torus, on $\mathbb{R}^2$. Compare also Baues-Goldman [12]. These structures were first discussed by Kuiper [94].

3.3 AUSLANDER'S CONJECTURE AND MILNOR'S QUESTION.

In [8], Auslander asserted that every discrete subgroup $\Gamma < \mathsf{Aff}(\mathsf{A}^n)$ acting properly and cocompactly on A^n is virtually solvable. This was his approach to proving Chern's conjecture that *the Euler characteristic of a compact affine manifold vanishes* in the case the manifold is complete. The general theory described in Section 3.2 implies that if Γ is virtually polycyclic, then up to a finite covering $M = \Gamma\backslash\mathsf{A}^n$ has a particularly tractable algebraic structure as a *solvmanifold,* a homogeneous space of a 1-connected solvable Lie group by a lattice.

Namely, M identifies with the quotient $\Gamma\backslash G$, where G is a crystallographic hull. The simply transitive affine action of G on A^n identifies $M = \Gamma\backslash\mathsf{A}^n$ with $\Gamma\backslash G$. Furthermore, M admits the finite covering space $(\Gamma \cap G^0)\backslash\mathsf{A}^n$, which identifies with the solvmanifold $(\Gamma \cap G^0)\backslash G^0$. This gives a satisfying picture of virtually polycyclic affine crystallographic groups generalizing Bieberbach's theorem. See Grunewald-Segal [79] for more details.

Unfortunately, Auslander's proof is incomplete. His assertion that *every affine crystallographic group is virtually polycyclic* remains unsolved and, following Fried-Goldman [61], has been called the *Auslander conjecture.* It is one of the fundamental open questions in the theory of affine manifolds. The main result of [61] is the proof of this conjecture in dimension 3.

Vanishing of the Euler characteristic of a complete compact affine manifold was later proved by Kostant-Sullivan [93] independently of Auslander's conjecture.

3.3.1 *Proper affine actions of* F_2

Affine geometry is significantly more complicated than Euclidean geometry in that discrete groups of affine transformations need not act properly. Suppose

$$\Gamma \overset{\rho}{\hookrightarrow} \mathsf{Isom}(\mathsf{E}^n)$$

defines a faithful isometric action of Γ. This action is properly discontinuous (that is, *proper* with respect to the discrete topology on Γ) if and only if the image of Γ is a discrete subgroup of $\mathsf{Isom}(\mathsf{E}^n)$ (that is, ρ is a *discrete embedding*). However, if ρ is only affine (that is, the linear part $\mathsf{L}(\Gamma)$ is not assumed to lie in $\mathsf{O}(n)$), then discrete embeddings do not necessarily define proper actions.

Milnor realized that the assumption of compactness in Auslander's conjecture was not necessary to raise an interesting question. Tits [124] proved that every subgroup Γ of $\mathsf{Aff}(\mathsf{A}^n)$ of affine transformations is either

- virtually solvable or
- contains a subgroup isomorphic to a two-generator free group.

If Γ is also assumed to be discrete, then the first condition of virtual solvability can be strengthened to virtual polycyclicity.

Milnor then asked whether proper affine actions exist when Γ is a two-generator free group F_2. Nonexistence implies Auslander's conjecture, which would result in a satisfying structure theory generalizing the Bieberbach theory. Attacking this question requires a criterion for solvability.

Evidently, Γ is virtually solvable if and only if the *Zariski* closure $\overline{\rho(\Gamma)}^{\mathsf{Zar}}$ of $\rho(\Gamma)$ in $\mathsf{Aff}(\mathsf{A})$ is virtually solvable. Since Zariski closed subgroups have finitely many connected components in the classical topology, the identity component (in the classical topology) $\left(\overline{\rho(\Gamma)}^{\mathsf{Zar}}\right)^0$ is a connected solvable closed (Lie) subgroup, which has finite index in $\overline{\rho(\Gamma)}^{\mathsf{Zar}}$. In turn, this is equivalent to its linear part

$$G := \mathscr{L}\left(\left(\overline{\rho(\Gamma)}^{\mathsf{Zar}}\right)^0\right)$$

being a connected solvable closed subgroup of $\mathsf{GL}(\mathsf{V})$.

Recall the *Levi decomposition*: a connected Lie group is the semidirect product of a maximal normal solvable connected subgroup, called its *radical*, by a semisimple subgroup, called its *semisimple part* or its *Levi subgroup*. In particular, a group is solvable if and only if its semisimple part is trivial.

Summarizing:

PROPOSITION 3.1. *Let* $\Gamma \hookrightarrow \mathsf{Aff}(\mathsf{A})$ *be an affine representation. The following conditions are equivalent:*

- Γ *is virtually solvable.*
- $\mathsf{L}(\Gamma)$ *is virtually solvable.*
- $\overline{\mathsf{L}(\Gamma)}^{\mathsf{Zar}}$ *is virtually solvable.*
- *The identity component* $\left(\overline{\mathsf{L}(\Gamma)}^{\mathsf{Zar}}\right)^0$ *is solvable.*
- *The semisimple part of* $\left(\overline{\mathsf{L}(\Gamma)}^{\mathsf{Zar}}\right)^0$ *is trivial.*

This raises the question of which groups can arise as semisimple parts of $\left(\overline{\mathsf{L}(\Gamma)}^{\mathsf{Zar}}\right)^0$ for a proper affine action $\Gamma \xrightarrow{\rho} \mathsf{Aff}(\mathsf{A})$ where Γ is *not* virtually solvable. Following Smilga, we say that a closed connected subgroup $G < \mathsf{GL}(\mathsf{V})$ is *Milnor* if no such proper affine deformation with

$$\left(\overline{\mathsf{L}(\Gamma)}^{\mathsf{Zar}}\right)^0 = G$$

exists. By Tits [124], we can replace Γ in the above definition by the two-generator free group F_2.

DEFINITION 3.2. *Let* $\rho : G \to \mathsf{GL}(\mathsf{V})$ *be a linear representation of an algebraic group* G *on a vector space* V. *Then* $G \ltimes_\rho \mathsf{V}$ *acts affinely on* V. *We call* ρ *Milnor if there does not exist a subgroup* $\Gamma < G \ltimes_\rho \mathsf{V}$ *that is isomorphic to a nonabelian free group, has linear part* $\mathsf{L}(\Gamma) < G$ *Zariski dense, and acts properly discontinuously on* V.

A solvable subgroup is (trivially) Milnor. Similarly, Bieberbach's structure theorem implies compact groups are Milnor. Thus Milnor's question can be rephrased as whether non-Milnor subgroups exist.

In fact, even many nonsolvable subgroups, for example $\mathsf{GL}(\mathsf{V})$ and $\mathsf{SL}(\mathsf{V})$, are easily seen to be Milnor from the following (see Proposition 3.4).

LEMMA 3.3. *Suppose that* $\Gamma \xrightarrow{\rho} \mathsf{Aff}(\mathsf{A})$ *defines a free action on* A. *Then every element of* $\overline{\mathsf{L}(\Gamma)}^{\mathsf{Zar}}$ *has* 1 *as an eigenvalue.*

This lemma was first used by Kostant-Sullivan [93] in the proof that the Euler characteristic of a compact complete affine manifold vanishes.

Lemma 3.3 follows from two elementary observations:

- If $g \in \mathsf{Aff}(\mathsf{A})$ acts freely on A, then $\mathsf{L}(g) \in \mathsf{GL}(\mathsf{V})$ has 1 as an eigenvalue.
- The condition that $A \in \mathsf{GL}(\mathsf{V})$ has 1 as an eigenvalue—namely, that $\det(A - \mathbb{I}) = 0$—is a polynomial condition on A and thus passes to the Zariski closure.

Summarizing:

PROPOSITION 3.4. *Suppose that* $G < \mathsf{GL}(\mathsf{V})$ *is not Milnor. Then every element of* G *has* 1 *as an eigenvalue.*

3.3.2 Complete affine 3-manifolds

Fried–Goldman [61] classifies which connected semisimple subgroups G can arise as semisimple parts of $\overline{\mathsf{L}(\Gamma)}^{\mathsf{Zar}}$ when $\dim \mathsf{V} = 3$. It follows from their work that the only connected semisimple subgroup $G < \mathsf{GL}(3, \mathbb{R})$ that is not Milnor is $G = \mathsf{SO}(2, 1)^0$. We recall the argument here.

By an easy calculation, the only connected semisimple subgroups of $\mathsf{GL}(3, \mathbb{R})$ are (up to conjugacy):

- $\mathsf{SL}(3, \mathbb{R})$,
- $\mathsf{SO}(3)$,
- $\mathsf{SL}(2, \mathbb{R})$, and
- $\mathsf{SO}(2, 1)^0$,

embedded in the standard ways. We have already excluded the case $G = \mathsf{SL}(3, \mathbb{R})$. The case $G = \mathsf{SO}(3)$ is excluded by the Bieberbach theorems (since $\mathsf{L}(\Gamma)$ must be a finite group, G is trivial).

Suppose $G = \mathsf{SL}(2, \mathbb{R})$. Then $\mathsf{L}(\Gamma)$ can be conjugated into one of these forms:

$$\begin{bmatrix} * & * & * \\ 0 & * & * \\ 0 & * & * \end{bmatrix}, \quad \begin{bmatrix} * & 0 & 0 \\ * & * & * \\ * & * & * \end{bmatrix}$$

The condition that these matrices have 1 as an eigenvalue implies that the $(1,1)$ entry equals 1.

In the first case, the vector field $\partial/\partial x$ is a Γ-invariant parallel vector field that descends to a parallel vector field on M. In the second case, the 1-form dx is a Γ-invariant parallel 1-form that descends to a parallel 1-form on M.

These cases are eliminated as follows. A parallel 1-form can be perturbed to have rational periods and integrates to give a fibration of M over S^1 with fibers closed complete affine 2-manifolds, from which the virtual solvability follows by the two-dimensional case. In the case of a parallel vector field ξ, the Zariski density implies the existence of two elements γ_1, γ_2, which generate a nonabelian free group and correspond to closed orbits of the flow of ξ. These closed orbits are hyperbolic in the sense of hyperbolic dynamics, but their stable manifolds intersect (by lifting them to A), which is a contradiction. See Fried-Goldman [61] for further details.

Finally, consider the most interesting case—namely, $G = \mathsf{SO}(2,1)^0$. Margulis's breakthrough [102, 103] may be restated that $\mathsf{SO}(2,1)^0$ is *not* Milnor. Thus $\mathsf{SO}(2,1)^0$ was the first example of a non-Milnor group. Suppose $M^3 = \Gamma\backslash\mathsf{A}^3$ is a complete affine 3-manifold whose fundamental group Γ is nonsolvable. By the above, $\overline{\mathsf{L}(\Gamma)}^{\mathsf{Zar}}$ is (conjugate to) $\mathsf{SO}(2,1)^0$. Hence, the $\mathsf{O}(2,1)$-invariant inner product on V^3 defines a flat Lorentzian metric on A^3 invariant under the action of Γ. When equipped with this metric, we denote the affine space by $\mathsf{E}^{2,1}$. Hence, $M^3 = \Gamma\backslash\mathsf{E}^{2,1}$ inherits a *flat Lorentzian structure* from the $\mathsf{SO}(2,1)$-invariant Lorentzian inner product on V. Such a complete flat Lorentzian 3-manifold M^3 is called a *Margulis spacetime*.

3.4 THE ASSOCIATED HYPERBOLIC SURFACE.

PROPOSITION 3.5 (Fried-Goldman [61]). *Suppose that $\Gamma < \mathsf{Isom}(\mathsf{E}^{2,1})$ is discrete and acts properly on $\mathsf{E}^{2,1}$. Either Γ is virtually polycyclic or $\Gamma \xrightarrow{\mathsf{L}} \mathsf{O}(2,1)$ is an isomorphism of Γ onto a discrete subgroup of $\mathsf{SO}(2,1) < \mathsf{O}(2,1)$.*

Selberg [116] proved that every finitely generated matrix group contains a torsion-free subgroup of finite index (compare Raghunathan [113], corollary 6.13). Thus Γ contains a finite index subgroup containing no elliptic elements. Henceforth, we restrict to torsion-free discrete subgroups.

Hyperbolic geometry enters here, as $\mathsf{SO}(2,1)$ is the isometry group of H^2 and every discrete subgroup of $\mathsf{SO}(2,1)$ acts properly on H^2. The quotient

$$\Sigma^2 := \mathsf{L}(\Gamma)\backslash\mathsf{H}^2$$

is a complete hyperbolic surface. Since M^3 and Σ^2 are both quotients of contractible spaces by proper and free actions of Γ, M^3 and Σ^2 are homotopy equivalent. We call Σ^2 the *hyperbolic surface associated to* M^3 and M^3 an *affine deformation* of Σ^2.

3.4.1 Margulis spacetimes are not closed

Note that M^3 cannot be compact by the following cohomological dimension argument. If M were compact, then

$$2 = \mathsf{dim}(\Sigma) \geq \mathsf{cd}\big(\mathsf{L}(\Gamma)\big) = \mathsf{cd}(\Gamma) = \mathsf{dim}(M^3) = 3.$$

This contradiction completes the proof of Auslander's conjecture in dimension 3.

Using similar arguments, Goldman–Kamishima [71] proved Auslander's conjecture for flat Lorentzian manifolds (linear holonomy in $\mathsf{O}(n,1)$), and Grunewald–Margulis [78] proved Auslander's conjecture for affine deformations for which the linear holonomy lies in other rank 1 subgroups. See also Tomanov [125, 126].

Around 1990, Mess proved the following:

THEOREM 3.6 (Mess [105]).
The fundamental group of a closed surface admits no proper affine action on A^3.

In particular, Σ is not a closed surface and Γ_0 is not a uniform lattice. In fact, Γ must be a free group: every Margulis spacetime is an affine deformation of a *noncompact* complete hyperbolic surface Σ.

In 1999, Goldman-Margulis [76] gave alternate proofs of Theorem 3.6; compare the discussion in Section 7.2. Later, Labourie [95] and Danciger-Zhang [43] generalized Mess's theorem to show that for a certain class of linear surface group representations, called *Hitchin representations*, affine deformations are never proper. On the other hand, proper affine actions by surface groups do exist by recent work of Danciger–Guéritaud–Kassel; see Theorem 7.10. (Compare [41, 121].)

3.4.2 Affine deformations of hyperbolic surfaces

Sections 4–7 discuss the geometry and classification of Margulis spacetimes. To facilitate this discussion, we use Proposition 3.5 to recast these questions in a more convenient form.

Suppose that $M^3 = \Gamma\backslash \mathsf{E}^{2,1}$ is a Margulis spacetime, where $\Gamma < \mathsf{Isom}^+(\mathsf{E}^{2,1})$ is a discrete subgroup acting properly on $\mathsf{E}^{2,1}$. By Proposition 3.5, we can assume the linear holonomy group $\mathsf{L}(\Gamma)$ is a Fuchsian subgroup $\Gamma_0 < \mathsf{SO}(2,1)$.

Fix Γ_0 and consider Γ as an affine deformation of Γ_0. Affine deformations of Γ_0 are determined by the translational part $\Gamma_0 \xrightarrow{\mathsf{u}} \mathbb{R}^{2,1}$, and we denote the affine deformation determined by the cocycle u by Γ_{u}. In particular, the zero cocycle determines $\mathsf{L}(\Gamma)$, so the notation for $\Gamma_0 = \mathsf{L}(\Gamma)$ is consistent. Translational conjugacy classes of affine deformations form the vector space $\mathsf{H}^1(\Gamma_0, \mathbb{R}^{2,1})$, which has dimension $3(r-1)$ if Γ_0 is a free group of rank $r > 1$.

More geometrically, consider M^3 to be an affine deformation of the hyperbolic surface Σ. Then identify $\mathsf{H}^1(\Gamma_0, \mathbb{R}^{2,1})$ with the cohomology $\mathsf{H}^1(\Sigma, \mathsf{V})$, where V denotes the local system (flat vector bundle) over Σ determined by the linear holonomy homomorphism

$$\pi_1(\Sigma) \xrightarrow{\cong} \Gamma_0 < \mathsf{Isom}(\mathbb{R}^{2,1}) = \mathsf{O}(2,1).$$

The main goal now becomes determining which elements of the vector space $\mathsf{H}^1(\Gamma_0, \mathbb{R}^{2,1})$ determine *proper affine deformations*. This was foreshadowed by Milnor [107], where he proposed a possible way of constructing proper affine actions of non–virtually solvable groups:

> *"Start with a free discrete subgroup of* $\mathsf{O}(2,1)$ *and add translation components to obtain a group of affine transformations which acts freely. However it seems difficult to decide whether the resulting group action is properly discontinuous."*

In retrospect, these are the only ways of constructing such actions in dimension 3.

4 Construction of Margulis spacetimes

We turn now to a direct construction of Margulis spacetimes via fundamental domains bounded by piecewise linear surfaces called *crooked planes*. While Margulis's original examples were constructed from a dynamical point of view (we defer discussion of his original proof until Section 5.3), crooked fundamental domains bring a geometric perspective to the subject. Their introduction by Drumm in 1990 launched a classification program for Margulis spacetimes that was completed recently by Danciger-Guéritaud-Kassel (see Section 7.7).

Continuing with the discussion of the previous section, we consider affine deformations of Fuchsian subgroups $\Gamma_0 < \mathsf{SO}(2,1)$. Associate to Γ_0

a complete hyperbolic surface $\Sigma = \Gamma_0 \backslash \mathsf{H}^2$. In order for an affine deformation to have a chance at being proper, Σ must be non-compact, and we assume this going forward. The section will give a brief overview of the fundamentals of crooked geometry leading to a discussion of Drumm's theorem that every complete noncompact hyperbolic surface Σ admits a proper affine deformation as a Margulis spacetime. This requires models for both the hyperbolic plane H^2 and the three-dimensional Lorentzian affine space, which we call *Minkowski space* $\mathsf{E}^{2,1}$. (Geometrically, $\mathsf{E}^{2,1}$ is characterized as the unique simply connected, geodesically complete, flat Lorentzian manifold.) We motivate the discussion by relating three-dimensional Lorentzian geometry to two-dimensional hyperbolic geometry.

After discussing these models, we introduce *crooked half-spaces* to build fundamental polyhedra for Margulis spacetimes. Our exposition follows Burelle-Charette-Drumm-Goldman [20] as modified by Danciger-Guéritaud-Kassel [40]. Schottky's classical construction of hyperbolic surfaces (now called *ping pong*) extends to crooked geometry, giving a geometric construction of Margulis spacetimes. This was first developed by Drumm [52, 53, 54]. For more details and background, see [55], [26], and [81].

Whereas in the initial examples of Margulis the topology of the quotients is unclear, Margulis spacetimes that have a crooked fundamental polyhedron are topologically equivalent to solid handlebodies and thus *topologically tame*. The tameness of all Margulis spacetimes is discussed at the end of Section 4.4.

4.1 THE GEOMETRY OF H^2 AND $E^{2,1}$.

This introductory section describes the geometry of the hyperbolic plane and its relation to the Lorentzian geometry of Minkowski 3-space. In particular, we discuss the basic geometric objects needed to build hyperbolic surfaces and their extensions to Minkowski space. Then we discuss the classical theory of Schottky groups, which, in the next section, we extend to proper affine deformations of Fuchsian groups.

4.1.1 *The projective model for* H^2

Start with the familiar model of H^2 as the upper half-plane in $\mathbb{C}$, consisting of $x + iy \in \mathbb{C}$ with $y > 0$ with the Poincaré metric. The group $\mathsf{PSL}(2,\mathbb{R})$ acts on H^2 by linear fractional transformations, comprises all orientation-preserving isometries, and is the identity component of $\mathsf{Isom}(\mathsf{H}^2)$. The complement of $\mathsf{PSL}(2,\mathbb{R})$ in $\mathsf{Isom}(\mathsf{H}^2)$ is the other connected component, comprised of orientation-reversing isometries.

A natural model for the Lorentzian vector space $\mathbb{R}^{2,1}$ is the set of *Killing vector fields* on H^2, or the Lie algebra $\mathfrak{sl}(2,\mathbb{R})$ of $\mathsf{PSL}(2,\mathbb{R})$. The Lie algebra $\mathfrak{sl}(2,\mathbb{R})$ is identified with the set of traceless 2×2 real matrices, and the action of $\mathsf{Isom}(\mathsf{H}^2)$ on $\mathfrak{sl}(2,\mathbb{R})$ is by Ad, the adjoint representation (see Section 7.1). The (indefinite) inner product on $\mathfrak{sl}(2,\mathbb{R})$ is defined by

$$\mathbf{v}\cdot\mathbf{w} := \frac{1}{2}\mathsf{tr}(\mathbf{vw});$$

this is 1/8 the Killing form on $\mathfrak{sl}(2,\mathbb{R})$. The basis

$$\mathbf{x}_1 = \begin{bmatrix} 0 & 1 \\ 1 & 0 \end{bmatrix}, \quad \mathbf{x}_2 = \begin{bmatrix} 1 & 0 \\ 0 & -1 \end{bmatrix}, \quad \mathbf{x}_3 = \begin{bmatrix} 0 & -1 \\ 1 & 0 \end{bmatrix},$$

is *Lorentzian-orthonormal* in the sense that

$$\mathbf{x}_1\cdot\mathbf{x}_1 = \mathbf{x}_2\cdot\mathbf{x}_2 = 1, \quad \mathbf{x}_3\cdot\mathbf{x}_3 = -1,$$

and $\mathbf{x}_i\cdot\mathbf{x}_j = 0$ for $i \neq j$.

A natural model for H^2 is one of the two components of the quadric

$$\mathbf{u}\cdot\mathbf{u} = -1.$$

A natural isometry from the upper half-plane $\{x+iy \in \mathbb{C} \mid y > 0\}$ with the Poincaré metric to the Lie algebra $\mathfrak{sl}(2,\mathbb{R}) \longleftrightarrow \mathbb{R}^{2,1}$ with the above inner product is:

$$\begin{array}{ccccc} \mathsf{H}^2 & \longrightarrow & \mathfrak{sl}(2,\mathbb{R}) & \longleftrightarrow & \mathbb{R}^{2,1} \\ x+iy & \longmapsto & \frac{1}{y}\begin{bmatrix} x & -(x^2+y^2) \\ 1 & -x \end{bmatrix} & \longleftrightarrow & \frac{1}{2y}\begin{bmatrix} 1-x^2-y^2 \\ 2x \\ 1+x^2+y^2 \end{bmatrix}. \end{array}$$

Here the vector on the right-hand side represents the coordinates with respect to the basis $\mathbf{x}_1, \mathbf{x}_2, \mathbf{x}_3$ of $\mathbb{R}^{2,1}$:

$$\frac{1}{y}\begin{bmatrix} x & -(x^2+y^2) \\ 1 & -x \end{bmatrix} = \frac{1-x^2-y^2}{2y}\,\mathbf{x}_1 + \frac{x}{y}\,\mathbf{x}_2 + \frac{1+x^2+y^2}{2y}\,\mathbf{x}_3.$$

Note that the Lie algebra $\mathfrak{sl}(2,\mathbb{R})$ comes naturally equipped with an orientation, defined in terms of the bracket. Indeed we define the ordered basis $(\mathbf{x}_1, \mathbf{x}_2, \mathbf{x}_3)$ to be positive since $[\mathbf{x}_1, \mathbf{x}_2] = +2\mathbf{x}_3$. This orientation of $\mathfrak{sl}(2,\mathbb{R})$ is naturally associated with an orientation of H^2—namely, the orientation for which $\mathbf{x}_3$ is an infinitesimal rotation in the positive direction. Note that the orientation-reversing isometries of the Lorentzian structure on $\mathfrak{sl}(2,\mathbb{R})$ flip the sign of the Lie bracket. The adjoint action of an orientation-reversing isometry of H^2 preserves the Lie bracket and hence the orientation of $\mathfrak{sl}(2,\mathbb{R})$; however, it exchanges H^2 with the other component of the quadric $\mathbf{u}\cdot\mathbf{u}=-1$.

In relativistic terminology, a vector $\mathbf{v}\in\mathbb{R}^{2,1}\setminus\{\mathbf{0}\}$ is called *spacelike* if $\mathbf{v}\cdot\mathbf{v}$ is positive, *timelike* if it is negative, *null* or *lightlike* if it is zero. A spacelike (respectively, timelike) vector $\mathbf{v}$ is *unit-spacelike* (respectively, *unit-timelike*) if and only if $\mathbf{v}\cdot\mathbf{v}=1$ (respectively, $\mathbf{v}\cdot\mathbf{v}=-1$). A Killing vector field $\xi\in\mathfrak{sl}(2,\mathbb{R})$ is spacelike (respectively, null, timelike) if and only if it generates a hyperbolic (respectively, parabolic, elliptic) one-parameter group of isometries.

The set of null vectors (including $\mathbf{0}$) is a cone, called the *light cone* and denoted $\mathcal{N}$. The set $\mathcal{N}\setminus\mathbf{0}$ has two components, or *nappes*.

Choosing a preferred nappe is equivalent to choosing a *time orientation*. For example, we choose lightlike vectors with $v_3>0$ to be *future* pointing and lightlike vectors with $v_3<0$ to be *past pointing*. The connected components of timelike vectors are similarly defined to be future and past pointing. One model for H^2, already described above, is the space of *future-pointing* unit-timelike vectors.

An equivalent model for H^2 is the subset of the projective space $\mathsf{P}(\mathbb{R}^{2,1})$ comprised of timelike lines, with $\partial\mathsf{H}^2$ the set of null lines. Spacelike vectors $\mathbf{w}\in\mathbb{R}^{2,1}$ determine geodesics and half-planes in H^2 in the projective model as follows:

$$\mathfrak{h}_{\mathbf{w}}:=\{\mathbf{v}\in\mathsf{H}^2\mid\mathbf{v}\cdot\mathbf{w}>0\}$$

is the open halfplane defined by $\mathbf{w}$. The boundary $\partial\mathfrak{h}_{\mathbf{w}}=\mathsf{H}^2\cap\mathbf{w}^{\perp}$ is the geodesic corresponding to $\mathbf{w}$. The orientation of H^2 together with $\mathfrak{h}_{\mathbf{w}}$ determines a natural orientation on $\partial\mathfrak{h}_{\mathbf{w}}$. That is, unit-spacelike vectors in $\mathbb{R}^{2,1}=\mathfrak{sl}(2,\mathbb{R})$ correspond to *oriented geodesics* in H^2. Note that as a Killing vector field, $\mathbf{w}$ is an infinitesimal translation along $\partial\mathfrak{h}_{\mathbf{w}}$ in the positive direction. Generally, the Killing field on H^2 associated to $\mathbf{w}\in\mathbb{R}^{2,1}$ is given explicitly in terms of the Lorentzian cross product on $\mathbb{R}^{2,1}$, which is just the Lie bracket $[\cdot,\cdot]$ on $\mathfrak{sl}(2)$, by restricting the vector field $x\mapsto[\mathbf{w},x]$ to the hyperboloid of timelike future-pointing vectors.

4.1.2 *Cylinders and fundamental slabs*

A basic hyperbolic surface is a *hyperbolic cylinder*, arising as the quotient $\Sigma:=\langle A\rangle\backslash\mathsf{H}^2$, where $A\in\mathsf{Isom}(\mathsf{H}^2)$ is an isometry that is *hyperbolic*, meaning

A leaves invariant a unique geodesic l_A along which it translates by a distance ℓ_A. The image $\langle A\rangle\backslash l_A$ in $\langle A\rangle\backslash \mathsf{H}^2$ is a closed geodesic in Σ of length ℓ_A. The scalar invariant ℓ_A completely describes the isometry type of Σ.

One can build fundamental domains for the action of $\langle A\rangle$ as follows. Choose any geodesic $l_0 \in \mathsf{H}^2$ meeting l_A in a point $a_0 \in \mathsf{H}^2$, and let $\mathfrak{h}_0$ be the half-plane bounded by l_0 containing $A(a_0)$. Then $A(\mathfrak{h}_0) \subset \mathfrak{h}_0$, and the complement

$$\Delta := \mathfrak{h}_0 \setminus A(\mathfrak{h}_0)$$

is a fundamental domain for the cyclic group $\langle A\rangle$ acting on H^2. If $\mathbf{w}$ is unit-spacelike and $\mathfrak{h}_0 = \mathfrak{h}_{\mathbf{w}}$ is the open half-plane defined by $\mathbf{w}$ as above, then the fundamental domain takes the form

$$\Delta = \mathfrak{h}_{\mathbf{w}} \cap \mathfrak{h}_{-A(\mathbf{w})},$$

and we call Δ a *fundamental slab* for $\langle A\rangle$.

4.1.3 *Affine deformations of cylinders*

Recall the *Minkowski 3-space,* $\mathsf{E}^{2,1}$—namely, the complete 1-connected flat Lorentzian manifold in dimension 3. Equivalently, $\mathsf{E}^{2,1}$ is an affine space whose underlying vector space is equipped with a Lorentzian inner product. As above, we model the underlying Lorentzian vector space $\mathbb{R}^{2,1}$ on the Lie algebra $\mathfrak{sl}(2,\mathbb{R})$ of Killing vector fields on H^2.

The group of *linear* orientation preserving isometries $\mathsf{Isom}(\mathbb{R}^{2,1})$ equals the special orthogonal group $\mathsf{SO}(2,1) \cong \mathsf{Isom}(\mathsf{H}^2)$. Its identity component $\mathsf{Isom}^+(\mathsf{H}^2)$, comprising the orientation preserving isometries of H^2, is naturally identified via the adjoint representation with $\mathsf{PSL}(2,\mathbb{R})$. A hyperbolic element $A \in \mathsf{PSL}(2,\mathbb{R})$ pointwise fixes one spacelike line, and this line contains exactly two unit-spacelike vectors that are negatives of each other.

In terms of Killing vector fields, the line fixed by A is just the infinitesimal centralizer of A. Indeed,

$$A = \exp\left(\frac{\ell_A}{2}\mathbf{w}_A\right) \sim \begin{bmatrix} e^{\ell_A/2} & 0 \\ 0 & e^{-\ell_A/2} \end{bmatrix},$$

where $\mathbf{w}_A$ is one of the two unit-spacelike generators of this line, and $\ell_A > 0$ is the translation length of A in H^2. Since $\ell_A > 0$, $A \mapsto \mathbf{w}_A$ is well-defined and equivariant under the action of $\mathsf{Isom}^+(\mathsf{H}^2)$, in the sense that $\mathbf{w}_{BAB^{-1}} = B\mathbf{w}_A$, for any $B \in \mathsf{Isom}^+(\mathsf{H}^2)$.

Let $g \in \mathsf{Isom}(\mathsf{E}^{2,1})$ be an affine deformation of A. That is,

$$g(p) = A(p) + \mathsf{u},$$

where the vector u is the *translational part* of g. There is a unique g-invariant line, denoted $\mathsf{Axis}(g)$, parallel to the fixed line $\mathbb{R}\mathbf{w}_A$ of the linear part A of g. The line $\mathsf{Axis}(g)$ inherits a natural orientation induced from $\mathbf{w}_A$.

The restriction of g to $\mathsf{Axis}(g)$ is a translation, and the *signed* displacement along this spacelike geodesic is a scalar defined by

$$\alpha(g) = \mathbf{w}_{\mathsf{L}(g)} \cdot \mathsf{u}(g). \tag{4.4}$$

Here, $\mathsf{L}(g) := A$ is the linear part of g and $\mathsf{u}(g)$ is the translational part of g, as in Equation (4.1). Clearly, g acts freely if and only if $\alpha(g) \neq 0$.

DEFINITION 4.1. *The scalar quantity $\alpha(g)$ is called the Margulis invariant of g.*

Through a translational change of coordinates, the origin may be located on $\mathsf{Axis}(g)$ so that

$$\mathsf{u}(g) = \alpha(g)\mathbf{w}_{\mathsf{L}(g)}.$$

By an orientation preserving linear change of coordinates, the linear part $A = \mathsf{L}(g)$ diagonalizes and g takes the form:

$$g(x) = \begin{bmatrix} e^{\ell_A} & 0 & 0 \\ 0 & e^{-\ell_A} & 0 \\ 0 & 0 & 1 \end{bmatrix} x + \begin{bmatrix} 0 \\ 0 \\ \alpha(g) \end{bmatrix}.$$

4.1.4 The role of orientation

We note that the definition of the vector $\mathbf{w}_A$, and hence of the Margulis invariant $\alpha(g)$, depends on more than just the structure of $\mathbb{R}^{2,1} \longleftrightarrow \mathfrak{sl}(2,\mathbb{R})$ as a Lorentzian vector space. It depends, more specifically, on the Lie algebra structure, where the operation of the *Lorentzian cross product* is determined entirely by the Lorentzian structure and orientation. Margulis's original work does not use the Lie algebra $\mathfrak{sl}(2,\mathbb{R})$. There the definition of α is given directly in terms of the Lorentzian structure and a choice of orientation, as follows. The positive direction $\mathbf{w}_A$ of the 1-eigenspace of $\mathsf{L}(g)$ is the one making the basis $(\mathbf{w}^+, \mathbf{w}^-, \mathbf{w}_A)$ positive, where $\mathbf{w}^+, \mathbf{w}^-$ denote representatives of the attracting and repelling eigenlines of $\mathsf{L}(g)$ that have negative inner product $\mathbf{w}^+ \cdot \mathbf{w}^- < 0$.

4.1.5 Parallel slabs

Affine lines parallel to $\mathsf{Axis}(g)$ describe a g-invariant foliation, and the foliation $\mathscr{F}$ by planes orthogonal to these lines is a g-invariant two-dimensional foliation defined by the g-equivariant orthogonal projection

$$\mathsf{E}^{2,1} \xrightarrow{\Pi} \mathsf{Axis}(g) \longleftrightarrow \mathbb{R}.$$

In particular, since g acts by translation by $\alpha(g)$ on $\mathsf{Axis}(g)$, the preimage $\Pi^{-1}\Big[0, |\alpha(g)|\Big]$ of the closed interval

$$\Big[0, |\alpha(g)|\Big] \subset \mathbb{R}$$

is a fundamental domain for $\langle g\rangle$. Since the faces of this fundamental domain are the parallel hyperplanes $\Pi^{-1}(0)$ and $\Pi^{-1}|\alpha(g)|$, we call these fundamental domains *parallel slabs.*

4.1.6 Schottky groups and ping pong

Having discussed actions of the infinite cyclic group $\mathbb{Z}$ on both H^2 and $\mathsf{E}^{2,1}$, we now turn to nonabelian free groups. We recall Schottky's [115] construction of discrete free groups acting on the hyperbolic plane.

For brevity, let us focus on the two-generator case.

Suppose $A_1, A_2 \in \mathsf{PSL}(2,\mathbb{R})$ are hyperbolic elements with respective translation axes l_1 and l_2. Let $\mathbf{w}_1$ and $\mathbf{w}_2$ be unit spacelike vectors associated to half-spaces $\mathfrak{h}_{\mathbf{w}_1}, \mathfrak{h}_{\mathbf{w}_2}$, such that for each $i=1,2$, the boundary of $\mathfrak{h}_{\mathbf{w}_i}$ crosses l_i in H^2 and satisfies $A_i \cdot \mathfrak{h}_{\mathbf{w}_i} \subset \mathfrak{h}_{\mathbf{w}_i}$ so that a fundamental domain for the cyclic group $\langle A_i\rangle$ is given by $\Delta_i = \mathfrak{h}_{\mathbf{w}_i} \cap \mathfrak{h}_{-A_i\mathbf{w}_i}$, as in Section 4.1.2.

PROPOSITION 4.2 (Schottky). *Suppose that the four half-spaces*

$$\mathfrak{h}_{-\mathbf{w}_1}, \mathfrak{h}_{A_1\mathbf{w}_1}, \mathfrak{h}_{-\mathbf{w}_2}, \mathfrak{h}_{A_2\mathbf{w}_2} \tag{4.5}$$

are pairwise disjoint. Then A_1 and A_2 generate a discrete free subgroup $\Gamma = \langle A_1, A_2\rangle < \mathsf{PSL}(2,\mathbb{R})$.

Let us sketch the proof of this well-known fact. Consider the polygon

$$\Delta = \mathfrak{h}_{\mathbf{w}_1} \cap \mathfrak{h}_{-A_1\mathbf{w}_1} \cap \mathfrak{h}_{\mathbf{w}_2} \cap \mathfrak{h}_{-A_2\mathbf{w}_2} \subset \mathsf{H}^2,$$

which is bounded by four disjoint lines. Then the image of Δ under any nontrivial reduced word w in $A_1, A_1^{-1}, A_2, A_2^{-1}$ lies in one of the four half-spaces in Equation (4.5). Indeed, observe the following relations:

$$A_1 \cdot (\mathfrak{h}_{-\mathsf{w}_1})^c = A_1 \cdot \overline{\mathfrak{h}_{\mathsf{w}_1}} = \overline{\mathfrak{h}_{A_1\mathsf{w}_1}}$$

$$A_1^{-1} \cdot (\mathfrak{h}_{A_1\mathsf{w}_1})^c = A_1^{-1} \cdot \overline{\mathfrak{h}_{-A_1\mathsf{w}_1}} = \overline{\mathfrak{h}_{-\mathsf{w}_1}}$$

$$A_2 \cdot (\mathfrak{h}_{-\mathsf{w}_2})^c = A_2 \cdot \overline{\mathfrak{h}_{\mathsf{w}_2}} = \overline{\mathfrak{h}_{A_2\mathsf{w}_2}}$$

$$A_2^{-1} \cdot (\mathfrak{h}_{A_2\mathsf{w}_2})^c = A_2^{-1} \cdot \overline{\mathfrak{h}_{-A_2\mathsf{w}_2}} = \overline{\mathfrak{h}_{-\mathsf{w}_2}}$$

Then by induction on the length of the reduced word w, the image of Δ (the "ping pong ball") under the action of w lies:

- inside $\mathfrak{h}_{A_1\mathsf{w}_1}$ if the first (that is, leftmost) letter of w is A_1;
- inside $\mathfrak{h}_{-\mathsf{w}_1}$ if the first letter is A_1^{-1};
- inside $\mathfrak{h}_{A_2\mathsf{w}_2}$ if the first letter is A_2; or
- inside $\mathfrak{h}_{-\mathsf{w}_2}$ if the first letter is A_2^{-1}.

This proves the proposition.

Since Γ is discrete it acts properly on H^2; however, Δ might not be a fundamental domain for the action. Indeed, in some cases,

$$\bigcup_{\gamma \in \Gamma} \gamma \cdot \overline{\Delta}$$

is a proper subset of H^2. It is always possible, however, to choose a polyhedron Δ as above that is a fundamental domain.

The disjointness of the half-spaces in Equation (4.5) is essential in this construction.

However, in affine space at most *two* half-spaces can be disjoint. Hence, a ping pong construction using affine half-spaces does not work in affine geometry. Nonetheless, Schottky fundamental domains in $\mathsf{A}^3 = \mathsf{E}^{2,1}$ do exist and are constructed from *crooked half-spaces.*

4.2 CROOKED GEOMETRY. Milnor [107] essentially proposed building proper actions of free groups by combining proper actions of cyclic groups. However, deciding whether multiple proper actions by cyclic groups generate a proper action of the free product is quite delicate. As we observed in the previous section, hyperplanes, which are perhaps the most natural separating surfaces in affine geometry, are not well suited for a Schottky-style construction of fundamental domains for free groups. Observe that, by contrast to Euclidean geometry, in our Lorentzian setting the linear part of an affine transformation dominates the translational part for "most" points. Hence building fundamental polyhedra adapted more to the linear part than the translational part seems preferable.

4.2.1 Crooked planes and crooked half-spaces

In [52], Drumm introduced so-called crooked planes in order to build fundamental domains for proper affine actions of nonabelian free groups. A crooked plane disconnects $\mathsf{E}^{2,1}$ into two regions, called crooked half-spaces. Unlike a linear plane, a crooked plane has a distinguished point, called the *vertex*. In particular, crooked planes are not homogeneous. Drumm's original construction was given purely in terms of Lorentzian geometry. Here, however, we make use of the identification $\mathbb{R}^{2,1} \cong \mathfrak{sl}(2,\mathbb{R})$ and define them in terms of Killing fields on the hyperbolic plane, following Danciger-Guéritaud-Kassel [40].

Let $\mathbf{w} \in \mathfrak{sl}(2,\mathbb{R})$ be a spacelike unit vector, let

$$\ell = \ell_{\mathbf{w}} = \mathbf{w}^{\perp} \cap \mathsf{H}^2$$

be the oriented geodesic associated to $\mathbf{w}$, and let $\mathbf{w}^+$ and $\mathbf{w}^-$ be future-oriented lightlike vectors respectively representing the forward and backward endpoints $[\mathbf{w}^+], [\mathbf{w}^-]$ of $\ell_{\mathbf{w}}$ in $\partial\mathsf{H}^2$. Here we think of the ideal boundary $\partial\mathsf{H}^2$ as the projectivized null cone in $\mathbb{R}^{2,1} = \mathfrak{sl}(2,\mathbb{R})$. We first define the crooked plane $\mathcal{C}(\mathbf{0},\ell)$ with vertex the origin $\mathbf{0}$. The crooked plane $\mathcal{C}(\mathbf{x},\ell)$ with vertex $\mathbf{x}$ is just the translate $\mathcal{C}(\mathbf{0},\ell) + \mathbf{x}$.

The *crooked plane* $\mathcal{C}(\mathbf{0},\ell)$ is the union of three linear pieces, a *stem* and two *wings*, described as follows. See Figure 4.2.

- The stem is the closure of the collection of all elliptic Killing fields whose fixed point in H^2 lies on $\ell_{\mathbf{w}}$.
- The wing associated to the forward endpoint $[\mathbf{w}^+]$ of $\ell_{\mathbf{w}}$ is the union of the parabolic Killing fields $\mathbb{R}\mathbf{w}^+$ that fix $[\mathbf{w}^+]$ and the hyperbolic Killing fields for which $[\mathbf{w}^+]$ is a *repelling* fixed point. This wing meets the stem along the hinge $\mathbb{R}\mathbf{w}^+$.
- Similarly, the wing associated to $[\mathbf{w}^-]$ is the union of the parabolic Killing fields $\mathbb{R}\mathbf{w}^-$ fixing $[\mathbf{w}^-]$ and the hyperbolic Killing fields for which $[\mathbf{w}^-]$ is a repelling fixed point. This wing meets the stem along the hinge $\mathbb{R}\mathbf{w}^-$.

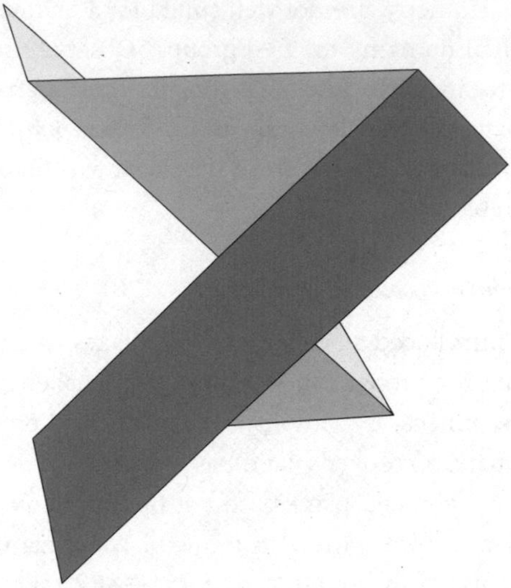

Figure 4.2. A crooked plane

- The line $\mathbb{R}\mathbf{w}$, called the *spine,* lies in $\mathcal{C}(\mathbf{0},\mathbf{w})$ and crosses the stem perpendicularly at $\mathbf{0}$. The positive ray $\mathbb{R}^+\mathbf{w}$ lies in the wing associated to $[\mathbf{w}^-]$ and the negative ray $\mathbb{R}^-\mathbf{w}$ lies in the wing associated to $[\mathbf{w}^+]$.

More succinctly, the crooked plane $\mathcal{C}(\mathbf{0},\ell)$ is the collection of all Killing fields with a *nonattracting fixed point* on the closure

$$\overline{\ell} = \ell \cup \{[\mathbf{w}^+],[\mathbf{w}^-]\}$$

of ℓ in $\overline{\mathsf{H}^2}$.

A crooked plane $\mathcal{C}(\mathbf{0},\ell)$ divides $\mathfrak{sl}(2,\mathbb{R})$ into two components, called crooked half-spaces. The *crooked half-space* $\mathcal{H}(\mathbf{0},\ell)$ is the collection of Killing vector fields with a nonattracting fixed point contained in the closure $\overline{\mathfrak{h}_{\mathbf{w}}} \subset \overline{\mathsf{H}^2}$ of the positive half-plane $\mathfrak{h}_{\mathbf{w}}$ bounded by $\ell_{\mathbf{w}}$. Note that

$$\mathcal{H}(\mathbf{0},\ell) \cup \mathcal{H}(\mathbf{0},-\ell) = \mathfrak{sl}(2,\mathbb{R}),$$
$$\mathcal{H}(\mathbf{0},\ell) \cap \mathcal{H}(\mathbf{0},-\ell) = \mathcal{C}(\mathbf{0},\ell) = \mathcal{C}(\mathbf{0},-\ell),$$

where $-\ell = \ell_{-\mathbf{w}}$ is the same geodesic ℓ but with the opposite orientation.

More generally, the crooked plane $\mathcal{C}(\mathbf{x},\mathbf{w})$ and crooked half-space $\mathcal{H}(\mathbf{x},\mathbf{w})$ with vertex $\mathbf{x}$ are obtained by translating by $\mathbf{x}$:

$$\mathcal{C}(\mathbf{x},\mathbf{w}) := \mathbf{x} + \mathcal{C}(\mathbf{0},\mathbf{w}),$$
$$\mathcal{H}(\mathbf{x},\mathbf{w}) := \mathbf{x} + \mathcal{H}(\mathbf{0},\mathbf{w}).$$

4.2.2 Crooked ping pong

The following Lorentzian ping pong lemma was proved in [53].

LEMMA 4.3. *Let* $\Gamma = \langle \gamma_1, \gamma_2, \ldots, \gamma_n \rangle$ *be a group in* $\mathsf{Isom}(\mathsf{E})$ *and* $\{\mathcal{H}_{\pm 1}, \mathcal{H}_{\pm 2}, \ldots, \mathcal{H}_{\pm n}\}$ *be* $2n$ *disjoint crooked half-spaces such that*

$$\gamma_i(\mathcal{H}_{-i}) = \overline{\mathsf{E} \setminus \mathcal{H}_{+i}}.$$

Then Γ *is a free group that acts properly on* E, *with fundamental domain*

$$\Delta := \overline{\mathsf{E} \setminus \cup_{i=1}^{n} (\mathcal{H}_{-i} \cup \mathcal{H}_{+i})}.$$

In particular, the quotient $\Gamma \backslash \mathsf{E}^{2,1}$ *is homeomorphic to an open solid handlebody.*

The conditions in the lemma immediately imply that the groups satisfying these conditions act properly on a subset of E. The difficult part of the proof is to demonstrate that

$$\mathsf{E} = \Gamma(\Delta) := \bigcup_{\gamma \in \Gamma} \gamma \Delta.$$

See Drumm [52, 53, 54], Charette-Goldman [30], and Danciger-Guéritaud-Kassel [40, Lemma 7.6].

Using Lemma 4.3, Drumm proved the following:

THEOREM 4.4.
Every finitely generated free discrete subgroup of $\mathsf{SO}(2,1)$ *admits a proper affine deformation with a fundamental domain bounded by crooked planes.*

4.3 DISJOINTNESS OF CROOKED HALF-SPACES AND PLANES.

The application of Lemma 4.3 requires crooked planes to be disjoint. We now give a criterion for disjointness, originally due to Drumm–Goldman [58] and later conceptually clarified by Burelle–Charette–Drumm–Goldman [20].

Consider a set of pairwise disjoint geodesics $\{\ell_1, \ell_2, \ldots, \ell_n\}$ in H^2 that bound a common region. The geodesics can be oriented consistently so that the interiors of the crooked half-spaces $\mathcal{H}(\mathbf{0}, \ell_i)$ are disjoint. All of the crooked half-spaces meet at the origin $\mathbf{0}$, and pairs of the corresponding crooked planes, boundaries of the crooked half-spaces, may share a wing.

Translations $\mathbf{u}_i$ exist for which the sets $\{\mathcal{H}(\mathbf{u}_i, \ell_i)\}$ are pairwise disjoint. This situation is exactly the one described in Lemma 4.3. To this end, define the following:

DEFINITION 4.5. *For an oriented geodesic* ℓ, *the (open) stem quadrant* $Q(\ell)$ *is the open quadrant of the plane containing the stem of* $C(\mathbf{0}, \ell)$ *that lies inside the interior of* $\mathcal{H}(\mathbf{0}, \ell)$.

Figure 4.3. A crooked plane and a translation of the crooked plane by a vector in the stem quadrant

The stem quadrant $\mathcal{Q}(\ell)$ is composed of spacelike vectors and bounded by two null rays inside the plane that contains the stem of $\mathcal{C}(\mathbf{0},\ell)$. In the Lie algebra interpretation, the spacelike vectors in $\mathcal{Q}(\ell)$ are hyperbolic Killing vector fields whose invariant geodesics are perpendicular to ℓ and point into the interior of the half-space defined by ℓ. The null rays on the boundary are the parabolic Killing vector fields whose fixed points are the endpoints of ℓ.

Stem quadrants were defined in [20] and used to show the following:

LEMMA 4.6. *$\mathcal{H}(\mathbf{u},\ell)\subset\mathcal{H}(\mathbf{0},\ell)$ if $\mathbf{u}\in\mathcal{Q}(\ell)$. Furthermore, $\mathcal{H}(\mathbf{u}_1,\ell_1)$ and $\mathcal{H}(\mathbf{u}_2,\ell_2)$ are disjoint if and only if $\mathbf{u}_1-\mathbf{u}_2\in\mathcal{Q}(\ell_1)-\mathcal{Q}(\ell_2)$.*

See Figure 4.3.

In particular, start with the collection of crooked half-spaces $\{\mathcal{H}(\mathbf{0},\ell_i)\}$ whose interiors are disjoint. Translate each crooked half-space in a stem quadrant direction $\mathbf{u}_i \in \mathcal{Q}(\ell_i)$ to create a collection $\{\mathcal{H}(\mathbf{u}_i, \ell_i)\}$ of disjoint crooked half-spaces as in Lemma 4.3.

Burelle-Charette-Drumm-Goldman [20] introduce foliations by crooked planes; Burelle-Francoeur [21] show that every crooked slab admits a foliation by crooked planes, answering a question raised by Charette-Kim [32].

4.4 TAMENESS. A natural question, in a direction converse to Theorem 4.4, is whether every Margulis spacetime arises from a crooked polyhedron—that is, whether Drumm's construction gives all Margulis spacetimes. This question, first asked by Drumm–Goldman [57], motivated much of the recent work on Margulis spacetimes. This *Crooked Plane Conjecture* was established by Danciger–Guéritaud–Kassel [10, 12] in general, following earlier work for two generator groups by Charette-Drumm-Goldman [27, 29]. See Section 7.6 for a discussion of these ideas.

This has the following purely topological consequence:

THEOREM 4.7.
A complete affine 3-manifold with fundamental group Γ free of rank r is homeomorphic to a handlebody of genus r.

Theorem 4.7 is the analog of the *Marden Conjecture* for hyperbolic 3-manifolds, proved independently by Agol [6] and Calegari-Gabai [23], which implies that every complete hyperbolic 3-manifold with free fundamental group is homeomorphic to an open solid handlebody. There are two proofs of Theorem 4.7 due independently to Choi-Drumm-Goldman and to Danciger-Guéritaud-Kassel, which do not use crooked planes and which preceded the resolution of the Crooked Plane Conjecture.

Choi–Goldman [35] proved Theorem 4.7 in the case that Σ has compact convex core (i.e., the linear holonomy group is convex cocompact). This was later extended by Choi–Drumm–Goldman [34] to include the case that Σ has cusps. The proof involves compactifying a Margulis spacetime M^3 with convex cocompact linear holonomy, as an $\mathbb{RP}^3$-manifold with geodesic (ideal) boundary. The boundary is an $\mathbb{RP}^2$-manifold obtained by grafting annuli to two copies of Σ along its boundary, as in Goldman [67] and Choi [33]. The boundary $\mathbb{RP}^2$ surface is naturally the quotient of a domain in the projective sphere at infinity for $\mathsf{E}^{2,1}$. Given that the Γ action on $\mathsf{E}^{2,1}$ and on this domain at infinity are both proper, the difficulty lies in proving that the Γ-action on the union is also proper. This is accomplished by using the dynamics of the lifted geodesic flow as in Goldman–Labourie–Margulis [73] and the fact that the linear holonomy group Γ_0 acts on $\overline{\mathsf{H}^2}$ as a *convergence group*. When Γ_0 is no longer convex cocompact (but still finitely generated), then the proof requires a detailed technical analysis of the geometry near a cusp.

From a different point of view, Danciger–Guéritaud–Kassel [39, 42] proved (Proposition 7.5) that any Margulis spacetime M is fibered in affine (timelike) lines over the associated surface Σ. This also gives a proof of Theorem 4.7. See Section 7 for further discussion.

In another direction, Frances [60] defines an ideal boundary for Margulis spacetimes, using the action on the (Lorentzian) conformal compactification $\mathsf{Ein}^{2,1}$ of Minkowski space, sometimes called the *Einstein Universe*. This extends the local conformal Lorentzian geometry of $\mathsf{E}^{2,1}$ in the same way that the conformal geometry of S^n extends conformal Euclidean geometry on E^n. The Einstein Universe is diffeomorphic to the mapping torus of the antipodal map on S^2. Its automorphism group is the projective orthogonal group $\mathsf{PO}(3,2)$. (Compare [10, 69].)

The main result is that the action extends to the conformal boundary in much the same way that actions of discrete isometry groups on hyperbolic (n-1)-space extends to its ideal boundary S^n. Frances defines a *limit set* Λ such that Γ acts properly discontinuously on the complement $\mathsf{Ein}^{2,1} \setminus \Lambda$ and describes a compactification for the quotient $\big(\mathsf{Ein}^{2,1} \setminus \Lambda\big)/\Gamma$ (which is not a manifold).

5 The Margulis spectrum

5.1 THE MARKED SIGNED LORENTZIAN LENGTH SPECTRUM.

The *marked length spectrum* of a hyperbolic surface Σ is an important invariant, which determines the isometry type of Σ. Recall that this is the function

$$\pi_1(\Sigma) \xrightarrow{\ell_\Sigma} \mathbb{R}_{\geq 0},$$

which associates to the homotopy class of a based loop γ the infimum of the lengths of loops (freely) homotopic to γ. When Σ is closed, then $\ell(\gamma)$ equals the length of the closed geodesic in Σ homotopic to γ; in particular $\ell(\gamma) > 0$. In general, γ has parabolic holonomy if and only if $\ell(\gamma) = 0$.

This function is part of a general construction defined on the group $\mathsf{Isom}(\mathsf{H}^2)$. The *geodesic displacement function*

$$\mathsf{Isom}(\mathsf{H}^2) \xrightarrow{\ell} \mathbb{R}_{\geq 0} \qquad (4.6)$$

associates to g the infimum $d(p, g(p))$, where $p \in \mathsf{H}^2$. If g is elliptic or parabolic, then $\ell(g) = 0$. If g is hyperbolic, then $\ell(g)$ equals the length of the shortest closed geodesic in the cylinder $\mathsf{H}^2/\langle g \rangle$, as in Section 4.1.2.

If $M^3 = \Gamma \backslash \mathsf{E}^{2,1}$ is a Margulis spacetime with associated hyperbolic surface $\Sigma \sim M^3$, then each homotopy class of closed curve $\gamma \in \Gamma$ with nonparabolic holonomy is represented by a unique spacelike geodesic whose Lorentzian length is $|\alpha(\gamma)|$, where α is defined by Equation (4.4) in Section 4.1.3. More generally, the function

$$\pi_1(M) \xrightarrow{\alpha} \mathbb{R} \tag{4.7}$$

is an important invariant of M^3 called the *marked Lorentzian length spectrum* and is analogous to the marked length spectrum ℓ_Σ for the associated hyperbolic surface. As we shall see in the next subsection, the signs of the Margulis invariants $\alpha(\gamma)$ also play a central role in the theory.

5.2 PROPERTIES OF THE MARGULIS INVARIANT. Margulis defined the function α in [102] and [103]. Recall from Section 4.1.3 that if $g \in \mathsf{Isom}^+(\mathsf{E}^{2,1})$ is an orientation preserving Lorentzian isometry with $\mathscr{L}(g)$ hyperbolic, then g leaves invariant a unique spacelike line $\mathsf{Axis}(g)$, which carries a natural orientation induced from the orientation of $\mathsf{E}^{2,1}$ (see Section 4.1.3). Furthermore, the restriction of g to $\mathsf{Axis}(g)$ is a translation by a multiple $\alpha(g)\mathbf{w}_g$, where $\mathbf{w}_g$ is the unit-spacelike vector parallel to $\mathsf{Axis}(g)$ determined by the orientation of $\mathsf{Axis}(g)$. The *sign* of g is defined as the sign of $\alpha(g) \in \mathbb{R}$ (positive, negative, or zero).

Margulis's invariant has the following important properties:

LEMMA 5.1. *Suppose* $g \in \mathsf{Isom}^+(\mathsf{E}^{2,1})$ *with* $\mathsf{L}(g)$ *hyperbolic.*

(1) $\alpha(g) = 0$ *if and only if g has a fixed point.*
(2) $\alpha(g) = (g(p) - p) \cdot \mathbf{w}_g$ *for any* $p \in \mathsf{E}$.
(3) $\alpha(g) = \alpha(\eta g \eta^{-1})$ *for any* $\eta \in \mathsf{Isom}(\mathsf{E}^{2,1})$.
(4) $\alpha(g^n) = |n|\alpha(g)$ *for* $n \neq 0$.

While the definition of $\alpha(\gamma)$ provides the conceptual meaning of the Margulis invariant, Lemma 5.1.(2) is a useful formula for its computation. Lemma 5.1.(4) implies that the sign of a power is independent of the exponent and, in particular,

$$\alpha(\gamma^{-1}) = \alpha(\gamma). \tag{4.8}$$

The four properties of Lemma 5.1 are elementary. In contrast, the following *Opposite Sign Lemma* is deep, playing an important role in characterizing proper affine deformations (Theorem 6.1).

THEOREM 5.2 (Opposite Sign Lemma).
If g, h are isometries with hyperbolic linear part, with opposite signs—that is, $\alpha(g)\alpha(h) \leq 0$—then $\langle g, h\rangle$ does not act properly on A^3.

Abels's survey paper [1] provides a detailed proof of Margulis's Opposite Sign Lemma, along the lines of the original proof in [102] and [103].

5.3 MARGULIS'S ORIGINAL CONSTRUCTION. The Margulis invariant is also key in his original construction of proper affine deformations of free discrete groups in $\Gamma_0 < \mathsf{SO}(2,1)$.

To that end, first define the *hyperbolicity* of a hyperbolic element $g \in \mathsf{SO}(2,1)$ as the Euclidean distance

$$d(S^2 \cap \langle g^+\rangle, S^2 \cap \langle g^-\rangle),$$

where S^2 is the Euclidean unit sphere and $\langle g^+\rangle$ (respectively $\langle g^-\rangle$) is the attracting (respectively repelling) eigenline for g. Hyperbolicity is related to the distance of a fixed basepoint $0 \in \mathsf{H}^2$ to the invariant geodesic $l_g \subset \mathsf{H}^2$ of g. Call an element ϵ*-hyperbolic* if its hyperbolicity is greater than ϵ.

Moreover, two elements $g, h \in \mathsf{SO}(2,1)$ are said to be ϵ-transverse if they are ϵ-hyperbolic and

$$d(S^2 \cap \langle g^\pm\rangle, S^2 \cap \langle h^\pm\rangle) > \epsilon.$$

Margulis showed that, for any two ϵ-hyperbolic, ϵ-transverse elements $g, h \in \mathsf{SO}^+(2,1)$ that are "sufficiently contracting" (this basically means that their largest eigenvalues are sufficiently large), we have

$$\alpha(gh) \approx \alpha(g) + \alpha(h). \tag{4.9}$$

Now consider a free, two-generator discrete group $\Gamma_0 < \mathsf{SO}^+(2,1)$ whose limit set Λ is not all of $\partial\mathsf{H}^2$ (equivalently, Γ_0 is not a lattice). Then there exists

$$\eta \in \mathsf{SO}^+(2,1)/\Gamma_0$$

so that every element in the coset $\eta\Gamma_0$ is ϵ-hyperbolic. (In particular η is ϵ-hyperbolic, with attracting fixed point outside of Λ.) Then using Equations (4.8) and (4.9), Margulis showed that, for an affine deformation Γ whose translational parts of the generators satisfy a suitable condition, $|\alpha(\eta\gamma)|$ grows roughly like the word length of γ for $\gamma \in \Gamma$. Once the hyperbolicity is bounded below by ϵ, the Margulis invariant $\alpha(\eta\gamma)$ controls the minimum Euclidean distance $\eta\gamma$ moves any point. For any compact $K \subset \mathsf{E}^{2,1}$,

$$\{\gamma \in \Gamma \mid \eta\gamma(K) \cap K \neq \emptyset\}$$

is finite. This implies that Γ acts properly on $\mathsf{E}^{2,1}$. For further details, compare Drumm-Goldman [56].

5.4 LENGTH SPECTRUM RIGIDITY.

The *marked length spectrum* of a hyperbolic structure on a surface Σ is the map that assigns to each free homotopy class $[\gamma]$ of loop, the length $\ell(\gamma)$ of the unique closed geodesic in that homotopy class. Regarding $\mathsf{Isom}^+(\mathsf{H}^2) = \mathsf{PSL}(2,\mathbb{R})$, suppose $\pi_1(\Sigma) \xrightarrow{\rho_0} \mathsf{PSL}(2,\mathbb{R})$ is the holonomy representation of the hyperbolic structure on Σ. Then $\ell(\gamma)$ relates to the *character* of ρ_0 by:

$$\mathsf{tr}\big(\rho_0(\gamma)\big) = \pm 2\cosh\left(\frac{\ell\big(\rho_0(\gamma)\big)}{2}\right).$$

Hence, a hyperbolic structure on Σ is determined by its length spectrum, simply because the holonomy representation ρ_0 is determined by its character. This is a general algebraic fact about irreducible linear representations; see, for example, Goldman [68] for a general proof. For details on this question, see Abikoff [5]. More recently Otal [109] and Croke [37] proved marked length spectrum rigidity for surfaces of *variable negative curvature,* where the algebraic methods are unavailable. For length spectrum rigidity for locally symmetric spaces, see Kim [89, 88] and Cooper–Delp [36].

Now we discuss to what extent the marked Lorentzian length spectrum determines the isometry type of a Margulis spacetime. As a consequence of Theorem 5.2, either the $\alpha(g)$ are all positive or all negative. By changing the orientation of $\mathsf{E}^{2,1}$, we may assume they are all positive.

Suppose M^3 is a Margulis spacetime whose associated (complete) hyperbolic surface Σ has a compact convex core. (In this case the holonomy group $\mathsf{L}(\Gamma)$ of Σ is said to be *convex cocompact.*) As in Section 5.1, every element of $\mathsf{L}(\Gamma)\setminus\{1\}$ is hyperbolic and every closed curve in M^3 is freely homotopic to a unique closed geodesic in M^3. The absolute value $|\alpha(\gamma)|$ equals the *Lorentzian length* of this closed geodesic in M^3. Thus the function

$$\pi_1(M^3) \xrightarrow{\alpha\circ\rho} \mathbb{R}$$

represents the analogous *marked Lorentzian length spectrum* of M^3.

THEOREM 5.3.

Consider two affine deformations ρ, ρ' *of* F_n *with the same convex cocompact representation as linear part. Suppose that* $\alpha \circ \rho = \alpha \circ \rho'$. *Then* ρ *and* ρ' *are conjugate in* $\mathsf{Isom}(\mathsf{E})$.

This was proved by Drumm-Goldman [59] for $n=2$, to which we shall refer. We give below the modifications needed to prove this for general $n > 2$. Charette-Drumm [25] proved the stronger statement without the assumption that ρ and ρ' have the same linear part, only assuming that $\alpha \circ \rho = \alpha \circ \rho'$. See also Kim [90] and Ghosh [65].

Assume inductively the result for all free groups of rank at most n, where $n \geq 2$. For $\mathsf{F}_{n+1} = \langle x_1, x_2, \ldots, x_{n+1} \rangle$, consider the three n-generator subgroups:

$$\begin{aligned} S_1 &= \langle x_2, x_3, x_4, \ldots, x_{n+1} \rangle \\ S_2 &= \langle x_1, x_3, x_4, \ldots, x_{n+1} \rangle \\ S_3 &= \langle x_1, x_2, x_4, \ldots, x_{n+1} \rangle \end{aligned}$$

In the following we will only be concerned with the generators x_1, x_2, x_3, which do not occur inside every such subgroup. Without loss of generality, we may choose the generators x_1, x_2, x_3 so that the 1-eigenspaces of the linear parts $\mathsf{L}(x_1), \mathsf{L}(x_2)$, and $\mathsf{L}(x_3)$ do not have a nontrivial linear dependence. If this is not the case, we simply replace x_1 by $x_2 x_1 x_2^{-1}$ and the assumption will hold.

We consider two representations ρ, ρ' of F_{n+1} and their restrictions to S_i for $i = 1, 2, 3$. Denote the translational parts of ρ and ρ' by

$$\mathsf{u}, \mathsf{u}' \in \mathsf{Z}^1(\mathsf{F}_{n+1}, \mathbb{R}^{2,1})$$

respectively. Let $i = 1, 2$, or 3. Since the Margulis invariants α, α' agree, their restrictions to the n-generator subgroup S_i also agree. Thus the restrictions of u and u' to S_i are cohomologous in $\mathsf{Z}^1(S_i, \mathbb{R}^{2,1})$. That is, there exists $a_i \in \mathbb{R}^{2,1}$ so that

$$\mathsf{u}'(\gamma) - \mathsf{u}(\gamma) = \delta(a_i)(\gamma) = a_i - \mathsf{L}(\gamma) a_i \tag{4.10}$$

for $\gamma \in S_i$.

We show that the vector $a_2 - a_3$ lies in the fixed line

$$\mathsf{Fix}\big(\mathsf{L}(x_1)\big) = \mathsf{Ker}\big(\mathbb{I} - \mathsf{L}(x_1)\big).$$

Apply (4.10) to $\gamma = x_1$ and $i = 2, 3$:

$$a_2 - \mathsf{L}(x_1) a_2 = \mathsf{u}'(x_1) - \mathsf{u}(x_1) = a_3 - \mathsf{L}(x_1) a_3,$$

from which follows

$$a_2 - a_3 = \mathsf{L}(x_1)(a_2 - a_3)$$

as claimed. Similarly, $a_3 - a_1 \in \mathsf{Fix}(\mathsf{L}(x_2))$ and $a_1 - a_2 \in \mathsf{Fix}(\mathsf{L}(x_3))$. By our assumption above, the three lines $\mathsf{Fix}(\mathsf{L}(x_i))$ for $i = 1, 2, 3$ are not coplanar, so in particular they form a direct sum decomposition of $\mathbb{R}^3$. Observing that

$$(a_2 - a_3) + (a_3 - a_1) + (a_1 - a_2) = 0,$$

we deduce that the vectors $a_2 - a_3$, $a_3 - a_1$, $a_1 - a_2$ must each be zero. Thus the vectors $a_1 = a_2 = a_3$ are all equal, and (4.10) holds over the entire group F_{n+1}.

5.5 FURTHER REMARKS ON THE MARGULIS LENGTH SPECTRUM. Charette–Goldman [31] proved an analog of McShane's identity [104], a relation on the marked length spectrum for hyperbolic punctured tori.

If a discrete group Γ of affine isometries with hyperbolic linear part acts properly on E, then the Margulis invariants $\alpha(\gamma)$ are either all positive or all negative. In general, infinitely many positivity conditions are needed to ensure properness (but see Section 6.2 for the two examples of Σ_0 where only finitely many conditions suffice). Charette [24] found a sequence of affine deformations ρ_n of a two-generator Fuchsian group Γ_0 with the following property: for any given integer n,

- $\alpha(\rho_n(\gamma)) > 0$ for all $\gamma \in \Gamma_0$ with word length less than n; and
- $\alpha(\rho_n(\gamma')) < 0$ for some $\gamma' \in \Gamma_0$.

Using *strip deformations*, Minsky [74] explicitly showed there exist free groups with convex cocompact linear part with the property that the Margulis invariants of all elements have one sign but that do not act properly on E. See the discussion of Theorem 7.7 in Section 7.6.

The sign of an affine deformation is undefined for elliptic affine transformations. Charette–Drumm–Goldman [28] extended Margulis's sign to *parabolic* affine transformations. A parabolic element γ of $\mathsf{SO}(2,1)$ fixes no spacelike vectors and no closed geodesic has holonomy γ. Charette and Drumm find a subspace of null vectors fixed by γ with a natural orientation and extend the *sign* of the Margulis invariant to γ. Lemma 5.1 can be adapted to parabolic transformations. Moreover, Theorem 5.2 extends to the case where either or both transformations are parabolic, using this extension of Margulis's invariant.

6 Diffusing the Margulis invariant

In this section we describe the extension of Margulis's marked Lorentzian length spectrum to the space of geodesic currents and state the properness criterion of Goldman–Labourie–Margulis, which leads to a description of the deformation space of Margulis spacetimes associated to a given hyperbolic surface.

6.1 NORMALIZING THE MARGULIS INVARIANT.

The (signed) Margulis invariant and the geodesic length function enjoy the same homogeneity (Lemma 5.1 (4)):

$$\ell(\gamma^n) = |n|\ell(\gamma)$$
$$\alpha(\gamma^n) = |n|\alpha(\gamma)$$

Thus the quotient

$$\Gamma_0 \xrightarrow{\widehat{\alpha}} \mathbb{R}$$
$$\gamma \longmapsto \frac{\alpha(\gamma)}{\ell(\gamma)}$$

is constant on cyclic subgroups of $\Gamma \cong \pi_1(\Sigma)$.

Cyclic hyperbolic subgroups of $\pi_1(\Sigma)$ correspond to closed geodesics on Σ. Closed geodesics on Σ correspond to periodic trajectories of the geodesic flow Φ on the unit tangent bundle $U\Sigma$ and hence determine Φ-invariant probability measures on $U\Sigma$ supported on the velocity vector field of the closed geodesic.

Recall that a *geodesic current* on Σ is a Φ-invariant probability measure on $U\Sigma$. See Bonahon [17]. The convex set $\mathfrak{C}(\Sigma)$ of all geodesic currents is equipped with the weak-* topology. It is compact if Σ has compact convex core. Geodesic currents corresponding to closed geodesics are dense in $\mathfrak{C}(\Sigma)$.

For a fixed affine deformation ρ of a Fuchsian representation ρ_0, the above function $\widehat{\alpha}$ extends to a continuous map

$$\mathfrak{C}(\Sigma) \xrightarrow{\widehat{\alpha}} \mathbb{R}.$$

Moreover, if we let the $\rho = \rho_{[\mathbf{u}]}$ vary over the space $\mathsf{H}^1(\Gamma_0, \mathbb{R}^{2,1})$ of affine deformations, then we have the following:

THEOREM 6.1 (Goldman-Labourie-Margulis [73]).
Fix a hyperbolic surface Σ with holonomy representation ρ_0.

- *There exists a continuous map*

$$\mathsf{H}^1(\Gamma_0, \mathbb{R}^{2,1}) \times \mathfrak{C}(\Sigma) \xrightarrow{\Psi} \mathbb{R}$$

such that for a fixed affine deformation ρ corresponding to $[u] \in \mathsf{H}^1(\Gamma_0, \mathbb{R}^{2,1})$ and an element $\gamma \in \Gamma$ corresponding to an Φ-invariant probability measure μ,

$$\Psi([\mathbf{u}], \mu) = \widehat{\alpha}_\rho(\gamma)$$

as above. Furthermore this function is bi-affine with respect to the linear structure on $\mathsf{H}^1(\Gamma_0, \mathbb{R}^{2,1})$ and the affine structure on $\mathfrak{C}(\Sigma)$.
- *The affine deformation $\rho = \rho_{[\mathbf{u}]}$ is proper if and only if the image $\Psi\Big(\{\mathbf{u}\} \times \mathfrak{C}(\Sigma)\Big)$ is bounded away from 0.*

Since probability measures supported on periodic trajectories are dense, properness is equivalent to bounding $\widehat{\alpha}(\gamma) = \alpha(\gamma)/\ell(\gamma)$ away from zero. In particular, Minsky's [74] construction of nonproper affine deformations has the property that every element has the same sign, but there is a sequence of elements whose normalized Margulis invariants approach zero.

Theorem 6.1 immediately implies the Opposite Sign Lemma (Theorem 5.2) as follows. Suppose that $\alpha(\gamma_1) < 0 < \alpha(\gamma_2)$. Let μ_i denote the invariant probability measure corresponding to γ_i. Then

$$\widehat{\alpha}(\mu_1) = \frac{\alpha(\gamma_1)}{\ell(\gamma_1)} < 0 < \frac{\alpha(\gamma_2)}{\ell(\gamma_2)} = \widehat{\alpha}(\mu_2).$$

Since $\mathfrak{C}(\Sigma)$ is convex, a continuous path μ_t (for $1 \le t \le 2$) joins μ_1 to μ_2. Continuity of $\widehat{\alpha}$ and the Intermediate Value Theorem imply that $\widehat{\alpha}(\mu_t) = 0$ for some $1 < t < 2$ and $\mu_t \in \mathfrak{C}(\Sigma)$. By Theorem 6.1, the affine deformation is not proper.

Let us briefly contrast the properness criterion of Theorem 6.1 with the properness criterion of Benoist [14] and Kobayashi [92] for reductive homogeneous spaces. In the setting of reductive homogeneous spaces G/H, properness of the action of a discrete group $\Gamma < G$ is characterized by the behavior of the Cartan projection (singular values) of Γ, specifically that the Cartan projection of Γ goes away from the Cartan projection of H. There is no known analogue of this simple criterion in nonreductive settings, such as Minkowski geometry $\mathsf{E}^{2,1}$. In Minkowski geometry, the Margulis invariant of an element is less like a Cartan projection and more like an infinitesimal Jordan projection (eigenvalues). The work of Danciger-Guéritaud-Kassel [39] interprets an action of a group Γ on Minkowski space $\mathsf{E}^{2,1}$ as an infinitesimal action on the three-dimensional anti–de Sitter space, the model for constant negative curvature

Lorentzian geometry in dimension 3. Note that anti–de Sitter space is a reductive homogeneous space, so the Benoist-Kobayashi properness criterion applies. However, Kassel [87] and Guéritaud-Kassel [82] give a different properness criterion in terms of uniform behavior of the Jordan projections. The new proof of Theorem 6.1 given in Danciger-Guéritaud-Kassel [39] interprets the uniform behavior of the Margulis invariant as an infinitesimal analogue of the Guéritaud-Kassel properness criterion in anti–de Sitter geometry. See the discussion in Section 7.3.

6.2 CLASSIFICATION OF MARGULIS SPACETIMES. Theorem 4.4 implies that proper affine deformations exist whenever $\Sigma = \Gamma_0 \backslash \mathsf{H}^2$ is noncompact. Another consequence of Theorem 6.1 is a determination of the deformation space of Margulis spacetimes as a convex domain. Theorem 6.1 implies that the space of *all* proper affine deformations of Γ_0 equals the subspace of $\mathsf{H}^1(\Gamma_0, \mathbb{R}^{2,1})$ comprised of $[\mathbf{u}]$ such that $\Phi([\mathbf{u}], \mu)$ is either always positive or always negative for all $\mu \in \mathfrak{C}(\Sigma)$.

The positive affine deformations, those $[\mathbf{u}]$ for which $\Phi([\mathbf{u}], \mu) > 0$ for all $\mu \in \mathfrak{C}(\Sigma)$, form an open and convex cone. This cone is the interior of the intersection over $\gamma \in \Gamma$ of the set of half-spaces defined by $\alpha_{[\mathbf{u}]}(\gamma) > 0$. In fact, it suffices to take this intersection over γ corresponding to *simple* loops on Σ. See Goldman-Labourie-Margulis-Minsky [74] and Danciger-Guéritaud-Kassel [39]. Indeed, properness is implied by positivity (or negativity) of the Margulis invariant $\Psi([\mathbf{u}], \mu)$ over all *measured laminations* μ (those currents with self-intersection zero).

Figures 4.4 and 4.5 depict the deformation space for four hyperbolic surfaces Σ representing the four different topological types for which $\chi(\Sigma) = -1$, or equivalently $\pi_1(\Sigma) \cong \mathsf{F}_2$. Although in these cases

$$\dim \mathsf{H}^1(\Gamma_0, \mathbb{R}^{2,1}) = 3,$$

we may projectivize the set and draw the image of this cone in the two-dimensional projective space $\mathsf{P}(\mathsf{H}^1(\Gamma_0, \mathbb{R}^{2,1}))$. The lines drawn in these pictures are defined by $\alpha(\gamma) = 0$, where γ is a *primitive* element of F_2 (that is, an element living in a free basis of F_2). However, only in the case of the one-holed torus do the primitive elements correspond to simple nonseparating loops.

6.3 DYNAMICAL IDEAS FROM THE PROOF OF THEOREM 6.1. Let us sketch some ideas from Goldman-Labourie-Margulis [73]. The basic idea of the properness criterion in Theorem 6.1 is to translate properness of the discrete group action into a question of properness of a continuous flow.

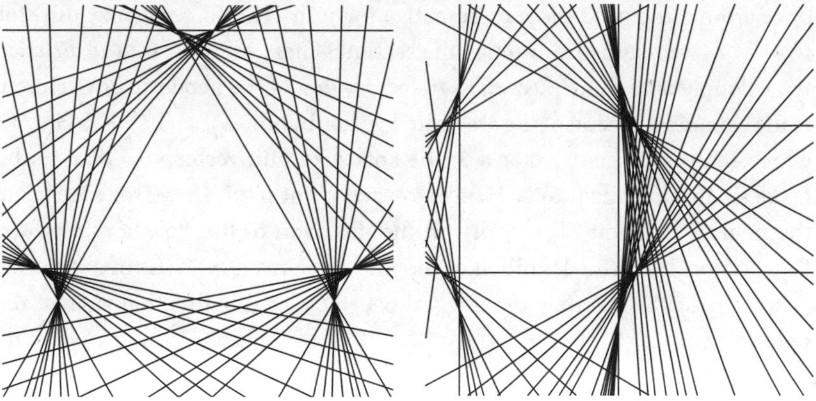

Figure 4.4. Deformation spaces for the three-holed sphere and two-holed cross-surface (projective plane)

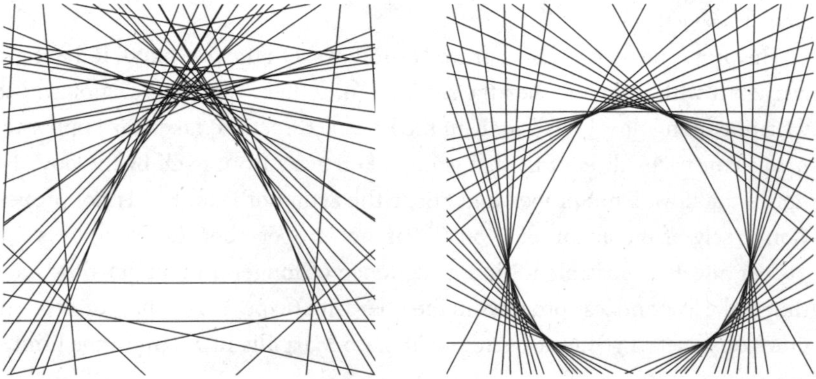

Figure 4.5. Deformation spaces for the one-holed Klein bottle and one-holed torus

Consider the flat affine bundle $\mathsf{E}_\rho \to U\Sigma$, defined as the quotient $(\mathsf{E}^{2,1} \times U\mathsf{H}^2)/\Gamma$, where Γ acts by $\rho = \rho_{[\mathbf{u}]}$ in the $\mathsf{E}^{2,1}$ factor and by ρ_0 on the $U\mathsf{H}^2$ factor. The geodesic flow φ_t on $U\Sigma$ lifts, via the flat connection, to a flow Φ_t on E_ρ. Then the properness of the ρ action of Γ on $\mathsf{E}^{2,1}$ is equivalent to properness of the flow Φ_t on E_ρ. To determine properness of this flow, it suffices to consider only the recurrent part $U_{rec}\Sigma$ of the unit tangent bundle $U\Sigma$.

Let us describe how to extend the Margulis invariant function to the space of currents. Let $s\colon U\Sigma \to \mathsf{E}_\rho$ be a smooth section. Using the flat connection, one may measure how much s changes along a path in the unit tangent bundle. The Margulis invariant $\alpha(\gamma)$ of an element $\gamma \in \Gamma$ is equal to the amount the section s changes in the direction of the translation axis $\mathsf{Axis}(\rho(\gamma))$ after going once around the geodesic representative for γ in $U\Sigma$. To generalize this, for $u \in U_{rec}\Sigma$, we may ask how much the section s changes in the *neutral direction* $\nu(u)$ along (some finite piece of) a trajectory $\Phi_t u$ of the geodesic flow. Here ν is the canonical section of the associated vector bundle $V_\rho = (\mathbb{R}^{2,1} \times U\mathsf{H}^2)/\Gamma$, which maps a tangent vector u to the spacelike unit vector $\nu(u)$ dual to the geodesic in Σ, tangent to u. Hence, to each point $u \in U\Sigma$, we may associate the neutral variation $\langle d_{\varphi_t} s, \nu \rangle$ of s in the direction of the flow, a real valued function on $U\Sigma$. The definition of the Margulis invariant $\widehat{\alpha}(\mu)$ of a geodesic current μ on Σ is simply the integral over U_{rec} of this function against the measure μ:

$$\widehat{\alpha}(\mu) := \int_{U_{rec}} \langle d_{\varphi_t} s, \nu \rangle d\mu.$$

It can be shown that the definition does not depend on the choice of section s. Indeed, the functional $\mu \mapsto \widehat{\alpha}(\mu)$ is continuous and on the dense subset of currents μ_γ that are supported on a closed geodesic γ, the definition gives $\widehat{\alpha}(\mu_\gamma) = \widehat{\alpha}(\gamma)$.

The properness criterion is proved as follows. On the one hand, if there is a current μ with $\widehat{\alpha}(\mu) = 0$, then the section s (or rather a modified section whose variation along flow lines is only in the neutral direction) takes the support of μ to a compact subset of E_ρ that is not taken away from itself by the flow Φ_t; hence the flow is not proper, and hence the action of Γ on $\mathsf{E}^{2,1}$ is not proper. Conversely, if the action of Γ on $\mathsf{E}^{2,1}$ is not proper, then the flow Φ_t is not proper and it is possible to find a sequence of longer and longer flow lines that make less and less progress in the fiber with respect to the flat connection. One constructs a geodesic current with zero Margulis invariant in the limit.

6.4 DYNAMICAL STRUCTURE OF MARGULIS SPACETIMES.

While we do not explore the ideas here, we mention some further work on the dynamical structure of Margulis spacetimes.

In [72], Goldman and Labourie show that the union of closed geodesics in a Margulis spacetime with convex cocompact linear holonomy is dense in the projection of the *nonwandering set* for the geodesic flow. (Goldman and Labourie call the projections of the nonwandering orbits *recurrent.*) In other words, a Margulis spacetime with convex cocompact linear holonomy group has a compact "dynamical core." This is completely analogous to the behavior of nonwandering orbits for the geodesic flow for a convex cocompact hyperbolic surface.

Further, Ghosh [62] proves an Anosov property for the geodesic flow in the dynamical core of a Margulis spacetime. From that he constructs a *pressure metric* on the moduli space [64], analogous to the pressure metric defined in higher Teichmüller theory constructed by Bridgeman-Canary-Labourie-Sambarino [19, 18].

7 Affine actions and deformations of geometric structures

The finer structure of the classification of Margulis spacetimes is deeply tied to the theory of *infinitesimal deformations* of hyperbolic surfaces. The connection comes from the low-dimensional coincidence that the standard representation of $\mathsf{SO}(2,1)$ acting on $\mathbb{R}^{2,1}$ is isomorphic to the *adjoint representation* of $\mathsf{SO}(2,1)$ acting on its Lie algebra $\mathfrak{so}(2,1)$. In general, if G is a Lie group, affine actions with linear part in the adjoint representation of G are in direct correspondence with infinitesimal deformations of representations into G, which in many cases correspond to deformations of geometric structures modeled on some homogeneous space of G. The dynamics of the affine action is often closely related to the geometry of the associated deformation of geometric structures. For the case of $G = \mathsf{SO}(2,1)$ corresponding to Margulis spacetimes, Danciger–Guéritaud–Kassel [39] showed that properness of the affine deformation is equivalent to the condition that the associated deformation of hyperbolic structure is *contracting*, in a sense to be made precise later in this section.

7.1 AFFINE ACTIONS AND DEFORMATION THEORY.

For the moment, let us work in the general setting that G is an arbitrary finite-dimensional Lie group. Recall that the Lie algebra $\mathfrak{g}$ of G is the Lie subalgebra of $\mathsf{Vec}(G)$ consisting of *right-invariant vector fields* on G. These vector fields generate flows by left-multiplication by one-parameter subgroups of G. The action of G on itself by left-multiplication induces a (left-) action on $\mathsf{Vec}(G)$. Since left- and right-multiplication commute, the left-action on $\mathsf{Vec}(G)$ preserves $\mathfrak{g} < \mathsf{Vec}(G)$. The resulting action is the adjoint representation

$$G \xrightarrow{\mathsf{Ad}} \mathsf{Aut}(\mathfrak{g}).$$

The linear action of G on $\mathfrak{g}$ extends to an affine action of the semi-direct product $G \ltimes_{\mathsf{Ad}} \mathfrak{g}$, where $\mathfrak{g}$ acts by translations:

$$\mathbf{v} \xmapsto{(g,u)} \mathsf{Ad}(g)\mathbf{v} + \mathsf{u}. \tag{4.11}$$

The adjoint action of G on $\mathfrak{g}$ preserves the Killing form $B(\cdot,\cdot)$. If G is semisimple, which will be the case in all of our applications, B is a nondegenerate symmetric bilinear form of indefinite signature (p,q). Hence the affine action of $G \ltimes_{\mathsf{Ad}} \mathfrak{g}$ on $\mathfrak{g}$ is by isometries of a flat pseudo-Riemannian metric of signature (p,q). This gives a map

$$G \ltimes_{\mathsf{Ad}} \mathfrak{g} \xrightarrow{\Phi_G} \mathsf{Isom}(\mathsf{E}^{p,q}), \tag{4.12}$$

which maps G into the stabilizer of a point, a copy of $\mathsf{SO}(p,q)$, and maps $\mathfrak{g}$ to the translation subgroup $\mathbb{R}^{p,q}$.

Affine actions of the form (4.11) closely relate to infinitesimal deformations of geometric structures and representations. The affine group $G \ltimes_{\mathsf{Ad}} \mathfrak{g}$ is naturally isomorphic to the total space of the tangent bundle $\mathsf{T}G$ of the Lie group G, under the map that associates $g \in G$ and $\mathsf{u} \in \mathfrak{g}$ to the evaluation $\mathsf{u}_g \in \mathsf{T}_g G$ of u at g:

$$\mathsf{T}G \cong G \ltimes_{\mathsf{Ad}} \mathfrak{g}. \tag{4.13}$$

Given a representation $\rho_0 : \Gamma \to G$ of a discrete group Γ in G, an *infinitesimal deformation* of ρ_0 is a homomorphic lift ρ to the tangent bundle $\mathsf{T}G$:

$$\begin{array}{ccc} & & \mathsf{T}G \\ & \overset{\rho}{\nearrow} & \Big\downarrow \Pi_G \\ \Gamma & \xrightarrow{\rho_0} & G \end{array}$$

Infinitesimal deformations arise naturally as tangent vectors to paths in the analytic set $\mathsf{Hom}(\Gamma, G)$. Indeed, if $\rho_t \in \mathsf{Hom}(\Gamma, G)$ is a smooth path, then

$$\rho(\gamma) := \frac{\mathrm{d}}{\mathrm{d}t}\bigg|_{t=0} \rho_t(\gamma) \in \mathsf{T}_{\rho_0(\gamma)} G \tag{4.14}$$

defines an infinitesimal deformation. Using the isomorphism (4.13), an infinitesimal deformation ρ of a fixed representation $\rho_0 \in \mathsf{Hom}(\Gamma, G)$ is efficiently described as a cocycle

$$\mathsf{u} \in \mathsf{Z}^1(\Gamma, \mathfrak{g}_{\mathsf{Ad}\rho_0}),$$

where $\mathfrak{g}_{\mathsf{Ad}\rho_0}$ denotes the Γ-module defined by the composition

$$\Gamma \xrightarrow{\rho_0} G \xrightarrow{\mathsf{Ad}} \mathsf{Aut}(\mathfrak{g}).$$

(Compare Raghunathan [113], Section 6.)

We refer to cocycles in $\mathsf{Z}^1(\Gamma, \mathfrak{g}_{\mathsf{Ad}\rho_0})$ as *deformation cocycles*. When ρ is the derivative of a conjugation path

$$\rho_t(\cdot) = g_t \rho_0(\cdot) g_t^{-1},$$

where g_t is a smooth path in G based at the identity, the associated cocycle u is the coboundary δv, where $v \in \mathfrak{g}$ extends the tangent vector

$$\left.\frac{d}{dt}\right|_{t=0} g_t \in \mathsf{T}_e(G).$$

The set $\mathsf{B}^1(\Gamma, \mathfrak{g}_{\mathsf{Ad}\rho_0})$ of such coboundaries makes up the *infinitesimal conjugations*, or trivial infinitesimal deformations. The cohomology group $\mathsf{H}^1(\Gamma, \mathfrak{g}_{\mathsf{Ad}\rho_0})$ describes the equivalence classes of infinitesimal deformations up to infinitesimal conjugation. For further details, see Sikora [117] or Labourie [97].

7.2 MARGULIS INVARIANTS AND LENGTH FUNCTIONS.

For the remainder of this section (except Section 7.9) we consider the specific case when

$$G := \mathsf{Isom}(\mathsf{H}^2) \cong \mathsf{PGL}(2, \mathbb{R}) \cong \mathsf{SO}(2,1),$$
$$\mathfrak{g} := \mathfrak{sl}(2, \mathbb{R}) \cong \mathfrak{so}(2,1),$$

and $\Gamma_0 \overset{\rho_0}{\hookrightarrow} G$ is the inclusion of a finitely generated, torsion-free, discrete subgroup corresponding to the hyperbolic surface $\Sigma = \Gamma_0 \backslash \mathsf{H}^2$.

Since the adjoint action of G on $\mathfrak{g}$ is isomorphic to the standard representation of $\mathsf{SO}(2,1)$ on $\mathbb{R}^{2,1}$, the action of $\mathsf{Isom}^+(\mathsf{E}^{2,1})$ on $\mathsf{E}^{2,1}$ identifies with the affine action of $G \ltimes_{\mathsf{Ad}} \mathfrak{g}$ on the Lie algebra $\mathfrak{g}$—in other words, the map

$$G \ltimes_{\mathsf{Ad}} \mathfrak{g} \xrightarrow{\Phi_G} \mathsf{Isom}^+(\mathsf{E}^{2,1})$$

from (4.12) is an isomorphism. In particular, a cocycle in

$$\mathsf{Z}^1(\Gamma_0, \mathbb{R}^{2,1}_{\rho_0}) \cong \mathsf{Z}^1(\Gamma_0, \mathfrak{so}(2,1)_{\mathsf{Ad}})$$

corresponds both to an affine deformation Γ_{u} of Γ_0 and to an infinitesimal deformation of the representation $\Gamma_0 \overset{\rho_0}{\hookrightarrow} G$. By the Ehresmann-Weil-Thurston principle (see Goldman [70]), infinitesimal deformations of ρ_0 correspond to infinitesimal deformations of the hyperbolic structure on Σ.

Goldman-Margulis [76] observed the first key entry in the dictionary between the dynamics of the affine action and the geometry of the associated infinitesimal deformation.

Recall the geodesic displacement function $G \xrightarrow{\ell} \mathbb{R}_{\geq 0}$ defined in Equation (4.6). Its restriction to the hyperbolic elements of G (an open subset) is a smooth function, whose differential we denote by $\mathsf{T}G \xrightarrow{\mathrm{d}\ell} \mathbb{R}$.

LEMMA 7.1 ([76]). *Let*

$$(g, \mathsf{u}) \in G \ltimes_{\mathsf{Ad}} \mathfrak{g} = \mathsf{T}G$$

be an infinitesimal deformation of the hyperbolic element $g \in G$, *and let* $\Phi_G(g, \mathsf{u})$ *be the corresponding orientation preserving affine isometry of* $\mathsf{E}^{2,1}$. *Then the Margulis invariant (see Section 5.2) equals the derivative of the length. That is,*

$$\alpha\big(\Phi_G(g, \mathsf{u})\big) = \mathrm{d}\ell(g, u).$$

Let $\Gamma_0 < G$ be a fixed convex cocompact subgroup. Recall the extension of the Margulis invariant to the space of currents from Theorem 6.1:

$$\mathsf{H}^1(\Gamma_0, \mathbb{R}^{2,1}) \times \mathfrak{C}(\Sigma) \xrightarrow{\Psi} \mathbb{R}.$$

Lemma 7.1 implies that this function is exactly the differential of the length function for geodesic currents. More specifically, to each geodesic current μ on $\Sigma = \Gamma_0 \backslash \mathsf{H}^2$ is associated a length function ℓ_μ on the space $\mathsf{Hom}_{cc}(\Gamma_0, G)$ of convex cocompact representations. The map $\mu \mapsto \ell_\mu$ taking currents to continuous G-invariant functions on $\mathsf{Hom}_{cc}(\Gamma_0, G)$ is continuous. Density of geodesic currents corresponding to closed geodesics implies that $\ell_\mu(\cdot)$ may be approximated in terms of the usual length functions $\ell_\gamma(\cdot)$ for $\gamma \in \Gamma_0$, defined by:

$$\ell_\gamma(\rho) := \ell\big(\rho(\gamma)\big).$$

Lemma 7.1 implies that under the identification

$$\mathsf{H}^1(\Gamma_0, \mathbb{R}^{2,1}) \cong \mathsf{H}^1(\Gamma_0, \mathfrak{g}_{\mathsf{Ad}}),$$

the diffused Margulis invariant function Ψ is precisely the map

$$\mathsf{H}^1(\Gamma_0, \mathfrak{g}_{\mathsf{Ad}}) \times \mathfrak{C}(\Sigma) \xrightarrow{\mathrm{d}\ell} \mathbb{R}$$

taking a cohomology class $[\mathsf{u}]$ of infinitesimal deformations and a current $\mu \in \mathfrak{C}(\Sigma)$ to the derivative $\mathrm{d}\ell_\mu(\mathsf{u})$ of the length of μ in the u direction. Therefore, the Goldman–Labourie–Margulis properness criterion (Theorem 6.1) may be restated:

PROPOSITION 7.2. *Let $\Gamma_0 < G$ be a convex cocompact subgroup and* $\mathsf{u} \in \mathsf{Z}^1(\Gamma_0, \mathfrak{g}_{\mathsf{Ad}\rho_0})$ *a cocycle defining an infinitesimal deformation of the inclusion* $\Gamma_0 \xrightarrow{\rho_0} G$. *Then the corresponding affine action* $\Phi_G(\rho_0, \mathsf{u})$ *is properly discontinuous if and only if* $d\ell_\mu(\mathsf{u}) \neq 0$ *for all geodesic currents* $\mu \in \mathfrak{C}(\Sigma)$.

By exchanging u with $-\mathsf{u}$ (which gives an affine equivalent action), we may assume that $d\ell_\nu(\mathsf{u}) \leq 0$ for some current $\nu \in \mathfrak{C}(\Sigma)$. Hence, since the space of currents is connected, the condition that $d\ell_\mu(\mathsf{u}) \neq 0$ for all $\mu \in \mathfrak{C}(\Sigma)$ is equivalent to the condition that $d\ell_\mu(\mathsf{u}) < 0$ for all $\mu \in \mathfrak{C}(\Sigma)$. Equivalently,

$$\sup_{\gamma \in \Gamma \setminus \{e\}} \frac{d\ell_\gamma(\mathsf{u})}{\ell(\gamma)} < 0. \tag{4.15}$$

In other words, all closed geodesics on Σ become uniformly shorter under the infinitesimal deformation.

Mess's Theorem 3.6, which states that any affine deformation of a cocompact surface group $\Gamma_0 < G$ fails to be proper, follows easily from Proposition 7.2. Indeed, suppose that $\Sigma_0 = \Gamma_0 \backslash \mathsf{H}^2$ is a closed surface. There exists a geodesic current μ_{Γ_0}, the *Liouville current* associated to Σ_0, whose length in any hyperbolic structure Σ is minimized for $\Sigma = \Sigma_0$. Hence $d\ell_{\mu_{\Gamma_0}}(\mathsf{u}) = 0$ for any infinitesimal deformation u.

In fact, a slightly stronger statement is true: any nontrivial infinitesimal deformation u of a closed hyperbolic surface must increase the lengths of some closed geodesics while decreasing the lengths of others. A hint as to why that should be true is that the area of a closed hyperbolic surface of genus $g \geq 2$ is constant, equal to $4\pi(g-1)$ by the Gauss–Bonnet formula. Thus, if a deformation contracts in some directions, then it should stretch/lengthen in other directions.

The same basic idea underpins Thurston's theory of the Lipschitz metric on Teichmüller space [123]. This metric measures distance between two hyperbolic structures on a closed surface according to the minimum Lipschitz constant of Lipschitz maps between the two structures. Guéritaud-Kassel [87, 82] extended Thurston's theory to finite-type hyperbolic surfaces, as well as higher dimensional hyperbolic manifolds. An application is a properness criterion in the setting of $G \times G$ acting on G by right- and left-multiplication, where $G = \mathsf{SO}(2,1)^0$ is the identity component of the isometry group of hyperbolic 2-space H^2.

This theory is the starting point for the work of Danciger–Guéritaud–Kassel [39] on Margulis spacetimes, so we digress briefly to explain it.

7.3 CONTRACTING DEFORMATIONS AND PROPER ACTIONS ON LIE GROUPS.

Consider the identity component $G = \mathsf{SO}(2,1)^0$ of $\mathsf{Isom}(\mathsf{H}^2)$ and the action of $G \times G$ on G by *right- and left-multiplication:*

$$(g_0, g_1) \cdot h := g_1 h g_0^{-1}. \tag{4.16}$$

Thanks to the exceptional isomorphism $\mathfrak{so}(2,2) \simeq \mathfrak{so}(2,1) \oplus \mathfrak{so}(2,1)$, the action of $G \times G$ on G then models three-dimensional Lorentzian geometry of constant negative curvature, also known as *anti–de Sitter (AdS) geometry.*

Consider a discrete (geometrically finite, nonelementary) embedding $\Gamma \xrightarrow{\rho_0} G$ defining a proper action on H^2 whose quotient

$$\Sigma_0 = \rho_0(\Gamma)\backslash \mathsf{H}^2$$

is a complete hyperbolic surface. Consider a second discrete embedding $\rho_1 : \Gamma \hookrightarrow G$. Then via (4.16), the pair (ρ_0, ρ_1) defines an action of Γ on G. Such an action is not necessarily properly discontinuous: for instance, if

$$\rho_1 = \mathsf{Inn}(h) \circ \rho_0$$

for $h \in G$, then $(\rho_0, \rho_1)(\gamma)$ fixes h for all $\gamma \in \Gamma$.

Here is the properness criterion for such $G \times G$ actions on G, due to Guéritaud–Kassel [87, 82]. Say that ρ_1 is a *contracting Lipschitz deformation* of ρ_0 if and only if there exists a Lipschitz map

$$\mathsf{H}^2 \xrightarrow{f} \mathsf{H}^2$$

with Lipschitz constant $\mathrm{Lip}(f) < 1$ that is (ρ_0, ρ_1)-equivariant—that is,

$$f \circ \rho_0(\gamma) = \rho_1(\gamma) \circ f \tag{4.17}$$

for all $\gamma \in \Gamma$.

THEOREM 7.3 (Guéritaud–Kassel [87, 82]).
Up to switching the roles of ρ_0 and ρ_1, the (ρ_0, ρ_1)-action of Γ on G is proper if and only if ρ_1 is a contracting Lipschitz deformation of ρ_0.

Note that if ρ_1 is also injective and discrete with quotient $\Sigma_1 = \rho_1(\Gamma)\backslash \mathsf{H}^2$, then f corresponds to a Lipschitz deformation of hyperbolic surfaces $\Sigma_0 \to \Sigma_1$. To see why the existence of such a map f suffices for properness, observe that the (ρ_0, ρ_1) action on G projects equivariantly down to the ρ_0 action on H^2 via the *fixed point map*

$$\mathsf{g} \mapsto \mathsf{Fix}(\mathsf{g}^{-1} \circ f),$$

which is well defined by the contraction property. Proper discontinuity of the action on the base H^2 then implies proper discontinuity on G.

Further generalizing Thurston's theory, Guéritaud–Kassel also studied the relationship between the optimal Lipschitz constant over all (ρ_0, ρ_1)-equivariant maps and the factor by which translation lengths are stretched in ρ_1 compared with ρ_0. In particular, when ρ_0 is convex cocompact, a contracting Lipschitz (ρ_0, ρ_1)-equivariant map exists if and only if

$$\sup_{\gamma \in \Gamma \backslash \{e\}} \frac{\ell(\rho_1(\gamma))}{\ell(\rho_0(\gamma))} < 1. \tag{4.18}$$

Observe the similarity with the properness criterion Proposition 7.2. Indeed, Condition (4.15) may be viewed as the infinitesimal version of (4.18). The similarity exemplifies a more fundamental principle at work: the affine action of $G \ltimes_{\mathsf{Ad}} \mathfrak{g}$ on $\mathfrak{g}$ is the *infinitesimal analogue* of the action by right-and left-multiplication of $G \times G$ on G. This guiding principle led Danciger–Guéritaud–Kassel to develop an infinitesimal analogue of the theory of Lipschitz contraction and proper actions. The next sections will dive into that theory and its consequences for the structure and classification of Margulis spacetimes.

One consequence, in line with the guiding principle above, is the following geometric transition statement: *every Margulis spacetime is the rescaled limit of a family of collapsing AdS spacetimes.* For the sake of brevity, we do not give details here; see [39, Theorem 1.4]. For a different survey that develops more thoroughly the parallel between Margulis spacetimes and complete AdS 3-manifolds, see Guéritaud [81].

Let us also mention that this geometric transition result has recently been generalized to the setting of the affine group $\mathsf{SO}(2n+2, 2n+1) \ltimes \mathbb{R}^{4n+3}$, seen as an infitesimal analogue of the reductive group $\mathsf{SO}(2n+2, 2n+2)$ (Danciger, Guéritaud, and Kassel's result is the case $n = 1$). Of course for arbitrary n the exceptional isomorphism then no longer applies, so the general case no longer fits into the framework of $G \ltimes_{\mathsf{Ad}} \mathfrak{g}$ and $G \times G$. This generalization has been done independently in Danciger and Zhang [43] on the one hand and Ghosh [63] on the other hand.

7.4 DEFORMATION VECTOR FIELDS AND INFINITESIMAL CONTRACTION. Let us return now to the setting of Margulis spacetimes. Let $G = \mathsf{SO}(2,1)$ and let $\mathfrak{g} = \mathfrak{so}(2,1)$ be its Lie algebra. Fix a discrete faithful representation $\rho_0 : \Gamma \hookrightarrow G$ of the finitely generated, torsion-free, group Γ determining a hyperbolic surface $\Sigma_0 = \big(\rho_0(\Gamma)\big)\backslash\mathsf{H}^2$. Let $\mathsf{u} \in \mathsf{Z}^1(\Gamma, \mathfrak{g}_{\mathsf{Ad}\rho_0})$ be a cocycle tangent to a smooth deformation path $\rho_t \in \mathsf{Hom}(\Gamma, G)$ based at ρ_0, as in (4.14). Assume further that ρ_0 is convex cocompact, so that for small $t \geq 0$, the representations ρ_t are also discrete and faithful and determine a family of hyperbolic surfaces $\Sigma_t := \big(\rho_t(\Gamma)\big)\backslash\mathsf{H}^2$. These hyperbolic surfaces may be organized into a smoothly varying family of hyperbolic structures on a single surface $\Sigma := \Sigma_0$ by finding a smoothly varying family of *developing maps* $f_t : \mathsf{H}^2 \to \mathsf{H}^2$ that satisfy the following conditions:

- $f_0 = \mathrm{id}$
- f_t is a homeomorphism for all t
- f_t is (ρ_0, ρ_t)-equivariant:

$$f_t \circ \rho_0(\gamma) = \rho_t(\gamma) \circ f_t \qquad (4.19)$$

These conditions ensure that for each t, f_t descends to a homeomorphism $\Sigma \to \Sigma_t$ that becomes "close to the identity" as $t \to 0$. Consider the tangent vector field $X \in \mathsf{Vec}(\mathsf{H}^2)$ to the deformation of developing maps, defined by

$$X(p) := \left.\frac{\mathrm{d}}{\mathrm{d}t}\right|_{t=0} f_t(p).$$

This vector field satisfies an equivariance condition coming from taking the derivative of Condition (4.19).

Before stating the equivariance condition, observe that the group G acts on the space $\mathsf{Vec}(\mathsf{H}^2)$ of all vector fields on H^2. Indeed, $\mathsf{H}^2 = G/K$ is the space of right cosets of the maximal compact subgroup $K < G$, and so $\mathsf{Vec}(\mathsf{H}^2)$ identifies with the subspace of right-K-invariant vector fields in $\mathsf{Vec}(G)$. The left action of G on $\mathsf{Vec}(G)$ determines an action of G on $\mathsf{Vec}(\mathsf{H}^2)$. Thinking of the Lie algebra $\mathfrak{g}$ as the space of right-G-invariant vector fields on G as in Section 7.1, we have a natural embedding $\mathfrak{g} \hookrightarrow \mathsf{Vec}(\mathsf{H}^2)$. The image of $\mathfrak{g}$ in $\mathsf{Vec}(\mathsf{H}^2)$ is precisely the space of *Killing vector fields*—that is, those vector fields whose flow preserves the hyperbolic metric. We will denote the image of $\mathsf{u} \in \mathfrak{g}$ in $\mathsf{Vec}(\mathsf{H}^2)$ again by u.

The group Γ acts on $\mathsf{Vec}(\mathsf{H}^2)$ via the representation $\rho_0 : \Gamma \hookrightarrow G$. Differentiating Condition (4.19) yields that

$$X - \rho_0(\gamma) \cdot X = \mathsf{u}(\gamma) \tag{4.20}$$

holds for all $\gamma \in \Gamma$; in other words, while X is not invariant under the ρ_0-action of Γ, it differs from any translate by a Killing vector field determined by the deformation cocycle u. A vector field satisfying this condition is called u-*equivariant* or *automorphic*. We observe the following:

PROPOSITION 7.4. *Let $X \in \mathsf{Vec}(\mathsf{H}^2)$ be a u-equivariant vector field. Then the coset $X - \mathfrak{g}$ is an affine subspace of $\mathsf{Vec}(\mathsf{H}^2)$ invariant under the ρ_0-action of Γ. Furthermore, the action of Γ on $X - \mathfrak{g}$,*

$$X - \xi \overset{\gamma}{\longmapsto} X - \mathsf{u}(\gamma) - \mathsf{Ad}(\rho_0(\gamma))\xi, \tag{4.21}$$

identifies with the affine action $\Phi_G(\rho_0, \mathsf{u})$ of Γ on $\mathsf{E}^{2,1}$.

Each element $X - \xi \in X - \mathfrak{g}$ satisfies the equivariance property (4.20), but for a different cocycle—namely, the cocycle

$$\gamma \mapsto \mathsf{u}(\gamma) + \mathsf{Ad}(\rho_0(\gamma))\xi - \xi,$$

which is cohomologous to u. The affine space $X - \mathfrak{g}$ bijectively corresponds to the cohomology class $[\mathsf{u}] \subset \mathsf{Z}^1(\Gamma, \mathfrak{g}_{\mathsf{Ad}\rho_0})$. Note that we insist on writing $X - \mathfrak{g}$ rather than $X + \mathfrak{g}$ so that the action, as written in (4.21), matches that of $\Phi_G(\rho_0, \mathsf{u})$ in (4.12).

Properness of the affine action of Γ on $X - \mathfrak{g}$ may be expressed in terms of an infinitesimal version of the Lipschitz contraction condition of Section 7.3. Suppose the maps

$$\mathsf{H}^2 \xrightarrow{f_t} \mathsf{H}^2$$

above are K_t-Lipschitz with Lipschitz constant

$$K_t = 1 + kt + O(t^2)$$

converging smoothly to 1 as $t \to 0$. In this case, the deformation vector field X satisfies an infinitesimal version of Lipschitz, which Danciger–Guéritaud–Kassel [39] call *k-lipschitz*, with lower case "*l*": for all $x \neq y$ in H^2,

$$d'_X(x, y) \leq k d(x, y), \tag{4.22}$$

where $k \in \mathbb{R}$ is a constant and

$$d'_X(x,y) := \left.\frac{\mathrm{d}}{\mathrm{d}t}\right|_{t=0} d\left(f_t(x), f_t(y)\right) \tag{4.23}$$

is the rate at which the vector field X pushes the points x and y away from each other. Note that this depends only on X, not on the particular path of maps f_t.

For any Killing vector field $\xi \in \mathfrak{g}$, the family $\exp(-t\xi) \circ f_t$ is also K_t-lipschitz. The corresponding deformation vector field $X - \xi$ is then also k-lipschitz for the same constant k. Thus, the entire affine space $X - \mathfrak{g}$ consists of k-lipschitz vector fields. Properness of the action on $X - \mathfrak{g}$ occurs in the case that these vector fields are *contracting*, that is, $k < 0$.

PROPOSITION 7.5 (Danciger–Guéritaud–Kassel [39]). *Suppose X is k-lipschitz for some $k < 0$. Then the affine space $X - \mathfrak{g}$ admits a Γ-equivariant fibration*

$$X - \mathfrak{g} \xrightarrow{\mathcal{Z}} \mathsf{H}^2.$$

In particular, Γ acts properly on the affine space $X - \mathfrak{g}$ with quotient a complete affine three-manifold M. The quotient map

$$M := \Gamma \backslash (X - \mathfrak{g}) \longrightarrow \Gamma_0 \backslash \mathsf{H}^2 = \Sigma$$

is an affine line bundle over Σ with total space M.

The proof of Proposition 7.5 is straightforward, following the same "contracting fixed point" idea from Section 7.3:

Proof. Fix a point $p \in \mathsf{H}^2$. For any Killing vector field $\xi \in \mathfrak{g}$, the vector field $X - \xi$ is also k-lipschitz. For sufficiently large $R > 0$ (depending on $\|(X - \xi)(p)\|$ and k), the vector field $X - \xi$ points inward along $\partial B_R(p)$. Thus, $X - \xi$ has a zero z inside $B_R(p)$, by the well-known vector field analogue of the Brouwer fixed point theorem.

Furthermore, z is the *unique* zero of $X - \xi$: by the contraction property (4.22), $X - \xi$ pushes any point $x \neq z$ closer to z. Hence, define

$$\mathcal{Z}(X - \xi) := z$$

to be this unique zero. $\mathcal{Z}$ is a continuous map intertwining the affine action of Γ on $X - \mathfrak{g}$ with the ρ_0-action of Γ on H^2. Since the ρ_0-action on the base H^2 is properly discontinuous, the action on $X - \mathfrak{g}$ is also properly discontinuous.

In fact, $\mathcal{Z}$ is a fibration. The fiber $\mathcal{Z}^{-1}(p)$ over $p \in \mathsf{H}^2$ consists of all vector fields $X - \xi$ vanishing at p. In particular, every tangent vector $\mathbf{v} \in \mathsf{T}_p\mathsf{H}^2$ is the value of a Killing vector field $\xi_{\mathbf{v}}$ at p. Now take $\mathbf{v} = X(p)$. The vector field $X - \xi_{\mathbf{v}}$ vanishes at p, so $\mathcal{Z}(X - \xi_{\mathbf{v}}) = p$. Thus $\mathcal{Z}^{-1}(p) \neq \emptyset$. If $\xi_1, \xi_2 \in \mathfrak{g}$ and

$$\mathcal{Z}(X - \xi_1) = \mathcal{Z}(X - \xi_2) = p,$$

then $\xi_1 - \xi_2$ is a Killing vector field that vanishes at p. Therefore, two elements of $\mathcal{Z}^{-1}(p)$ differ by an element of the infinitesimal stabilizer of the point p, a copy of $\mathfrak{so}(2)$ inside of $\mathfrak{g}$, which is an affine line of negative (timelike) signature for the Killing form. $\square$

Danciger–Guéritaud–Kassel show that *all* Margulis spacetimes arise from contracting infinitesimal deformations. The following theorem was proved in the case that ρ_0 is convex cocompact in [39] and in the general case in [42].

THEOREM 7.6 (Danciger–Guéritaud–Kassel).
Consider a discrete embedding $\Gamma \xrightarrow{\rho_0} G$ of a finitely generated, nonelementary group Γ and a deformation cocycle $\Gamma \xrightarrow{\mathsf{u}} \mathfrak{g}$. Suppose that $d\ell_\gamma(\mathsf{u}) \leq 0$ for at least one $\gamma \in \Gamma$. Then the affine action $\Phi_G(\rho_0, \mathsf{u})$ of Γ on $\mathsf{E}^{2,1}$ is properly discontinuous if and only if there exists a (ρ, u)-equivariant vector field that is k-lipschitz for some $k < 0$. In particular, any proper affine action of a nonabelian free group Γ on $\mathbb{R}^3$ is conjugate to one as in Proposition 7.5.

Note that the assumption that $d\ell_\gamma(\mathsf{u}) \leq 0$ for a least one $\gamma \in \Gamma$ is satisfied by either u or $-\mathsf{u}$, and that the affine actions $\Phi_G(\rho_0, \mathsf{u})$ and $\Phi_G(\rho_0, -\mathsf{u})$ are conjugate by an orientation reversing affine transformation. The proof of Theorem 7.6 follows the same strategy as the work of Guéritaud-Kassel [82] discussed in Section 7.3. The key point is that if the infimum $k_{\min}$ of lipschitz constants for (ρ_0, u)-equivariant vector fields is nonnegative, then any (ρ_0, u)-equivariant vector field realizing $k_{\min}$ must infinitesimally stretch (the lift of) a geodesic lamination in the convex core of Σ at a rate precisely equal to $k_{\min}$.

However, contrary to the setting of Lipschitz maps, the Arzelà-Ascoli compactness theorem does *not* hold for lipschitz vector fields. Indeed, the limit of a (bounded) sequence of k-lipschitz vector fields is not necessarily a vector field but instead a convex set valued section of the tangent bundle called a *convex field*. Much technical care is needed in adapting the arguments of [82] to this setting.

7.5 TAMENESS OF MARGULIS SPACETIMES. The topology of a Margulis spacetime may be read off from Theorem 7.6 and Proposition 7.5. By Theorem 7.6, every proper affine action of a free group Γ on $\mathsf{E}^{2,1}$ comes from a contracting infinitesimal deformation of a non-compact hyperbolic surface Σ as in Proposition 7.5. The quotient Margulis spacetime $M = \Gamma\backslash\mathsf{E}^{2,1}$ is an affine line bundle over the surface Σ. This implies the topological tameness of M (Theorem 4.7). See the discussion in Section 4.4.

7.6 THE MODULI SPACE OF MARGULIS SPACETIMES: STRIP DEFORMATIONS. Fix a discrete embedding ρ_0 of a free group Γ into $G = \mathsf{SO}(2,1)$. It follows from Proposition 7.2 or Theorem 7.6 that the set of cohomology classes $[\mathsf{u}] \in \mathsf{H}^1(\Gamma_0, \mathfrak{g}_{\mathsf{Ad}})$ of infinitesimal deformations of ρ_0 for which the affine action $\Phi_G(\rho_0, \mathsf{u})$ is proper is an open cone in $\mathsf{H}^1(\Gamma_0, \mathfrak{g}_{\mathsf{Ad}})$, sometimes called the *admissible cone* or *cone of proper deformations*. The admissible cone is the disjoint union of a properly convex cone and its negative. One convex component contains the infinitesimal deformations that uniformly contract the geometry of the surface $\Sigma_0 = \rho_0(\Gamma)\backslash\mathsf{H}^2$ in the sense of Equation (4.15) and Theorem 7.6. The other component contains the infinitesimal deformations that uniformly lengthen. The projectivization of the admissible cone will be denoted $\mathsf{adm}(\rho_0)$; it is the moduli space of Margulis spacetimes associated to a fixed hyperbolic surface Σ_0, considered up to affine equivalence.

In [40] and [42], Danciger-Guéritaud-Kassel give a combinatorial parameterization of $\mathsf{adm}(\rho_0)$ in terms of the *arc complex* of Σ_0 in a similar spirit to Penner's cell decomposition of the decorated Teichmüller space of a punctured surface [111]. The parameterization realizes each contracting deformation $[\mathsf{u}] \in \mathsf{H}^1(\Gamma_0, \mathfrak{g}_{\mathsf{Ad}})$ as an *infinitesimal strip deformation*. The following construction goes back to Thurston [123] (see also Papadopolous-Théret [110]). Let us assume the hyperbolic surface Σ_0 has no cusp, so that all infinite ends are funnels—that is, ρ_0 is convex cocompact (this assumption was present in [40] but removed in [42]). Starting with the hyperbolic surface Σ_0,

- Choose a collection of disjoint nonisotopic properly embedded geodesic arcs $\alpha_1, \ldots, \alpha_r \subset \Sigma_0$.
- For each $1 \le i \le r$ choose an arc α_i' disjoint from, but very close to, α_i (in particular isotopic to α_i) so that α_i and α_i' bound a *strip* in Σ_0—that is, a region isometric to the region between two ultraparallel geodesics in the hyperbolic plane H^2. Let p_i and p_i' be the points on α_i and α_i' respectively with minimal distance. The geodesic segment $[p_i, p_i']$ is called the *waist* of the strip.

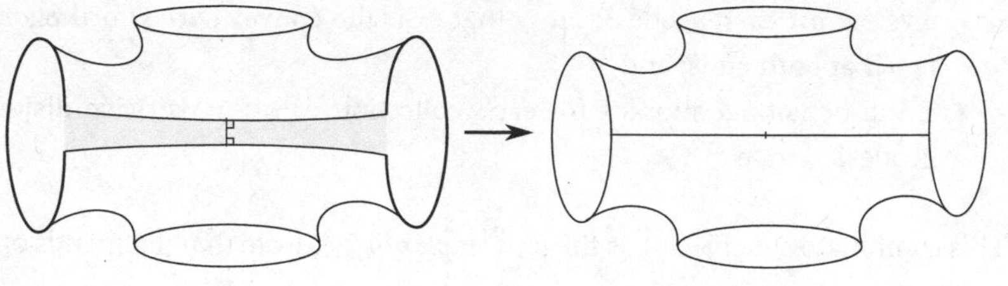

Figure 4.6. A strip deformation along a single arc

Then, for each i, delete the strip bounded by α_i and α_i' (we assume the strips are disjoint), and glue α_i to α_i' by the isometry that identifies p_i to p_i'. The result is a hyperbolic surface $\Sigma_1 = \rho_1(\Gamma)\backslash \mathsf{H}^2$ equipped with a natural 1-Lipschitz map $\Sigma_0 \xrightarrow{f} \Sigma_1$ (which collapses the strips). The holonomy representation $\Gamma \xrightarrow{\rho_1} G$ (defined here up to conjugation) is a new representation of Γ, which we call a *strip deformation* of ρ_0. See Figure 4.6. One can show that, even though f is only 1-Lipschitz, in fact if the arcs $\alpha_1, \ldots, \alpha_r$ cut Σ_0 into disks, then the Lipschitz constant may be improved to < 1 by deforming f; in particular, (4.18) holds.

Now consider a family ρ_t of strip deformations of ρ_0, as follows. For each $1 \le i \le r$, let α_i' move closer and closer to the fixed arc α_i in such a way that the endpoint $p_i \in \alpha_i$ of the waist remains constant and the width

$$d(p_i, p_i') = w_i t + O(t^2)$$

tends to zero at some linear rate $w_i \in \mathbb{R}^+$. The cohomology class $[\mathsf{u}] \in \mathsf{H}^1(\Gamma_0, \mathfrak{g}_{\mathsf{Ad}})$ of the derivative of the path ρ_t is called an *infinitesimal strip deformation* of ρ_0. The points $p_i \in \alpha_i$ are called the *waists* and the coefficients w_i are called the *widths* of the infinitesimal strip deformation. Note that if the arcs cut Σ_0 into disks, then every closed curve must cross the arcs a number of times roughly proportional to its length, which should make plausible the fact that lengths of closed curves are decreasing at a uniform rate as in (4.15); hence $\mathsf{u} \in \mathsf{adm}(\rho_0)$ in this case.

Danciger-Guéritaud-Kassel [40] proved that every contracting infinitesimal deformation u is realized by an infinitesimal strip deformation. The realization becomes unique if further requirements are put on the strips. For example, let us require that each α_i crosses the boundary of the convex core $\Omega \subset \Sigma_0$ at a right angle and that p_i is the midpoint of $\Omega \cap \alpha_i$. Strip deformations of this type are naturally organized into an abstract simplicial complex $\overline{\mathcal{X}}$, with:

- a vertex for each geodesic arc α that exits the convex core Ω orthogonal to $\partial\Omega$ at both ends; and
- a k-dimensional simplex for each collection of $k+1$ pairwise disjoint geodesic arcs $\alpha_1, \ldots, \alpha_{k+1}$.

This combinatorial object $\overline{\mathcal{X}}$ is the *arc complex* of Σ_0. Note that it depends only on the topology of Σ_0.

Consider the map

$$\overline{\mathcal{X}} \xrightarrow{\textbf{Strip}} \mathsf{H}^1(\Gamma_0, \mathfrak{g}_{\mathsf{Ad}})$$

defined as follows. Write any element $x \in \overline{\mathcal{X}}$ as a formal weighted sum of arcs

$$x = w_1\alpha_1 + \cdots + w_{k+1}\alpha_{k+1}$$

with each $w_i > 0$ and $\sum w_i = 1$. Then define $\textbf{Strip}(x)$ to be the infinitesimal strip deformation for the arcs $\alpha_1, \ldots, \alpha_{k+1}$, where for $1 \le i \le k+1$, the waist of the infinitesimal strip at α_i is the midpoint of $\alpha_i \cap \Omega$ and the width is w_i. Denote by $\mathcal{X}$ the subset of $\overline{\mathcal{X}}$ obtained by removing all open faces corresponding to collections of arcs that fail to cut the surface into disks.

Penner [111] showed that $\mathcal{X}$ is homeomorphic to a ball of dimension 1 smaller than the dimension of the Fricke-Teichmüller space of complete hyperbolic structures on Σ_0. The map **Strip** sends $\mathcal{X}$ into the contracting half of the admissible cone in $\mathsf{H}^1(\Gamma, \mathfrak{g}_{\mathsf{Ad}\rho_0})$. The projectivization of the restriction of **Strip**, denoted

$$\mathcal{X} \xrightarrow{\mathsf{Strip}} \mathsf{adm}(\rho_0),$$

is then a map between balls of the same dimension. The main theorem of [40] (extended in [42] to when Σ_0 may have cusps), is the following:

THEOREM 7.7.
$\mathcal{X} \xrightarrow{\mathsf{Strip}} \mathsf{adm}(\rho_0)$ *is a homeomorphism.*

The proof has two parts: first, Strip is a local homeomorphism, and second, Strip is proper. Both are nontrivial, but let us comment only on the second. Consider a sequence x_n going to infinity in $\mathcal{X}$. There are two ways this can happen. First, it could be that, up to subsequence, x_n converges in $\overline{\mathcal{X}}$ to a point $x_\infty \in \overline{\mathcal{X}} \setminus \mathcal{X}$, which is supported on arcs whose complement includes a subsurface of nontrivial topology. The limit

$$[\mathsf{u}_\infty] = [\textbf{Strip}(x_\infty)]$$

of the projective classes $[\mathsf{u}_n] = \mathsf{Strip}(x_n)$ is a projective class of infinitesimal deformations leaving unchanged the lengths of closed curves in this subsurface; hence $[\mathsf{u}_\infty] \in \partial\mathsf{adm}(\rho_0)$. Consider second the case that x_n diverges even in $\overline{\mathcal{X}}$. Then the supporting arcs of x_n become more and more complicated and, after taking a subsequence, converge in the Hausdorff sense to (up to twice as many) geodesic arcs $\beta_1, \ldots, \beta_s$, which are no longer properly embedded but rather accumulate in one direction around a geodesic lamination Λ in the convex core Ω. The limit $[\mathsf{u}_\infty]$ of the strip deformations $[\mathsf{u}_n] = \mathsf{Strip}(x_n)$ should be thought of as a strip deformation for which the waists of the strips are infinitely deep in the lamination Λ; in other words, $[\mathsf{u}_\infty]$ is obtained by removing (infinitesimal) parabolic strips, each of whose thickness goes to zero as the strip winds closer and closer to Λ. The lengths of longer and longer closed curves $\gamma \in \Gamma$ that travel very close to Λ are affected (proportionally) less and less by $[\mathsf{u}_\infty]$, showing that uniform contraction (4.15) fails for $[\mathsf{u}_\infty]$, so that again $[\mathsf{u}_\infty] \in \partial\mathsf{P}(\mathsf{adm}(\rho_0))$. Thus Strip is proper.

Note that Minsky [74] used strip deformations with parabolic strips to show that there exist affine deformations of a one-holed torus that are not proper but for which the Margulis spectrum is positive. See the discussion in Section 5.5.

7.7 STRIP DEFORMATIONS AND CROOKED PLANES.

One consequence of Theorem 7.7 is the resolution of the *Crooked Plane Conjecture*; see Section 4.2.

COROLLARY 7.8. Consider a discrete embedding $\Gamma \xhookrightarrow{\rho_0} G$ of a free group Γ of rank $r \geq 1$ and a deformation cocycle $\Gamma \xrightarrow{\mathsf{u}} \mathfrak{g}$. Suppose that the affine action $\Phi_G(\rho_0, \mathsf{u})$ of Γ on $\mathsf{E}^{2,1}$ is properly discontinuous. Then there exists a fundamental domain in $\mathsf{E}^{2,1}$ bounded by $2r$ pairwise disjoint crooked planes.

Before explaining the proof, we make a quick note about more general fundamental domains. Recall from Proposition 7.5 that any (ρ_0, u)-equivariant vector field Y that is k-lipschitz for some $k < 0$ determines a Γ-equivariant fibration

$$\mathsf{E}^{2,1} \xrightarrow{\mathcal{Z}} \mathsf{H}^2.$$

If $\Delta \subset \mathsf{H}^2$ is a fundamental domain, then $\mathcal{Z}^{-1}\Delta \subset \mathsf{E}^{2,1}$ is a fundamental domain. The surfaces bounding $\mathcal{Z}^{-1}\Delta$ are ruled by affine lines but do not have any other particularly nice structure and are far from canonical. Indeed much freedom exists in choosing Y. However, Theorem 7.7 implies u is realized uniquely as an infinitesimal strip deformation.

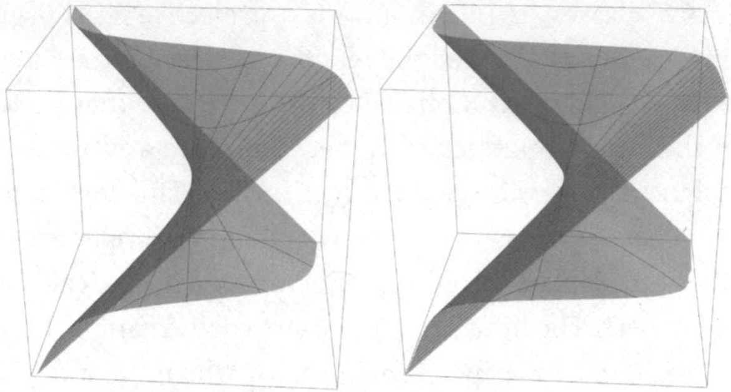

Figure 4.7. The preimage of an arc $\widetilde{\alpha} \subset \mathsf{H}^2$ under the fibration determined by a k-lipschitz vector field Y, as Y converges to an infinitesimal strip deformation X along $\widetilde{\alpha}$, with $d'_Y(p,q) \to 0$ for all $p, q \in \widetilde{\alpha}$ as $Y \to Y_\infty$. The limit is a crooked plane.

As described below, an infinitesimal strip deformation is a (ρ_0, u)-equivariant piecewise Killing vector field X on H^2. The vector field X is discontinuous along a $\rho_0(\Gamma)$-invariant collection $\widetilde{\mathscr{A}}$ of pairwise disjoint geodesic arcs—namely, the lifts of the arcs $\mathscr{A} = \{\alpha_1, \ldots, \alpha_r\}$ supporting the strip deformation. Although X is only 0-lipschitz, it is sufficiently contractive to define a singular version of the fibration from Proposition 7.5. The surfaces in $\mathsf{E}^{2,1}$ that lift arcs $\widetilde{\alpha} \in \widetilde{\mathscr{A}}$ of the strip deformation are precisely crooked planes! Indeed, crooked planes are seen in the limit of the fibrations for k-lipschitz vector fields Y converging to X, with $k \to 0^-$. See Figure 4.7. Here is the precise recipe for finding crooked planes from strip deformation data.

First, we describe in more detail the (ρ_0, u)-equivariant vector field X associated to the strip deformation realizing X. The connected components of the complement of $\alpha_1 \cup \cdots \cup \alpha_r$ in Σ_0 are each homeomorphic to a disk (if the collection $\mathscr{A}$ of arcs is maximal, each component is a hyper-ideal triangle). We denote the set of these components by $\mathscr{T}$. The lift to H^2 of $\mathscr{T}$ is denoted $\widetilde{\mathscr{T}}$; its elements are the *tiles* of a $\rho_0(\Gamma)$-invariant tiling of H^2. Then,

- The restriction of X to each of the tiles $\Delta \in \widetilde{\mathscr{T}}$ is a Killing field $\xi_\Delta \in \mathfrak{g}$.
- If two tiles $\Delta, \Delta' \in \widetilde{\mathscr{T}}$ are adjacent along an arc $\widetilde{\alpha} \in \widetilde{\mathscr{A}}$, then the relative motion of Δ with respect to Δ'—namely, the difference

$$\psi_{\Delta,\Delta'} := \xi_\Delta - \xi_{\Delta'} \in \mathfrak{g}$$

—is an infinitesimal translation along an axis orthogonal to $\widetilde{\alpha}$ in the direction of Δ'. If $\widetilde{\alpha}$ is a lift of α_i, then the axis of $\psi_{\Delta,\Delta'}$ intersects $\widetilde{\alpha}$

at the lift of the waist $p_i \in \alpha_i$ and the velocity of the translation is equal to the width w_i, as defined in Section 7.6. Note $\psi_{\Delta,\Delta'} = -\psi_{\Delta',\Delta}$.

- Since X must be discontinuous along $\widetilde{\alpha}$, we define X along $\widetilde{\alpha}$ to agree with the Killing field
$$\mathbf{v}_{\widetilde{\alpha}} := \frac{(\xi_\Delta + \xi_{\Delta'})}{2},$$
which is the average of the Killing fields associated to the adjacent tiles Δ and Δ'. Think of $\mathbf{v}_{\widetilde{\alpha}}$ as the infinitesimal motion of the arc $\widetilde{\alpha}$ under the deformation.

Each arc $\widetilde{\alpha} \in \widetilde{\mathscr{A}}$ together with its infinitesimal motion $\mathbf{v}_{\widetilde{\alpha}}$ determines a crooked plane,

$$\mathcal{C}_{\widetilde{\alpha}} := \mathcal{C}(\mathbf{v}_{\widetilde{\alpha}}, \widetilde{\alpha}).$$

Here we identify $\mathsf{E}^{2,1}$ with the (affine space of the) Lie algebra $\mathfrak{g}$, as in Section 4.2, and recall that for ℓ a geodesic in H^2 and $\mathbf{v} \in \mathfrak{g}$ a Killing vector field, the crooked plane $\mathcal{C}(\mathbf{v}, \ell) \subset \mathfrak{g}$ is the collection of Killing fields $\mathbf{w} \in \mathfrak{g}$ such that $\mathbf{w} - \mathbf{v}$ has a nonattracting fixed point on the closure $\overline{\ell}$ of ℓ in $\overline{\mathsf{H}^2}$. Equipping ℓ with a transverse orientation, the closed crooked half-space $\mathcal{H}(\mathbf{v}, \ell) \subset \mathfrak{g}$ is the collection of Killing fields $\mathbf{w} \in \mathfrak{g}$ such that $\mathbf{w} - \mathbf{v}$ has a nonattracting fixed point on the closure in $\overline{\mathsf{H}^2}$ of the positive half-space h_ℓ bounded by ℓ.

LEMMA 7.9. *Let $\widetilde{\alpha}, \widetilde{\alpha}' \in \widetilde{\mathscr{A}}$ and endow each arc with a transverse orientation so that the positive half-space of $\widetilde{\alpha}$ is contained in that of $\widetilde{\alpha}'$. Then*

$$\mathcal{H}(\mathbf{v}_{\widetilde{\alpha}}, \widetilde{\alpha}) \subset \operatorname{Int}\left(\mathcal{H}(\mathbf{v}_{\widetilde{\alpha}'}, \widetilde{\alpha}')\right). \tag{4.24}$$

Proof. First, consider the case that $\widetilde{\alpha}, \widetilde{\alpha}' \in \widetilde{\mathscr{A}}$ are two distinct arcs on the boundary of a common tile Δ''. Let Δ (respectively Δ') denote the tile on the other side of $\widetilde{\alpha}$ (respectively $\widetilde{\alpha}'$) from Δ''. Then the vertices of the crooked planes $\mathcal{C}_{\widetilde{\alpha}}$ and $\mathcal{C}_{\widetilde{\alpha}'}$ may be written:

$$\mathbf{v}_{\widetilde{\alpha}} = \xi_{\Delta''} + (1/2)\psi_{\Delta,\Delta''}, \qquad \mathbf{v}_{\widetilde{\alpha}'} = \xi_{\Delta''} + (1/2)\psi_{\Delta',\Delta''}.$$

Hence the crooked half-space $\mathcal{H}_{\widetilde{\alpha}} := \mathcal{H}(\mathbf{v}_{\widetilde{\alpha}}, \widetilde{\alpha})$ is obtained from $\mathcal{H}(\xi_{\Delta''}, \widetilde{\alpha})$ by translating in the direction $(1/2)\psi_{\Delta,\Delta''}$; similarly $\mathcal{H}_{\widetilde{\alpha}'} := \mathcal{H}(\mathbf{v}_{\widetilde{\alpha}'}, \widetilde{\alpha}')$ is obtained from $\mathcal{H}(\xi_{\Delta''}, \widetilde{\alpha}')$ by translating in the direction $(1/2)\psi_{\Delta',\Delta''}$.

In fact, the two crooked half-spaces, $\mathcal{H}(\xi_{\Delta''}, \widetilde{\alpha})$ and $\mathcal{H}(\xi_{\Delta''}, \widetilde{\alpha}')$, are nested and their bounding crooked planes meet only at the vertex:

$$\mathcal{H}(\xi_{\Delta''}, \widetilde{\alpha}) \subset \operatorname{Int}\left(\mathcal{H}(\xi_{\Delta''}, \widetilde{\alpha}')\right) \cup \{\xi_{\Delta''}\}. \tag{4.25}$$

The key observation is Lemma 4.6. If $\mathbf{w} \in \mathfrak{g}$ is an infinitesimal translation along an axis orthogonal to $\widetilde{\alpha}$ and pushes toward the negative side of $\widetilde{\alpha}$, then affine translation by $\mathbf{w}$ pushes the crooked half-space $\mathcal{H}(\mathbf{0}, \widetilde{\alpha})$ inside of itself. In particular,

$$
\begin{aligned}
\mathcal{H}_{\widetilde{\alpha}} &= \mathbf{v}_{\widetilde{\alpha}} + \mathcal{H}(\mathbf{0}, \widetilde{\alpha}) = \xi_{\Delta''} + (1/2)\psi_{\Delta,\Delta''} + \mathcal{H}(\mathbf{0}, \widetilde{\alpha}) \\
&\subset \xi_{\Delta''} + \mathcal{H}(\mathbf{0}, \widetilde{\alpha}) = \mathcal{H}(\xi_{\Delta''}, \widetilde{\alpha}).
\end{aligned} \tag{4.26}
$$

Similarly, $\mathcal{H}(\xi_{\Delta''}, \widetilde{\alpha}') + (1/2)\psi_{\Delta'',\Delta'} \subset \mathcal{H}(\xi_{\Delta''}, \widetilde{\alpha}')$, and hence:

$$
\begin{aligned}
\mathcal{H}(\xi_{\Delta''}, \widetilde{\alpha}') &\subset \mathcal{H}(\xi_{\Delta''}, \widetilde{\alpha}') - (1/2)\psi_{\Delta'',\Delta'} \\
&= \mathcal{H}(\xi_{\Delta''}, \widetilde{\alpha}') + (1/2)\psi_{\Delta',\Delta''} = \mathcal{H}(\mathbf{v}_{\widetilde{\alpha}'}, \widetilde{\alpha}').
\end{aligned} \tag{4.27}
$$

So (4.24) follows from (4.25), (4.26), and (4.27) upon observing that the vertex $\xi_{\Delta''}$ is not contained in $\mathcal{H}_{\widetilde{\alpha}}$.

Now a simple inductive argument shows that (4.24) indeed holds for *any* pair of arcs $\widetilde{\alpha}, \widetilde{\alpha}'$ oriented so that the positive halfspace of $\widetilde{\alpha}$ is contained in that of $\widetilde{\alpha}'$. □

Observe that by Lemma 7.9, the crooked planes in the collection

$$
\left\{ \mathcal{C}_{\widetilde{\alpha}} := \mathcal{C}(\mathbf{v}_{\widetilde{\alpha}}, \widetilde{\alpha}) : \widetilde{\alpha} \in \widetilde{\mathscr{A}} \right\} \tag{4.28}
$$

are pairwise disjoint. Further, the half-spaces bounded by these crooked planes obey the same inclusion relations that hold for half-planes in H^2 bounded by the corresponding arcs. To find a fundamental domain in $\mathsf{E}^{2,1} \cong \mathfrak{g}$ bounded by disjoint crooked planes, one simply chooses the crooked planes associated to a subset of arcs of $\widetilde{\mathscr{A}}$ that bound a fundamental domain for the action on H^2. This proves Corollary 7.8.

In the same spirit of Section 7.3, there is a parallel theory of strip deformations and crooked planes in the setting of three-dimensional anti–de Sitter geometry; see [38] and [69].

7.8 TWO-GENERATOR GROUPS.

We now focus on the special case that the free group Γ has rank 2, corresponding to Euler characteristic $\chi(\Sigma_0) = -1$. There are four possible topological types for Σ_0. In each case, the arc complex $\overline{\mathcal{X}}$ is two-dimensional, but the combinatorics is quite different across the cases (see Figure 4.8). This results in a substantially different picture of the (projectivized) cone of proper deformations $\mathsf{adm}(\rho_0)$, depending on the

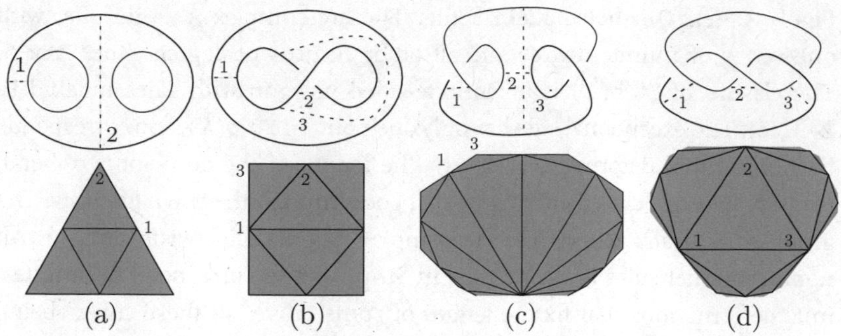

Figure 4.8. Four surfaces of small complexity (*top*) and their arc complexes, mapped under Strip to the closure of $\mathsf{adm}(\rho_0)$ in an affine chart of $\mathsf{P}(\mathsf{H}^1(\Gamma, \mathfrak{g}_{\mathrm{Ad}\rho_0}))$ (*bottom*). Some arcs are labeled by arabic numerals. (*Source:* [40].)

topology of Σ_0 (see again Figure 4.8). We describe the qualitative behavior in each of the four cases below in the language of Theorem 7.7. However, we remark that the understanding of $\mathsf{adm}(\rho_0)$ in the rank 2 case, in particular each description below, predates Theorem 7.7. Charette–Drumm–Goldman [27, 28, 29] described a tiling of $\mathsf{adm}(\rho_0)$ according to which isotopy classes of crooked planes embed disjointly in the associated Margulis spacetime. From this, they deduced the Crooked Plane Conjecture, Corollary 7.8, in the rank 2 case. The relationship between $\mathsf{adm}(\rho_0)$ and the arc complex $\overline{\mathcal{X}}$ of Σ_0 is already apparent in this work, which was an important precursor to Theorem 7.7 and Corollary 7.8.

Figure 4.8(a): Three-holed sphere. The arc complex $\overline{\mathcal{X}}$ has 6 vertices, 9 edges, 4 faces. Its image $\mathsf{Strip}(\overline{\mathcal{X}})$ is a triangle whose sides stand in natural bijection with the three boundary components of the convex core of Σ_0: an infinitesimal deformation u of ρ_0 lies in a side of the triangle if and only if it fixes the length of the corresponding boundary component, to first order. The set $\mathsf{adm}(\rho) = \mathsf{Strip}(\mathcal{X})$ is the interior of the triangle. See also the left part of Figure 4.4.

Figure 4.8(b): Two-holed projective plane. The arc complex $\overline{\mathcal{X}}$ has 8 vertices, 13 edges, 6 faces. Its image $\mathsf{Strip}(\overline{\mathcal{X}})$ is a quadrilateral. The horizontal sides of the quadrilateral correspond to infinitesimal deformations u that fix the length of a boundary component. The vertical sides correspond to infinitesimal deformations that fix the length of one of the two simple closed curves running through the half-twist. The set $\mathsf{adm}(\rho_0) = \mathsf{Strip}(\mathcal{X})$ is the interior of the quadrilateral. See also the right part of Figure 4.4.

Figure 4.8(c): One-holed Klein bottle. The arc complex $\overline{\mathcal{X}}$ is infinite, with one vertex of infinite degree and all other vertices of degree either 2 or 5. The closure of $\mathsf{Strip}(\overline{\mathcal{X}})$ is an infinite-sided polygon with sides indexed in $\mathbb{Z} \cup \{\infty\}$. The exceptional side has only one point in $\mathsf{Strip}(\overline{X})$, and corresponds to infinitesimal deformations that fix the length of the only nonperipheral, two-sided simple closed curve γ, which goes through the two half-twists. The group $\mathbb{Z}$ naturally acts on the arc complex $\overline{\mathcal{X}}$ via Dehn twists along γ. All nonexceptional sides are contained in $\mathsf{Strip}(\overline{\mathcal{X}})$ and correspond to infinitesimal deformations that fix the length of some curve, all these curves being related by some power of the Dehn twist along γ. The set $\mathsf{adm}(\rho_0) = \mathsf{Strip}(\mathcal{X})$ is the interior of the polygon. See also the left part of Figure 4.5.

Figure 4.8(d): One-holed torus. The arc complex $\overline{\mathcal{X}}$ is infinite, with all vertices of infinite degree; it is known as the *Farey triangulation*. The arcs are parameterized by $\mathbb{P}^1(\mathbb{Q})$. The closure of $\mathsf{Strip}(\overline{\mathcal{X}})$ contains infinitely many segments in its boundary. These segments, also indexed by $\mathbb{P}^1(\mathbb{Q})$, are in natural correspondence with the simple closed curves. However, the boundary is not the union of these segments; there are additional points corresponding to deformations for which the length of every curve decreases but the length of some lamination remains constant.

The structure of the boundary of $\mathsf{adm}(\rho_0)$ in this case was described in Guéritaud [80] and Goldman-Labourie-Minsky-Margulis [74]. See also the right part of Figure 4.5. For more details on the affine deformations of nonorientable surfaces, compare also Goldman-Laun [75] and Laun's thesis [99].

7.9 BEYOND FREE GROUPS: RIGHT-ANGLED COXETER GROUPS.

The existence of proper affine actions by nonabelian free groups suggests the possibility that other finitely generated groups that are not virtually solvable might also admit proper affine actions. However, in the more than 30 years since Margulis's discovery, very few examples have appeared. In particular, until recently, all known examples of word hyperbolic groups acting properly by affine transformations on $\mathbb{R}^n$ were virtually free groups. To conclude this section, we summarize further work of Danciger-Guértaud-Kassel [41] that generalizes the ideas of Section 7.4 to give many new examples, both word hyperbolic and not.

THEOREM 7.10 ([41, Theorem 1.1]).
Any right-angled Coxeter group on k generators admits proper affine actions on $\mathbb{R}^{k(k-1)/2}$.

A right-angled Coxeter group Γ is a finitely presented group of the form

$$\Gamma = \langle s_1, \ldots, s_g \mid (s_i s_j)^{m_{ij}} = 1, \forall\, 1 \leq i, j \leq g \rangle,$$

where $m_{ii} = 1$, that is, each generator s_i is an involution, and $m_{ij} = m_{ji} \in \{2, \infty\}$ for $i \neq j$, meaning two distinct generators either commute ($m_{ij} = 2$) or have no relation ($m_{ij} = \infty$). Some examples come from *reflection groups* in hyperbolic space. Indeed, the group generated by reflections in the faces of a right-angled polyhedron in H^n is a right-angled Coxeter group. Though simple to define, right-angled Coxeter groups have a rich structure and contain many interesting subgroups. As a corollary to Theorem 7.10, the following groups admit proper affine actions:

- the fundamental group of any closed orientable surface of negative Euler characteristic;
- any right-angled Artin group, see [44];
- any virtually special group, see [83];
- any Coxeter group (not just right-angled), see [84];
- any cubulated word hyperbolic group, using Agol's virtual specialness theorem [7];
- therefore, all fundamental groups of closed hyperbolic 3-manifolds, using [114, 86]: see [16];
- the fundamental groups of many other 3-manifolds, see [128, 101, 112].

Januszkiewicz–Świątkowski [85] found word hyperbolic right-angled Coxeter groups of arbitrarily large virtual cohomological dimension. See also [108] for another construction. Hence, another consequence of Theorem 7.10 is the following:

COROLLARY 7.11. *There exist proper affine actions by word hyperbolic groups of arbitrarily large virtual cohomological dimension.*

The Auslander Conjecture is equivalent to the statement that a group acting properly by affine transformations on $\mathbb{R}^n$ is either virtually solvable or has virtual cohomological dimension $< n$. In the examples from Theorem 7.10, the dimension $n = k(k-1)/2$ of the affine space grows quadratically in the number of generators k, while the virtual cohomological dimension of the Coxeter group acting is naively bounded above by k. Hence, Corollary 7.11 is far from giving counterexamples to the Auslander Conjecture.

The affine actions from Theorem 7.10 come from infinitesimal deformations of representations into a Lie group G as in Section 7.1, for G an indefinite orthogonal group. Indeed, a right-angled Coxeter group Γ on k generators (say, infinite and irreducible) admits explicit families of discrete *reflection group* embeddings

$$\Gamma \xrightarrow{\rho} \mathsf{O}(p, q+1) =: G$$

for $k = p+q+1$, which have long been studied by Tits, Vinberg, and others. The strategy from Section 7.4 of ensuring properness of the affine action from contraction of the deformation works well when $q=0$. In that case, each representation ρ acts by reflections in the walls of a right-angled polytope Δ_ρ in hyperbolic space H^p. For two such representations ρ, ρ', natural (ρ, ρ')-equivariant maps f are described explicitly by mapping Δ_ρ to $\Delta_{\rho'}$ projectively, walls-to-walls, and extending equivariantly by reflections. Deformations ρ' for which the maps f are Lipschitz contracting are found, roughly, by pushing the walls of Δ_ρ closer together. The derivative of an appropriate path of such contracting Lipschitz deformations, for ρ' smoothly converging to ρ, gives a (ρ_0, u)-equivariant contracting lipschitz vector field and hence a proper affine action $\Phi_G(\rho_0, \mathsf{u})$ by the argument given in Proposition 7.5.

It should be noted that, still in the case $q=0$, the dimension of the representations, and hence of the corresponding affine actions, may sometimes be reduced: if for some $n \geq 2$, Γ admits an action on H^n generated by reflections in some polytope Δ, then a contracting deformation as above may be found in H^{n+c-1} if the faces of Δ may be colored with c colors so that neighboring faces have different color (see [41, Proposition 6.1]). For example, if Γ is the group generated by reflections in a right-angled $2m$-gon in the hyperbolic plane (for $m \geq 3$), then we may take $c=2$. There exists a path of deformations of this reflection group into the isometry group of $\mathsf{H}^3 = \mathsf{H}^{2+2-1}$ for which tangent vectors to the path give proper affine actions in dimension $6 = \dim(\mathfrak{so}(3,1))$. Note that in this example, Γ contains surface subgroups of finite index.

The general case of Theorem 7.10 requires indefinite orthogonal groups of higher $\mathbb{R}$-rank—that is, $q > 0$; indeed, not all right-angled Coxeter groups may be realized as reflection groups in some hyperbolic space. Here, one could attempt the contraction strategy of Section 7.4 in the higher rank Riemannian symmetric space X of G. However, the most natural space in which to see the geometry of the Tits–Vinberg representations $\rho : \Gamma \to G$ is in a *pseudo-Riemannian symmetric space*—namely, the pseudo-Riemannian analogue $\mathsf{H}^{p,q} \subset \mathbb{R}\mathsf{P}^{k-1}$ of H^p in signature (p, q). Indeed, as above, the ρ-action of Γ is by reflections in the walls of a natural fundamental domain, a certain polytope $\Delta_\rho \subset \mathsf{H}^{p,q}$, and natural (ρ, ρ')-equivariant maps f are defined by taking Δ_ρ projectively to $\Delta_{\rho'}$, walls-to-walls. Further, since the "distances" in $\mathsf{H}^{p,q}$ are computed by a simple cross-ratio formula, similar to H^p (in the projective model), the "contraction" properties of the maps f are easy to check locally in the fundamental domain Δ_ρ. Theorem 7.10 is proved by employing a version of the contraction strategy from the H^p setting, adjusted and reinterpreted appropriately to work in the pseudo-Riemannian space $\mathsf{H}^{p,q}$. Despite the obvious hurdle that $\mathsf{H}^{p,q}$ is not a metric space, enough structure survives for this approach to work. One key observation is that $\rho(\Gamma)$-orbits in $\mathsf{H}^{p,q}$ escape only in *spacelike* (that is, positive) directions, in which their growth resembles that of actions on H^p.

8 Higher dimensions

8.1 NON-MILNOR REPRESENTATIONS. Margulis's original work can be reinterpreted as the discovery of the first known non-Milnor representation (see Definition 3.2)—namely, the standard representation of $\mathsf{SO}(2,1)$ on $\mathbb{R}^3$. We now discuss the question of identifying Milnor and non-Milnor representations in higher dimensions. Recall Proposition 3.4: if ρ does not have the property that every element acts with 1 as an eigenvalue, it is automatically Milnor. Observe, as in Section 7, that the standard representation of $\mathsf{SO}(2,1)$ is isomorphic to the adjoint representation of $\mathsf{SO}(2,1)$ and, more generally, that the adjoint representation of any semisimple Lie group G has the property that every element of infinite order acts with 1 as an eigenvalue.

THEOREM 8.1 (Smilga [119]).
For every semisimple real linear Lie group G, the adjoint representation is non-Milnor whenever G is not compact.

Note that the proper affine actions by right-angled Coxeter groups of Section 7.9 have linear part in adjoint representations of special orthogonal groups.

In a different direction, the work of Abels-Margulis-Soifer [2, 3] definitively settles the case of the standard representation of the special orthogonal groups $\mathsf{SO}(p,q)$ on $\mathbb{R}^{p,q}$:

THEOREM 8.2 (Abels-Margulis-Soifer [2, 3]).
Let $p \geq q$. Then the standard representation of $\mathsf{SO}(p,q)$ on $\mathbb{R}^{p+q}$ is

(1) Milnor if
- $p-q\neq 1$ *or*
- $p-q=1$ *and* p *is odd; or*

(2) non-Milnor if $p-q=1$ *and* p *is even.*

Observe that when $p-q$ is even, it is not the case that every element of $\mathsf{SO}(p,q)$ has 1 as an eigenvalue and Proposition 3.4 implies that $\mathsf{SO}(p,q)$ is Milnor. The general case, however, involves the detailed analysis from Margulis's original argument. The case when $q=p-1$ is the most interesting. Then a Margulis invariant α may be defined for elements with regular linear holonomy and the Opposite Sign Lemma holds. If p is odd, then $\alpha(\gamma)=-\alpha(\gamma^{-1})$, and hence no proper affine actions of F_2 exist. However, if p is even, then $\alpha(\gamma)=\alpha(\gamma^{-1})$ as in Lemma 5.1(4), so there is no obvious sign obstruction to proper affine actions. Indeed, Margulis's construction may be generalized in this case.

Smilga [118] constructs fundamental domains for the proper actions of F_2 with linear part in $\mathsf{SO}(p,p-1)$ with $p=2k+2$ even. They are bounded by hypersurfaces inspired by the crooked planes of Section 4.2, but these hypersurfaces are curved rather than piecewise flat. Burelle-Treib [22], on the other hand, have found a generalization of crooked planes to $\mathsf{SO}(2k+2,2k+1)$ constructed by flat hypersurfaces, which give rise to fundamental domains for the action of the group on the sphere (quotient of $\mathbb{R}^{4k+3}\backslash\{0\}$ by positive scalars) minus the limit set. The Burelle-Treib construction may very likely be extended to fundamental polyhedra in the affine space. In the same setting, Ghosh-Treib [66] proved an analogue of the Goldman-Labourie-Margulis properness criterion 6.1. Following Ghosh's earlier work [62] on the dynamical structure of Margulis spacetimes, they interpret proper affine actions as an extension of Labourie's Anosov representations [96] to the non-reductive context.

Smilga gave a sufficient condition for an irreducible representation of a semisimple group to be non-Milnor, which is conjectured [121] to be *necessary* as well.

Let G be a semisimple real Lie group with Lie algebra $\mathfrak{g}$. Choose in $\mathfrak{g}$ a *Cartan subspace* $\mathfrak{a}$ and a system Σ^+ of positive restricted roots. Recall that $\mathfrak{a}$ is a maximal abelian subalgebra of $\mathfrak{g}$ consisting of hyperbolic elements, and a *restricted root* is an element $\alpha\in\mathfrak{a}^*$ such that the restricted root space

$$\mathfrak{g}^{\alpha}:=\{Y\in\mathfrak{g}\mid\forall X\in\mathfrak{a},\ [X,Y]=\alpha(X)Y\}$$

is nonzero. Restricted roots form a root system Σ; a system of positive roots Σ^+ is a subset of Σ contained in a half-space, such that

$$\Sigma=\Sigma^+\sqcup-\Sigma^+.$$

Let $A := \exp(\mathfrak{a})$, and let L be the centralizer of $\mathfrak{a}$ in G. The *longest element* of the *restricted Weyl group* $W := N_G(A)/Z_G(A)$ is the unique element w_0 such that $w_0(\Sigma^+) = \Sigma^-$. We choose some representative $\tilde{w}_0 \in G$ of this element. (Compare [91].)

THEOREM 8.3 (Smilga [121]).
Suppose that $G \xrightarrow{\rho} \mathsf{GL}(\mathsf{V})$ *is an irreducible representation such that*

(i) $\forall l \in L,\ \rho(l) \cdot v = v$ *and*
(ii) $\rho(\tilde{w}_0) \cdot v \neq v$

for some $v \in \mathsf{V}$. *Then* ρ *is non-Milnor.*

Le Floch and Smilga [100] have classified such ρ when G is split.

Since $L \supset A$, the first condition implies that the *restricted weight space associated to* 0

$$\mathsf{V}^A = \{v \in \mathsf{V} \mid \forall a \in A,\ \rho(a) \cdot v = v\}$$

is nonzero. Equivalently, every element of $\rho(G)$ has 1 as an eigenvalue, consistent with Proposition 3.4.

The proofs of Theorems 8.2(1), 8.1 and 8.3 all follow the same basic template as Margulis's original proof (see Section 5.3), although the more general proofs are more complicated.

The main idea is to decompose the representation space V as a direct sum of three subspaces

$$\mathsf{V} = \mathsf{V}^{>} \oplus \mathsf{V}^{=} \oplus \mathsf{V}^{<}$$

and then construct a "generalized Schottky group" Γ_ℓ in $\rho(G)$ so that every element $\gamma \in \Gamma_\ell$, conjugated by a suitable map, preserves all three spaces and has very large eigenvalues on $\mathsf{V}^{>}$, very small eigenvalues on $\mathsf{V}^{<}$, and eigenvalues "close to 1" on $\mathsf{V}^{=}$. Moreover, there is then a further decomposition

$$\mathsf{V}^{=} = \mathsf{V}^{t} \oplus \mathsf{V}^{r}$$

such that every element of Γ_ℓ has a conjugate that stabilizes V^r and *fixes* V^t. A crucial point is that $\mathsf{V}^t \neq 0$ (this comes from Condition (i), as it turns out to be precisely the subspace of fixed points of L).

Now let A be the affine space corresponding to the vector space V. Then every affine map g with linear part in Γ_ℓ, when conjugated by a suitable map, preserves the decomposition of A into three subspaces

$$\mathsf{A} = \mathsf{V}^{>} \oplus \mathsf{A}^{=} \oplus \mathsf{V}^{<}$$

(where $\mathsf{A}^=$ is an affine subspace of A parallel to $\mathsf{V}^=$). Its restriction to $\mathsf{A}^=$ is then a sort of generalized "screw-displacement": it preserves the directions parallel to V^t and V^r and acts by pure translation along A^t. We call such affine transformations *quasi-translations,* and we call the translation vector along V^t of the (suitable conjugate of) g the *Margulis invariant* $M(g)$.

In particular cases, these constructions can be simplified. For $G = \mathsf{SO}(2,1)$ acting on $\mathbb{R}^{2,1}$, the space V^t has dimension 1, so that the vector Margulis invariant $M(g)$ reduces to the classical (scalar) Margulis invariant $\alpha(g)$; the space V^r is trivial, so that quasi-translations reduce to just translations.

Now in the general case, the analog of Formula (4.8) is

$$M(g^{-1}) = \rho(\tilde{w}_0) \cdot M(g). \tag{4.29}$$

Proving this turns out to be straightforward. Most of the effort goes into proving an analog of (4.9): for elements g and h in G that are "regular," "sufficiently transverse," and "sufficiently contracting,"

$$M(gh) \approx M(g) + M(h). \tag{4.30}$$

We then conclude in the same way as Margulis: we combine (4.30) with (4.29) to prove that, for a suitable choice of the translation parts of the generators, Margulis invariants of large elements of the group grow unboundedly. More precisely, we prescribe these translation parts in such a way that the Margulis invariants of the generators become equal to (some sufficiently large multiple of) the vector v supplied by the hypotheses of Theorem 8.3. Here Condition (ii) is crucial, as it allows the Margulis invariants of both the generators and their inverses to go in the same direction.

We conjecture the converse of Theorem 8.3, which generalizes part (1) of Theorem 8.2. We have the following partial result.

Say that an irreducible representation $G \xrightarrow{\rho} \mathsf{GL}(\mathsf{V})$ is *nonswinging* if and only if ρ has no nonzero w_0-invariant weight. In particular if G has no simple factor of type $A_{n\geq 2}$, D_{2n+1}, or E_6, then $w_0 = -\mathbb{I}$ and every representation is nonswinging.

THEOREM 8.4 ([120]).
Let G be a semisimple Lie group. Furthermore, suppose

- *the group G is split;*
- *ρ is a nonswinging irreducible representation; and*
- *ρ does not satisfy the hypotheses of Theorem 8.3.*

Then ρ is Milnor.

8.2 AUSLANDER'S CONJECTURE IN DIMENSION AT MOST 6.

THEOREM 8.5 (Abels-Margulis-Soifer [4]).
Let $n \leq 6$. Suppose $\Gamma < \mathsf{Aff}(\mathsf{A}^n)$ is a discrete subgroup acting properly and cocompactly on A^n. Then Γ is virtually solvable.

In the case that $n \leq 5$, this result follows from Fried-Goldman [61] for $n \leq 3$ and independently Tomanov [127] and Abels-Margulis-Soifer [4] for $n = 4, 5$. (As Abels-Margulis-Soifer point out in [4], an earlier version of Tomanov's work contained a gap, which was subsequently filled in [127].) Furthermore Tomanov [127] proposed a suggestive generalization of Auslander's conjecture to arbitrary algebraic groups of mixed type, and proved this stronger statement for $n \leq 5$:

CONJECTURE 8.6. *Let G be a real algebraic group, and suppose that $H < G$ contains a maximal reductive subgroup of G. Suppose that $\Gamma < G$ acts crystallographically on G/H (that is, Γ is a discrete subgroup, acts properly on G/H, and the quotient $\Gamma \backslash G/H$ is compact). Then Γ is virtually polycyclic.*

For another perspective and possible attack on Auslander's conjecture, compare Labourie [98].

Here are some components of the proof of Auslander's conjecture in low dimensions. For convenience, consider the equivalent formulation: Assume $\Gamma < \mathsf{Aff}(\mathsf{A}^n)$ is a discrete subgroup acting properly on A^n and Γ is not virtually solvable. Then $\Gamma \backslash \mathsf{A}^n$ is not compact.

As in Proposition 3.1, consider the semisimple part S of the identity component $G = \left(\overline{\mathsf{L}(\Gamma)}^{\mathsf{Zar}}\right)^0$ of the Zariski closure. Write $S = S_1 \cdots S_k$ as an almost direct product of simple Lie groups. The first key ingredient is the following theorem.

THEOREM 8.7 (Soifer [122], Tomanov [126, 127]).
If each factor S_i has real rank ≤ 1, then Γ is not cocompact.

Hence assume that at least one simple factor S_1 has real rank ≥ 2. By a dynamical argument, for certain S satisfying the hypotheses, Γ does not act properly. In other cases, a cohomological dimension argument (as in Section 3.4.1) rules out cocompactness. What remains is a handful of interesting cases requiring some more sophisticated arguments.

Most interesting is the standard (six-dimensional) representation of $G = \mathsf{SO}(2,1) \times \mathsf{SL}(3,\mathbb{R})$ on $\mathbb{R}^{2,1} \oplus \mathbb{R}^3$. Since G is not Milnor, proper affine deformations of non–virtually solvable discrete subgroups of G exist. Since $\mathsf{cd}(\Gamma) \leq \mathsf{dim}(G/K) = 7$, where Γ is a torsion-free discrete subgroup of G and K is the maximal compact subgroup of G, cohomological dimension does not obstruct the existence of an affine crystallographic action on A^6. In this case, Abels-Margulis-Soifer define a Margulis invariant $\alpha(g)$ for elements $g \in \Gamma$ whose linear part $\mathsf{L}(g)$ is regular enough. It is essentially the Margulis invariant of the $\mathsf{SO}(2,1) \ltimes \mathbb{R}^{2,1}$ part of g. Then they prove the Opposite Sign Lemma in this setting. Note that the dynamics in this setting is more complicated than in the setting of Margulis spacetimes. One sign that things are more complicated is that the attracting subspace for a regular element $\mathsf{L}(g)$ and for its inverse $\mathsf{L}(g)^{-1}$ have different dimension. To conclude the proof in this case, they show that cocompactness implies that some subset of an orbit escapes in a timelike direction of the $\mathbb{R}^{2,1}$ factor and derive a contradiction to the Opposite Sign Lemma.

References

[1] Herbert Abels, *Properly discontinuous groups of affine transformations: a survey*, Geom. Dedicata **87** (2001), no. 1–3, 309–333. MR 1866854

[2] Herbert Abels, Gregory A. Margulis, and Gregory A. Soifer, *On the Zariski closure of the linear part of a properly discontinuous group of affine transformations*, J. Differential Geom. **60** (2002), no. 2, 315–344. MR 1938115

[3] ———, *The linear part of an affine group acting properly discontinuously and leaving a quadratic form invariant*, Geom. Dedicata **153** (2011), 1–46. MR 2819661

[4] ———, *The Auslander conjecture for dimension less than 7*, arXiv:1211.2525

[5] William Abikoff, *The real analytic theory of Teichmüller space*, Lecture Notes in Mathematics, vol. 820, Springer, Berlin, 1980. MR 590044

[6] Ian Agol, *Tameness of hyperbolic 3-manifolds*, arXiv:math.GT/0405568, 2004.

[7] ———, *The virtual Haken conjecture*, Doc. Math. **18** (2013), 1045–1087, with an appendix by Ian Agol, Daniel Groves, and Jason Manning. MR 3104553

[8] L. Auslander, *The structure of complete locally affine manifolds*, Topology **3** (1964), no. suppl. 1, 131–139. MR 0161255

[9] L. Auslander and L. Markus, *Flat Lorentz* 3*-manifolds*, Mem. Amer. Math. Soc. No. **30** (1959), 60. MR 0131842 (24 #A1689)

[10] Thierry Barbot, Virginie Charette, Todd Drumm, William M. Goldman, and Karin Melnick, *A primer on the* $(2+1)$ *Einstein universe*, Recent developments in pseudo-Riemannian geometry, ESI Lect. Math. Phys., Eur. Math. Soc., Zürich, 2008, pp. 179–229. MR 2436232

[11] Oliver Baues, *Varieties of discontinuous groups*, Crystallographic groups and their generalizations (Kortrijk, 1999), Contemp. Math., vol. 262, Amer. Math. Soc., Providence, RI, 2000, pp. 147–158. MR 1796130 (2001i:58013)

[12] Oliver Baues and William M. Goldman, *Is the deformation space of complete affine structures on the 2-torus smooth?*, Geometry and dynamics, Contemp. Math., vol. 389, Amer. Math. Soc., Providence, RI, 2005, pp. 69–89. MR 2181958 (2006j:57066)

[13] Yves Benoist, *Une nilvariété non affine*, J. Differential Geom. **41** (1995), no. 1, 21–52. MR 1316552

[14] ———, *Actions propres sur les espaces homogènes réductifs*, Ann. of Math. (2) **144** (1996), no. 2, 315–347. MR 1418901

[15] Yves Benoist and Karel Dekimpe, *The uniqueness of polynomial crystallographic actions*, Math. Ann. **322** (2002), no. 3, 563–571. MR 1895707

[16] Nicolas Bergeron and Daniel T. Wise, *A boundary criterion for cubulation*, Amer. J. Math. **134** (2012), no. 3, 843–859. MR 2931226

[17] Francis Bonahon, *The geometry of Teichmüller space via geodesic currents*, Invent. Math. **92** (1988), no. 1, 139–162. MR 931208

[18] Martin Bridgeman, Richard Canary, François Labourie, and Andres Sambarino, *The pressure metric for Anosov representations*, Geom. Funct. Anal. **25** (2015), no. 4, 1089–1179. MR 3385630

[19] Martin Bridgeman, Richard Canary, and Andrés Sambarino, *An introduction to pressure metrics for higher Teichmüller spaces*, Ergodic Theory Dynam. Systems **38** (2018), no. 6, 2001–2035. MR 3833339

[20] Jean-Philippe Burelle, Virginie Charette, Todd A. Drumm, and William M. Goldman, *Crooked halfspaces*, Enseign. Math. **60** (2014), no. 1–2, 43–78. MR 3262435

[21] Jean-Philippe Burelle and Dominik Francoeur, *Foliations between crooked planes in 3-dimensional Minkowski space*, Internat. J. Math. **30** (2019), no. 1, 1950004, 7. MR 3916271

[22] Jean-Philippe Burelle and Nicolaus Treib, *Schottky presentations of positive representations*, arXiv:1807.05286, 2018.

[23] Danny Calegari and David Gabai, *Shrinkwrapping and the taming of hyperbolic 3-manifolds*, J. Amer. Math. Soc. **19** (2006), no. 2, 385–446. MR 2188131

[24] Virginie Charette, *Non-proper affine actions of the holonomy group of a punctured torus*, Forum Math. **18** (2006), no. 1, 121–135. MR 2206247

[25] Virginie Charette and Todd A. Drumm, *Strong marked isospectrality of affine Lorentzian groups*, J. Differential Geom. **66** (2004), no. 3, 437–452. MR 2106472

[26] ———, *Complete Lorentzian 3-manifolds*, Geometry, groups and dynamics, Contemp. Math., vol. 639, Amer. Math. Soc., Providence, RI, 2015, pp. 43–72. MR 3379819

[27] Virginie Charette, Todd A. Drumm, and William M. Goldman, *Affine deformations of a three-holed sphere*, Geom. Topol. **14** (2010), no. 3, 1355–1382. MR 2653729

[28] ———, *Finite-sided deformation spaces of complete affine 3-manifolds*, J. Topol. **7** (2014), no. 1, 225–246. MR 3180618

[29] ———, *Proper affine deformations of the one-holed torus*, Transform. Groups **21** (2016), no. 4, 953–1002. MR 3569564

[30] Virginie Charette and William M. Goldman, *Affine Schottky groups and crooked tilings*, Crystallographic groups and their generalizations (Kortrijk, 1999), Contemp. Math., vol. 262, Amer. Math. Soc., Providence, RI, 2000, pp. 69–97. MR 1796126

[31] ———, *McShane-type identities for affine deformations*, Ann. Inst. Fourier (Grenoble) **67** (2017), no. 5, 2029–2041. MR 3732683

[32] Virginie Charette and Youngju Kim, *Foliations of Minkowski* $2+1$ *spacetime by crooked planes*, Internat. J. Math. **25** (2014), no. 9, 1450088, 25. MR 3266531

[33] Suhyoung Choi, *The convex and concave decomposition of manifolds with real projective structures*, Mém. Soc. Math. Fr. (N.S.) (1999), no. 78, vi+102. MR 1779499 (2001j:57030)

[34] Suhyoung Choi, Todd A. Drumm, and William M. Goldman, *Tameness of Margulis spacetimes with parabolics*, Forum Math. (to appear).

[35] Suhyoung Choi and William Goldman, *Topological tameness of Margulis spacetimes*, Amer. J. Math. **139** (2017), no. 2, 297–345. MR 3636632

[36] Daryl Cooper and Kelly Delp, *The marked length spectrum of a projective manifold or orbifold*, Proc. Amer. Math. Soc. **138** (2010), no. 9, 3361–3376. MR 2653965

[37] Christopher B. Croke, *Rigidity for surfaces of nonpositive curvature*, Comment. Math. Helv. **65** (1990), no. 1, 150–169. MR 1036134

[38] Jeffrey Danciger, François Guéritaud, and Fanny Kassel, *Fundamental domains for free groups acting on anti–de Sitter 3-space*, Math. Res. Lett. **23** (2016), no. 3, 735–770. MR 3533195

[39] ———, *Geometry and topology of complete Lorentz spacetimes of constant curvature*, Ann. Sci. Éc. Norm. Supér. (4) **49** (2016), no. 1, 1–56. MR 3465975

[40] ———, *Margulis spacetimes via the arc complex*, Invent. Math. **204** (2016), no. 1, 133–193. MR 3480555

[41] ———, *Proper affine actions for right-angled Coxeter groups*, Duke Math. J. **169** (2020), no. 12, 2231–2280.

[42] ———, *Margulis spacetimes with parabolic elements*, in preparation.

[43] Jeffrey Danciger and Tengren Zhang, *Affine actions with Hitchin linear part*, Geom. Funct. Anal. **29** (2019), no. 5, 1369–1439. MR 4025516

[44] Michael W. Davis and Tadeusz Januszkiewicz, *Right-angled Artin groups are commensurable with right-angled Coxeter groups*, J. Pure Appl. Algebra **153** (2000), no. 3, 229–235. MR 1783167

[45] Karel Dekimpe, *Polynomial structures and the uniqueness of affinely flat infra-nilmanifolds*, Math. Z. **224** (1997), no. 3, 457–481. MR 1439202

[46] ———, *Polycyclic-by-finite groups: from affine to polynomial structures*, Groups St. Andrews 1997 in Bath, I, London Math. Soc. Lecture Note Ser., vol. 260, Cambridge Univ. Press, Cambridge, 1999, pp. 219–236. MR 1676619

[47] ———, *Solvable Lie algebras, Lie groups and polynomial structures*, Compositio Math. **121** (2000), no. 2, 183–204. MR 1757881

[48] ———, *Affine and polynomial structures on virtually 2-step solvable groups*, Comm. Algebra **29** (2001), no. 11, 4965–4988. MR 1856924

[49] ———, *Polynomial crystallographic actions on the plane*, Geom. Dedicata **93** (2002), 47–56. MR 1934685

[50] Karel Dekimpe and Paul Igodt, *Polynomial structures on polycyclic groups*, Trans. Amer. Math. Soc. **349** (1997), no. 9, 3597–3610. MR 1422895

[51] ———, *Polynomial alternatives for the group of affine motions*, Math. Z. **234** (2000), no. 3, 457–485. MR 1774093

[52] Todd A. Drumm, *Fundamental polyhedra for Margulis space-times*, ProQuest LLC, Ann Arbor, MI, 1990, Thesis (Ph.D.)–University of Maryland, College Park. MR 2638637

[53] ———, *Fundamental polyhedra for Margulis space-times*, Topology **31** (1992), no. 4, 677–683. MR 1191372

[54] ———, *Linear holonomy of Margulis space-times*, J. Differential Geom. **38** (1993), no. 3, 679–690. MR 1243791

[55] ———, *Lorentzian geometry*, Geometry, topology and dynamics of character varieties, Lect. Notes Ser. Inst. Math. Sci. Natl. Univ. Singap., vol. 23, World Sci. Publ., Hackensack, NJ, 2012, pp. 247–280. MR 2987620

[56] Todd A. Drumm and William M. Goldman, *Complete flat Lorentz 3-manifolds with free fundamental group*, Internat. J. Math. **1** (1990), no. 2, 149–161. MR 1060633

[57] ———, *Crooked planes*, Electron. Res. Announc. Amer. Math. Soc. **1** (1995), no. 1, 10–17. MR 1336695

[58] ———, *The geometry of crooked planes*, Topology **38** (1999), no. 2, 323–351. MR 1660333

[59] ———, *Isospectrality of flat Lorentz 3-manifolds*, J. Differential Geom. **58** (2001), no. 3, 457–465. MR 1906782

[60] Charles Frances, *The conformal boundary of Margulis space-times*, C. R. Math. Acad. Sci. Paris **336** (2003), no. 9, 751–756. MR 1989275

[61] David Fried and William M. Goldman, *Three-dimensional affine crystallographic groups*, Adv. in Math. **47** (1983), no. 1, 1–49. MR 689763

[62] Sourav Ghosh, *Anosov structures on Margulis spacetimes*, Groups Geom. Dyn. **11** (2017), no. 2, 739–775. MR 3668058

[63] ———, *Avatars of Margulis invariants and proper actions*, arXiv:1812.03777, 2018.

[64] ———, *The pressure metric on the Margulis multiverse*, Geom. Dedicata **193** (2018), 1–30. MR 3770278

[65] ———, *Margulis multiverse: Infinitesimal rigidity, pressure form and convexity*, arXiv:1907.12348, 2019.

[66] Sourav Ghosh and Nicolaus Treib, *Affine Anosov representations and proper actions*, arXiv:1711.09712, 2017.

[67] William M. Goldman, *Projective structures with Fuchsian holonomy*, J. Differential Geom. **25** (1987), no. 3, 297–326. MR 882826

[68] ———, *Trace coordinates on Fricke spaces of some simple hyperbolic surfaces*, Handbook of Teichmüller theory, vol. II, IRMA Lect. Math. Theor. Phys., vol. 13, Eur. Math. Soc., Zürich, 2009, pp. 611–684. MR 2497777

[69] ———, *Crooked surfaces and anti-de Sitter geometry*, Geom. Dedicata **175** (2015), 159–187. MR 3323635

[70] ———, *Flat affine, projective and conformal structures on manifolds: a historical perspective*, Geometry in History (S. Dani and A. Papadopoulos, eds.), Springer, Inc., Cham, 2019.

[71] William M. Goldman and Yoshinobu Kamishima, *The fundamental group of a compact flat Lorentz space form is virtually polycyclic*, J. Differential Geom. **19** (1984), no. 1, 233–240. MR 739789 (85i:53064)

[72] William M. Goldman and François Labourie, *Geodesics in Margulis spacetimes*, Ergodic Theory Dynam. Systems **32** (2012), no. 2, 643–651. MR 2901364

[73] William M. Goldman, François Labourie, and Gregory Margulis, *Proper affine actions and geodesic flows of hyperbolic surfaces*, Ann. of Math. (2) **170** (2009), no. 3, 1051–1083. MR 2600870

[74] William M. Goldman, François Labourie, Gregory Margulis, and Yair Minsky, *Geodesic laminations and proper affine actions*, in preparation.

[75] William M. Goldman and Gregory Laun, *Affine Coxeter extensions of the two-holed projective plane*, arXiv:1511.05228, 2015.

[76] William M. Goldman and Gregory A. Margulis, *Flat Lorentz 3-manifolds and cocompact Fuchsian groups*, Crystallographic groups and their generalizations (Kortrijk, 1999), Contemp. Math., vol. 262, Amer. Math. Soc., Providence, RI, 2000, pp. 135–145. MR 1796129

[77] Walter Helbig Gottschalk and Gustav Arnold Hedlund, *Topological dynamics*, American Mathematical Society Colloquium Publications, vol. 36, Amer. Math, Society. Providence, RI, 1955. MR 0074810

[78] Fritz Grunewald and Gregori Margulis, *Transitive and quasitransitive actions of affine groups preserving a generalized Lorentz-structure*, J. Geom. Phys. **5** (1988), no. 4, 493–531 (1989). MR 1075720

[79] Fritz Grunewald and Dan Segal, *On affine crystallographic groups*, J. Differential Geom. **40** (1994), no. 3, 563–594. MR 1305981

[80] François Guéritaud, *Lengthening deformations of singular hyperbolic tori*, Ann. Fac. Sci. Toulouse Math. (6) **24** (2015), no. 5, 1239–1260. MR 3485334

[81] ———, *On Lorentz spacetimes of constant curvature*, Geometry, groups and dynamics, Contemp. Math., vol. 639, Amer. Math. Soc., Providence, RI, 2015, pp. 253–269. MR 3379833

[82] François Guéritaud and Fanny Kassel, *Maximally stretched laminations on geometrically finite hyperbolic manifolds*, Geom. Topol. **21** (2017), no. 2, 693–840. MR 3626591

[83] Frédéric Haglund and Daniel T. Wise, *Special cube complexes*, Geom. Funct. Anal. **17** (2008), no. 5, 1551–1620. MR 2377497

[84] ———, *Coxeter groups are virtually special*, Adv. Math. **224** (2010), no. 5, 1890–1903. MR 2646113

[85] Tadeusz Januszkiewicz and Jacek Świątkowski, *Hyperbolic Coxeter groups of large dimension*, Comment. Math. Helv. **78** (2003), no. 3, 555–583. MR 1998394

[86] Jeremy Kahn and Vladimir Markovic, *Immersing almost geodesic surfaces in a closed hyperbolic three manifold*, Ann. of Math. (2) **175** (2012), no. 3, 1127–1190. MR 2912704

[87] Fanny Kassel, *Quotients compacts d'espaces homogènes réels ou p-adiques*, Thesis (Ph.D.)–Université Paris-Sud 11, 2009.

[88] Inkang Kim, *Ergodic theory and rigidity on the symmetric space of non-compact type*, Ergodic Theory Dynam. Systems **21** (2001), no. 1, 93–114. MR 1826662

[89] ———, *Marked length rigidity of rank one symmetric spaces and their product*, Topology **40** (2001), no. 6, 1295–1323. MR 1867246

[90] ———, *Affine action and Margulis invariant*, J. Funct. Anal. **219** (2005), no. 1, 205–225. MR 2108366

[91] Anthony W. Knapp, *Lie groups beyond an introduction*, second ed., Progress in Mathematics, vol. 140, Birkhäuser, Boston, 2002. MR 1920389

[92] Toshiyuki Kobayashi, *Criterion for proper actions on homogeneous spaces of reductive groups*, J. Lie Theory **6** (1996), no. 2, 147–163. MR 1424629

[93] B. Kostant and D. Sullivan, *The Euler characteristic of an affine space form is zero*, Bull. Amer. Math. Soc. **81** (1975), no. 5, 937–938. MR 375341 (51 #11536)

[94] N. H. Kuiper, *Sur les surfaces localement affines*, Géométrie différentielle. Colloques Internationaux du Centre National de la Recherche Scientifique, Strasbourg, 1953, Centre National de la Recherche Scientifique, Paris, 1953, pp. 79–87. MR 0060288

[95] François Labourie, *Fuchsian affine actions of surface groups*, J. Differential Geom. **59** (2001), no. 1, 15–31. MR 1909247

[96] ———, *Anosov flows, surface groups and curves in projective space*, Invent. Math. **165** (2006), no. 1, 51–114. MR 2221137

[97] ———, *Lectures on representations of surface groups*, Zurich Lectures in Advanced Mathematics, Eur. Math. Soc. (EMS), Zürich, 2013. MR 3155540

[98] ———, *Entropy and affine actions for surface groups*, arXiv:1908.00599, 2019.

[99] Gregory Laun, *Fundamental domains for proper affine actions of Coxeter groups in three dimensions*, ProQuest LLC, Ann Arbor, MI, 2016, Thesis (Ph.D.)–University of Maryland, College Park. MR 3553574

[100] Bruno Le Floch and Ilia Smilga, *Action of Weyl group on zero-weight space*, C. R. Math. Acad. Sci. Paris **356** (2018), no. 8, 852–858. MR 3851538

[101] Yi Liu, *Virtual cubulation of nonpositively curved graph manifolds*, J. Topol. **6** (2013), no. 4, 793–822. MR 3145140

[102] G. A. Margulis, *Free completely discontinuous groups of affine transformations*, Dokl. Akad. Nauk SSSR **272** (1983), no. 4, 785–788. MR 722330

[103] ———, *Complete affine locally flat manifolds with a free fundamental group*, Journal of Soviet Mathematics **36** (1987), no. 1, 129–139.

[104] Greg McShane, *Simple geodesics and a series constant over Teichmuller space*, Invent. Math. **132** (1998), no. 3, 607–632. MR 1625712

[105] Geoffrey Mess, *Lorentz spacetimes of constant curvature*, Geom. Dedicata **126** (2007), 3–45. MR 2328921

[106] John Milnor, *Hilbert's problem 18: on crystallographic groups, fundamental domains, and on sphere packing*, Mathematical developments arising from Hilbert problems (Proc. Sympos. Pure Math., vol. XXVIII, Northern Illinois Univ., De Kalb, Ill., 1974), Amer. Math. Soc., Providence, RI, 1976, pp. 491–506. MR 0430101

[107] ———, *On fundamental groups of complete affinely flat manifolds*, Advances in Math. **25** (1977), no. 2, 178–187. MR 454886

[108] Damian Osajda, *A construction of hyperbolic Coxeter groups*, Comment. Math. Helv. **88** (2013), no. 2, 353–367. MR 3048190

[109] Jean-Pierre Otal, *Le spectre marqué des longueurs des surfaces à courbure négative*, Ann. of Math. (2) **131** (1990), no. 1, 151–162. MR 1038361

[110] Athanase Papadopoulos and Guillaume Théret, *Shortening all the simple closed geodesics on surfaces with boundary*, Proc. Amer. Math. Soc. **138** (2010), no. 5, 1775–1784. MR 2587462

[111] R. C. Penner, *The decorated Teichmüller space of punctured surfaces*, Comm. Math. Phys. **113** (1987), no. 2, 299–339. MR 919235

[112] Piotr Przytycki and Daniel T. Wise, *Mixed 3-manifolds are virtually special*, J. Amer. Math. Soc. **31** (2018), no. 2, 319–347. MR 3758147

[113] M. S. Raghunathan, *Discrete subgroups of Lie groups*, Ergebnisse der Mathematik und ihrer Grenzgebiete Band 68, Springer-Verlag, New York-Heidelberg, 1972. MR 0507234

[114] Michah Sageev, *Ends of group pairs and non-positively curved cube complexes*, Proc. London Math. Soc. (3) **71** (1995), no. 3, 585–617. MR 1347406

[115] F. Schottky, *Ueber die conforme Abbildung mehrfach zusammenhängender ebener Flächen*, J. Reine Angew. Math. **83** (1877), 300–351. MR 1579739

[116] Atle Selberg, *On discontinuous groups in higher-dimensional symmetric spaces*, Contributions to function theory (Internat. Colloq. Function Theory, Bombay, 1960), Tata Institute of Fundamental Research, Bombay, 1960, pp. 147–164. MR 0130324

[117] Adam S. Sikora, *Character varieties*, Trans. Amer. Math. Soc. **364** (2012), no. 10, 5173–5208. MR 2931326

[118] Ilia Smilga, *Fundamental domains for properly discontinuous affine groups*, Geom. Dedicata **171** (2014), 203–229. MR 3226793

[119] ———, *Proper affine actions on semisimple Lie algebras*, Ann. Inst. Fourier (Grenoble) **66** (2016), no. 2, 785–831. MR 3477891

[120] ———, *Construction of Milnorian representations*, Geom. Dedicata **206** (2020), 55–73.

[121] ———, *Proper affine actions: a sufficient criterion*, Math. Ann. (to appear), arXiv: 1612.08942v4.

[122] G. A. Soifer, *Affine crystallographic groups*, Algebra and analysis (Irkutsk, 1989), Amer. Math. Soc. Transl. Ser. 2, vol. 163, Amer. Math. Soc., Providence, RI, 1995, pp. 165–170. MR 1331393

[123] William P. Thurston, *Minimal stretch maps between hyperbolic surfaces*, arXiv:math.GT/9801039, 1998.

[124] J. Tits, *Free subgroups in linear groups*, J. Algebra **20** (1972), 250–270. MR 0286898

[125] George Tomanov, *On a conjecture of L. Auslander*, C. R. Acad. Bulgare Sci. **43** (1990), no. 2, 9–12. MR 1062291

[126] ———, *The virtual solvability of the fundamental group of a generalized Lorentz space form*, J. Differential Geom. **32** (1990), no. 2, 539–547. MR 1072918

[127] ———, *Properly discontinuous group actions on affine homogeneous spaces*, Tr. Mat. Inst. Steklova **292** (2016), Algebra, Geometriya i Teoriya Chisel, 268–279, reprinted in Proc. Steklov Inst. Math. **292** (2016), no. 1, 260–271. MR 3628466

[128] Daniel T. Wise, *Research announcement: the structure of groups with a quasiconvex hierarchy*, Electron. Res. Announc. Math. Sci. **16** (2009), 44–55. MR 2558631

5

TSACHIK GELANDER, YAIR GLASNER, AND GREGORY SOIFER

MAXIMAL SUBGROUPS OF COUNTABLE GROUPS: A SURVEY

To our friend, teacher, and colleague Grisha Margulis with great admiration. While proving the most remarkable theorems in the field as well as many other fantastic results, Margulis invented techniques and developed ideas that inspired so many mathematicians. This essay describes one such idea and its many ramifications as it allowed us and others to solve problems and extend the theory.

Abstract. This essay surveys the works of Margulis-Soifer on maximal subgroups and its many ramifications.

1 Introduction

This paper is a survey on the works [MS77, MS79, MS81] on maximal subgroups in finitely generated linear groups, and the works that followed it [GG08, GG13b, GG13a, Kap03, Iva92, HO16, GM16, AGS14, Sf90, Sf98, Per05, AKT16, FG18, GS17, Mei95] concerning maximal subgroups of infinite index in linear groups as well as in various other groups possessing a suitable geometry or dynamics.

1.1 THE MARGULIS-SOIFER THEOREM. The original motivation came from the following question of Platonov:

QUESTION 1.1. Does $SL_n(\mathbb{Z}), n \geq 3$ admit a maximal subgroup of infinite index?

Yair Glasner. Department of Mathematics. Ben-Gurion University of the Negev, P.O.B. 653, Be'er Sheva 84105, Israel
yairgl@math.bgu.ac.il

Tsachik Gelander. Department of Mathematics. Weizmann Institute of Science, Rehovot 76100, Israel
tsachik.gelander@weizmann.ac.il

Gregory Soifer. Department of Mathematics. Bar-Ilan University, Ramat-Gan, 5290002 Israel
soifer@math.biu.ac.il

The research of Glasner and Gelander was partially funded by Israel Science Foundation grant ISF 2919/19.

In [MS77], [MS79], and [MS81] this question was answered positively. Moreover these papers clarified the existence question of infinite index maximal subgroups for all finitely generated linear groups:

THEOREM 1.2 ([MS77, MS79, MS81]).
A finitely generated linear group admits a maximal subgroup of infinite index if and only if it is not virtually solvable.

The proof of theorem 1.2 is inspired by Tits's proof of the classical Tits alternative [Tit72]. Recall that Tits proved that a finitely generated linear group Γ that is not virtually solvable admits a free subgroup. In 1.2 it is shown that in fact Γ admits a profinitely dense free subgroup F. By Zorn's lemma F is contained in a maximal proper subgroup M of Γ. Since F is profinitely dense, so is M, and therefore $[\Gamma : M]$ must be infinite. The details of the proof, however, are quite involved, especially in the case where Γ is not Zariski connected.

1.2 PRIMITIVITY. Every subgroup H of a group Γ corresponds to a transitive action of Γ-namely, the action on the coset space Γ/H. The group H is maximal if and only if this coset action is primitive in the sense of the following:

DEFINITION 1.3. An action of a group Γ on a set X is *primitive* if $|X| > 1$ and there are no Γ-invariant equivalence relations on X apart from the two trivial ones.[1] An action is called *quasiprimitive* if every normal subgroup acts either trivially or transitively. A group is *primitive* or *quasiprimitive* if it admits a faithful primitive or quasiprimitive action on a set.

In particular Γ has a maximal subgroup of infinite index if and only if it admits a primitive permutation action on an infinite set. Primitive actions form the basic building blocks of the theory of permutation groups. A lot of research was dedicated to the study of finite primitive groups (cf. [AS85], [KL88], and [DM96]). The papers [MS77], [MS79], and [MS81] opened a door to the study of permutation representation of infinite linear groups.

The transition to permutation theoretic terminology suggests shifting the attention from infinite primitive groups to the study of groups admitting

[1]The trivial equivalence relations are those with a unique equivalence class or with singletons as equivalence classes. When $|X| = 2$, one should also require that the action is not trivial.

a faithful primitive action. This leads us to phrase the following guideline question.

QUESTION 1.4. Characterize the countable primitive groups.

Due to the method developed in [MS77], [MS79], and [MS81] a satisfactory answer is within reach for many natural families of groups. This brings us to the following definition, which will turn out to be central to our discussion:

DEFINITION 1.5. A countable group Γ is called *of almost simple type* if

- if it contains no nontrivial finite normal subgroups and
- $M, N \lhd \Gamma$ with $[M, N] = \langle e \rangle$, then either $M = \langle e \rangle$ or $N = \langle e \rangle$. In particular Γ contains no nontrivial abelian normal subgroups.

As a direct consequence of Theorem 1.2 one can prove the following:

THEOREM 1.6.
An infinite finitely generated linear group Γ is primitive if and only if it is of almost simple type.

The permutation representation viewpoint also suggests natural properties that are stronger than primitivity.

DEFINITION 1.7. An action $\Gamma \curvearrowright X$ is called 2-transitive if the induced action on pairs of distinct points $G \curvearrowright X \times X \setminus \mathrm{diag}(X)$ is transitive. An action $G \curvearrowright X$ is k-transitive if it is transitive on k-tuples of distinct points and is highly transitive if it is k-transitive for every k. A group will be called *highly transitive* if it admits a faithful, highly transitive action.

Every 2-transitive action is primitive. Indeed if $\sim$ is a Γ invariant equivalence relation on X and if $x \sim y$ for some $x \neq y \in X$, then 2-transitivity readily implies that any two points are equivalent.

1.3 THE CHARACTERIZATION OF COUNTABLE PRIMITIVE LINEAR GROUPS.

Theorem 1.2 was generalized to the setting of countable, but not necessarily finitely generated, linear groups.

THEOREM 1.8 ([GG08]).
Any countable linear nontorsion group of almost simple type is primitive. In fact such a group admits uncountably many nonequivalent faithful primitive actions.

Any countable linear nontorsion group that is not virtually solvable has uncountably many maximal subgroups of infinite index.

In the zero characteristic case, as well as in the finitely generated case, the theorem remains valid without the assumption that the group is nontorsion. In positive characteristic, we need this assumption for our proof. In fact, as in the finitely generated case, the proof of Theorem 1.8 actually establishes a stronger statement: the existence of a free subgroup that is contained in a maximal subgroup. This stronger statement fails for torsion groups like $\mathrm{PSL}_2(\overline{\mathbb{F}_7})$, where $\overline{\mathbb{F}_7}$ is the algebraic closure of the field $\mathbb{F}_7$, of seven elements. Note, however, that $\mathrm{PSL}_2(\overline{\mathbb{F}_7})$ does not violate Theorem 1.8 because it is primitive, and in fact it even admits a faithful 3-transitive action on the projective line $\mathbb{P}\overline{F_7}$.

Another difference that stands out between this theorem and its finitely generated counterpart is the lack of the converse direction. The missing implication is actually the easy direction of Theorem 1.2. But, upon leaving the realm of finitely generated groups, it fails. An easy example is the 2-transitive action of the solvable group $\Gamma =$

$$\left\{ \begin{pmatrix} a & b \\ 0 & 1 \end{pmatrix} \,\middle|\, a \in \mathbf{Q}^*, b \in \mathbf{Q} \right\} \text{ on the invariant set } \left\{ \begin{pmatrix} x \\ 1 \end{pmatrix} \in \mathbf{Q}^2 \,\middle|\, x \in \mathbf{Q} \right\}.$$

This action can also be identified as the natural affine action of the semidirect product $\mathbf{Q}^* \ltimes \mathbf{Q} \curvearrowright \mathbf{Q}$.

DEFINITION 1.9. Let $\Gamma = \Delta \ltimes M$ be a semidirect product. The natural affine (or standard) action of Γ is the action $\Gamma \curvearrowright M$ in which M acts on itself by left translations and Δ acts on M by the conjugation

$$m \cdot x = mx, \ \forall m \in M, \qquad \delta \cdot x = \delta x \delta^{-1}, \ \forall \delta \in \Delta.$$

As it turns out, this example is quite indicative, as can be seen from the following two theorems.

THEOREM 1.10.
Let Γ be a primitive countable group that is not of almost simple type. Then Γ splits as a semidirect product $\Gamma = \Delta \ltimes M$, and the given primitive action is equivalent to its natural affine action on M.

In particular it follows that the faithful primitive action is unique in this case. In fact it is even unique among all faithful quasiprimitve actions of this group. Of course the same group might admit additional primitive actions that are not faithful. Take, for example, the group $SL_2(\mathbf{Q}) \ltimes \mathbf{Q}^2$. Its natural action on $\mathbf{Q}^2$ is 2-transitive—hence the unique primitive faithful action of this group by Theorem 1.10. However, this group admits a quotient that is of almost simple type $SL_2(\mathbf{Q})$. Thus by Theorem 1.8 it does admit uncountably many primitive actions factoring through this quotient.

The semidirect products whose natural action is primitive/faithful are easily classified.

THEOREM 1.11.
The affine action of a semidirect product $\Gamma = \Delta \ltimes M \curvearrowright M$ is faithful iff $Z_M(\Delta) = \{\delta \in \Delta \mid [\delta, m] = e \ \forall m \in M\} = \langle e \rangle$. This action is primitive if and only if the only subgroups of M that are normalized by Δ are M itself and the trivial group $\langle e \rangle$.

DEFINITION 1.12. Let $\Gamma = \Delta \ltimes M$ be a countable, semidirect product whose natural action is primitive and faithful—that is, such that M is characteristically simple and admits no nontrivial Δ-invariant subgroups and $Z_M(\Delta) = \langle e \rangle$. Then, if in addition M is abelian, Γ is called *primitive of affine type*; if M is nonabelian, Γ is called *primitive of diagonal type.*

Combining Theorem 1.8 with the elementary classification of primitive groups that are not of almost simple type yields a characterization of countable linear primitive groups subject to the additional assumption that the groups in question contain at least one element of infinite order. We like to think of this theorem as a rough generalization of the Aschbacher-O'Nan-Scott theorem (see [AS85], [DM96]) to the setting of countable linear groups.

THEOREM 1.13.
A countable nontorsion linear group Γ is primitive if and only if one of the following mutually exclusive conditions holds.

- *Γ is primitive of almost simple type.*
- *Γ is primitive of affine type.*
- *Γ is primitive of diagonal type.*

In the affine and the diagonal cases the group admits a unique faithful quasiprimitive action. In the almost simple case the group admits uncountably many nonisomorphic faithful primitive actions.

REMARK 1.14. For a finitely generated group Γ, only the first possibility can occur in the above theorem.

REMARK 1.15. In all the cases under consideration, it is shown that a group is primitive if and only if it is quasiprimitive.

EXAMPLE 1.16. Let Σ be any simple countable linear group that is not torsion. For example one can take $\Sigma = \mathrm{PSL}_2(\mathbf{Q})$. Now consider the two groups $\Gamma_1 = \Sigma \times \Sigma$ and $\Gamma_2 = (\mathbf{Z}/2\mathbf{Z}) \ltimes (\Sigma \times \Sigma)$. Despite the clear similarity between these two groups (one being an index two subgroup in the other), their respective permutation representation theories are quite different. Γ_1 is primitive of diagonal type and hence it admits a unique faithful primitive action. Moreover this action is very explicit; it is the action $\Gamma_1 \curvearrowright \Sigma$ given by $(\gamma_1, \gamma_2) \cdot \sigma = \gamma_1 \sigma \gamma_2^{-1}$. On the contrary the group Γ_2 is primitive of almost simple type and hence admits uncountably many, nonisomorphic faithful primitive actions. Yet, we do not have a good explicit description for any of these actions.

EXAMPLE 1.17. Let $\Gamma = \mathrm{PSL}_n(\mathbf{Q})$. This group admits a very explicit faithful primitive action—namely, its action on the projective line $\mathbb{P}(\mathbf{Q}^n)$. When $n = 2$ this action is not only primitive but also 3-transitive. Being a group of almost simple type, the above theorem yields uncountably many other nonisomorphic primitive permutation representations. Again, we do not have any explicit descriptions of these actions.

Section 2 is dedicated to the classification of countable primitive groups that are not of almost simple type. It deals with general countable groups and uses only soft, group theoretic arguments. Section 3 is dedicated to the linear case and the proof of Theorem 1.13.

1.4 THE VARIETY OF MAXIMAL SUBGROUPS. Since the construction of maximal subgroups of infinite index in [MS77], [MS79], and [MS81], it is expected that there should be examples of such maximal subgroups of various different natures. In particular, in the latter paper the existence of uncountably many maximal subgroups in any finitely generated non-virtually-solvable linear group was established. However, as the proof is nonconstructive and relies on the axiom of choice, it is highly nontrivial to lay one's hands on specific properties of the resulting groups.

In many special cases one can find examples of maximal subgroups in the same group that are very different from each other. Example 1.17 describes actions of $\mathrm{PSL}_n(Q)$, some of which have Zariski dense stabilizers and others not. For $\mathrm{SL}_2(\mathbf{Q})$ one even has a 3-transitive action with a solvable point stabilizer. In another direction, Section 5 constructs highly transitive faithful actions for any nontorsion group of almost simple type $\Gamma < \mathrm{SL}_2(k)$, where k is any local field. Many of these groups also admit actions that are not highly transitive. These examples and many more come to show that there is probably a whole zoo of maximal subgroups out there that we are only starting to see.

To some extent, a benchmark example is the group $\mathrm{SL}_n(\mathbf{Z})$, $n \geq 3$, the same group appearing in the original question of Platonov. These groups are very rigid in nature, and it is quite possible that a good understanding of the family of maximal subgroups here would shed light on the general case.

The first step would be to show that, indeed, maximal subgroups $\Delta \leq \mathrm{SL}_n(\mathbf{Z})$ of different nature do exist. As of today, little can be said about the intrinsic algebraic structure of Δ. Instead, one is led to focus on the way it sits inside $\mathrm{SL}_n(\mathbf{Z})$. Two points of view that are natural to consider are

- the associated permutation representation $\Gamma \curvearrowright \Gamma/\Delta$ and
- the action of Δ on the associated projective space $\mathbb{P}^{n-1}(\mathbf{R})$.

The following results were established in [GM16]:

THEOREM 1.18.
Let $n \geq 3$. There are $2^{\aleph_0}$ infinite index maximal subgroups in $\mathrm{SL}_n(\mathbf{Z})$.

THEOREM 1.19.
Let $n \geq 3$. There exists a maximal subgroup Δ of $\mathrm{SL}_n(\mathbf{Z})$ that does not have a dense orbit in $\mathbb{P}^{n-1}(\mathbf{R})$. In particular, the limit set of Δ (in the sense of [CG00]) is nowhere dense.

THEOREM 1.20.
Let $n \geq 3$. There exists an infinite index maximal subgroup M of $\mathrm{PSL}_n(\mathbf{Z})$ and an element $g \in \mathrm{PSL}_n(\mathbf{Z})$ such that $M \bigcap gMg^{-1} = \{id\}$.

THEOREM 1.21.
Let $n \geq 3$. There exists a primitive permutation action of $\mathrm{SL}_n(\mathbf{Z})$ that is not 2-transitive.

REMARK 1.22. These theorems remain true also for $\mathrm{SL}_n(\mathbf{Q})$ instead of $\mathrm{SL}_n(\mathbf{Z})$.

REMARK 1.23. Recall that Theorem 1.2 is much more general. It holds for any finitely generated non-virtually-solvable linear group Γ. However the proof of the last results rely on special properties of $SL_n(\mathbf{Z})$, $n \geq 3$. In particular one important ingredient here is the beautiful result of Venkataramana about commuting unipotents, Theorem 4.8. Another ingredient is the result of Conze and Guivarc'h, Theorem 4.12. Some of these results can be extended to the class of arithmetic groups of higher $\mathbf{Q}$-rank.

1.5 HIGHLY TRANSITIVE ACTIONS. Over the years many authors generalized the results and the methods of [MS77], [MS79], and [MS81]. In section 5 we describe two major directions that eventually converged together in a very nice way. The first direction involves implementing the methods of [MS81] to various *linear-like* settings. Notable examples include the works of Kapovich [Kap03] for subgroups of hyperbolic groups and Ivanov [Iva92] on mapping class groups. More specifically than just linear-like, these examples exhibit boundary dynamics closer in nature to that of subgroups of rank-1 Lie groups. From the group theoretic point of view, this has the effect that often after ruling out obvious obstructions, all groups in question are of almost simple type and primitive.

The second direction involves construction of k-transitive or highly transitive permutation representations, which are a priori harder to construct. Originally the predominant feeling was that highly transitive groups are much rarer. However, over a period spanning a few decades, wider and more elaborate constructions of highly transitive groups were given. Some notable papers in this direction are [McD77], [Dix90], [Hic92], [Gun92], [Kit12], [MS13], [Cha12], [GG13a], and [HO16]. The paper of Hull and Osin establishes the following:

THEOREM 1.24 ([HO16, theorem 1.2]).
Any countable acylindrically hyperbolic group with no finite normal subgroups is highly transitive

The family of acylindrically hyperbolic groups is very wide, encompassing within it most groups that could be considered as *rank-1*, even in a weak sense of the word. Thus, while Theorem 1.24 does not imply Theorem 1.2, it ties together the strings, establishing a very strong form of this theorem for a wide range of rank-1 examples, thereby generalizing the theorems mentioned above. To emphasize this we quote from their paper a few specific situations where their theorem applies.

- A countable group relatively hyperbolic to a collection of proper subgroups is highly transitive iff it is not virtually cyclic and has no finite normal subgroups.
- $\mathrm{Mod}(\Sigma_{g,n,p})$, the mapping class group of a compact orientable surface, of genus g, p-punctures, and n-boundary components, is highly transitive if and only if $n=0$ and $3g+p\geq 5$.
- $\mathrm{Out}(F_n)$ is highly transitive iff $n\geq 3$.
- $\pi_1(M)$, where M is a compact irreducible 3-manifold, is highly transitive iff it is not virtually solvable and M is not Seifert fibered.
- A right-angled Artin group is highly transitive if and only if it is noncyclic and directly indecomposable.

We can summarize all of the above by saying that in each and every one of these situations, a subgroup is highly transitive if and only if it is of almost simple type.

Inspired by the work of Hull and Osin, and applying the theory of linear groups, we establish the following result.

THEOREM 1.25.
Let k be a local field, and let $\Gamma < \mathrm{SL}_2(k)$ *be a center-free unbounded countable group. Then the following conditions are equivalent for* Γ.

(1) Γ *is of almost simple type.*
(2) Γ *is Zariski dense.*
(3) Γ *is not virtually solvable.*
(4) Γ *is highly transitive.*

We note that this is the only result in this survey that has not been proved before.

From a group theoretic point of view we use Hull-Osin's characterization of the point stabilizers in highly transitive actions. From the geometric point of view, we use the topological dynamics of the boundary action, instead of the small cancellation-type methods used in [HO16].

1.6 OTHER GEOMETRIC SETTINGS. As seen so far the ideas behind the proof of Theorem 1.2 have been substantially generalized to many linear-like settings, by which we refer very loosely to the situation where there is a group action with a rich enough proximal dynamic. The outcome of the theory in many of these settings is that all groups of almost simple type are primitive. Sometimes, notably in negatively curved type settings, much stronger transitivity properties are established by similar techniques.

In section 6 we offer just a glimpse, without proofs, into some fascinating works outside the linear-like setting. Here we encounter completely different methods and different types of behavior.

2 Faithful primitive actions of countable groups

2.1 NECESSARY CONDITIONS AND GROUP TOPOLOGIES ON Γ.

The goal of this section is to establish Theorems 1.10 and 1.11 from the introduction. A more detailed version of these, which appears here as Proposition 2.1, can be thought of as a summary of those implications in Theorem 1.13, which hold for general countable groups without any additional assumptions. It provides the classification of countable primitive groups into three disjoint classes: affine, diagonal, and almost simple. The structure of primitive groups of affine or diagonal type is well understood. But the structure of primitive groups of almost simple type remains mysterious; this is the class of groups for which Question 1.4 is the most interesting. We start with the proof of Theorem 1.11, characterizing primitivity and faithfulness for the standard actions of semidirect products.

Proof of Theorem 1.11. Consider the natural action of a semidirect product $\Gamma = \Delta \ltimes M \curvearrowright M$. One easily verifies that $\Delta = \mathrm{Stab}_\Gamma(\{e\})$; hence the kernel of the action is $Z_M(\Delta) = \{\delta \in \Delta \mid \delta m \delta^{-1} = m, \ \forall m \in M\}$, which proves the first statement. If $\langle e \rangle \neq K \lneqq M$ is normalized by Δ, then $\Delta < \Delta \ltimes K < \Gamma$, so Δ is not maximal and the standard action fails to be primitive. Conversely, assume that Δ fails to be a maximal, with $\Delta \lneqq \Sigma \lneqq \Gamma$ an intermediate subgroup. Using the unique product decomposition $\Gamma = \Delta M$, we find that $\langle e \rangle \neq \Sigma \bigcap M$ is a nontrivial proper subgroup of M normalized by Δ. □

This gives rise to a very explicit description of primitive groups of both affine and diagonal type. It follows that in a primitive group of affine or diagonal type the normal subgroup M must be characteristically simple. When M is abelian and countable this means that M is the additive group of a countable vector space—that is, either $M \cong \mathbb{F}_p^\infty$ or $M \cong \mathbf{Q}^n$, $n \in \mathbf{N} \bigcup \{\infty\}$. In this case Δ can be identified with an irreducible subgroup of $\mathrm{GL}(M)$.

When M is nonabelian it is center-free, so the natural map $\iota : M \to \mathrm{Inn}(M) < \mathrm{Aut}(M)$ is injective. Since there are no nontrivial Δ-invariant subgroups, $\Delta \bigcap \mathrm{Inn}(M)$ is either trivial or equal to $\mathrm{Inn}(M)$. In the first possibility

Δ commutes with $\mathrm{Inn}(M)$ contradicting the fact that there are no Δ-invariant subgroups in M; hence $\mathrm{Inn}(M) < \Delta$. Recall that we identified the action of M with its left action on itself. Now that we have realized the inner automorphisms of M as a subgroup of Δ, we can distinguish another subgroup of Γ, which is isomorphic to M and commutes with it—namely, the action of M on itself from the right:

$$N = \{\iota(m^{-1})m \mid m \in M\} < \Gamma.$$

This is of course the source of the name *diagonal type*. Indeed we have constructed here the diagonal action $M \times M \curvearrowright M$ described in Example 1.16 as a subaction of any primitive group of diagonal type. We turn to the proof of Proposition 2.1, which is a more detailed version of Theorem 1.10 from the introduction.

PROPOSITION 2.1. *Let $\Gamma \curvearrowright \Omega$ be a faithful primitive action of a countably infinite group on a set. Fix a basepoint $\omega_0 \in \Omega$ and let $\Delta = \Gamma_{\omega_0}$ be its stabilizer. Then every nontrivial normal subgroup of Γ is infinite. Moreover Γ falls into precisely one of the following three categories:*

(1) Either Γ is primitive of almost simple type.
(2) Or Γ is primitive of affine type. In this case $\Gamma = \Delta \ltimes V$, with V a countable vector space over a prime field F (possibly $F = \mathbf{Q}$) and Δ an irreducible subgroup of $\mathrm{GL}_F(V)$. The given action is equivalent to the standard affine action of Γ on V.
(3) Or Γ is primitive of diagonal type. In this case $\Gamma = \Delta \ltimes M$ with M a nonabelian, characteristically simple group. $\mathrm{Inn}(M) < \Delta < \mathrm{Aut}(M)$ *and M has no nontrivial subgroups that are invariant under the Δ-action. Again the given action in this case is equivalent to the standard action of Γ on M.*

Conversely if Γ falls into categories (2) or (3), then its natural action is faithful and primitive.

Proof. As the action is faithful, Ω must be infinite, so it is absurd that a finite normal subgroup would act transitively. If Γ fails to be of almost simple type, then we can find $\langle e\rangle \neq M, N \lhd \Gamma$ with $[M,N] = \langle e\rangle$. Both normal subgroups must act transitively on Ω since the action is primitive and faithful. But $M_{n\omega_0} = nM_{\omega_0}n^{-1} = M_{\omega_0}$, $\forall n \in N$ so that the stabilizer M_{ω_0} will fix all of $N\omega_0 = \Omega$, which implies $M_{\omega_0} = \langle e\rangle$ by the faithfulness of the action. Thus M acts regularly on Ω. Identifying Ω with M via the orbit map $m \mapsto m \cdot \omega_0$, it is now routine to verify that $\Gamma = \Delta \ltimes M$ and that the given action of Γ on Ω is equivalent to the standard affine action of this semidirect product. The question whether the group is of affine or diagonal type now depends only on whether M is abelian or not. The detailed description of the structure of the group in these two cases follows from our discussion above.

Let us note that, just like M, the group $K = M \bigcap N$ is a normal subgroup commuting with N. So, either $K = \langle e \rangle$ or K too acts regularly by the same argument, in which case $M = N$. These two cases, of course, correspond to the diagonal and the affine cases respectively. □

This theorem highlights the family of *almost simple groups* for which Question 1.4 is interesting. In all other cases the existence question of a faithful primitive action is easily resolved. Proposition 2.1 is very similar in its structure to our main Theorem 1.13. It basically summarizes all the "easy" implications of that theorem that do not require the linearity assumption. The remaining implication does not work without the linearity, as the following example shows. Section 6 will be dedicated to a more comprehensive treatment of such examples.

THEOREM 2.2 ([Per05]).
The first Grigorchuk group is an example of a group that is of almost simple type but does not admit any infinite primitive action.

2.2 THE NORMAL TOPOLOGY. We like to think of the definition of *groups of almost simple type* in topological terms. Let

$$\mathcal{N}(\Gamma) = \{\langle e \rangle \neq N \lhd \Gamma\}$$

be the collection of all nontrivial normal subgroups. If Γ is of almost simple type, every $N \in \mathcal{N}(\Gamma)$ is infinite, and this family is closed under intersections. Clearly it is invariant under conjugation. Thus it forms a basis of identity neighborhoods for a group topology on Γ.

DEFINITION 2.3. The *normal topology* τ_N on a countable group of almost simple type Γ is the topology obtained by taking $\mathcal{N}(\Gamma)$ as a basis of open neighborhoods of the identity.

PROPOSITION 2.4. *The normal topology is second countable. It is finer*[2] *than the profinite topology. It is Hausdorff if and only if* $\mathrm{Ncore}(\Gamma) = \bigcap_{N \in \mathcal{N}(\Gamma)} N = \langle e \rangle$.

[2] Possibly the two topologies are equal; this happens exactly when the group admits the Margulis normal subgroup property.

Proof. We denote by $\langle\langle g\rangle\rangle = \langle \gamma g \gamma^{-1} \mid \gamma \in \Gamma\rangle \lhd \Gamma$ the normal subgroup of Γ generated by the conjugacy class of the element $g \in \Gamma$. The countable collection $\{\langle\langle g\rangle\rangle \mid g \in G\}$ forms a basis of identity neighborhoods for the normal topology. Thus the topological group (Γ, τ_N) is first (and consequently also second) countable. All the rest of the statements are obvious. □

In [GG13b, definition 2.3] we gave a different definition for linear groups of almost simple type. The following lemma shows that in the specific case of countable linear groups these two definitions agree.

LEMMA 2.5. *Let Γ be a countable linear group. Then Γ is of almost simple type if and only if there exists a faithful linear representation $\phi : \Gamma \to \mathrm{GL}_n(k)$, with k algebraically closed, such that the Zariski closure $G = \left(\overline{\phi(\Gamma)}^Z\right)$ satisfies $G^{(0)} = H^m$ where H is a simple, center-free, algebraic group and Γ acts transitively by conjugation on these m simple factors. Moreover, for any such representation, if $\langle e\rangle \neq N \lhd \Gamma$, then $G^{(0)} < \left(\overline{\phi(N)}^Z\right)$.*

Proof. Assume that Γ is linear of almost simple type. Realize $\Gamma < \mathrm{GL}_n(k)$ as a linear group, over an algebraically closed field k, and let $G = \overline{\Gamma}^Z$ be the Zariski closure. We may assume, without loss of generality, that G is semisimple. If not, we divide out by the solvable radical of G. Since Γ has no nontrivial abelian normal subgroups, it also has no nontrivial solvable normal subgroups and it will still map injectively into the semisimple quotient. Next we may assume without loss of generality that G is adjoint upon replacing it by its image under the adjoint representation. Since Γ has no finite or abelian normal subgroups, it maps injectively into this new group too. Note that now $G^{(0)} = H_1 \times H_2 \times \cdots \times H_l$ is a direct product of simple, center-free groups. Γ acts on $G^{(0)}$ by conjugation. We claim that this action must permute these l-simple factors. Fix $\gamma \in \Gamma$ and $1 \leq i, j \leq l$; by simplicity of these factors there are only two options: either $\gamma H_i \gamma^{-1} = H_j$ or $\gamma H_i \gamma^{-1} \bigcap H_j = \langle e\rangle$. When the second of the two options holds, we have $[\gamma H_i \gamma^{-1}, H_j] = e$. Since $\gamma H_i \gamma^{-1}$ cannot commute with the whole group, there must be a (necessarily unique) j such that $\gamma H_i \gamma^{-1} = H_j$. Let us denote the permutation representation thus obtained by $\gamma H_i \gamma^{-1} = H_{\pi(\gamma,i)}$. After rearranging we can rewrite the decomposition of the connected component into simple factors as follows: $G^{(0)} = G_1 \times \cdots \times G_k$, where each G_i is the direct power of simple factors $G_i = H_i^{m_i}$ that are permuted transitively by Γ. This gives rise to an injective map $\phi : \Gamma \to \mathrm{Aut}(G_1) \times \cdots \times \mathrm{Aut}(G_k)$, and since we assumed that Γ is subdirect irreducible we can find a factor i_0 such that the map $\phi_{i_0} : \Gamma \to \mathrm{Aut}(G_{i_0})$ is already injective. This is our desired quotient.

Conversely, assume that Γ admits such a linear representation. If $\langle e\rangle \neq N \lhd \Gamma$, then $W = \left(\overline{\phi(N)}^{Z}\right) \lhd G$. If N were either finite or abelian, W would be the same, contradicting the fact that G does not admit finite or abelian normal subgroups.

By assumption N is infinite so that W has positive dimension and $W \bigcap G^{(0)} \neq \langle e\rangle$. This means that $W \bigcap H \neq \langle e\rangle$ for at least one of the simple factors $G^{(0)} = H^m$. By simplicity W actually contains this simple factor, and by transitivity of the Γ action on the factors $W > G^{(0)}$. This establishes the last statement.

Finally, assume that $N_1 \bigcap N_2 = \langle e\rangle$ for two normal subgroups $\langle e\rangle \neq N_1, N_2 \lhd \Gamma$. Then $[N_1, N_2] = \langle e\rangle$. Passing to the Zariski closure $W_i = \overline{N_i}^{Z}$, we have $[W_1, W_2] = \langle e\rangle$. But by the previous paragraph this implies that $G^{(0)}$ is abelian, which is absurd. This concludes the proof of the lemma. □

DEFINITION 2.6. Let Γ be a group of almost simple type. A subgroup $\Delta < \Gamma$ is called *prodense* if it is dense in the normal topology.

DEFINITION 2.7. The core of a subgroup $\Delta < \Gamma$ is the normal subgroup $\mathrm{Core}_\Gamma(\Delta) = \bigcap_{\gamma \in \Gamma} \gamma \Delta \gamma^{-1}$. The subgroup Δ is called *core-free* if $\mathrm{Core}_\Gamma(\Delta) = \langle e\rangle$.

Note that $\mathrm{Core}_\Gamma(\Delta)$ is exactly the kernel of the permutation representation $\Gamma \curvearrowright \Gamma/\Delta$. Consequently this action is faithful if and only if Δ is core-free. We leave the verification of the following easy lemma to the reader.

LEMMA 2.8. *The following conditions are equivalent for a subgroup Δ in a group Γ of almost simple type:*

- *Δ is prodense.*
- *$\Delta N = \Gamma$ for every $\langle e\rangle \neq N \lhd \Gamma$.*
- *$\phi(\Delta) = \phi(\Gamma)$ whenever $\phi : \Gamma \to H$ is a homomorphism with nontrivial kernel.*

In particular prodense subgroups are always core-free. Clearly every prodense subgroup is profinitely dense.

Such subgroups play a central role in the strategy for construction of maximal subgroups. The idea is that if $\Delta < \Gamma$ is a prodense (respectively profinitely dense subgroup), then any larger subgroup still has the same property. In particular if $\Delta < M < \Gamma$ is a proper maximal subgroup containing Δ, then M too must be prodense (respectively profinitely dense). This in turn guarantees that M is core-free (respectively of infinite index). In order to make sure that Δ is contained in some proper maximal subgroup, we use the following condition.

DEFINITION 2.9. A subgroup $\Delta < \Gamma$ is called *cofinitely generated* if $\Gamma = \langle \Delta, F \rangle$ for some finite $F \subset \Gamma$.

If $\Delta < \Gamma$ is cofinitely generated, let F be a finite subset as in the above definition. Zorn's lemma yields a subgroup $\Delta < M < \Gamma$, which is maximal with respect to the condition that $F \not\subset M$. Any strictly larger subgroup will contain $\Gamma = \langle M, F \rangle$, so M must be a maximal subgroup. To summarize the discussion:

PROPOSITION 2.10. *Let Γ be any group.*

(1) *Γ contains a maximal subgroup of infinite index if and only if it contains a profinitely dense cofinitely generated subgroup.*
(2) *Γ contains a core-free maximal subgroup if and only if it contains a prodense cofinitely generated subgroup.*

3 Linear groups

For countable linear groups, there is almost a complete answer to Question 1.4. The missing part of the puzzle is the case of amenable countable linear groups of almost simple type. We give a more or less complete description of what is known, starting with some preliminaries.

3.1 PROJECTIVE TRANSFORMATIONS OVER VALUATION FIELDS.

In this section we shall review some definitions and results from [BG03] and [BG07] regarding the dynamical properties of projective transformations that we shall use in the proof.

Let k be a local field and $\|\cdot\|$ the standard norm on k^n—that is, the standard Euclidean norm if k is Archimedean and $\|x\| = \max_{1 \le i \le n} |x_i|$, where $x = \sum x_i e_i$ when k is non-Archimedean and $(e_1, \ldots, e_n)$ is the canonical basis of k^n. This norm extends in the usual way to $\Lambda^2 k^n$. We define the *standard metric* on $\mathbb{P}(k^n)$ by $d(\overline{v}, \overline{v}) = \frac{\|v \wedge w\|}{\|v\| \|w\|}$, where $\overline{v}$ denotes the projective point corresponding to $v \in k^n$. $\|v \wedge w\|$ can be thought of as the area of the parallelogram defined by the vectors v, w—that is, $\|v\|$ times the distance of w to the line spanned by v. Unless otherwise specified all our notation will refer to this metric (see also [Tit72]). In particular $(A)_\epsilon := \{x \in \mathbb{P}(k^n) \mid d(x, A) < \epsilon\}$

will denote the open ϵ-neighborhood and $[A]_\epsilon := \{x \in \mathbb{P}(k^n) \mid d(x, A) \leq \epsilon\}$ the closed neighborhood of a set $A \subset \mathbb{P}(k^n)$. With respect to this metric, every projective transformation is bi-Lipschitz on $\mathbb{P}(k^n)$.

DEFINITION 3.1. For $\epsilon \in (0, 1)$, we call a projective transformation $g \in \mathrm{PGL}_n(k)$ *ϵ-contracting* if $g\left(\mathbb{P}(k^n) \setminus (\overline{H})_\epsilon\right) \subset (\overline{v})_\epsilon$ for some point $\overline{v} \in \mathbb{P}(k^n)$ and projective hyperplane $\overline{H} < \mathbb{P}(K^n)$, which are referred to as an attracting point and a repelling hyperplane for g. We say that g is *ϵ-very contracting* if g and g^{-1} are ϵ-contracting. A projective transformation $g \in \mathrm{PGL}_n(k)$ is called (r, ϵ)-*proximal* $(r > 2\epsilon, > 0)$ if it is ϵ-contracting with respect to some attracting point $\overline{v} \in \mathbb{P}(k^n)$ and repelling hyperplane $\overline{H}$, such that $d(\overline{v}, \overline{H}) \geq r$. A projective transformation g is called (r, ϵ)-*very proximal* if both g and g^{-1} are (r, ϵ)-proximal. Finally, g is simply called *proximal* (respectively *very proximal*) if it is (r, ϵ)-proximal (respectively (r, ϵ)-very proximal) for some $r > 2\epsilon, > 0$.

The attracting point $\overline{v}$ and repelling hyperplane $\overline{H}$ of an ϵ-contracting transformation are not uniquely defined. Yet, if g is proximal with good enough parameters, then we have the following natural choice for an attracting point and a repelling hyperplane:

LEMMA 3.2. ([BG07, lemma 3.1]). *Let k be a local field and $\epsilon \in (0, \frac{1}{4})$. There exist two constants $c_1, c_2 \geq 1$ such that if $g \in \mathrm{PGL}_n(k)$ is an (r, ϵ)-proximal transformation with $r \geq c_1\epsilon$ and associated attracting point $\overline{v}$ and repelling hyperplane $\overline{H}$, then g fixes a unique point $\overline{v}_g \in (\overline{v})_\epsilon$ and a unique projective hyperplane $\overline{H}_g \subset \left(\overline{H}\right)_\epsilon$. Moreover, if $r \geq c_1\epsilon^{2/3}$, then the positive powers g^n, $n \geq 1$, are $(r - 2\epsilon, (c_2\epsilon)^{\frac{n}{3}})$-proximal transformations with respect to these same $\overline{v}_g$ and $\overline{H}_g$. The constants c_1, c_2 may depend on the local field k, but they become only better when passing to a finite extension field.*

REMARK 3.3. In what follows, whenever we add the article *the* (or *the canonical*) to an attracting point and repelling hyperplane of a proximal transformation g, we shall mean the fixed point $\overline{v}_g$ and fixed hyperplane $\overline{H}_g$ obtained in Lemma 3.2.

Moreover, when r and ϵ are given, we shall denote by $A(g)$, $R(g)$ the ϵ-neighborhoods of $\overline{v}_g$, $\overline{H}_g$ respectively. In some cases, we shall specify different attracting and repelling sets for a proximal element g. In such a case we shall denote them by $\mathcal{A}(g)$, $\mathcal{R}(g)$ respectively. This means that

$$g\big(\mathbb{P}(k^n) \setminus \mathcal{R}(g)\big) \subset \mathcal{A}(g).$$

If g is very proximal and we say that $\mathcal{A}(g), \mathcal{R}(g), \mathcal{A}(g^{-1}), \mathcal{R}(g^{-1})$ are specified attracting and repelling sets for g, g^{-1}, then we shall always require additionally that

$$\mathcal{A}(g) \bigcap (\mathcal{R}(g) \bigcup \mathcal{A}(g^{-1})) = \mathcal{A}(g^{-1}) \bigcap (\mathcal{R}(g^{-1}) \bigcup \mathcal{A}(g)) = \emptyset.$$

DEFINITION 3.4. If g, h are very proximal elements with given associated repelling and attracting neighborhoods, we will say that g *is dominated by h* if $\mathcal{R}(g^\eta) \subset \mathcal{R}(h^\eta), \mathcal{A}(g^\eta) \subset \mathcal{A}(h^\eta)$ for $\eta \in \pm 1$.

These notions depend not only on the elements themselves but also on the specific choice of attracting and repelling neighborhoods, but when these neighborhoods are clear, we will suppress them from the notation.

Definition 3.1 is stated in terms of the topological dynamics of the action of a single projective transformation $g \in \mathrm{PGL}_n(k)$ on the projective space $\mathbb{P}(k^n)$. A fundamental idea to the whole theory, which was fully developed in [BG03] and [BG07], is that contraction can also be alternatively expressed in metric or in algebraic terms: metrically, in terms of the Lipschitz constant of g on an open set away from the repelling hyperplane; algebraically, in terms of the singular values of the corresponding matrix. The equivalence between the different notions is quantitative.

LEMMA 3.5. *Let F be a local field, c_1, c_2 the constants given in Lemma 3.2, and $g \in \mathrm{GL}_n(F)$. Denote by $[g]$ the image of g in $\mathrm{PGL}_n(F)$ and by $g = kak'$ its Cartan decomposition, with $a = diag(a_1, a_2, \ldots, a_n), a_1 \geq a_2 \geq \ldots a_n > 0$. Set $H = \overline{\mathrm{Span}\{k'^{-1}(e_i)\}_{i=2}^n}, p = \overline{ke_1}$. Then there exists a constant $c > 0$ depending only on the field such that for any $0 < \epsilon < \frac{1}{4}$ we have the following:*

- *If $\frac{a_1}{a_2} \geq \frac{1}{\epsilon^2}$, then $[g]$ is ϵ-contracting with (H, a) as a repelling hyperplane and an attracting point. Moreover $[g]$ is $\frac{\epsilon^2}{r^2}$ Lipschitz outside the r-neighborhood of H.*
- *Assume that the restriction of $[g]$ to some open set $O \subset \mathbb{P}(F^n)$ is ϵ-Lipschitz; then $\frac{a_1}{a_2} \geq \frac{1}{2\epsilon}$ and $[g]$ is $c\sqrt{\epsilon}$-contracting.*
- *If $[g]$ is (r, ϵ)-contracting for $r > c_1\epsilon$, then it is $c\frac{\epsilon^2}{d^2}$-Lipschitz outside the d-neighborhood of the repelling hyperplane.*

Using proximal elements, one constructs free groups with the following variant of the classical ping pong lemma.

LEMMA 3.6. *Suppose that* $\{g_i\}_{i\in I} \subset \mathrm{PGL}_n(k)$ *is a set of very proximal elements, each associated with some given attracting and repelling sets for itself and for its inverse. Suppose that for any* $i \neq j$, $i,j \in I$, *the attracting set of* g_i *(respectively of* g_i^{-1}*) is disjoint from both the attracting and repelling sets of both* g_j *and* g_j^{-1}. *Then the* g_i*'s form a free set—that is, they are free generators of a free group.*

This lemma calls for several important definitions that will play a central role in this essay. Our general setting is somewhat more general than usual, because we work with countable linear groups that are not necessarily finitely generated.

Let (K,ν) be a complete valued field and $\Gamma < \mathrm{PGL}_n(K)$ a countable group. For every $\Delta < \Gamma$ let us denote by $(k(\Delta),\nu(\Delta))$ the closed subfield that is generated by the matrix coefficients of elements in Δ. Even though K itself will typically not be a local field, we do assume that $(k(\Delta),\nu(\Delta))$ is a local field whenever Δ is finitely generated.

DEFINITION 3.7. A finite list of elements $\{g_i \in \Gamma\}_{i\in I}$ will be called a *ping pong* or a *Schottky* tuple, and the group that they generate $\Delta = \langle g_i \mid i \in I\rangle$ will be called a *Schottky subgroup* if they satisfy the conditions of Lemma 3.6 on $\mathbb{P}(k^n)$ for every intermediate local field $k(\Delta) < k < K$. Such a tuple, as well as the group it generates, will be referred to as *spacious Schottky* if there exists an additional element $g \in \Gamma$ such that $\{g_i \mid i \in I\} \bigcup \{g\}$ is still a Schottky tuple. Finally we will call $\Delta < \Gamma$ *locally Schottky* or *locally spacious Schottky* if every finitely generated subgroup of Δ is such. We denote by $X = X(\Gamma)$ the collection of locally spacious Schottky subgroups of Γ. When Γ is an abstract group, with an action on a projective space given by a representation $\rho : \Gamma \to \mathrm{PGL}_n(k)$, we will denote by $X^\rho(\Gamma)$ the collection of subgroups whose image under ρ is locally spacious Schottky.

One example of a locally spacious Schottky group is just an *infinite Schottky group*. By definition this is a subgroup with a given infinite generating set $\Delta = \langle g_i \mid i \in I\rangle$ such that for every finite $J \subset I$ the tuple $\{g_i \mid i \in J\}$ is a ping pong tuple, and hence $\Delta_J = \langle g_j \mid j \in J\rangle$ plays ping pong on $\mathbb{P}(k_J)$ with $k_J = k(\Delta_J)$. The local fields $\{k_J\}$ and corresponding projective spaces $\{\mathbb{P}(k_J)\}$ form a direct system inside K and $\mathbb{P}(K)$ respectively. Though note that in our current setting there is no canonical ping pong playground where the generators play all together. The following lemma is easy; we leave its proof to the reader.

LEMMA 3.8. *Let* $\Delta = \langle \eta_1, \eta_2, \ldots, \eta_m, \zeta \rangle \in X(\Gamma)$ *be a spacious Schottky subgroup. Let* $\{e \neq w_j \mid j \in \mathbf{N}\}$ *be any ordered set of nontrivial elements, possibly containing repetitions, in the group* $\langle \eta_1, \eta_2, \ldots, \eta_m \rangle$. *Then* $\{\zeta^{-i} w_i \zeta^{-i} \mid i \in \mathbf{N}\}$ *is an infinite Schottky tuple.*

DEFINITION 3.9. Let K be any field. A pair $\overline{\alpha} = \left(\overline{H}^+, \overline{v}^-\right)$, with $\overline{H}^+ < \mathbb{P}(K^n)$ a hyperplane and $\overline{v}^- \in \overline{H}^+$ a point, will be called a *minimal flag* or an *M-flag* for short. We denote by $\mathbb{M}(K^n)$ the (projective) variety of all such M-flags. We will say that two M-flags $\overline{\alpha}_1, \overline{\alpha}_2$ *touch each other* if either $\overline{v}_1^- \in \overline{H}_2^+$ or $\overline{v}_2^- \in \overline{H}_1^+$. A collection of M-flags $\left\{\alpha_i = \left(\overline{H}_i^+, \overline{v}_i^-\right) \mid i \in I\right\}$ is said to be *in general position* if the following conditions are satisfied:

- $\bigcap_{i \in J} \overline{H}_i^+ = \emptyset$ for every $J \subset I$ with $|J| = n$.
- $\mathrm{Span}\{\overline{v}_j^- \mid j \in J\} := \overline{\mathrm{Span}\left\{v_j^- \mid j \in J\right\}} = \mathbb{P}(K^n)$ for every $J \subset I$ with $|J| = n$.

Note that the plus and minus superscripts indices in this definition do not really play any role; they are only there in anticipation of the following:

DEFINITION 3.10. If k is a local field and $g \in \mathrm{PGL}_n(k)$ is a very proximal element, then we can associate with it an M-flag $\alpha(g) = \left(\overline{H}_g, \overline{v}_{g^{-1}}\right)$. We will say that a ping pong tuple $(g_1, g_2, \ldots, g_m)$ is *in general position* if both $\{\overline{\alpha}(g_i) \mid 1 \leq i \leq m\}$ and $\{\overline{\alpha}(g_i^{-1}) \mid 1 \leq i \leq m\}$ are.

LEMMA 3.11. *If* $\{\alpha_i \in \mathbb{M}(K^n) \mid 1 \leq i \leq m\}$ *is in general position with* $m \geq 2n-1$, *then no* $\alpha \in \mathbb{M}(K^n)$ *can touch simultaneously all of the M-flags* $\{\alpha_i,\ 1 \leq i \leq m\}$.

Proof. Let $I = \{1, 2, \ldots, 2n-1\}$ and assume that $\alpha_i = \left(\overline{H}_i^+, \overline{v}_i^-\right)$ and $\alpha = \left(\overline{H}^+, \overline{v}^-\right)$. Set $J = \{i \in I \mid \overline{v}^- \notin \overline{H}_i^+\}$. Since $\overline{v}^- \in \bigcap_{i \in I \setminus J} \overline{H}_i^+ \neq \emptyset$, the first condition in the general position assumption implies that $|J| \geq n$. Now the second condition implies that $\mathrm{Span}\left\{\overline{v}_j^- \mid j \in J\right\} = \mathbb{P}(K^n)$ so that $\overline{v}_{i_0}^- \notin \overline{H}^+$ for some $i_0 \in J$. Thus the two M-flags α, α_{i_0} fail to touch and the lemma is proved. □

DEFINITION 3.12. A linear representation $\rho : \Gamma \to \mathrm{GL}_n(k)$ is called *strongly irreducible* if one of the following equivalent conditions holds:

- $\rho(\Gamma)$ does not preserve a finite collection of nontrivial proper subspaces.
- $\rho(\Delta)$ is irreducible for every finite index subgroup $\Delta < \Gamma$.
- $\left(\overline{\rho(\Gamma)}^Z\right)^{(0)}$ is irreducible.

We leave the verification of the equivalence to the reader. Strongly irreducible representations of groups of almost simple type offer a lot of flexibility, a fact that we attempt to capture in the next two lemmas.

LEMMA 3.13. *Let* $\Gamma < \mathrm{GL}_n(K)$ *be a group of almost simple type,* $G = \overline{\Gamma}^Z$ *its Zariski closure with* $G^{(0)} = H^m$ *as in Lemma 2.5. Let* $\rho: G \to \mathrm{GL}_m(K)$ *be a strongly irreducible representation defined over K. Then* $\rho(N)$ *is irreducible for every* $\langle e \rangle \neq N \lhd \Gamma$*. In particular, given a vector* $0 \neq v \in K^m$ *and a hyperplane* $V < K^m$ *there is some* $n \in N$ *such that* $\rho(n)(v) \notin V$.

Proof. By the equivalence of the conditions in the previous definition, $\rho(G^{(0)})$ is irreducible. By Lemma 2.5, $\overline{N}^Z > G^{(0)}$ so that $\rho(N)$ is irreducible as well. Since $\rho(N)$ is irreducible we have

$$\mathrm{Span}\{\rho(n)v \mid n \in N\} = K^m,$$

so this set cannot be contained in the proper subspace V. □

LEMMA 3.14. *Let k be a local field,* $m \in \mathbf{N}$*, and* $\Gamma < \mathrm{GL}_n(k)$ *a strongly irreducible group of almost simple type. Assume that* Γ *contains a very proximal element g. Then* Γ *contains a Schottky tuple* $(\eta_1, \eta_2, \ldots, \eta_m)$ *in general position. More generally if* $(\gamma_1, \gamma_2, \ldots, \gamma_l)$ *is a spacious Schottky tuple, then one can find a spacious Schottky tuple* $(\eta_1, \eta_2, \ldots, \eta_m)$ *in general position, such that* $(\gamma_1, \gamma_2, \ldots, \gamma_l, \eta_1, \eta_2, \ldots, \eta_m)$ *is still spacious Schottky.*

Proof. We start with the first statement—namely, with the special case $l = 0$. Assume by induction that we already have a spacious Schottky tuple $\{\beta_i \mid i \in I\}$ whose corresponding attracting points and repelling hyperplanes satisfy all of the conditions implied so far:

- $\dim\left(\bigcap_{j \in J} \overline{H}_j^+\right) = n - |J| - 1, \ \forall J \subset I$
- $\dim(\mathrm{Span}\{\overline{v}_i^- \mid j \in J\}) = |J| - 1, \ \forall J \subset I, |J| \leq n$
- $\overline{v}_i^\eta \notin \overline{H}_j^\epsilon$ whenever either $i \neq j$ or $\epsilon = \eta$

Here *dim* denotes projective dimension, and we have adopted the convention that a negative dimension corresponds to an empty set. As the basis of the induction we can take β_1 to be any conjugate of g.

Now let us fix a finite set of points $A \subset \mathbb{P}(k^n)$ such that

$$\{\overline{v}_i^\eta \in A, \mid i \in I, \eta \in \pm 1\} \subset A$$

and also $A \cap \left(\bigcap_{j \in J} \overline{H}_j^{\eta} \right) \neq \emptyset$ for every subset $J \subset I$ with $|J| < n$ and every $\eta \in \{\pm 1\}$. Since Γ is strongly irreducible we can find an element $\gamma \in \Gamma$ subject to the following two conditions:

- $A \cap \gamma \overline{H}_g^{\epsilon} = \emptyset$, $\forall \epsilon \in \pm 1$.
- $\gamma \overline{v}_g^{-} \notin \bigcup_{\substack{J \subset I \\ |J| < n}} \mathrm{Span}\{\overline{v}_j^{-} \mid j \in J\}$.

Setting $\beta_{p+1} = \gamma g \gamma^{-1}$ we have $\overline{H}_{\beta_{p+1}}^{\epsilon} = \gamma \overline{H}_g^{\epsilon}$ and $\overline{v}_{\beta_{p+1}}^{\epsilon} = \gamma \overline{v}_g^{\epsilon}$, $\forall \epsilon \in \pm 1$. After m such steps we obtain a collection of very proximal elements whose attracting points and repelling hyperplanes are subject to all of the desirable conditions mentioned above. We obtain the desired ping pong tuple by setting $\eta_i = \beta_i^N$ for a high enough value of N.

Now assume $l \neq 0$ and let g be the proximal element such that $(\gamma_1, \gamma_2, \ldots, \gamma_l, g)$ is Schottky. Let us construct $(\eta_1', \eta_2', \ldots, \eta_n')$ as a spacious Schottky tuple in general position, just as we did in the previous paragraph. By taking care we may assume that these were constructed in such a way that $(g^n, \eta_1', \eta_2', \ldots, \eta_n')$ is still Schottky. Now setting $\eta_i = g \eta_i' g^{-1}$ we obtain a Schottky tuple in general position $(\eta_1, \eta_2, \ldots, \eta_n)$ subject to the additional condition that $\mathcal{A}(\eta_i^{\pm}) \subset \mathcal{A}(g), \mathcal{R}(\eta_i^{\pm}) \subset \mathcal{R}(g)$. This implies that $(\gamma_1, \gamma_2, \ldots, \gamma_l, \eta_1, \eta_2, \ldots, \eta_l)$ is as required. $\qquad \square$

The main ingredient in the method we use for generating free subgroups is a projective representation whose image contains contracting elements and acts strongly irreducibly. The following theorem is a particular case of Theorem 4.3 from [BG07]. Note that a similar statement also appeared earlier in [MS81].

THEOREM 3.15.

Let F be a field and $\mathbb{H}$ an algebraic F-group for which the connected component $\mathbb{H}^{\circ}$ is not solvable, and let $\Psi < \Gamma < \mathbb{H}$ be Zariski dense subgroups with Ψ finitely generated and Γ countable. Assume that Ψ contains at least one element of infinite order. Then we can find a valued field (K, v), an embedding $F \hookrightarrow K$, an integer n, and a strongly irreducible projective representation $\rho : \mathbb{H}(K) \to \mathrm{PGL}_n(K)$ defined over K with the following properties.

(1) $(k(\Delta), v(\Delta))$ is a local field for every finitely generated subgroup $\Delta < \Gamma$.

(2) There exists an element $\psi \in \Psi$, such that $\rho(\psi)$ is a very proximal element on $\mathbb{P}(k^n)$ for some parameters (ϵ, r) satisfying the stronger condition appearing in Lemma 3.2 for every intermediate local field $k(\Psi) < k < K$.

Where as above, $(k(\Delta), \nu(\Delta))$ denotes the closed subfield of K generated by all matrix coefficients of $\rho(\Delta)$.

3.2 PRIMITIVITY FOR LINEAR GROUPS OF ALMOST SIMPLE TYPE.

THEOREM 3.16.

Let Γ be a countable linear group of almost simple type, containing at least one element of infinite order. Then Γ is primitive.

Let $\mathbb{H} = \overline{\Gamma}^Z$ be the Zariski closure. By dimension considerations we can find a finitely generated subgroup $\Psi < \Gamma$ with the same Zariski closure. Obviously Ψ contains an element of infinite order. We now apply Theorem 3.15 and fix once and for all the data that this theorem yields. In fact we will just identify Γ with its image under this representation $\rho(\Gamma)$ in order to avoid cumbersome notation. Once we have fixed this representation we will denote by $X(\Gamma) = X^\rho(\Gamma)$ the collection of all subgroups whose image under this fixed representation is locally spacious Schottky. With all this notation in place Theorem 3.16 will follow from the following, slightly more general statement.

THEOREM 3.17.

Any finitely generated spacious Schottky subgroup $D < \Gamma$ is contained in a maximal core-free subgroup $D < M < \Gamma$.

Proof. By Proposition 2.10 it suffices to construct a subgroup $D < \Delta < \Gamma$ that is prodense and cofinitely generated. As a feature of the proof, the group Δ we construct will be spacious Schottky, and in particular $\Delta \in X(\Gamma)$.

Let $(\delta_1, \delta_2, \ldots, \delta_m)$ be a Schottky generating set for the subgroup D. Let $D < D' < \Gamma$ be a larger (not necessarily Schottky) finitely generated subgroup that has the same Zariski closure as Γ. Letting $k = k(D')$ be the local field generated by the matrix coefficients of D', this implies that $\rho(\Gamma) \bigcap \mathrm{GL}_n(k)$ is strongly irreducible. Appealing to Lemma 3.14 we can construct a spacious Schottky tuple $(\delta_1, \delta_2, \ldots, \delta_m, \sigma_1, \ldots, \sigma_{2n-1}, \zeta)$ with $\Sigma = \langle \sigma_1, \ldots, \sigma_2, \ldots, \sigma_{2n-1}, \zeta \rangle$, a Schottky subgroup whose generators are in general position, as in Definition 3.9. The last element is denoted differently just because it will play a separate role in the proof. We set $\Delta_0 := D, k_0 = k(\langle \Delta_0, \Sigma \rangle)$ and in $\mathbb{P}(k_0^n)$ we denote by $\overline{H}_i^\pm = \overline{H}_{\sigma_i}^\pm, \overline{v}_i^\pm = \overline{v}_{\sigma_i}^\pm, \overline{H}^\pm = \overline{H}_\zeta^\pm, \overline{v}^\pm = \overline{v}_\zeta^\pm$ the attracting points and repelling hyperplanes of the corresponding ping pong game, all defined over k_0. We did not name the attracting points and repelling neighborhoods of the δ_i's, as we will not refer to them explicitly.

Let us use odd indices $\{\gamma_1\langle\langle n_1\rangle\rangle, \gamma_3\langle\langle n_3\rangle\rangle, \gamma_5\langle\langle n_5\rangle\rangle, \ldots\}$ to list all cosets of all these normal subgroups that are generated by one, nontrivial, conjugacy class. Since any nontrivial normal subgroup contains one of these, the collection of these cosets forms a basis for the normal topology on Γ. So a subgroup intersecting all of them nontrivially will be prodense. Similarly we use even indices $(\Sigma\theta_2\Sigma, \Sigma\theta_4\Sigma, \ldots)$ to enumerate the nontrivial double cosets of Σ in Γ. A subgroup $\Delta < \Gamma$ that has a nontrivial intersection with all of these double cosets will be cofinitely generated by virtue of the fact that $\Gamma = \langle \Delta, \Sigma\rangle$.

We will construct an infinitely generated Schottky group

$$\Delta = \langle \delta_1, \delta_2, \ldots, \delta_m, \eta_1, \eta_2, \ldots\rangle$$

such that $\eta_i \in \gamma_i\langle\langle n_i\rangle\rangle$ for every odd i and $\eta_i \in \Sigma\theta_i\Sigma$ for every even i. This will be done by induction on i, with $\Delta_i := \langle D, \eta_1, \eta_2, \ldots, \eta_i\rangle$ the group constructed at the ith step and $k_i = k(\langle \Delta_i, \Sigma\rangle)$ the corresponding local subfield of K. We obtain a sequence of local fields $k_0 < k_1 < k_2 < \ldots < K$ with corresponding, direct sequence of projective spaces $\mathbb{P}(k_0^n) \subset \mathbb{P}(k_1^n) \subset \mathbb{P}(k_2^n) \subset \ldots \subset \mathbb{P}(K^n)$. The generators of Δ_i will form a ping pong tuple in their action on $\mathbb{P}(k_i^n)$. By Lemma 3.5, the canonical attracting and repelling neighborhoods of the ping pong players η_i will be of the form $\left(\overline{H}_{\eta_i}^{\pm}\right)_{\epsilon_i}, \left(\overline{v}_{\eta_i}^{\pm}\right)_{\epsilon_i}$. These are all defined over k_i, in the sense that $v_i \in k_i^n$, H_i has a basis consisting of $(n-1)$ vectors in k_i^n and ϵ is determined by the singular values of η_i that are inside k_i. Thus the same elements will form a ping pong tuple also on $\mathbb{P}(k_j)$ for any $j > i$. The attracting points and repelling hyperplanes in the extended vector space are obtained by an extension of scalars and ϵ_i remains the same.

Assume we constructed $\Delta_\ell = \langle \delta_1, \ldots, \delta_m, \eta_1, \eta_2, \ldots, \eta_{\ell-1}\rangle < \mathrm{PGL}_n(k_{\ell-1})$, satisfying

- $\eta_i \in \gamma_i\langle\langle n_i\rangle\rangle$ if i is odd, and $\eta_i \in \Sigma\theta_i\Sigma$ if i is even; and
- $\eta_i = \zeta^i q_i \zeta^{-i}$ with q_i very proximal and dominated by some nontrivial element in $\langle \sigma_1, \sigma_2, \ldots, \sigma_{2n-1}\rangle < \Sigma$ for every $1 \le i < \ell$.

The second condition guarantees that $\Delta_{\ell-1}$ is Schottky by Lemma 3.8.

When ℓ is odd, we are looking for an element $\eta_\ell \in \gamma_\ell\langle\langle n_\ell\rangle\rangle$. We extend scalars to K, but by abuse of notation we will identify $L_i^{\pm} < k_i^n$ with their scalar extensions $L_i^{\pm} \otimes_{k_{\ell-1}} K < K^n$ and similarly for $w_i^{\pm} \in k_{\ell-1}^n \subset K^n$. By Lemma 3.13 we find $n_1, n_2, n_3 \in \langle\langle n_\ell\rangle\rangle$ such that $n_1^{\epsilon}\overline{v}_1^{+} \notin \overline{H}_1^{-}, n_2\gamma_\ell\overline{v}^{+} \notin \overline{H}^{-}, n_2^{-1}\overline{v}^{+} \notin \gamma\overline{H}^{-}, n_3^{\epsilon}\overline{v}_1^{-} \notin \overline{H}_1^{+}$, $\forall \epsilon \in \{\pm 1\}$. Let $k_\ell = k(\langle \Delta_{\ell-1}, \Sigma, n_1, n_2, n_3\rangle)$ and consider the element

$$\eta_\ell = \zeta^\ell q_\ell \zeta^{-\ell} = \zeta^\ell \sigma_1^N n_3 \sigma_1^{-N} \zeta^{-\ell} n_2 \gamma_\ell \zeta^\ell \sigma_1^{-N} n_1 \sigma_1^N \zeta^{-\ell} \in \gamma_\ell \langle\langle n_\ell \rangle\rangle,$$

where the auxiliary element q_ℓ is also defined by the above equation. Following the dynamics of the action of this element on $\mathbb{P}(k_\ell^n)$ we see that, for N large enough, q_ℓ is very proximal and dominated by σ_1. Thus by Lemma 3.8 the tuple $(\delta_1, \delta_2, \ldots, \delta_m, \eta_1, \eta_2, \ldots, \eta_\ell) = (\delta_1, \delta_2, \ldots, \delta_m, \zeta q_1 \zeta^{-1}, \ldots, \zeta^\ell q_\ell \zeta^{-\ell})$ is Schottky. Group theoretically it is easy to verify that η_ℓ thus defined belongs to the desired coset $\gamma_\ell \langle\langle n_\ell \rangle\rangle$.

Next, assume that ℓ is even. Now our goal is to construct a ping pong player of the form $\eta_\ell = \zeta^\ell q_\ell \zeta^{-\ell} \in \Sigma \theta_\ell \Sigma$. By Lemma 3.11 we can find some $i = i_\ell$ such that the M-flag $\alpha(\sigma_i)$ does not touch the M-flag $\theta_\ell \alpha(\sigma_1^{-1})$. Explicitly this means that $\overline{v}_i^- \notin \theta_\ell \overline{H}_1^-$ and $\theta_\ell \overline{v}_1^+ \notin \overline{H}_i^+$. These are exactly the conditions needed in order to ensure that, for a high enough value of $n = n_l$, the element $q_\ell := \sigma_i^n \theta_l \sigma_1^n$ is dominated by the element $\sigma_i^n \sigma_1^n \in \Sigma$. Now set $\eta_\ell := \zeta^l \sigma_i^n \theta_\ell \sigma_1^n \zeta^{-\ell} \in \Sigma \theta_\ell \Sigma$. This concludes the even step of the induction and completes the proof. □

4 Counting maximal subgroups of $\mathrm{SL}_n(\mathbb{Z})$

When restricting the attention to $\mathrm{SL}_n(\mathbb{Z})$, $n \geq 3$, one can make use of the arithmetic structure and the abundance of unipotent elements to produce $2^{\aleph_0}$ different maximal subgroups. We follow the argument of [GM16].

4.1 PROJECTIVE SPACE. Let $n \geq 3$. By $\mathbb{P}(k^n)$ we denote the $(n-1)$–dimensional real projective space endowed with the metric defined in Section 3.1. For every $0 \leq k \leq n-1$, the set $\mathbb{L}_k$ of k-dimensional subspaces of $\mathbb{P}(k^n)$ can be endowed with the metric defined by

$$\mathrm{dist}_{\mathbb{L}_k}(L_1, L_2) := \max\{d(k^n)(x, L_i) \mid x \in L_{3-i} \forall\, 1 \leq i \leq 2\}$$

for every $L_1, L_2 \in \mathbb{L}_k$. Note that $\mathbb{L}_k$ is naturally homeomorphic to the Grassmannian $\mathrm{Gr}(k+1, \mathbf{R}^n)$.

4.2 UNIPOTENT ELEMENTS.

DEFINITION 4.1 (rank-1 unipotent elements). We say that a unipotent element u has rank 1 if $\mathrm{rank}(u - I_n) = 1$. The point $p_u \in \mathbb{P}(k^n)$ that is induced by the Euclidean line $\{ux - x \mid x \in \mathbf{R}^n\}$ is called the point of attraction of u. The

$(n-2)$–dimensional subspace $L_u \subseteq \mathbb{P}(k^n)$ induced by the Euclidean $(n-1)$–dimensional space $\{x \in \mathbf{R}^n \mid ux = x\}$ is called the fixed hyperplane of u. The set of rank-1 unipotent elements in $SL_n(\mathbf{Z})$ is denoted by $\mathcal{U}$.

The following two lemmas follow directly from the definition of $\mathcal{U}$ and are stated for future reference.

LEMMA 4.2 (structure of unipotent elements). *The set $\mathcal{U}$ can be divided into equivalence classes in the following way: $u, v \in \mathcal{U}$ are equivalent if there exist nonzero integers r and s such that $u^s = v^r$. The map $u \mapsto (p_u, L_u)$ is a bijection between equivalence classes in $\mathcal{U}$ and the set of pairs (p, L) where $p \in \mathbb{P}(k^n)$ is a rational point and $L \subseteq \mathbb{L}_{n-2}$ is an $(n-2)$–dimensional rational subspace that contains p.*

LEMMA 4.3 (dynamics of unipotent elements). *Let $u \in \mathcal{U}$. For every $\varepsilon > 0$ and every $\delta > 0$ there exists a constant c such that if $m \geq c$ and $v = u^m$, then $v^k(x) \in (p_u)_\varepsilon$ for every $x \in \mathbb{P}(k^n) \setminus (L_u)_\delta$ and every $k \neq 0$. Note that the previous lemma implies that $p_u = p_v$ and $L_u = L_v$.*

4.3 SCHOTTKY SYSTEMS.

DEFINITION 4.4. Assume that $\mathcal{S}$ is a nonempty subset of $\mathcal{U}$ and $\mathcal{A} \subseteq \mathcal{R}$ are closed subsets of $\mathbb{P}(k^n)$. We say that $\mathcal{S}$ is a Schottky set with respect to the attracting set $\mathcal{A}$ and the repelling set $\mathcal{R}$ and call the triple $(\mathcal{S}, \mathcal{A}, \mathcal{R})$ a Schottky system if for every $u \in \mathcal{S}$ there exist two positive numbers $\delta_u \geq \epsilon_u$ such that the following properties hold:

(1) $u^k(x) \in (p_u)_{\varepsilon_u}$ for every $x \in \mathbb{P}(k^n) \setminus (L_u)_{\delta_u}$ and every $k \neq 0$;
(2) If $u \neq v \in \mathcal{S}$ then $(p_u)_{\varepsilon_u} \bigcap (L_v)_{\delta_v} = \emptyset$;
(3) $\bigcup_{u \in \mathcal{S}} (p_u)_{\varepsilon_u} \subseteq \mathcal{A}$; and
(4) $\bigcup_{u \in \mathcal{S}} (L_u)_{\delta_u} \subseteq \mathcal{R}$.

DEFINITION 4.5. The Schottky system $(\mathcal{S}, \mathcal{A}, \mathcal{R})$ is said to be profinitely dense if $\mathcal{S}$ generates a profinitely dense subgroup of $\mathrm{SL}_n(\mathbf{Z})$. We say that the Schottky system $(\mathcal{S}_+, \mathcal{A}_+, \mathcal{R}_+)$ contains the Schottky system $(\mathcal{S}, \mathcal{A}, \mathcal{R})$ if $\mathcal{S}_+ \supseteq \mathcal{S}$, $\mathcal{A}_+ \supseteq \mathcal{A}$, and $\mathcal{R}_+ \supseteq \mathcal{R}$.

LEMMA 4.6. *Let $(\mathcal{S}, \mathcal{A}, \mathcal{R})$ be a Schottky system. Assume that $[p]_\epsilon \bigcap \mathcal{A} = \emptyset$ and $[L]_\delta \bigcap \mathcal{R} = \emptyset$ where $\delta \geq \epsilon > 0$ and p is a rational point that is continued in a rational subspace $L \in \mathbb{L}_{n-2}$. Denote $\mathcal{A}_+ = \mathcal{A} \bigcup [p]_\epsilon$ and $\mathcal{R}_+ = \mathcal{R} \bigcup [L]_\delta$. Then there* *exists* $v \in \mathcal{U}$ *with* $p = p_v$, $L = L_v$ *such that* $(\mathcal{S}_+, \mathcal{A}_+, \mathcal{R}_+)$ *is a Schottky system that contains* $(\mathcal{S}, \mathcal{A}, \mathcal{R})$ *where* $\mathcal{S}_+ := \mathcal{S} \bigcup \{v\}$.

Proof. Lemma 4.2 implies that there exists $u \in \mathcal{U}$ such that $p_u = p$ and $L_u = L$. Lemma 4.3 implies that there exists $m \geq 1$ such that $v := u^m$ satisfies the required properties. □

The following lemma is a version of the well-known ping pong lemma:

LEMMA 4.7 (ping pong). *Let* $(\mathcal{S}, \mathcal{A}, \mathcal{R})$ *be a Schottky system. Then the natural homomorphism* $*_{u \in \mathcal{S}} \langle u \rangle \to \langle \mathcal{S} \rangle$ *is an isomorphism.*

An important ingredient for our methods is the following beautiful result:

THEOREM 4.8 (Venkataramana [Ven87]).
Let Γ *be a Zariski-dense subgroup of* $\mathrm{SL}_n(\mathbf{Z})$. *Assume that* $u \in \mathcal{U} \bigcap \Gamma$, $v \in \Gamma$ *is unipotent and* $\langle u, v \rangle \simeq \mathbf{Z}^2$. *Then* Γ *has finite index in* $\mathrm{SL}_n(\mathbf{Z})$. *In particular, if* Γ *is profinitely dense, then* $\Gamma = \mathrm{SL}_n(\mathbf{Z})$.

Note that if $u, v \in \mathrm{SL}_n(\mathbf{Z}) \bigcap \mathcal{U}$ and $p_u = p_v$, then $(u-1)(v-1) = (v-1)(u-1) = 0$ and in particular $uv = vu$. Thus we get the following lemma:

LEMMA 4.9. *Let* $g \in \mathrm{SL}_n(\mathbf{Z})$ *and* $u_1, u_2 \in \mathcal{U}$. *Assume that* $p_{u_2} = g p_{u_1}$ *and* $L_{u_2} \neq g L_{u_1}$. *Then* $\langle u_1, g^{-1} u_2 g \rangle \simeq \mathbf{Z}^2$.

LEMMA 4.10. *Assume that g is an element of* $\mathrm{SL}_n(\mathbf{Z})$, $(\mathcal{S}, \mathcal{A}, \mathcal{R})$ *is a profinitely dense Schottky system,* $\delta \geq \epsilon > 0$, p_1 *and* p_2 *are rational points, and* L_1 *and* L_2 *are rational* $(n-2)$*–dimensional subspaces such that the following conditions hold:*

- *(1)* $([p_1]_\epsilon \bigcup [p_2]_\epsilon) \bigcap \mathcal{R} = \emptyset$ *and* $([L_1]_\delta \bigcup [L_2]_\delta) \bigcap \mathcal{A} = \emptyset$;
- *(2)* $[p_1]_\epsilon \bigcap [L_2]_\delta = \emptyset$ *and* $[p_2]_\epsilon \bigcap [L_1]_\delta = \emptyset$; *and*
- *(3)* $p_1 = g p_2$ *and* $L_1 \neq g L_2$.

Denote $\mathcal{A}_+ = \mathcal{A} \bigcup [p_1]_\epsilon \bigcup [p_2]_\epsilon$ *and* $\mathcal{R}_+ = \mathcal{R} \bigcup [L_1]_\delta \bigcup [L_2]_\delta$. *Then there exists a set* $\mathcal{S}_+ \supseteq \mathcal{S}$ *such that* $(\mathcal{S}_+, \mathcal{A}_+, \mathcal{R}_+)$ *is a Schottky system that contains* $(\mathcal{S}, \mathcal{A}, \mathcal{R})$ *and* $\langle \mathcal{S}_+, g \rangle = \mathrm{SL}_n(\mathbf{Z})$.

Proof. For every $1 \leq i \leq 2$ choose $u_i \in \mathcal{U}$ such that $p_{u_i} = p_i$ and $L_{u_i} = L_i$. Lemma 4.9 implies that $\langle u_1, g^{-1} u_2 g \rangle \simeq \mathbf{Z}^2$. Lemma 4.3 implies that there exists $m \geq 1$

such that $(\mathcal{S}_+, \mathcal{A}_+, \mathcal{R}_+)$ is a Schottky system where $v_1 := u_1^m$, $v_2 := u_2^m$, and $\mathcal{S}_+ := \mathcal{S} \bigcup \{v_1, v_2\}$. Theorem 4.8 implies that $\langle \mathcal{S}_+, g\rangle = \mathrm{SL}_n(\mathbf{Z})$. □

DEFINITION 4.11. Let $1 \leq k \leq n$. A k-tuple $(p_1, \ldots, p_k)$ of projective points is called generic if $p_1, \ldots, p_k$ span a $(k-1)$–dimensional subspace of $\mathbb{P}(k^n)$. Note that the set of generic k-tuples of $\mathbb{P}(k^n)$ is an open subset of the product of k copies of the projective space; indeed it is even Zariski open.

THEOREM 4.12 (Conze-Guivarc'h, [CG00]).
Assume that $n \geq 3$ and that $\Gamma \leq \mathrm{SL}_n(\mathbf{R})$ is a lattice. Then Γ acts minimally on the set of generic $(n-1)$-tuples.

COROLLARY 4.13. *Assume that $n \geq 3$ and $\Gamma \leq \mathrm{SL}_n(\mathbf{R})$ is a lattice. For every $1 \leq i \leq 2$, let $p_i \in L_i \in \mathbb{L}_{n-2}$. Then for every positive number ε and δ there exists $g \in \Gamma$ such that $gp_1 \in (p_2)_\varepsilon$ and $gL_1 \in (L_2)_\delta$.*

The proof of the following proposition is based on the proof of the main result of [AGS14].

PROPOSITION 4.14. *Assume that $n \geq 3$ and $p \in L \in \mathbb{L}_{n-2}$. Then for every $\delta \geq \epsilon > 0$ there exists a finite subset $\mathcal{S} \subseteq \mathcal{U}$ such that $(\mathcal{S}, \mathcal{A}, \mathcal{R})$ is a profinitely dense Schottky system where $\mathcal{A} := [p]_\epsilon$ and $\mathcal{R} := [L]_\delta$.*

We will make use of the following well-known lemma:

LEMMA 4.15. *Let $n \geq 3$ and $\Gamma \leq \mathrm{SL}_n(\mathbf{Z})$ be a subgroup such that Γ projects on $\mathrm{SL}_n(\mathbf{Z}/4\mathbf{Z})$ and $\mathrm{SL}_n(\mathbf{Z}/p\mathbf{Z})$ for every odd prime p. Then Γ is profinitely dense in $\mathrm{SL}_n(\mathbf{Z})$.*

We sketch a proof that was shown to us by Chen Meiri:

Proof. For $1 \leq i \neq j \leq n$ and $a \in \mathbf{Z}$, let $E_{i,j}(a)$ be the matrix with 1 on the diagonal, a on the (i,j) entry, and zero elsewhere. For $a, b, m \in \mathbf{Z}$ we write $a \equiv_m b$ to indicate that a is equal to b modulo m. For every prime p and every $m \geq 1$, let $\pi_{p^m} : \mathrm{SL}_n(\mathbf{Z}) \to \mathrm{SL}_n(\mathbf{Z}/p^m\mathbf{Z})$ be the reduction map. The following two facts are straightforward:

FACT 4.16. Let p be an odd prime. If $A \equiv_p E_{i,j}(1)$, then for every $m \geq 1$, $A^{p^m} \equiv_{p^{m+1}} E_{i,j}(p^m)$.

FACT 4.17. If $A \equiv_4 E_{i,j}(2)$, then for every $m \geq 1$, $A^{2^m} \equiv_{2^{m+2}} E_{i,j}(2^{m+1})$.

CLAIM 4.18. Let p be an odd prime and $S \subseteq \mathrm{SL}_n(\mathbf{Z})$. If $\pi_p(S) \supseteq \pi_p(\{E_{i,j}(1) \mid 1 \leq i \neq j \leq n\})$, then for every $m \geq 1$, $\pi_{p^m}(\langle S \rangle) = \mathrm{SL}_n(\mathbf{Z}/p^m\mathbf{Z})$.

Proof. The proof is by induction on m. The case $m = 1$ is clear. Assume that the claim holds for some $m \geq 1$. By 4.16,

$$\pi_{p^{m+1}}(\langle S \rangle) \supseteq \{\pi_{p^{m+1}}(E_{i,j}(p^m)) \mid 1 \leq i \leq j \leq n\}.$$

The result follows since $\mathrm{SL}_n(\mathbf{Z}/p^m\mathbf{Z})/\mathrm{SL}_n(\mathbf{Z}/p^{m+1}\mathbf{Z})$ is isomorphic to $\mathfrak{sl}_n(\mathbf{Z}/p\mathbf{Z})$ and $\mathfrak{sl}_n(\mathbf{Z}/p\mathbf{Z})$ is spanned by $\{ge_{i,j}(1)g^{-1} \mid g \in \mathrm{SL}_n(\mathbf{Z}/p\mathbf{Z}),\ 1 \leq i \neq j \leq n\}$. □

Arguing similarly, one obtains the following:

CLAIM 4.19. Let $S \subseteq \mathrm{SL}_n(\mathbf{Z})$. If $\pi_4(S) \supseteq \pi_4(\{E_{i,j}(1) \mid 1 \leq i \neq j \leq n\})$ then for every $m \geq 2$, $\pi_{2^m}(\langle S \rangle) = \mathrm{SL}_n(\mathbf{Z}/2^m\mathbf{Z})$.

CLAIM 4.20. For all distinct primes $q, p_1, \ldots, p_k$ with q odd and every $m \geq 1$, the group Γ contains an element A such that $\pi_q(A) = \pi_q(E_{i,j}(1))$ and $\pi_{p_r^m}(A) = \pi_{p_r^m}(I_n)$ for every $1 \leq r \leq k$.

Proof. For every prime p, $\mathrm{PSL}_n(\mathbf{Z}/p\mathbf{Z})$ is simple. The Jordan-Holder theorem implies that Γ projects onto $\mathrm{SL}_n(\mathbf{Z}/q\mathbf{Z}) \times \prod_{1 \leq r \leq k} \mathrm{PSL}_n(\mathbf{Z}/p_r\mathbf{Z})$. Choose $1 \leq l \leq n$ distinct from i and j. Then Γ contains an element B such that $\pi_q(B) = \pi_q(E_{l,j}(1))$ and $\pi_{p_r}(B) = \pm\pi_{p_r}(I_n)$ for every $1 \leq r \leq k$. Choose $C \in \Gamma$ such that $\pi_q(C) = \pi_q(E_{i,l}(1))$. Denote $D := [C, B]$. Then, $\pi_q(D) = \pi_q(E_{i,j}(1))$ and $\pi_{p_r}(D) = \pi_{p_r}(I_n)$ for every $1 \leq r \leq k$. Choose t such that $t(p_1 \cdots p_k)^{m-1} \equiv_q 1$. Then $A = D^t$ is the required element. □

In order to complete the proof of Lemma 4.15, note that in view of the congruence subgroup property it is enough to prove that for every distinct prime $p_1, \ldots, p_k$ and every $m \geq 2$, Γ projects onto $\prod_{1 \leq r \leq k} \mathrm{SL}_n(\mathbf{Z}/p_r^m)$. Claim 4.19 implies that for every $m \geq 1$, Γ projects onto $\mathrm{SL}_n(\mathbf{Z}/2^m\mathbf{Z})$. The result follows from Claims 4.18 and 4.20. □

Proof of Proposition 4.14. We recall some facts about Zariski-dense and profinitely dense subgroups. For a positive integer $d \geq 2$, let $\pi_d : \mathrm{SL}_n(\mathbf{Z}) \to \mathrm{SL}_n(\mathbf{Z}/d\mathbf{Z})$ be the modulo-d homomorphism and denote $K_d := \ker \pi_d$.

(1) If $H \le SL_n(\mathbf{Z})$ and $\pi_p(H) = SL_n(\mathbf{Z}/p\mathbf{Z})$ for some odd prime p, then H is Zariski dense [Wei96, Lub99].
(2) The strong approximation theorem of Weisfeiler [Wei84] and Nori [Nor87] implies that if a subgroup H of $SL_n(\mathbf{Z})$ is Zariski dense, then there exists some positive integer q such that $\pi_d(H) = SL_n(\mathbf{Z}/d\mathbf{Z})$ whenever $\gcd(q,d) = 1$.

Fix $\delta \ge \epsilon > 0$ and set $\mathcal{A} := [p]_\epsilon$ and $\mathcal{R} := [L]_\delta$. For every $1 \le i \le 2n^2 - n$, fix a point p_i belonging to an $(n-2)$–dimensional subspace L_i and positive numbers $\delta_i \ge \epsilon_i > 0$ such that the following two conditions hold:

(1) $\bigcup_{1 \le i \le 2n^2 - n} (p_i)_{\varepsilon_i} \subseteq \mathcal{A}$ and $\bigcup_{1 \le i \le 2n^2 - n} (L_i)_{\delta_i} \subseteq \mathcal{R}$.
(2) For every $1 \le i \ne j \le 2n^2 - n$, $(p_i)_{\varepsilon_i} \bigcap (L_j)_{\delta_j} = \emptyset$.

For every $1 \le i \ne j \le n$, let $e_{i,j} \in SL_n(\mathbf{Z})$ be the matrix with 1 on the diagonal and on the (i,j) entry and zero elsewhere, and let $e_1, \ldots, e_{n^2-n}$ be an enumeration of the $e_{i,j}$'s. Denote the exponent of $SL_n(\mathbf{Z}/3\mathbf{Z})$ by t. If $g_1, \ldots, g_{n^2-n} \in K_3$ and $k_1, \ldots, k_{n^2-n}$ are positive integers, then $\pi_3(H_1) = SL_n(\mathbf{Z}/3\mathbf{Z})$, where $u_i := g_i e_i^{tk_i+1} g_i^{-1}$ and $H_1 := \langle u_i \mid 1 \le i \le n^2 - n \rangle$. Note that for every $u \in \mathcal{U}$ and $g \in SL_n(\mathbf{Z})$, $p_{gug^{-1}} = gp_u$ and $L_{gug^{-1}} = gL_u$. Thus, Lemma 4.3 and Corollary 4.13 imply that it is possible to choose g_i's and k_i's such that

(3) $u_i^k(x) \in (p_i)_{\varepsilon_i}$ for every $1 \le i \le n^2 - n$, every $x \notin (L_i)_{\delta_i}$, and every $k \ne 0$.

In particular, $\{u_1, \ldots, u_{n^2-n}\}$ is a Schottky set with respect to $\mathcal{A}$ and $\mathcal{R}$, which generates a Zariski-dense subgroup H_1.

The strong approximation theorem implies that there exists some positive integer q such that $\pi_d(H_1) = SL_n(\mathbf{Z}/d\mathbf{Z})$ whenever $\gcd(q,d) = 1$. Denote the exponent of $SL_n(\mathbf{Z}/q^2\mathbf{Z})$ by r. As before, there exist $g_{n^2-n+1}, \ldots, g_{2n^2-2n} \in K_{q^2}$ and positive integers $k_{n^2-n+1}, \ldots, k_{2n^2-2n}$ such that the elements of the form $u_i := g_i e_i^{rk_i+1} g_i^{-1}$ satisfy

(4) $\pi_{q^2}(H_2) = SL_n(\mathbf{Z}/q^2\mathbf{Z})$ where $H_2 := \langle u_i \mid n^2 - n + 1 \le i \le 2n^2 - 2n \rangle$; and
(5) $u_i^k(x) \in (p_i)_{\varepsilon_i}$ for every $n^2 - n + 1 \le i \le 2n^2 - 2n$, every $x \notin (L_i)_{\delta_i}$, and every $k \ne 0$.

Denote $\mathcal{S} := \{u_1, \ldots, u_{2n^2-2n}\}$. Lemma 4.15 implies that $\pi_d(\langle \mathcal{S} \rangle) = SL_n(\mathbf{Z}/d\mathbf{Z})$ for every $d \ge 1$. Thus, $(\mathcal{S}, \mathcal{A}, \mathcal{R})$ is the required profinitely dense Schottky system. □

Zorn's lemma implies that every proper subgroup H of $\mathrm{SL}_n(\mathbf{Z})$ is contained in a maximal subgroup M (since $\mathrm{SL}_n(\mathbf{Z})$ is finitely generated, an increasing union of proper subgroups is a proper subgroup). If H is profinitely dense then so is M; hence M should have infinite index. Thus, Theorem 1.18 follows from the following proposition:

THEOREM 4.21.

Let $n \geq 3$. There exist $2^{\aleph_0}$ infinite-index profinitely dense subgroups of $\mathrm{SL}_n(\mathbf{Z})$ such that the union of any two of them generates $\mathrm{SL}_n(\mathbf{Z})$.

Proof. For every nonnegative integer i, fix a rational point p_i belonging to a rational $(n-2)$–dimensional subspace L_i and two numbers $\delta_i \geq \epsilon_i > 0$ such that $[p_i]_{\varepsilon_i} \bigcap [L_j]_{\delta_j} = \emptyset$ for every $i \neq j$. Let $\mathcal{A}$ and $\mathcal{R}$ be the closures of $\bigcup_{i \geq 0} (p_i)_{\epsilon_i}$ and $\bigcup_{i \geq 0} (L_i)_{\delta_i}$ respectively. Proposition 4.14 implies that there exists a finite subset $\mathcal{S}_0 \subseteq \mathcal{U}$ such that $(\mathcal{S}_0, \mathcal{A}_0, \mathcal{R}_0)$ is a profinitely dense Schottky system where $\mathcal{A}_0 = [p_0]_{\varepsilon_0}$ and $\mathcal{R}_0 = [L_0]_{\delta_0}$. Lemmas 4.2 and 4.3 imply that for every $i \geq 1$ there are $u_{i,1}, u_{i,2} \in \mathcal{U}$ such that

(1) $p_i = p_{u_{i,1}} = p_{u_{i,2}}$ and $L_{u_{i,1}} \neq L_{u_{i,2}} \subseteq (L_i)_{\delta_i}$ (hence, $\langle u_{i,1}, u_{i,2} \rangle \cong \mathbf{Z}^2$); and
(2) $u_{i,j}^k(x) \in (p_i)_\varepsilon$ for every $1 \leq j \leq 2$, every $x \notin (L_i)_{\delta_i}$, and every $k \neq 0$.

For every function f from the positive integers to $\{0, 1\}$, the set $\mathcal{S}_f := \mathcal{S}_0 \bigcup \{u_{i,f(i)|i \geq 1}\}$ is a Schottky set with respect to the attracting set $\mathcal{A}$ and the repelling set $\mathcal{R}$. If f and g are distinct functions, then $\mathcal{S}_f \bigcup \mathcal{S}_g$ contains $\{u_{i,1}, u_{i,2}\}$ for some $i \geq 1$, so Theorem 4.8 implies that $\langle \mathcal{S}_f \bigcup \mathcal{S}_g \rangle = \mathrm{SL}_n(\mathbf{Z})$. □

5 Higher transitivity in negative curvature settings

Our goal in this section is to prove Theorem 1.25.

5.1 PRECISE PING PONG DYNAMICS. The proof will proceed via the topological dynamics of the action of Γ on $\mathbb{P} := \mathbb{P}(k^2)$ and on the limit set $L = L(\Gamma) \subset \mathbb{P}$ (see Lemma 5.1). By a neighborhood of a point p we mean any set containing p in its interior. By a *fundamental domain* for the action of Γ on an open invariant subset $Y \subset \mathbb{P}$, we will always refer to an open subset $O \subset Y$ satisfying (a) $\gamma O \bigcap O = \emptyset$, $\forall \gamma \in \Gamma \setminus \{1\}$, (b) $Y \subset \bigcup_{\gamma \in \Gamma} \gamma \overline{O}$, and (c) ∂O has an empty interior.

Let $\gamma \in \mathrm{SL}_2(k)$ be a very proximal element. In our current rank-1 setting there is a new symmetry between the attracting and repelling neighborhoods:

the repelling hyperplane $\overline{H}_{\gamma}^{+}$ reduces to a single point and coincides with the attracting point of the inverse $\overline{v}_{\gamma}^{-}$. To emphasize this we slightly change the notation. We say that $\Omega_{\gamma}^{\pm} \subset \mathbb{P}$ are attracting and repelling neighborhoods for a very proximal element γ on $\mathbb{P}$ if they are closed, disjoint neighborhoods of the attracting and repelling points $\overline{v}_{\gamma}^{\pm} \in \Omega_{\gamma}^{\pm}$ satisfying $\gamma\left(L \setminus \Omega^{-}\right) \subset \Omega^{+}$, or equivalently $\gamma^{-1}\left(L \setminus \Omega^{+}\right) \subset \Omega^{-}$. We will further call such attracting and repelling neighborhoods *precise* if $O = L \setminus (\Omega^{+} \bigcup \Omega^{-})$ is a fundamental domain for the action of the cyclic group $\langle \gamma \rangle$ on $L \setminus \{\overline{v}^{+}, \overline{v}^{-}\}$.

LEMMA 5.1. *Let k be a local field and $\Gamma < \mathrm{SL}_2(k)$ a center-free unbounded countable group. Assume Γ fixes neither a point nor a pair of points in $\mathbb{P}(k^2)$. Then*

(1) *Γ contains a very proximal element.*
(2) *There is a unique, minimal, closed, Γ-invariant subset $L = L(\Gamma) \subset \mathbb{P}(k^2)$. Moreover $L(\Gamma)$ is perfect as a topological space.*
(3) *The collection $\{(\overline{v}_{\gamma}^{+}, \overline{v}_{\gamma}^{-}) \mid \gamma \in \Gamma,$ very proximal$\}$ is dense in $L(\Gamma)^2$.*
(4) *If $e \neq \gamma \in \Gamma$, then $\mathrm{Supp}(\gamma) = \{x \in L(\Gamma) \mid \gamma x \neq x\}$ is an open dense subset of $L(\Gamma)$.*

Proof. Since Γ is unbounded it contains elements with an unbounded ratio between their singular values. Thus by Lemma 3.5, for every $\epsilon > 0$, we can arrange for $g \in \Gamma$ such that both g, g^{-1} are ϵ-contractions. Let $h \in \Gamma$ be such that $h\overline{v}_g^{+} \neq v_g^{-}$; then hg^n will be very proximal for a high enough value of n.

Let g be a very proximal element with attracting and repelling points $\overline{v}^{\pm}$. We know that $\lim_{n\to\infty} g^n(x) = \overline{v}^{+}, \forall x \neq \overline{v}^{-}$, and if $h\overline{v}^{-} \neq \overline{v}^{-}$, then $\lim_{n\to\infty} g^n h(\overline{v}^{-}) = \overline{v}^{+}$. So $L := \Gamma(\overline{v}_1^{+})$ is contained in every closed Γ-invariant set, proving the first sentence of (2).

Now let $U^{\pm} \subset L(\Gamma)$ be two (relatively) open subsets. As the action of Γ on $L(\Gamma)$ is clearly minimal we have elements $h^{\pm} \in \Gamma$ such that $h^{\pm}\overline{v}^{\pm} \in U^{\pm}$, respectively. The element $q = h^{+}g^{n}(h^{-})^{-1}$ will be very proximal with attracting and repelling points $\overline{v}_q^{\pm} \in U^{\pm}$. This proves (3). If q, g are two very proximal elements with different attracting and repelling points, then $\{q^n \overline{v}_g^{+} \mid n \in \mathbf{Z}\}$ is an infinite set of points contained in $L(\Gamma)$. Thus $L(\Gamma)$ is an infinite compact minimal Γ-space and hence perfect. This concludes (2). Finally, (4) follows from the fact that $|\mathrm{Fix}(\gamma)| < 3$ for every $e \neq \gamma \in \Gamma$. □

LEMMA 5.2. *Any very proximal element $\gamma \in \Gamma \leq \mathrm{SL}_2(k)$ admits precise attracting and repelling neighborhoods $\Omega_{\gamma}^{\pm} \subset L(\Gamma)$ in its action on $L(\Gamma)$.*

Proof. Given attracting and repelling neighborhoods $\Omega_1^{\pm} \subset L(\Gamma)$, we replace them by precise neighborhoods by setting $\Omega^- := \overline{\Omega_1^-}$ and $\Omega^+ := \overline{\gamma(L(\Gamma) \setminus \Omega^-)}$. It is clear that $\Omega^{\pm}$ thus defined are closed and that the sets $\{\gamma^k O \mid k \in \mathbf{Z}\}$ are pairwise disjoint where $O = L(\Gamma) \setminus (\Omega^- \bigcup \Omega^+)$. Given any $x \in L(\Gamma) \setminus \{\overline{v}_\gamma^+, \overline{v}_\gamma^-\}$, let $k \in \mathbf{Z}$ be the largest number such that $\gamma^k x \notin \Omega^+$. Then $\gamma^k x$ is neither in Ω^+ nor in Ω^-, since $\gamma^{k+1} x \in \Omega^+$. It follows that $\gamma^k x \in O$. □

In this rank-1 setting we can obtain a more precise version of the ping pong lemma (3.6).

LEMMA 5.3 (precise ping pong lemma). *Let $S = \{\gamma_1, \gamma_2, \ldots, \gamma_N\} \subset \mathrm{SL}_2(k)$ be a collection of $N \geq 2$ very proximal elements. Set $\Delta = \langle S \rangle$. Suppose that $\{\Omega_i^{\pm} \subset \mathbb{P}\}_{i=1}^N$ are pairwise disjoint, precise attracting and repelling neighborhoods for the γ_i action on $\mathbb{P}$. Set $\Omega_i = \Omega_i^+ \bigcup \Omega_i^-$ and $\Omega = \bigcup_{i=1}^N \Omega_i$ and assume that $O = \mathbb{P} \setminus \Omega$ is nonempty. Then,*

(1) S freely generates a free group Δ.
(2) There is a unique minimal closed Δ-invariant subset $L = L(\Delta) \subset \mathbb{P}$.
(3) There is a Δ-equivariant homeomorphism $\ell : \partial\Delta \to L$.
(4) ΔO is open dense and L is nowhere dense in $\mathbb{P}$.
(5) O is a fundamental domain for the action $\Delta \curvearrowright \mathbb{P} \setminus L$.

Proof. Statement (1) is the standard ping pong lemma, as proved, for example, in [Gel15, proposition 3.4].

Let us denote by $\overline{v_i}^{\pm} \in \mathbb{P}$ the attracting and repelling points for γ_i. We know that $\lim_{n\to\infty} \gamma_1^n(x) = \overline{v}_1^+, \forall x \neq \overline{v}_1^-$, and $\lim_{n\to\infty} \gamma_1^n \gamma_2(\overline{v}_1^-) = \overline{v}_1^+$. Set $L := \overline{\Delta(\overline{v}_1^+)}$. Then L is contained in every closed Δ-invariant set proving (2). Note that L contains every attracting or repelling point for every $e \neq \delta \in \Delta$.

We identify $\partial\Delta$ (the geometric boundary of the free group Δ) with infinite reduced words in $S \sqcup S^{-1}$. If $\xi \in \partial\Delta$, let $\xi(k) \in \Delta$ denote its k-prefix. Now define recursively a map $\ell : \Delta \setminus \{e\} \to \mathrm{Cl}(\mathbb{P})$ of Δ into the space of closed subsets of X by setting $\ell(s^\epsilon) = \Omega_s^\epsilon$ for $s \in S, \epsilon \in \{\pm 1\}$ and $\ell(s^\epsilon w) = s^\epsilon \ell(w)$ whenever $w \in \Delta$ is represented by a reduced word that does not start with $s^{-\epsilon}$. The ping pong dynamics yield two properties that are easy to verify: (a) $\ell(w) \subset \ell(v)$ whenever the reduced word representing v is a prefix of that representing w and (b) $\gamma\ell(w) \subset \ell(\gamma w)$ whenever $\gamma, w \in \Delta$ and w is represented by a long enough word in the generators.

Now we extend this definition to $\partial\Delta$ by setting $\ell(\xi) = \bigcap_{k\in\mathbf{N}} \ell(\xi(k))$ for every $\xi \in \partial\Delta$. $\ell(\xi)$ is non-empty by the finite intersection property. Using the metric contraction properties of the very proximal elements $\gamma_i^{\pm}$ given in Lemma 3.5, we verify that $\ell(\xi(k)) \in \mathbb{P}$ is a single point. Thus ℓ defines a point map, which by abuse of notation we will still denote by $\ell : \partial\Delta \to \mathbb{P}$. By property (a) above, this map is injective; by property (b) it is Δ invariant. Assume that $\xi_n \to \xi$ is $\partial\Delta$. By the definition of the standard topology on $\partial\Delta$, this just means that for every $m \in \mathbf{N}$ the m prefixes $\{\xi_n(m) \mid n \in \mathbf{Z}\}$ eventually stabilize and are equal to $\xi(m)$. Consequently, $\ell(\xi_n) \in \ell(\xi(m))$ for every n large enough, which immediately implies continuity; because both spaces are compact and metric ℓ is a homeomorphism onto its image. But the image is a minimal Δ set and hence equal to $L(\Delta)$ by (2). This concludes the proof of (3).

Set $O_i = \mathbb{P} \setminus \Omega_i$ and note that $O = \bigcap_{i=1}^N O_i$. By our assumption O_i is a fundamental domain for the action of $\langle \gamma_i \rangle$ on $\mathbb{P} \setminus \{\overline{v}_i^{\pm}\}$. Since the sets Ω_i are disjoint, $\overline{O} = \bigcap_{i=1}^N \overline{O_i}$. It is clear from the ping pong dynamics that the Δ translates of this set are disjoint. Thus to demonstrate (4) we take $x \in \mathbb{P} \setminus L(\Delta)$ and produce some element $\delta \in \Delta$ such that $\delta x \in \overline{O}$. We achieve this by an inductive procedure setting $x_0 := x$ and defining a sequence of points $x_0, x_1, \ldots \subset \Delta x_0$, stopping on the first time we hit $\overline{O}$. Thus if $x_m \in \overline{O}$, we are finished. If not, there is a unique index $1 \le i_m \le N$ such that $x_m \notin \overline{O}_{i_m}$. But $x_m \notin \{\overline{v}_{i_m}^{\pm}\} \subset L(\Delta)$ so we can find some $n_m \in \mathbf{Z}$ so that $\gamma_{i_m}^{-n_m} x_m \in \overline{O}_{i_m}$. This inductive procedure must terminate after finitely many steps. Indeed it is easy to verify by induction that $\gamma_0^{n_0}\gamma_1^{n_1} \cdot\cdots\cdot \gamma_m^{n_m}$ is a reduced word and that $x \in \ell(\gamma_0^{n_0}\gamma_1^{n_1} \cdot\cdots\cdot \gamma_m^{n_m})$. If the procedure never terminates we will obtain an infinite reduced word $\xi = \gamma_0^{n_0}\gamma_1^{n_1} \cdot\cdots\cdot \gamma_m^{n_m} \cdots \in \partial\Delta$ with $x \in \ell(\xi)$, contradicting our assumption that $x \notin L(\Delta)$.

Finally it follows directly from the above that ΔO is a dense open subset contained in $X \setminus L(\Delta)$, proving (5). □

5.2 POSSIBLE PARTIAL PERMUTATIONS. A *possible partial permutation* is a triplet of the form $\phi = (m, \alpha, \beta)$ with $m \in \mathbf{N}$, $\alpha = (a_1, a_2, \ldots, a_m)$, $\beta = (b_1, b_2, \ldots, b_m) \in \Gamma^m$. A possible partial permutation is called *special* if $a_1 = b_1 = e$. We will use the notation $\phi = (m(\phi), \alpha(\phi), \beta(\phi))$ to emphasize the data that is associated with a given possible partial permutation ϕ. Denote by $\mathrm{PPP} = \mathrm{PPP}(\Gamma)$ the set of all possible partial permutations of Γ, and by PPP^0 the collection of special ones. A given $\phi \in \mathrm{PPP}$ is said to be *legitimate* modulo a subgroup $\Delta < \Gamma$ if both α, β give rise to m distinct elements

$$\alpha\Delta = \{\alpha_i\Delta \mid 1 \le i \le m\}, \ \beta\Delta = \{\beta_i\Delta \mid 1 \le i \le m\} \subset \Gamma/\Delta.$$

Such a legitimate $\phi \in \mathrm{PPP}$ defines a partial map on Γ/Δ, which we denote by the same letter $\phi : \alpha\Delta \to \beta\Delta$, given by $\phi(\alpha_i\Delta) = \beta_i\Delta$, $1 \le i \le m$. Finally if there exists an element $\gamma \in \Gamma$ such that the partial map ϕ is the restriction of the quasiregular action $\gamma : \Gamma/\Delta \to \Gamma/\Delta$, we will say that ϕ is *realized by* γ. With this terminology in place note the following characterization of highly transitive actions in terms of the properties of a stabilizer of a point.

LEMMA 5.4. *Let $\Delta < \Gamma$ be a subgroup and $\psi : \Gamma \to \mathrm{Sym}(\Gamma/\Delta)$ the corresponding transitive action. Then the following conditions are equivalent:*

- *ψ is highly transitive (i.e., has a dense image).*
- *Every possible partial permutation $\phi \in \mathrm{PPP}(\Gamma)$ that is legitimate modulo Δ is realized by some $\gamma \in \Gamma$.*
- *Every special possible partial permutation $\phi \in \mathrm{PPP}^0(\Gamma)$ that is legitimate modulo Δ is realized by some $\delta \in \Delta$.*

DEFINITION 5.5. A subgroup $\Delta < \Gamma$ satisfying the equivalent conditions of Lemma 5.4 will be called *co-highly transitive* or *co-ht* for short.

5.3 HIGH TRANSITIVITY PROOFS.

Proof of Theorem 1.25. The equivalence of conditions (1), (2), and (3) follows directly from Lemma 2.5 combined with the fact that every connected proper algebraic subgroup of SL_2 is solvable. That condition (4) implies condition (1) follows from Proposition 2.1 combined with the fact that primitive groups of both affine and diagonal type are never highly transitive. Indeed, it is proven in that proposition that these groups are semidirect products of the from $\Delta \ltimes M$ and that their unique primitive action is the standard affine action $\Gamma \curvearrowright M$. If this action were to be highly transitive, it would follow that the conjugation action of Δ on $M \setminus \{e\}$ is highly transitive, which is absurd for any group. Thus we remain with our main task for this section: to construct a co-ht core-free subgroup Δ in Γ whenever $\Gamma < SL_2(k)$ is of almost simple type. We construct such a subgroup that is infinitely generated Schottky coming from a ping pong game on $Y = L(\Gamma) \subset \mathbb{P}(k^2)$. The limit set of Γ, given by Lemma 5.1(2).

Let $\mathrm{PPP}^0 = \{\phi_2, \phi_4, \phi_6, \ldots\}$ be an enumeration of all special possible partial permutations of Γ. Let $\{\gamma_1\langle\langle n_1\rangle\rangle, \gamma_3\langle\langle n_3\rangle\rangle, \gamma_5\langle\langle n_5\rangle\rangle, \ldots\}$ be an enumeration of all cosets of all the normal subgroups that are generated by a nontrivial conjugacy class. We construct, by induction on ℓ, an increasing sequence of precise, spacious Schottky groups $\Delta_\ell = \langle \delta_1, \delta_2, \ldots, \delta_{N_\ell}\rangle$ subject to the following properties:

- $(\delta_1, \delta_2, \ldots, \delta_{N_\ell})$ is a precise spacious ping pong tuple on $L(\Gamma)$.
- For ℓ even, if ϕ_ℓ is legitimate modulo $\Delta_{\ell-1}$, then it is realized modulo Δ_ℓ (by some element of Δ_ℓ).
- For ℓ odd, $\Delta_\ell \bigcap \gamma_\ell \langle\langle n_\ell \rangle\rangle \neq \emptyset$.

Setting $\Delta = \bigcup_\ell \Delta_\ell = \langle \delta_1, \delta_2, \ldots \rangle$, the condition imposed on the odd steps guarantees that Δ will be prodense, and in particular core-free. The even steps ensure that it is co-ht, by virtue of Lemma 5.4. Hence the coset action $\Gamma \curvearrowright \Gamma/\Delta$ will be faithful and highly transitive.

The odd steps of the induction were already treated, in greater generality, in the proof of Theorem 3.16. We will not repeat the argument here but turn directly to the even steps. Fix an even ℓ. For the sake of better legibility, we will often omit ℓ from the notation. For example, we will denote $\phi = \phi_\ell = (m, \alpha, \beta)$. If $\phi = \phi_\ell$ is illegitimate modulo $\Delta_{\ell-1}$, then it would definitely be illegitimate modulo any larger subgroup. In this case we just declare $\Delta_\ell = \Delta_{\ell-1}$. From here on we suppose that ϕ_ℓ is legitimate modulo $\Delta_{\ell-1}$.

By our induction assumption, $\Delta_{\ell-1}$ is generated by a precise ping pong tuple $\Delta_{\ell-1} = \langle \delta_1, \delta_2, \ldots, \delta_{N_{\ell-1}} \rangle$. We will find $\gamma \in \Gamma$ such that $\{\delta_1, \delta_2, \ldots, \delta_N, \gamma, b_2^{-1}\gamma a_2, \ldots, b_m^{-1}\gamma a_m\}$ still constitutes a precise ping pong tuple. Setting Δ_ℓ to be the group generated by these elements, after renaming them appropriately[3], we verify that ϕ_n is now realized by $\gamma \in \Delta_\ell$ modulo Δ_ℓ.

Let $\{\Omega_i^\pm\}_{i=1 \ldots N_{\ell-1}}$ be precise attracting and repelling points for the generators of $\Delta_{\ell-1}$ and $O = L(\Gamma) \setminus \bigcup_{i=1}^{N_{\ell-1}} (\Omega_i^- \cup \Omega_i^+)$ the fundamental domain, given by Lemma 5.3, for the action $\Delta_{\ell-1} \curvearrowright L(\Gamma) \setminus L(\Delta_{\ell-1})$. Since, by assumption, $\Delta_{\ell-1}$ is spacious Schottky inside Γ, the set $L(\Gamma) \setminus L(\Delta_{\ell-1})$ is nonempty. Hence by item (4) in that same lemma, $O \subset L(\Gamma)$ is relatively open and dense.

Consider the sets

$$R = \bigcap_{i=1}^{m} a_i^{-1} \Delta O, \quad A = \bigcap_{i=1}^{m} b_i^{-1} \Delta O \subset L(\Gamma).$$

By the Baire category theorem both sets are open and dense in $L(\Gamma)$. In particular we can find two points $a \in A$, $r \in R$. Let

$$\{\theta_j \mid 1 \leq j \leq m\}, \{\eta_j \mid 1 \leq j \leq m\}$$

be the unique elements of $\Delta_{\ell-1}$ satisfying $\theta_j^{-1} a_j^{-1} r \in O$ and $\eta_j^{-1} b_j^{-1} a \in O$, $\forall 1 \leq j \leq m$. Thus the sets

[3] That is, setting $\delta_{N_{\ell-1}+1} = \gamma, \delta_{N_{\ell-1}+2} = b_2^{-1}\gamma a_2, \ldots, \delta_{N_{\ell-1}+m} = b_m^{-1}\gamma a_m$ and setting $N_\ell = N_{\ell-1} + m$.

$$R_1 := \{x \in R \mid \theta_j^{-1} a_j^{-1} x \in O, \quad \forall 1 \leq j \leq m\} \text{ and}$$
$$A_1 := \{y \in A \mid \eta_j^{-1} b_j^{-1} y \in O, \quad \forall 1 \leq j \leq m\}$$

are open and nonempty.

By Lemma 5.1(3) we can find a very proximal element $\gamma \in \Gamma$ with repelling and attracting points $(\overline{v}^-, \overline{v}^+)$ subject to the following open conditions:

(1) $\overline{v}^- \in R_1, \overline{v}^+ \in A_1$
(2) $\overline{v}^- \in \mathrm{Supp}(a_k \theta_k \theta_j^{-1} a_j^{-1})$ and $\overline{v}^+ \in \mathrm{Supp}(b_k \eta_k \eta_j^{-1} b_j^{-1})$, $\forall 1 \leq j \neq k \leq m$
(3) $b_k \eta_k \theta_j^{-1} a_j^{-1} \overline{v}^- \neq \overline{v}^+, \qquad \forall 1 \leq j, k \leq m$

For the second conditions we used Lemma 5.1(3) together with our assumption that ϕ is legitimate modulo $\Delta_{\ell-1}$ to ensure that $a_k \theta_k \theta_j^{-1} a_j^{-1} \neq e$ and $b_k \eta_k \eta_j^{-1} b_j^{-1} \neq e$.

The above choices guarantee that

$$\{r_j := \theta_j^{-1} a_j^{-1} \gamma^-, a_j := \eta_j^{-1} b_j^{-1} \gamma^+ \mid 1 \leq j \leq m\}$$

are $2m$ distinct points inside O. We propose a precise ping pong generating set for Δ_ℓ of the form

$$\left\{\delta_1, \delta_2, \ldots, \delta_N, \gamma^k, (b_2\eta_2)^{-1}\gamma^k(a_2\theta_2), \ldots, (b_m\eta_m)^{-1}\gamma^k(a_m\theta_m)\right\}.$$

If we can adjust the parameter k so that these are indeed a precise ping pong tuple, all the desired properties hold. In particular $\gamma^k \in \Gamma$ would be the element realizing ϕ.

But if $\Omega^\pm$ are repelling and attracting neighborhoods for γ^k, then

$$r_j \in \theta_j^{-1} a_j^{-1} \Omega^-, a_j \in \eta_j^{-1} b_j^{-1} \Omega^+ \tag{5.1}$$

serve as repelling and attracting neighborhoods for $(b_j\eta_j)^{-1}\gamma^k(a_j\theta_j)$. By setting k large enough we can make the neighborhoods $\Omega^\pm$ arbitrarily small. Using Lemma 5.2 we can impose the condition that these sets are precise pairwise disjoint and contained in O. This completes the proof of the theorem. □

6 Groups that are not linear-like

The proof of Theorem 1.2 started with the construction of a profinitely dense subgroup, which then passed to a maximal subgroup containing it. We have seen by now many variants on this idea. One common feature of all constructions described so far is that the profinitely dense subgroup is free and constructed by various ping pong games á la Tits. Consequently all primitive groups that we have encountered so far are *large* in the sense that they contain a nonabelian free subgroup. One notable exception is some of the primitive groups of affine and diagonal type that we encountered.

This is a feature of the methods we have used so far. There are many primitive groups of almost simple type that do not contain free subgroups. In fact there are even linear examples. Take the group $PSL_n(K)$, where K is any countable locally finite field. One could take $K = \overline{F_p}$ to be the algebraic closure of F_p. These are all primitive groups of almost simple type, even though they are locally finite. In the case $n = 2$ the group $PSL_n(K)$ is even 3-transitive by virtue of its standard action on the projective line $\mathbb{P}(K)$. The goal of this section is to give a short survey, devoid of proofs, on some other results that are of different nature than the ones considered in previous sections.

6.1 SOME GROUPS OF SUBEXPONENTIAL GROWTH. It is easy to verify that the Grigorchuk group G is of almost simple type, though it is not stably so in the sense that many of its finite index subgroups fail to be of almost simple type. In fact $G_n < G$, the stabilizer in G of the nth level of the tree, splits as the direct product of 2^n subgroups.

Grigorchuk asked whether all maximal subgroups of the first Grigorchuk group are of finite index. This question was answered positively by Pervova:

THEOREM 6.1 ([Per05]).
Every maximal subgroup of the Grigorchuk group is of finite index.

Thus we have an example of a residually finite, finitely generated group of almost simple type admitting no maximal subgroup of infinite index. The results of Pervova were later generalized in [AKT16] to encompass a much larger family of groups. However, there are groups of subexponential growth that admit infinite index maximal subgroups, as shown by the following remarkable theorem of Francoeur and Garrido. The theorem deals with a family of groups that they coin *Šunić groups*, as they were defined by Zoran Šunić in [Š07]. We refer the reader to one of these articles for their precise definition, remarking here only that it is a family of finitely generated groups of subexponential growth acting on the binary rooted tree.

THEOREM 6.2 ([FG18]).
Let G be a nontorsion Šunić group that is not isomorphic to the infinite dihedral group, acting on the rooted binary tree. Then G admits countably many maximal subgroups, out of which only finitely many are of finite index.

Other examples of groups admitting countably many maximal subgroups of infinite index include Tarski monsters (that have only countably many subgroups to begin with) as well as some examples of affine type constructed by Hall in [Hal59].

6.2 THOMPSON'S GROUP **F**. The abelianization F/F' of the Thompson group is isomorphic to $\mathbf{Z}^2$. Any nontrivial normal subgroup of F contains the commutator; thus F is clearly of almost simple type. A subgroup $\Delta < F$ is profinitely dense if and only if it maps onto the abelianization, so it is not surprising that F contains many infinite index maximal subgroups.

In [Sav15] and [Sav10] Savchuk proves that all the orbits of the natural action of F on the interval $(0,1)$ are primitive. Otherwise put, $F_\nu < F$ is a maximal subgroup of infinite index for every $\nu \in (0,1)$. A notable fact is that many of these maximal subgroups are even finitely generated. Savchuk shows in particular that F_ν is finitely generated whenever $\nu \in \mathbf{Z}[1/2]$ and that it fails to be finitely generated whenever ν is irrational.

All of these facts are not difficult to verify. Svachuk asked whether Thompson's group F contains maximal infinite index subgroups that fail to fix a point. In a beautiful paper, Golan and Sapir [GS17] answer this question positively by constructing many very interesting examples of maximal subgroups. In particular they prove the following:

THEOREM 6.3 ([GS17]).
The Jones group $\vec{F} < F$, constructed in [Jon17], *is maximal of infinite index.*

This is the same group constructed by Vaughn Jones for establishing connections between link theory and Thompson's group F. It was shown by Brown in [Bro87] that the group $\vec{F}$ is isomorphic to the *triadic* version of the group F itself—namely, the group of all orientation preserving homeomorphisms of the interval $[0,1]$ that are piecewise linear with slopes that are powers of three and finitely many nondifferentiability points, all of which are contained in $\mathbf{Z}[1/3]$.

In addition to this one example, which is of particular interest due to the fact that this group is finitely generated and of independent interest, Golan and Sapir provide general methods of constructing a large variety of infinite index maximal subgroups in F.

6.3 LOCALLY FINITE SIMPLE GROUPS. By a famous theorem of Schur, every periodic linear group is locally finite. Thus, in view of Theorem 1.8 and the discussion following it, among the countable linear groups, locally finite groups of almost simple type are the only ones for which Question 1.4 is still open. Within this family, locally finite simple groups constitute an interesting special case.

In the emerging theory of locally finite simple groups (see, e.g., [HSBB95]), linear groups also play a special role. In fact it is quite customary in this subject to sort the locally finite simple groups into four families of increasing complexity: (a) finite, (b) linear, (c) finitary linear, and (d) the general case:

DEFINITION 6.4. A group Γ is called *finitary linear* if there exists a vector space V over a field F and an embedding $\Gamma < \mathrm{GL}(V)$ with the property that $\mathrm{Im}(\gamma - I)$ is a finite-dimensional subspace of V for every $\gamma \in \Gamma$.

In view of all this we find the following theorem of Meierfrankenfeld extremely interesting. His theorem gives a very strong solution to Question 1.4 exactly in the most complicated class of locally finite simple groups:

THEOREM 6.5 ([Mei95, theorem B]).
Let Γ be a locally finite simple group that is not finitary linear. Then every finite subgroup $D < \Gamma$ is contained in some proper maximal subgroup.

Note that since Γ here is simple, every maximal subgroup is automatically core-free.

6.4 NONESSENTIALLY FREE HOMEOMORPHISM GROUPS OF THE CIRCLE AND OF ∂T. Here we wish to highlight a very recent theorem of Le Boudec and Matte Bon. We view this theorem as one of the only nontrivial obstructions currently known to a group being highly transitive.

THEOREM 6.6 ([LBMB]).
Let $\Gamma < \mathrm{Homeo}(S^1)$. Assume that the action of Γ on S^1 is proximal, minimal, and not topologically free. Assume that distinct points in S^1 have distinct stabilizers. Then every faithful, 3-transitive action of G on a set is conjugate to its given action on one of the orbits within S^1.

In the same paper the authors prove a very similar theorem for group actions on a regular locally finite tree. This enables them to bound the transitivity degree of many groups. For example, for the natural index two extension $T^{\pm}$ of Thompson's group T (generated by T and a reflection of the circle), they show that it admits a 3-transitive action on a set (the set of dyadic points on the circle) but does not admit a 4-transitive action.

References

[AGS14] Menny Aka, Tsachik Gelander, and Gregory A. Soifer. Homogeneous number of free generators. *J. Group Theory*, 17(4):525–539, 2014.

[AKT16] Theofanis Alexoudas, Benjamin Klopsch, and Anitha Thillaisundaram. Maximal subgroups of multi-edge spinal groups. *Groups Geom. Dyn.*, 10(2):619–648, 2016.

[AS85] M. Aschbacher and L. Scott. Maximal subgroups of finite groups. *J. Algebra*, 92(1):44–80, 1985.

[BG03] E. Breuillard and T. Gelander. On dense free subgroups of Lie groups. *J. Algebra*, 261(2):448–467, 2003.

[BG07] E. Breuillard and T. Gelander. A topological Tits alternative. *Ann. of Math. (2)*, 166(2):427–474, 2007.

[Bro87] Kenneth S. Brown. Finiteness properties of groups. In *Proceedings of the Northwestern conference on cohomology of groups (Evanston, Ill., 1985)*, volume 44, pages 45–75, 1987.

[Cha12] Vladimir V. Chaynikov. *Properties of hyperbolic groups: Free normal subgroups, quasiconvex subgroups and actions of maximal growth*. ProQuest LLC, Ann Arbor, MI, 2012. Thesis (Ph.D.)–Vanderbilt University.

[CG00] J.-P. Conze and Y. Guivarc'h. Limit sets of groups of linear transformations. *Sankhyā Ser. A*, 62(3):367–385, 2000. Ergodic theory and harmonic analysis (Mumbai, 1999).

[Dix90] John D. Dixon. Most finitely generated permutation groups are free. *Bull. London Math. Soc.*, 22(3):222–226, 1990.

[DM96] John D. Dixon and Brian Mortimer. *Permutation groups*, volume 163 of *Graduate Texts in Mathematics*. Springer-Verlag, New York, 1996.

[FG18] Dominik Francoeur and Alejandra Garrido. Maximal subgroups of groups of intermediate growth. *Adv. Math.*, 340:1067–1107, 2018.

[GG13a] Shelly Garion and Yair Glasner. Highly transitive actions of $\mathrm{Out}(F_n)$. *Groups Geom. Dyn.*, 7(2):357–376, 2013.

[Gel15] Tsachik Gelander. Convergence groups are not invariably generated. arXiv: 1407.7226, 2015.

[GG08] Tsachik Gelander and Yair Glasner. Countable primitive groups. *Geom. Funct. Anal.*, 17(5):1479–1523, 2008.

[GG13b] Tsachik Gelander and Yair Glasner. An Aschbacher-O'Nan-Scott theorem for countable linear groups. *J. Algebra*, 378:58–63, 2013.

[GM16] T. Gelander and C. Meiri. Maximal subgroups of $\mathrm{SL}(n, \mathbb{Z})$. *Transform. Groups*, 21(4):1063–1078, 2016.

[GS17] Gili Golan and Mark Sapir. On subgroups of R. Thompson's group *F*. *Trans. Amer. Math. Soc.*, 369(12):8857–8878, 2017.

[Gun92] Steven V. Gunhouse. Highly transitive representations of free products on the natural numbers. *Arch. Math. (Basel)*, 58(5):435–443, 1992.

[Hal59] P. Hall. On the finiteness of certain soluble groups. *Proc. London Math. Soc. (3)*, 9:595–622, 1959.

[HSBB95] B. Hartley, G. M. Seitz, A. V. Borovik, and R. M. Bryant, editors. *Finite and locally finite groups*, volume 471 of *NATO Advanced Science Institutes Series C: Mathematical and Physical Sciences*. Kluwer Academic Publishers Group, Dordrecht, 1995.

[Hic92] K. K. Hickin. Highly transitive Jordan representations of free products. *J. London Math. Soc. (2)*, 46(1):81–91, 1992.

[HO16] Michael Hull and Denis Osin. Transitivity degrees of countable groups and acylindrical hyperbolicity. *Israel J. Math.*, 216(1):307–353, 2016.

[Iva92] Nikolai V. Ivanov. *Subgroups of Teichmüller modular groups*, volume 115 of *Translations of Mathematical Monographs*. American Mathematical Society, Providence, RI, 1992. Translated from the Russian by E. J. F. Primrose and revised by the author.

[Jon17] Vaughan Jones. Some unitary representations of Thompson's groups F and T. *J. Comb. Algebra*, 1(1):1–44, 2017.

[Kap03] Ilya Kapovich. The Frattini subgroup for subgroups of hyperbolic groups. *J. Group Theory*, 6(1):115–126, 2003.

[Kit12] Daniel Kitroser. Highly-transitive actions of surface groups. *Proc. Amer. Math. Soc.*, 140(10):3365–3375, 2012.

[KL88] Peter B. Kleidman and Martin W. Liebeck. A survey of the maximal subgroups of the finite simple groups. *Geom. Dedicata*, 25(1-3):375–389, 1988. Geometries and groups (Noordwijkerhout, 1986).

[LBMB] Adrien Le Boudec and Nicolás Matte Bon. Triple transitivity and non-free actions in dimension one. *Preprint 2021.*

[Lub99] Alexander Lubotzky. One for almost all: Generation of $\mathrm{SL}(n, p)$ by subsets of $\mathrm{SL}(n, \mathbf{Z})$. In *Algebra, K-theory, groups, and education (New York, 1997)*, volume 243 of *Contemporary Mathematics*, pages 125–128. American Mathematical Society, Providence, RI, 1999.

[MS77] G. A. Margulis and G. A. Soifer. A criterion for the existence of maximal subgroups of infinite index in a finitely generated linear group. *Dokl. Akad. Nauk SSSR*, 234(6):1261–1264, 1977.

[MS79] G. A. Margulis and G. A. Soifer. Nonfree maximal subgroups of infinite index of the group $\mathrm{SL}_n(\mathbf{Z})$. *Uspekhi Mat. Nauk*, 34(4(208)):203–204, 1979.

[MS81] G. A. Margulis and G. A. Soifer. Maximal subgroups of infinite index in finitely generated linear groups. *J. Algebra*, 69(1):1–23, 1981.

[McD77] T. P. McDonough. A permutation representation of a free group. *Quart. J. Math. Oxford Ser. (2)*, 28(111):353–356, 1977.

[Mei95] U. Meierfrankenfeld. Non-finitary locally finite simple groups. In *Finite and locally finite groups (Istanbul, 1994)*, volume 471 of *NATO Advanced Science Institutes Series C: Mathematical and Physical Sciences*, pages 189–212. Kluwer Academic Publishers, Dordrecht, 1995.

[MS13] Soyoung Moon and Yves Stalder. Highly transitive actions of free products. *Algebr. Geom. Topol.*, 13(1):589–607, 2013.

[Nor87] Madhav V. Nori. On subgroups of $\mathrm{GL}_n(\mathbf{F}_p)$. *Invent. Math.*, 88(2):257–275, 1987.

[Per05] E. L. Pervova. Maximal subgroups of some non locally finite *p*-groups. *Internat. J. Algebra Comput.*, 15(5-6):1129–1150, 2005.

[Sav10] Dmytro Savchuk. Some graphs related to Thompson's group *F*. In *Combinatorial and geometric group theory*, Trends in Mathematics, pages 279–296. Birkhäuser/Springer Basel AG, Basel, 2010.

[Sav15] Dmytro Savchuk. Schreier graphs of actions of Thompson's group *F* on the unit interval and on the Cantor set. *Geom. Dedicata*, 175:355–372, 2015.

[Sf90] G. A. Soifer. Maximal subgroups of infinite index of lattices of semisimple Lie groups. *Dokl. Akad. Nauk SSSR*, 311(3):536–539, 1990.

[Sf98] G. A. Soifer. Structure of infinite index maximal subgroups of $\mathrm{SL}_n(\mathbf{Z})$. In *Lie groups and ergodic theory (Mumbai, 1996)*, volume 14 of *Tata Institute of Fundamental Research Studies in Mathematics*, pages 315–324. Tata Institute of Fundamental Research, Bombay, 1998.

[Š07] Zoran Šunić. Hausdorff dimension in a family of self-similar groups. *Geom. Dedicata*, 124:213–236, 2007.

[Tit72] J. Tits. Free subgroups in linear groups. *J. Algebra*, 20:250–270, 1972.

[Ven87] T. N. Venkataramana. Zariski dense subgroups of arithmetic groups. *J. Algebra*, 108(2):325–339, 1987.

[Wei84] Boris Weisfeiler. Strong approximation for Zariski-dense subgroups of semisimple algebraic groups. *Ann. of Math. (2)*, 120(2):271–315, 1984.

[Wei96] Thomas Weigel. On the profinite completion of arithmetic groups of split type. In *Lois d'algèbres et variétés algébriques (Colmar, 1991)*, volume 50 of *Travaux en Cours*, pages 79–101. Hermann, Paris, 1996.

PART III

Expanders, representations, spectral theory

6

YVES BENOIST AND TOSHIYUKI KOBAYASHI

TEMPERED HOMOGENEOUS SPACES II

Dedicated to G. Margulis

Abstract. Let G be a connected semisimple Lie group with finite center and H a connected closed subgroup. We establish a geometric criterion that detects whether the representation of G in $L^2(G/H)$ is tempered.

1 Introduction

This essay is the sequel to our paper [3] dealing with harmonic analysis on homogeneous spaces G/H of semisimple Lie groups G. In the first paper, we studied the regular representation of G in $L^2(G/H)$ when both G and H are semisimple groups. The main result of [3] is a geometric criterion that detects whether the representation of G in $L^2(G/H)$ is tempered. The aim of the present paper is to extend this geometric criterion to the whole generality that H is an arbitrary closed connected subgroup.

In this introduction we will discuss the following questions:

- Why to care about tempered representations of semisimple Lie groups?
- What is our temperedness criterion for the homogeneous space G/H?
- What are the main ideas and ingredients in the proof of this criterion?

ACKNOWLEDGMENTS. The authors are grateful to the IHES and to the University of Tokyo for its support through the GCOE program. The second author was partially supported by Grant-in-Aid for Scientific Research (A) (25247006) JSPS.

YVES BENOIST. CNRS-Université Paris-Saclay, Orsay, France
yves.benoist@math.u-psud.fr
TOSHIYUKI KOBAYASHI. Graduate School of Mathematical Sciences and Kavli IPMU-University of Tokyo, Komaba, Japan
toshi@ms.u-tokyo.ac.jp

1.1 TEMPERED REPRESENTATIONS. Let G be a semisimple Lie group with finite center and K a maximal compact subgroup of G. Understanding unitary representations of G in Hilbert spaces is a major topic of research since the beginning of the 20th century. Its history includes the works of Cartan, Weyl, Gelfand, Harish-Chandra, Helgason, Langlands, Vogan, and many others. The main motivations came from quantum physics, analysis, and number theory.

Among unitary representations of G a smaller class called *tempered representations* plays a crucial role. Let us recall why they are so useful.

- By definition tempered representations are those that are weakly contained in the regular representation of G in $L^2(G)$. Therefore a unitary representation of G is tempered if and only if its disintegration into irreducible unitary representations involves only tempered representations.
- Tempered representations are those for which K-finite matrix coefficients belong to $L^{2+\varepsilon}(G)$ for all $\varepsilon > 0$. This definition does not look so enlightening, but equivalently these matrix coefficients are bounded by an explicit multiple of an explicit spherical function Ξ; see [8].
- Classification of irreducible tempered representations of G was accomplished by Knapp and Zuckerman in [16], while nontempered irreducible unitary representations have not yet been completely understood.
- Tempered representations are a cornerstone of the Langlands classification of admissible irreducible representations of G in [20]; see also [15].
- Irreducible tempered representations π can be characterized in term of the *leading exponents*: these exponents must be a positive linear combination of negative roots; see [15, chapter 8].
- One can also characterize them in terms of the *distribution character of* π: this character must be a tempered distribution on G; see [15, Chapter 12].
- Tempered representations are closed under induction, restriction, tensor product, and direct integral of unitary representations.
- The Kirilov-Kostant orbit methods work fairly well for tempered representations; see [13], for example.

1.2 THE REGULAR REPRESENTATION IN $L^2(G/H)$. One of the most studied representations of G is the natural unitary representations of G in $L^2(G/H)$ where H is a closed subgroup of G. When H is unimodular, the space G/H is implicitly endowed with a G-invariant measure and G acts naturally by translation on L^2 functions. When H is not assumed to be unimodular, the representation of G in $L^2(G/H)$ might involve an extra factor (see Equation 2.2) and is nothing but the (unitarily) induced representation $Ind_H^G(1)$.

The disintegration of $L^2(G/H)$ is sometimes called the *Plancherel formula* or L^2-harmonic analysis on G/H.

- For instance, Harish-Chandra's celebrated Plancherel formula [12] deals with the case where $H = \{e\}$.
- Another case that attracted a lot of interest in the late 20th century is the case where G/H is a symmetric space for which the disintegration of $L^2(G/H)$ is proved up to the classification of (singular) discrete series representations for symmetric spaces [9, 27].
- Even when H is not unimodular, the regular representation of G in $L^2(G/H)$ is still interesting. For instance, when H is a parabolic subgroup of G, this representation is known to be a finite sum of irreducible representations of G. This follows from Bruhat's theory for a minimal parabolic and is duc to Harish-Chandra in general (see [15, proposition 8.4]).
- The decomposition of the tensor product (*fusion rule*) is sometimes equivalent to the Plancherel formula for G/H where H is not necessarily unimodular (see section 5.3).

We refer to [3, introduction] for more remarks on the historical developments of the disintegration of the regular representation of G in $L^2(G/H)$. Getting a priori information on this disintegration was one of the main motivations in our search for such a general criterion. To the best of our knowledge there does not yet exist any general theorem involving simultaneously all these regular representations in $L^2(G/H)$ with H connected. Our temperedness criterion below seems to be the first one in that direction.

In this series of papers, we address the following question: *What kind of unitary representations occur in the disintegration of $L^2(G/H)$?* More precisely, *when are all of them tempered?*

As noted in [3], this question has not been completely solved even for reductive symmetric spaces G/H because the Plancherel formula involves a delicate algebraic problem on discrete series representations for (sub)symmetric spaces with singular infinitesimal characters (see Example 5.5).

We give a geometric necessary and sufficient condition on G/H under which all these irreducible unitary representations in the Plancherel formula are tempered, or equivalently under which the regular representation of G in

$L^2(G/H)$ is tempered. This criterion was first discovered in [3] in the special case where H is a reductive subgroup of G.

In the present essay we extend this criterion to any closed subgroup H with finitely many connected components.

Formally the extended criterion is exactly the same as for H reductive. Here is a short way to state our criterion (see Theorem 2.9).

$$(1.1) \qquad L^2(G/H) \text{ is tempered } \iff \rho_{\mathfrak{h}}(Y) \leq \rho_{\mathfrak{g}/\mathfrak{h}}(Y) \text{ for all } Y \in \mathfrak{h}.$$

Here $\mathfrak{g}$ and $\mathfrak{h}$ are the Lie algebras of G and H, and, for an $\mathfrak{h}$-module V and $Y \in \mathfrak{h}$, the quantity $\rho_V(Y)$ is half the sum of the absolute values of the real part of the eigenvalues of Y in V (see section 2.3). We note that our criterion holds beyond the (real) spherical case ([18]; cf. [28] for a non-Archimedean field) so that the disintegration of $L^2(G/H)$ may involve infinite multiplicities. We will give an explicit example of calculations of functions ρ_V in Corollary 5.8.

Our criterion in Equation (1.1) also detects whether $L^2(X)$ is tempered for any real algebraic G-variety X: $L^2(X)$ is unitarily equivalent to the direct integral of the regular representations for generic orbits, and one just has to check (1.1) at almost all orbits.

1.3 STRATEGY OF PROOF.

The proof of Equation (1.1) relies on the uniform decay of matrix coefficients as in [3], but the main techniques are different.

To avoid any confusion we will sometimes say *G-tempered* for *tempered as a representation of G*.

Dealing with nonunimodular subgroups H and dealing with the finitely many components of H will not be a problem because we prove in Corollary 3.3 the equivalence

$$\begin{aligned} L^2(G/H) \text{ is tempered } &\iff L^2(G/[H,H]) \text{ is tempered} \\ &\iff L^2(G/H_e) \text{ is tempered,} \end{aligned}$$

where $[H,H]$ is the derived subgroup, which is always unimodular, and where H_e is the identity component of H that is always connected! Therefore the temperedness of $L^2(G/H)$ depends only on $\mathfrak{h}$ and we can assume that $\mathfrak{h} = [\mathfrak{h}, \mathfrak{h}]$. Then the homogeneous space G/H admits a G-invariant Radon measure *vol* and, according to Corollary 2.7, the temperedness of $L^2(G/H)$ means that the compact subsets C of G/H intersect their translates gC in a set whose volume is bounded by a multiple of the Harish-Chandra function Ξ:

$$\text{vol}(g\,C \cap C) \leq M_C\, \Xi(g) \quad \text{for all } g \text{ in } G. \tag{1.2}$$

To prove the direct implication, $L^2(G/H)$ tempered $\Longrightarrow \rho_{\mathfrak{h}} \leq \rho_{\mathfrak{g}/\mathfrak{h}}$, we will just estimate in Proposition 3.7 the volume in Equation (1.2) when C is a small neighborhood of the base point of the space G/H.

As in [3], the converse implication, $\rho_{\mathfrak{h}} \leq \rho_{\mathfrak{g}/\mathfrak{h}} \Longrightarrow L^2(G/H)$ tempered, is much harder to prove. Using again Corollary 3.3, we can also assume that both G and H are Zariski connected algebraic groups. We proceed by induction on the dimension of G. We introduce in Lemma-Definition 4.1 two intermediate subgroups F and P:

$$H \subset F \subset P \subset G. \tag{1.3}$$

The group P is a parabolic subgroup of minimal dimension that contains H. We write $P = LU$ and $H = SV$, where U is the unipotent radical of P, L is a maximal reductive subgroup of P, $V \subset U$ is the unipotent radical of H, and $S \subset L$ is a maximal semisimple subgroup of H. The group F is given by $F := SU$.

When P is equal to G, the group H is semisimple and we apply the main result of [3]. We now assume that P is a proper parabolic subgroup of G. Let Z be the homogeneous space $Z = F/H \simeq U/V$ endowed with the natural F action. We denote

- by τ the regular representation of F in $L^2(F/H)$,
- by π the regular representation of P in $L^2(P/H)$, and
- by Π the regular representation of G in $L^2(G/H)$.

Let Z_0 be the same space $Z_0 = U/V$ but endowed with another F action, where U acts trivially and where S acts by conjugation.

We denote

- by τ_0 the regular representation of F in $L^2(Z_0)$,
- by π_0 the regular representation of P in $L^2(P \times_F Z_0)$, and
- by Π_0 the regular representation of G in $L^2(G \times_F Z_0)$.

Note that both the G-manifolds $X := G/H \simeq G \times_F Z$ and $X_0 := G \times_F Z_0$ have a G-equivariant fiber bundle structure over the same space G/F with fiber $Z \simeq Z_0$, but the unitary representations of G in $L^2(X)$ and in $L^2(X_0)$ are different. We want to study the representation of G in $L^2(X)$, whereas the induction hypothesis will give us information on $L^2(X_0)$. This is why we will need to compare in Equation (1.5) the volumes in the fibers Z and Z_0.

More precisely, the induction hypothesis combined with a reformulation of our criterion in Proposition 3.8 and a simple computation in Lemma 4.2 tell us that the representation π is L-tempered. If we knew that π were a P-tempered representation it would be easy to conclude, using Lemma 2.3, that the representation $\Pi = \mathrm{Ind}_P^G \pi$ is G-tempered. However, what we know is the temperedness of π_0 and $\mathrm{Ind}_P^G \pi_0$, and Corollary 2.7 gives us, for any compact subset C_0 of $G \times_F Z_0$, a bound:

$$(1.4) \qquad \mathrm{vol}(g\, C_0 \cap C_0) \leq M_{C_0}\, \Xi(g) \quad \text{for all g in } G.$$

In order to deduce Equation (1.2) from (1.4), we first focus on the representations τ in $L^2(Z)$ and τ_0 in $L^2(Z_0)$. We prove in Proposition 4.4 that for every compact subset D of the fiber $Z = F/H$, a uniform estimate of $\mathrm{vol}(fD \cap D)$ with respect to translates of the element $f \in F$ by elements of the unipotent subgroup U. Namely, there exists a compact subset $D_0 \subset Z_0$ such that

$$(1.5) \qquad \mathrm{vol}(fD \cap D) \leq \mathrm{vol}(fD_0 \cap D_0) \quad \text{for all } f \text{ in } F.$$

Since $\Pi = \mathrm{Ind}_F^G \tau$ and $\Pi_0 = \mathrm{Ind}_F^G \tau_0$, we deduce from (1.5) in Proposition 4.9 that for every compact subset $C \subset G/H \simeq G \times_F Z$, there exists a compact subset $C_0 \subset G \times_F Z_0$ such that

$$(1.6) \qquad \mathrm{vol}(g\, C \cap C) \leq \mathrm{vol}(g\, C_0 \cap C_0) \quad \text{for all g in } G.$$

And (1.2) follows from (1.4) and (1.6). This ends the sketch of the proof of the criterion (1.1). The details are explained in subsequent sections.

1.4 ORGANIZATION. Here is the organization of the paper.

In section 2 we recall the basic definition and state precisely our criterion.

In section 3 we collect the parts of the proof that do not involve the intermediate parabolic subgroup P. It includes the proof for the necessity of the inequality $\rho_{\mathfrak{h}} \leq \rho_{\mathfrak{g}/\mathfrak{h}}$ and a formulation of the temperedness criterion for $\mathrm{Ind}_H^G(L^2(V))$ (Theorem 3.6).

In section 4 we introduce the intermediate subgroups F and P in Equation (1.3) and detail how they work to conclude the proof for the hard part—that is, the proof that the inequality $\rho_{\mathfrak{h}} \leq \rho_{\mathfrak{g}/\mathfrak{h}}$ is sufficient.

In section 5 we give a few examples that illustrate the efficiency of our criterion.

1.5 DEDICATION. The notion of tempered representations plays an important role in quite a few papers of G. Margulis. As we have seen tempered representations are one of the main tools to obtain uniform estimates for coefficients of unitary representations. Margulis found many very different and very ingenious applications of these uniform estimates all along his mathematical career—for instance, to the construction of expanders in [22], to the nonexistence of compact quotients of homogeneous spaces in [23], to a pointwise ergodic theorem for probability measure preserving actions of semisimple Lie groups in [24], to a local rigidity phenomenon for actions of higher rank lattices in [11], to an effective rate of equidistribution of closed orbits of semisimple Lie groups in finite volume spaces in [10], to a uniform estimate on the smallest integral base change between two equivalent nonsingular integral quadratic forms in [21], and more.

One inspiration for Theorem 2.9 came from lemma 6.5.4 of [10], which says that for a connected semisimple Lie group G with no compact factor and a proper closed connected subgroup H, the representation of G in $L^2(G/H)$ always has a spectral gap

We are proud to dedicate this paper to G. Margulis.

2 Definition and main result

We collect in this chapter a few well-known facts on tempered representations, on almost L^2 representations, and on uniform decay of matrix coefficients.

2.1 REGULAR AND INDUCED REPRESENTATIONS. We first recall the construction of the regular representations and the induced representations.

2.1.1 Regular representations

Let G be a separable locally compact group, X be a separable locally compact space endowed with a continuous action of G, and ν_X be a Radon measure on X.

When the G action preserves the measure ν_X, one has a natural unitary representation λ_X of G in the Hilbert space $L^2(X) := L^2(X, \nu_X)$ called the *regular representation* and given by

$$(\lambda_X(g)\varphi)(x) = \varphi(g^{-1}x) \quad \text{for } g \text{ in } G,\ \varphi \text{ in } L^2(X) \text{ and } x \text{ in } X.$$

When the class of the measure ν_X is G-invariant, one still has a natural unitary representation π of G in the Hilbert space $L^2(X) := L^2(X, \nu_X)$ also called the *regular representation.* The formula will involve the Radon-Nikodym cocycle $c(g,x)$, which is defined for all g in G and for ν_X-almost all x in X by the equality

$$g_*\nu_X = c(g^{-1},x)\,\nu_X. \tag{2.1}$$

The regular representation of G in $L^2(X)$ is then given by

$$(\lambda_X(g)\varphi)(x) = c(g^{-1},x)^{1/2}\,\varphi(g^{-1}x) \text{ for } g \text{ in } G, \varphi \text{ in } L^2(X), \text{ and } x \text{ in } X. \tag{2.2}$$

2.1.2 Induced representations

Assume now that X is a homogeneous space G/H, where H is a closed subgroup of G. One can choose a G-invariant Radon measure on G/H if and only if the modular function of G coincides on H with that of H,

$$\Delta_G(h) = \Delta_H(h) \text{ for all } h \text{ in } H.$$

In general there always exists a measure ν on G/H whose class is G-invariant, and the regular representation of G in $L^2(G/H)$ is the induced representation of the trivial representation of H,

$$\lambda_{G/H} = \mathrm{Ind}_H^G(\mathbf{1}).$$

More generally, for any unitary representation π of H, one defines the (unitarily) induced representation $\Pi := \mathrm{Ind}_H^G(\pi)$ in the following way. The projection

$$G \longrightarrow G/H$$

is a principal bundle with structure group H. We fix a G-equivariant Borel measurable trivialization of this principal bundle

$$G \simeq G/H \times H, \tag{2.3}$$

which sends relatively compact subsets to relatively compact subsets. The action of G by left multiplication through this trivialization can be read as

$$g\,(x,h) = (gx, \sigma(g,x)h) \quad \text{for all } g \in G,\ x \in G/H, \text{ and } h \in H,$$

where $\sigma : G \times G/H \to H$ is a Borel measurable cocycle.

The space of the representation Π is the space $\mathcal{H}_\Pi := L^2(G/H; \mathcal{H}_\pi)$ of $\mathcal{H}_\pi$-valued L^2 functions on G/H, and the action of G is given, for g in G, ψ in $\mathcal{H}_\Pi$, x in G/H, by

$$(\Pi(g)\psi)(x) = c(g^{-1}, x)^{1/2}\, \pi(\sigma(g, g^{-1}x))\psi(g^{-1}x),$$

where c is again the Radon-Nikodym cocycle (2.1).

2.1.3 *Induced actions*

When the closed subgroup H of G is acting continuously on a locally compact space Z, one can define the induced action of G on the fibered space

$$G \times_H Z := (G \times Z)/H,$$

where the quotient is taken for the right H-action $(g, z)h = (gh, h^{-1}z)$ and where the G-action is given by $g_0(g, z) = (g_0 g, z)$, for all g_0, g in G, z in Z, and h in H. Using Equation (2.3), one gets a G-equivariant Borel measurable trivialization of this fibered space

$$G \times_H Z \simeq G/H \times Z.$$

Through this identification, the G-action is given by

$$g(x, z) = (gx, \sigma(g, x)z) \quad \text{for all } g \in G,\ x \in G/H, \text{ and } z \in Z.$$

When the H-action preserves the class of a measure ν_Z on Z, the G-action preserves the class of the measure $\nu_X := \nu \otimes \nu_Z$. In this case the regular representation of G in $L^2(G \times_H Z)$ is unitarily equivalent to the induced representation of the regular representation of H in Z:

(2.4) $$L^2(G \times_H Z) \simeq \mathrm{Ind}_H^G(L^2(Z)) \quad \text{as unitary representations of } G.$$

2.2 DECAY OF MATRIX COEFFICIENTS.

We now recall the control of the matrix coefficients of tempered representations of a semisimple Lie group.

2.2.1 *Tempered representations*

Let G be a locally compact group and π be a unitary representation of G in a Hilbert space $\mathcal{H}_\pi$. All representations π of G will be assumed to be continuous—that is, the map $G \to \mathcal{H}_\pi$, $g \mapsto \pi(g)v$ is continuous for all v in $\mathcal{H}_\pi$. The notion of tempered representation is due to Harish-Chandra.

DEFINITION 2.1. The unitary representation π is said to be tempered or G-tempered if π is weakly contained in the regular representation λ_G of G in $L^2(G)$—that is, if every matrix coefficient of π is a uniform limit on every compact subset of G of a sequence of sums of matrix coefficients of λ_G.

We refer to [1, appendix F] for more details on weak containments.

REMARK 2.2. The notion of temperedness is stable by passage to a finite index subgroup G' of G—that is, a unitary representation π of G is tempered if and only if π is tempered as a representation of G'.

This notion is also preserved by induction.

LEMMA 2.3. *Let G be a locally compact group, H be a closed subgroup of G, and π be a unitary representation of H. If π is H-tempered, then the induced representation $\mathrm{Ind}_H^G(\pi)$ is G-tempered.*

Proof. Since the H-representation π is weakly contained in the regular representation λ_H of H, the G-representation $\mathrm{Ind}_H^G(\pi)$ is weakly contained in the regular representation $\lambda_G = \mathrm{Ind}_H^G(\lambda_H)$ and hence is G-tempered. □

REMARK 2.4. (1) When G is amenable, according to the Hulanicki-Reiter theorem in [1, theorem G.3.2], every unitary representation of G is tempered. (2) When G is a product of two closed subgroups $G = SZ$ with Z central, *a unitary representation π of G is G-tempered if and only if it is S-tempered.* Indeed the regular representation of G in $L^2(G)$ is clearly S-tempered. Conversely, we want to prove that any unitary representation π of G that is S-tempered is also G-tempered. We can assume that π is G-irreducible. The action of Z in this representation is given by a unitary character χ, and π is weakly contained in the representation $\mathrm{Ind}_Z^G \chi$. Since χ is Z-tempered, this representation is G-tempered.

2.2.2 *Matrix coefficients*

Let now G be a semisimple Lie group (always implicitly assumed to be real Lie groups with finitely many connected components and whose identity component has finite center).

DEFINITION 2.5. A unitary representation π of G is said to be almost L^2 if there exists a dense subset $\mathcal{D} \subset \mathcal{H}_\pi$ for which the matrix coefficients $g \mapsto \langle \pi(g)v_1, v_2 \rangle$ are in $L^{2+\varepsilon}(G)$ for all $\varepsilon > 0$ and all v_1, v_2 in $\mathcal{D}$.

We fix a maximal compact subgroup K of G. Let Ξ be the Harish-Chandra spherical function on G (see [8]). By definition, Ξ is the matrix coefficient of a normalized K-invariant vector v_0 of the spherical unitary principal representation $\pi_0 = \mathrm{Ind}_{P_{\min}}^G(\mathbf{1}_{P_{\min}})$, where $P_{\min}$ is a minimal parabolic subgroup of G. That is,

$$\Xi(g) = \langle \pi_0(g)v_0, v_0 \rangle \quad \text{for all } g \text{ in } G. \tag{2.5}$$

Since $P_{\min}$ is amenable, the representation π_0 is G-tempered. Moreover, the function Ξ belongs to $L^{2+\varepsilon}(G)$ for all $\varepsilon > 0$ (see [15, proposition 7.15]). We will need the following much more precise version of this fact.

PROPOSITION 2.6 (Cowling, Haagerup, Howe [8]). *Let G be a connected semisimple Lie group with finite center and π be a unitary representation of G. The following are equivalent:*

(1) The representation π is tempered.
(2) The representation π is almost L^2.
(3) For every K-finite vectors v, w in $\mathcal{H}_\pi$, and for every g in G, one has

$$|\langle \pi(g)v, w \rangle| \le \Xi(g)\|v\|\|w\|(\dim\langle Kv \rangle)^{1/2}(\dim\langle Kw \rangle)^{1/2}.$$

See [8, theorems 1 and 2 and corollary]. See also [14], [25], and [26] for other applications of Proposition 2.6.

For regular representations this proposition becomes the following:

COROLLARY 2.7. *Let G be a connected semisimple Lie group with finite center and X a locally compact space endowed with a continuous action of G preserving a Radon measure* vol. *The regular representation of G in $L^2(X)$ is tempered if and only if, for any compact subset C of X, the function $g \mapsto \mathrm{vol}(gC \cap C)$ belongs to $L^{2+\varepsilon}(G)$ for all $\varepsilon > 0$.*

In this case, when C is K-invariant, one has

$$\mathrm{vol}(gC \cap C) \le \mathrm{vol}(C)\,\Xi(g) \quad \text{for all } g \text{ in } G. \tag{2.6}$$

Recall that the notation gC denotes the set $gC := \{gx \in X : x \in C\}$.

Proof. Note that a compact subset C of X is always included in a K-invariant compact subset C_0, that the function $\mathbf{1}_{C_0}$ is a K-invariant vector in $L^2(X)$, and that

$$\langle \lambda_X(g)\mathbf{1}_{C_0}, \mathbf{1}_{C_0}\rangle = \mathrm{vol}(g\, C_0 \cap C_0) \quad \text{for all } g \text{ in } G.$$

Note also that the functions $\mathbf{1}_C$ span a dense subspace in $L^2(X)$. □

2.3 THE FUNCTION ρ_V. We now define the functions $\rho_{\mathfrak{h}}$ and $\rho_{\mathfrak{g}/\mathfrak{h}}$ occurring in the temperedness criterion, explain how to compute them and emphasize their geometric meaning.

When H is a Lie group we denote by the corresponding gothic letter $\mathfrak{h}$ the Lie algebra of H. Let V be a real finite-dimensional representation of H. For an element Y in $\mathfrak{h}$, we consider the eigenvalues of Y in V (more precisely in the complexification $V_{\mathbb{C}}$), and we denote by V_+, V_0, and V_- the largest vector subspaces of V on which the real part of all the eigenvalues of Y are respectively positive, zero, and negative. One has the decomposition $V = V_+ \oplus V_0 \oplus V_-$. We define the nonnegative functions ρ_V^+ and ρ_V on $\mathfrak{h}$ by

$$\rho_V^+(Y) := \mathrm{Tr}(Y|_{V_+}),$$
$$\rho_V(Y) := \tfrac{1}{2}\, \rho_V^+(Y) + \tfrac{1}{2}\, \rho_V^+(-Y),$$

where Tr denotes the trace of a matrix. Note that one has the equality $\mathrm{Tr}(Y|_{V_-}) = -\rho_V^+(-Y)$.

By definition, one always has the equality $\rho_V(-Y) = \rho_V(Y)$. Moreover, when the action of H on V is volume preserving, one has the equality

$$\rho_V(Y) = \rho_V^+(Y).$$

The function called ρ_V in [3, section 3.1] is what we call now ρ_V^+. It coincides with our ρ_V since in [3] we only need to consider volume preserving actions.

Since this function $\rho_V\colon \mathfrak{h} \to \mathbb{R}_{\geq 0}$ plays a crucial role in our criterion, we begin by a few trivial but useful comments, which make it easy to compute when dealing with examples. To simplify these comments, we assume that H is an algebraic subgroup of $\mathrm{GL}(V)$. Let $\mathfrak{a} = \mathfrak{a}_{\mathfrak{h}}$ be a maximal split abelian Lie subalgebra of $\mathfrak{h}$—that is, the Lie subalgebra of a maximal split torus A of H. Any element Y in $\mathfrak{h}$ admits a unique Jordan decomposition $Y = Y_e + Y_h + Y_n$ as a sum of three commuting elements of $\mathfrak{h}$, where Y_e is a semisimple matrix with imaginary eigenvalues, Y_h is a semisimple matrix with real eigenvalues, and Y_n is a nilpotent matrix (see, for instance, [19, 2.1]). Moreover there exists an element λ_Y in $\mathfrak{a}$ that is H conjugate to Y_h. Then one has the equality

$$\rho_V(Y) = \rho_V(\lambda_Y) \text{ for all } Y \text{ in } \mathfrak{h}.$$

This equality tells us that the function ρ_V is completely determined by its restriction to $\mathfrak{a}$.

This function $\rho_V \colon \mathfrak{a} \to \mathbb{R}_{\geq 0}$ is continuous and is piecewise linear—that is, there exist finitely many convex polyhedral cones that cover $\mathfrak{a}$ and on which ρ_V is linear. Indeed, let P_V be the set of weights of $\mathfrak{a}$ in V and, for all α in P_V, let $m_\alpha := \dim V_\alpha$ be the dimension of the corresponding weight space. Then one has the equality

$$\rho_V(Y) = \tfrac{1}{2} \sum_{\alpha \in P_V} m_\alpha |\alpha(Y)| \text{ for all } Y \text{ in } \mathfrak{a}. \tag{2.7}$$

For example, when $\mathfrak{h}$ is semisimple and $V = \mathfrak{h}$ via the adjoint action, our function $\rho_{\mathfrak{h}}$ is equal on each positive Weyl chamber $\mathfrak{a}_+$ of $\mathfrak{a}$ to the sum of the corresponding positive roots—that is, to twice the usual ρ linear form. For other representations V, the maximal convex polyhedral cones on which ρ_V is linear are most often much smaller than the Weyl chambers. Explicit computations of the functions ρ_V will be given in section 5.

The geometric meaning of this function ρ_V is given by the following elementary Lemma as in [3, proposition 3.6].

LEMMA 2.8. *Let $V = \mathbb{R}^d$. Let $\mathfrak{a}$ be an abelian split Lie subalgebra of* End(V) *and C be a compact neighborhood of* 0 *in V. Then there exist constants $m_C > 0$, $M_C > 0$ such that*

$$m_C e^{-\rho_V(Y)} \leq e^{-\mathrm{Trace}(Y)/2} \,\mathrm{vol}(e^Y C \cap C) \leq M_C e^{-\rho_V(Y)} \text{ for all } Y \in \mathfrak{a}.$$

Such a factor $e^{-\mathrm{Trace}(Y)/2}$ occurs in computing the matrix coefficient of the vector $\mathbf{1}_C$ in the regular representation $L^2(V)$ when the action on V does not preserve the volume. Here *vol* denotes the volume with respect to the Lebesgue measure on V. The proof of Lemma 2.8 goes similarly to that of [3, proposition 3.6], which deals with the case where the action is volume preserving.

2.4 TEMPEREDNESS CRITERION FOR $L^2(G/H)$. We can now state precisely our temperedness criterion.

Let G be a semisimple Lie group and H a closed subgroup of G. Let $\mathfrak{g}$ and $\mathfrak{h}$ be the Lie algebras of G and H. The temperedness criterion for the regular representation of G in $L^2(G/H)$ will involve the functions $\rho_{\mathfrak{h}}$ and $\rho_{\mathfrak{g}/\mathfrak{h}}$ for the H-modules $V = \mathfrak{h}$ and $V = \mathfrak{g}/\mathfrak{h}$.

THEOREM 2.9.
Let G be a connected semisimple Lie group with finite center and H a closed connected subgroup of G. Then, one has the equivalence

$$L^2(G/H) \text{ is } G\text{-tempered} \iff \rho_{\mathfrak{h}} \leq \rho_{\mathfrak{g}/\mathfrak{h}}.$$

REMARK 2.10. The assumption that G and H are connected is not very important. As we shall explain in Corollary 3.3, Theorem 2.9 is still true when G and H have finitely many connected components as soon as the identity component G_e has finite center.

REMARK 2.11. When H is algebraic and $\mathfrak{a}$ is a maximal abelian split Lie subalgebra of $\mathfrak{h}$, inequality $\rho_{\mathfrak{h}} \leq \rho_{\mathfrak{g}/\mathfrak{h}}$ holds on $\mathfrak{h}$ if and only if it holds on $\mathfrak{a}$.

REMARK 2.12. When H is a minimal parabolic subgroup of G the representation of G in $L^2(G/H)$ is tempered because the group H is amenable. Our criterion is easy to check in this case since the functions $\rho_{\mathfrak{h}}$ and $\rho_{\mathfrak{g}/\mathfrak{h}}$ are equal. This example explains why, when H is nonunimodular, our temperedness criterion involves the functions ρ_V instead of the functions ρ_V^+.

3 Preliminary proofs

In this section we state a useful reformulation of Theorem 2.9 and prove the direct implication in Theorem 2.9.

3.1 THE HERZ MAJORATION PRINCIPLE. We first explain how to reduce the proof of Theorem 2.9 to the case where both G and H are algebraic and how to deal with groups having finitely many connected components.

PROPOSITION 3.1. *Let G be a semisimple Lie group with finitely many components such that the identity component G_e has finite center and $H' \subset H$ two closed subgroups of G.*

(1) *If $L^2(G/H)$ is G-tempered, then $L^2(G/H')$ is G-tempered.*
(2) *The converse is true when H' is normal in H and H/H' is amenable (for instance, finite, compact, or abelian).*

LEMMA 3.2. *Let G be a semisimple Lie group with finitely many connected components such that G_e has finite center, and let H be a closed subgroup of G. If the regular representation in $L^2(G/H)$ is G-tempered, then the induced representation $\Pi = \mathrm{Ind}_H^G(\pi)$ is also G-tempered for any unitary representation π of H.*

Proof of Lemma 3.2. This classical lemma is called the *Herz majoration principle* (see [2, chapter 6]). We recall the short argument since it will be very useful in Proposition 4.9. We use freely the notation of section 2. For a function φ in the space $L^2(G/H, \mathcal{H}_\pi)$ of the induced representation $\Pi = \mathrm{Ind}_H^G(\pi)$, we denote by $|\varphi|$ the function in the space $L^2(G/H)$ of the regular representation $\Pi_0 = \mathrm{Ind}_H^G(\mathbf{1})$ given by $|\varphi|(x) := \|\varphi(x)\|$ for x in G/H. The space $\mathcal{D}$ of bounded functions with compact support is dense in $L^2(G/H, \mathcal{H}_\pi)$. For φ and ψ in $\mathcal{D}$, one can compute the matrix coefficients

$$\langle \Pi(g)\varphi, \psi\rangle = \int_{G/H} c(g^{-1}, x)^{1/2} \langle \pi(\sigma(g, g^{-1}x))\varphi(g^{-1}x), \psi(x)\rangle \, d\nu(x),$$

$$|\langle \Pi(g)\varphi, \psi\rangle| \le \int_{G/H} c(g^{-1}, x)^{1/2} \|\varphi(g^{-1}x)\| \, \|\psi(x)\| \, d\nu(x)$$

$$\le \langle \Pi_0(g)|\varphi|, |\psi|\rangle.$$

Since Π_0 is tempered, these matrix coefficients belong to $L^{2+\varepsilon}(G)$ for all $\varepsilon > 0$. Therefore the representation Π is almost L^2 and hence is G-tempered by Proposition 2.6. □

Proof of Proposition 3.1.

(1) This follows from Lemma 3.2 applied to the regular representation π of H in $L^2(H/H')$.
(2) Since H/H' is amenable, the trivial representation of H is weakly contained in the regular representation of H in $L^2(H/H')$. Therefore, inducing to G, the regular representation of G in $L^2(G/H)$ is weakly contained in the regular representation of G in $L^2(G/H')$ and hence is G-tempered. □

The following corollary tells us that the temperedness of $L^2(G/H)$ depends only on the Lie algebras $\mathfrak{g}$, $\mathfrak{h}$ and does not change if we replace $\mathfrak{h}$ by its derived Lie algebra $[\mathfrak{h}, \mathfrak{h}]$.

COROLLARY 3.3. *Let G be a semisimple Lie group with finitely many connected components such that G_e has a finite center Z_G, and let H be a closed subgroup with finitely many connected components. Then the following are equivalent:*
(i) $L^2(G/H)$ is G-tempered $\Longleftrightarrow$ (ii) $L^2(G_e/H_e)$ is G_e-tempered $\Longleftrightarrow$
(iii) $L^2(G/HZ_G)$ is G/Z_G-tempered $\Longleftrightarrow$ (iv) $L^2(G/[H,H])$ is G-tempered

Proof. This follows from Proposition 3.1 and Remark 2.2 since the quotients H/H_e, HZ_G/H, and $H/[H,H]$ are amenable groups. □

REMARK 3.4. Corollary 3.3 is useful to reduce the proof of Theorem 2.9 to the case where both G and H are algebraic groups.

Indeed, every semisimple Lie algebra $\mathfrak{g}$ is the Lie algebra of an algebraic group: the group $\mathrm{Aut}(\mathfrak{g})$. Therefore, using *(i)* ⇔ *(ii)* ⇔ *(iii)*, we can assume that G is algebraic.

Moreover, by Chevalley's *théorie des repliques* in [6], for any closed subgroup H of an algebraic group G, there exists two algebraic subgroups H_1 and H_2 of G whose Lie algebras satisfy

$$\mathfrak{h}_1 \subset \mathfrak{h} \subset \mathfrak{h}_2 \text{ and } \mathfrak{h}_1 = [\mathfrak{h}, \mathfrak{h}] = [\mathfrak{h}_2, \mathfrak{h}_2].$$

Therefore, using *(i)* ⇔ *(iv)*, we can assume that H is an algebraic subgroup.

REMARK 3.5. Since the group $[H,H]$ is unimodular, Corollary 3.3 is also useful to reduce the proof of Theorem 2.9 to the case where H is unimodular.

3.2 A STRENGTHENING OF THE MAIN THEOREM.

Theorem 2.9 will be proven by induction on the dimension of G. This induction process forces us to prove simultaneously an apparently stronger theorem that involves $L^2(V)$-valued sections over G/H associated to a finite-dimensional H-module V.

THEOREM 3.6.
Let G be an algebraic semisimple Lie group, H an algebraic subgroup of G, and V a real finite-dimensional algebraic representation of H. Then, one has the equivalence:

$$\mathrm{Ind}_H^G(L^2(V)) \text{ is } G\text{-tempered} \quad \Longleftrightarrow \quad \rho_{\mathfrak{h}} \leq \rho_{\mathfrak{g}/\mathfrak{h}} + 2\,\rho_V.$$

Again, we only need to check this inequality on a maximal split abelian Lie subalgebra $\mathfrak{a}$ of $\mathfrak{h}$. Note also that, by Remark 3.4, Theorem 2.9 is the special case of Theorem 3.6 where $V = \{0\}$.

3.3 THE DIRECT IMPLICATION. We first prove the direct implication in Theorems 2.9 and 3.6.

From now on, we will set $\mathfrak{q} := \mathfrak{g}/\mathfrak{h}$.

PROPOSITION 3.7. *Let G be an algebraic semisimple Lie group, H an algebraic subgroup of G; and V an algebraic representation of H. If the representation* $\Pi = \mathrm{Ind}_H^G(L^2(V))$ *is G-tempered, then one has* $\rho_{\mathfrak{h}} \leq \rho_{\mathfrak{q}} + 2\rho_V$.

Proof. By Equation (2.4) this representation Π is also the regular representation of the G-space $X := G \times_H V$. Let A be a maximal split torus of H and $\mathfrak{a}$ be the Lie algebra of A. We choose an A-invariant decomposition $\mathfrak{g} = \mathfrak{h} \oplus \mathfrak{q}_0$ and small closed balls $B_0 \subset \mathfrak{q}_0$ and $B_V \subset V$ centered at 0. We can see B_V as a subset of X, and the map

$$B_0 \times B_V \longrightarrow G \times_H V, \qquad (u, v) \mapsto \exp(u)v$$

is a homeomorphism onto its image C. Since Π is tempered, one has a bound as in Equation (2.6),

$$\langle \Pi(g) 1_C, 1_C \rangle \leq M_C \, \Xi(g) \ \text{ for all } g \text{ in } G. \tag{3.1}$$

We will exploit this bound for elements $g = e^Y$ with Y in $\mathfrak{a}$. In our coordinate system (3.1), we can choose the measure ν_X to coincide with the Lebesgue measure on $\mathfrak{q}_0 \oplus V$. Taking into account the Radon-Nykodim derivative and the A-invariance of $\mathfrak{q}_0$, one computes

$$\langle \Pi(e^Y) 1_C, 1_C \rangle \geq e^{-Tr_{\mathfrak{q}_0}(Y)/2} e^{-Tr_V(Y)/2} \mathrm{vol}_{\mathfrak{q}_0}(e^Y B_0 \cap B_0) \, \mathrm{vol}_V(e^Y B_V \cap B_V),$$

and therefore, using Lemma 2.8, one deduces

$$\langle \Pi(e^Y) 1_C, 1_C \rangle \geq m_C \, e^{-\rho_{\mathfrak{q}}(Y)} e^{-\rho_V(Y)} \ \text{ for all } Y \text{ in } \mathfrak{a}. \tag{3.2}$$

Combining Equations (3.1) and (3.2) with known bounds for the spherical function Ξ as in [15, proposition 7.15], one gets, for suitable positive constants d, C,

$$\frac{m_C}{M_C} e^{-\rho_{\mathfrak{q}}(Y) - \rho_V(Y)} \leq \Xi(e^Y) \leq M_0 \, (1 + \|Y\|)^d e^{-\rho_{\mathfrak{g}}(Y)/2} \ \text{ for all } Y \text{ in } \mathfrak{a}.$$

Therefore one has $\rho_{\mathfrak{g}} \leq 2\,\rho_{\mathfrak{q}} + 2\,\rho_V$, and hence $\rho_{\mathfrak{h}} \leq \rho_{\mathfrak{q}} + 2\,\rho_V$ as required. □

3.4 EQUIVALENCE OF THE MAIN THEOREMS. We have already noticed that Theorem 2.9 is a special case of Theorem 3.6. We explain now why Theorem 3.6 is a consequence of Theorem 2.9.

PROPOSITION 3.8. *Let G be an algebraic semisimple Lie group. If the conclusion of Theorem 2.9 is true for all algebraic subgroups H of G, then the conclusion of Theorem 3.6 is also true for all algebraic subgroups H of G.*

The proof relies on the following lemma.

LEMMA 3.9. *Let H be a Lie group and V a finite-dimensional representation of H. Let $v \in V$ be a point whose orbit Hv has maximal dimension and H_v be the stabilizer of v in H. Then the action of H_v on $V/\mathfrak{h}v$ is trivial.*

Proof of Lemma 3.9. Assume by contradiction that there exist Y in $\mathfrak{h}$ and w in V such that the vector Yw does not belong to $\mathfrak{h}v$.

Choose a complementary subspace $\mathfrak{m}$ of $\mathfrak{h}_v$ in $\mathfrak{h}$ so that $\mathfrak{h} = \mathfrak{h}_v \oplus \mathfrak{m}$. Choose also a point $v_\varepsilon = v + \varepsilon w$ near v. For ε small, the tangent space $\mathfrak{h}\, v_\varepsilon$ to the orbit $H\, v_\varepsilon$ contains both the subspace $\mathfrak{m}\, v_\varepsilon$, which is near $\mathfrak{m}\, v = \mathfrak{h}\, v$, and the vector $\varepsilon^{-1} Yv_\varepsilon = Yw$. Therefore, for ε small, one has the inequality $\dim \mathfrak{h}v_\varepsilon > \dim \mathfrak{h}v$, which gives us a contradiction. □

Proof of Proposition 3.8. We assume that $\rho_{\mathfrak{h}} \leq \rho_{\mathfrak{q}} + 2\,\rho_V$, and we want to prove, using Theorem 2.9, that the regular representation of G in $L^2(G \times_H V)$ is tempered. Since the action is algebraic, there exists a Borel measurable subset $T \subset V$ that meets each of these H-orbits in exactly one point. Let ν_V be a probability measure on V with positive density and ν_T be the probability measure on $T \simeq H\backslash V$ given as the image of ν_V. One has a direct integral decomposition of the regular representation

$$L^2(G \times_H V) = \int_T^{\oplus} L^2(G/H_v)\, d\nu_T(v),$$

where H_v is the stabilizer of v in H. Since the direct integral of tempered representations is tempered, we only need to prove that, for ν_T-almost all v in T,

$$L^2(G/H_v) \text{ is } G\text{-tempered.} \tag{3.3}$$

Our assumption implies that

$$\rho_{\mathfrak{h}}(Y) \le \rho_{\mathfrak{q}}(Y) + 2\,\rho_V(Y) \text{ for all } Y \text{ in } \mathfrak{h}_v. \tag{3.4}$$

For ν_T-almost all v in T, the orbit Hv has maximal dimension; hence, by Lemma 3.9, the action of $\mathfrak{h}$ on the quotient $V/\mathfrak{h}v$ is trivial, and therefore one has the equality

$$\rho_V(Y) = \rho_{\mathfrak{h}}(Y) - \rho_{\mathfrak{h}_v}(Y) \text{ for all } Y \text{ in } \mathfrak{h}_v. \tag{3.5}$$

Combining (3.4) and (3.5), one gets, for ν_T-almost all v in T,

$$2\,\rho_{\mathfrak{h}_v}(Y) \le \rho_{\mathfrak{q}}(Y) + \rho_{\mathfrak{l}}(Y) - \rho_{\mathfrak{g}}(Y) \text{ for all } Y \text{ in } \mathfrak{h}_v,$$

which can be rewritten as the temperedness criterion $\rho_{\mathfrak{h}_v}(Y) \le \rho_{\mathfrak{q}_v}(Y)$ for $L^2(G/H_v)$ in Theorem 2.9 and hence proves Equation (3.3). □

4 Using parabolic subgroups

The aim of this section is to prove the converse implication in Theorem 2.9. As we have seen in Remarks 3.4 and 3.5, we can assume that G is a Zariski connected algebraic group and that H is a Zariski connected algebraic subgroup such that $\mathfrak{h} = [\mathfrak{h}, \mathfrak{h}]$.

The proof relies on the presence of two nice intermediate subgroups,

$$H \subset F \subset P \subset G.$$

4.1 THE INTERMEDIATE SUBGROUPS. We first explain the construction of these intermediate subgroups, F and P.

Let G be an algebraic semisimple Lie group and H a Zariski connected algebraic subgroup of G such that $\mathfrak{h} = [\mathfrak{h}, \mathfrak{h}]$.

LEMMA-DEFINITION 4.1. *We fix a parabolic subgroup P of G of minimal dimension that contains H and denote by U the unipotent radical of P. There exists a reductive subgroup $L \subset P$ such that $P = LU$ and $H = (L \cap H)(U \cap H)$. Moreover the group $S := L \cap H$ is semisimple and the group $V := U \cap H$ is the unipotent radical of H. We denote by F the group $F = SU$.*

Proof. The group $V := U \cap H$ is a unipotent normal subgroup of H. The quotient $S' := H/V$ is a Zariski connected subgroup of the reductive group P/U, which is not contained in any proper parabolic subgroup of P/U. Therefore, by [5, section 8.10] this group S' is reductive. Since $\mathfrak{h} = [\mathfrak{h}, \mathfrak{h}]$, this group S' is semisimple and there exists a semisimple subgroup $S \subset H$ such that $H = SV$. Since S is semisimple, the group V is the unipotent radical of H. Since maximal reductive subgroups L of P are U-conjugate, one can choose L containing S and therefore one has $S = L \cap H$. □

The following two lemmas will be useful in our induction process.

LEMMA 4.2. *With the notation of Definition 4.1, the following two functions on* $\mathfrak{s}$ *are equal:*

$$\rho_{\mathfrak{g}/\mathfrak{h}} - \rho_{\mathfrak{h}} = \rho_{\mathfrak{l}/\mathfrak{s}} + 2\,\rho_{\mathfrak{u}/\mathfrak{v}} - \rho_{\mathfrak{s}}. \tag{4.1}$$

Proof. Since $\rho_{\mathfrak{g}/\mathfrak{p}} = \rho_{\mathfrak{u}}$, one has the equalities of functions on $\mathfrak{s}$,

$$\rho_{\mathfrak{g}/\mathfrak{h}} = \rho_{\mathfrak{u}} + \rho_{\mathfrak{l}/\mathfrak{s}} + \rho_{\mathfrak{u}/\mathfrak{v}} \quad \text{and} \quad \rho_{\mathfrak{h}} = \rho_{\mathfrak{s}} + \rho_{\mathfrak{v}}.$$ □

LEMMA 4.3. *Let $P = LU$ be a real algebraic group that is a semidirect product of a reductive subgroup L and its unipotent radical U. Let π_0 be a unitary representation of P that is L-tempered and trivial on U. Then the representation π_0 is also P-tempered.*

Proof. The weak containment $\pi_0 \prec L^2(L)$ as unitary representations of L implies the weak containment $\pi_0 \prec L^2(P/U)$ as unitary representations of P because U acts trivially on both sides. Since U is amenable, the trivial representation of U is U-tempered; therefore by Lemma 2.3 the regular representation of P in $L^2(P/U)$ is P-tempered, and π_0 is also P-tempered. □

4.2 BOUNDING VOLUME OF COMPACT SETS. The proof of Theorem 2.9 relies on a control of the volume of the intersection of translates of compact sets in $X = G/H$. We first explain how to bound such volumes in $Z = F/H$. This bound is quite general.

PROPOSITION 4.4. *Let $F = SU$ be a real algebraic group that is a semidirect product of a reductive subgroup S and its unipotent radical U. Let $H = SV$ be an algebraic subgroup of F containing S where $V = U \cap H$. Let Z be the F space $Z = F/H = U/V$ endowed with a U-invariant Radon measure. Then for every compact* *subset $D \subset Z$, there exists a compact subset $D_0 \subset Z$ such that for all $s \in S$ and $u \in U$, one has*

$$\mathrm{vol}(suD \cap D) \leq \mathrm{vol}(sD_0 \cap D_0). \tag{4.2}$$

Here is the reformulation of Proposition 4.4 that we will use later.

DEFINITION 4.5. Let Z_0 be the same space $Z_0 = U/V$ as Z but endowed with another F-action where U acts trivially and where S acts by conjugation.

COROLLARY 4.6. *Same notation as in Proposition 4.4. Then for every compact subset $D \subset Z$, there exists a compact subset $D_0 \subset Z_0$ such that for every $f \in F$, one has*

$$\mathrm{vol}(fD \cap D) \leq \mathrm{vol}(fD_0 \cap D_0). \tag{4.3}$$

The proof of Proposition 4.4 is by induction on the dimension of Z. It relies only on geometric arguments and uses no representation theory.

Before studying the proof of Proposition 4.4 the reader could as an exercise focus on the following very simple example where $Z = \mathbb{R}^2$ is the affine 2-plane and F is the group of affine bijections

$$\begin{pmatrix} a & r \\ 0 & b \end{pmatrix} \begin{pmatrix} t \\ s \end{pmatrix} .$$

that preserve the horizontal foliation. In this case, S is the two-dimensional group of diagonal matrices and U is the three-dimensional Heisenberg group. The proof for this example relies on the same ideas while being very concrete.

Proof of Proposition 4.4.

Case 1: S is a split torus.

We denote by C the center of U and $C_V = V \cap C$. Let W be the closed subgroup $W := VC \subset U$. The projection

$$Z = U/V \longrightarrow Z' := U/W$$

is a principal bundle of group $C_W := C/C_V = W/V$. According to Lemma 4.7, there exists a continuous trivialization of this principal bundle

$$Z \simeq Z' \times C_W \tag{4.4}$$

such that the action of U and S through this trivialization can be read as

$$su\,(z', c) = (suz',\ sc + sc_0(u, z')) \tag{4.5}$$

for all $s \in S$, $u \in U$, $z' \in Z'$, $c \in C_W$, where c_0 is a continuous cocycle $c_0 : U \times Z' \to C_W$. We fix three compatible invariant measures vol_Z, $\mathrm{vol}_{Z'}$, and vol on Z, Z', and C_W.

We start with a compact set $D \subset Z$. Through the trivialization in Equation (4.4), this set D is included in a product of two compact sets $D' \subset Z'$ and $B \subset C_W$,

$$D \subset D' \times B,$$

where B is a symmetric convex set in the group C_W seen as a real vector space. By the induction hypothesis, there exists a compact set $D'_0 \subset Z'$ that satisfies the bound of Equation (4.2) for D', such that

$$\mathrm{vol}_{Z'}(suD' \cap D') \leq \mathrm{vol}_{Z'}(sD'_0 \cap D'_0) \ \text{ for all } s \in S,\ u \in U. \tag{4.6}$$

We compute using Equation (4.5) and Lemma 4.8, for all $s \in S$ and $u \in U$,

$$\begin{aligned} \mathrm{vol}_Z(suD \cap D) &\leq \int_{suD' \cap D'} \mathrm{vol}((sB + s\,c_0(u, (su)^{-1}z')) \cap B)\, dz' \\ &\leq \int_{suD' \cap D'} \mathrm{vol}(sB \cap B)\, dz', \end{aligned}$$

where dz' also denotes the U-invariant measure on Z'. Hence, using (4.6), we go on with

$$\begin{aligned} \mathrm{vol}_Z(suD \cap D) &\leq \mathrm{vol}_{Z'}(suD' \cap D')\, \mathrm{vol}(sB \cap B) \\ &\leq \mathrm{vol}_{Z'}(sD'_0 \cap D'_0)\, \mathrm{vol}(sB \cap B) \\ &= \mathrm{vol}_Z(sD_0 \cap D_0), \end{aligned}$$

where D_0 is the compact subset of Z given by $D_0 := D'_0 \times B$.

Case 2: S is a reductive group.

This general case will be deduced from the first case. Indeed, any reductive group admits a Cartan decomposition $S = K_s A_s K_s$, where K_s is a maximal compact subgroup of S and A_s is a maximal split torus of S. We start with a compact set D of Z. According to the first case, there exists a K_S-invariant compact set $D_0 \subset Z$ such that, for all $a \in A_S$ and $u \in U$, one has

$$\mathrm{vol}(auK_S D \cap K_S D) \leq \mathrm{vol}(aD_0 \cap D_0).$$

Therefore, for all s in S and u in U, writing $s = k_1 a k_2$ with k_1, k_2 in K_S and a in A_S, one has

$$\begin{aligned} \mathrm{vol}(suD \cap D) &\leq \mathrm{vol}(a(k_2 u k_2^{-1})k_2 D \cap k_1^{-1} D) \\ &\leq \mathrm{vol}(aD_0 \cap D_0) \\ &= \mathrm{vol}(sD_0 \cap D_0), \end{aligned}$$

as required. □

In the proof of Proposition 4.4, we have used the following two lemmas.

LEMMA 4.7. *Let U be a unipotent group, $V \subset U$ a unipotent subgroup, C be the center of U, $W := VC$, and $C_V := C \cap V$. Let $S \subset \mathrm{Aut}(U)$ be a split torus that preserves V. Then there exists a continuous trivialization of the U-equivariant principal bundle $U/V \to U/W$ with structure group C/C_V,*

$$U/V \simeq U/W \times C/C_V,$$

such that the action of U and S through this trivialization can be read as

$$su(y, c) = (suy,\ sc + sc_0(u, y))$$

for all $u \in U$, $s \in S$, $y \in U/W$, $c \in C/C_V$, where c_0 is a continuous cocycle $c_0 : U \times U/W \to C/C_V$.

Proof of Lemma 4.7. These claims are a variation of a classical result of Chevalley-Rosenlicht (see, for instance, [7, theorem 3.1.4]). The proof relies on the existence of "an adapted basis in a nilpotent Lie algebra." Here is a sketch of proof of these claims.

As usual, let $\mathfrak{u}$, $\mathfrak{v}$, $\mathfrak{c}$, and $\mathfrak{w}$ be the Lie algebras of the groups U, V, C, and W. Let I be the ordered set $I = \{1, \ldots, n\}$, where $n = \dim \mathfrak{u}$. We fix a basis $(e_i)_{i \in I}$ of $\mathfrak{u}$, such that

- for every $i \geq 1$, the vector space spanned by the e_j for $j \geq i$ is an ideal;
- for every $i \geq 1$, the line $\mathbb{R}e_i$ is invariant by S;
- there exists a subset $I_V \subset I$ such that $\mathfrak{v}$ is spanned by e_i for $i \in I_V$;
- there exists a subset $I_C \subset I$ such that $\mathfrak{c}$ is spanned by e_i for $i \in I_C$; and
- the Lie algebra $\mathfrak{w}$ is spanned by e_i for $i \in I_W := I_C \cup I_V$.

Then, the map

$$\Psi: \ \mathbb{R}^I \longrightarrow U, \quad (t_i)_{i\in I} \mapsto \prod_{i\in I} \exp(t_i e_i),$$

where the product is performed using the order on I, is a diffeomorphism, and one has

$$\Psi(\mathbb{R}^{I_V}) = V, \ \Psi(\mathbb{R}^{I_C}) = C, \ \text{and} \ \Psi(\mathbb{R}^{I_W}) = W.$$

Setting $J_V := I \setminus I_V$ and $J_W := I \setminus I_W$, the map

$$\Psi_V: \ \mathbb{R}^{J_V} \longrightarrow U/V, \quad (t_i)_{i\in J_V} \mapsto \prod_{i\in J_V} \exp(t_i e_i)V$$

is also a diffeomorphism, and the restriction of this map to the subset $\mathbb{R}^{J_W}$ gives an S-equivariant section of the bundle $U/V \to U/W$. □

Here is the second basic lemma used in the proof of Proposition 4.4.

LEMMA 4.8. *Let B, B' be two symmetric convex sets of $\mathbb{R}^d$; then one has*

$$\text{vol}((B+v)\cap B')) \leq \text{vol}(B\cap B') \ \text{ for all } v \in \mathbb{R}^d.$$

Proof. By the Brunn-Minkowski inequality (see [4, section 11]), the map $v \mapsto \text{vol}((B+v)\cap B')^{1/d}$ is concave on the convex set $B' - B$ and hence achieves its maximum value at $v = 0$. □

4.3 MATRIX COEFFICIENTS OF INDUCED REPRESENTATIONS.

We now explain how to control the volume of the intersection of translates of compact sets in the G-space $X = G/H$ with those in $X_0 := G \times_F Z_0$.

PROPOSITION 4.9. *Let G be an algebraic semisimple Lie group and H a Zariski connected algebraic subgroup such that $\mathfrak{h} = [\mathfrak{h}, \mathfrak{h}]$. Let $P = LU$, $F = SU$, and $H = SV$ be the groups introduced in Definition 4.1. Let Z_0 be the F-space introduced in*

Definition 4.5 and X_0 the G-space $X_0 := G \times_F Z_0$. Then, for every compact subset $C \subset G/H$, there exists a compact subset $C_0 \subset X_0$ such that

$$\mathrm{vol}(g\,C \cap C) \leq \mathrm{vol}(g\,C_0 \cap C_0) \quad \text{for all } g \text{ in } G. \tag{4.7}$$

In Proposition 4.9 the assumption $\mathfrak{h} = [\mathfrak{h}, \mathfrak{h}]$ can be removed but the conclusion in Equation (4.7) becomes slightly more technical when there is no G-invariant measure on G/H. Indeed, when $\mathfrak{h} \neq [\mathfrak{h}, \mathfrak{h}]$, one has to replace the bound (4.7) by a bound of K-finite matrix coefficients of the induced representation $\Pi = \mathrm{Ind}_F^G(L^2(F/H))$ thanks to K-finite matrix coefficients of the induced representation $\Pi_0 = \mathrm{Ind}_F^G(L^2(Z_0))$.

Proof of Proposition 4.9. The projection

$$G \to X' := G/F$$

is a G-equivariant principal bundle with structure group F. As in section 2.1, we fix a Borel measurable trivialization of this principal bundle

$$G \simeq X' \times F, \tag{4.8}$$

which sends relatively compact subsets to relatively compact subsets. The action of G by left-multiplication through this trivialization can be read as

$$g\,(x', f) = (gx', \sigma_F(g, x')f) \quad \text{for all } g \in G,\ x' \in X', \text{ and } f \in F,$$

where $\sigma_F \colon G \times X' \to F$ is a Borel measurable cocycle. This trivialization in Equation (4.8) induces a trivialization of the associated bundles

$$X = G \times_F Z \simeq X' \times Z,$$
$$X_0 = G \times_F Z_0 \simeq X' \times Z_0.$$

We start with a compact set C of X. Through the first trivialization, this compact set is included in a product of two compact sets $C' \subset X'$ and $D \subset Z$:

$$C \subset C' \times D \tag{4.9}$$

We denote by $D_0 \subset Z_0$ the compact set given by Corollary 4.6 and we compute using Equation (4.3) for g in G,

$$\begin{aligned}\mathrm{vol}_X(g\,C \cap C) &\leq \int_{gC' \cap C'} \mathrm{vol}_Z(\sigma_F(g, g^{-1}x')D \cap D)\, dx' \\ &\leq \int_{gC' \cap C'} \mathrm{vol}_{Z_0}(\sigma_F(g, g^{-1}x')D_0 \cap D_0)\, dx' \\ &\leq \mathrm{vol}_{X_0}(g\,C_0 \cap C_0),\end{aligned}$$

where dx' is a G-invariant measure on X' and C_0 is a compact subset of $X_0 \simeq X' \times Z_0$, which contains $C' \times D_0$. □

4.4 PROOF OF THE TEMPEREDNESS CRITERION. We conclude the proof of Theorem 2.9.

Proof of the converse implication in Theorem 2.9. We prove it by induction on the dimension of G. By Remarks 3.4 and 3.5, we can assume that G is a Zariski connected semisimple algebraic group and that H is a Zariski connected algebraic subgroup such that $\mathfrak{h} = [\mathfrak{h}, \mathfrak{h}]$. Let

$$H = SV \subset F = SU \subset P = LU \subset G$$

be the groups introduced in Definition 4.1. Let $Z_0 = U/V$ be the F-space introduced in Definition 4.5 and X_0 be the G-space $X_0 = G \times_F Z_0$.

When P is equal to G, the group H is semisimple and we apply [3, theorem 3.1]. We now assume that P is a proper parabolic subgroup of G.

By assumption one has $\rho_{\mathfrak{h}} \leq \rho_{\mathfrak{g}/\mathfrak{h}}$ on $\mathfrak{h}$. Therefore, by Lemma 4.2, one has the inequality on $\mathfrak{s}$,

$$\rho_{\mathfrak{s}} \leq \rho_{\mathfrak{l}/\mathfrak{s}} + 2\,\rho_{\mathfrak{u}/\mathfrak{v}}. \tag{4.10}$$

We introduce the regular representation π_0 of P in $L^2(P \times_F Z_0)$, which is unitarily equivalent to $\mathrm{Ind}_F^P(L^2(\mathfrak{u}/\mathfrak{v}))$ by the isomorphism in Equation (2.4). As a representation of L, one has

$$\pi_0|_L = \mathrm{Ind}_S^L(L^2(\mathfrak{u}/\mathfrak{v})).$$

Using our induction hypothesis on the dimension of G to the derived subgroup of L, Proposition 3.8 and Remark 2.4 tell us that the representation π_0

is L-tempered by Equation (2.4). Therefore, by Lemma 4.3, the representation π_0 is P-tempered. The regular representation Π_0 of G in $L^2(G \times_F Z_0)$ is unitarily equivalent to $\Pi_0 = \mathrm{Ind}_P^G(\pi_0)$ because $\Pi_0 \simeq \mathrm{Ind}_P^G(\mathrm{Ind}_F^P(L^2(\mathfrak{u}/\mathfrak{v})))$. Now, Lemma 2.3 implies that this representation Π_0 is G-tempered. Therefore, by Corollary 2.7, for any K-invariant compact subset C_0 of $G \times_F Z_0$, one has a bound

$$\mathrm{vol}(g\, C_0 \cap C_0) \leq \mathrm{vol}(C_0)\, \Xi(g) \quad \text{for all } g \text{ in } G.$$

Hence, by Proposition 4.9, for any compact subset C of G/H, one also has a bound

$$\mathrm{vol}(g\, C \cap C) \leq M_C\, \Xi(g) \quad \text{for all } g \text{ in } G.$$

Again by Corollary 2.7, this tells us that the representation of G in $L^2(G/H)$ is G-tempered. □

5 Examples

The criterion given in Theorem 2.9 allows us to easily detect for a given homogeneous space G/H whether the unitary representation of a semisimple Lie group G in $L^2(G/H)$ is tempered or not. We collect in this chapter a few examples, omitting the details of the computational verifications.

5.1 EXAMPLES OF TEMPERED HOMOGENEOUS SPACES. We first recall a few examples extracted from [3] where H is reductive.

EXAMPLE 5.1. $L^2(SL(p+q,\mathbb{R})/SO(p,q))$ is always tempered.
$L^2(SL(2m,\mathbb{R})/Sp(m,\mathbb{R}))$ is never tempered.
$L^2(SL(m+n,\mathbb{C})/SL(m,\mathbb{C}) \times SL(n,\mathbb{C}))$ is tempered iff $|m-n| \leq 1$.
$L^2(SO(m+n,\mathbb{C})/SO(m,\mathbb{C})) \times SO(n,\mathbb{C}))$ is tempered iff $|m-n| \leq 2$.
$L^2(Sp(m+n,\mathbb{C})/Sp(m,\mathbb{C}) \times Sp(n,\mathbb{C}))$ is tempered iff $m=n$.

EXAMPLE 5.2. Let $n = n_1 + \cdots + n_r$ with $n_1 \geq \cdots \geq n_r \geq 1$, $r \geq 2$.
$L^2(SL(n,\mathbb{R})/\prod SL(n_i,\mathbb{R}))$ is tempered iff $2n_1 \leq n+1$.
$L^2(Sp(n,\mathbb{R})/\prod Sp(n_i,\mathbb{R}))$ is tempered iff $2n_1 \leq n$.
Let $p = p_1 + p_2$, $q = q_1 + q_2$ with $p_1, p_2, q_1, q_2 \geq 1$.
$L^2(SO(p,q)/SO(p_1,q_1) \times SO(p_2,q_2))$ is tempered iff $|p_1 + q_1 - p_2 - q_2| \leq 2$.

EXAMPLE 5.3. Let G be an algebraic semisimple Lie group and K a maximal compact subgroup. $L^2(G_{\mathbb{C}}/K_{\mathbb{C}})$ is $G_{\mathbb{C}}$-tempered iff G is quasisplit.

REMARK 5.4. A way to justify this last example is to notice that our criterion $2\rho_{\mathfrak{k}_{\mathbb{C}}} \leq \rho_{\mathfrak{g}_{\mathbb{C}}}$ means that *the trivial K-type is a small K-type of G* in the terminology of Vogan's paper [29, definition 6.1] (see also Knapp's book [15, chapter 15]) and to use the following equivalences due to Vogan in the same paper [29, theorem 6.4]:

$$G \text{ has a small } K\text{-type} \Longleftrightarrow \text{the trivial } K\text{-type is small} \Longleftrightarrow G \text{ is quasisplit}$$

Here is a delicate example for semisimple symmetric spaces.

EXAMPLE 5.5. Let $G/H := Sp(2,1)/Sp(1) \times Sp(1,1)$. The Plancherel formula [9, 27] tells us that both the continuous part and a "generic portion" of the discrete part of $L^2(G/H)$ are tempered; however, our criterion in Equation (1.1) tells us that $L^2(G/H)$ is nontempered because $\rho_{\mathfrak{h}}(Y) = \frac{3}{2}\rho_{\mathfrak{q}}(Y) > \rho_{\mathfrak{q}}(Y)$ if Y is a nonzero hyperbolic element of $\mathfrak{h}$. In fact, the discrete part of $L^2(G/H)$ consists of Harish-Chandra's discrete series representations, say $\pi_n (n = 1, 2, \ldots)$, and two more nonvanishing representations π_0 and π_{-1} in the coherent family, where π_0 is still tempered but π_{-1} is nontempered ([17, theorem 1]).

Here is another direct application of our criterion in Equation (1.1) where H is not anymore assumed to be reductive.

COROLLARY 5.6. *Let G be an algebraic semisimple Lie group and H an algebraic subgroup.*

(1) *If the representation of $G_{\mathbb{C}}$ in $L^2(G_{\mathbb{C}}/H_{\mathbb{C}})$ is tempered, then the representation of G in $L^2(G/H)$ is tempered.*
(2) *The converse is true if H contains a maximal torus that is split.*

5.2 SUBGROUPS OF $SL(N,\mathbb{R})$. We now explain how to check our criterion in Equation (1.1) with a very concrete example.

In Table 6.1, we specify criterion (1.1) when $G = \mathrm{SL}(\mathbb{R}^p \oplus \mathbb{R}^q)$ and H is a subgroup of G normalized by the group $\mathrm{SL}(\mathbb{R}^p) \times \mathrm{SL}(\mathbb{R}^q)$.

In Table 6.2, we specify criterion (1.1) when $G = \mathrm{SL}(\mathbb{R}^p \oplus \mathbb{R}^q \oplus \mathbb{R}^r)$ and H is a subgroup of G normalized by the group $\mathrm{SL}(\mathbb{R}^p) \times \mathrm{SL}(\mathbb{R}^q) \times \mathrm{SL}(\mathbb{R}^r)$. Note that in these two tables, the center of the diagonal blocks is not important by Corollary 3.3.

REMARK 5.7. It is rather easy to guess the inequalities in Table 6.2. Here is the heuristic recipe: There is one inequality for each nonidentity diagonal

Table 6.1. The criterion $\rho_{\mathfrak{h}} \leq \rho_{\mathfrak{g}/\mathfrak{h}}$ when $G = \mathrm{SL}(p+q, \mathbb{R})$

$H_1: \begin{pmatrix} * & 0 \\ 0 & I \end{pmatrix}$	$H_2: \begin{pmatrix} * & * \\ 0 & I \end{pmatrix}$	$H_3: \begin{pmatrix} * & * \\ 0 & * \end{pmatrix}$	$H_4: \begin{pmatrix} * & 0 \\ 0 & * \end{pmatrix}$
$p \leq q+1$	$p=1$	$p=q=1$	$p \leq q+1$ $q \leq p+1$

Table 6.2. The criterion $\rho_{\mathfrak{h}} \leq \rho_{\mathfrak{g}/\mathfrak{h}}$ when $G = \mathrm{SL}(p+q+r, \mathbb{R})$

$H_1: \begin{pmatrix} * & 0 & * \\ 0 & I & 0 \\ 0 & 0 & I \end{pmatrix}$	$H_2: \begin{pmatrix} I & 0 & * \\ 0 & * & 0 \\ 0 & 0 & I \end{pmatrix}$	$H_3: \begin{pmatrix} I & * & * \\ 0 & * & 0 \\ 0 & 0 & I \end{pmatrix}$	$H_4: \begin{pmatrix} * & * & * \\ 0 & * & * \\ 0 & 0 & * \end{pmatrix}$
$p \leq q+1$	$q < p+r+1$	$q \leq r+1$	$p = q-r-1$
$H_5: \begin{pmatrix} * & 0 & 0 \\ 0 & * & 0 \\ 0 & 0 & I \end{pmatrix}$	$H_6: \begin{pmatrix} * & 0 & * \\ 0 & * & 0 \\ 0 & 0 & I \end{pmatrix}$	$H_7: \begin{pmatrix} * & * & * \\ 0 & * & 0 \\ 0 & 0 & I \end{pmatrix}$	$H_8: \begin{pmatrix} * & 0 & * \\ 0 & I & 0 \\ 0 & 0 & * \end{pmatrix}$
$p \leq q+r+1$ $q \leq p+r+1$	$p \leq q+1$ $q \leq p+r+1$	$p=1$ $q \leq r+1$	$p \leq q+1$ $r \leq q+1$
$H_9: \begin{pmatrix} I & * & * \\ 0 & * & 0 \\ 0 & 0 & * \end{pmatrix}$	$H_{10}: \begin{pmatrix} * & 0 & 0 \\ 0 & * & 0 \\ 0 & 0 & * \end{pmatrix}$	$H_{11}: \begin{pmatrix} * & 0 & * \\ 0 & * & 0 \\ 0 & 0 & * \end{pmatrix}$	$H_{12}: \begin{pmatrix} * & * & * \\ 0 & * & 0 \\ 0 & 0 & * \end{pmatrix}$
$q \leq r+1$ $r \leq q+1$	$p \leq q+r+1$ $q \leq p+r+1$ $r \leq p+q+1$	$p \leq q+1$ $q \leq p+r+1$ $r \leq q+1$	$p=1$ $q \leq r+1$ $r \leq q+1$

block. The left-hand side of this inequality is given by the size of this diagonal block, while the right-hand side can be guessed by looking at the size of the zero blocks on the right and on the top of it.

We will explain the proof for just the group $H = H_{11}$ in Table 6.2. The other cases are similar.

COROLLARY 5.8. *Let* $G = \mathrm{SL}(p+q+r, \mathbb{R})$ *and* H *the subgroup of matrices*

$$\begin{pmatrix} \alpha & 0 & z \\ 0 & \beta & 0 \\ 0 & 0 & \gamma \end{pmatrix}$$

with $\alpha \in \mathrm{GL}(p, \mathbb{R})$, $\beta \in \mathrm{GL}(q, \mathbb{R})$, $\gamma \in \mathrm{GL}(r, \mathbb{R})$, $z \in \mathrm{M}(p, r; \mathbb{R})$. *Then* $L^2(G/H)$ *is* G*-tempered if and only if* $p \leq q+1$, $q \leq p+r+1$, $r \leq q+1$.

Proof of Corollary 5.8. We denote by $\mathfrak{a}$ the Lie algebra of diagonal matrices

$$\mathfrak{a} = \{Y = (x, y, z) \in \mathbb{R}^p \oplus \mathbb{R}^q \oplus \mathbb{R}^r \mid \mathrm{Trace}(Y) = 0\}.$$

We only need to check the criterion in Equation (1.1) on the chamber

$$\mathfrak{a}_+ = \{Y = (x, y, z) \in \mathfrak{a} \mid \ x,\ y, \text{ and } z \text{ have nondecreasing coordinates}\}.$$

We recall that $\mathfrak{q} = \mathfrak{g}/\mathfrak{h}$ and we compute for $Y \in \mathfrak{a}_+$,

$$\rho_{\mathfrak{h}}(Y) = \textstyle\sum_{i=1}^{p} a_i x_i + \sum_{j=1}^{q} b_j y_j + \sum_{k=1}^{r} c_k z_k\,,$$

where $a_i := 2i - p - 1,\ b_j := 2j - q - 1,\ c_k := 2k - r - 1,$ and

$$\rho_{\mathfrak{q}}(Y) = \textstyle\sum_{i,j} |x_i - y_j| + \sum_{j,k} |y_j - z_k|.$$

Assume first that the criterion $\rho_{\mathfrak{h}} \le \rho_{\mathfrak{q}}$ is satisfied on $\mathfrak{a}$. It is then also satisfied on $\mathbb{R}^{p+q+r}$. Applying it successively to the three vectors $Y = e_p$, $Y = e_{p+q}$, and $Y = e_{p+q+r}$ of the standard basis $e_1, \ldots, e_{p+q+r}$ of $\mathbb{R}^{p+q+r}$, one gets successively the three inequalities $p \le q+1$, $q \le p+r+1$, and $r \le q+1$.

Assume now that these three inequalities are satisfied. Note that

$$\begin{aligned}\rho_{\mathfrak{q}}(Y) &\ge \textstyle\sum_{a_i > b_j}(x_i - y_j) + \sum_{b_j > a_i}(y_j - x_i) + \sum_{b_j > c_k}(y_j - z_k) + \sum_{c_k > b_j}(z_k - y_j)\\ &= \textstyle\sum_{i=1}^{p} \ell_i x_i + \sum_{j=1}^{q} m_j y_j + \sum_{k=1}^{r} n_k z_k,\end{aligned}$$

where

$$\begin{aligned}\ell_i &= |\{j \mid b_j < a_i\}| - |\{j \mid b_j > a_i\}|\\ m_j &= |\{i \mid a_i < b_j\}| - |\{i \mid a_i > b_j\}| + |\{k \mid c_k < b_j\}| - |\{k \mid c_k > b_j\}|\\ n_k &= |\{j \mid b_j < c_k\}| - |\{j \mid b_j > c_k\}|.\end{aligned}$$

Since $p \le q+1$, one has $\ell_i = a_i$ for all $1 \le i \le p$.
Since $r \le q+1$, one has $n_k = c_k$ for all $1 \le k \le r$.
For $1 \le j \le q$, one has $m_{q+1-j} = -m_j$ and, when $j > q/2$,

$$m_j = \min(b_j, p) + \min(b_j, r).$$

Since $q \le p+r+1$, one has $m_j \ge b_j$ for all $j > q/2$. Then using the fact that the y_j's are nondecreasing functions of j, one gets, for Y in $\mathfrak{a}_+$,

$$\rho_{\mathfrak{q}}(Y) - \rho_{\mathfrak{h}}(Y) = \textstyle\sum_{j=1}^{q} (m_j - b_j) y_j \ge 0.$$

This proves that the criterion $\rho_{\mathfrak{h}} \le \rho_{\mathfrak{q}}$ is satisfied. □

Some of the subgroups in Table 6.2 appear naturally in analyzing the tensor product representations of $SL(n, \mathbb{R})$ as below.

5.3 TENSOR PRODUCT OF NONTEMPERED REPRESENTATIONS.

Suppose Π and Π' are unitary representations of G. The tensor product representation $\Pi \otimes \Pi'$ is tempered if Π or Π' is tempered. In contrast, $\Pi \otimes \Pi'$ may be and may not be tempered when both Π and Π' are nontempered.

For instance, let $n = n_1 + \cdots + n_k$ be a partition, and we consider the (degenerate) principal series representation $\Pi_{n_1,\cdots,n_k} := \mathrm{Ind}_{P_{n_1,\cdots,n_k}}^{G}(1)$ of $G = SL(n, \mathbb{R})$, where $P_{n_1,\cdots,n_k}$ is the standard parabolic subgroup with Levi subgroup $S(GL(n_1, \mathbb{R}) \times \cdots \times GL(n_k, \mathbb{R}))$. Then $\Pi_{n_1,\cdots,n_k}$ is tempered iff $k = n$ and $n_1 = \cdots = n_k = 1$. Here are some examples of the temperedness criterion in Equation (1.1) applied to the tensor product of two such representations.

PROPOSITION 5.9. *Let $0 \le k, l \le n$, and $a+b+c = n$.*

(1) $\Pi_{k,n-k} \otimes \Pi_{n-l,l}$ *is tempered iff* $|k-l| \le 1$ *and* $|k+l-n| \le 1$.
(2) $\Pi_{a,b,c} \otimes \Pi_{b+c,a}$ *is tempered iff* $\max(b,c) - 1 \le a \le b+c+1$.
(3) $\Pi_{a,b,c} \otimes \Pi_{c,b,a}$ *is tempered iff* $2\max(a,b,c) \le n+1$.

Proof. For any parabolic subgroups P and P' of G, there exists an element $w \in G$ such that PwP' is open dense in G, and thus the tensor product $\mathrm{Ind}_P^G(1) \otimes \mathrm{Ind}_{P'}^G(1)$ is unitarily equivalent to the regular representation in $L^2(G/H)$ by the Mackey theory, where $H = w^{-1}Pw \cap P'$. In the above cases, we have the following unitary equivalences:

$$\Pi_{k,n-k} \otimes \Pi_{n-l,l} \simeq L^2(G/H_{12}) \ \text{ with } (p,q,r) = (|k-l|, \min(k,l), n-\max(k,l)),$$

$$\Pi_{a,b,c} \otimes \Pi_{b+c,a} \simeq L^2(G/H_{11}) \ \text{ with } (p,q,r) = (b,a,c),$$

$$\Pi_{a,b,c} \otimes \Pi_{c,b,a} \simeq L^2(G/H_{10}) \ \text{ with } (p,q,r) = (a,b,c),$$

whence Proposition 5.9 follows from Table 6.2 in section 5.2. □

References

[1] B. Bekka, P. de la Harpe, and A. Valette. *Kazhdan's Property (T)*. Camb. Univ. Press, 2008.

[2] B. Bekka and Y. Guivarc'h. On the spectral theory of groups of affine transformations of compact nilmanifolds. *Ann. Sci. Éc. Norm. Supér.*, 48:607–645, 2015.

[3] Y. Benoist and T. Kobayashi. Tempered reductive homogeneous spaces. *J. Eur. Math. Soc.*, 17:3015–3036, 2015.

[4] T. Bonnesen and W. Fenchel. *Theory of Convex Bodies*. Springer, 1974.

[5] N. Bourbaki. *Lie Groups and Lie Algebras. Chapters 7–9*. Springer, 2005.

[6] C. Chevalley. Algebraic Lie algebras. *Ann. of Math.*, 48:91–100, 1947.

[7] L. Corwin and F. Greenleaf. *Representations of Nilpotent Lie Groups and Their Applications*. Camb. Univ. Press, 1990.

[8] M. Cowling, U. Haagerup, and R. Howe. Almost L^2 matrix coefficients. *J. Reine Angew. Math.*, 387:97–110, 1988.

[9] P. Delorme. Formule de Plancherel pour les espaces symétriques réductifs. *Ann. of Math.*, 147:417–452, 1998.

[10] M. Einsiedler, G. Margulis, and A. Venkatesh. Effective equidistribution for closed orbits of semisimple groups on homogeneous spaces. *Invent. Math.*, 177(1):137–212, 2009.

[11] D. Fisher and G. Margulis. Local rigidity of affine actions of higher rank groups and lattices. *Ann. of Math. (2)*, 170(1):67–122, 2009.

[12] Harish-Chandra. Harmonic analysis on real reductive groups. III. *Ann. of Math.*, 104:117–201, 1976.

[13] B. Harris and T. Weich. Wave front sets of reductive lie group representations iii. *Adv. Math.*, 313:176–236, 2017.

[14] R. Howe and E.-C. Tan. *Nonabelian Harmonic Analysis*. Universitext. Springer, 1992.

[15] A. Knapp. *Representation Theory of Semisimple Groups*. Princeton Landmarks in Mathematics. Princeton University Press, 2001.

[16] A. Knapp and G. Zuckerman. Classification of irreducible tempered representations of semisimple groups. *Ann. of Math.*, 116:389–455, 1982.

[17] T. Kobayashi. Singular unitary representations and discrete series for indefinite Stiefel manifolds $U(p,q;F)/U(p-m,q;F)$. *Mem. Amer. Math. Soc.*, 95(462), 1992.

[18] T. Kobayashi and T. Oshima. Finite multiplicity theorems for induction and restriction. *Adv. Math.*, 248:921–944, 2013.

[19] B. Kostant. On convexity, the Weyl group and the Iwasawa decomposition. *Ann. Sci. École Norm. Sup.*, 6:413–455, 1973.

[20] R. Langlands. *On the Classification of Irreducible Representations of Real Algebraic Groups*. Volume 31 of *Math. Surv. Mon.*, pages 101–170. AMS, 1989.

[21] H. Li and G. Margulis. Effective estimates on integral quadratic forms: Masser's conjecture, generators of orthogonal groups, and bounds in reduction theory. *Geom. Funct. Anal.*, 26(3):874–908, 2016.

[22] G. Margulis. Explicit constructions of expanders. *Problemy Peredači Informacii*, 9(4):71–80, 1973.

[23] G. Margulis. Existence of compact quotients of homogeneous spaces, measurably proper actions, and decay of matrix coefficients. *Bull. Soc. Math. France*, 125(3):447–456, 1997.

[24] G. Margulis, A. Nevo, and E. Stein. Analogs of Wiener's ergodic theorems for semisimple Lie groups. II. *Duke Math. J.*, 103(2):233–259, 2000.

[25] A. Nevo. Spectral transfer and pointwise ergodic theorems for semi-simple Kazhdan groups. *Math. Res. Lett.*, 5:305–325, 1998.

[26] H. Oh. Uniform pointwise bounds for matrix coefficients of unitary representations and applications. *Duke Math. J.*, 113:133–192, 2002.

[27] T. Oshima. A method of harmonic analysis on semisimple symmetric spaces. In *Algebraic Analysis, Vol. II*, pages 667–680. Academic Press, 1988.

[28] Y. Sakellaridis and A. Venkatesh. Periods and harmonic analysis on spherical varieties. *Astérisque*, 396:viii+360, 2017.

[29] D. Vogan, Jr. The algebraic structure of the representation of semisimple Lie groups. I. *Ann. of Math.*, 109:1–60, 1979.

7

EMMANUEL BREUILLARD AND ALEXANDER LUBOTZKY

EXPANSION IN SIMPLE GROUPS

Dedicated to Grisha Margulis with admiration and affection

Abstract. Two short seminal papers of Margulis used Kazhdan's property (T) to give, on the one hand, explicit constructions of expander graphs and to prove, on the other hand, the uniqueness of some invariant means on compact simple Lie groups. These papers opened a rich line of research on expansion and spectral gap phenomena in finite and compact simple groups. In this essay we survey the history of this area and point out a number of problems that are still open.

1 Introduction

Grisha Margulis has the Midas touch: whatever he touches becomes gold. It seems that he did not have a particular interest in combinatorics, but in the early seventies events of life brought him to work at the Institute for Information Transmission in Moscow, where he became aware of the concept of expander graphs. Such graphs were known to exist at the time only by counting considerations (à la Erdős's random graph theory), but because of their importance in computer science, explicit constructions were very desirable. Margulis noticed that such explicit constructions could be made using the (new at the time) Kazhdan property (T) from representation theory of semisimple Lie groups and their discrete subgroups. His short paper [Mar73] opened a new area of research with a wealth of remarkable achievements.

A similar story happened with Margulis's contribution to the so-called Ruziewicz problem. Namely, must every rotation invariant finitely additive measure on the sphere S^n be equal to the Lebesgue measure? It had been known for a long time that the answer is "no" for $n = 1$, but Margulis [Mar80] (as well as Sullivan [Sul81]) showed, again using Kazhdan's property (T), that the answer is "yes" for $n \geq 4$.

Emmanuel Breuillard. DPMMS, University of Cambridge, UK
emmanuel.breuillard@maths.cam.ac.uk

Alexander Lubotzky. Einstein Institute of Mathematics, Hebrew University
alex.lubotzky@mail.huji.ac.il

These two seemingly unrelated topics are actually very much connected. This was explained in detail in [Lub94]. We will repeat it in a nutshell in Section 2 and give a brief historical description. Both directions of research led to a problem of the following type:

Problem. *Let G be a simple nonabelian finite (respectively, compact Lie) group. For a finite symmetric, that is, $S = S^{-1}$, subset S of G, consider*

$$\Delta^S = \sum_{s \in S} s$$

as an operator on $\mathbf{L}^2(G)$, where $sf(x) := f(s^{-1}x)$. It is easy to see that its largest eigenvalue is $k - |S|$ (with the constant functions being the eigenspace). It has multiplicity one if and only if S generates (respectively, generates topologically) G. Find S with spectral gap, that is, for which the second largest eigenvalue of Δ^S is bounded away from $|S|$.

This problem has many variants. Do we take S optimal (best case scenario), worst (worst case scenario), or random? Do we want $k = |S|$ to be fixed? Is the "bounded away" uniform? In what: The generators? All groups? All generators?

A quite rich theory has been developed around these questions, which grew out from the above two papers of Margulis. The goal of this essay is to describe this story and to point out several problems that are still open.

In Section 2, we will give some more history and show how the central problem we study here is related to expanders and to the Ruziewicz problem.

In Section 3, we describe the numerous developments the expansion problem for finite simple groups has had in the last decade or so. This led to amazing connections with additive combinatorics, diophantine geometry, Hilbert's fifth problem, and more. It also led to a new noncommutative sieve method with some remarkable applications. These subjects have been discussed in a number of books and surveys [Lub94, Tao15, Lub12, Bre16, Bre15, Bre14c], so we do not cover them here.

In section 4, we will turn our attention to the compact simple Lie groups. Here much less is known but an important connection has been made between the spectral gap problem and a certain noncommutative diophantine problem. We discuss this and state some further open problems.

Finally, in section 5 we discuss another direction, which has recently received renewed interest from questions in quantum computation (golden gates). Now, one looks not only for topological generators $S := \{g_1, \ldots, g_k\}$ in

G with spectral gap but also for an algorithm that will enable us to find for every $g \in G$ a short word w in $g_1, \dots, g_k$ such that $w(g_1, \dots, g_k)$ is very close to g.

This essay, which only illustrates a small part of Margulis's influence on modern mathematics, is dedicated to Grisha with admiration and affection. He has been a personal and professional inspiration for both of us.

ACKNOWLEDGMENTS. The authors are grateful to Ori Parzanchevski, Nicolas de Saxcé, and Peter Sarnak for useful comments. We also thank the referee for his careful reading. The first author acknowledges support from the ERC (grant no. 617129). The second author acknowledges support from the ERC, the NSF, and the BSF.

2 Expanders and invariant means

A family of finite k-regular graphs $X_i = (V_i, E_i)$ is called *an expanding family*, if there exists $\epsilon > 0$ such that for every i and every subset $Y \subset V_i$ with $|Y| \leq |V_i|/2$,

$$|\partial Y| \geq \epsilon |Y|,$$

where $\partial Y = E(Y, \overline{Y})$ is the set of edges going out from Y to its complement $\overline{Y}$.

Margulis made the following seminal observation, which connected expanders and representation theory:

PROPOSITION 2.1 (Margulis). *Let Γ be a group generated by a finite set S with $S = S^{-1}$ and $|S| = k$. Assume that Γ has Kazhdan property (T); then the family of finite k-regular Cayley graphs $Cay(\Gamma/N; S)$, where N runs over the finite index normal subgroups of Γ, forms an expanding family.*

Let us give a sketch of proof: property (T) means that the trivial representation is an isolated point in the unitary dual of Γ, the space of irreducible unitary representations of Γ up to equivalence endowed with the Fell topology. In concrete terms for $\Gamma = \langle S \rangle$ as above, it says that there exists an $\epsilon' > 0$, such that whenever Γ acts unitarily on a Hilbert space $\mathcal{H}$ via a (not necessarily irreducible) unitary representation ρ without a nonzero fixed vector, for every vector $v \neq 0$ in $\mathcal{H}$, there exists $s \in S$ such that

$$\|\rho(s)v - v\| \geq \epsilon' \|v\|. \tag{1}$$

In our situation, let Y be a subset of Γ/N—that is, a subset of vertices in $Cay(\Gamma/N, S)$. Let f be the function f in $L^2(\Gamma/N)$ defined by

$$f(y) = |\overline{Y}|$$

if $y \in Y$ and

$$f(y) = -|Y|$$

if $y \in \overline{Y}$, where $\overline{Y}$ is the complement of Y.

Then $f \in L_0^2(\Gamma/N)$, that is, $\sum_{y \in V} f(y) = 0$. Now, Γ acts unitarily by left translations on $L_0^2(\Gamma/N)$, which, as a representation, is a direct sum of nontrivial irreducible representations. We may thus apply (1) and deduce that there exists $s \in S$ such that

$$\|\rho(s)f - f\| \geq \epsilon' \|f\|. \tag{2}$$

Spelling out the meaning of f, and noting that f is essentially the (normalized) characteristic function of Y, one sees that

$$|sY \triangle Y| \geq \epsilon'' |Y|,$$

which implies the desired result. See [Lub94] for the full argument with the constants involved.

This fundamental argument gave a lot of families of finite (simple) groups that are expanding families. Every *mother group* Γ with property (T) gives rise to a family of expanders.

For example, for every $n \geq 3$, $\Gamma_n = \mathrm{SL}_n(\mathbb{Z})$ has (T) by Kazhdan's theorem [Kaz67]. Fix a finite symmetric set S of generators in Γ. One deduces that the family $\{Cay(\mathrm{PSL}_n(p), S); p \text{ prime}\}$ is a family of expanders. Naturally this raises the question whether all $\mathrm{PSL}_n(p)$ (all n all p) or even all nonabelian finite simple groups can be made into an expanding family simultaneously. This will be discussed in section 3. Meanwhile, let us give it another interpretation. The well-known result of Alon, Milman, and others (see [Lub94] for detailed history) gives the connection between expanders and the spectral gap. Let us formulate it in the context of finite groups.

PROPOSITION 2.2. *Let $\{G_i\}_{i \in I}$ be a family of finite groups with symmetric generating sets S_i with $|S_i| = k$ for every i. The following are equivalent:*

(1) *$\{Cay(G_i; S_i)\}_{i \in I}$ forms an expanding family of k-regular graphs.*

(2) There exists $\epsilon' > 0$ such that all eigenvalues of

$$\Delta^{S_i} = \sum_{s \in S_i} s$$

acting on $L_0^2(G_i)$ are at most[1] *$k - \epsilon'$.*

Let us now move to the invariant mean problem (a.k.a. the Ruziewicz problem) and we will see that the same spectral gap issue comes up. The problem to start with was formulated for the spheres S^n on which $G = SO(n+1)$ acts. But it generalizes naturally to the group G itself and actually to every compact group, so we will formulate it in this generality.

Let G be a compact group. An invariant mean from $L^\infty(G)$ to $\mathbb{R}$ is a linear functional m satisfying for every $f \in L^\infty(G)$,

- $m(f) \geq 0$ if $f \geq 0$,
- $m(1_G) = 1$, and
- $m(g.f) = m(f)$ for all $g \in G$,

where 1_G is the constant function equal to 1 on G, and for $g \in G$, $g.f(x) = f(g^{-1}x)$. Integration against the Haar measure, or Haar integration, is such an invariant mean. It is the only such if we assume in addition that m is countably additive. The question is whether this is still true also among the finitely additive invariant means.

THEOREM 2.3 (Rosenblatt [Ros81]).
Let G be a compact group, $S = S^{-1}$ a finite symmetric set in G with $|S| = k$, and $\Gamma = \langle S \rangle$ the subgroup generated by S. The following are equivalent:

- *The Haar integration is the unique Γ-invariant mean on $L^\infty(G)$.*
- *There exists $\epsilon' > 0$ such that all the eigenvalues of $\Delta^S = \sum_{s \in S} s$ acting on $L_0^2(G)$ are at most $k - \epsilon'$.*

It is not surprising now that if $\Gamma = \langle S \rangle \subset G$ is a dense subgroup with property (T) (i.e., Γ has (T) as an abstract discrete group), then every Γ-invariant mean (and hence G-invariant mean) of G is equal to the Haar integration.

[1] This is a one-sided condition, i.e., for each eigenvalue $\lambda \leq k - \epsilon'$. However, it is also equivalent to the two-sided condition $|\lambda| \leq k - \epsilon'$, provided the G_i's do not have an index two subgroup disjoint from S_i; see [BGGT15, appendix E] and [Bis19].

This was the way Margulis [Mar80] and Sullivan [Sul81] solved the Ruziewicz problem for S^n (and $SO(n+1)$), $n \geq 4$, to start with.

Note that when G is finite, then the spectral gap property is not so interesting for a single group G; it just says that S generates G. For an infinite compact group, S generates G topologically if and only if k has multiplicity one—that is, all other eigenvalues are less than k. But we want them (there are infinitely many of them!) to be bounded away from k by ϵ'. So the question is of interest even for a single group G.

Let us mention here a result that connects the two topics directly:

THEOREM 2.4 (Shalom [Sha97]).
Let $\Gamma = \langle S \rangle$, $S = S^{-1}$, $|S| = k$ be a finitely generated group and $G = \widehat{\Gamma}$ its profinite completion. Then the Haar integration is the only Γ-invariant mean on G if and only if the family $\{Cay(\Gamma/N; S); N \lhd \Gamma, |\Gamma/N| < \infty\}$ forms an expanding family.

In what follows, if G is a finite or compact group and $S = S^{-1}$ is a subset of G with $|S| = k$, we will say that S is *ϵ-expanding* if all eigenvalues of $\Delta^S = \sum_{s \in S} s$ acting on $L_0^2(G)$ are at most $k - \epsilon$. Sometimes we simply say *expanding*, omitting the ϵ, when we talk about an infinite group G or about an infinite collection of G's with the same ϵ.

In the case of finite groups, if all eigenvalues of Δ^S on $L_0^2(G)$ are, in absolute value, either k or at most $2\sqrt{k-1}$, we say that S is a Ramanujan subset of G.

In this case $Cay(G, S)$ is a Ramanujan graph [LPS88, Lub94] and $2\sqrt{k-1}$ is the best possible bound one can hope for (for an infinite family of finite groups) by the well-known Alon-Boppana result (see [Sar90, (Section 3) 3.2.7]). This notion extends naturally to subsets S of an infinite compact group G. Also, here $2\sqrt{k-1}$ is the best possible bound (even for a single such group G) because $2\sqrt{k-1}$ is the rate of exponential growth of the number of closed paths of length n based at a point in the k-valent tree.

In sections 3 and 4, we will describe what is known about expanding sets in finite simple groups, and in compact simple Lie groups. Very little is known about the existence of Ramanujan subsets, and we will raise there some questions.

3 Expansion in finite simple groups

In this section we are interested in expanding subsets of size k in finite simple groups. Abelian groups cannot give rise to expanders, (see [LW93]), so when we say *simple*, we always mean noncommutative simple groups. We will divide our discussion into three subsections: best, random, and worst case generators.

3.1 BEST CASE GENERATORS.

The classification of the finite simple groups can be used to show that every such group is generated by two elements. In our context it is natural to ask if all finite simple groups are uniformly expanding. As mentioned in section 2, it was shown at an early stage that for fixed $n_0 \geq 3$, $\{\mathrm{SL}_{n_0}(p)\}_{p\in\mathcal{P}}$ is an expanding family when p runs over the set $\mathcal{P}$ of all primes, using the generators of the mother group $\mathrm{SL}_n(\mathbb{Z})$. But what about the family $G_n(p_0) := \mathrm{SL}_n(p_0)$ when this time p_0 is fixed and n varies?

In [LW93] it was shown that the family $\{G_n(p_0)\}_{n\geq 2}$ is not expanding with respect to *some* set of generators of bounded size (see section 3.3; for $\{G_{n_0}(p)\}_{p\in\mathcal{P}}$ this is still an open problem!). This was deduced there by embedding a finitely generated amenable group as a dense subgroup of $\prod_n \mathrm{SL}_n(p_0)$, something which is impossible in $\prod_p \mathrm{SL}_{n_0}(p)$. It has been suggested that maybe *bounded rank* (i.e., n_0 fixed) groups behave differently regarding expansion than unbounded rank (i.e., p_0 fixed and $n \to +\infty$). This still might be the case regarding worst case generators. A combination of works of Kassabov [Kas07a, Kas07b], Lubotzky [Lub11b], Nikolov [Nik07] (see [KLN06]), and Breuillard-Green-Tao [BGT11b] gives us the following:

THEOREM 3.1.
There exist $k \in \mathbb{N}$ and $\epsilon > 0$ such that every non-abelian finite simple group G has a subset $S = S^{-1}$ of size k such that $Cay(G; S)$ is an ϵ-expander.

The breakthrough for the proof of Theorem 3.1 was the paper of Kassabov [Kas07b] in which he broke the barrier of bounded rank to show that $\{\mathrm{SL}_{3n}(p); p \in \mathcal{P}, n \in \mathbb{N}\}$ is an expanding family. Rather than describing the exact historical development (which can be found in [KLN06]), let us give the conceptual explanation.

In [EJZ10] it is shown that $\Gamma = E_d(\mathbb{Z}\langle x_1, \ldots, x_\ell\rangle)$ has property (T) for every $d \geq 3$ and $\ell \in \mathbb{N}$, where $\mathbb{Z}\langle x_1, \ldots, x_\ell\rangle$ is the free noncommutative ring on ℓ free variables and $E_d(R)$, for a ring R, is the group of $d \times d$ matrices over R generated by the elementary matrices $\{I + re_{i,j}; 1 \leq i \neq j \leq d, r \in R\}$. Now, for every prime power q and every $n \in \mathbb{N}$, the matrix ring $M_n(\mathbb{F}_q)$ is 2-generated as a ring—that is, $\mathbb{Z}\langle x_1, x_2\rangle$ can be mapped onto $M_n(\mathbb{F}_q)$. This implies that $\Gamma = E_3(\mathbb{Z}\langle x_1, x_2\rangle)$ can be mapped onto $E_3(M_n(\mathbb{F}_q)) \simeq \mathrm{SL}_{3n}(\mathbb{F}_q)$ and hence by Margulis's original result—that is, Proposition 2.1—$\{SL_{3n}(\mathbb{F}_q)\}_{n\geq 1}$ is an expanding family.

Let us take the opportunity to observe that this can be used to answer (positively!) a question asked in [LW93]; it was asked there whether it is possible to have an infinite compact group K containing two finitely generated dense subgroups A and B such that A is amenable and B has property (T). If K is a compact Lie group, this is impossible because the Tits alternative forces A, and hence G and B, to have a solvable subgroup of finite index, but a (T) group that is also amenable must be finite. On the other hand, it was shown in that paper that the compact group $\prod_{n\geq 3} SL_n(\mathbb{F}_p)$ does contain a finitely generated amenable dense subgroup. Hence its quotient $K := \prod_{n\geq 1} SL_{3n}(\mathbb{F}_p)$ also has such an A. But from the previous paragraph, we see that K also has a (T) subgroup B. Indeed the diagonal image of $\Gamma = E_3(\mathbb{Z}\langle x_1, x_2\rangle)$ has (T) and must be dense in K because it maps onto each of the nonisomorphic quasi-simple groups $SL_{3n}(\mathbb{F}_p)$ (Goursat's lemma).

Now let us move ahead with expanders. An easy lemma shows that if a group is a bounded product of expanding groups, then it is also expanding (for different k and ϵ). Nikolov [Nik07] showed that when the rank is large enough, every finite simple group of Lie type is a bounded product of the groups treated by Kassabov, thus extending the result for all high rank. But what about lower rank and first of all SL_2?

Let us observe first that one cannot hope for a proof of the Margulis/Kassabov kind for the groups $SL_2(q) = SL_2(\mathbb{F}_q)$. In fact there is no mother group Γ with property (T) that is mapped onto $SL_2(q)$ for infinitely many prime powers q. Indeed, if such a group Γ exists, then by some standard ultraproduct argument (or elementary algebraic geometry; see, e.g., [LMS93]), Γ has an infinite representation into $SL_2(F)$ for some algebraically closed field F. However, this is impossible as every property (T) subgroup of $SL_2(F)$ must be finite (see, e.g., [Lub94, Theorem 3.4.7] for the proof when $\mathrm{char}(F) = 0$, but the same argument works in positive characteristic: any action of a (T) group on a Bruhat-Tits tree or hyperbolic space must fix a point, so it must have compact closure in all field completions; see also [dlHV89, chapter 6, Proposition 26]).

So a different argument is needed; for p prime, it has been deduced from Selberg's theorem ($\lambda_1 \geq \frac{3}{16}$) that $\{SL_2(p)\}_{p\in\mathcal{P}}$ are expanding; see, for example, [Gam02] and [Bre14b]. Similar reasoning (using Drinfeld instead of Selberg) gives a similar result for $\{SL_2(p_0^{\ell})\}$, where p_0 is fixed and $\ell \in \mathbb{N}$; see [Mor94]. But how to handle them together? This was done by Lubotzky [Lub11b] using a very specific construction of Ramanujan graphs (and Ramanujan complexes).

That construction, in [LSV05b], made $G = \mathrm{SL}_2(p^\ell)$ into a $(p+1)$–regular Ramanujan graph using a set of $p+1$ generators of the following type: $\{tct^{-1}; t \in T\}$, where c is a specific element in $G = \mathrm{SL}_2(p^\ell)$ and T is a nonsplit torus in $H = \mathrm{SL}_2(p)$. Now by Selberg as above, H is expanding with respect to 2 generators (and their inverses)—say,

$$a = \begin{pmatrix} 1 & 1 \\ 0 & 1 \end{pmatrix} \text{ and } b = \begin{pmatrix} 1 & 0 \\ 1 & 1 \end{pmatrix};$$

by the previous sentence, G is so with respect to one H-conjugate orbit of $c \in G$. From this, one deduces that G is an expander with respect to $\{a^{\pm 1}, b^{\pm 1}, c^{\pm 1}\}$ (see [Lub11b] for details and [KLN06] for an exposition). In fact, it is also shown there that one can use the more general Ramanujan complexes constructed in [LSV05a] and [LSV05b] to give an alternative proof to Kassabov's result for $\mathrm{SL}_n(q)$, all n, and all q simultaneously.

Anyway, once we have $\mathrm{SL}_2(q)$ at our disposal, all finite simple groups are bounded products of $\mathrm{SL}_n(q)$ (all n, all q) except for two families that still need special treatment for the expanding problem and for proving Theorem 3.1.

One family is the family of Suzuki groups; these finite simple groups (which are characterized by the fact that they are the only finite simple groups whose order is not divisible by 3; see [Gla77]), do not contain copies of $PSL_2(\mathbb{F}_q)$ and resist all the above methods. They were eventually resolved by Breuillard-Green-Tao [BGT11b] by random methods, so we postpone their treatment to section 3.2.

Last but not least is the most important family of finite simple groups, $Alt(n)$. They do contain copies of groups of Lie type, but one can show that they are *not* bounded products of such. So a new idea was needed here; what Kassabov [Kas07a] did is to consider first n's of the form $n = d^6$ when $d = 2^k - 1$ for some $k \in \mathbb{N}$. The fact that $\mathrm{SL}_3(\mathbb{Z}\langle x_1, x_2\rangle)$ has property (T) implies that the direct product $\mathrm{SL}_{3k}(\mathbb{F}_2)^{d^5}$ is an expanding family, and he embedded this group into $Alt(n)$ in six different ways. The product of these six copies is still not the full $Alt(n)$, but (borrowing an idea from Roichman [Roi96]) he treated separately representations of $Alt(n)$ corresponding to partitions $\lambda = (\lambda_1 \geq \ldots \geq \lambda_\ell)$ of n with $\lambda_1 \leq n - d^{5/4}$ and all the others. The first were treated by appealing to results on "normalized character values," and the second were treated collectively by giving their sum a combinatorial meaning and showing that the action of a bounded product of six copies of that model mixes in a few steps. The reader is referred to the full details of this ingenious proof in [Kas07a] or to the exposition in [KLN06].

All in all, the knowledge on best case expansion in finite simple groups is in pretty good shape, as Theorem 3.1 shows, and certainly better than what we will describe in the other five topics. Still, some natural problems arise here:

PROBLEMS 3.1.

(a) Is there a discrete group with property (T) that is mapped onto all finite simple groups of large rank? Or on all $Alt(n)$? By the computer-assisted recent breakthrough in [KNO], we now know that the group of (order preserving) automorphisms of the free group F_5 has property (T). This group is known to surject onto $Alt(n)$ for infinitely many n's; see [Gil77]. See also [Lub11a], [KKN], and [Nit] for the even more recent extension of [KNO] to $Aut(F_k)$ for all $k \geq 4$.

(b) The proof of Theorem 3.1 described above gives an explicit set of generators in all cases except the Suzuki groups. It would be of interest to cover this case. Theorem 3.1 also gives a certain fixed number k of generators, which is bounded but larger than two. One hopes to get a proof with smaller sets of generators (perhaps two). This is especially of interest for $Alt(n)$.

(c) We discussed expanding families—that is, the eigenvalues are bounded away from $k = |S|$. What about Ramanujan families—that is, families of groups G_i with $|S_i| = k$ such that all nontrivial eigenvalues are bounded by $2\sqrt{k-1}$. As of now, only subfamilies of $\{SL_2(p^\ell); p \text{ prime}, \ell \in \mathbb{N}\}$ are known to have such generators; see [LPS88] and [Mor94]. What about $SL_3(p)$? $Alt(n)$? In [P20] Parzanchevski defines Ramanujan directed graphs. Strangely enough, while it is not known how to turn many finite simple groups into Ramanujan graphs, he manages in [P18] to turn them into Ramanujan directed graphs!

3.2 RANDOM GENERATORS. A well-known result, proved by Dixon in [Dix69] for the symmetric groups, Kantor-Lubotzky [KL90] for the classical groups, and Liebeck-Shalev [LS95] for the exceptional ones, asserts that for every $m \geq 2$, randomly chosen m elements of a finite simple group G generate G. This means that

$$Prob\big((x_1, \ldots, x_m) \in G^m; \langle x_1, \ldots, x_m \rangle = G\big) \longrightarrow_{|G| \to +\infty} 1.$$

The basic question of this section is whether they form expanders—namely, is there $\epsilon > 0$ such that

$$Prob\big(Expd(G, m, \epsilon)\big) \longrightarrow_{|G| \to +\infty} 1, \tag{3}$$

where

$$Expd(G, m, \epsilon) := \{(x_1, \ldots, x_m) \in G^m;\ Cay(G; S) \text{ is } \epsilon\text{-expanding for } S = \{x_1^{\pm 1}, \ldots, x_m^{\pm 1}\}?$$

This is still widely open. The best result as of now is the following:

THEOREM 3.3 (Breuillard-Guralnick-Green-Tao [BGGT15]).
For each $m \geq 2$ and $r \geq 1$, there is $\epsilon > 0$ such that (3) *holds for all finite simple groups G of rank at most r.*

The rate of convergence in (3) is even polynomial in $|G|^{-1}$. In particular this holds for the groups $G = \mathrm{PSL}_n(q)$ when the rank $n-1$ is bounded and q goes to infinity. It also includes the family of Suzuki groups, thus completing Theorem 3.1 by showing the existence of some expanding Cayley graph; see [BGT11b] for this special family, a case that was not covered by the Kassabov-Lubotzky-Nikolov methods. The case of $\mathrm{PSL}_2(p)$ in the above theorem was first established by Bourgain and Gamburd in [BG08b].

The method of proof here, pioneered in [Hel08] and [BG08b] for the family of groups $\{\mathrm{PSL}_2(p)\}_p$, is based on the classification of approximate subgroups of G (see Theorem 3.4 below), an important statistical lemma in arithmetic combinatorics (the so-called Balog-Szemerédi-Gowers lemma), and a crucial property of characters of finite simple groups: the smallest degree of their nontrivial irreducible characters is at least $|G|^{\delta}$, where $\delta > 0$ depends only on the rank of G.

This property, going back to Frobenius for $\mathrm{PSL}_2(p)$, was established in full generality in a classic paper by Landazuri-Seitz [LS74]. It was used by Sarnak-Xue [SX91] and in Gamburd's thesis [Gam02] in the closely related context of spectral gap estimates for the Laplacian on hyperbolic surfaces. It also plays an important role in various combinatorial questions regarding finite groups. It was coined *quasi-randomness* by Gowers [Gow08]. In particular it implies what is now called *Gowers's trick*: namely, the fact that given any finite subset A of a finite simple group, we have $AAA = G$, that is, every element of G can be written as the product of three elements from A, provided $|A| > |G|^{1-\delta}$, where $\delta > 0$ depends only on the rank of G. See [BNP08, Bre14a] for proofs of this fact.

The approximate groups mentioned above are by definition subsets A of G such that AA can be covered by a bounded number of translates of A. This bound, say, K, determines the quality of the K-approximate subgroup. With this definition, 1-approximate subgroups of G are simply genuine subgroups of G. The classification of subgroups of finite simple groups is a vast subject, which of course is part of the Classification of Finite Simple Groups (CFSG). Starting with Jordan's 19th-century theorem that finite subgroups of $GL_n(\mathbb{C})$ are bounded-by-abelian [Jor78, Bre12] and Dickson's early 20th-century classification of subgroups of $PSL_2(q)$ [Dic58], it climaxes with the Larsen-Pink theorem [LP11], which gives a CFSG-free classification of subgroups of finite linear groups, saying in essence that they are close to being given by the $\mathbb{F}_q$ points of some algebraic group. Regarding approximate groups, the main result is as follows:

THEOREM 3.4 (Classification of approximate subgroups). *Let G be a finite simple group, $A \subset G$ a generating subset, and $K \geq 1$. If $AA \subset XA$ for some $X \subset G$ with $|X| \leq K$, then either $|A| \leq CK^C$ or $|G|/|A| \leq CK^C$, where C depends only on the rank of G. Moreover, for all generating subsets $A \subset G$,*

$$|AAA| \geq \min\{|G|, |A|^{1+\delta}\}$$

for some $\delta > 0$ depending only on the rank of G.

For $PSL_2(p)$ and more generally in rank 1, the above result can be established by elementary methods based on the sum-product theorem à la Bourgain-Katz-Tao [BKT04]. This was proved by Helfgott for $PSL_2(p)$ [Hel08] and generalized by Dinai [Din11] to $PSL_2(q)$ for all q. In high rank, new ideas were required and although some mileage had been achieved by Helfgott [Hel11] and Gill-Helfgott [GH10] for $PSL_3(p)$ and $PSL_n(p)$, the solution came after Hrushovski [Hru12] proved a very general qualitative version of the above theorem based on a model-theoretic generalization of the Larsen-Pink theorem and ideas from geometric stability theory. The result in the form above was finally proved independently in [PS16] by Pyber-Szabó (all groups) and in [BGT11a] by Breuillard-Green-Tao (who had initially only announced it for Chevalley groups) without using any model theory but rather more down-to-earth algebraic geometry in positive characteristic. See [Bre15] for an exposition.

We now briefly explain the link between Theorems 3.4 and 3.3 following the strategy first developed in [BG08b]. We refer the reader to the expository paper [Bre15] and to the book [Tao15] for further details. In order to get a spectral gap, it is enough to show that the probability that the simple random walk on the Cayley graph returns to the identity in $O(\log|G|)$ steps is close to $1/|G|$. The Cayley graph is assumed to have large girth (see below Problem 3.2(c)), so during the first $c\log|G|$ steps (c is a small constant), the random walk evolves on a tree and we understand it very well (via Kesten's [Kes59] theorem in particular). The main point is then to establish further decay of this return probability between $c\log|G|$ steps and $C\log|G|$ steps (C is a large constant). Here the main tool, the *ℓ^2-flattening lemma* of Bourgain-Gamburd, is a consequence of the celebrated additive-combinatorial Balog-Szemerédi-Gowers lemma (see [TV10]). It implies that decay takes place at some rate (of order $\exp(-n^{\alpha})$ for some small $\alpha > 0$), provided the random walk does not accumulate on an approximate subgroup of G. Theorem 3.4 then kicks in and allows one to reduce the proof to showing that the random walk does not accumulate on subgroups of G. This last step, which is in fact the main part of [BGGT15], is straightforward in rank 1 but requires several new ideas in high rank, in particular the existence of so-called *strongly dense* free subgroups of simple algebraic groups in positive characteristic, proved in [BGGT12] for this purpose.

Finally, we mention that the above method also produces expander Cayley graphs for finite groups that are no longer simple (for example, $SL_d(\mathbb{Z}/n\mathbb{Z})$ when n is no longer assumed to be prime). The generators of these Cayley graphs can be chosen to be the reduction modulo n of a set of generators in a fixed Zariski dense subgroup in $SL_d(\mathbb{Z})$. This is the subject of *super strong approximation*, for which we refer the reader to the works of Bourgain, Varjú, and Salehi-Golsefidy [Var12, SaVa12, BV12], and its many applications—in particular, to the affine sieve [BGS10, SaSa13] (see also the surveys [Sal14], [Bre15]).

PROBLEMS 3.2.

(a) The corresponding problem for the family of alternating groups $Alt(n)$ with n growing to infinity is still wide open. Is there $\epsilon > 0$ and $m \geq 2$ such that Equation (3) holds when $G = Alt(n)$? Can this even happen with probability bounded away from 0? How about $G = Sym(n)$ the full symmetric group? Looking instead at the random Schreier graphs $Sch(X_{n,r}; S)$ of $Sym(n)$ obtained by the action on the set $X_{n,r}$ of r-tuples of n elements, it is well-known that (3) holds for some $\epsilon = \epsilon(r) > 0$ when r is fixed (e.g., see [Lub94], [FJR$^+$98]), while one would need $r = n$ to get the full Cayley graph of $Sym(n)$. Nevertheless, a conjecture of Kozma and Puder ([PP20, Conjecture 1.8]) asserts that for every generating set S, the spectral gap of the Cayley graph $Cay(Sym(n); S)$ ought to be entirely governed by that of the Schreier graph $Sch(X_{n,4}; S)$ with $r = 4$. This conjecture, if true, would imply that random Cayley graphs of $Alt(n)$ are expanders.

(b) Being an expander implies that the diameter of the Cayley graph is logarithmic in the size of the graph; see [Lub94]. However, when the rank of the finite simple groups goes to infinity, such as for the family $Alt(n)$, we do not even know whether or not the diameter of a random k-regular Cayley graph can be bounded logarithmically in the size of G. In the case of $Alt(n)$, however, poly logarithmic bounds have been established (see [BH06], [SP12], [HSZ15]). See also [EJ20] for the case of classical groups of high rank.

(c) Girth lower bounds are also relevant to the problem. For finite simple groups of bounded rank, it is known that m randomly chosen elements generate Cayley graphs with girth at least $c \log |G|$, where $c > 0$ depends only on m and on the rank of G. In other words, the group's presentation has no relation of length $< c \log |G|$; see [GHS$^+$09]. As pointed out above this was used in the proof of Theorem 3.3. However, logarithmic girth lower bounds when the rank of the groups is allowed to go to infinity are still completely open, even for $Alt(n)$.

(d) What about Ramanujan graphs? Numerical evidence [LR93] hints that random Cayley graphs of $PSL_2(p)$ may not be Ramanujan. However, it is plausible that they are in fact *almost Ramanujan*, in the sense that for each $\epsilon > 0$ with very high probability as $p \to +\infty$, all nontrivial eigenvalues are bounded by $2\sqrt{k-1} + \epsilon$. See [RS17] where an upper bound on the number of exceptional eigenvalues is established and numerics are given. The same could be said of the family of alternating groups $Alt(n)$ (and perhaps even of the full family of all finite simple groups). Partial evidence in this direction is provided by Friedman's proof of Alon's conjecture [Fri08] that the Schreier graphs of $Alt(n)$ acting on n elements are almost Ramanujan (see also [Pud15], [Bor15], [BC19]).

3.3 WORST CASE GENERATORS. The family of finite simple groups $Alt(n)$ was shown (see section 3.1) to be a family of expanders with respect to some choice of generators, but it is not with respect to others: for example, take $\tau = (1, 2, 3)$ and $\sigma = (1, 2, \ldots, n)$ if n is odd and $\sigma = (2, \ldots, n)$ if n is even. Then $Cay(Alt(n); \{\tau^{\pm 1}, \sigma^{\pm 1}\})$ are not expanders (see [Lub94]).

A similar kind of argumentation can be performed for every family of finite simple groups $\{G_i\}_{i \in I}$ of Lie type with unbounded Lie rank. In [Som15] it was shown that for each such family there is a generating set S_i of G_i of size at most 10, such that the sequence of graphs $\{Cay(G_i, S_i)\}_{i \in I}$ is not expanding.

By contrast we have the following conjecture (see [Bre14c]):

CONJECTURE 3.6. *Given $r \in \mathbb{N}$ and $k \in \mathbb{N}$, there exists an $\epsilon = \epsilon(r,k) > 0$ such that for every finite simple group of rank $\leq r$ and every set of generators S of size $|S| \leq k$, $Cay(G, S)$ is an ϵ-expander.*

Some evidence toward this conjecture is provided by the following result:

THEOREM 3.7 ([BG10]).
There exists a set of primes $\mathcal{P}_1$ of density 1 among all primes satisfying the following: there exists $\epsilon > 0$ such that if $p \in \mathcal{P}_1$ and x, y are two generators of $\mathrm{SL}_2(p)$, then for $S = \{x^{\pm 1}, y^{\pm 1}\}$, $Cay(\mathrm{SL}_2(p), S)$ is an ϵ-expander.

The proof uses the uniform Tits alternative proved in [Bre11] as well as the same Bourgain-Gamburd method used in the proof of Theorem 3.3. The uniform Tits alternative in combination with the effective Nullstellensatz is used to show that for most primes p, the probability of return to the identity (or even to any proper subgroup) of the simple random walk on $\mathrm{SL}_2(p)$ after $n = \log p$ steps is at most p^{-c} for some fixed $c > 0$ independent of the generating set. This in turn implies a spectral gap via the Bourgain-Gamburd method and Theorem 3.4.

4 Expansion in compact simple Lie groups

In this section we are interested in expanding subsets of size k in compact simple Lie groups. Here again, we will divide our discussion into three subsections: best, random, and worst case (topological) generators.

4.1 BEST CASE GENERATORS. Here the question is to find a topological generating set with spectral gap:

THEOREM 4.1 (Margulis, Sullivan, Drinfeld).
Every simple compact Lie group contains a finite topological generating set with spectral gap.

Every simple compact Lie group G not locally isomorphic to $\mathrm{SO}(3)$ contains a countable dense subgroup with Kazhdan's property (T). Indeed one can find an irreducible high rank arithmetic lattice in a product $G \times H$, where H is a certain noncompact semisimple Lie group, and project it to G; see [Mar91, III.5.7]. Any finite generating subset S of this countable (T) group Γ will provide an example of a topological generating set of G with a spectral gap (in particular, the conditions of Theorems 2.3 and 2.4 will hold). These observations were made by Margulis [Mar80] and Sullivan [Sul81].

However, the case of SO(3) (and its double cover SU(2)) is exceptional: it does not contain any countable infinite group with property (T) (see [dlHV89, Chapter 6, Proposition 26]). So it seems much harder to find a topological generating set with spectral gap. Nevertheless this was achieved by Drinfeld shortly after the work of Margulis and Sullivan in a one-page paper [Dri84]. Kazhdan's original proof that lattices in high rank simple Lie groups have property (T) is representation theoretic by nature. It uses heavily the fact that the discrete group is a lattice, so that one can induce unitary representations from the lattice to the ambient Lie group and thus reduce the problem to a good understanding of the representation theory of the Lie group. Drinfeld's idea is similar: the countable dense subgroup of G he uses arises from the group of invertible elements in a quaternion algebra defined over $\mathbb{Q}$, which ramifies at the real place (so that the associated Lie group is locally isomorphic to SO(3)). But the tools to establish the spectral gap are much more sophisticated: namely, the Jacquet-Langlands correspondence is used to reduce the question to spectral gap estimates for Hecke operators associated to irreducible $PGL_2(\mathbb{Q}_p)$ representations arising from automorphic representations on the adelic space $L^2(PGL_2(\mathbb{A})/PGL_2(\mathbb{Q}))$. These estimates follow either from the work of Deligne on the Ramanujan-Peterson conjectures [Del74] or from earlier estimates due to Rankin [Ran39]. We refer to Drinfeld's original paper and to the book [Lub94] for the details of this argument.

These methods produce some specific (topological) generating sets arising from generators of a lattice in a bigger group. One can be very explicit and write down concrete matrices for the generators. See [Lub94], [Sar90], and [CdV89]. Lubotzky-Phillips-Sarnak [LPS87] pushed this to produce many examples of families of topological generators of SO(3) with optimal spectral gap (i.e., Ramanujan). The set S consisting of the three rotations of angle $\arccos(-\frac{3}{5})$ around the coordinate axes of $\mathbb{R}^3$ and their inverses provides such an example.[2]

In these examples the quality of the gap deteriorates as the dimension tends to infinity. In [Sar90, section 2.4], Sarnak gives an inductive construction

---

[2]These generators are called V gates in the quantum computing literature; see section 5.

starting with a set S of size k in $SO(n)$ with spectral gap at least ϵ, which produces a new set S' in $SO(n+1)$ of size $2k$ with spectral gap at least $\epsilon/2k$. However, the following is still open:

PROBLEM 4.1. Does there exist $\epsilon > 0$ and $k > 0$ and for each n a symmetric set of k topological generators of $SO(n)$ with all eigenvalues $< k - \epsilon$ (i.e., a spectral gap that is uniform in n)?

4.2 RANDOM CASE. Here the situation is wide open. Sarnak [Sar90, p. 58] asks the question whether for a generic pair of rotations a, b in $SO(3)$ the corresponding set $S = \{a, b, a^{-1}, b^{-1}\}$ has a spectral gap. This question is still open regardless of whether *generic* is understood in the sense of Lebesgue measure or in the sense of Baire category.

Regarding the latter, an interesting observation was made in [LPS86]: if G is a compact simple Lie group, then for a Baire generic family of (topological) generating sets S of size k, generating a free subgroup of G, the Laplace operator Δ^S on $L^2_0(G)$ has infinitely many exceptional eigenvalues (i.e., eigenvalues of size $> 2\sqrt{k-1}$); see the end of section 4.

It is also worth mentioning that generically a k-tuple of elements in G generates a free subgroup. This is true both in the sense of Baire category and in the sense of measure (see, e.g., [GK03], [BG03], [Aou11]).

Another interesting observation in the random (with respect to Lebesgue) situation was made by Fisher [Fis06]. He observed that the property of an m-tuple $(a_1, \ldots, a_m)$ in G^m to have some nonzero spectral gap (i.e., $S = \{a_1^{\pm 1}, \ldots, a_m^{\pm 1}\}$ has a spectral gap) is invariant under the group of automorphisms of the free group F_m. Indeed if S has a spectral gap, so does any other generating subset of the group $\langle S \rangle$ generated by S. Now the action of $Aut(F_m)$ on G^m is known to be ergodic when $m \geq 3$ by a result of Goldman [Gol07] (case of $SU(2)$) and Gelander [Gel08] (general G); see also [Lub11a] for this and general background on $Aut(F_m)$ and its actions. Hence we have the following zero-one law:

THEOREM 4.3 (Fisher [Fis06]).
Let G be a simple compact Lie group. If $m \geq 3$, then either Lebesgue almost all m-tuples have a spectral gap, or Lebesgue almost no m-tuple has a spectral gap.

In the next paragraph we discuss the new method introduced by Bourgain and Gamburd to establish a spectral gap.

4.3 WORST CASE. Even worse: we do not know even a single example of a finite topological generating set of SO(3) (or any compact simple Lie group) without a spectral gap. Nevertheless a breakthrough took place a decade ago when Bourgain and Gamburd produced many more examples of topological generators with spectral gap. They showed in [BG08a] that every topological generating set all of whose matrix entries are algebraic numbers (i.e., in $SU(2, \overline{\mathbb{Q}})$) have a spectral gap. This has now been generalized, first to $SU(d)$ by Bourgain-Gamburd themselves [BG12], then to arbitrary compact simple Lie groups by Benoist and Saxcé [BdS16]:

THEOREM 4.4 (Bourgain-Gamburd, Benoist-Saxcé).
Let G be a compact simple Lie group and $S \subset G$ be a finite symmetric subset generating a dense subgroup such that $\mathrm{tr}(Ad(s))$ is an algebraic number for every $s \in S$. Then S has a spectral gap.

Note that the best case examples mentioned above and produced by Margulis, Sullivan, and Drinfeld have algebraic entries (property (T) groups have an algebraic trace field by rigidity) and so they fall in the class of subsets handled by the above theorem. The converse, however, is not true: in Theorem 4.4 the subgroups generated by S are usually not lattices in any Lie group, and although they can be made discrete under the usual geometric embedding looking at the different places of the trace field, they will only be *thin subgroups* there—that is, Zariski-dense of infinite covolume.

The proof of Theorem 4.4 is inspired from the above-mentioned method Bourgain and Gamburd first pioneered for the family of finite simple groups $PSL_2(p)$, but it is much more involved. It still contains a significant combinatorial input in that instead of the growth properties of triple products of finite subsets as in Theorem 3.4, one needs to consider the growth under triple products of δ-separated sets and thus consider *discretized approximate groups*. The cardinality of a finite set A is replaced by the δ-discretized cardinality $\mathcal{N}_\delta(A)$, which is the minimum number of balls of radius δ needed to cover A. This setting was explored by Bourgain in his proof that there is no nontrivial Borel subring of the reals with positive Hausdorff dimension [B03] culminating with *Bourgain's discretized sum-product theorem* and later by Bourgain-Gamburd in [BG08a]. In his thesis Saxcé was able to prove the suitable analogue of Theorem 3.4 in the context of discretized sets in compact Lie groups.

THEOREM 4.5 (Saxcé's product theorem [dS15]).
Let G be a simple compact Lie group and $\delta > 0$. For every $\kappa > 0$ and $\sigma > 0$ there is $\epsilon > 0$ such that for every set $A \subset G$ that is (a) of intermediate size (i.e., $\mathcal{N}_\delta(A) =$

$\delta^{-\alpha}$ for $\alpha \in [\sigma, \dim G - \sigma]$), (b) κ-non-concentrated (i.e., $\mathcal{N}_\rho(A) \geq \delta^\epsilon \rho^{-\kappa}$ for all $\rho \geq \delta$), and (c) ϵ-away from subgroups (i.e., for every proper closed subgroup H of G there is $a \in A$ with $d(a, H) \geq \delta^\epsilon$), we have

$$\mathcal{N}_\delta(AAA) \geq \mathcal{N}_\delta(A)^{1+\epsilon}.$$

An interesting consequence of this theorem (proved in [dS17]) is that sets of positive Hausdorff dimension have a bounded covering number: namely, given $\sigma > 0$, there is $p \in \mathbb{N}$ such that for any typologically generating Borel subset A of G with Hausdorff dimension at least σ, $A^p = G$.

As in the case of finite groups of Lie type, the spectral gap in Theorem 4.4 is established by showing the fast equidistribution of the simple random walk on G induced by S. A similar combinatorial argument based on the Balog-Szemerédi-Gowers lemma shows that the fast equidistribution must take place unless the walk is stuck in a δ-discretized approximate group and hence (applying Theorem 4.5) in the neighborhood of some closed subgroup. It only remains to show that the random walk cannot spend too much time close to any subgroup. This is where the algebraicity assumption comes in. In fact, as shown in [BdS16], it is enough to know that the probability of being exponentially close to a closed subgroup is uniformly exponentially small.

DEFINITION 4.6 (Weak diophantine property). A finite symmetric subset S of a compact simple Lie group G (with bi-invariant metric $d(\cdot,\cdot)$) is said to be *weakly diophantine* if there are $c_1, c_2 > 0$ such that for all large enough n and every proper closed subgroup $H \leq G$, we have

$$Prob_{\{w;|w|=n\}}\left(d(w, H) < e^{-c_1 n}\right) \leq e^{-c_2 n},$$

where the probability is taken uniformly over the k^n words w of length $|w| = n$ in the alphabet S.

Note that the presence of a spectral gap gives a rate of equidistribution of the random walk. In particular, it easily implies the weak diophantine property. But we now have the following:

THEOREM 4.7 (see [BdS16]).
For a finite symmetric subset S the weak diophantine property and the existence of a spectral gap are equivalent.

If S has algebraic entries (or if the trace field generated by $\mathrm{tr}(Ad(s))$ is a number field), then it is well-known and easy to show that it satisfies a *strong diophantine property*—namely, there is $c_1 > 0$ such that $d(w, 1) > e^{-c_1|w|}$ for every word w in S not equal to the identity. Benoist and Saxcé verify that it must also satisfy the weak diophantine property.

The uniform Tits alternative yields a weaker version of the weak diophantine property (where $e^{-c_1 n}$ is replaced by $e^{-n^{c_1}}$), which holds for *every* topological generating set S; see [Bre11].

It has been conjectured in [GJS99] and [Gam04, section 4.2] that the strong diophantine property ought to hold for Lebesgue almost every S (it does not hold for every S; for example, it fails if S contains a rotation whose angle mod π is a Liouville number).

Finally, we propose a stronger spectral gap conjecture:

CONJECTURE 4.8. *Let G be a simple compact Lie group and $k \geq 4$. There is $\epsilon > 0$ such that for every symmetric set S of size k generating a dense subgroup of G, Δ^S has only finitely many eigenvalues outside the interval $[-k+\epsilon, k-\epsilon]$.*

It is easy to see that not all eigenvalues can be contained in a proper subinterval: for example, if the generators are close to the identity in G, then there will be many eigenvalues close to the maximal eigenvalue k. Partial evidence for this conjecture is supported by the fact that the analogous statement (even without exceptional eigenvalues) does hold, with a uniform ϵ, for the action of S on the regular representation $\ell^2(\langle S \rangle)$ of the abstract group $\langle S \rangle$, as follows from the uniform Tits alternative; see [Bre11].

What about Ramanujan topological generating sets? As mentioned above Lubotzky-Phillips-Sarnak produced such examples in [LPS88]. In [LPS86, Theorem 1.4], however, they observed that generic (in the sense of Baire) generators are in general not Ramanujan. In this vein the following is still open:

PROBLEM 4.3. Is being *asymptotically Ramanujan* a Baire generic property? Namely, is there a countable intersection Ω of dense open subsets of G^k such that for every $\epsilon > 0$ and every k-tuple of symmetric generators $S \in \Omega$, there are only finitely many eigenvalues of Δ^S outside the interval $[-2\sqrt{k-1} - \epsilon, 2\sqrt{k-1} + \epsilon]$? The same question can be asked for almost every k-tuple of generators in the sense of Lebesgue measure, and in this case the argument for the zero-one law of Theorem 4.3 no longer applies.

5 Navigation and golden gates

5.1 NAVIGATION IN FINITE SIMPLE GROUPS. One of the most important applications of expander graphs is their use for the construction of communication networks. Let us imagine n microprocessors working simultaneously within a supercomputer. Ideally we would like to have them all connected to each other, but this would require $\Omega(n^2)$ connections, which is not feasible. Expander graphs give a replacement that can be implemented with $O(n)$ connections and with reasonable performance. But this also requires a navigation algorithm, which will find a short path between any two vertices of the graph. It is easy to see that ϵ-expander k-regular graphs X have diameter bounded by $C\log_{k-1}(|X|)$, where C depends only on the expansion coefficient ϵ. In section 3, we showed that there exist k, ϵ such that every (nonabelian) finite simple group G has a symmetric set of generators S of size at most k such that $Cay(G, S)$ is an ϵ-expander, and so there is C such that $\mathrm{diam}(Cay(G, S)) \leq C\log_{k-1}(|G|)$. But the proof that provided these generators did not offer an algorithm to find a path between two given points of length at most $C\log_{k-1}(|G|)$. This is still open:

PROBLEM 5.1.a. Find $k \in \mathbb{N}$ and $\epsilon, C > 0$ such that every nonabelian finite simple group has a symmetric set of generators of size at most k, for which $Cay(G, S)$ is an ϵ-expander and there exists a polynomial time (i.e., polynomial in $\log_{k-1}|G|$) algorithm that expresses any given element in G as a word in S of length at most $C\log_{k-1}(|G|)$.

In [BKL89] a set S of size 14 was presented (for almost all the finite simple groups) for which $\mathrm{diam}(Cay(G, S)) = O(\log|G|)$ with an absolute implied constant, even though these Cayley graphs were not uniform expanders.

The case when G is the alternating group $Alt(n)$ (or the symmetric group $Sym(n)$) is of special interest: we know a set of generators S that would give expanders (see [Kas07a] or [KNO]), but they come with no navigation. On the other hand, [BKL89] gives such a navigation algorithm but not expanders.

Another case of special interest is the family of groups $\mathrm{PSL}_2(\mathbb{F}_p)$, where p runs overs the primes. For this family,

$$S = \left\{ \begin{pmatrix} 1 & \pm 1 \\ 0 & 1 \end{pmatrix}, \begin{pmatrix} 1 & 0 \\ \pm 1 & 1 \end{pmatrix} \right\}$$

gives rise to expanders, but the best navigation algorithm with this generating set is due to Larsen [Lar03]. However, his probabilistic algorithm gives words of length $O((\log p)^2)$ rather than the desired $O(\log p)$. Here is a baby version of this problem:

PROBLEM 5.1.b. Find an algorithm to express

$$\begin{pmatrix} 1 & \frac{p+1}{2} \\ 0 & 1 \end{pmatrix} = \begin{pmatrix} 1 & 1 \\ 0 & 1 \end{pmatrix}^{\frac{p+1}{2}}$$

as a word of length $O(\log p)$ using

$$\begin{pmatrix} 1 & \pm 1 \\ 0 & 1 \end{pmatrix} \quad \text{and} \quad \begin{pmatrix} 1 & 0 \\ \pm 1 & 1 \end{pmatrix}.$$

Let us mention that if one allows to add an extra generator—say,

$$t = \begin{pmatrix} 2 & 0 \\ 0 & \frac{1}{2} \end{pmatrix},$$

then this is easily done. Denote by u_x the unipotent matrix

$$\begin{pmatrix} 1 & x \\ 0 & 1 \end{pmatrix}.$$

For $b = 1, \dots, p-1$, write $b = \sum_{i=0}^{r} a_i 4^i$ with $r \leq \log_4(p)$ and $0 \leq a_i \leq 3$. Then, since $tu_x t^{-1} = u_{4x}$, we get $u_b = u_{a_0} t u_{a_1} t^{-1} \cdot \ldots \cdot t^r u_{a_r} t^{-r}$. This means that u_b is a word of length $O(r) = O(\log p)$ using only the letters u_1 and t. A similar trick for the lower unipotent matrices plus the observation that every matrix in $\mathrm{SL}_2(p)$ is a product of at most four upper and lower unipotent matrices shows the following:

PROPOSITION 5.3. *Problem 5.1.a. has an affirmative answer for the family of groups* $\{PSL_2(p); p \text{ prime}\}$.

The navigation algorithm in $\mathrm{PSL}_2(q)$, with respect to the $p+1$ generators provided by the LPS Ramanujan graph, received some special attention [TZ08, PLQ08, Sar19, Sar17] as it has been suggested that these Cayley graphs can be used to construct efficient hash functions. It turned out that this problem is intimately related to some deep problems in number theory asking for solutions of some diophantine equations. Some of these problems are

NP-complete and some are solved in polynomial time! These works give a probabilistic polynomial time algorithm to navigate $G = \mathrm{PSL}(2, q)$ (with respect to the $p+1$ generators of the LPS Ramanujan graphs) finding a path of length at most $(3 + o(1)) \log_p |G|$ between any two points, while the typical distance between two vertices chosen at random in the graph is $(1 + o(1)) \log_p |G|$. But already this implies that these Cayley graphs are not a good choice for hash functions. On the other hand, finding the shortest path between any two points is NP-complete!

5.2 NAVIGATION IN SIMPLE COMPACT GROUPS. Let G be a compact group with bi-invariant metric d, where our main interest will be compact simple Lie groups with the metric induced by the Riemannian structure. In this case the analogous question to those discussed in section 5.1 for finite simple groups is of interest even for a single group G and has the following form.

PROBLEM 5.2. Find a finite symmetric subset S of G of size k, which generates a dense subgroup Γ of G, and find an algorithm that given $\epsilon > 0$ and $g \in G$ provides a word w of short length in S with $d(w, g) < \epsilon$.

What do we mean by *short*? Let μ_ϵ be the volume of a ball of radius ϵ in G. We normalize the volume so that G has volume 1. We want to cover G by balls of radius ϵ around the words of length $\leq \ell$ in S. The number T of such words satisfies $|T| \leq k^\ell$, and so the best we can hope for is $\ell \leq O_k(\log \frac{1}{\mu_\epsilon})$. For simple Lie groups, $\mu_\epsilon \sim c\epsilon^{\dim G}$, so we can hope for $\ell \leq O_{k,G}(\log \frac{1}{\epsilon})$. Ideally we would like also to have an efficient algorithm that, when ϵ and $g \in G$ are given, will find $w \in \Gamma$ that is ϵ-close to g and will express w as a word of length $O(\log(\frac{1}{\epsilon}))$ in the elements of S. This problem for the group $\mathrm{PU}(n)$ (especially $\mathrm{PU}(2)$, but also for larger n) is of fundamental importance in Quantum Computing. The elements of Γ are usually called the *gates*, and optimal gates are *golden gates* (see [NC00], [PS18], and the references therein for more on this). We will not go in this direction here but will just mention that the classical Solovay-Kitaev algorithm works for general gates (i.e., a subset S as before) but gives a word w that is of polylogarithmic length $\log(\frac{1}{\epsilon})^{O(1)}$, while the spectral methods to be discussed briefly below work for special choices of gates but give w of smaller length, sometimes even with an almost optimal implicit constant. We refer the reader to Varjú's work [Var13] for the best-known polylogarithmic estimates for general gates in all compact simple Lie groups.

Problem 5.2 has two parts, and each one is nontrivial: (a) Given $g \in G$, find $w \in \Gamma$ that is short and ϵ-close to g, and (b) Express w as an explicit short word (circuit) in terms of S.

The work of Ross and Selinger [RS16] essentially gave a solution to both parts for the case $G = PU(2)$. They observed that every $g \in G$ can be written as a product of three diagonal matrices and showed how to solve the problem for each diagonal matrix. For both (a) and (b), they used the group $\Gamma = PU_2(\mathbb{Z}[\sqrt{2}][\frac{1}{\sqrt{2}}])$ (which is the first factor projection of the corresponding arithmetic lattice in $PU_2(\mathbb{R}) \times PGL_2(K)$, where K is the degree 2 extension of the field $\mathbb{Q}_2$ of 2-adic numbers associated to the prime $\sqrt{2}$).

The work of Parzanchevski and Sarnak [PS18] gives a conceptual explanation for this and a vast generalization. They find a number of groups Γ that are suitable to achieve this goal: all the Γ's are arithmetic lattices, which appear naturally as lattices in $PU(2) \times PGL_2(K)$, when K is a local non Archimedean field (see [Lub94] for a thorough explanation of this). The projection of Γ to $PU(2)$ gives the desired dense subgroup. But the more interesting point is that the discrete projection to $PGL_2(K)$ and the action of Γ on its associated Bruhat-Tits tree gives the navigation algorithm that solves part (b) of the problem. Some special choices of such Γ's gives *super golden gates*, which are essentially optimal.

The work of Evra and Parzanchevski [EP18] takes the story a step further by studying the analogous problem for $PU(3)$. This time, this is done via arithmetic discrete subgroups Γ of $PU(3) \times PGL_3(\mathbb{Q}_p)$. Again the projection to $PU(3)$ gives the desired dense subgroup of $PU(3)$, while the projection to the other factor gives an action of Γ on the Bruhat-Tits building, which enables one to also solve the navigation problem (in spite of not being a tree). The reader is referred to [EP18] for this emerging beautiful theory and for more open questions.

References

[Aou11] Richard Aoun. Random subgroups of linear groups are free. *Duke Math. J.*, 160(1):117–173, 2011.

[BH06] László Babai and Thomas P. Hayes. The probability of generating the symmetric group when one of the generators is random. *Publ. Math. Debrecen*, 69(3): 271–280, 2006.

[BKL89] László Babai, William M. Kantor, and Alexander Lubotzky. Small-diameter Cayley graphs for finite simple groups. *Eur. J. Combin.*, 10(6):507–522, 1989.

[BNP08] László Babai, Nikolay Nikolov, and László Pyber. Product growth and mixing in finite groups. In *Proceedings of the Nineteenth Annual ACM-SIAM Symposium on Discrete Algorithms*, pages 248–257. ACM, New York, 2008.

[BdS16] Yves Benoist and Nicolas de Saxcé. A spectral gap theorem in simple Lie groups. *Invent. Math.*, 205(2):337–361, 2016.

[Bis19] Arindam Biswas. On a cheeger type inequality in Cayley graphs of finite groups. *Eur. J. Combin.*, 81:298–308, 2019.

[Bor15] Charles Bordenave. A new proof of Friedman's second eigenvalue theorem and its extension to random lifts. *Ann. Sci. Éc. Norm. Supér.*, 53(4):1393–1440, 2020.

[BC19] Charles Bordenave and Benoît Collins. Eigenvalues of random lifts and polynomial of random permutations matrices. *Ann. of Math. (2)* 190(3):811–875, 2019.

[B03] Jean Bourgain. On the Erdős-Volkmann and Katz-Tao ring conjectures. *Geom. Funct. Anal.*, 13(2):334–365, 2003.

[BG08a] Jean Bourgain and Alexander Gamburd. On the spectral gap for finitely generated subgroups of $SU(2)$. *Inven. Math.*, 171:83–121, 2008.

[BG08b] Jean Bourgain and Alexander Gamburd. Uniform expansion bounds for Cayley graphs of $\mathrm{SL}_2(\mathbb{F}_p)$. *Ann. of Math. (2)*, 167(2):625–642, 2008.

[BG12] Jean Bourgain and Alexander Gamburd. A spectral gap theorem in $\mathrm{SU}(d)$. *J. Eur. Math. Soc. (JEMS)*, 14(5):1455–1511, 2012.

[BGS10] Jean Bourgain, Alexander Gamburd, and Peter Sarnak. Affine linear sieve, expanders, and sum-product. *Invent. Math.*, 179(3):559–644, 2010.

[BKT04] Jean Bourgain, Nets Katz, and Terence Tao. A sum-product estimate in finite fields, and applications. *Geom. Funct. Anal.*, 14(1):27–57, 2004.

[BV12] Jean Bourgain and Péter Varjú. Expansion in $SL_d(\mathbf{Z}/q\mathbf{Z})$, q arbitrary. *Invent. Math.*, 188(1):151–173, 2012.

[Bre11] Emmanuel Breuillard. Heights on SL_2 and free subgroups. In *Geometry, rigidity, and group actions*, Chicago Lectures in Math., pages 455–493. Univ. Chicago Press, Chicago, IL, 2011.

[Bre12] Emmanuel Breuillard. An exposition of Jordan's original proof of his theorem on finite subgroups of $GL_n(\mathbb{C})$. *Preprint available on the author's website*, 2012.

[Bre14a] Emmanuel Breuillard. A brief introduction to approximate groups. In *Thin groups and superstrong approximation*, volume 61 of *Math. Sci. Res. Inst. Publ.*, pages 23–50. Cambridge Univ. Press, Cambridge, 2014.

[Bre14b] Emmanuel Breuillard. Expander graphs, property (τ) and approximate groups. In *Geometric group theory*, volume 21 of *IAS/Park City Math. Ser.*, pages 325–377. Amer. Math. Soc., Providence, RI, 2014.

[Bre14c] Emmanuel Breuillard. Diophantine geometry and uniform growth of finite and infinite groups. In *Proceedings of the 2014 ICM, Seoul, Vol. III*, pages 27–50. Kyung Moon Sa, Seoul, 2014.

[Bre15] Emmanuel Breuillard. Approximate subgroups and super-strong approximation. In *Groups St Andrews 2013*, volume 422 of *London Math. Soc. Lecture Note Ser.*, pages 1–50. Cambridge Univ. Press, Cambridge, 2015.

[Bre16] Emmanuel Breuillard. Lectures on approximate groups and Hilbert's 5th problem. In *Recent trends in combinatorics*, volume 159 of *IMA Vol. Math. Appl.*, pages 369–404. Springer, Cham, 2016.

[BG10] Emmanuel Breuillard and Alex Gamburd. Strong uniform expansion in $\mathrm{SL}(2,p)$. *Geom. Funct. Anal.*, 20(5):1201–1209, 2010.

[BG03] Emmanuel Breuillard and Tsachik Gelander. On dense free subgroups of Lie groups. *J. Algebra*, 261(2):448–467, 2003.

[BGGT12] Emmanuel Breuillard, Ben Green, Robert Guralnick, and Terence Tao. Strongly dense free subgroups of semisimple algebraic groups. *Israel J. Math.*, 192(1):347–379, 2012.

[BGGT15] Emmanuel Breuillard, Ben Green, Robert Guralnick, and Terence Tao. Expansion in finite simple groups of Lie type. *J. Eur. Math. Soc. (JEMS)*, 17(6):1367–1434, 2015.

[BGT11a] Emmanuel Breuillard, Ben Green, and Terence Tao. Approximate subgroups of linear groups. *Geom. Funct. Anal.*, 21(4):774–819, 2011.

[BGT11b] Emmanuel Breuillard, Ben Green, and Terence Tao. Suzuki groups as expanders. *Groups Geom. Dyn.*, 5(2):281–299, 2011.

[CdV89] Yves Colin de Verdière. Distribution de points sur une sphère (d'après Lubotzky, Phillips et Sarnak). *Astérisque*, 177–178(Exp. No. 703):83–93, 1989. Séminaire Bourbaki, Vol. 1988/89.

[Del74] Pierre Deligne. La conjecture de Weil. I. *Inst. Hautes Études Sci. Publ. Math.*, (43):273–307, 1974.

[Dic58] Leonard Eugene Dickson. *Linear groups: With an exposition of the Galois field theory*. With an introduction by W. Magnus. Dover Publications, Inc., New York, 1958.

[Din11] Oren Dinai. Growth in SL_2 over finite fields. *J. Group Theory*, 14(2):273–297, 2011.

[Dix69] John D. Dixon. The probability of generating the symmetric group. *Math. Z.*, 110:199–205, 1969.

[Dri84] Vladimir G. Drinfeld. Finitely-additive measures on S^2 and S^3, invariant with respect to rotations. *Funktsional. Anal. i Prilozhen.*, 18(3):77, 1984.

[EJ20] Sean Eberhard and Urban Jezernik. Babai's conjecture for high-rank classical groups with random generators. 2020. Preprint arXiv:2005.09990.

[EJZ10] Mikhail Ershov and Andrei Jaikin-Zapirain. Property (T) for noncommutative universal lattices. *Invent. Math.*, 179(2):303–347, 2010.

[EP18] Shai Evra and Ori Parzanchevski. Ramanujan complexes and Golden Gates in PU(3). Preprint arXiv:1810.04710.

[Fis06] David Fisher. Out(F_n) and the spectral gap conjecture. *Int. Math. Res. Not.*, Art. ID 26028, 2006.

[Fri08] Joel Friedman. A proof of Alon's second eigenvalue conjecture and related problems. *Mem. Amer. Math. Soc.*, 195(910):viii+100, 2008.

[FJR+98] Joel Friedman, Antoine Joux, Yuval Roichman, Jacques Stern, and Jean-Pierre Tillich. The action of a few permutations on r-tuples is quickly transitive. *Random Structures Algorithms*, 12(4):335–350, 1998.

[Gam02] Alex Gamburd. On the spectral gap for infinite index "congruence" subgroups of $SL_2(\mathbf{Z})$. *Israel J. Math.*, 127:157–200, 2002.

[Gam04] Alexander Gamburd. Expander graphs, random matrices and quantum chaos. In *Random walks and geometry*, pages 109–140. Walter de Gruyter, Berlin, 2004.

[GHS+09] Alexander Gamburd, Shlomo Hoory, Mehrdad Shahshahani, Aner Shalev, and Bálint Virág. On the girth of random Cayley graphs. *Random Structures Algorithms*, 35(1):100–117, 2009.

[GJS99] Alexander Gamburd, Dmitry Jacobson, and Peter Sarnak. Spectra of elements in the group ring of SU(2). *J. Eur. Math. Soc. (JEMS)*, 1(1):51–85, 1999.

[GK03] Paul M. Gartside and Robin W. Knight. Ubiquity of free subgroups. *Bull. London Math. Soc.*, 35(5):624–634, 2003.

[Gel08] Tsachik Gelander. On deformations of F_n in compact Lie groups. *Israel J. Math.*, 167:15–26, 2008.

[GH10] Nick Gill and Harald Helfgott. Growth in solvable subgroups of $\mathrm{GL}_r(\mathbb{Z}/p\mathbb{Z})$. *Math. Ann.* 360(1–2):157–208, 2014.

[Gil77] Robert Gilman. Finite quotients of the automorphism group of a free group. *Canad. J. Math.*, 29(3):541–551, 1977.

[Gla77] George Glauberman. *Factorizations in local subgroups of finite groups. Regional Conference Series in Mathematics, No. 33.* Amer. Math. Soc., Providence, RI, 1977.

[Gol07] William M. Goldman. An ergodic action of the outer automorphism group of a free group. *Geom. Funct. Anal.*, 17(3):793–805, 2007.

[Gow08] William Timothy Gowers. Quasirandom groups. *Combin. Probab. Comput.*, 17(3):363–387, 2008.

[dlHV89] Pierre de la Harpe and Alain Valette. La propriété (T) de Kazhdan pour les groupes localement compacts (avec un appendice de Marc Burger). *Astérisque*, (175):158, 1989.

[Hel08] Harald Helfgott. Growth and generation in $\mathrm{SL}_2(\mathbb{Z}/p\mathbb{Z})$. *Ann. of Math. (2)*, 167(2):601–623, 2008.

[Hel11] Harald Helfgott. Growth in $\mathrm{SL}_3(\mathbb{Z}/p\mathbb{Z})$. *J. Eur. Math. Soc. (JEMS)*, 13(3):761–851, 2011.

[HSZ15] Harald Helfgott, Ákos Seress, and Andrzej Zuk. Random generators of the symmetric group: Diameter, mixing time and spectral gap. *J. Algebra*, 421:349–368, 2015.

[Hru12] Ehud Hrushovski. Stable group theory and approximate subgroups. *J. Amer. Math. Soc.*, 25(1):189–243, 2012.

[Jor78] Camille Jordan. Mémoire sur les équations différentielles linéaires à intégrale algébrique. *J. Reine Angew. Math.*, (84):89–215, 1878.

[KKN] Marek Kaluba, David Kielak, and Piotr W. Nowak. On property (T) for $Aut(F_n)$ and $SL_n(\mathbb{Z})$. To appear in *Ann. of Maths.*

[KNO] Marek Kaluba, Piotr W. Nowak, and Narutaka Ozawa. $Aut(F_5)$ has property (T). *Math. Ann.*, 375(3–4):1169–1191, 2019.

[KL90] William M. Kantor and Alexander Lubotzky. The probability of generating a finite classical group. *Geom. Dedicata*, 36(1):67–87, 1990.

[Kas07a] Martin Kassabov. Symmetric groups and expander graphs. *Invent. Math.*, 170(2):327–354, 2007.

[Kas07b] Martin Kassabov. Universal lattices and unbounded rank expanders. *Invent. Math.*, 170(2):297–326, 2007.

[KLN06] Martin Kassabov, Alexander Lubotzky, and Nikolay Nikolov. Finite simple groups as expanders. *Proc. Natl. Acad. Sci. USA*, 103(16):6116–6119, 2006.

[Kaz67] David Kazhdan. On the connection of the dual space of a group with the structure of its closed subgroups. *Funkcional. Anal. i Priložen.*, 1:71–74, 1967.

[Kes59] Harry Kesten. Symmetric random walks on groups. *Trans. Amer. Math. Soc.*, 92:336–354, 1959.

[LR93] John Lafferty and Daniel Rockmore. Numerical investigation of the spectrum for certain families of Cayley graphs. In *Expanding graphs (Princeton, NJ, 1992)*, volume 10 of *DIMACS Ser. Discrete Math. Theoret. Comput. Sci.*, pages 63–73. Amer. Math. Soc., Providence, RI, 1993.

[LS74] Vicente Landazuri and Gary M. Seitz. On the minimal degrees of projective representations of the finite Chevalley groups. *J. Algebra*, 32:418–443, 1974.

[Lar03] Michael Larsen. Navigating the Cayley graph of $SL_2(\mathbb{F}_p)$. *Int. Math. Res. Not.*, 27:1465–1471, 2003.

[LP11] Michael J. Larsen and Richard Pink. Finite subgroups of algebraic groups. *J. Amer. Math. Soc.*, 24(4):1105–1158, 2011.

[LS95] Martin W. Liebeck and Aner Shalev. The probability of generating a finite simple group. *Geom. Dedicata*, 56(1):103–113, 1995.

[Lub94] Alexander Lubotzky. *Discrete groups, expanding graphs and invariant measures*, volume 125 of *Progress in mathematics*. Birkhäuser Verlag, Basel, 1994. With an appendix by Jonathan D. Rogawski.

[Lub11a] Alexander Lubotzky. Dynamics of Aut(F_N) actions on group presentations and representations. In *Geometry, rigidity, and group actions*, Chicago Lectures in Math., pages 609–643. Univ. Chicago Press, Chicago, IL, 2011.

[Lub11b] Alexander Lubotzky. Finite simple groups of Lie type as expanders. *J. Eur. Math. Soc. (JEMS)*, 13(5):1331–1341, 2011.

[Lub12] Alexander Lubotzky. Expander graphs in pure and applied mathematics. *Bull. Amer. Math. Soc. (N.S.)*, 49(1):113–162, 2012.

[LMS93] Alexander Lubotzky, Avinoam Mann, and Dan Segal. Finitely generated groups of polynomial subgroup growth. *Israel J. Math.*, 82(1–3):363–371, 1993.

[LPS86] Alexander Lubotzky, Ralph Phillips, and Peter Sarnak. Hecke operators and distributing points on the sphere. I. *Comm. Pure Appl. Math.*, 39(S, suppl.):S149–S186, 1986. *Frontiers of the mathematical sciences: 1985* (New York, 1985).

[LPS87] Alexander Lubotzky, Ralph Phillips, and Peter Sarnak. Hecke operators and distributing points on S^2. II. *Comm. Pure Appl. Math.*, 40(4):401–420, 1987.

[LPS88] Alexander Lubotzky, Ralph Phillips, and Peter Sarnak. Ramanujan graphs. *Combinatorica*, 8(3):261–277, 1988.

[LSV05a] Alexander Lubotzky, Beth Samuels, and Uzi Vishne. Explicit constructions of Ramanujan complexes of type $\tilde{A}_d$. *Eur. J. Combin.*, 26(6):965–993, 2005.

[LSV05b] Alexander Lubotzky, Beth Samuels, and Uzi Vishne. Ramanujan complexes of type $\tilde{A}_d$. *Israel J. Math.*, 149:267–299, 2005. Probability in mathematics.

[LW93] Alexander Lubotzky and Benjamin Weiss. Groups and expanders. In *Expanding graphs (Princeton, NJ, 1992)*, volume 10 of *DIMACS Ser. Discrete Math. Theoret. Comput. Sci.*, pages 95–109. Amer. Math. Soc., Providence, RI, 1993.

[Mar73] Gregory A. Margulis. Explicit constructions of expanders. *Problemy Peredači Informacii*, 9(4):71–80, 1973.

[Mar80] Gregory A. Margulis. Some remarks on invariant means. *Monatsh. Math.*, 90(3):233–235, 1980.

[Mar91] Gregory A. Margulis. *Discrete subgroups of semisimple Lie groups*, volume 17 of *Ergebnisse der Mathematik und ihrer Grenzgebiete*. Springer-Verlag, Berlin, 1991.

[Mor94] Moshe Morgenstern. Existence and explicit constructions of $q+1$ regular Ramanujan graphs for every prime power q. *J. Combin. Theory Ser. B*, 62(1):44–62, 1994.

[NC00] Michael A. Nielsen and Isaac L. Chuang. *Quantum computation and quantum information*. Cambridge Univ. Press, Cambridge, 2000.

[Nik07] Nikolay Nikolov. A product of decomposition for the classical quasisimple groups. *J. Group Theory*, 10(1):43–53, 2007.

[Nit] Martin Nitsche, Computer proofs for Property (T), and SDP duality. arXiv.2009.05134, 2020.

[P18] Ori Parzanchevski. Optimal generators for matrix groups. 2018 in preparation.

[P20] Ori Parzanchevski. Ramanujan graphs and digraphs. In *Analysis and geometry on graphs and manifolds* (LMS Lecture Notes 461), M. Keller, D. Lenz and R. Wojciechowski Eds. Cambridge Univ. Press, Cambridge, 2020.

[PP20] Ori Parzanchevski and Doron Puder. Aldous' spectral gap conjecture for normal sets. *Trans. Amer. Math. Soc.*, 373(10):7067–7086, 2020.

[PS18] Ori Parzanchevski and Peter Sarnak. Super-golden-gates for $PU(2)$. *Adv. Math.*, 327:869–901, 2018.

[PLQ08] Christophe Petit, Kristin Lauter, and Jean-Jacques Quisquater. Full cryptanalysis of LPS and Morgenstern hash functions. In *Security and cryptography for networks*, Rafail Ostrovsky, Roberto De Prisco, and Ivan Visconti, Eds, pp. 263–277. Springer Berlin, Heidelberg, 2008.

[Pud15] Doron Puder. Expansion of random graphs: New proofs, new results. *Invent. Math.*, 201(3):845–908, 2015.

[PS16] László Pyber and Endre Szabó. Growth in finite simple groups of Lie type. *J. Amer. Math. Soc.*, 29(1):95–146, 2016.

[Ran39] Robert A. Rankin. Contributions to the theory of Ramanujan's function $\tau(n)$ and similar arithmetical functions. I. *Proc. Cambridge Philos. Soc.*, 35:351–372, 1939.

[RS17] Igor Rivin and Naser T. Sardari. Quantum chaos on random Cayley graphs of $SL(2, p)$. *Exp. Math.*, 28(3):328–341, 2019.

[Roi96] Yuval Roichman. Upper bound on the characters of the symmetric groups. *Invent. Math.*, 125(3):451–485, 1996.

[Ros81] Joseph Rosenblatt. Uniqueness of invariant means for measure-preserving transformations. *Trans. Amer. Math. Soc.*, 265(2):623–636, 1981.

[RS16] Neil J. Ross and Peter Selinger. Optimal ancilla-free Clifford + T approximation of z-rotations. *Quantum Inf. Comput.*, 16(11–12):901–953, 2016.

[Sal14] Alireza Salehi-Golsefidy. Affine sieve and expanders. In *Thin groups and superstrong approximation*, volume 61 of *Math. Sci. Res. Inst. Publ.*, pages 325–342. Cambridge Univ. Press, Cambridge, 2014.

[SaSa13] Alireza Salehi-Golsefidy and Peter Sarnak. The affine sieve. *J. Amer. Math. Soc.*, 26(4):1085–1105, 2013.

[SaVa12] Alireza Salehi-Golsefidy and Péter Varjú. Expansion in perfect groups. *Geom. Funct. Anal.*, 22(6):1832–1891, 2012.

[Sar17] Naser T. Sardari. Complexity of strong approximation on the sphere. 2017. Preprint arXiv:1703.02709.

[Sar19] Naser T. Sardari. Diameter of Ramanujan graphs and random Cayley graphs. *Combinatorica*, 39(2):427–446, 2019.

[Sar90] Peter Sarnak. *Some applications of modular forms*. Cambridge Tracts in Mathematics. Cambridge Univ. Press, Cambridge, 1990.

[SX91] Peter Sarnak and Xiao Xi Xue. Bounds for multiplicities of automorphic representations. *Duke Math. J.*, 64(1):207–227, 1991.

[dS15] Nicolas de Saxcé. A product theorem in simple Lie groups. *Geom. Funct. Anal.*, 25(3):915–941, 2015.

[dS17] Nicolas de Saxcé. Borelian subgroups of simple Lie groups. *Duke Math. J.*, 166(3):573–604, 2017.

[SP12] Jan-Christoph Schlage-Puchta. Applications of character estimates to statistical problems for the symmetric group. *Combinatorica*, 32(3):309–323, 2012.

[Sha97] Yehuda Shalom. Expanding graphs and invariant means. *Combinatorica*, 17(4):555–575, 1997.

[Som15] Gábor Somlai. Non-expander Cayley graphs of simple groups. *Comm. Algebra*, 43(3):1156–1175, 2015.

[Sul81] Dennis Sullivan. For $n > 3$ there is only one finitely additive rotationally invariant measure on the n-sphere defined on all Lebesgue measurable subsets. *Bull. Amer. Math. Soc. (N.S.)*, 4(1):121–123, 1981.

[Tao15] Terence Tao. *Expansion in finite simple groups of Lie type*, volume 164 of *Graduate Studies in Mathematics*. Amer. Math. Soc., Providence, RI, 2015.

[TV10] Terence Tao and Van H. Vu. *Additive combinatorics*, volume 105 of *Cambridge Studies in Advanced Mathematics*. Cambridge Univ. Press, Cambridge, 2010.

[TZ08] Jean-Pierre Tillich and Gilles Zémor. Collisions for the LPS expander graph hash function. *Advances in cryptology—EUROCRYPT* 2008, pages 254–269. Lecture Notes in Comput. Sci., 4965. Springer, Berlin, 2008.

[Var12] Péter Varjú. Expansion in $SL_d(\mathcal{O}_K/I)$, I square-free. *J. Eur. Math. Soc.* 14(1):273–305, 2012.

[Var13] Péter Varjú. Random walks in compact groups. *Doc. Math.*, 18:1137–1175, 2013.

8

ANDERS KARLSSON

ELEMENTS OF A METRIC SPECTRAL THEORY

Dedicated to Margulis, with admiration

Abstract. This essay discusses a general method for spectral-type theorems using metric spaces instead of vector spaces. Advantages of this approach are that it applies to genuinely nonlinear situations and also to random versions. Metric analogs of operator norm, spectral radius, eigenvalue, linear functional, and weak convergence are suggested. Applications explained include generalizations of the mean ergodic theorem, the Wolff-Denjoy theorem, and Thurston's spectral theorem for surface homeomorphisms.

1 Introduction

In one line of development of mathematics, considerations progressed from concrete functions, to vector spaces of functions, and then to abstract vector spaces. In parallel, the standard operations, such as derivatives and integrals, were generalized to the abstract notions of linear operators, linear functionals, and scalar products. The study of the category of topological vector spaces and continuous linear maps is basically what is now called functional analysis. Dieudonné wrote that if one were to reduce the complicated history of functional analysis to a few keywords, the emphasis should fall on the evolution of two concepts: *spectral theory* and *duality* [Di81]. Needless to say, as most often is the case, the abstract general study does not supersede the more concrete considerations in every respect. In the context of analysis, one can compare the two different points of view in the excellent texts [L02] and [StS11].

The metric space axioms were born out of the same development; see the historical note in [Bo87] or [Di81]. In the present essay, I would like to argue for another step: from normed vector spaces to metric spaces (and their

Anders Karlsson. Section de mathématiques, Université de Genève, 2-4 Rue du Lièvre, Case Postale 64, 1211 Genève 4, Suisse
anders.karlsson@unige.ch and

Matematiska institutionen, Uppsala universitet, Box 256, 751 05 Uppsala, Sweden
anders.karlsson@math.uu.se

This work was supported in part by the Swiss NSF.

generalizations), and bounded linear operators to semicontractions. This could be called *metric functional analysis*, or in view of the particular focus here, a *metric spectral theory*. Indeed we will in the metric setting discuss a spectral principle and duality in the form of metric functionals. This is motivated by situations that are genuinely nonlinear, but there is also an interest in the metric perspective even in the linear case. The latter can be exemplified by a well-known classical instance: for many questions in the study of groups of 2×2 real matrices, it is easier to employ their (associated) isometric action on the hyperbolic plane, which is indeed a metric and not a linear space, instead of the linear action on $\mathbb{R}^2$. The isometric action of $PSL_2(\mathbb{R})$ is by fractional linear transformations preserving the upper half-plane. This generalizes to $n \times n$ matrices and the associated symmetric space.

Geometric group theory is a subject that has influenced the development of metric geometry during the last few decades. Gromov has been the leading person in this subject with his many and diverse contributions. Originally he found some inspiration from combinatorial group theory and the Mostow-Margulis rigidity theory (for example, the Gromov product appeared in Lyndon's work, Mostow introduced the crucial notion of quasi-isometry, and Margulis noted that one can argue in terms of word metrics in this context of quasi-isometries and boundary maps).

There is another strand of metric geometry sometimes called the *Ribe program*; see Naor's recent ICM plenary lecture [N18] for some history and appropriate references. Bourgain wrote already in 1986 [B86] in this context that "the notions from local theory of normed spaces are determined by the metric structure of the space and thus have a purely metrical formulation. The next step consists in studying these metrical concepts in general metric spaces in an attempt to develop an analogue of the linear theory." The present text suggests something similar, yet rather different. The properties of the Banach spaces and metric spaces studied in the Ribe program are rather subtle; in contrast, we are here much more basic and in particular motivated by understanding distance preserving self-maps. This latter topic we see as a kind of metric spectral theory with consequences within several areas of mathematics: geometry, topology, group theory, ergodic theory, probability, complex analysis, operator theory, fixed point theory, and more.

We consider metric spaces (X, d), at times with the symmetry axiom removed, and the corresponding morphisms, here called semicontractions (in contrast to bi-Lipschitz maps in the context of Bourgain, Naor, and others). A map f between two metric spaces is a *semicontraction* if distances are not increased—that is, for any two points x and y, it holds that

$$d(f(x), f(y)) \leq d(x, y).$$

Synonyms are 1-Lipschitz or nonexpansive maps.

It is reasonable to wonder whether in such a general setting there could be anything worthwhile to uncover. One useful general fact is well-known: the *contraction mapping principle*. The abstract statement appeared in Banach's thesis, but some version might have been used before (for the existence and uniqueness of solutions to certain ordinary differential equations). In this essay I will suggest a complement to this principle, which basically appeared in [Ka01] and that is applicable more generally than the contraction mapping principle since isometries are included.

The objective here is to discuss metric space analogs of the linear concepts

- linear functionals and weak topology,
- operator norm and spectral radius, and
- eigenvalues and Lyapunov exponents,

and then show how these metric notions can be applied. At the center for applications is, as already indicated, a complement to the contraction mapping principle—namely, a *spectral principle* [Ka01, GV12], its ergodic theoretic generalization [KaM99, KaL11, GK15] (see also [G18]), and a special type of metrics that could be called spectral metrics [T86, Ka14].

Here is an example: Let M be an oriented closed surface of genus $g \geq 2$. Let $\mathcal{S}$ denote the isotopy classes of simple closed curves on M not isotopically trivial. For a Riemannian metric ρ on M, let $l_\rho(\beta)$ be the infimum of the length of curves isotopic to β. In a seminal preprint from 1976 [T88], Thurston could show the following consequence (the details are worked out in [FLP79, "Théorème Spectrale"]):

Theorem 1.1 ([T88, theorem 5]).
For any diffeomorphism f of M, there is a finite set $1 \leq \lambda_1 < \lambda_2 < \cdots < \lambda_K$ of algebraic integers such that for any $\alpha \in \mathcal{S}$ there is a λ_i such that for any Riemannian metric ρ,

$$\lim_{n\to\infty} l_\rho(f^n\alpha)^{1/n} = \lambda_i.$$

The map f is isotopic to a pseudo–Anosov map iff $K=1$ and $\lambda_1 > 1$.

This is analogous to a simple statement for linear transformations A in finite dimensions: given a vector v there is an associated exponent λ (absolute value of an eigenvalue), such that

$$\lim_{n\to\infty} \|A^n v\|^{1/n} = \lambda.$$

To spell out the analogy: diffeomorphism f instead of a linear transformation A, a length instead of a norm, and a curve α instead of a vector v. Below we will show how to get the top exponent, even for a random product of homeomorphisms, using our metric ideas and a lemma in Margulis's and my paper [KaM99]. This is a different approach than [Ka14]. To get all the exponents (without their algebraic nature) requires some additional arguments; see [H16].

One of the central notions in the present text is that of a Busemann function or metric functional. This notion appears implicitly in classical mathematics, with Poisson and Eisenstein, and is by now recognized by many people as a fundamental tool. In differential geometry, see the discussion in Yau's survey [Y11]. Busemann functions play a crucial role in the Cheeger-Gromoll splitting theorem for manifolds with nonnegative Ricci curvature. The community of researchers of nonpositive curvature also has frequently employed Busemann functions. For example, it has been noted by several people that the horofunction boundary (metric compactification) is the right notion when generalizing Patterson-Sullivan measures; see, for example, [CDST18] for a recent contribution. In my work with Ledrappier, we used this notion without knowing anything about the geometry of the Cayley graphs, in particular without any curvature assumption. Related to this, with a view toward another approach to Gromov's polynomial growth theorem, see [TY16]. There are many other instances one could mention, but still, it seems that the notion of a Busemann function remains a bit off the mainstream, instead of taking its natural place dual to geodesics.

A note on terminology: When I had a choice, or need, to introduce a word for a concept, I sometimes followed Serge Lang's saying that terminology should (ideally) be functorial with respect to the ideas. Hence I use *metric functional* for a variant of the notion of horofunction usually employed and introduced by Gromov, generalizing an older concept due to Busemann, in turn extending a notion in complex analysis (and also from Martin boundary theory). While some people do not like this, I thought it could avoid confusion to have different terms for different concepts, even when, or precisely because, these are variants of each other. In addition to being functorial in the ideas, *metric functional* also sounds more basic and fundamental as a notion than *horofunction* does. Indeed, the present essay tries to argue for the analogy with the linear case and the basic importance of the metric concept of horofunctions or metric functionals. See [Ka19] for a metric Hahn-Banach theorem.

For the revision of this text I thank the referee, David Fisher, Erwann Aubry, Thomas Haettel, Massimo Picardello, Marc Peigné, and especially Armando Gutiérrez for comments.

2 Functionals

2.1 LINEAR THEORY. For vector spaces E, *lines*

$$\gamma : \mathbb{R} \to \mathrm{E}$$

are of course fundamental objects, as are their dual objects, the *linear functionals*

$$\phi : E \to \mathbb{R}.$$

In the case of normed vector spaces the existence of continuous linear functionals relies in general on Zorn's lemma via the Hahn-Banach theorem. It is an abstraction of integrals. The sublevel sets of ϕ define half-spaces. The description of these functionals is an important aspect of the theory; see, for example, the section entitled "The Search for Continuous Linear Functionals" in [Di81].

2.2 METRIC THEORY. For metric spaces X, *geodesic lines*

$$\gamma : \mathbb{R} \to X$$

are fundamental. The map γ is here an isometric embedding. (Note that *geodesic lines* has two meanings: in differential geometry they are locally distance minimizing, while in metric geometry they are most often meant to be globally distance minimizing. The concepts coincide lifted to contractible universal covering spaces.) Now we will discuss what the analog of linear functionals should be; that is, some type of maps

$$h : X \to \mathbb{R}.$$

Observation 2.1. Let X be a real Hilbert space. Take a vector v with $\|v\| = 1$ and consider

$$\lim_{t\to\infty} \|tv - y\| - \|tv\| = \lim_{t\to\infty} \sqrt{(tv-y, tv-y)} - t = \lim_{t\to\infty} \frac{(tv-y, tv-y) - t^2}{\sqrt{(tv-y, tv-y)} + t}$$

$$= \lim_{t\to\infty} \frac{t\left(-2(y,v) + (y,y)/t\right)}{t\left(\sqrt{1 - 2(y,v)/t + (y,y)/t^2} + 1\right)} = -(y,v).$$

In this way one can recover the scalar product from the norm differently than from the polarization identity.

In an *analytic continuation of ideas,* as it were, one is then led to the next observation (which maybe is not how Busemann was thinking about this):

Observation 2.2 (Busemann). Let γ be a geodesic line (or just a ray $\gamma : \mathbb{R}_+ \to X$). Then the following limit exists:

$$h_\gamma(y) = \lim_{t\to\infty} d(\gamma(t), y) - d(\gamma(t), \gamma(0)).$$

The reason for the existence of the limit for each y is that the sequence in question is bounded from below and monotonically decreasing (thanks to the triangle inequality); see [BGS85] and [BrH99].

Example 2.3. The open unit disk of the complex plane admits the Poincaré metric in its infinitesimal form

$$ds = \frac{2\,|dz|}{1-|z|^2}.$$

This gives a model for the hyperbolic plane, and moreover it is fundamental in the way that every holomorphic self-map of the disk is a semicontraction in this metric; this is the content of the Schwarz-Pick lemma. The Busemann function associated to the ray from 0 to the boundary point ζ—in other words, $\zeta \in \mathbb{C}$ with $|\zeta| = 1$—is

$$h_\zeta(z) = \log \frac{|\zeta - z|^2}{1-|z|^2}.$$

These functions appear (in disguise) in the Poisson integral representation formula and in the Eisenstein series.

We can take one more step, which will be parallel to the construction of the Martin boundary in potential theory. This specific metric idea might have come from Gromov around 1980 (except that he considers another topology—an important point for us here).

Let (X, d) be a metric space (perhaps without the symmetric axiom for d satisfied; this point is discussed in [W14] and [GV12]). Let

$$\Phi : X \to \mathbb{R}^X$$

be defined via

$$x \mapsto h_x(\cdot) := d(\cdot, x) - d(x_0, x).$$

This is a continuous injective map. The functions h and their limits are called *metric functionals*. In view of Observation 2.2.2, Busemann functions are examples of metric functionals and are easily seen as not being of the form h_x, with $x \in X$. Even though geodesics may not exist, metric functionals always exist. Note that like in the linear case functionals are normalized to be 0 at the origin: $h(x_0) = 0$.

Every horofunction (i.e., uniform limit on bounded subsets of functions h_x as x tends to infinity) is a metric functional, and every Busemann function is a metric functional. On the other hand, in general it is a well-recognized fact that not every horofunction is a Busemann function (such spaces could perhaps be called nonreflexive) and not every Busemann function is a horofunction; some artificial counterexamples showing this can be thought of:

Example 2.4. Take one ray $[0, \infty]$ that will be geodesic, then add an infinite number of points at distance 1 to the point 0 and distance 2 to each other. Then at each point n on the ray, connect it to one of the points around 0 with a geodesic segment of length $n - 1/2$. This way $h_\gamma(y) = \lim_{t\to\infty} d(\gamma(t), y) - d(\gamma(t), \gamma(0))$ still of course converges for each y but not uniformly. Hence the Busemann function h_γ is a metric functional but not a horofunction.

As already stated, to any geodesic ray from the origin there is an associated metric functional (Busemann function); compare this with the situation in the linear theory that the fundamental Hahn-Banach theorem addresses. In the metric category the theory of injective metric spaces considers when semicontractions (1-Lipschitz maps) defined on a subset can be extended; see [La13] and references therein. See also [Ka19]. The real line is injective, which means that for any subset A of a metric space B and semicontraction $f : A \to \mathbb{R}$ there is an extension of f to $B \to \mathbb{R}$ without increasing the Lipschitz constant—for example,

$$\bar{f}(b) := \sup_{a\in A} \big(f(a) - d(a, b)\big)$$

or

$$\bar{f}(b) := \inf_{a\in A} \big(f(a) + d(a, b)\big).$$

It would require a lengthy effort to survey all the purposes horofunctions have served in the past. Two instances can be found in differential geometry: in nonnegative curvature, the Cheeger-Gromoll theorem, and in nonpositive curvature, the Burger-Schroeder-Adams-Ballmann theorem. In my experience, many people know of one or a few applications, but few have an overview of all the applications. Other applications are found below or in papers listed in the bibliography; for example, let us mention a recent Furstenberg-type formula for the drift of random walks on groups [CLP17], in part building on [KaL06] and [KaL11]. It is also the case that the last two decades have seen identifications and understanding of horofunctions for various classes of metric spaces.

3 Weak convergence and weak compactness

3.1 LINEAR THEORY. One of the main uses for continuous linear functionals is to define weak topologies that have compactness properties even when the vector space is of infinite dimension (the Banach-Aloglu theorem); see [L02].

3.2 METRIC THEORY. We will now discuss how the definition of metric functionals on a metric space will provide the metric space with a weak topology for which the closure is compact. There have been other, more specific efforts to achieve this in special situations. Maybe the first one for trees can be found in Margulis's paper [Ma81]; see also [CSW93] for another approach, [Mo06] for a discussion in nonpositive curvature, and then [GV12] for the general method taken here.

Let X be a set. By a *hemi-metric* on X we mean a function

$$d : X \times X \to \mathbb{R}$$

such that $d(x,y) \leq d(x,z) + d(z,y)$ for every $x, y, z \in X$ and $d(x,y) = 0$ if and only if $x = y$. (The latter axiom can be satisfied by passing to a quotient space.) In other words, we do not insist that d is symmetric (one could symmetrize it) or positive. For more discussion about such metrics, see [GV12] and [W14]. One way to proceed is to consider

$$D(x,y) := \max\{d(x,y), d(y,x)\},$$

which clearly is symmetric but also positive (see [GV12]), so an honest metric. One can take the topology on X from D.

For a weak topology there are a couple of alternative definitions, but we proceed as follows. As defined in the previous section, let

$$\Phi : X \to \mathbb{R}^X,$$

defined via

$$x \mapsto h_x(\cdot) := d(\cdot, x) - d(x_0, x).$$

This is a continuous injective map. By the triangle inequality we note that

$$-d(x_0, y) \leq h_x(y) \leq d(y, x_0).$$

A consequence of this in view of Tychonoff's theorem is that with the pointwise (= product) topology, the closure $\overline{\Phi(X)}$ is compact. In general this is not a compactification in the strict and standard sense that the space sits as an open dense subset in a compact Hausdorff space, but it is convenient to still call it a compactification; for a discussion about this terminology, see [Si15, 6.5].

Example 3.1. This has by now been studied for a number of classes of metric spaces: nonpositively curved spaces [BGS85, BrH99], Gromov hyperbolic spaces ([BrH99], or more recent and closer to our consideration is [MT18]), Banach spaces [W07, Gu17, Gu18], Teichmüller spaces (see [Ka14] for references in particular to Walsh), Hilbert metrics [W14, W18, LN12], Roller boundary of CAT(0)-cube complex (due to Bader-Guralnick; see [FLM18]), and symmetric spaces of noncompact type equipped with Finsler metrics [KL18].

Let me introduce some terminology. We call $\overline{\Phi(X)}$ the *metric compactification* (the term was also coined for proper geodesic metric spaces by Rieffel in a paper on operator algebras and noncommutative geometry) and denote it by $\overline{X}$, even though this is a bit abusive, since the topology of X itself might be different. The closure that is usually considered starting from Gromov (see [BGS85, BrH99]) is to take the topology of uniform convergence on bounded sets (note that uniform convergence on compact sets is in the present context equivalent to our pointwise convergence), and following [BrH99] we call this the *horofunction bordification*. For proper geodesic spaces the two notions coincide.

Example 3.2. A simple useful example is the following metric space, which I learned from Uri Bader. Consider longer and longer finite closed intervals $[0, n]$ all glued to a point x_0 at the point 0. This becomes a countable (metric) tree that is unbounded but contains no infinite geodesic ray. By virtue of being a tree it is CAT(0). It is easy to directly verify that there are no limits in terms of the topology of uniform convergence on bounded subsets. Alternatively, one can see this less directly since for CAT(0) spaces every horofunction is a Busemann function, but there are no (infinite) geodesic rays. So there are no horofunctions in the usual sense; the horofunction bordification is empty; no points are added. The metric compactification also does not add any new points, but new topology is such that every unbounded subsequence converges to h_{x_0}. This shows in particular that there are minor inaccuracies in [BrH99, 8.15 exercises] and [GV12, remark 14].

Let us discuss some more terminology: We call, as said above, the elements in $\overline{\Phi(X)}$ *metric functionals*. We call *horofunctions* those that arise from unbounded subsequences via the strong topology, that of uniform convergence on bounded subsets. The metric functionals coming from geodesic rays, via Busemann's observation above, are called *Busemann functions*. As observed, not every Busemann function is a horofunction, and vice versa.

In my opinion these examples show the need for a precise and new terminology, instead of just using the word *horofunction* for all these concepts, with its precise definition depending on the context.

Moreover, we attempt to distinguish further between various classes of metric functionals. We have *finite metric functionals* and *metric functionals at infinity*. The latter are those functions that have $-\infty$ as their infimum; the former are those metric functionals that have a finite infimum. Busemann functions are always at infinity. The tree example above shows that even an unbounded sequence can converge to a finite metric functional. (What can easily be shown, though, is that every metric functional at infinity can only be reached via an unbounded sequence). An example of a metric functional from an unbounded sequence that has finite infimum is the $h_{\infty,0} \equiv 0$ in the Hilbert space example in the next section.

One can have metrically improper metric functionals with infinite infimum. For the finite metric functionals, we suggest moreover that the ones coming from points $x \in X$, h_x are *internal (finite) metric functionals* and the complement of these are the *exotic (finite) metric functionals*. Examples of the latter are provided by the Hilbert space proposition in the next section (their existence is needed since we claim to obtain a compact space in which the Hilbert space sits). For related division of metric functionals in the context of Gromov hyperbolic spaces, see [MT18].

Example 3.3. Here is a simple illustration of how the notion of metric functionals interacts with Gromov hyperbolicity. Let h be a metric functional (Busemann function) defined by a sequence y_m belonging to a geodesic ray from x_0. Assume that x_n is a sequence such that $h(x_n) < 0$ and $x_n \to \infty$. Then

$$2\left(x_n, y_m\right) = d(x_n, x_0) + d(y_m, x_0) - d(x_n, y_m) > d(x_n, x_0)$$

for any n with m sufficiently large in view of $0 > h(x_n) = \lim_{m\to\infty} d(y_m, x_n) - d(y_m, x_0)$. So for each n we can find a sufficiently large m such that this inequality holds along this subsequence $(x_n, y_m) \to \infty$, showing that the two sequences hence converge to one and the same point of the Gromov boundary. For more on metric functionals for (nonproper) Gromov hyperbolic spaces, we refer to [MT18].

4 Examples: Banach spaces

4.1 LINEAR THEORY. The set of continuous linear functionals forms a new normed vector space, called the *dual space*, with norm

$$\|f\| = \sup_{v \neq 0} \frac{|f(v)|}{\|v\|}.$$

4.2 METRIC THEORY. The weak compactification and the horofunctions of Banach spaces introduce a new take on a part of classical functional analysis, especially as they have a similar role as continuous linear functionals. Two features stand out: first, the existence of these new functionals do not need any Hahn-Banach theorem, which in general is based on Zorn's lemma; second, the horofunctions are always convex and sometimes linear. Horofunctions interpolate between the norm ($h_0(x) = \|x\|$) and linear functionals. More precise statements now follow.

Proposition 4.1. *Let E be a normed vector space. Every function $h \in \overline{E}$ is convex; that is, for any $x, y \in X$, one has*

$$h\left(\frac{x+y}{2}\right) \leq \frac{1}{2}h(x) + \frac{1}{2}h(y).$$

Proof. Note that for $z \in E$, one has

$$\begin{aligned} h_z((x+y)/2) &= \left\|(x+y)/2 - z\right\| - \|z\| = \frac{1}{2}\left\|x - z + y - z\right\| - \|z\| \\ &\leq \frac{1}{2}\|x - z\| + \frac{1}{2}\left\|y - z\right\| - \|z\| = \frac{1}{2}h_z(x) + \frac{1}{2}h_z(y). \end{aligned}$$

This inequality passes to any limit point of such h_z. □

Furthermore, as Busemann noticed in the context of geodesic spaces, any vector v gives rise to a horofunction via

$$h_{\infty v}(x) = \lim_{t \to \infty} \|x - tv\| - t\|v\|.$$

Often this is a norm 1 linear functional; it happens precisely when $v/\|v\|$ is a smooth point of the unit sphere [W07, Gu17, Gu18].

Note that in this case one has, in addition to the convexity, that $h_{\infty v}(\lambda x) = \lambda h_{\infty v}(x)$ for scalars λ, and so $h_{\infty v}$ is a homogeneous sublinear function. By the Hahn-Banach theorem we have a norm 1 linear functional ψ associated to unit vector v for which $\psi(v) = 1$ and such that $\psi \leq h_{\infty v}$.

Proposition 4.2. *Let H be a real Hilbert space with scalar product $(\cdot,\cdot)$. The elements of $\overline{H}$ are parametrized by $0 < r < \infty$ and vectors $v \in H$ with $\|v\| \leq 1$, and the element corresponding to $r = 0$, $v = 0$. When $\|v\| = 1$,*

$$h_{r,v}(y) = \|y - rv\| - r,$$

and for general v,

$$h_{r,v}(y) = \sqrt{\|y\|^2 - 2(y, rv) + r^2} - r.$$

In addition there is $h_0(y) := h_{0,0}(y) = \|y\|$ and the $r = \infty$ cases

$$h_{\infty,v}(y) = -(y, v),$$

where $v \in H$ with $\|v\| \leq 1$. A sequence (t_i, v_i) with $\|v_i\| = 1$ converges to $h_{r,v}$ iff $t_i \to r \in (0, \infty]$ and $v_i \to v$ in the standard weak topology or to h_0 iff $t_i \to 0$.

Proof. In order to identify the closure we look at vectors $tv \in H$ where we have normalized so that $\|v\| = 1$. By weak compactness we may assume that a sequence $t_i v_i$ (or net) clusters at some radius r and some limit vector v in the weak topology with $\|v\| \leq 1$. In the case $r < \infty$ we clearly get the functions

$$h_{r,v}(y) = \sqrt{r^2(1 - \|v\|^2) + \|y - rv\|^2} - r,$$

which after developing the norms give the functions in the proposition. Note that in case $t \to 0$ the function is just h_0 independently of v.

In the case $t_i \to \infty$ we have the following calculation:

$$h_{\infty,v}(y) = \lim_{i\to\infty} \sqrt{(t_i v_i - y, t_i v_i - y)} - t = \lim_{i\to\infty} \frac{(t_i v_i - y, t_i v_i - y) - t^2}{\sqrt{(t_i v_i - y, t_i v_i - y)} + t}$$

$$= \lim_{i\to\infty} \frac{t_i\left(-2(y, v) + (y, y)/t_i\right)}{t_i\left(\sqrt{1 - 2(y, v)/t_i + (y, y)/t_i^2} + 1\right)} = -(y, v).$$

It is rather immediate that the functions described are all distinct, which means that for convergent sequences both t_i and v_i must converge (with the trivial exception of when $t_i \to 0$). □

We have in this way compactified Hilbert spaces. To illustrate the relation with the (linear) weak topology, consider an ON basis $\{e_n\}$. It is a first example of the weak topology that $e_n \rightharpoonup 0$ weakly; likewise does the sequence $\lambda_n e_n$ for any sequence of scalars $0 < \lambda_n < 1$. In $\overline{H}$ it is true that $e_n \to h_{1,0}$, but $\lambda_n e_n$ does not necessarily converge. On the other hand, $n \cdot e_1$ does not converge weakly as $n \to \infty$ but $n \cdot e_1 \to h_{\infty,e_1}(\cdot) = -(\cdot, e_1)$ in $\overline{H}$.

For L^p spaces we refer to [W07], [Gu17], [Gu18], and [Gu19]. An interesting detail that Gutiérrez showed is that the function identically equal to zero is not a metric functional for ℓ^1. He also observed how a famous fixed point–free example of Alspach must fix a metric functional.

5 Basic spectral notions

5.1 LINEAR THEORY. Let E be a normed vector space and $A: E \to E$ a bounded (or continuous) linear map (operator). One defines the *operator norm*

$$\|A\| = \sup_{v \neq 0} \frac{\|Av\|}{\|v\|}.$$

A basic notion is the *spectrum* that is a closed nonempty set of complex numbers. As Beurling and Gelfand observed, its radius can be calculated by

$$\rho(A) = \lim_{n \to \infty} \left\| A^n \right\|^{1/n},$$

called the *spectral radius* of A. (The existence of the limit comes from a simple fact, known as the Fekete lemma, in view of the submultiplicative property of the norm; see [L02, 17.1]). One has the obvious inequality

$$\rho(A) \leq \|A\|.$$

In many important cases there is in fact an equality here, such as for normal operators, which includes all unitary and self-adjoint operators.

For a given vector v one may ask for the existence of

$$\lim_{n \to \infty} \left\| A^n v \right\|^{1/n}.$$

Such considerations are called *local spectral theory*. In infinite dimensions this limit may not exist when the spectral theory fails. In finite dimensions the limit exists, as is clear from the Jordan normal form. A counterexample can be given in ℓ^2 where A is a combination of a shift and a diagonal operator, having two exponents each alternating in longer and longer stretches, making the behavior seem different for various periods of n. See, for example, [Sc91] for details.

When A^n is replaced by a random product of operators, an ergodic cocycle, then Oseledets multiplicative ergodic theorem asserts that these limits, called Lyapunov exponents, exist a.e.

5.2 METRIC THEORY. Let (X, d) be a metric space and $f: X \to X$ a semicontraction (i.e., a 1-Lipschitz map). One defines the *minimal displacement*

$$d(f) = \inf_x d(x, f(x)).$$

Like in hyperbolic geometry, or for nonpositively curved spaces [BGS85], one can classify semicontractions of a metric space as follows:

- *Elliptic* if $d(f) = 0$ and the infimum is attained (i.e., there is a fixed point)
- *Hyperbolic* if $d(f) > 0$ and the infimum is attained
- *Parabolic* if the minimum is not attained.

Usually the parabolic maps are the more complicated. It might also be useful to divide semicontractions according to whether all orbits are bounded or all orbits are unbounded, and in the latter case whether all orbits tend to infinity. For example, a circle rotation is hyperbolic and bounded. In this general context we again recommend [G18] for examples and a simpler proof of Calka's theorem, which asserts that for proper metric spaces, unbounded orbits necessarily tend to infinity.

Another basic associated number is the *translation number* (or *drift* or *escape rate*)

$$\tau(f) = \lim_{n\to\infty} \frac{1}{n} d(x, f^n(x)).$$

Notice that this number is independent of x because by the 1-Lipschitz property any two orbits stay on bounded distance from each other. This number exists by the Fekete lemma in view of the subadditivity coming from the triangle inequality and the 1-Lipschitz property. It also has the tracial property $\tau(fg) = \tau(gf)$, as is simple to see.

One has the obvious inequality

$$\tau(f) \leq d(f).$$

In important cases one has equality, especially under nonpositive curvature: for isometries see [BGS85], and for the most general version see [GV12]. In view of the fact that holomorphic maps preserve Kobayashi pseudo-distances, one can study the corresponding invariants and ask when equality holds:

Problem 5.1. For holomorphic self-maps f, when do we have equality $\tau(f) = d(f)$ in the Kobayashi pseudo-distance?

This has been studied by Andrew Zimmer and is analogous to operators when the spectral radius equals the norm.

The following fact is a spectral principle [Ka01] that is analogous to the discussion about the local spectral theory. Note that in contrast to the linear case, it holds in all situations. The first statement can also be thought of as a weak spectral theorem or weak Jordan normal form. (For comparison, there is a stronger version in [GV12] for a restricted class of metric spaces.)

Theorem 5.2 (metric spectral principle [Ka01]).
Given a semicontraction $f:(X,d)\to(X,d)$ with drift τ, there exists $h\in\overline{X}$ such that

$$h(f^k(x_0)) \leq -\tau k$$

for all $k>0$, and for any $x\in X$,

$$\lim_{k\to\infty} -\frac{1}{k}h(f^k(x)) = \tau.$$

Proof. Given a sequence $\epsilon_i \searrow 0$ we set $b_i(n) = d(x_0, f^n(x_0)) - (l-\epsilon_i)n$. Since these numbers are unbounded in n for each fixed i, we can find a subsequence such that $b_i(n_i) > b_i(m)$ for any $m < n_i$. We have for any $k \geq 1$ and i that

$$\begin{aligned}
d(f^k(x_0), f^{n_i}(x_0)) - d(x_0, f^{n_i}(x_0)) &\leq d(x_0, f^{n_i-k}x_0) - d(x_0, f^{n_i}x_0) \\
&= b_i(n_i-k) + (l-\epsilon_i)(n_i-k) - b_i(n_i) - (l-\epsilon_i)n_i \\
&\leq -(l-\epsilon_i)k.
\end{aligned}$$

By compactness, there is a limit point h of the sequence $d(\cdot, f^{n_i}(x_0)) - d(x_0, f^{n_i}(x_0))$ in $\overline{X}$. Passing to the limit in the above inequality gives

$$h(f^k(x_0)) \leq -lk$$

for all $k > 0$. Finally, the triangle inequality

$$d(x, f^k(x)) + d(f^k(x), z) \geq d(x, z)$$

implies that

$$h(f^k(x_0)) \geq -d(x_0, f^k(x_0)).$$

From this, the second statement in the theorem follows in view of the fact that changing x_0 to x only is a bounded change since f is 1-Lipschitz:

$$\left| d(x_0, f^k(x)) - d(x_0, f^k(x_0)) \right| \leq \max \left\{ d(f^k(x), f^k(x_0)), d(f^k(x_0), f^k(x)) \right\}$$
$$\leq \max \{ d(x, x_0), d(x_0, x) \} . \qquad \square$$

Example 5.3. The classical instance of this is the Wolff-Denjoy theorem in complex analysis. This is thanks to Pick's version of the Schwarz lemma, which asserts that every holomorphic map of the unit disk to itself is 1-Lipschitz with respect to the Poincaré metric ρ. It says that given a holomorphic self-map of the disk, either there is a fixed point or there is a point on the boundary circle that attracts every orbit. From basic hyperbolic geometry one can deduce this from our theorem. Wolff also considered horodisks but may not have discussed lengths τ, which here equal $\inf_{z \in D} \rho(z, f(z))$, as follows, for example, from [GV12].

In the isometry case, in the same way, looking at times for which the orbit is closer to the origin than all future orbit points, one can show that there exists a metric functional h such that

$$h(f^{-n} x_0) \geq \tau_{f^{-1}} \cdot n$$

for all $n \geq 1$.

6 Application: Extensions of the mean ergodic theorem

In 1931, in response to a famous hypothesis in statistical mechanics, von Neumann used spectral theory to establish that for unitary operators U,

$$\frac{1}{n} \sum_{k=0}^{n-1} U^k g \to Pg,$$

where P is the projection operator onto the U invariant elements in the Hilbert space in question. Carleman showed this independently at the same time (or before), and a nice proof of a more general statement (for U with $\|U\| \leq 1$) was found by F. Riesz, inspired by Carleman's method. Such a convergence statement is known not to hold in general for all Banach spaces, in the sense that there is no strong convergence of the average. On the other hand, let $f(w) = Uw + v$; then we have

$$f^n(0) = \sum_{k=0}^{n-1} U^k v.$$

If $\|U\| \leq 1$, then f is semicontractive and Theorem 5.2 applies, and it does so for *any* Banach space.

In other words, the theorem is weak enough to always hold. On the other hand, when the situation is better—for example when we are studying transformation of a Hilbert space—then the weak convergence can be upgraded to a stronger statement, thanks to knowledge about the metric functionals. Here is an example:

Let U and f be as above acting on a real Hilbert space. Theorem 5.2 applied to f hands us a metric functional h, for which

$$\frac{1}{n} h\left(\sum_{k=0}^{n-1} U^k v\right) \to -\tau,$$

where as before τ is the growth rate of the norm of the ergodic average. Either $\tau = 0$ and we have

$$\frac{1}{n}\sum_{k=0}^{n-1} U^k v \to 0,$$

or else we need to have that h is a metric functional at infinity (because h must be unbounded from below; see Proposition 4.2); in fact, it must be of the form $h(x) = -(x, w)$ with $\|w\| = 1$ (since τ is the growth of the norm that h applied to the orbit matches). It is a well-known simple fact that if we have a sequence of points x_n in a Hilbert space and a vector w with norm $\|w\| \leq 1$, such that $(x_n, w) \to 1$ and $\|x_n\| \to 1$, then necessarily $x_n \to w$ and $\|w\| = 1$. These are the details for the current situation:

$$\left\|\frac{1}{n}\sum_{k=0}^{n-1} U^k v - \tau w\right\|^2 = \left\|\frac{1}{n}\sum_{k=0}^{n-1} U^k v\right\|^2 - 2\left(\frac{1}{n}\sum_{k=0}^{n-1} U^k v, \tau w\right)$$
$$+ \|\tau w\|^2 \to \tau^2 - 2\tau^2 + \tau^2 = 0$$

This finishes the proof of the classical mean ergodic theorem.

7 Spectral metrics

At the moment I do not see an appropriate axiomatization for the type of metrics that will be useful. Here is an informal description; precise definitions will follow in the particular situations studied later. We will have a group of transformations, with elements denoted f or g, of a space. This space has objects denoted α with some sort of length l; the set or subset of these objects should be invariant under the transformation, and we define

$$d(f,g) = \log \sup_{\alpha} \frac{l(g^{-1}\alpha)}{l(f^{-1}\alpha)}.$$

The triangle inequality is automatic from the supremum, as is the invariance. The function d separates f and g if the set of α's is sufficiently extensive. On the other hand, this distance is not necessarily symmetric. If desired it can be symmetrized in a couple of trivial ways.

Example 7.1. Define a hemi-metric between two linear operators A and B of a real Hilbert space H:

$$d(A, B) = \log \sup_{v \neq 0} \frac{\|B^t v\|}{\|A^t v\|}.$$

(Here t denotes the transpose.) Note that we may take the supremum over the vectors that have unit length, and we see that there is the obvious connection to the operator norm

$$d(I, A) = \log \|A^t\| = \log \|A\|,$$

where I denotes the identity operator.

Here is an example of classical and very useful metrics:

Example 7.2. Metrics on the Teichmüller space of a surface,

$$d(x, y) = \log \sup_{\alpha \in \mathcal{S}} \frac{l_y(\alpha)}{l_x(\alpha)},$$

where x and y denote different equivalence classes of metrics (or complex structures) on a fixed surface, $\mathcal{S}$ is the set of nontrivial isotopy classes of simple closed curves, and l could denote various notions of length, depending on the choice whether the metric is asymmetric. See the next section for more details and applications.

Here is another possibility:

Example 7.3. Taken from ([DKN18]). Given two intervals I, J and a C^1-map $g: I \to J$, which is a diffeomorphism onto its image, the distortion coefficient is defined by

$$K(g; I) := \sup_{x,y \in I} \left| \log \left(\frac{g'(x)}{g'(y)} \right) \right|.$$

This is subadditive under composition, and $K(g, I) = K(g^{-1}, g(I))$.

Other examples of such metrics include the Hilbert, Funk, and Thompson metrics on cones [LN12], the Kobayashi pseudo-metric in the complex category, Hofer's metric on symplectomorphisms [Gr07], and the Lipschitz metric on outer space.

8 Application: Surface homeomorphisms

Let Σ be a surface of finite type. Let $\mathcal{S}$ be the set of nontrivial isotopy classes of simple closed curves on Σ. One denotes by $l_x(\alpha)$ the infimal length of curves in the class of α in the metric x. The metric x can be considered to be a point in the Teichmüller space $\mathcal{T}$ of Σ and hence a hyperbolic metric; the length will be realized on a closed geodesic. Thurston introduced the following asymmetric metric on $\mathcal{T}$:

$$L(x, y) = \log \sup_{\alpha \in \mathcal{S}} \frac{l_y(\alpha)}{l_x(\alpha)}.$$

Thurston's seminal work provided a sort of Jordan normal form for mapping classes of diffeomorphisms of Σ and deduced from this the existence of Lyapunov exponents, or eigenvalues as it were. A different approach was proposed in [Ka14]. In this section we will use the metrics directly, without metric functionals explicitly. We will use a lemma in a paper by Margulis and me [KaM99], which was substantially sharpened in [GK15].

Let (Ω, ρ) be a measure space with $\rho(\Omega) = 1$, and let $T: \Omega \to \Omega$ be an ergodic measure preserving map. We consider a measurable map $\omega \mapsto f_\omega$, where f_ω are homeomorphisms of Σ (or more generally semicontractions of $\mathcal{T}$). We assume the appropriate measurability and integrability assumptions. We form $Z_n(\omega) := f_\omega \circ f_{T\omega} \circ \cdots \circ f_{T^{n-1}\omega}$. Let

$$a(n,\omega) = L(x_0, Z_n(\omega)x_0),$$

which is a subadditive (sub-)cocycle that by the subadditive ergodic theorem

$$a(n,\omega)/n$$

converges for a.e. ω to a constant that we denote by τ. Given a sequence of ϵ_i tending to 0, proposition 4.2. in [KaM99] implies that a.e. there is an infinite sequence of n_i and numbers K_i such that

$$a(n_i,\omega) - a(n_i - k, T^k\omega) \geq (\tau - \epsilon_i)k$$

for all $K_i \leq k \leq n_i$. Moreover, we may assume that $(\tau - \epsilon_i)n_i \leq a(n_i,\omega) \leq (\tau + \epsilon_i)n_i$ for all i.

We will now use a property of L established in [LRT12] (that was not used in [Ka14]). Namely, there is a finite set of curves $\mu = \mu_{x_0}$ such that

$$L(x_0, y) = \log \sup_{\alpha \in \mathcal{S}} \frac{l_y(\alpha)}{l_{x_0}(\alpha)} \asymp \log \max_{\alpha \in \mu} \frac{l_y(\alpha)}{l_{x_0}(\alpha)}$$

up to an additive error.

Now by the pigeonhole principle refine n_i such that there is one curve α_1 in μ that realizes the maximum for each $y = Z_{n_i}(\omega)x_0$; in other words,

$$l_{Z_{n_i}x_0}(\alpha_1) \asymp \exp(n_i(\tau \pm \epsilon_i)).$$

Given the way n_i were selected, we have

$$-\log \sup_{\alpha \in \mathcal{S}} \frac{l_{Z_n x_0}(\alpha)}{l_{Z_k x_0}(\alpha)} \geq -a(n_i - k, T^k\omega) \geq (\tau - \epsilon_i)k - a(n_i,\omega).$$

(The first inequality is an equality in case the maps are isometries and not merely semicontractions.) It follows, like in [Ka14], that

$$l_{Z_k x_0}(\alpha_1) \geq l_{Z_{n_i}x_0}(\alpha_1) e^{-a(n_i,\omega)} e^{(\tau - \epsilon_i)k}.$$

Since no length of a curve can grow faster than $e^{\tau k}$, we get

$$l_{Z_k x_0}(\alpha_1)^{1/k} \to e^{\tau}.$$

In other words, the top Lyapunov exponents exist in this sense. For the other exponents in the independent, identically distributed case, we refer to Horbez [H16] and, in the general ergodic setting, to a forthcoming joint paper with

Horbez. The purpose of this section was to show a different technique to get such results using spectral metrics and subadditive ergodic theory. For a similar statement but with the complex notion of extremal length and using metric functionals, see [GK15].

9 Conclusion

9.1 A BRIEF DISCUSSION OF EXAMPLES OF METRICS. The hyperbolic plane was discovered (rather late) as a consequence of the inquiries on the role of the parallel axiom in Euclidean geometry. At that time it was probably considered a curiosity, but it has turned out to be a basic example connected to an enormous amount of mathematics. In particular it is often the first example in the following list of metric spaces (for references see [Gr07], [Ka05], [GK15]).

- L^2 metrics: The fundamental group of a Riemannian manifold acts by isometry on the universal covering space. In geometric group theory, it is important to have isometric actions on CAT(0) spaces—for example, CAT(0)-cube complexes.
- Symmetric space–type metric spaces: Extending the role of the hyperbolic plane for 2×2 matrices and the moduli of two-dimensional tori, there are the Riemannian symmetric spaces. These have recently also been considered with Finsler metrics. Other extensions are Teichmüller spaces, outer space, spaces of Riemannian metrics on which homeomorphisms or diffeomorphisms have induced isometric actions, and invertible bounded operators on spaces of positive operators.
- Hyperbolic metrics: The most important notion is Gromov hyperbolic spaces, appearing in infinite group theory (Cayley-Dehn; see below), the curve complex (nonlocally compact!), and similar complexes coming from topology and group theory, and Hilbert and Kobayashi metrics in the next item.
- L^∞ metrics. Again generalizing the hyperbolic plane and the positivity aspect of spaces of metrics are cones and convex sets with metrics of Hilbert metric–type. In complex analysis in one or several variables, we have pseudo-metrics of a similar type, generalizing the Poincaré metric, the maximal one being the Kobayashi pseudo-metric. The operator norm, Hofer's metric, and Thurston's asymmetric metric are further examples. Roughly speaking, these are the metrics referred to above as *spectral metrics*, and the natural maps in question in all these examples are semicontractions.

- L^1 metrics: Cayley-Dehn graphs are associated with groups and a generating set; the group itself acts on the graph by automorphisms, which amount to isometries with respect to the word *metric*.

9.2 FURTHER DIRECTIONS. Horbez in [H16] extended [Ka14] to give all exponents in the independent, identically distributed case, thus in particular recovering Thurston's theorem (except for the algebraic nature of the exponents), and implemented the same scheme for the outer automorphisms group via an intricate study of the Culler-Vogtmann outer space, in particular its metric functionals. The paper by Gaubert-Vigeral [GV12], which in particular establishes, with another method, a strengthening of the metric spectral principle above in case the metrics admit a combing of nonpositive curvature, contains further references to examples of hemi-metrics and semi-contractions arising in areas such as game theory and optimal control. Other directions could be

- Symplectomorphisms and Hofer's metric.
- Reproving some statements for invertible linear transformations or compact operators using the asymmetric metric above.
- Diffeomorphisms of manifolds. There are several suggestions for spectral metrics here; see, for instance, Navas's preprint [Na18] on distortion of one-dimensional diffeomorphisms.
- The subject of Kalman filters via the metric approach of Bougerol and others, giving rise to semicontractions; see [Wo07].

In the works of Cheeger and collaborators on differentiability of functions on metric spaces, (see [Ch99], [Ch12]), the notion of a generalized linear function appears. In [Ch99] Cheeger connects this to Busemann functions; on the other hand he remarks in [Ch12] that nonconstant functions do not exist for most spaces. Perhaps it remains to investigate how metric functionals relate to this subject.

References

[BGS85] Ballmann, Werner; Gromov, Mikhael; Schroeder, Viktor. Manifolds of nonpositive curvature. Progress in Mathematics, 61. Birkhäuser Boston, Inc., Boston, MA, 1985.

[BrH99] Bridson, Martin R.; Haefliger, André. Metric spaces of non-positive curvature. Grundlehren der Mathematischen Wissenschaften [Fundamental Principles of Mathematical Sciences], 319. Springer-Verlag, Berlin, 1999.

[Bo87] Bourbaki, N. Topological vector spaces, chapters 1–5. Springer Verlag, Berlin, 1987.

[B86] Bourgain, J. The metrical interpretation of superreflexivity in Banach spaces. Israel J. Math. 56 (1986), no. 2, 222–230.

[CLP17] Carrasco, Matias; Lessa, Pablo; Paquette, Elliot. A Furstenberg type formula for the speed of distance stationary sequences, https://arxiv.org/abs/1710.00733

[CSW93] Cartwright, Donald I.; Soardi, Paolo M.; Woess, Wolfgang. Martin and end compactifications for non-locally finite graphs. Trans. Amer. Math. Soc. 338 (1993), no. 2, 679–693.

[Ch99] Cheeger, J. Differentiability of Lipschitz functions on metric measure spaces. Geom. Funct. Anal. 9 (1999), no. 3, 428–517.

[Ch12] Cheeger, Jeff. Quantitative differentiation: a general formulation. Comm. Pure Appl. Math. 65 (2012), no. 12, 1641–1670.

[Cl18] Claassens, Floris. The horofunction boundary of the infinite dimensional hyperbolic space. Geom. Dedicata 207 (2020), 255–263.

[CDST18] Coulon, R.; Dougall, R.; Schapira, B.; Tapie, S. Twisted Patterson-Sullivan measures and applications to amenability and coverings, https://hal.archives-ouvertes.fr/hal-01881897

[DKN18] Deroin, Bertrand; Kleptsyn, Victor; Navas, Andrés. On the ergodic theory of free group actions by real-analytic circle diffeomorphisms. Invent. Math. 212 (2018), no. 3, 731–779.

[Di81] Dieudonné, Jean. History of functional analysis. North-Holland Mathematics Studies, 49. Notas de Matemática [Mathematical Notes], 77. North-Holland Publishing Co., Amsterdam-New York, 1981.

[FLP79] Fathi, A.; Laudenbach, F.; and Poénaru, V. *Travaux de Thurston sur les surfaces.* Astérisque, 66–67. Société Mathématique de France, Paris, 1979.

[FLM18] Fernós, Talia; Lécureux, Jean; Mathéus, Frédéric. Random walks and boundaries of CAT(0) cubical complexes. Comment. Math. Helv. 93 (2018), no. 2, 291–333.

[GV12] Gaubert, S.; Vigeral, G. A maximin characterisation of the escape rate of non-expansive mappings in metrically convex spaces, Math. Proc. Cambridge Phil. Soc. 152, Issue 2 (2012), 341–363.

[G18] Gouëzel, S. Subadditive cocycles and horofunctions. Proceedings of the International Congress of Mathematicians–Rio de Janeiro 2018. Vol. III. Invited lectures, 1933–1947, World Sci. Publ., Hackensack, NJ, 2018.

[GK15] Gouëzel, S.; Karlsson, A. Subadditive and multiplicative ergodic theorems, J. Eur. Math. Soc. (JEMS) 22 (2020), no. 6, 1893–1915.

[Gr07] Gromov, Misha. Metric structures for Riemannian and non-Riemannian spaces. Based on the 1981 French original. With appendices by M. Katz, P. Pansu, and S. Semmes. Translated from French by Sean Michael Bates. Reprint of the 2001 English edition. Modern Birkhäuser Classics. Birkhäuser Boston, Inc., Boston, MA, 2007.

[Gu17] Gutiérrez, Armando W. The horofunction boundary of finite-dimensional ℓ_p spaces. Colloq. Math. 155 (2019), no. 1, 51–65.

[Gu18] Gutiérrez, Armando W. On the metric compactification of infinite-dimensional ℓ_p spaces. Canad. Math. Bull. 62 (2019), no. 3, 491–507

[Gu19] Gutiérrez, Armando W. Characterizing the metric compactification of L_p spaces by random measures. Ann. Funct. Anal. 11 (2020), no. 2, 227–243.

[H16] Horbez, Camille. The horoboundary of outer space, and growth under random automorphisms, Ann. Scient. Ec. Norm. Sup. (4) 49 (2016), no. 5, 1075–1123.

[KL18] Kapovich, M.; Leeb, B. Finsler bordifications of symmetric and certain locally symmetric spaces. Geometry and Topology, 22 (2018) 2533–2646.

[Ka01] Karlsson, Anders. Non-expanding maps and Busemann functions. Ergodic Theory Dynam. Systems 21 (2001), no. 5, 1447–1457.

[Ka05] Karlsson, Anders. On the dynamics of isometries. Geom. Topol. 9 (2005), 2359–2394.

[Ka14] Karlsson, Anders. Two extensions of Thurston's spectral theorem for surface diffeomorphisms, Bull. London Math. Soc. 46 (2014), no. 2, 217–226.

[Ka19] Karlsson, Anders. Hahn-Banach for metric functionals and horofunctions. J. Funct. Anal. 281 (2021), no. 2, 109030.

[KaL06] Karlsson, Anders; Ledrappier, François. On laws of large numbers for random walks. Ann. Probab. 34 (2006), no. 5, 1693–1706.

[KaL11] Karlsson, Anders; Ledrappier, François. Noncommutative ergodic theorems. *Geometry, rigidity, and group actions*, 396–418, Chicago Lectures in Math., Univ. Chicago Press, Chicago, IL, 2011.

[KaM99] Karlsson, Anders; Margulis, Gregory A. A multiplicative ergodic theorem and nonpositively curved spaces. Comm. Math. Phys. 208 (1999), no. 1, 107–123.

[La13] Lang, Urs. Injective hulls of certain discrete metric spaces and groups, J. Topol. Anal. 5 (2013), 297–331.

[L02] Lax, Peter D. *Functional analysis*, Wiley, New York, 2002

[LN12] Lemmens, Bas; Nussbaum, Roger. *Nonlinear Perron-Frobenius theory.* Cambridge Tracts in Mathematics, 189. Cambridge University Press, Cambridge, 2012.

[LRT12] Lenzhen, Anna; Rafi, Kasra; Tao, Jing. Bounded combinatorics and the Lipschitz metric on Teichmüller space. Geom. Dedicata 159 (2012), 353–371.

[MT18] Maher, Joseph; Tiozzo, Giulio. Random walks on weakly hyperbolic groups. J. Reine Angew. Math. 742 (2018), 187–239.

[Ma81] Margulis, G. A. On the decomposition of discrete subgroups into amalgams. Selected translations. Selecta Math. Soviet. 1 (1981), no. 2, 197–213.

[Mo06] Monod, Nicolas. Superrigidity for irreducible lattices and geometric splitting. J. Amer. Math. Soc. 19 (2006), no. 4, 781–814.

[N18] Naor, Assaf. Metric dimension reduction: a snapshot of the Ribe program. Proceedings of the International Congress of Mathematicians–Rio de Janeiro 2018. Vol. I. Plenary lectures, 759–837, World Sci. Publ., Hackensack, NJ, 2018.

[Na18] Navas, Andrés. On conjugates and the asymptotic distortion of 1-dimensional C^{1+bv} diffeomorphisms, https://arxiv.org/pdf/1811.06077.pdf

[Sc91] Schaumlöffel, Kay-Uwe. Multiplicative ergodic theorems in infinite dimensions. Lyapunov exponents (Oberwolfach, 1990), 187–195, Lecture Notes in Math., 1486, Springer, Berlin, 1991.

[Si15] Simon, Barry. *Operator theory: a comprehensive course in analysis, part 4*, AMS, Providence, RI, 2015.

[StS11] Stein, Elias M.; Shakarchi, Rami. *Functional analysis: introduction to further topics in analysis.* Princeton University Press, Princeton, NJ, 2011

[T86] Thurston, W. Minimal stretch maps between hyperbolic surfaces. Preprint, arXiv:math GT/9801039, 1986.

[T88] Thurston, William P. On the geometry and dynamics of diffeomorphisms of surfaces. Bull. Amer. Math. Soc. (N.S.) 19 (1988), no. 2, 417–431.

[TY16] Tointon, Matthew C. H.; Yadin, Ariel. Horofunctions on graphs of linear growth. C. R. Math. Acad. Sci. Paris 354 (2016), no. 12, 1151–1154.

[W07] Walsh, Cormac. The horofunction boundary of finite-dimensional normed spaces. Math. Proc. Cambridge Philos. Soc., 142 (2007), no. 3, 497–507.

[W14] Walsh, Cormac. The horoboundary and isometry group of Thurston's Lipschitz metric. Handbook of Teichmüller theory. Vol. IV, 327–353, IRMA Lect. Math. Theor. Phys., 19, Eur. Math. Soc., Zürich, 2014.

[W18] Walsh, Cormac. Hilbert and Thompson geometries isometric to infinite-dimensional Banach spaces. Ann. Inst. Fourier (Grenoble) 68 (2018), no. 5, 1831–1877.

[Wo07] Wojtkowski, Maciej P. Geometry of Kalman filters. J. Geom. Symmetry Phys. 9 (2007), 83–95.

[Y11] Yau, Shing-Tung. Perspectives on geometric analysis. Geometry and analysis, no. 2, 417–520, Adv. Lect. Math. (ALM), 18, Int. Press, Somerville, MA, 2011.

PART IV

Homogeneous dynamics

9

VICTOR BERESNEVICH* AND DMITRY KLEINBOCK†

QUANTITATIVE NONDIVERGENCE AND DIOPHANTINE APPROXIMATION ON MANIFOLDS

Dedicated to G. A. Margulis, with admiration

Abstract. The goal of this survey is to discuss the quantitative nondivergence estimate on the space of lattices and present a selection of its applications. The topics covered include extremal manifolds, Khintchine-Groshev-type theorems, rational points lying close to manifolds, and badly approximable points on manifolds. The main emphasis is on the role of the quantitative nondivergence estimate in the aforementioned topics within the theory of Diophantine approximation; therefore this paper should not be regarded as a comprehensive overview of the area.

1 Quantitative nondivergence estimate and its origins

1.1 BACKGROUND. The main purpose of this survey is to discuss a particular strand of fruitful interactions between Diophantine approximation and the methods of homogeneous dynamics. The focus will be on the technique/estimate developed in [KM98] by Margulis and the second-named author, which is commonly known by the name of *quantitative nondivergence* (QnD). Before considering any quantitative aspects of the theory, it will be useful to explain the meaning of *nondivergence* of sequences and maps in the space

$$X_k := \mathrm{SL}_k(\mathbb{R})/\mathrm{SL}_k(\mathbb{Z})$$

of real unimodular lattices. As is well-known, the quotient topology induced from $\mathrm{SL}_k(\mathbb{R})$ makes this space noncompact. Naturally, a nondivergent sequence in X_k is then defined by requiring that it keeps returning into some compact set. To give this narrative description more rigor it is convenient to

VICTOR BERESNEVICH. Department of Mathematics. University of York, Heslington, York, YO10 5DD, England
victor.beresnevich@york.ac.uk

DMITRY KLEINBOCK. Department of Mathematics. Brandeis University, Waltham, MA 02454 USA
kleinboc@brandeis.edu

*ORCID: 0000-0002-1811-9697 (also known as V. Berasnevich).
†ORCID: 0000-0002-9418-5020. Supported in part by NSF grants DMS-1600814 and DMS-1900560.

use Mahler's compactness theorem and the function

$$\delta : X_k \to \mathbb{R}_+,$$

which assigns the length of the shortest nonzero vector to a given lattice. Thus,

$$\delta(\Lambda) := \inf \left\{ \|\mathbf{v}\| : \mathbf{v} \in \Lambda \setminus \{\mathbf{0}\} \right\} \quad \text{for every} \quad \Lambda \in X_k.$$

Mahler's Compactness Theorem [Mah46] states that "a subset S of X_k is relatively compact if and only if there exists $\varepsilon > 0$ such that $\delta(\Lambda) \geq \varepsilon$ for all $\Lambda \in S$." Thus, a sequence of lattices is nondivergent if and only if for a suitably chosen $\varepsilon > 0$ the sequence contains infinitely many elements in the (compact) set

$$\mathcal{K}_\varepsilon := \left\{ \Lambda \in X_k : \delta(\Lambda) \geq \varepsilon \right\}. \tag{1.1}$$

The choice of the norm $\|\cdot\|$ does not affect Mahler's theorem. For simplicity we shall stick to the supremum norm: $\|\mathbf{v}\| = \max_{1 \leq i \leq k} |v_i|$ for $\mathbf{v} = (v_1, \ldots, v_k)$.

Similarly, given a continuous map

$$\phi : [0, +\infty) \to X_k,$$

we will say that $\phi(x)$ is *nondivergent* (as $x \to +\infty$) if there exists $\varepsilon > 0$ such that $\phi(x) \in \mathcal{K}_\varepsilon$ for arbitrarily large x.

The development of the QnD estimate in [KM98] was preceded by several important nonquantitative results instigated by Margulis [Mar71] regarding the orbits of one-parameter unipotent flows. The main result of [Mar71] verifies that if $\{u_x\}_{x \in \mathbb{R}}$ is a one-parameter subgroup of $\mathrm{SL}_k(\mathbb{R})$ consisting of unipotent matrices, then $\phi(x) = u_x \Lambda$ is nondivergent for any $\Lambda \in X_k$. Several years later Dani [Dan79] strengthened Margulis's result by showing that such orbits return into a suitably chosen compact set with positive frequency. To be more precise, Dani proved that there are $0 < \varepsilon, \eta < 1$ such that for any interval $[0, t] \subset [0, +\infty)$, one has that

$$\lambda \left\{ x \in [0, t] : \phi(x) \notin \mathcal{K}_\varepsilon \right\} < \eta\, t, \tag{1.2}$$

where λ stands for Lebesgue measure on $\mathbb{R}$. Subsequently Dani [Dan86] improved his result by showing that under a mild additional constraint on $\{u_x\}_{x \in \mathbb{R}}$, the parameter $\eta > 0$ in (1.2) can be made arbitrarily small, in which case, of course, ε has to be chosen appropriately small. Later Shah [Sha94] generalized Dani's result to polynomial maps ϕ that are not necessarily orbits of some subgroups of $SL_k(\mathbb{R})$. It has to be noted that the nondivergence theorem of Margulis was used as an ingredient in his proof of arithmeticity of nonuniform lattices in semisimple Lie groups of higher rank [Mar75] and that subsequent qualitative nondivergence estimates—in particular, Dani's result in [Dan86]—were an important part of various significant developments of the time, such as Ratner's celebrated theorems [Rat94]. The essence of the quantitative nondivergence estimate obtained in [KM98] is basically an explicit dependence of η on ε in Equation (1.2). More to the point, it is applicable to a very general class of maps ϕ of several variables that do not have to be polynomial, let alone the orbits of unipotent subgroups. In the next subsection we give the precise formulation of the QnD estimate.

1.2 THE QUANTITATIVE NONDIVERGENCE ESTIMATE.

First we recall some notation and definitions. Given a ball $B = B(x_0, r) \subset \mathbb{R}^d$ centered at x_0 of radius r and $c > 0$, by cB we will denote the ball $B(x_0, cr)$. Throughout, λ_d will denote Lebesgue measure on $\mathbb{R}^d$. Given an open subset $U \subset \mathbb{R}^d$ and real numbers $C, \alpha > 0$, a function $f : U \to \mathbb{R}$ is called *(C, α)-good on U* if for any ball $B \subset U$ for any $\varepsilon > 0$:

$$\lambda_d\big(\{x \in B : |f(x)| < \varepsilon\}\big) \leq C \left(\frac{\varepsilon}{\sup_{x\in B}|f(x)|}\right)^{\alpha} \lambda_d(B).$$

Finally, given $\mathbf{v}_1, \ldots, \mathbf{v}_r \in \mathbb{R}^k$, the number $\|\mathbf{v}_1 \wedge \ldots \wedge \mathbf{v}_r\|$ will denote the supremum norm of the exterior product $\mathbf{v}_1 \wedge \ldots \wedge \mathbf{v}_r$ with respect to the standard basis of $\bigwedge^r(\mathbb{R}^k)$. Up to sign the coordinates of $\mathbf{v}_1 \wedge \ldots \wedge \mathbf{v}_r$ can be computed as all the $r \times r$ minors of the matrix composed of the coordinates of $\mathbf{v}_1, \ldots, \mathbf{v}_r$ in the standard basis. Recall that the norm $\|\mathbf{v}_1 \wedge \ldots \wedge \mathbf{v}_r\|$ is equivalent to the r-dimensional volume of the parallelepiped spanned by $\mathbf{v}_1, \ldots, \mathbf{v}_r$, which is precisely the Euclidean norm of $\mathbf{v}_1 \wedge \ldots \wedge \mathbf{v}_r$.[1]

THEOREM 1.1 (quantitative nondivergence estimate [KM98, Theorem 5.2]).
Let $k, d \in \mathbb{N}$, $C, \alpha > 0$, $0 < \rho \leq 1/k$, a ball B in $\mathbb{R}^d$ and a function $h : 3^k B \to GL_k(\mathbb{R})$ be given. Assume that for any linearly independent collection of integer vectors $\mathbf{a}_1, \ldots, \mathbf{a}_r \in \mathbb{Z}^k$,

[1] In Diophantine approximation, if $\mathbf{v}_1, \ldots, \mathbf{v}_r$ is a basis of $\mathbb{Z}^k \cap V$, where $V = \mathrm{Span}_{\mathbb{R}}(\mathbf{v}_1, \ldots, \mathbf{v}_r)$, the Euclidean norm of $\mathbf{v}_1 \wedge \ldots \wedge \mathbf{v}_r$ is known as the *height* of the linear rational subspace V of $\mathbb{R}^k$; see [Sch91].

(1) *the function* $x \mapsto \|h(x)\mathbf{a}_1 \wedge \ldots \wedge h(x)\mathbf{a}_r\|$ *is* (C,α)*-good on* $3^k B$, *and*

(2) $\sup_{x\in B} \|h(x)\mathbf{a}_1 \wedge \ldots \wedge h(x)\mathbf{a}_r\| \geq \rho$.

Then for any $\varepsilon > 0$,

$$\lambda_d\big(\{x \in B : \delta\big(h(x)\mathbb{Z}^k\big) < \varepsilon\}\big) \;\leq\; kC(3^d N_d)^k \left(\frac{\varepsilon}{\rho}\right)^{\alpha} \lambda_d(B), \tag{1.3}$$

where N_d *is the Besicovitch constant.*

We note that in [Kle08] the second-named author established a version of the QnD estimate where the norm $\|\cdot\|$ is made to be Euclidean, not supremum. This made it possible to remove the condition $\rho < 1/k$ and at the same time replace condition (2) above by a weaker condition:

(3) $\sup_{x\in B} \|h(x)\mathbf{a}_1 \wedge \ldots \wedge h(x)\mathbf{a}_r\| \geq \rho^r$.

The latter has been especially useful for studying Diophantine approximation on affine subspaces. See [Kle10a] for a detailed exposition of the proof of the refined version of Theorem 1.1 established in [Kle08].

The estimate in Equation (1.3) is amazingly general. However, in many applications, analyzing conditions (1) and (2)/(3) represents a substantial, often challenging, task. In the case when h is analytic—that is, every entry of h is a real analytic function of several variables—condition (1) always holds for some C and α; see [KM98] for details. For the rest of this survey we shall mainly discuss developments in the theory of Diophantine approximation where the QnD estimate played an important, if not crucial, role. Naturally, we begin with the application of the QnD estimate that motivated its discovery.

2 The Baker-Sprindžuk conjecture and extremality

In this section we explain the role of QnD in establishing the Baker-Sprindžuk conjecture—a combination of two prominent problems in the theory of Diophantine approximation on manifolds, one due to A. Baker [Bak75] and the other due to V. G. Sprindžuk [Spr80]. The Baker-Sprindžuk conjecture is not merely a combination of two disjoint problems. Indeed, the origin of both problems lies in a single conjecture of Mahler [Mah32]—an important problem in the theory of transcendence posed in 1932 and proved by Sprindžuk in 1964 [Spr69]. Mahler's conjecture/Sprinzŭk's theorem states that for any $n \in \mathbb{N}$ and any $\varepsilon > 0$ for almost every real number x the inequality

$$(2.1) \qquad |p + q_1 x + q_2 x^2 + \cdots + q_n x^n| < \|\mathbf{q}\|^{-n(1+\varepsilon)}$$

has only finitely many solutions $(p, \mathbf{q}) \in \mathbb{Z} \times \mathbb{Z}^n$, where $\mathbf{q} = (q_1, \ldots, q_n)$ and, as before, $\|\mathbf{q}\| = \max_{1 \leq i \leq n} |q_i|$. Note that, by Dirichlet's theorem or Minkowski's theorem for systems of linear forms, if $\varepsilon \leq 0$, then (2.1) has infinitely many integer solutions $(p, \mathbf{q})$ for any $x \in \mathbb{R}$. The condition $\varepsilon > 0$ in Mahler's conjecture is therefore sharp; that does not mean, however, that one cannot improve on it! The improvements may come about when one replaces the right-hand side of (2.1) with a different function of $\mathbf{q}$. One such improvement was conjectured by A. Baker [Bak75], who proposed that the statement of Mahler's conjecture had to be true if the right-hand side of (2.1) was replaced by

$$\Pi_+(\mathbf{q})^{-(1+\varepsilon)}, \qquad \text{where } \Pi_+(\mathbf{q}) := \prod_{\substack{i=1 \\ q_i \neq 0}}^{n} |q_i|.$$

Clearly this leads to a stronger statement than Mahler's conjecture since

$$(2.2) \qquad \Pi_+(\mathbf{q}) \leq \|\mathbf{q}\|^n.$$

The essence of replacement of $\|\mathbf{q}\|^n$ with $\Pi_+(\mathbf{q})$ is to make the error of approximation depend on the size of each coordinate of $\mathbf{q}$ rather than on their maximum.

In another direction Sprindžuk [Spr80] proposed a generalization of Mahler's conjecture by replacing the powers of x in (2.1) with arbitrary analytic functions of real variables that together with 1 are linearly independent over $\mathbb{R}$. The two conjectures (of Baker and Sprindžuk) can be merged in an obvious way to give what is known as the Baker-Sprindžuk conjecture:

CONJECTURE 2.1 (Baker-Sprindžuk). *Let $f_1, \ldots, f_n : U \to \mathbb{R}$ be real analytic functions defined on a connected open set $U \subset \mathbb{R}^d$. Suppose that $1, f_1, \ldots, f_n$ are linearly independent over $\mathbb{R}$. Then for any $\varepsilon > 0$ and for almost every $x \in U$, the inequality*

$$|p + q_1 f_1(x) + \cdots + q_n f_n(x)| < \Pi_+(\mathbf{q})^{-1-\varepsilon}$$

has only finitely many solutions $(p, \mathbf{q}) \in \mathbb{Z} \times \mathbb{Z}^n$.

At the time these conjectures were posed each seemed intractable, and for a long while only limited partial results were known. Indeed, both conjectures remained open for $n \geq 4$ almost until the turn of the millennium, when they were solved in [KM98] as an elegant application of the quantitative nondivergence estimate (Theorem 1.1). The ultimate solution applies to the wider class of nondegenerate maps (defined below) that are not necessarily analytic.

Nondegeneracy. Let $\mathbf{f} = (f_1, \dots, f_n) : U \to \mathbb{R}^n$ be a map defined on an open subset U of $\mathbb{R}^d$. Given a point $x_0 \in U$, we say that $\mathbf{f}$ is *ℓ-nondegenerate at x_0* if $\mathbf{f}$ is ℓ times continuously differentiable on some sufficiently small ball centered at x_0 and the partial derivatives of $\mathbf{f}$ at x_0 of orders up to ℓ span $\mathbb{R}^n$. The map $\mathbf{f}$ is called *nondegenerate* at x_0 if it is ℓ-nondegenerate at x_0 for some $\ell \in \mathbb{N}$; $\mathbf{f}$ is called nondegenerate almost everywhere (in U) if it is nondegenerate at almost every $x_0 \in U$ with respect to Lebesgue measure. The nondegeneracy of differentiable submanifolds of $\mathbb{R}^n$ is defined via their parameterization(s). Note that a real analytic map $\mathbf{f}$ defined on a connected open set is nondegenerate almost everywhere if and only if $1, f_1, \dots, f_n$ are linearly independent over $\mathbb{R}$.

With the definition of nondegeneracy in place we are now ready to state the following flagship result of [KM98] that solved the Baker-Sprindžuk conjecture in full generality, not only in the analytic case but also for arbitrary nondegenerate maps.

THEOREM 2.2 ([KM98], Theorem A).
Let $\mathbf{f} = (f_1, \dots, f_n)$ be a map defined on an open subset U of $\mathbb{R}^d$ that is nondegenerate almost everywhere. Then for any $\varepsilon > 0$, for almost every $x \in U$ the inequality

$$|p + q_1 f_1(x) + \cdots + q_n f_n(x)| < \Pi_+(\mathbf{q})^{-1-\varepsilon} \tag{2.3}$$

has only finitely many solutions $(p, \mathbf{q}) \in \mathbb{Z} \times \mathbb{Z}^n$.

It is worth making further comments on the terminology used around the Baker-Sprindžuk conjecture. The point $\mathbf{y} \in \mathbb{R}^n$ is called *very well approximable (VWA)* if for some $\varepsilon > 0$ the inequality

$$|p + q_1 y_1 + \cdots + q_n y_n| < \|\mathbf{q}\|^{-n(1+\varepsilon)} \tag{2.4}$$

has infinitely many solutions $(p, \mathbf{q}) \in \mathbb{Z}^{n+1}$. The point $\mathbf{y}$ that is not VWA is often referred to as *extremal*. The point $\mathbf{y} \in \mathbb{R}^n$ is called *very well multiplicatively approximable (VWMA)* if for some $\varepsilon > 0$ the inequality

$$|p + q_1 y_1 + \cdots + q_n y_n| < \Pi_+(\mathbf{q})^{-1-\varepsilon} \tag{2.5}$$

has infinitely many solutions $(p, \mathbf{q}) \in \mathbb{Z}^{n+1}$. The point $\mathbf{y}$ that is not VWMA is often referred to as *strongly extremal*. As one would expect from a well-set terminology we have that

$$\mathbf{y} \text{ is strongly extremal} \quad \Longrightarrow \quad \mathbf{y} \text{ is extremal.}$$

This is due to Equation (2.2).

Similarly, if μ is a measure on $\mathbb{R}^n$, one says that μ is extremal or strongly extremal if so is μ-a.e. point of $\mathbb{R}^n$. The same goes for subsets of $\mathbb{R}^n$ carrying naturally defined measures. For example, the notion of *almost all* for points lying on a differentiable submanifold in $\mathbb{R}^n$ can be defined in several equivalent ways. Perhaps the simplest is to fix a parameterization $\mathbf{f}: U \to \mathbb{R}^n$ (possibly restricting $\mathcal{M}$ to a local coordinate chart) and consider the pushforward $\mathbf{f}_* \lambda d$ of Lebesgue measure on $\mathbb{R}^d$. Then a subset S of $\mathbf{f}(U)$ is null if and only if $\lambda_d(\mathbf{f}^{-1}(S)) = 0$. Now Theorem 2.2 can be rephrased as follows: *almost every point of any nondegenerate*[2] *submanifold $\mathcal{M}$ of $\mathbb{R}^n$ is strongly extremal*, or alternatively *any nondegenerate submanifold $\mathcal{M}$ of $\mathbb{R}^n$ is strongly extremal*.

Sketch of the proof of Theorem 2.2 (for full details see [KM98]). Define the following $(n+1) \times (n+1)$ matrix:

$$u_{\mathbf{f}(x)} = \begin{pmatrix} 1 & \mathbf{f}(x) \\ 0 & I_n \end{pmatrix}. \tag{2.6}$$

For $\mathbf{t} = (t_1, \ldots, t_n) \in \mathbb{Z}^n_{\geq 0}$ define the following $(n+1) \times (n+1)$ diagonal matrix:

$$g_{\mathbf{t}} = \begin{pmatrix} e^t & & & \\ & e^{-t_1} & & \\ & & \ddots & \\ & & & e^{-t_n} \end{pmatrix}, \qquad \text{where } t = t_1 + \cdots + t_n.$$

Given a solution $(p, \mathbf{q}) \in \mathbb{Z}^{n+1}$ to Equation (2.3), one defines $r = \Pi_+(\mathbf{q})^{-\varepsilon/(n+1)}$ and the smallest nonnegative integers t_i such that

$$e^{-t_i} \max\{1, |q_i|\} \leq r.$$

Observe that $e^{t_i} < r^{-1} e \max\{1, |q_i|\}$. Then, by (2.3),

$$e^t |p + q_1 f_1(x) + \cdots + q_n f_n(x)| < \left(e^n r^{-n} \Pi_+(\mathbf{q})\right) \Pi_+(\mathbf{q})^{-1-\varepsilon} = e^n r.$$

---

[2]Here we say that $\mathcal{M}$ is nondegenerate if $\mathbf{f}$ is nondegenerate at λ_d-almost every point of U.

Also, an elementary computation shows that $r < e^{-t\gamma} e^{n\gamma}$ with $\gamma = \varepsilon/(n+1+n\varepsilon)$. As a result we have that

$$\delta(g_{\mathbf{t}} u_{\mathbf{f}(x)} \mathbb{Z}^{n+1}) < e^n r < e^{n(1+\gamma)} e^{-\gamma t}. \tag{2.7}$$

Clearly, if for some $x \in U$ (2.3) holds for infinitely many $\mathbf{q}$, then (2.7) holds for infinitely many $\mathbf{t} \in \mathbb{Z}^n_{\geq 0}$. Then the obvious line of proof of Theorem 2.2 is to demonstrate that the sum of measures

$$\sum_{\mathbf{t}} \lambda_d \{x \in B : \delta(g_{\mathbf{t}} u_{\mathbf{f}(x)} \mathbb{Z}^{n+1}) < e^{n(1+\gamma)} e^{-\gamma t}\}$$

converges, where B is a sufficiently small ball centered at an arbitrary point $x_0 \in U$ such that $\mathbf{f}$ is nondegenerate at x_0. Indeed, the nondegeneracy condition placed on $\mathbf{f}$ justifies the restriction of x to B while the Borel-Cantelli Lemma from probability theory ensures that for almost all x, Equation (2.7) holds only finitely often subject to the convergence of the above series.

The following are the remaining steps in the proof:

- Take $h(x) = g_{\mathbf{t}} u_{\mathbf{f}(x)}$;
- Verify conditions (1) and (2) of Theorem 1.1 for suitably chosen balls B;
- Conclude that

$$\lambda_d(\{x \in B : \delta(g_{\mathbf{t}} u_{\mathbf{f}(x)} \mathbb{Z}^{n+1}) < e^{n(1+\gamma)} e^{-\gamma t}\}) \leq \text{Const} \cdot e^{-\gamma \alpha t}; \textit{ and}$$

- Observe that for each t there are no more than t^{n-1} integer n-tuples $\mathbf{t}$ such that $t_1 + \cdots + t_n = t$ and that

$$\sum_{t=0}^{\infty} t^{n-1} e^{-\gamma \alpha t} < \infty,$$

thus completing the proof. □

It should be noted that the above sketch of proof is missing the details of verifying conditions (1) and (2) of Theorem 1.1. In particular, this requires explicit calculations of actions of h on discrete subgroups of $\mathbb{Z}^{n+1}$ and understanding why certain maps are (C, α)-good. Details can be found in [KM98].

2.1 FURTHER REMARKS ON THE BAKER-SPRINDŽUK CONJECTURE AND EXTREMALITY. The ideas of [KM98] have been taken to a whole new level in [KLW04] by identifying a large class of so-called *friendly measures* that are strongly extremal. Examples include measures supported on a large class of fractal sets (the Cantor ternary set, the Sierpinski gasket, the attractors of certain iterated function systems) and their pushforwards by nondegenerate maps. See also the work of Das, Fishman, Simmons, and Urbański [DFSU18, DFSU21], where they introduced an even larger class of *weakly quasi-decaying measures* to which the QnD method applies.

Theorem 2.2 was generalized in [Kle03] to affine subspaces of $\mathbb{R}^n$ (lines, hyperplanes, etc.) that satisfy a natural Diophantine condition, as well as to submanifolds of such affine subspaces. Affine subspaces and their submanifolds represent a very natural (if not the only natural) example of degenerate submanifolds of $\mathbb{R}^n$. One striking consequence of [Kle03] is the following criterion for analytic submanifolds: *an analytic submanifold $\mathcal{M}$ of $\mathbb{R}^n$ is (strongly) extremal if and only if the smallest affine subspace of $\mathbb{R}^n$ that contains $\mathcal{M}$ is (strongly) extremal.* See also [Kle08] and [Kle10b] for an extension of this inheritance principle to arbitrary (possibly nonextremal) affine subspaces of $\mathbb{R}^n$. Other natural generalizations of Theorem 2.2 include Diophantine approximation on complex analytic manifolds [Kle04], Diophantine approximation in positive characteristics [Gho07], and in $\mathbb{Q}_p$ (p-adic numbers) and more generally Diophantine approximation on submanifolds in the product of several real and p-adic spaces [KT07].

The Baker-Sprindžuk conjecture deals with small values of one linear form of integer variables. This is a special case of the more general framework of systems of several linear forms in which the notions of extremal and strongly extremal matrices are readily available. Given an $n \times m$ matrix X with real entries, one says that X is extremal (not VWA) if and only if for any $\varepsilon > 0$,

$$\|\mathbf{q}X - \mathbf{p}\|^m < \|\mathbf{q}\|^{-n(1+\varepsilon)} \tag{2.8}$$

holds for at most finitely many $(\mathbf{p}, \mathbf{q}) \in \mathbb{Z}^m \times \mathbb{Z}^n$. The choice of the norm $\|\cdot\|$ does not affect the notion, but again for simplicity we choose the supremum norm: $\|\mathbf{q}\| = \max_{1 \le i \le n} |q_i|$ for $\mathbf{q} = (q_1, \ldots, q_n)$. Similarly, one says that X is strongly extremal (not VWMA) if and only if for any $\varepsilon > 0$,

$$\Pi(\mathbf{q}X - \mathbf{p}) < \Pi_+(\mathbf{q})^{-(1+\varepsilon)} \tag{2.9}$$

holds for at most finitely many $(\mathbf{p}, \mathbf{q}) \in \mathbb{Z}^m \times \mathbb{Z}^n$. Here $\Pi(\mathbf{y}) = \prod_{j=1}^m |y_j|$ for $\mathbf{y} = (y_1, \ldots, y_m)$ and, as before, $\Pi_+(\mathbf{q}) = \prod_{i=1,\ q_i \neq 0}^n |q_i|$ for $\mathbf{q} = (q_1, \ldots, q_n)$.

The QnD estimate can be applied to establish (strong) extremality of submanifolds of the space of matrices; however, conditions (1) and (2) of Theorem 1.1 are more difficult to translate into a natural and practically checkable definition of nondegeneracy. Indeed, identifying natural generalizations of the notion of nondegeneracy for matrices has been an active area of recent research; see [KMW10] and [BKM15]. In the case of analytic submanifolds of the space of matrices, the goal has been attained in [ABRS18], with the notion of *constraining pencils* in the space of matrices replacing affine hyperplanes of $\mathbb{R}^n$. See [DS19] for a solution to the matrix version of Baker's Conjecture proposed by the second-named author in [Kle10b].

3 Khintchine-Groshev-type results

The theory of extremality discussed in the previous section deals with Diophantine inequalities with the right-hand side written as the function

$$\psi_\varepsilon(h) = h^{-1-\varepsilon} \tag{3.1}$$

of either $\Pi_+(\mathbf{q})$ or $\|\mathbf{q}\|^n$; see Equations (2.8) and (2.9). Of course, there are other choices for the height function of $\mathbf{q}$. For instance, in the case of the so-called *weighted* Diophantine approximation, one uses the quasi-norm defined by

$$\|\mathbf{q}\|_{\mathbf{r}} = \max_{1\le i\le n} |q_i|^{1/r_i}\,, \tag{3.2}$$

where $\mathbf{r} = (r_1, \ldots, r_n) \in \mathbb{R}^n_{>0}$ is an n-tuple of weights that satisfy the condition

$$r_1 + \cdots + r_n = 1. \tag{3.3}$$

In this section we discuss the refinement of the theory of extremality that involves replacing the specific function ψ_ε given by (3.1) with an arbitrary (monotonic) function ψ, akin to the classical results of Khintchine [Khi24, Khi26]. Below we state Khintchine's theorem in the one-dimensional case. Given a function $\psi : \mathbb{N} \to \mathbb{R}_+$, let

$$\mathcal{A}(\psi) := \{x \in [0,1] \colon |qx - p| < \psi(q) \text{ for infinitely many } (p,q) \in \mathbb{Z}\times\mathbb{N}\}.$$

THEOREM 3.1 (Khintchine's theorem).

$$\lambda_1\big(\mathcal{A}(\psi)\big) = \begin{cases} 0 & \textit{if} \quad \sum_{h=1}^{\infty} \psi(h) < \infty\,, \\ 1 & \textit{if} \quad \sum_{h=1}^{\infty} \psi(h) = \infty \ \textit{ and } \psi \textit{ is nonincreasing.} \end{cases}$$

This beautiful finding has been generalized in many ways, and the theory for independent variables is now in a very advanced state; see, for instance, [BBDV09] and [BV10]. The generalization of Khintchine's theorem to systems of linear forms was first established by Groshev [Gro38]. In the modern-day theory, various generalizations of Khintchine's theorem to Diophantine approximation on manifolds are often called Khintchine-type or Groshev-type or Khintchine-Groshev-type results. We will not define precise meanings of these words as there is some inconsistency in their use across the literature, although some good effort to harmonize the terminology was made in the monograph [BD99].

It has to be noted that the convergence case of Theorem 3.1 is a relatively simple consequence of the Borel-Cantelli Lemma. In the case of Diophantine approximation on manifolds this is no longer the case, and establishing convergence Khintchine-Groshev-type results for manifolds leads to a major challenge—indeed, even in the special case associated with extremality. In this section we shall describe the role played by the QnD estimate in addressing this major challenge—namely, in establishing convergence Khintchine-Groshev-type refinements of Theorem 2.2 that were proved in [BKM01]. The key result of [BKM01] reads as follows.

THEOREM 3.2 (see [BKM01]).
Let $\mathbf{f} = (f_1, \dots, f_n)$ be a map defined on an open subset U of $\mathbb{R}^d$ that is nondegenerate almost everywhere. Let $\Psi : \mathbb{Z}^n \to \mathbb{R}_+$ be any function such that

$$\Psi(q_1, \dots, q_i, \dots, q_n) \le \Psi(q_1, \dots, q_i', \dots, q_n) \text{ if } |q_i| \ge |q_i'| \text{ and } q_i q_i' > 0. \tag{3.4}$$

Suppose that

$$\sum_{\mathbf{q} \in \mathbb{Z}^n} \Psi(\mathbf{q}) < \infty. \tag{3.5}$$

Then for almost every $x \in U$ the inequality

$$|p + q_1 f_1(x) + \cdots + q_n f_n(x)| < \Psi(\mathbf{q}) \tag{3.6}$$

has only finitely many solutions $(p, \mathbf{q}) \in \mathbb{Z}^{n+1}$.

Observe that $\Psi(\mathbf{q}) = \Pi_+(\mathbf{q})^{-1-\varepsilon}$ for $\varepsilon > 0$ satisfies the conditions of Theorem 3.2, and thus Theorem 3.2 is a true generalization of Theorem 2.2.

Prior to describing the ideas of the proof of Theorem 3.2, we formally state the following three corollaries: standard, weighted, and multiplicative Khintchine-Groshev-type results.

COROLLARY 3.3. *Let* $\mathbf{f}$ *be as in Theorem 3.2 and let* $\psi : \mathbb{R}_+ \to \mathbb{R}_+$ *be any monotonic function. Suppose that*

$$\sum_{h=1}^{\infty} \psi(h) < \infty. \tag{3.7}$$

Then for almost every $\mathbf{x} \in U$ *the inequality*

$$|p + q_1 f_1(\mathbf{x}) + \cdots + q_n f_n(\mathbf{x})| < \psi(\|\mathbf{q}\|^n) \tag{3.8}$$

has only finitely many solutions $(p, \mathbf{q}) \in \mathbb{Z}^{n+1}$.

This Khintchine-Groshev-type theorem, a direct generalization of Sprindžuk's conjecture discussed in the previous section, is in fact a partial case of the following, more general, weighted version.

COROLLARY 3.4. *Let* $\mathbf{f}$ *be as in Theorem 3.2,* $\psi : \mathbb{R}_+ \to \mathbb{R}_+$ *be any monotonic function, and* $\mathbf{r} = (r_1, \ldots, r_n) \in \mathbb{R}^n_{>0}$ *be an n-tuple satisfying Equation* (3.3). *Suppose that* (3.7) *holds. Then for almost every* $\mathbf{x} \in U$ *the inequality*

$$|p + q_1 f_1(\mathbf{x}) + \cdots + q_n f_n(\mathbf{x})| < \psi(\|\mathbf{q}\|_{\mathbf{r}}) \tag{3.9}$$

has only finitely many solutions $(p, \mathbf{q}) \in \mathbb{Z}^{n+1}$.

Recall again that Corollary 3.3 is a special case of Corollary 3.4. Indeed, all one has to do to see it is to set $\mathbf{r} = (\frac{1}{n}, \ldots, \frac{1}{n})$. Corollary 3.3 was established in [Ber02] using an approach that does not rely on the QnD estimate. However, without new ideas, that approach does not seem to be possible to extend to the weighted case, let alone multiplicative approximation (the next corollary), where the QnD estimate has proven to be robust.

COROLLARY 3.5. *Let* $\mathbf{f}$ *be as in Theorem 3.2 and* $\psi : \mathbb{R}_+ \to \mathbb{R}_+$ *be any monotonic function. Suppose that*

$$\sum_{h=1}^{\infty} (\log h)^{n-1} \psi(h) < \infty. \tag{3.10}$$

Then for almost every $x \in U$ the inequality

$$|p + q_1 f_1(x) + \cdots + q_n f_n(x)| < \psi\big(\Pi_+(\mathbf{q})\big) \tag{3.11}$$

has only finitely many solutions $(p, \mathbf{q}) \in \mathbb{Z}^{n+1}$.

Sketch of the proof of Theorem 3.2 (for full details see [BKM01]). The proof again uses the QnD estimate (or rather an appropriate generalization of Theorem 1.1). However, this time the QnD estimates are not directly applicable to get the required result. The reason is that α in Equation (1.3) is not matching the heuristic expectation; in fact it can hardly match it. The idea of the proof, which goes back to Bernik's paper [Ber89],[3] is to separate two independent cases as described below.

Case 1: Fix a small $\delta > 0$ and consider the set of $x \in U$ such that Equation (3.6) is satisfied simultaneously with the following condition on the gradient

$$\big\|\nabla\big(q_1 f_1(x) + \cdots + q_n f_n(x)\big)\big\| \geq \|\mathbf{q}\|^{0.5+\delta} \tag{3.12}$$

for infinitely many $(p, \mathbf{q}) \in \mathbb{Z}^{n+1}$.

Case 2: Fix a small $\delta > 0$ and consider the set of $x \in U$ such that (3.6) is satisfied simultaneously with the opposite condition on the gradient.

$$\big\|\nabla\big(q_1 f_1(x) + \cdots + q_n f_n(x)\big)\big\| < \|\mathbf{q}\|^{0.5+\delta} \tag{3.13}$$

for infinitely many $(p, \mathbf{q}) \in \mathbb{Z}^{n+1}$.

Clearly, Theorem 3.2 would follow if one could show that the set $x \in U$ under consideration in each of the two cases is null. So how does splitting into two cases help?

Regarding Case 2, note that the two conditions of Equations (3.4) and (3.5) imposed on Ψ imply that

$$\Psi(\mathbf{q}) \leq \Pi_+(\mathbf{q})^{-1}$$

when $\|\mathbf{q}\|$ is sufficiently large. Hence it suffices to show that the set of $x \in U$ such that

$$\begin{cases} |p + q_1 f_1(x) + \cdots + q_n f_n(x)| < \Pi_+(\mathbf{q})^{-1} \\ \big\|\nabla\big(q_1 f_1(x) + \cdots + q_n f_n(x)\big)\big\| < \|\mathbf{q}\|^{0.5+\delta} \end{cases} \tag{3.14}$$

[3] It is worth mentioning that in [Ber89], Bernik essentially proved Corollary 3.3 in the case $f_i(x) = x^i$.

for infinitely many $(p, \mathbf{q}) \in \mathbb{Z}^{n+1}$ is null. For $\delta < 0.5$ this effectively brings us back to an extremality problem, this time for matrices, which is dealt with using the QnD estimate in a similar way to the proof of Theorem 2.2, albeit with much greater technical difficulties in verifying conditions (1) and (2) of Theorem 1.1 (or rather an appropriate generalization of Theorem 1.1).

Regarding Case 1, the presence of the extra condition—namely, that of Equation (3.12), leads to the following two key properties, which can be verified provided U is of a sufficiently small (fixed) diameter.

Separation property: There is a constant $c > 0$ such that for any $\mathbf{q} \in \mathbb{Z}^n$ with sufficiently large $\|\mathbf{q}\|$ for any $x \in U$ satisfying (3.12), the inequality

$$|p + q_1 f_1(x) + \cdots + q_n f_n(x)| < \frac{c}{\|\nabla(q_1 f_1(x) + \cdots + q_n f_n(x))\|} \tag{3.15}$$

can hold for at most one integer value of p.

Measure comparison property: For some constant $C > 0$ for any $\mathbf{q} \in \mathbb{Z}^n$ with sufficiently large $\|\mathbf{q}\|$ and any integer p,

$$\begin{aligned} &\lambda_d(\{x \in U : (3.6) \ \& \ (3.12) \text{ hold}\}) \\ &\quad \le C\Psi(\mathbf{q})\lambda_d(\{x \in U : (3.6) \ \& \ (3.15) \text{ hold}\}). \end{aligned} \tag{3.16}$$

The separation property implies that for a fixed $\mathbf{q}$,

$$\sum_{p \in \mathbb{Z}} \lambda_d(\{x \in U : (3.6) \ \& \ (3.15) \text{ hold}\}) \le \lambda_d(U).$$

Putting this together with Equation (3.16) and summing over $\mathbf{q}$ gives

$$\sum_{\substack{(p,\mathbf{q}) \in \mathbb{Z}^{n+1} \\ \mathbf{q} \neq \mathbf{0}}} \lambda_d(\{x \in U : (3.6) \ \& \ (3.12) \text{ hold}\}) \le C\lambda_d(U) \sum_{\mathbf{q} \in \mathbb{Z}^n} \Psi(\mathbf{q}) < \infty.$$

It remains to apply the Borel-Cantelli Lemma to complete the proof. □

REMARK 3.6. The above sketch proof is a significantly simplified version of the full proof presented in [BKM01], which is far more effective. The effective elements of the proof are stated as two independent results—Theorems 1.3 and 1.4 in [BKM01]. These underpin a range of further interesting applications of the QnD estimate, which we shall touch on in later sections.

3.1 FURTHER REMARKS ON KHINTCHINE-GROSHEV-TYPE RESULTS. Similarly to the theory of extremality, Khintchine-Groshev-type results are not limited to nondegenerate manifolds and have been extended to affine subspaces of $\mathbb{R}^n$; see, for instance, [Gho05], [Gho10], and [Gho11]. Khintchine-Groshev-type results in p-adic and more generally S-arithmetic settings received their attention too; see, for instance, [MSG09] and [MSG12]. Another remarkable application of the QnD estimate initially discovered in [Ber05] for polynomials and then extended in [BD09] to nondegenerate curves enables one to remove the monotonicity constrain on ψ from Corollary 3.3.

The state of the art for Diophantine approximation of matrices is far less satisfactory, where we do not have sufficiently general convergence Khintchine-Groshev-type results. Partial results include matrices with independent columns; see [BBB17]. The key difficulty likely lies within the case of simultaneous Diophantine approximation, which boils down to counting rational points near manifolds, and will be discussed in section 5.

Theorem 3.2 implies that, under the convergence assumption of Equation (3.5), for almost every $x \in U$ there exists a constant $\kappa > 0$ such that

$$|p + q_1 f_1(x) + \cdots + q_n f_n(x)| \geq \kappa \Psi(\mathbf{q}) \tag{3.17}$$

for all $(p, \mathbf{q}) \in \mathbb{Z}^{n+1}$ with $\mathbf{q} \neq \mathbf{0}$. Clearly, the constant κ depends on x. Let $\mathcal{B}(\mathbf{f}, \Psi; \kappa)$ denote the set of $x \in U$ satisfying the above condition for the same κ. Thus, $\mathcal{B}(\mathbf{f}, \Psi; \kappa)$ is the set of all $x \in U$ such that (3.17) holds for all $(p, \mathbf{q}) \in \mathbb{Z}^{n+1}$ with $\mathbf{q} \neq \mathbf{0}$. The conclusion of Theorem 3.2 exactly means that $\lambda_d(U \setminus \mathcal{B}(\mathbf{f}, \Psi; \kappa)) \to 0$ as $\kappa \to 0^+$. In general, the measure of $U \setminus \mathcal{B}(\mathbf{f}, \Psi; \kappa)$ is positive. It is therefore of interest to understand how fast it is converging to zero. Motivated by applications in network information theory, this question was explicitely posed in [Jaf10]. The answer was provided in [ABLVZ16, Theorem 3] and reads as follows:

$$\lambda_d(U \setminus \mathcal{B}(\mathbf{f}, \Psi; \kappa)) \leq \delta \lambda_d(U)$$

for any

$$\kappa \leq \min \left\{ \kappa_0,\ C_0 \delta \left(\sum_{\mathbf{q} \in \mathbb{Z}^n \setminus \{\mathbf{0}\}} \Psi(\mathbf{q}) \right)^{-1},\ C_1 \delta^{d(n+1)(2l-1)} \right\},$$

where l is a parameter characterizing the nondegeneracy of $\mathbf{f}$ and κ, C_0, C_1 are positive (explicitly computable) constants that depend only on $\mathbf{f}$. This quantitative version of the Khintchine-Groshev theorem for manifolds has been generalized to affine subspaces [GG19].

Given that Khintchine's theorem (Theorem 3.1) treats both convergence and divergence, the question of establishing divergence counterparts to the convergence statements of this section is very natural. In this respect we now state the following known result.

THEOREM 3.7.
Let $\mathbf{f}$ *be as in Theorem 3.2,* $\psi:\mathbb{R}_+\to\mathbb{R}_+$ *be any monotonic function, and* $\mathbf{r}=(r_1,\ldots,r_n)\in\mathbb{R}^n_{>0}$ *be an n-tuple satisfying* (3.3). *Suppose that*

$$\sum_{h=1}^{\infty}\psi(h)=\infty. \tag{3.18}$$

Then for a.e. $x\in U$ *the inequality* (3.9) *has infinitely many solutions* $(p,\mathbf{q})\in\mathbb{Z}^{n+1}$.

Motivated by a Khintchine-Groshev generalization of Mahler's conjecture, Theorem 3.7 was first established in [Ber99] for equal weights $r_1=\cdots=r_n$ and functions $f_i(x)=x^i$ of one variable. Subsequently the method was generalised in [BBKM02] to arbitrary nondegenerate maps and later to arbitrary weights [BBV13]. Theorem 3.7 provides the divergence counterpart to Corollaries 3.3 and 3.4. Establishing the divergence counterpart to Corollary 3.5 (the multiplicative case) remains a challenging open problem even in dimension $n=2$:

PROBLEM 3.8. Let $\mathbf{f}$ be as in Theorem 3.2 and $\psi:\mathbb{R}_+\to\mathbb{R}_+$ be any monotonic function. Suppose that

$$\sum_{h=1}^{\infty}(\log h)^{n-1}\psi(h)=\infty. \tag{3.19}$$

Prove that for almost every $x\in U$ Equation (3.11) holds for infinitely many $(p,\mathbf{q})\in\mathbb{Z}^{n+1}$.

In the light of the recent progress on a version of Problem 3.8 for simultaneous rational approximations for lines made in [BHV20], [Cho18], [CY], and [CT19], it would also be very interesting to investigate Problem 3.8 when $\mathbf{f}$ is a linear/affine (and hence degenerate) map from $\mathbb{R}^d$ into $\mathbb{R}^n$ $(1\le d<n)$. Note that the case $d=n$ of Problem 3.8 follows from a result of Schmidt [Sch64].

We conclude this section by one final comment: the proof of Theorem 3.7 is also underpinned by the QnD estimate. At first glance this may seem rather counter intuitive since the QnD estimate deals with upper bounds while Theorem 3.7 is all about lower bounds. One way or another, this is the case, and we shall return to explaining the role of the QnD estimate in establishing Theorem 3.7 in the next section, within the more general context of Hausdorff measures.

4 Hausdorff measures and dimension

It this section we discuss another refinement of the theory of extremality that aims at understanding the Hausdorff dimension of exceptional sets. The basic question is as follows: *Given a nondegenerate submanifold* $\mathcal{M}$ *of* $\mathbb{R}^n$ *and* $\varepsilon > 0$ *(not necessarily small), what is the Hausdorff dimension of the set of* $\mathbf{y} \in \mathcal{M}$ *such that Equation* (2.4) *holds for infinitely many* $(p, \mathbf{q}) \in \mathbb{Z}^{n+1}$? The same question can be posed in the multiplicative setting of Equation (2.5), for weighted Diophantine approximation (when $\|\mathbf{q}\|^n$ is replace by $\|\mathbf{q}\|_{\mathbf{r}}$ given by (3.2)), and for Diophantine approximation of matrices such as in (2.8) and (2.9).

The background to this question lies with the classical results of Jarník [Jar29] and Besicovitch [Bes34] stated below in the one-dimensional case.

THEOREM 4.1 (Jarník-Besicovitch Theorem).
Let $\varepsilon > 0$ *and* $\psi_\varepsilon(x) = x^{-1-\varepsilon}$. *Then*

$$\dim \mathcal{A}(\psi_\varepsilon) = \frac{2}{2+\varepsilon}.$$

This fundamental result tells us exactly how the size of $\mathcal{A}(\psi_\varepsilon)$ get smaller as we increase ε and thus make the approximation function ψ_ε decrease faster. As with Khintchine's theorem, the Jarník-Besicovitch theorem has been generalized in many ways, and the theory for independent variables is now in a very advanced state; see, for instance, [BV06], [BBDV09], [BV10], and [AB18]. It has to be noted that showing the upper bound $\dim \mathcal{A}(\psi_\varepsilon) \le \frac{2}{2+\varepsilon}$ is a relatively simple consequence of the so-called Hausdorff-Cantelli Lemma [BD99], an analogue of the Borel-Cantelli Lemma. However, in the case of Diophantine approximation on manifolds, establishing upper bounds for manifolds leads to a major challenge that is still very much open (see Problem 4.4). On the contrary, the lower bounds have been obtained in reasonable generality. The main purpose of this section is to exhibit the role played by the QnD estimate in obtaining the lower bounds.

As before, $\mathbf{f} : U \to \mathbb{R}^n$, and $\mathbf{r} = (r_1, \ldots, r_n)$ is an n-tuple of positive numbers satisfying Equation (3.3). Given $s > 0$, $\mathcal{H}^s$ will denote the s-dimensional Hausdorff measure. Let $\delta > 0$, $H > 1$, and

$$\Phi_{\mathbf{r}}(H,\delta) := \left\{ x \in U : \exists\, \mathbf{q} \in \mathbb{Z}^n \setminus \{0\} \text{ such that } \left\{ \begin{array}{l} |p+\mathbf{q}\cdot\mathbf{f}(x)| < \delta H^{-1} \\ \|\mathbf{q}\|_{\mathbf{r}} \leq H \end{array} \right. \right\}.$$

Also, given a function $\psi : \mathbb{R}_+ \to \mathbb{R}_+$, let $\mathcal{W}(\mathbf{f}, U, \mathbf{r}, \psi)$ be the set of $x \in U$ such that (3.9) has infinitely many solutions $(p, \mathbf{q}) \in \mathbb{Z}^{n+1}$. The following homogeneous version of one of the main results from [BBV13] works as a black box to proving divergence Khintchine-Groshev-type results (such as Theorem 3.7) and lower bounds for Hausdorff dimension.

THEOREM 4.2 ([BBV13], Theorem 3).
Let $\mathbf{f} : U \to \mathbb{R}^n$ be a C^2 map on an open subset U of $\mathbb{R}^d$, and let $\mathbf{r} = (r_1, \ldots, r_n)$ be an n-tuple of positive numbers satisfying (3.3). Suppose that for almost every point $x_0 \in U$ there is an open neighborhood $V \subset U$ of x_0 and constants $0 < \delta, \omega < 1$ such that for any ball $B \subset V$ we have that

$$\lambda_d\big(\Phi_{\mathbf{r}}(H,\delta) \cap B\big) \leq \omega \lambda_d(B) \tag{4.1}$$

for all sufficiently large H. Let $d-1 < s \leq d$ and $\psi : \mathbb{R}_+ \to \mathbb{R}_+$ be monotonic. Then

$$\mathcal{H}^s\big(\mathcal{W}(\mathbf{f}, U, \mathbf{r}, \psi)\big) = \mathcal{H}^s(U) \qquad \text{if} \qquad \sum_{\mathbf{q}\in\mathbb{Z}^n_{\neq 0}} \|\mathbf{q}\| \left(\frac{\psi(\|\mathbf{q}\|_{\mathbf{r}})}{\|\mathbf{q}\|}\right)^{s+1-d} = \infty.$$

The proof of this result makes use of *ubiquitous systems* as defined in [BDV06]; see also [Ber00], [BBKM02], and the survey [BBD02b] for the related notion of *regular systems*. The QnD estimate steps in when one wishes to apply Theorem 4.2—namely, to verify the condition of Equation (4.1). In particular, as was demonstrated in [BBV13], any nondegenerate map $\mathbf{f}$ satisfies this condition for any collection of weights $\mathbf{r}$. Upon taking $s = d$ one then verifies that

$$\text{Theorem 4.2} \quad \Longrightarrow \quad \text{Theorem 3.7.}$$

Another consequence of Theorem 4.2 is the following lower bound on the Hausdorff dimension of exceptional sets that contributes to resolving the problem outlined at the beginning of this section. For simplicity we only state the result for the case of equal weights: $\mathbf{r}_0 = (\frac{1}{n}, \ldots, \frac{1}{n})$.

COROLLARY 4.3 ([BBV13], Corollary 2). *Let $\mathbf{f}$ be as in Theorem 3.2, let $\psi : \mathbb{R}_+ \to \mathbb{R}_+$ be any monotonic function, and let*

$$\tau_\psi := \liminf_{t\to\infty} \frac{-\log\psi(t)}{\log t}.$$

Suppose that $n \leq \tau_\psi < \infty$. *Then*

$$\dim \mathcal{W}(\mathbf{f}, U, \mathbf{r}_0, \psi) \geq s := d - 1 + \frac{n+1}{n\tau_\psi + 1}. \tag{4.2}$$

The number τ_ψ is often referred to as the *lower order* of $1/\psi$ at infinity. It indicates the growth of the function $1/\psi$ near infinity. Naturally for $\psi_\varepsilon(t) = t^{-1-\varepsilon}$ we have that $\tau_{\psi_\varepsilon} = 1 + \varepsilon$. Estimate (4.2) was previously shown in [DD00] for arbitrary extremal submanifolds of $\mathbb{R}^n$. The additional benefit of Theorem 4.2 compared to (4.2) is that it allows one to compute the Hausdorff measure of $\mathcal{W}(\mathbf{f}, U, \mathbf{r}_0, \psi)$ at $s = \dim \mathcal{W}(\mathbf{f}, U, \mathbf{r}_0, \psi)$. It is believed that the lower bound given by (4.2) is exact for nondegenerate maps $\mathbf{f}$, at least in the analytic case. It is readily seen that to establish the desired equality for all ψ in question, it suffices to consider approximation functions ψ_ε only. Hence we have the following:

PROBLEM 4.4. Let $\mathbf{f} : U \to \mathbb{R}^n$ be an analytic nondegenerate map defined on a ball U in $\mathbb{R}^d$. Prove that for every $\varepsilon > 0$

$$\dim \mathcal{W}(\mathbf{f}, U, \mathbf{r}_0, \psi_\varepsilon) = d - 1 + \frac{n+1}{n+1+n\varepsilon}. \tag{4.3}$$

Problem 4.4 was established in full in the case $n = 2$ by R. C. Baker [Bak78]. For $n \geq 3$ it remains very much open. However, for the polynomial maps $\mathbf{f} = (x, \ldots, x^n)$, it was settled by Bernik in [Ber83] for arbitrary n.

4.1 FURTHER REMARKS ON HAUSDORFF MEASURES AND DIMENSION. Although establishing the upper bound in (4.3) remains a prominent open problem, it was shown in [BBD02a] that the QnD estimate can be used to resolve it for small ε—namely, for $0 < \varepsilon < 1/(4n^2 + 2n - 4)$ when $d = 1$. This seemingly inconsequential result turned out to have major significance in resolving open problems on badly approximable points on manifolds, which will be discussed in section 6.

We can refine Problem 4.4 in the spirit of Khintchine-Groshev-type results by asking to prove the following convergence counterpart to Theorem 4.2.

PROBLEM 4.5. Let $\mathbf{f} : U \to \mathbb{R}^n$ be a nondegenerate analytic map defined on a ball U in $\mathbb{R}^d$, $0 < s < d$, and $\mathbf{r} = (r_1, \ldots, r_n)$ an n-tuple of positive numbers satisfying Equation (3.3). Suppose that $\psi : \mathbb{R}_+ \to \mathbb{R}_+$ is any monotonic function.

Prove that

$$\mathcal{H}^s(\mathcal{W}(\mathbf{f}, U, \mathbf{r}, \psi)) = 0 \qquad \text{if} \qquad \sum_{\mathbf{q}\in\mathbb{Z}^n_{\neq 0}} \|\mathbf{q}\| \left(\frac{\psi(\|\mathbf{q}\|_{\mathbf{r}})}{\|\mathbf{q}\|}\right)^{s+1-d} < \infty.$$

For $n=2$, Problem 4.5 was resolved in [Hua18] for equal weights, but in higher dimensions it is still open even for $\mathbf{f}(x) = (x, \ldots, x^n)$, let alone nondegenerate maps. In fact, for $n=2$ the case of nonequal weights can be reduced to the case of equal weights. This can be shown by modifying the proof of Theorem 2 from [BB00].

5 Rational points near manifolds

Rational and integral points lying on or near curves and surfaces crop up in numerous problems in number theory and are often one of the principle objects of study (e.g., in analytic number theory, Diophantine approximation, Diophantine geometry). The goal of this section is to demonstrate the role of the QnD estimate in recent counting results on rational points lying close to manifolds [Ber12, BDV07, BZ10, BVVZ]. The motivation lies within the theory of simultaneous Diophantine on manifolds, which boils down to understanding the proximity of rational points $\mathbf{p}/q = (p_1/q, \ldots, p_n/q)$ to points $\mathbf{y} = (y_1, \ldots, y_n) \in \mathbb{R}^n$ restricted to a submanifold $\mathcal{M}$. Here $p_1, \ldots, p_n \in \mathbb{Z}$ and $q \in \mathbb{N}$ is the common denominator of the coordinates of $\mathbf{p}/q$. By Dirichlet's theorem, for every irrational point $\mathbf{y} \in \mathbb{R}^n$ there are infinitely many $\mathbf{p}/q \in \mathbb{Q}^n$ such that

$$\max_{1\le i\le n} \left| y_i - \frac{p_i}{q} \right| < q^{-1-1/n}.$$

This inequality can be rewritten in the form

$$\max_{1\le i\le n} |qy_i - p_i|^n < q^{-1}. \tag{5.1}$$

Understanding when the right-hand side of (5.1) can be replaced by $\psi_\varepsilon(q) = q^{-1-\varepsilon}$, $\varepsilon > 0$ (or even by a generic monotonic function ψ) is the subject matter of many classical problems and famous results. The basic question is about the solvability of

$$\max_{1\le i\le n} |qy_i - p_i|^n < \psi_\varepsilon(q) \tag{5.2}$$

in $(\mathbf{p}, q) \in \mathbb{Z}^n \times \mathbb{N}$ for arbitrarily large q. For example, celebrated Schmidt's subspace theorem states that for any algebraic $y_1, \ldots, y_n$ such that $1, y_1, \ldots, y_n$

are linearly independent over $\mathbb{Q}$ and any $\varepsilon > 0$, Equation (5.2) has only finitely many solutions $(\mathbf{p}, q) \in \mathbb{Z}^n \times \mathbb{N}$.

When the point $\mathbf{y}$ lies on a submanifold $\mathcal{M}$, (5.2) forces the rational point $\mathbf{p}/q$ to lie near $\mathcal{M}$. Hence, metric problems concerning (5.2) (e.g., Khintchine-type results, analogues of the Jarník-Besicovitch theorem) have resulted in significant interest in counting and understanding the distribution of rational points lying close to submanifolds of $\mathbb{R}^n$. The basic setup is as follows. Given $Q > 1$ and $0 < \psi < 1$, let

$$R_{\mathcal{M}}(Q, \psi) = \left\{(\mathbf{p}, q) \in \mathbb{Z}^n \times \mathbb{N} : 1 \leq q \leq Q, \ \mathrm{dist}(\mathbf{p}/q, \mathcal{M}) \leq \psi/q\right\}.$$

It is not difficult to work out the following heuristic

$$\#R_{\mathcal{M}}(Q, \psi) \asymp \psi^m Q^{d+1}, \tag{5.3}$$

where $d = \dim \mathcal{M}$, $m = n - d = \operatorname{codim} \mathcal{M}$, and # stands for the cardinality. Also, $\asymp$ means the simultaneous validity of two Vinogradov symbols $\ll$ and $\gg$, where $\ll$ means the inequality $\leq$ up to a positive multiplicative constant.

REMARK 5.1. This heuristic estimate has to be treated with caution as, for instance, the unit circle $y_1^2 + y_2^2 = 1$ will always contain at least $\textsc{const} \cdot Q$ rational points (given by Pythagorean triples) resulting in $\#R_{\mathcal{M}}(Q, \psi) \gg Q$ no matter how small ψ is. On the contrary, the circle $y_1^2 + y_2^2 = 3$ contains no rational points resulting in $\#R_{\mathcal{M}}(Q, \psi) = 0$ for large Q when $\psi = o(Q^{-1})$. Also, rational (affine) subspaces inherently contain many rational points and so any manifold that contains a rational subspace may break the heuristic with ease for moderately small ψ.

The following is the principal problem in this area; see [Ber12] and [Hua20].

PROBLEM 5.2. Show that (5.3) holds for any suitably curved compact differentiable submanifold $\mathcal{M}$ of $\mathbb{R}^n$ when $\psi \geq Q^{-1/m+\delta}$, where $\delta > 0$ is arbitrary, and $m = \operatorname{codim} \mathcal{M}$.

Ideally, it would be desirable to resolve this problem for all nondegenerate submanifolds of $\mathbb{R}^n$. The condition $\psi \geq Q^{-1/m+\delta}$ is pretty much optimal unless one imposes further constraints on the internal geometry of $\mathcal{M}$. For instance, to relax the condition on ψ, one has to exclude the manifolds that contain a rational subspace of dimension $d-1$ (when $d > 1$). In what follows we shall describe the role of the QnD estimate in establishing the lower bound for analytic nondegenerate manifolds.

THEOREM 5.3 ([Ber12], Corollary 1.5).
For any analytic nondegenerate submanifold $\mathcal{M} \subset \mathbb{R}^n$ of dimension d and codimension $m = n - d$, there exist constants $C_1, C_2 > 0$ such that

$$\#R_{\mathcal{M}}(Q, \psi) \geq C_1 \psi^m Q^{d+1} \tag{5.4}$$

for all sufficiently large Q and all real ψ satisfying

$$C_2 Q^{-1/m} < \psi < 1. \tag{5.5}$$

Sketch of the proof (for full details see [Ber12]). For simplicity we will assume that $\mathcal{M}$ is the image $\mathbf{f}(\mathcal{U})$ of a cube $\mathcal{U} \subset \mathbb{R}^d$ under a map $\mathbf{f}$ and that $\mathcal{M}$ is bounded. Further, without loss of generality, we will assume that $\mathbf{f}$ is of the *Monge form*—that is,

$$\mathbf{f}(x_1, \ldots, x_d) = \big(x_1, \ldots, x_d, f_1(x_1, \ldots, x_d), \ldots, f_m(x_1, \ldots, x_d)\big).$$

The rational points $\mathbf{p}/q$ give rise to the integer vectors $\mathbf{a} = (q, p_1, \ldots, p_n)$, which are essentially projective representations of $\mathbf{p}/q$. Define $\mathbf{y}(x) = \big(1, \mathbf{f}(x)\big)$, a projective representation of $\mathbf{f}(x)$. As is well-known, the distance of $\mathbf{p}/q$ from $\mathbf{f}(x)$ is comparable to the projective distance between them, defined as the sine of the acute angle between $\mathbf{a}$ and $\mathbf{y}(x)$. To make this angle $\ll \psi/q$ and thus ensure that $\mathbf{p}/q$ lies in $R_{\mathcal{M}}(Q, \psi)$, it is enough to verify that

$$|q| \leq Q \qquad \text{and} \qquad |\mathbf{g}_i(x) \cdot \mathbf{a}| \ll \psi \qquad (1 \leq i \leq n) \tag{5.6}$$

for any fixed collection $\mathbf{g}_1(x), \ldots, \mathbf{g}_n(x)$ of vector orthogonal to $\mathbf{y}(x)$ and such that $\|\mathbf{g}_i(x)\| \ll 1$ and $\|\mathbf{g}_1(x) \wedge \ldots \wedge \mathbf{g}_n(x)\| \asymp \|\mathbf{g}_1(x)\| \cdots \|\mathbf{g}_n(x)\|$. Clearly, (5.6) defines a convex body of $\mathbf{a} \in \mathbb{R}^{n+1}$, and one can potentially use Minkowski's theorem on convex bodies to find $\mathbf{a}$. However, for ψ much smaller than $Q^{-1/n}$, this is impossible—the volume of the body is too small. To overcome this difficulty, the convex body is expanded in the directions tangent to the manifolds written in the projective coordinates; see (5.7) below. For this purpose it is convenient to make the following choices for $\mathbf{g}_i(x)$. First of all, for $i = 1, \ldots, m$, the vectors $\mathbf{g}_i(x)$ are taken to be orthogonal to $\mathbf{y}(x), \partial\mathbf{y}(x)/\partial x_1, \ldots, \partial\mathbf{y}(x)/\partial x_d$, which are linearly independent for $\mathbf{f}$ of the Monge form. Next, for $i = m + 1, \ldots, n$, the vectors $\mathbf{g}_i(x)$ are taken to be orthogonal to $\mathbf{y}(x), \mathbf{g}_1(x), \ldots, \mathbf{g}_m(x)$.

We will need two auxiliary positive parameters $\widetilde{\psi}$ and $\widetilde{Q}$, which will be proportional to ψ and Q respectively. Now, for $\kappa > 0$ and a given x, we consider the convex body of $\mathbf{a} \in \mathbb{R}^{n+1}$ defined by

$$\begin{aligned} &|\mathbf{g}_i(x)\cdot \mathbf{a}| < \widetilde{\psi} \quad (1 \le i \le m), \\ &|\mathbf{g}_{m+j}(x)\cdot \mathbf{a}| < (\widetilde{\psi}^m \widetilde{Q})^{-\frac{1}{d}} \quad (1 \le j \le d), \\ &|q| \le \kappa \widetilde{Q}. \end{aligned} \tag{5.7}$$

For a suitably chosen constant $\kappa = \kappa_0$ dependent only on n and $\mathbf{f}$, this body is of sufficient volume to apply Minkowski's theorem. This results in the existence of an $\mathbf{a} \in \mathbb{Z}^{n+1} \setminus \{0\}$ satisfying (5.7). Thus the set

$$B(\widetilde{\psi}, \widetilde{Q}, \kappa) = \{x \in U : \exists\, \mathbf{a} \in \mathbb{Z}^{n+1}_{\neq 0} \text{ satisfying (5.7)}\}$$

coincides with all of U for $\kappa = \kappa_0$. Now suppose that $\kappa_1 < \kappa_0$ and

$$x \in U \setminus B(\widetilde{\psi}, \widetilde{Q}, \kappa_1) \qquad (= B(\widetilde{\psi}, \widetilde{Q}, \kappa_0) \setminus B(\widetilde{\psi}, \widetilde{Q}, \kappa_1)).$$

Then, the first two collections of inequalities in (5.7) are satisfied for some vector $\mathbf{a} = (q, p_1, \ldots, p_n) \in \mathbb{Z}^{n+1}$ such that

$$\kappa_1 \widetilde{Q} \le |q| \le \kappa_0 \widetilde{Q}.$$

The first set of inequalities in (5.7) keeps the rational point $\mathbf{p}/q$ at distance $\ll \varepsilon_1 = \widetilde{\psi}/\widetilde{Q}$ from the tangent plane to the manifold at $\mathbf{f}(x)$. The second set of inequalities in (5.7) keeps the rational point $\mathbf{p}'/q$, where $\mathbf{p}' = (p_1, \ldots, p_d)$ at distance $\ll \varepsilon_2 = (\widetilde{\psi}^m \widetilde{Q})^{-\frac{1}{d}} \widetilde{Q}^{-1}$ from x. Since the tangent plane deviates from the manifold quadratically, assuming that $\varepsilon_2^2 \le \varepsilon_1$, we conclude that the point $\mathbf{p}/q$ remains at distance $\ll \varepsilon_1 + \varepsilon_2^2 \ll \varepsilon_1$ from the manifold; see [Ber12, Lemma 4.3]. The proof of this uses nothing but the second-order Taylor's formula.

To sum up,

$$U \setminus B(\widetilde{\psi}, \widetilde{Q}, \kappa_1) \subset \bigcup_{(\mathbf{p},q) \in R_{\mathcal{M}}(c_1 \widetilde{Q}, c_2 \widetilde{\psi})} B(\mathbf{p}'/q, c_3 \varepsilon_2),$$

where $B(x, r)$ is a ball in $\mathbb{R}^d$ centered at x of radius r, and $c_1, c_2, c_3 > 0$ are some constants. Then

$$\begin{aligned} \lambda_d\big(U \setminus B(\widetilde{\psi}, \widetilde{Q}, \kappa_1)\big) &\ll \#R_{\mathcal{M}}(c_1 \widetilde{Q}, c_2 \widetilde{\psi})(c_3 \varepsilon_2)^d \\ &\asymp \#R_{\mathcal{M}}(c_1 \widetilde{Q}, c_2 \widetilde{\psi})\psi^{-m} Q^{-(d+1)}. \end{aligned} \tag{5.8}$$

At this point the QnD estimate is used to verify that for a suitably small constant κ_1, the set $B(\widetilde{\psi}, \widetilde{Q}, \kappa_1)$ has small measure—say, $\leq \frac{1}{2}\lambda_d(U)$. Thus $\lambda_d\big(U \setminus B(\widetilde{\psi}, \widetilde{Q}, \kappa_1)\big) \geq \frac{1}{2}\mu_d(U)$, and (5.8) implies the desired result on requiring that $\widetilde{Q} \leq Q/c_1$ and $\widetilde{\psi} \leq \psi/c_2$.

To finish this discussion, we shall show explicitly how (5.7) can be rewritten for the purpose of applying the QnD estimate. For simplicity we consider a nondegenerate planar curve $\mathcal{C} = \{(x, f(x) : x \in U\}$, where U is an interval, and so $d = m = 1$ and $n = 2$, and we restrict ourselves to the case when $\widetilde{\psi} \asymp \widetilde{Q}^{-1}$. The latter means that we are counting rational points closest to $\mathcal{C}$. Let $\widetilde{\psi}\widetilde{Q} = \kappa^{1/3}$, $e^t = \widetilde{\psi}^{-1}\kappa^{1/3}$, and $e^{-t} = \widetilde{Q}^{-1}\kappa^{-2/3}$.

Then, (5.7) can be replaced by

$$\delta(g_t G_x \mathbb{Z}^3) < \kappa^{1/3}, \tag{5.9}$$

where

$$G_x = \begin{pmatrix} f(x) - xf'(x) & f'(x) & -1 \\ x & -1 & 0 \\ 1 & 0 & 0 \end{pmatrix}$$

and

$$g_t = \begin{pmatrix} e^t & 0 & 0 \\ 0 & 1 & 0 \\ 0 & 0 & e^{-t} \end{pmatrix}.$$

Indeed, the first row of G_x is simply $\mathbf{g}_1(x)$ appearing in (5.7), and the second row of G_x is simply a multiple of $\mathbf{g}_2(x)$ appearing in (5.7). Thus, counting rational points closest to a planar curve as discussed above relies on finding an appropriately small constant $\kappa > 0$ such that the set of $x \in U$ satisfying (5.9) has measure at most, say, $\frac{1}{2}\lambda_1(U)$ for all sufficiently large t. To rephrase this, half of the curve $x \mapsto G_x\mathbb{Z}^3$ in X_3 has to remain in the compact set $\mathcal{K}_\varepsilon$ defined by Equation (1.1) with $\varepsilon = \kappa^{1/3}$ under the action by the g_t; that is,

$$\lambda_1(\{x \in U : g_t G_x \mathbb{Z}^3 \in \mathcal{K}_\varepsilon\}) \geq \tfrac{1}{2}\lambda_1(U) \qquad \text{for all sufficiently large } t. \tag{5.10}$$

□

5.1 FURTHER REMARKS ON RATIONAL POINTS NEAR MANIFOLDS.

When $d = 1$, it was shown in [Ber12, Theorem 7.1] that for analytic nondegenerate curves, Equation (5.5) can be relaxed to

$$C_2 Q^{-\frac{3}{2n-1}} < \psi < 1. \tag{5.11}$$

More recently, the condition of the analyticity was removed in [BVVZ] following a more careful and explicit application of the QnD estimate. In essence, the analytic case does not require us to deal with condition (1) of Theorem 1.1, the latter task being accomplished in [BVVZ]. For $d > 1$, removing the analyticity condition from Theorem 5.3 remains an open problem. In the case of planar curves, Problem 5.2 was solved for nondegenerate planar curves as a result of [Hux94], [BDV07], [BZ10], and [VV06]; see also asymptotic and inhomogeneous results in [BVV11], [Hua15], [Cho17], and [Gaf14]. Upper bounds in higher dimensions represent a challenging open problem, but see [BVVZ17], [BY], [Sim18], and [Hua20] for some recent results.[4]

For $n = 2$ ($d = 1$), the condition in Equation (5.11) does not actually improve on (5.5). In fact, (5.5) is optimal within the class of all nondegenerate hypersurfaces, and in particular nondegenerate planar curves; see Remark 5.1. In principle, the existence of rational points as opposed to counting does not require using the QnD; see [BLVV17].

Detecting rational points near planar curves closer than the limit set by the left-hand side of Equation (5.4) will require additional conditions on top of nondegeneracy and represents an interesting problem:

PROBLEM 5.4. Find reasonable conditions on a connected analytic curve C in $\mathbb{R}^2$ sufficient to satisfy

$$\liminf_{q\to\infty} q^2 \operatorname{dist}(C, \tfrac{1}{q}\mathbb{Z}^2) = 0. \tag{5.12}$$

Observe that for ellipses in $\mathbb{R}^2$, Problem 5.4 reduces to the Oppenheim conjecture (1929) remarkably proved by Margulis in 1986:

THEOREM 5.5 (Margulis, 1986).
Let Q be a nondegenerate indefinite quadratic form of three real variables, and suppose that Q is not a multiple of a form with rational coefficients. Then for any $\varepsilon > 0$ there exist nonzero integers a, b, and c such that

$$|Q(a, b, c)| < \varepsilon. \tag{5.13}$$

To see the link between Theorem 5.5 and Problem 5.4, first divide (5.13) through by c^2 and, using the fact that Q is a homogeneous polynomial of degree 2, obtain the following equivalent inequality:

---

[4] In particular, the upper bound of [BY] for rational points avoiding certain exceptional sets uses the QnD estimate to measure these exceptional sets and is sharp.

$$\left|Q\left(\frac{a}{c},\frac{b}{c},1\right)\right| < \frac{\varepsilon}{c^2}. \tag{5.14}$$

Since Q is indefinite, without loss of generality one can assume that

$$Q(x,y,1) = \widetilde{Q}(x-x_0, y-y_0) - r^2$$

for some positive definite quadratic form $\widetilde{Q}$ of two variables and some $r>0$. If necessary one can permute the variables a, b, and c to make sure this is the case. If $\mathcal{C}$ denotes the curve in $\mathbb{R}^2$ defined by the equation $Q(x,y,1)=0$, then an elementary check shows that (5.14) is equivalent to $\mathrm{dist}(\mathcal{C},(a/c,b/c)) \ll \varepsilon/c^2$ and also that $|c| \gg \max\{|a|,|b|\}$. Hence it becomes obvious that Theorem 5.5 is equivalent to (5.12) for the specific type of curves $\mathcal{C}$ in question. For instance, if $Q(x,y,z)=x^2+y^2-(rz)^2$ for some $r>0$, then $\mathcal{C}$ is the circle of radius r centered at the origin. In general, $\mathcal{C}$ is an ellipse.

Apparently, when attacking Problem 5.4 one has to appeal to an unbounded g_t-orbit of $G_x\mathbb{Z}^3$ as opposed to the bounded parts of this orbit appearing in (5.10), where g_t and G_x are the same as in (5.9). Indeed, assuming that $\mathcal{C} = \{(x,f(x)) : x \in U\}$ is bounded, it is a relatively simple task to verify that

$$(5.12) \quad \Longrightarrow \quad \{g_tG_x\mathbb{Z}^3 : x \in U, t \geq 0\} \text{ is unbounded in } X_3\,, \tag{5.15}$$

while the converse requires a slight tightening of the condition on the right by replacing U with any closed subset U' of the interior of U, in which case we have that

$$(5.12) \quad \Longleftarrow \quad \{g_tG_x\mathbb{Z}^3 : x \in U', t \geq 0\} \text{ is unbounded in } X_3. \tag{5.16}$$

The argument in support of (5.15) and (5.16) can be obtained on modifying the technique used for detecting rational points near manifolds that we discussed above and as detailed in [Ber12], [BDV07], [BZ10], and [BVVZ]. Of course, due to Margulis's theorem on the Oppenheim conjecture, Equation (5.12) and consequently the right-hand side of (5.15) hold for irrational ellipses.

Oppenheim's conjecture is only one example of problems on *small values of homogeneous polynomials at integral points.* Clearly, any problem of this ilk falls into the framework of *rational points near manifolds.* To give another example, which is of current interest and where the QnD estimate plays an important role, consider counting integral (irreducible) polynomials P of degree n and height $H(P) \leq Q$ with relatively small discriminant $D(P)$. Indeed, for polynomials of degree 2 the problem reduces to counting rational points near the parabola $y=x^2$; see [BBG16, section 2]. In general,

$D(P)$ can be written as a homogeneous polynomial $D(a_0, \ldots, a_n)$ of the coefficients $a_0, \ldots, a_n$ of $P = a_n x^n + \cdots + a_0$; the degree of D is $2n-2$. Thus, when $H(P) = \max\{|a_0|, \ldots, |a_n|\} \le Q$, we have that $|D(P)| \ll Q^{2n-2}$. This gives rise to the following:

PROBLEM 5.6. Let $n \ge 2$ be an integer and $v \in [0, n-1]$. Establish the asymptotic behavior (as $Q \to \infty$) of the number $N_n(Q)$ of integral irreducible polynomials P of degree n and height $H(P) \le Q$ satisfying the condition

$$0 < |D(P)| \ll Q^{2n-2-2v}. \tag{5.17}$$

The problem can be equally restated for monic polynomials $P = x^{n+1} + a_n x^n + \cdots + a_0$ of degree $n+1$.

It was shown in [BBG16] that

$$N_n(Q) \gg Q^{n+1-\frac{n+2}{n}v} \tag{5.18}$$

for any $v \in [0, n-1]$. Quite remarkably, the proof of (5.18) represents yet another application of the QnD estimate. To be more precise, establishing (5.18) uses counting irreducible polynomials P such that P and its derivatives have prescribed values at points x from a subset of $[-1/2, 1/2]$ of measure at least $1/2$; see [BBG10, Lemma 4]. The latter is proved by using the QnD estimate applied to the system

$$|P^{(i)}(x)| < \theta_i \qquad (0 \le i \le n)$$

for a suitable choice of positive parameters θ_i such that the product $\theta_0 \cdots \theta_n$ is a sufficiently small constant; see [BBG10, Lemma 1] or more generally [Ber12, Theorem 5.8]. In all likelihood (5.18) is sharp, but the complementary upper bound remains unknown except for $n=2$ [BBG16] and $n=3$ when $0 < v < 3/5$ [GKK14]. Very recently, in [DOS, Theorem 1.1], an upper bound for the number of monic irreducible polynomials of a fixed discriminant and height $H(P) \le Q$ has been established for arbitrary degrees ≥ 3. However, this recent upper bound seems to have enough room for further improvement, even for monic polynomials of degree 3; thus finding upper bounds within Problem 5.6 remains an almost entirely open challenge.

6 Badly approximable points on manifolds

The notion of badly approximable points in $\mathbb{R}^n$ comes about by reversing the inequalities in Dirichlet's theorem with a suitably small constant. Recall again, by Dirichlet's theorem, for every $\mathbf{y} = (y_1, \ldots, y_n) \in \mathbb{R}^n$ there are infinitely many $q \in \mathbb{N}$ such that

$$\max_{1 \leq i \leq n} |\langle q y_i \rangle|^n < q^{-1},$$

where $|\langle q y_i \rangle|$ is the distance from $q y_i$ to the nearest integer p_i. Thus, the point $\mathbf{y} \in \mathbb{R}^n$ is *badly approximable* if there exists a constant $c = c(\mathbf{y}) > 0$ such that

$$\max_{1 \leq i \leq n} |\langle q y_i \rangle|^n \geq c q^{-1} \tag{6.1}$$

for all $q \in \mathbb{N}$. More generally, given an n-tuple of weights $\mathbf{r} = (r_1, \ldots, r_n) \in \mathbb{R}^n_{\geq 0}$ normalized by Equation (3.3), the point $\mathbf{y} \in \mathbb{R}^n$ is called $\mathbf{r}$*-badly approximable* if there exists $c = c(\mathbf{y}) > 0$ such that

$$\max_{1 \leq i \leq n} |\langle q y_i \rangle|^{1/r_i} \geq c q^{-1} \tag{6.2}$$

for all $q \in \mathbb{N}$. Here, by definition, $|\langle q y_i \rangle|^{1/0} = 0$. In what follows, the set of $\mathbf{r}$-badly approximable points in $\mathbb{R}^n$ will be denoted by $\mathbf{Bad}(\mathbf{r})$. It is a well-known fact that $\mathbf{Bad}(\mathbf{r})$ is always of Lebesgue measure zero. Therefore, in Diophantine approximation, one is interested in understanding how small the sets $\mathbf{Bad}(\mathbf{r})$ really are by using, for example, Hausdorff dimension. More sophisticated problems arise when one considers the intersections of $\mathbf{Bad}(\mathbf{r})$ and restrictions to submanifolds of $\mathbb{R}^n$. This broad theme has been around for several decades and investigated in great depth; see [Dav64], [PV02], [KW05], [KTV06], [KW10], [Fis09], [Sch66], [Ber15], [BV14], [BPV11], [NS14], [Nes13], [ABV18], [An16], and [An13] among many dozens of other papers on the topic. There is also a natural link, known as Dani's correspondence [Dan85], between badly approximable points in $\mathbb{R}^n$ and bounded orbits of the lattices

$$\Lambda_{\mathbf{y}} = \begin{pmatrix} I_n & \mathbf{y} \\ 0 & 1 \end{pmatrix} \mathbb{Z}^{n+1},$$

where $\mathbf{y} \in \mathbb{R}^n$ is treated as a column and I_n is the identity matrix. According to Dani's correspondence, a point $\mathbf{y} \in \mathbb{R}^n$ is badly approximable if and only if $g_t \Lambda_{\mathbf{y}}$ $(t \geq 0)$ is bounded in the space of lattices X_{n+1}, where $g_t := \operatorname{diag}\{e^t, \ldots, e^t, e^{-nt}\}$. Later it was shown in [Kle98] that Dani's correspondence extends to Diophantine approximation with weights and for matrices. In particular, a point $\mathbf{y} \in \mathbb{R}^n$ is $\mathbf{r}$-badly approximable if and only if the trajectory $g_t \Lambda_{\mathbf{y}}$ $(t \geq 0)$ is bounded in X_{n+1}, where $g_t := \operatorname{diag}\{e^{tr_1}, \ldots, e^{tr_n}, e^{-t}\}$.

The purpose of this section is to expose the role of the QnD estimate in a recent proof given in [Ber15] that countable intersections of the sets **Bad**(**r**) restricted to a nondegenerate submanifold of $\mathbb{R}^n$ have full Hausdorff dimension. The following key result of [Ber15] will be the main subject of discussion in this section.

THEOREM 6.1 ([Ber15], Theorem 1).
Let $n, d \in \mathbb{N}$, W be a finite or countable collection of n-tuples $(r_1, \ldots, r_n) \in \mathbb{R}^n_{\geq 0}$ with $r_1 + \cdots + r_n = 1$. Assume that

$$\inf\{\tau(\mathbf{r}) : \mathbf{r} \in W\} > 0 \tag{6.3}$$

where

$$\tau(r_1, \ldots, r_n) = \min\{r_i > 0 : 1 \leq i \leq n\}.$$

Let $\mathcal{F}_n(B)$ be a finite collection of analytic nondegenerate maps defined on a ball $B \subset \mathbb{R}^d$. Then

$$\dim \bigcap_{\mathbf{f} \in \mathcal{F}_n(B)} \bigcap_{\mathbf{r} \in W} \mathbf{f}^{-1}\big(\mathbf{Bad}(\mathbf{r})\big) = d. \tag{6.4}$$

Sketch of the proof (for full details see [Ber15]). To begin with, one uses a transference principle to reformulate **Bad**(**r**) in terms of approximations by one linear form: $\mathbf{y} \in \mathbb{R}^n$ is in **Bad**(**r**) if and only if there exists $c > 0$ such that for any $H \geq 1$ the only integer solution $(p, q_1, \ldots, q_n)$ to the system

$$\begin{aligned} &|p + q_1 y_1 + \cdots + q_n y_n| < cH^{-1}, \\ &|q_i| < H^{r_i} \qquad (1 \leq i \leq n) \end{aligned} \tag{6.5}$$

is zero—that is, $p = q_1 = \cdots = q_n = 0$. Another simplification is that one can assume that $d = 1$—that is, it suffices to deal with curves. This is due to the existence of appropriate techniques for fibering analytic nondegenerate manifolds into nondegenerate curves and Marstrand's slicing lemma; see [Ber15].

For simplicity we will assume that $\#W = 1$, $\#\mathcal{F}_n(B) = 1$, and $B = [0, 1]$. Then Equation (6.4) becomes

$$\dim \underbrace{\{x \in [0, 1] : \mathbf{f}(x) \in \mathbf{Bad}(\mathbf{r})\}}_{S} = 1. \tag{6.6}$$

The basic idea is to construct a Cantor set

$$\mathcal{K} := \bigcap_{t=1}^{\infty} \mathcal{K}_{t+m}$$

starting from $\mathcal{K}_m = [0, 1]$, where m is a large integer, and fulfilling the condition

$$\mathcal{K}_{t+m} \subset \mathcal{K}_{t-1+m} \setminus \{x \in [0,1] : \delta(g_t u_{\mathbf{f}(x)} \mathbb{Z}^{n+1}) < \kappa\} \qquad \text{for } t \in \mathbb{N}, \tag{6.7}$$

where $\eta > 0$ is a suitably large constant, $g_t = \operatorname{diag}\{e^{\eta t}, e^{-\eta t r_1}, \ldots, e^{-\eta t r_n}\}$, and $u_{\mathbf{f}(x)}$ is the same as in Equation (2.6). By Dani's correspondence, or rather by its version from [Kle98], $\mathcal{K}$ is a subset of S defined in (6.6). The goal is thus to demonstrate that for any $\delta > 0$ there exists a suitably small $\kappa > 0$ such that

$$\dim \mathcal{K} \geq 1 - \delta. \tag{6.8}$$

The level sets $\mathcal{K}_t$ of $\mathcal{K}$ are made of small *building blocks*—closed subintervals of length R^{-t} with disjoint interiors, where the parameter R is a large positive integer. This requirement makes it easier to estimate the Hausdorff dimension of $\mathcal{K}$. Essentially, $\mathcal{K}_t$ is obtained from $\mathcal{K}_{t-1}$ by chopping up each building block of $\mathcal{K}_{t-1}$ into R equal pieces and then removing some of them. The building blocks that have to be removed are identified by requirement (6.7). Effectively, to achieve the dimension bound in (6.8) one has to show that we remove relatively little. How little is determined by a technical statement on Cantor sets originally obtained in [BPV11] and [BV11] and developed further in [Ber15] into a notion of *Cantor rich* sets. Cantor rich sets are closed under countable intersections, albeit there is a mild technical condition attached to intersections; see also [BHNS18] for a comparison of Cantor rich sets with other similar notions. It is the nature of Cantor rich sets that allowed us to assume that $\#W = 1$ and $\#\mathcal{F}_n(B) = 1$.

To accomplish the final goal one has to analyze the composition of the set

$$\{x \in [0,1] : \delta(g_t u_{\mathbf{f}(x)} \mathbb{Z}^{n+1}) < \kappa\}$$

—that is, the set removed in (6.7). This set is defined as the union over all the integer points $(p, q_1, \ldots, q_n)$ subject to $|q_i| < e^{\eta t r_i}$ $(1 \leq i \leq n)$ of all the solutions x to

$$|p + q_1 f_1(x) + \cdots + q_n f_n(x)| < \kappa e^{-\eta t}. \tag{6.9}$$

For a fixed $(p, q_1, \ldots, q_n)$, inequality (6.9) defines a finite collection of intervals. The number of these intervals is bonded by a constant depending on n and $\mathbf{f}$; however, the length of these intervals depends on the slope of the graph of the function $x \mapsto p + q_1 f_1(x) + \cdots + q_n f_n(x)$—that is, on

$$|q_1 f_1'(x) + \cdots + q_n f_n'(x)| \tag{6.10}$$

—and thus can vary hugely. It is convenient to combine together the intervals of similar size by sandwiching (6.10) between consecutive powers of a real number. Effectively, for some $\ell \in \mathbb{Z}$, one considers the system

$$\begin{aligned} &|p + q_1 f_1(x) + \cdots + q_n f_n(x)| < \kappa e^{-\eta t}, \\ &e^{\eta(\gamma t - \gamma' \ell)} \le |q_1 f_1'(x) + \cdots + q_n f_n'(x)| < e^{\eta(\gamma t - \gamma'(\ell - 1))}, \\ &|q_i| < e^{\eta t r_i} \qquad (1 \le i \le n). \end{aligned} \tag{6.11}$$

Since the maximum of (6.10) is $\ll e^{\eta t \gamma}$, where $\gamma = \max\{r_1, \ldots, r_n\}$, it suffices to assume that ℓ is nonnegative. The parameter γ' is used for convenience to eventually synchronize the (approximate) length of the intervals arising from (6.11) with that of building blocks of an appropriate level of $\mathcal{K}$. Indeed, for relatively small ℓ the intervals of x arising from (6.11) for a fixed $(p, q_1, \ldots, q_n)$ are of length

$$\asymp \kappa e^{-\eta t} e^{\eta(\gamma t - \gamma' \ell)}. \tag{6.12}$$

The proof uses a counting argument from the geometry of numbers to estimate the number of different points $(p, q_1, \ldots, q_n)$ that give rise to a nonempty set of x satisfying (6.11), and this estimate put together with (6.12) appears to be sufficient to make the Cantor rich sets work.

The problem remains in the case of relatively large ℓ. And this is precisely the case where the QnD estimate comes to the rescue. The idea is to consider the system

$$\begin{aligned} &|p + q_1 f_1(x) + \cdots + q_n f_n(x)| < \kappa e^{-\eta t}, \\ &|q_1 f_1'(x) + \cdots + q_n f_n'(x)| < e^{\eta t(\gamma - \varepsilon)}, \\ &|q_i| < e^{\eta t r_i} \qquad (1 \le i \le n), \end{aligned} \tag{6.13}$$

where ε is a fixed constant. In practice, ε can be chosen within the limits $1/n \le \varepsilon \le 2/n$. The solutions of (6.11) with $\ell \gg \varepsilon t$ will fall into the set S_t of solutions x to (6.13). Using the version of the QnD estimate from [BKM01], one verifies that the measure of S_t is

$$\ll e^{-t\alpha\varepsilon},$$

where α depends only on n. In fact, if we swell the set S_t up by placing a ball of radius

$$\Delta := \kappa e^{-\eta t}/e^{\eta t(\gamma-\varepsilon)}$$

around each point of S_t, the QnD estimate applies to this bigger set $\widehat{S}_t$. Due to its construction the set $\widehat{S}_t$ can be written as a disjoint union of intervals of length $\asymp \Delta$, while the total measure of these intervals is still $\ll e^{-t\alpha\varepsilon}$. Hence, one gets a bound on the number of the intervals, and this bound appears good enough to complete the proof. □

REMARK 6.2. The basic idea for treating Equation (6.13) that we described above evolved from the paper [BBD02a], which deals with a very special case of Problem 4.4 discussed in section 4.1. Indeed, the method of [BBD02a], which relies on the QnD estimate, can be easily modified to obtain the upper bound for the Hausdorff dimension,

(6.14)

$$\dim\left\{x : \begin{array}{l}\text{(6.13) has a nonzero solution } (p, q_1, \ldots, q_n)\\ \text{for infinitely many } t \in \mathbb{N}\end{array}\right\} \le 1 - c(\alpha, \varepsilon, n),$$

for some explicitly computable parameter $c(\alpha,\varepsilon,n) > 0$ depending only on α, ε, and n. Recall that within the proof of Theorem 6.1 the target set $\mathcal{K}$ given by Equation (6.8) is sought to satisfy (6.8) for arbitrarily small $\delta > 0$. In the case of (6.11) this goal is attained by taking η sufficiently large and κ sufficiently small. Now note that the estimate (6.14) is independent of η and κ. This means that when constructing the levels $\mathcal{K}_t$ of our Cantor set, the case (6.13) removes a set of dimension strictly smaller than $1-\delta$ as long as we impose the condition $0 < \delta < c(\alpha,\varepsilon,n)$ with δ the same as in (6.8).

6.1 FURTHER REMARKS ON BADLY APPROXIMABLE POINTS ON MANIFOLDS.

The technical condition of Equation (6.3) on the weights of approximation arises within the part of the proof of Theorem 6.1 that does not use the QnD. Introducing new ideas to this part, Lei Yang [Yan19] managed to remove (6.3) completely.

Theorem 6.1 has a straightforward consequence to real numbers badly approximable by algebraic numbers. These can be defined via small values of polynomials:

$$\mathcal{B}_n = \left\{\xi \in \mathbb{R} : \begin{array}{c}\exists\, c_1 = c_1(\xi, n) > 0 \text{ such that } |P(\xi)| \ge c_1 H(P)^{-n}\\ \text{for all nonzero } P \in \mathbb{Z}[x],\ \deg P \le n\end{array}\right\}.$$

As a consequence of Theorem 6.1 we have that for any natural number N and any interval I in $\mathbb{R}$,

$$\dim \bigcap_{n=1}^{N} \mathcal{B}_n \cap I = 1. \tag{6.15}$$

However, Theorem 6.1 leaves the following problem open: *show that* (6.15) *holds when* $N = \infty$. The generalization of Yang [Yan19] that removes condition (6.3) does not solve this problem. However, it has been resolved in [BNY1] on showing that the sets

$$\{x \in \mathbb{R} : (x, \ldots, x^n) \text{ is badly approximable}\}$$

are winning. Previously, this was shown in dimension $n = 2$ [ABV18]. More generally, it is shown in [BNY1] that for any $n \in \mathbb{N}$ and any n-tuple $\mathbf{r}$ of weights, the set of $\mathbf{r}$-badly approximable points on any nondegenerate analytic curve in $\mathbb{R}^n$ is absolute winning. We note that the results of [BNY1] represent yet another powerful application of the QnD, this time for fractal measures as established in [KLW04]. Another remarkable application of the QnD for fractal measures is the proof that the sets $\mathbf{Bad}(\mathbf{r})$ are hyperplane absolute winning, established in [BNY2].

We note that the above exposition is not a complete account of known applications of the QnD estimates. There is no doubt that many new exciting applications are still awaiting to be discovered!

References

[ABLVZ16] F. Adiceam, V. Beresnevich, J. Levesley, S. Velani and E. Zorin. Diophantine approximation and applications in interference alignment. *Adv. Math.*, 302:231–279, 2016.

[ABRS18] M. Aka, E. Breuillard, L. Rosenzweig and N. de Saxcé. Diophantine approximation on matrices and Lie groups. *Geom. Funct. Anal.*, 28(1):1–57, 2018.

[AB18] D. Allen and V. Beresnevich. A mass transference principle for systems of linear forms and its applications. *Compos. Math.*, 154(5):1014–1047, 2018.

[An13] J. An. Badziahin-Pollington-Velani's theorem and Schmidt's game. *Bull. Lond. Math. Soc.*, 45(4):721–733, 2013.

[An16] J. An. 2-dimensional badly approximable vectors and Schmidt's game. *Duke Math. J.*, 165(2):267–284, 2016.

[ABV18] J. An, V. Beresnevich and S. Velani. Badly approximable points on planar curves and winning. *Adv. Math.*, 324:148–202, 2018.

[BBV13] D. Badziahin, V. Beresnevich and S. Velani. Inhomogeneous theory of dual Diophantine approximation on manifolds. *Adv. Math.*, 232:1–35, 2013.

[BHNS18] D. Badziahin, S. Harrap, E. Nesharim and D. Simmons. Schmidt games and Cantor winning sets. arXiv:1804.06499, 2018.

[BPV11] D. Badziahin, A. Pollington and S. Velani. On a problem in simultaneous Diophantine approximation: Schmidt's conjecture. *Ann. of Math. (2)*, 174(3):1837–1883, 2011.

[BV11] D. Badziahin and S. Velani. Multiplicatively badly approximable numbers and generalised Cantor sets. *Adv. Math.*, 228(5):2766–2796, 2011.

[BV14] D. Badziahin and S. Velani. Badly approximable points on planar curves and a problem of Davenport. *Math. Ann.*, 359(3–4):969–1023, 2014.

[Bak75] A. Baker. Transcendental number theory. Cambridge Univ. Press, London–New York, 1975.

[Bak78] R. C. Baker. Dirichlet's theorem on Diophantine approximation. *Math. Proc. Cambridge Philos. Soc.*, 83:37–59, 1978.

[Ber99] V. Beresnevich. On approximation of real numbers by real algebraic numbers. *Acta Arith.*, 90(2):97–112, 1999.

[Ber00] V. Beresnevich. Application of the concept of regular systems of points in metric number theory. *Vestsī Nats. Akad. Navuk Belarusī Ser. Fīz.-Mat. Navuk*, (1):35–39, 140, 2000.

[Ber02] V. Beresnevich. A Groshev type theorem for convergence on manifolds. *Acta Math. Hungar.*, 94(1–2):99–130, 2002.

[Ber05] V. Beresnevich. On a theorem of V. Bernik in the metric theory of Diophantine approximation. *Acta Arith.*, 117(1):71–80, 2005.

[Ber12] V. Beresnevich. Rational points near manifolds and metric Diophantine approximation. *Ann. of Math. (2)*, 175(1):187–235, 2012.

[Ber15] V. Beresnevich. Badly approximable points on manifolds. *Invent. Math.*, 202(3):1199–1240, 2015.

[BB00] V. Beresnevich and V. I. Bernik. Baker's conjecture and Hausdorff dimension. *Publ. Math. Debrecen*, 56(3–4):263–269, 2000. Dedicated to Professor Kálmán Győry on the occasion of his 60th birthday.

[BBD02a] V. Beresnevich, V. I. Bernik and M. M. Dodson. On the Hausdorff dimension of sets of well-approximable points on nondegenerate curves. *Dokl. Nats. Akad. Nauk Belarusi*, 46(6):18–20, 124, 2002.

[BBD02b] V. Beresnevich, V. I. Bernik and M. M. Dodson. Regular systems, ubiquity and Diophantine approximation. In *A panorama of number theory or the view from Baker's garden (Zürich, 1999)*, pages 260–279. Cambridge Univ. Press, Cambridge, 2002.

[BBDV09] V. Beresnevich, V. I. Bernik, M. M. Dodson and S. Velani. Classical metric Diophantine approximation revisited. In *Analytic number theory*, pages 38–61. Cambridge Univ. Press, Cambridge, 2009.

[BBG10] V. Beresnevich, V. Bernik, F. Götze. The distribution of close conjugate algebraic numbers. *Compos. Math.*, 146(5):1165–1179, 2010.

[BBG16] V. Beresnevich, V. Bernik, F. Götze. Integral polynomials with small discriminants and resultants. *Adv. Math.*, 298:393–412, 2016.

[BBKM02] V. Beresnevich, V. I. Bernik, D. Kleinbock and G. A. Margulis. Metric Diophantine approximation: the Khintchine-Groshev theorem for nondegenerate manifolds. *Mosc. Math. J.*, 2(2):203–225, 2002. Dedicated to Yuri I. Manin on the occasion of his 65th birthday.

[BBB17] V. Beresnevich, N. Budarina and V. I. Bernik. Systems of small linear forms and diophantine approximation on manifolds. arXiv:1707.00371, 2017.

[BDV06] V. Beresnevich, D. Dickinson and S. Velani. Measure theoretic laws for lim sup sets. *Mem. Amer. Math. Soc.*, 179(846):x+91, 2006.

[BDV07] V. Beresnevich, D. Dickinson and S. Velani. Diophantine approximation on planar curves and the distribution of rational points. *Ann. of Math. (2)*, 166(2):367–426, 2007. With an Appendix II by R. C. Vaughan.

[BHV20] V. Beresnevich, A. Haynes and S. Velani. Sums of reciprocals of fractional parts and multiplicative Diophantine approximation. *Mem. Amer. Math. Soc.*, 263(1276):vii+77, 2020.

[BKM15] V. Beresnevich, D. Kleinbock and G. A. Margulis. Non-planarity and metric Diophantine approximation for systems of linear forms. *J. Théor. Nombres Bordeaux*, 27(1):1–31, 2015.

[BLVV17] V. Beresnevich, L. Lee, R. C. Vaughan and S. Velani. Diophantine approximation on manifolds and lower bounds for Hausdorff dimension. *Mathematika*, 63(3):762–779, 2017.

[BNY1] V. Beresnevich, E. Nesharim and L. Yang. Winning property of badly approximable points on curves. arXiv:2005.02128, 2020.

[BNY2] V. Beresnevich, E. Nesharim and L. Yang. Bad(w) is hyperplane absolute winning. *Geom. Funct. Anal.*, 31(1):1–33, 2021.

[BVV11] V. Beresnevich, R. C. Vaughan and S. Velani. Inhomogeneous Diophantine approximation on planar curves. *Math. Ann.*, 349(4):929–942, 2011.

[BVVZ17] V. Beresnevich, R. C. Vaughan, S. Velani and E. Zorin. Diophantine approximationon manifolds and the distribution of rational points: contributions to the convergence theory. *Int. Math. Res. Not. IMRN*, (10):2885–2908, 2017.

[BVVZ] V. Beresnevich, R. C. Vaughan, S. Velani and E. Zorin. Diophantine approximation on curves and the distribution of rational points: divergence theory. *Adv. Math.*, 388:107861, 2021.

[BV06] V. Beresnevich and S. Velani. A mass transference principle and the Duffin-Schaeer conjecture for Hausdorff measures. *Ann. of Math. (2)*, 164(3):971–992, 2006.

[BV10] V. Beresnevich and S. Velani. Classical metric Diophantine approximation revisited: the Khintchine-Groshev theorem. *Int. Math. Res. Not. IMRN*, (1):69–86, 2010.

[BY] V. Beresnevich and L. Yang. Khintchine's theorem and Diophantine approximation on manifolds. arXiv:2105.13872, 2021.

[BZ10] V. Beresnevich and E. Zorin. Explicit bounds for rational points near planar curves and metric Diophantine approximation. *Adv. Math.*, 225(6):3064–3087, 2010.

[Ber83] V. I. Bernik. An application of Hausdorff dimension in the theory of Diophantine approximation. *Acta Arith.*, 42(3):219–253, 1983. (In Russian; English transl. in *Amer. Math. Soc. Transl.*, 140:15–44, 1988.)

[Ber89] V. I. Bernik. The exact order of approximating zero by values of integral polynomials. *Acta Arith.*, 53(1):17–28, 1989.

[BD99] V. I. Bernik and M. M. Dodson. *Metric Diophantine approximation on manifolds*, volume 137 of *Cambridge tracts in mathematics*. Cambridge Univ. Press, Cambridge, 1999.

[BKM01] V. I. Bernik, D. Kleinbock and G. A. Margulis. Khintchine-type theorems on manifolds: the convergence case for standard and multiplicative versions. *Internat. Math.Res. Notices*, (9):453–486, 2001.

[Bes34] A. S. Besicovitch. Sets of fractional dimensions (IV): on rational approximation to real numbers. *J. London Math. Soc.*, 9(2):126–131, 1934.

[BD09] N. Budarina and D. Dickinson. Diophantine approximation on non-degenerate curves with non-monotonic error function. *Bull. Lond. Math. Soc.*, 41(1):137–146, 2009.

[Cho17] S. Chow. A note on rational points near planar curves. *Acta Arith.*, 177(4):393–396, 2017.

[Cho18] S. Chow. Bohr sets and multiplicative Diophantine approximation. *Duke Math. J.*, 167(9):1623–1642, 2018.

[CT19] S. Chow and N. Technau. Higher-rank Bohr sets and multiplicative diophantine approximation. *Compos. Math.* 155(11):2214–2233, 2019.

[CY] S. Chow and L. Yang. An effective Ratner equidistribution theorem for multiplicative Diophantine approximation on planar lines. arXiv:1902.06081, 2019.

[Dan79] S. G. Dani. On invariant measures, minimal sets and a lemma of Margulis. *Invent. Math.*, 51(3):239–260, 1979.

[Dan85] S. G. Dani. Divergent trajectories of flows on homogeneous spaces and Diophantine approximation. *J. Reine Angew. Math.*, 359:55–89, 1985.

[Dan86] S. G. Dani. On orbits of unipotent flows on homogeneous spaces. II. *Ergodic Theory Dynam. Systems*, 6(2):167–182, 1986.

[DFSU18] T. Das, L. Fishman, D. Simmons and M. Urbański. Extremality and dynamically defined measures, part I: Diophantine properties of quasi-decaying measures. *Selecta Math. (N.S.)*, 24(3):2165–2206, 2018.

[DFSU21] T. Das, L. Fishman, D. Simmons and M. Urbański. Extremality and dynamically defined measures, part II: measures from conformal dynamical systems. *Ergodic Theory Dynam. Systems*, 41(8):2311–2348, 2021.

[DS19] T. Das and D. Simmons. A proof of the matrix version of Baker's conjecture in Diophantine approximation. *Math. Proc. Cambridge Philos. Soc.*, 167(1):159–169, 2019.

[Dav64] H. Davenport. A note on Diophantine approximation. II. *Mathematika*, 11:50–58, 1964.

[DD00] H. Dickinson and M. M. Dodson. Extremal manifolds and Hausdorff dimension. *Duke Math. J.*, 101(2):271–281, 2000.

[DOS] R. Dietmann, A. Ostafe and I. E. Shparlinski. Discriminants of fields generated by polynomials of given height. arXiv:1909.00135, 2019.

[Fis09] L. Fishman. Schmidt's game on fractals. *Israel J. Math.*, 171:77–92, 2009.

[Gaf14] A. Gafni. Counting rational points near planar curves. *Acta Arith.*, 165(1):91–100, 2014.

[GG19] A. Ganguly and A. Ghosh. Quantitative Diophantine approximation on affine subspaces. *Math. Z.*, 292(3–4):923–935, 2019.

[GKK14] F. Götze, D. Kaliada and O. Kukso. The asymptotic number of integral cubic polynomials with bounded heights and discriminants. *Lith. Math. J.*, 54(2):150–165, 2014.

[Gho05] A. Ghosh. A Khintchine-type theorem for hyperplanes. *J. London Math. Soc. (2)*, 72(2):293–304, 2005.

[Gho07] A. Ghosh. Metric Diophantine approximation over a local field of positive characteristic. *J. Number Theory*, 124(2):454–469, 2007.

[Gho10] A. Ghosh. Diophantine approximation on affine hyperplanes. *Acta Arith.*, 144(2):167–182, 2010.

[Gho11] A. Ghosh. A Khintchine-Groshev theorem for affine hyperplanes. *Int. J. Number Theory*, 7(4):1045–1064, 2011.

[Gro38] A. V. Groshev. A theorem on a system of linear forms. *Dokl. Akad. Nauk SSSR*, 19:151–152, 1938. (In Russian.)

[Hua15] J.-J. Huang. Rational points near planar curves and Diophantine approximation. *Adv. Math.*, 274:490–515, 2015.

[Hua18] J.-J. Huang. Hausdorff theory of dual approximation on planar curves. *J. Reine Angew. Math.*, 740:63–76, 2018.

[Hua20] J.-J. Huang. The density of rational points near hypersurfaces. *Duke Math. J.* 169(11):2045–2077, 2020.

[Hux94] M. N. Huxley. The rational points close to a curve. *Ann. Scuola Norm. Sup. Pisa Cl. Sci. (4)*, 21(3):357–375, 1994.

[Jaf10] S. A. Jafar. Interference alignment: a new look at signal dimensions in a communication network. *Foundations and Trends in Communications and Information Theory*, 7(1), 2011.

[Jar29] V. Jarník. Diophantischen Approximationen und Hausdorffsches Mass. *Mat. Sbornik*, 36:371–382, 1929.

[Khi24] A. Khintchine. Einige Sätze über Kettenbrüche, mit Anwendungen auf die Theorie der Diophantischen Approximationen. *Math. Ann.*, 92(1–2):115–125, 1924.

[Khi26] A. Khintchine. Zur metrischen Theorie der diophantischen Approximationen. *Math. Z.*, 24(1):706–714, 1926.

[Kle98] D. Kleinbock. Flows on homogeneous spaces and Diophantine properties of matrices. *Duke Math. J.*, 95(1):107–124, 1998.

[Kle03] D. Kleinbock. Extremal subspaces and their submanifolds. *Geom. Funct. Anal.*, 13(2):437–466, 2003.

[Kle04] D. Kleinbock. Baker-Sprindžuk conjectures for complex analytic manifolds. In *Algebraic groups and arithmetic*, pages 539–553. Tata Inst. Fund. Res., Mumbai, 2004.

[Kle08] D. Kleinbock. An extension of quantitative nondivergence and applications to Diophantine exponents. *Trans. Amer. Math. Soc.*, 360(12):6497–6523, 2008.

[Kle10b] D. Kleinbock. An 'almost all versus no' dichotomy in homogeneous dynamics and Diophantine approximation. *Geom. Dedicata*, 149(1):205–218, 2010.

[Kle10a] D. Kleinbock. Quantitative nondivergence and its Diophantine applications. In *Homogeneous flows, moduli spaces and arithmetic*, pages 131–153. Clay Math. Proc., 10, Amer. Math. Soc., Providence, RI, 2010.

[KLW04] D. Kleinbock, E. Lindenstrauss and B. Weiss. On fractal measures and Diophantine approximation. *Selecta Math. (N.S.)*, 10(4):479–523, 2004.

[KM98] D. Kleinbock and G. A. Margulis. Flows on homogeneous spaces and Diophantine approximation on manifolds. *Ann. of Math. (2)*, 148(1):339–360, 1998.

[KMW10] D. Kleinbock, G. A. Margulis and J. Wang. Metric Diophantine approximation for systems of linear forms via dynamics. *Int. J. Number Theory*, 6(5):1139–1168, 2010.

[KT07] D. Kleinbock and G. Tomanov. Flows on *S*-arithmetic homogeneous spaces and applications to metric Diophantine approximation. *Comment. Math. Helv.*, 82(3):519–581, 2007.

[KW05] D. Kleinbock and B. Weiss. Badly approximable vectors on fractals. *Israel J. Math.*, 149:137–170, 2005. Probability in mathematics.

[KW10] D. Kleinbock and B. Weiss. Modified Schmidt games and Diophantine approximation with weights. *Adv. Math.*, 223(4):1276–1298, 2010.

[KTV06] S. Kristensen, R. Thorn and S. Velani. Diophantine approximation and badly approximable sets. *Adv. Math.*, 203(1):132–169, 2006.

[Mah32] K. Mahler. Über das Maßder Menge aller *S*-Zahlen. *Math. Ann.*, 106(1):131–139, 1932.

[Mah46] K. Mahler. On lattice points in *n*-dimensional star bodies. I. Existence theorems. *Proc. Roy. Soc. London. Ser. A.*, 187:151–187, 1946.

[Mar71] G. A. Margulis. The action of unipotent groups in a lattice space. *Mat. Sb. (N.S.)*, 86(128):552–556, 1971.

[Mar75] G. A. Margulis. Non-uniform lattices in semisimple algebraic groups. In *Lie groups and their representations (Proc. Summer School on Group Representations of the Bolyai János Math. Soc., Budapest, 1971)*, pages 371–553. Halsted, New York, 1975.

[MSG09] A. Mohammadi and A. Salehi Golsefidy. S-arithmetic Khintchine-type theorem. *Geom. Funct. Anal.*, 19(4):1147–1170, 2009.

[MSG12] A. Mohammadi and A. Salehi Golsefidy. Simultaneous Diophantine approximation in non-degenerate p-adic manifolds. *Israel J. Math.*, 188:231–258, 2012.

[Nes13] E. Nesharim. Badly approximable vectors on a vertical Cantor set. *Mosc. J. Comb. Number Theory*, 3(2):88–116, 2013. With an appendix by Barak Weiss and the author.

[NS14] E. Nesharim and D. Simmons. $\mathbf{Bad}(s,t)$ is hyperplane absolute winning. *Acta Arith.*, 164(2):145–152, 2014.

[PV02] A. Pollington and S. Velani. On simultaneously badly approximable numbers. *J. London Math. Soc. (2)*, 66(1):29–40, 2002.

[Rat94] M. Ratner. Invariant measures and orbit closures for unipotent actions on homogeneous spaces. *Geom. Funct. Anal.*, 4(2):236–257, 1994.

[Sch64] W. M. Schmidt. Metrical theorems on fractional parts of sequences. *Trans. Amer. Math. Soc.*, 110:493–518, 1964.

[Sch66] W. M. Schmidt. On badly approximable numbers and certain games. *Trans. Amer. Math. Soc.*, 123:178–199, 1966.

[Sch91] W. M. Schmidt. Diophantine approximations and Diophantine equations. In *Lecture notes in mathematics*, vol. 1467, Springer-Verlag, Berlin, 1991.

[Sha94] N. Shah. Limit distributions of polynomial trajectories on homogeneous spaces. *Duke Math. J.*, 75(3):711–732, 1994.

[Sim18] D. Simmons. Some manifolds of Khinchin type for convergence. *J. Théor. Nombres Bordeaux*, 30(1):175–193, 2018.

[Spr69] V. G. Sprindžuk. *Mahler's problem in metric number theory*. Translated from Russian by B. Volkmann. *Translations of mathematical monographs*, vol. 25, American Mathematical Society, Providence, RI, 1969.

[Spr80] V. G. Sprindžuk. Achievements and problems of the theory of Diophantine approximations. *Uspekhi Mat. Nauk*, 35(4(214)):3–68, 248, 1980.

[VV06] R. C. Vaughan and S. Velani. Diophantine approximation on planar curves: the convergence theory. *Invent. Math.*, 166(1):103–124, 2006.

[Yan19] L. Yang. Badly approximable points on manifolds and unipotent orbits in homogeneous spaces. *Geom. Funct. Anal.*, 29(4):1194–1234, 2019.

10

ALEX ESKIN AND SHAHAR MOZES

MARGULIS FUNCTIONS AND THEIR APPLICATIONS

To Grisha with our admiration

1 Definition and basic properties

1.1 MOTIVATION. In many cases, one wants to show that trajectories of some dynamical system spend most of the time in compact sets or more generally avoid on average a certain subset of the space. The construction of a Margulis function allows one to obtain remarkably sharp estimates of this type. The first construction is due to Margulis in [EMM98] to show quantitative recurrence for the action of $SO(p,q)$ on $SL(p+q,\mathbb{R})/SL(p+q,\mathbb{Z})$; this is used in the proof of the "quantitative Oppenheim conjecture". The difficulty in this problem is related to the complicated geometry of the noncompact part of the space. However, the method is remarkably versatile and has seen many other applications.

We now proceed with the formal definition and give examples later. The reader is encouraged to skip ahead to the examples as necessary.

Let X be the space where our dynamics takes place. First we need an averaging operator A. This is formally just a linear map from the space $C(X)$ of

Alex Eskin. Department of Mathematics. University of Chicago, Chicago, Illinois 60637, USA
eskin@math.uchicago.edu

Shahar Mozes. Institute of Mathematics. The Hebrew University of Jerusalem, Givat Ram, Jerusalem 9190401, Israel
mozes@math.huji.ac.il

Research of Eskin is partially supported by NSF grants DMS 1501126 and DMS 1800646 and by the Simons Foundation.
Research of Mozes is partially supported by ISF-Moked grants 2095/15 and 2919/19.

continuous functions on X to itself, where X is the space where the dynamics takes place. We always assume that A is a Markov operator—that is, that A takes nonnegative functions to nonnegative functions and takes the constant function 1 to itself.

Let Y be a possibly empty subset of X. If Y is not empty, we assume that it is invariant in the sense that if $h \in C(X)$ is supported on $X \setminus Y$, then so is Ah.

DEFINITION 1.1. A continuous function $f : X \to [1, \infty]$ is called a Margulis function for Y if the following hold:

(a) $f(x) = \infty$ if and only if $x \in Y$. For each $\ell > 0$, the set $\{x : f(x) \leq \ell\}$ is a compact subset of $X \setminus Y$.

(b) There exists $c < 1$ and $b < \infty$ such that for all $x \in X$,

$$(Af)(x) \leq cf(x) + b. \tag{1.1}$$

The continuity assumption on f is often modified; this will be mentioned below.

We now state an immediate consequence of the definition:

LEMMA 1.2. *Suppose $x \in X \setminus Y$. Then, there exists $N = N(x)$ such that for all $n > N$,*

$$(A^n f)(x) \leq \frac{2b}{1-c} < \infty. \tag{1.2}$$

The constant $N(x)$ depends only on $f(x)$ and can thus be chosen uniformly over the compact sets $\{x : f(x) \leq \ell\}$.

Proof. By iterating (1.1) we obtain

$$(A^n f) \leq c^n f(x) + c^{n-1} b + \cdots + cb + b \leq c^n f(x) + \frac{b}{1-c},$$

where for the last estimate we summed the geometric series. Now choose n so that $c^n f(x) < b/(1-c)$. □

2 Random walks

In this setting, a Margulis function is also called a Foster-Lyapunov (or drift) function and has been used extensively. See the book [MT09] for further references.

Suppose we are considering a random walk on X. This means that for each $x \in X$ we have a probability measure μ_x on X so that the probability of moving in one step of the random walk from x into some subset $E \subset X$ is $\mu_x(E)$. Now, for $h \in C(X)$, let

$$(Ah)(x) = \int h \, d\mu_x,$$

so A is the averaging operator with respect to one step of the random walk. Then A^n is the averaging operator with respect to n steps of the random walk, and we can write $A^n h = \int_X h \, d\mu_x^n$, where $\mu_x^n(E)$ is the probability of moving in n steps of the random walk into some set E.

LEMMA 2.1. *Suppose $Y \subset X$ and that a Margulis function f can be constructed for Y. Then, for any $\epsilon > 0$ there exists a compact subset F_ϵ of $X \setminus Y$ such that for any $x \in X \setminus Y$, for all sufficiently large n (depending on x and ϵ), $\mu_x^n(F_\epsilon) > 1 - \epsilon$.*

In particular, Lemma 2.1 shows that any weak-star limit μ_x^∞ of the measures μ_x^n is a probability measure satisfying $\mu_x^\infty(Y) = 0$.

Proof. The equation (1.2) has the interpretation that for any $x \in X$, for large enough n,

$$\int f \, d\mu_x^n \leq 2b/(1-c).$$

Now suppose $\epsilon > 0$ is given, and choose $\ell > \frac{2b}{(1-c)\epsilon}$. By Markov's inequality we have

$$\mu_x^n(\{x : f(x) > \ell\}) \leq \frac{2b}{(1-c)\ell} < \epsilon.$$

Thus, the μ_x^n measure of the compact set $\{x : f(x) \leq \ell\}$ is at least $1 - \epsilon$. □

The existence of a Margulis function also implies certain large deviation results; see, for example, §6.

3 Actions of semisimple groups

Suppose the space X admits a continuous action of a semisimple group G. For simplicity of presentation, we will assume in this section that $G = SL(2, \mathbb{R})$. For the case where $G = SO(p, q)$ see the original paper [EMM98].

Recall that G acts on the upper half plane $\mathbb{H}$ by Möbius transformations. It is convenient to write this action as

$$\begin{pmatrix} a & b \\ c & d \end{pmatrix} \cdot z = \frac{dz - b}{-cz + a}.$$

This is a right action of G, and the stabilizer of $i \in \mathbb{H}$ is $K = SO(2)$. Thus, $\mathbb{H}$ is canonically identified with $K\backslash G$. This action is by hyperbolic isometries; thus $d_{\mathbb{H}}(Kg_1 g, Kg_2 g) = d_{\mathbb{H}}(Kg_1, Kg_2)$ for all $g_1, g_2, g \in G$, where $d_{\mathbb{H}}$ is the hyperbolic metric on $\mathbb{H}$. We will also use this action to identify the unit tangent bundle $T^1(\mathbb{H})$ of $\mathbb{H}$ with G.

Let $a_t = \begin{pmatrix} e^t & 0 \\ 0 & e^{-t} \end{pmatrix}$, $r_\theta = \begin{pmatrix} \cos\theta & -\sin\theta \\ \sin\theta & \cos\theta \end{pmatrix}$. Then, under the identification of G with $T^1(\mathbb{H})$, left-multiplication by a_t on G corresponds to geodesic flow for time t on $T^1(\mathbb{H})$. In particular,

$$d_{\mathbb{H}}(Ka_t g, Kg) = t.$$

Also, since $Kr_\theta g = Kg$, $r_\theta g$ corresponds to the same point in $\mathbb{H}$ as g but with a different tangent vector. For $g \in G$, let

$$S_\tau(Kg) = \{Ka_\tau r_\theta g \ : \ 0 \le \theta < 2\pi\} \subset \mathbb{H}.$$

Then, $S_\tau(Kg)$ is the circle of radius τ around the point $Kg \in \mathbb{H}$.

Now suppose X is an arbitrary space with a continuous (left) $SL(2, \mathbb{R})$ action. For a function $h: X \to \mathbb{R}$, we can pull back h to a function h_x on $G \cong T^1(\mathbb{H})$. We then let the averaging operator A_τ be defined as

$$(3.1) \qquad (A_\tau h)(x) = \frac{1}{2\pi} \int_0^{2\pi} h(a_\tau r_\theta x)\, d\theta = \frac{1}{2\pi} \int_0^{2\pi} h_x(a_\tau r_\theta)\, d\theta.$$

We will usually take h to be invariant under the action of $K = SO(2) \subset G$. Using the identification of $K\backslash G$ with $\mathbb{H}$, we think of $(A_\tau h)(x)$ as the average of h over a circle of radius τ in the G orbit through x, or more precisely the average of h_x over $S_\tau(K)$, where K is the base point of $K\backslash G \cong \mathbb{H}$.

Suppose $Y \subset X$ is a G-invariant submanifold. (Again, $Y = \emptyset$ is allowed).

DEFINITION 3.1. A K-invariant function $f: X \to [1, \infty]$ is called a Margulis function for Y if it satisfies the following properties:

(a) There exists $\sigma > 1$ such that for all $0 \leq t \leq 1$ and all $x \in X$,

$$\sigma^{-1} f(x) \leq f(a_t x) \leq \sigma f(x). \tag{3.2}$$

(This holds if $\log f$ is uniformly continuous along the G-orbits).

(b) For every $c_0 > 0$ there exist $\tau > 0$ and $b_0 > 0$ such that for all $x \in X$,

$$A_\tau f(x) \leq c_0 f(x) + b_0.$$

(c) $f(x) = \infty$ if and only if $x \in Y$, and f is bounded on compact subsets of $X \setminus Y$. For any $\ell > 0$, the set $\overline{\{x : f(x) \leq \ell\}}$ is a compact subset of $X \setminus Y$.

LEMMA 3.2. *Suppose there exists a Margulis function f for Y. Then,*

(i) *For all $c < 1$ there exists $t_0 > 0$ (depending on σ and c) and $b > 0$ (depending only on b_0, c_0, and σ) such that for all $t > t_0$ and all $x \in X$,*

$$(A_t f)(x) \leq c f(x) + b.$$

(ii) *There exists $B > 0$ (depending only on c_0, b_0, and σ) such that for all $x \in X$, there exists $T_0 = T_0(x, c_0, b_0, \sigma)$ such that for all $t > T_0$,*

$$(A_t f)(x) \leq B.$$

(iii) *For every $\epsilon > 0$ there exists a compact subset F_ϵ of $X \setminus Y$ such that for all $x \in X$, there exists $T_0 = T_0(x, c_0, b_0, \sigma)$ such that for all $t > T_0$,*

$$|\{\theta \in [0, 2\pi) : a_t r_\theta x \in F_\epsilon\}| \geq 2\pi(1 - \epsilon).$$

For completeness, we include the proof of this lemma. It is essentially taken from [EMM98, section 5.3], specialized to the case $G = SL(2, \mathbb{R})$.

The basic observation is the following standard fact from hyperbolic geometry:

LEMMA 3.3. *There exist absolute constants $0 < \delta' < 1$ and $\delta > 0$ such that for any $p, q \in \mathbb{H}$, for any $t > 0$, for at least δ'-fraction of $z \in S_t(q)$ (with respect to the visual measure from q), we have*

$$d_{\mathbb{H}}(p, q) + t - \delta \leq d_{\mathbb{H}}(p, z) \leq d_{\mathbb{H}}(p, q) + t. \tag{3.3}$$

Using the identification of G with $T^1(\mathbb{H})$ we can restate Lemma 3.3 as follows:

COROLLARY 3.4. *There exist absolute constants $0 < \delta' < 1$ and $\delta > 0$ such that for any $t > 0$, any $s > 0$, and any $g \in G$, for at least δ'-fraction of $\phi \in [0, 2\pi]$,*

$$t + s - \delta \leq d_{\mathbb{H}}(Ka_t r_\phi a_s g, Kg) \leq t + s. \tag{3.4}$$

Proof. This is indeed Lemma 3.3 with $p = Kg$, $q = Ka_s g$ (so $d_{\mathbb{H}}(p, q) = s$). As ϕ varies, the points $Ka_t r_\phi a_s g$ trace out $S_t(q) = S_t(Ka_s g) \subset \mathbb{H}$. □

COROLLARY 3.5. *Suppose $f : X \to [1, \infty]$ is a K-invariant function satisfying (3.2). Then, there exists $\sigma' > 1$ depending only on σ such that for any $t > 0$, $s > 0$, and any $x \in X$,*

$$(A_{t+s} f)(x) \leq \sigma'(A_t A_s f)(x). \tag{3.5}$$

Outline of proof. Fix $x \in X$. For $g \in SL(2, \mathbb{R})$, let $f_x(g) = f(gx)$, and let

$$\tilde{f}_x(g) = \frac{1}{2\pi} \int_0^{2\pi} f(gr_\theta x)\, d\theta.$$

Then, $\tilde{f}_x : \mathbb{H} \to [1, \infty]$ is a spherically symmetric function—that is, $\tilde{f}_x(g)$ depends only on $d_{\mathbb{H}}(Kg, Ke)$, where e is the identity of G.

We have

$$(A_t A_s f)(x) = \frac{1}{2\pi} \int_0^{2\pi} \frac{1}{2\pi} \int_0^{2\pi} f(a_t r_\phi a_s r_\theta x)\, d\phi\, d\theta = \frac{1}{2\pi} \int_0^{2\pi} \tilde{f}_x(a_t r_\phi a_s). \tag{3.6}$$

By Corollary 3.4, for at least δ'-fraction of $\phi \in [0, 2\pi]$, (3.4) holds (with $g = e$). Then, by (3.2), for at least δ'-fraction of $\phi \in [0, 2\pi]$,

$$\tilde{f}_x(a_t r_\phi a_s) \geq \sigma_1^{-1} \tilde{f}_x(a_{t+s}),$$

where $\sigma_1 = \sigma_1(\sigma, \delta) > 1$. Plugging in to (3.6), we get

$$(A_t A_s f)(x) \geq (\delta' \sigma_1^{-1}) \tilde{f}_x(a_{t+s}) = (\delta' \sigma_1^{-1})(A_{t+s} f)(x),$$

as required. □

Proof of Lemma 3.2. By condition (b) of Definition 3.1 we can choose τ large enough in (b) so that c_0 is sufficiently small so that $\kappa \equiv c_0\sigma' < 1$, where σ' is as in Corollary 3.5. Then, for any $s \in \mathbb{R}$ and for all x,

$$\begin{aligned}(A_{s+\tau}f)(x) &\le \sigma' A_s(A_\tau f)(x) && \text{by (3.5)}\\ &\le \sigma' A_s(c_0 f(x) + b_0) && \text{by condition (b)}\\ &= \kappa(A_s f)(x) + \sigma' b_0 && \text{since } \sigma' c_0 = \kappa.\end{aligned}$$

Iterating this we get, for $n \in \mathbb{N}$,

$$(A_{n\tau}f)(x) \le \kappa^n f(x) + \sigma' b_0 + \kappa\sigma' b_0 + \cdots + \kappa^{n-1}\sigma' b_0 \le \kappa^n f(x) + B,$$

where $B = \frac{\sigma' b_0}{1-\kappa}$. Since $\kappa < 1$, $\kappa^n f(x) \to 0$ as $n \to \infty$. Therefore both (i) and (ii) follow for $t \in \tau\mathbb{N}$. The general case of both (i) and (ii) then follows by applying again condition (a). The derivation of (iii) from (ii) is the same as in the random walk case. □

As in the random walk setting, the existence of a Margulis function implies certain large deviation results; see section 6.

4 Construction of Margulis functions I: Easy cases

In this section, we construct Margulis functions in the simplest possible settings. A much more elaborate (and useful) construction is done in the next section.

We begin with the following elementary calculation:

LEMMA 4.1. *Fix $0 \le \delta < 1$. Then there exists a constant $c(\delta)$ such that for any $\tau > 0$ and any $v \in \mathbb{R}^2 - \{(0,0)\}$,*

$$\frac{1}{2\pi}\int_0^{2\pi} \frac{d\theta}{\|a_\tau r_\theta v\|^{1+\delta}} \le \frac{c(\delta)e^{-\tau(1-\delta)}}{\|v\|^{1+\delta}}. \tag{4.1}$$

Proof. By rescaling and rotating, we may assume that $v = (0,1)$. Then, the left-hand side of (4.1) becomes

$$\frac{1}{2\pi}\int_0^{2\pi} (e^{2\tau}\sin^2\theta + e^{-2\tau}\cos^2\theta)^{-(1+\delta)/2}\, d\theta.$$

We decompose $[0, 2\pi] = R_1 \cup R_2$, where $R_1 = \{\theta : e^{2\tau} \sin^2\theta \le e^{-2\tau}\cos^2\theta\}$ and R_2 is the set where the opposite inequality holds. Note that there exist absolute constants $0 < c_1 < c_2$ such that

$$c_1 e^{-2\tau} \le |R_1| \le c_2 e^{-2\tau}. \tag{4.2}$$

On R_1, the integrand is bounded by a constant multiple of $e^{\tau(1+\delta)}$. Hence, in view of (4.2), the integral over R_1 is $O(e^{-\tau(1-\delta)})$, as required. Now the integral over R_2 is bounded by

$$e^{-\tau(1+\delta)} \int_{R_2} |\sin\theta|^{-(1+\delta)}\, d\theta = O(e^{-\tau(1-\delta)}),$$

where in the last estimate we used (4.2). □

4.1 INTERPRETATION IN THE HYPERBOLIC UPPER HALF PLANE.

Given $g \in SL(2, \mathbb{R})$, we may write

$$g^{-1} = \begin{pmatrix} 1 & x \\ 0 & 1 \end{pmatrix} \begin{pmatrix} y^{1/2} & 0 \\ 0 & y^{-1/2} \end{pmatrix} \begin{pmatrix} \cos\theta & \sin\theta \\ -\sin\theta & \cos\theta \end{pmatrix}.$$

In view of our conventions at the beginning of section 3, $g \cdot i = x + iy$, and let $\phi(g) = x + iy$. Then, ϕ gives an identification between $SO(2)\backslash SL(2, \mathbb{R})$ and the hyperbolic upper half plane $\mathbb{H}$. Under this identification, the right-multiplication action of $SL(2, \mathbb{R})$ on $SO(2)\backslash SL(2, \mathbb{R})$ becomes action by Möbius transformations on $\mathbb{H}$.

Let $\beta : \mathbb{H} \to \mathbb{R}^+$ be defined by $\beta(x + iy) = y^{1/2}$. Note that in view of the definitions of β and ϕ,

$$\beta(\phi(g)) = \left\| g \begin{pmatrix} 1 \\ 0 \end{pmatrix} \right\|^{-1}.$$

Thus, Lemma 4.1 is equivalent to the following, which is well-known:

LEMMA 4.2. *Fix $0 \le \delta < 1$. Then there exists a constant $c(\delta)$ such that for any $\tau > 0$ and any $z \in \mathbb{H}$,*

$$\int_{S_\tau(z)} \beta^{1+\delta}(w)\, dm_z(w) \le c(\delta) e^{-\tau(1-\delta)} \beta(z)^{1+\delta}, \tag{4.3}$$

where $S_\tau(z)$ is the sphere in $\mathbb{H}$ centered at z and with radius τ, and the measure m_z is the normalized visual measure (from z) on the sphere $S_\tau(z)$.

(In other words, most of the measure of $S_\tau(z)$ is concentrated closer to the x axis than z). If, as in section 3, we let

$$(A_\tau h)(z) = \int_{S_\tau(z)} h \, dm_z, \tag{4.4}$$

then (4.3) may be rewritten as

$$(A_\tau \beta^{1+\delta})(z) \leq c(\delta) e^{-\tau(1-\delta)} \beta^{1+\delta}(z). \tag{4.5}$$

4.2 TAKING THE QUOTIENT BY $SL(2,\mathbb{Z})$.

Let $\alpha(z) = \sup_{\gamma \in SL(2,\mathbb{Z})} \beta(\gamma z)$.

LEMMA 4.3. *For any $0 \leq \delta < 1$ and any τ large enough depending on δ, the function $\alpha^{1+\delta}$ is a Margulis function for the averaging operator (4.4) on $X = \mathbb{H}/SL(2,\mathbb{Z})$ with $Y = \emptyset$.*

Proof. The property (a) of Definition 1.1 is immediate from the description of the fundamental domain of the action of $SL(2,\mathbb{Z})$ on $\mathbb{H}$. To show (1.2), fix $\tau > 0$ large enough so that $c_0 \equiv c(\delta) e^{-(1-\delta)\tau} < 1$.

Note that if $\operatorname{Im} z \geq 1$, then $\alpha(z) = \beta(z)$. Thus, if $\operatorname{Im} z$ is large enough so that $S(z,\tau) \subset \{x + iy \, : \, y \geq 1\}$, then, in view of (4.5), we have

$$(A_\tau \alpha^{1+\delta})(z) \leq c_0 \alpha^{1+\delta}(z).$$

If $\operatorname{Im} z$ is not large enough, then $\alpha(z) \leq C(\tau)$, and then for all $w \in S(z,\tau)$, $\alpha(z)^{1+\delta} \leq b(\delta,\tau)$, where $b(\delta,\tau)$ is some constant. Thus, in this case,

$$(A_\tau \alpha^{1+\delta})(z) \leq c_0 \alpha^{1+\delta}(z) + b(\delta,\tau).$$

Thus, for all $z \in \mathbb{H}$,

$$(A_\tau \alpha^{1+\delta})(z) \leq c_0 \alpha^{1+\delta}(z) + b(\delta,\tau),$$

and $c_0 < 1$. This verifies condition (b) of Definition 1.1. □

4.3 THE SPACE $SL(2,\mathbb{R})/SL(2,\mathbb{Z})$. The space $\mathcal{L}_2$ of unimodular lattices in $\mathbb{R}^2$ admits a transitive action by $SL(2,\mathbb{R})$, and the stabilizer of the square lattice is $SL(2,\mathbb{Z})$; thus $\mathcal{L}_2$ is isomorphic to the quotient space $SL(2,\mathbb{R})/SL(2,\mathbb{Z})$.

Note that the map ϕ is $SL(2,\mathbb{Z})$-equivariant. Let $d(L)$ denote the length of the shortest vector in the lattice L. From the definitions we see the following:

LEMMA 4.4. *For any $g \in SL(2,\mathbb{R})$,*

$$\alpha(\phi(g)) = d(g\mathbb{Z}^2)^{-1}.$$

Then, as a corollary of Lemma 4.3, we get the following:

LEMMA 4.5. *For any $0 \le \delta < 1$, the function $d^{-(1+\delta)}$ is a Margulis function (in the sense of Definition 3.1 with the averaging operator A_τ given by (3.1)) for the action of $SL(2,\mathbb{R})$ on $X = \mathcal{L}_2$, with $Y = \emptyset$.*

We now come full circle by indicating a direct proof of Lemma 4.5 (i.e., without thinking of the hyperbolic plane). Note that a unimodular lattice in $\mathbb{R}^2$ can have at most one (linearly independent) vector of length < 1 (otherwise the covolume is too small). If the shortest vector v of a lattice L is sufficiently short (depending on τ), then for all $\theta \in [0, 2\pi)$, $d(a_\tau r_\theta L) = \|a_\tau r_\theta v\|$. Then, by Lemma 4.1,

$$(A_\tau d^{-(1+\delta)})(L) \le c(\delta) e^{-(1-\delta)\tau} d^{-(1+\delta)}(L).$$

If not, then $d(L)^{-(1+\delta)} \le C(\delta, \tau)$, and then

$$(A_\tau d^{-(1+\delta)})(L) \le b(\delta, \tau) d^{-(1+\delta)}(L).$$

Then, in all cases, provided τ is large enough so that $c_0 \equiv c(\delta) e^{-(1-\delta)\tau} < 1$, we have

$$(A_\tau d^{-(1+\delta)})(L) \le c_0 d^{-(1+\delta)}(L) + b_0,$$

where $c_0 < 1$. Thus, (b) of Definition 3.1 holds. The condition (c) holds by Mahler compactness, and (a) follows immediately from the definitions. □

4.4 BALL AVERAGES. For $h: \mathbb{H} \to \mathbb{R}$, let $(B_\tau h)(z)$ denote the average of h over the ball $B(z, \tau)$ of radius τ centered at z with respect to the hyperbolic volume. Thus, B_τ is similar to A_τ but is doing ball averages instead of sphere averages. In view of hyperbolic geometry (and in particular the fact that most of the hyperbolic volume of a ball is concentrated near its outer radius) and the results for the sphere averages A_τ, we see that for all $0 \le \delta < 1$, assuming τ is sufficiently large depending on δ, we have for all $z \in \mathbb{H}$,

$$(B_\tau \alpha^{1+\delta})(z) \le c_0 \alpha^{1+\delta}(z) + b_0,$$

where $c_0 < 1$ and $b_0 = b_0(\delta, \tau)$.

4.5 PRODUCTS OF UPPER HALF PLANES. Suppose $X = \mathbb{H} \times \mathbb{H}$ is a product of two copies of the hyperbolic plane. We consider X with the supremum metric (i.e., distance on X is the supremum of the distances in the two factors). Then, the ball of radius τ in X is the product of the balls of radius τ in the two factors. Hence, if B_τ^X is the averaging operator over the ball in X of radius τ, then $B_\tau^X = B_\tau^1 B_\tau^2$, where B_τ^1 is the averaging operator over the ball of radius τ in the first factor, and B_τ^2 is the analogous thing in the second factor.

For $z = (z_1, z_2) \in X$, let $\alpha_1(z) = \alpha(z_1)$, $\alpha_2(z) = \alpha(z_2)$.

LEMMA 4.6. *Suppose* $0 \le \delta < 1$. *Let*

$$u(z) = \epsilon(\alpha_1(z)\alpha_2(z))^{1+\delta} + \alpha_1(z)^{1+\delta} + \alpha_2(z)^{1+\delta}.$$

Then (provided τ is large enough depending on δ) and ϵ is chosen sufficiently small depending on δ and τ, u is a Margulis function for the averages B_τ^X on X, with $Y = \emptyset$.

Proof. We have

$$\begin{aligned} B_\tau^X(\epsilon\alpha_1^{1+\delta}\alpha_2^{1+\delta}) &= \epsilon(B_\tau^1\alpha_1^{1+\delta})(B_\tau^2\alpha_2^{1+\delta}) \le \epsilon(c_0\alpha_1^{1+\delta} + b_0)(c_0\alpha_2^{1+\delta} + b_0) \\ &\le \epsilon c_0^2\alpha_1^{1+\delta}\alpha_2^{1+\delta} + \epsilon b_0\alpha_1^{1+\delta} + \epsilon b_0\alpha_2^{1+\delta} + \epsilon b_0^2. \end{aligned} \tag{4.6}$$

Also, for $i = 1, 2$,

$$B_\tau^X(\alpha_i^{1+\delta}) \le c_0\alpha_i^{1+\delta} + b_0.$$

Thus,

$$B_\tau^X u \le \epsilon c_0^2\alpha_1^{1+\delta}\alpha_2^{1+\delta} + (\epsilon b_0 + c_0)\alpha_1^{1+\delta} + (\epsilon b_0 + c_0)\alpha_2^{1+\delta} + \epsilon b_0^2 + 2b_0.$$

We now choose ϵ sufficiently small so that $c_1 \equiv \epsilon b_0 + c_0 < 1$. We get

$$B_\tau^X u \le c_1 u + b_1,$$

where $c_1 < 1$ and $b_1 = \epsilon b_0^2 + 2b_0$. This completes the proof. □

Similar constructions work for the product of any number of copies of $\mathbb{H}$, but the coefficients ϵ become more complicated. The Minsky product region theorem [Mi96] states that the geometry at infinity of Teichmüller space is similar to that of products of hyperbolic planes (with the supremum metric). In view of this, an analogue of the function u of Lemma 4.6 was used in [EMi11] to show that most closed geodesics return to a given compact set. A more refined version (which can deal with random geodesics on strata of quadratic or abelian differentials) was proved in [EMR12].

5 Construction of Margulis functions: $SL(n,\mathbb{R})/SL(n,\mathbb{Z})$

Let Δ be a lattice in $\mathbb{R}^n$. We say that a subspace L of $\mathbb{R}^n$ is Δ-*rational* if $L\cap\Delta$ is a lattice in L. For any Δ-rational subspace L, we denote by $d_\Delta(L)$ or simply by $d(L)$ the volume of $L/(L\cap\Delta)$. Let us note that $d(L)$ is equal to the norm of $u_1\wedge\cdots\wedge u_\ell$ in the exterior power $\bigwedge^\ell(\mathbb{R}^n)$ where $\ell=\dim L$, $(u_1,\cdots,u_\ell)$ is a basis over $\mathbb{Z}$ of $L\cap\Delta$, and the norm on $\bigwedge(\mathbb{R}^n)$ is induced from the Euclidean norm on $\mathbb{R}^n$. If $L=\{0\}$ we write $d(L)=1$. A lattice is Δ *unimodular* if $d_\Delta(\mathbb{R}^n)=1$. The space of unimodular lattices is canonically identified with $SL(n,\mathbb{R})/SL(n,\mathbb{Z})$.

Let us introduce the following notation:

$$\alpha_i(\Delta)=\sup\left\{\frac{1}{d(L)}\,\middle|\, L \text{ is a } \Delta\text{-rational subspace of dimension } i\right\},\quad 0\le i\le n,$$

$$\alpha(\Delta)=\max_{0\le i\le n}\alpha_i(\Delta). \tag{5.1}$$

The classical Mahler compactness theorem states that for any $M>0$, the set $\{\Delta\in SL(n,\mathbb{R})/SL(n,\mathbb{Z})\ :\ \alpha(\Delta)\le M\}$ is compact.

Let $G=SL(n,\mathbb{R})$, $\Gamma=SL(n,\mathbb{Z})$, $\hat{K}\cong SO(n)$ is a maximal compact subgroup of G, $H\cong SO(p,q)\subset G$ and $K=H\cap\hat{K}$ is a maximal compact subgroup of H.

For any K-invariant function f on G/Γ, let $(A_tf)(x)=\int_K f(a_tkx)\,dm(k)$, where m is the normalized Haar measure on K. Suppose $x\in G/\Gamma$ and the stabilizer of x in H is trivial. Then $K\backslash Hx$ is isomorphic to the symmetric space $K\backslash H$, with x corresponding to the origin. If rank $K\backslash H=1$, then $(A_tf)(x)$ can be interpreted as the average of f over the sphere of radius $2t$ centered at the origin in the symmetric space $K\backslash Hx$.

If $p\ge 3$ and $0<s<2$, or if $(p,q)=(2,1)$ or $(2,2)$ and $0<s<1$, it is shown in [EMM98, Lemma 5.6] that for any $c>0$ there exist $t>0$, and $\omega>1$ so that the functions α_i^s satisfy the following system of integral inequalities in the space of lattices:

$$A_t\alpha_i^s \le c_i\alpha_i^s + \omega^2 \max_{0<j\le\min(n-i,i)} \sqrt{\alpha_{i+j}^s\alpha_{i-j}^s}, \tag{5.2}$$

where A_t is the averaging operator $(A_tf)(\Delta) = \int_K f(a_tk\Delta)$ and $c_i \le c$. If $(p,q) = (2,1)$ or $(2,2)$ and $s=1$, then (5.2) also holds (for suitably modified functions α_i), but some of the constants c_i cannot be made smaller than 1. (The proof of [EMM98, Lemma 5.6] is a much more complicated version of the direct proof of Lemma 4.5 in section 4.)

In [EMM98, section 5.4] it is shown that if the α_i satisfy (5.2), then for any $\epsilon > 0$, the function $f = f_{\epsilon,s} = \sum_{0\le i\le n} \epsilon^{i(n-i)}\alpha_i^s$ satisfies the scalar inequality

$$A_tf \le cf + b, \tag{5.3}$$

where t, c, and b are constants. (This proof is a more complicated version of the proof of Lemma 4.6 in section 4.) If $c<1$, which occurs in the case $p\ge 3$, it follows that f is a Margulis function (for the case $Y=\emptyset$).

If $c=1$, which will occur in the $SO(2,1)$ and $SO(2,2)$ cases, then (5.3) implies that $(A_rf)(1)$ is growing at most linearly with the radius.

Throughout [EMM98] one considers the functions $\alpha(g)^s$ for $0<s<2$, even though for the application to quadratic forms one only needs $s=1+\delta$ for some $\delta>0$. This yields a better integrability result and is also necessary for the proof of the convergence results [EMM98, Theorem 3.4] and [EMM98, Theorem 3.5].

Even though the function f is not strictly speaking a Margulis function for the case $s=1$, $p=2$, $q=2$, it plays a key role in the analysis of the $(2,2)$ case of the quantitative Oppenheim conjecture in [EMM05].

6 Large deviation estimates

For simplicity we state the results for the $SL(2,\mathbb{R})$-action setting. For the random walk setting, see [Ath06, Theorem 1.2].

Let $C_\ell = \{x\in X : f(x) < \ell\}$. Let m denote the uniform measure on $SO(2) \subset SL(2,\mathbb{R})$. We refer to the trajectories of the group $\{a_t : t\in\mathbb{R}\}$ as "geodesics."

THEOREM 6.1 ([Ath06, Theorem 1.1]).

1. *For all* l *sufficiently large and all* $x \notin C_l$, *there are positive constants* $c_1 = c_1(l, x), c_2(l)$, *with*

$$m\{\theta : a_t r_\theta x \notin C_l, 0 \le t \le T\} \le c_1 e^{-c_2 T}$$

for all T *sufficiently large. That is, the probability that a random geodesic trajectory has not visited* C_l *by time* T *decays exponentially in* T.

2. *For all* l, S, T *sufficiently large and all* $x \in X$, *there are positive constants* $c_3 = c_3(S, l, x), c_4 = c_4(l)$, *with*

$$m\{\theta : a_t r_\theta x \notin C_l, S \le t \le S + T\} \le c_3 e^{-c_4 T}.$$

That is, the probability that a random geodesic trajectory does not enter C_l *in the interval* $[S, S+T]$ *decays exponentially in* T.

3. *Let* $x \in X$. *For any* $0 < \lambda < 1$, *there is* $l \ge 0$, *and* $0 < \gamma < 1$, *such that for all* T *sufficiently large (depending on all the above constants),*

$$m\{\theta : \frac{1}{T}|\{0 \le t \le T : a_t r_\theta x \notin C_l\}| > \lambda\} \le \gamma^T.$$

Result (3) above may be thought of as a large deviations result for the geodesics. Suppose μ_Q is an $SL(2, \mathbb{R})$-invariant measure on X (which we think of as the volume). While ergodicity guarantees that $\frac{1}{T}|\{0 \le t \le T : a_t x \in C_l\}| \to \mu_Q(C_l)$ for μ_Q-almost every $x \in X$, Theorem 6.1 gives explicit information for any $x \in X$ about the likelihood of bad trajectories starting in the set $SO(2)x$. Notice, however, this is *not* a traditional large deviations result, which estimates the probability of a deviation of any $\epsilon > 0$ from the ergodic average.

7 Other constructions and applications

7.1 HOMOGENEOUS DYNAMICS. Let G be a semisimple Lie group, and let Γ be a lattice in G. Suppose μ is a probability measure on G; then μ defines a random walk on G/Γ.

In [EMa05], provided that the group generated by the support of μ is Zariski dense in G, a Margulis function for this random walk (and $Y = \emptyset$) was constructed; in the case $G = SL(n, \mathbb{R})$ and $\Gamma = SL(n, \mathbb{Z})$, the function is in fact the same as the function in section 5.

In [BQ12], a Margulis function for this random walk (again with $Y = \emptyset$) was constructed under the weaker assumption that the Zariski closure of the group generated by the support of μ is semisimple. For a treatment of the case where Y is a closed orbit of some semisimple subgroup, see [BQ13].

In [GM10] a different Margulis function was used in conjunction with Fourier analysis to give polynomial error terms for the quantitative Oppenheim conjecture in at least five variables. This also gives an alternative proof of the definite case of the Oppenheim conjecture in five or more variables first proved in [G04].

A Margulis function (for $Y = \emptyset$) was constructed for the space of inhomogeneous lattices in [MM11] in order to prove the analogue of the quantitative Oppenheim conjecture for inhomogeneous quadratic forms.

In [HLM17], the construction of the Margulis function on the space of lattices was extended to the S-arithmetic case and used to prove the S-arithmetic version of the quantitative Oppenheim conjecture.

In [EK12] and [KKLM17], a modification of the Margulis function from [EMM98] was used to control the entropy contribution from the thin part of the space of lattices. In [Kh18], a Margulis function was used to study the Hausdorff dimension of the set of diverging trajectories of a diagonalizable element on the space of lattices.

7.2 TEICHMÜLLER DYNAMICS. The idea of Margulis functions has played a key role in Teichmüller dynamics. In [EMas01], a Margulis function for the action of $SL(2,\mathbb{R})$ for $Y = \emptyset$ on strata of Abelian or quadratic differentials has been constructed. The construction has some parallels to that of section 5. This function was used in [Ath06] to prove some exponential large deviation estimates for Teichmüller geodesic rays starting at a given point in the space. Athreya's results were later used in [AthF08] to control deviation of ergodic averages in almost all directions for a billiard flow in a rational polygon. The Margulis function of [EMas01] was later used in [AG13] in their proof of exponential decay of correlations for the Teichmüller geodesic flow. Building on the work of [EMas01] and [Ath06], a Margulis function for the same action but arbitrary $SL(2,\mathbb{R})$-invariant submanifolds Y was constructed in [EMM15]. Together with the measure classification theorem of [EMi18], this function played a key role in the proof that $SL(2,\mathbb{R})$-orbit closures are invariant submanifolds. This function is also used in many related results, such as [CE15].

A modified (and independently developed) version of the Margulis function technique was used in [AF07] to prove that the generic interval exchange transformation is weak mixing.

7.3 OTHER APPLICATIONS. Suppose μ is a probability measure on $SL(n,\mathbb{R})$. We may then consider random products of independent matrices, each with the distribution μ. We can then ask if the Lyapunov exponents $\lambda_i(\mu)$ of these random products depend continuously on μ. In [V14, Chapter 10] a new version of the Margulis function technique, due to Avila and Viana, which involves a modification of the natural averaging operator so that a Margulis function can be constructed, was used to show that in dimension 2, a natural continuity statement holds; namely, if $\mu_j \to \mu$ in the weak-star topology and also the support of μ_j tends to the support of μ in the Hausdorff topology, then for $i=1,2$, $\lambda_i(\mu_j) \to \lambda_i(\mu)$. (The assumption about the support is necessary; see [V14, Chapter 10] for a counterexample). A more complicated proof was given previously in [BV10] without use of Margulis functions.

The result of [V14, Chapter 10] was extended in [MV14] to the case of Markov processes. (For the case $n \geq 2$, see the next paragraph).

7.4 ADDITIVE MARGULIS FUNCTIONS. Suppose we have a decomposition of $X = C \cup D$, where $C \cap D = \emptyset$. Let A be an averaging operator. An additive Margulis function (relative to this decomposition) is a function ϕ: $X \to \mathbb{R}^+$ with the following properties:

(a) There exists a constant $\kappa_C > 0$ such that for $x \in C$,

$$(A\phi)(x) < \phi(x) - \kappa_C. \tag{7.1}$$

(b) There exists a constant $\kappa_D > 0$ such that for $x \in D$,

$$(A\phi)(x) < \phi(x) + \kappa_D. \tag{7.2}$$

Suppose f is a Margulis function, and choose $\Lambda > b/c$. Let

$$C = \{x \in X \,:\, f(x) > \Lambda\} \qquad D = \{x \in X \,:\, f(x) \leq \Lambda\}.$$

Then, it follows from Jensen's inequality that $\log f$ is an additive Margulis function relative to the decomposition $X = C \cup D$.

However, it is not true that if ϕ is an additive Margulis function, then e^ϕ is a multiplicative one. In fact, the inequality (1.1) is very sensitive to the "worst case behavior" of e^ϕ on the support of the measure μ_x defining A; on the other hand, the inequalities (7.1) and (7.2) depend more on the "average case" behavior of ϕ. Because of this effect, it is often much easier to construct an additive Margulis function than a multiplicative one. (In fact we do not know how to construct a useful multiplicative Margulis function in the setting of [V14, Chapter 10] beyond the case $n=2$).

Additive Margulis functions are useful because of the following:

LEMMA 7.1. *Suppose ϕ is an additive Margulis function for A relative to the decomposition $X=C\cup D$, and suppose η is a measure on X with $\int_X \phi\, d\eta < \infty$. Suppose also $\int_X (A\phi)(x)\, d\eta(x) \geq \int_X \phi(x)\, d\eta(x)$ (for example, this holds if η is A invariant). Then,*

$$\eta(D) \geq \frac{\kappa_C}{\kappa_C+\kappa_D}\eta(X). \tag{7.3}$$

Proof. We have

$$\int_X \phi(x)\, d\eta(x) \leq \int_X (A\phi)(x)\, d\eta(x) < \int_X \phi(x)\, d\eta(x) - \kappa_C\eta(C) + \kappa_D\eta(D).$$

Thus, $-\kappa_C\eta(C)+\kappa_D\eta(D) > 0$, which implies (7.3). □

This circle of ideas was used in [AEV] in order to extend the results on continuity of Lyapunov exponents in [V14, Chapter 10] to arbitrary dimensions, and was also used in [BBB15] in a nonlinear setting.

8 Comparison to other techniques

In the homogeneous dynamics setting, there is another technique for proving results similar in flavor to what can be obtained using Margulis functions. For the case $Y=\emptyset$, this originates with the paper [Mar71] and was further developed in [Dan84] and [Dan86]. These ideas were used in many of the foundational papers in homogeneous dynamics such as [DM89], [DM90], and [Ra91]. For other Y, the key result is the linearization technique of [DM93] (in which in particular the asymptotically exact lower bounds for the quantitative Oppenheim conjecture were proved). An abstract framework for these methods in terms of "(C,α)-good" functions defined in [EMS97] is developed in [KMar98]. These techniques (and in particular the framework in [KMar98]) have numerous applications to Diophantine approximations and other areas, which are beyond the scope of this survey.

The "(C,α)-good" techniques rely essentially on the variants of the polynomial nature of the unipotent flow and have limited applicability outside of homogeneous dynamics. (Exceptions are [MW02] and [MW14], where the authors manage to obtain results on nondivergence in the Teichmüller dynamics setting using essentially polynomial techniques). In the homogeneous setting, one usually obtains sharper estimates if one manages to construct a Margulis function; for example, the quantitative Oppenheim conjecture cannot be proved by (C, α)-good techniques since the estimates one obtains that way are too weak. (This is in fact the original motivation for Margulis functions). However, a construction of a Margulis function is not always possible, for example, for the action of a single unipotent. This is related to the fact that (1.1) has to hold for *all* $x \in X$. This can be easier to do if one considers additive Margulis functions instead, but then the results are even weaker than what is obtained by (C, α)-good methods. In general, (nonadditive) Margulis functions are an extremely powerful tool, but in many cases their construction is a difficult engineering challenge.

References

[Ath06] J. Athreya. Quantitative recurrence and large deviations for Teichmüller geodesic flow. *Geom. Dedicata* **119** (2006), 121–140.

[AthF08] J. Athreya, G. Forni. Deviation of ergodic averages for rational polygonal billiards. *Duke Math. J.* **144** (2008), no. 2, 285–319.

[AEV] A. Avila, A. Eskin, M. Viana. Continuity of Lyapunov exponents of random matrix products. *In preparation.*

[AF07] A. Avila, G. Forni. Weak mixing for interval exchange transformations and translation flows. *Ann. of Math.* (2) **165** (2007), no. 2, 637–664.

[AG13] A. Avila, S. Gouëzel. Small eigenvalues of the Laplacian for algebraic measures in moduli space, and mixing properties of the Teichmüller flow. *Ann. of Math.* (2) **178** (2013), no. 2, 385–442.

[BBB15] L. Backes, A. Brown, C. Butler. Continuity of Lyapunov exponents for cocycles with invariant holonomies, 2015, arXiv:1507.08978.

[BQ12] Y. Benoist, J-F. Quint. Random walks on finite volume homogeneous spaces. *Invent. Math.* **187** (2012), no. 1, 37–59.

[BQ13] Y. Benoist, J-F. Quint. Stationary measures and invariant subsets of homogeneous spaces (III). *Ann. of Math* (2) **178** (2013), no. 3, 1017–1059.

[BV10] C. Bocker, M. Viana. Continuity of Lyapunov exponents for 2D random matrices, 2010, arXiv:1012.0872.

[CE15] J. Chaika, A. Eskin. Every flat surface is Birkhoff and Oseledets generic in almost every direction. *J. Mod. Dyn.* **9** (2015), 1–23.

[Dan79] S. G. Dani. On invariant measures, minimal sets and a lemma of Margulis. *Invent. Math.* **51** (1979), 239–260.

[Dan81] S. G. Dani. Invariant measures and minimal sets of horoshperical flows. *Invent. Math.* **64** (1981), 357–385.

[Dan84] S. G. Dani. On orbits of unipotent flows on homogeneous spaces. *Ergod. Theor. Dynam. Syst.* **4** (1984), 25–34.

[Dan86] S. G. Dani. On orbits of unipotent flows on homogenous spaces II. *Ergod. Theor. Dynam. Syst.* **6** (1986), 167–182.

[DM89] S. G. Dani, G. A. Margulis. Values of quadratic forms at primitive integral points. *Invent. Math.* **98** (1989), 405–424.

[DM90] S. G. Dani, G. A. Margulis. Orbit closures of generic unipotent flows on homogeneous spaces of $SL(3,\mathbb{R})$. *Math. Ann.* **286** (1990), 101–128.

[DM91] S. G. Dani, G. A. Margulis. Asymptotic behaviour of trajectories of unipotent flows on homogeneous spaces. *Indian. Acad. Sci. J.* **101** (1991), 1–17.

[DM93] S. G. Dani, G. A. Margulis. Limit distributions of orbits of unipotent flows and values of quadratic forms. In *I. M. Gelfand Seminar*, pp. 91–137, Amer. Math. Soc., Providence, RI, 1993.

[EK12] Manfred Einsiedler, Shirali Kadyrov. Entropy and escape of mass for $\mathrm{SL}_3(\mathbb{Z})\backslash\mathrm{SL}_3(\mathbb{R})$. *Israel J. Math.*, 190:253–288, 2012.

[EMa05] A. Eskin, G. Margulis. Recurrence properties of random walks on finite volume homogeneous manifolds. *In Random walks and geometry*, pp. 431–444, Walter de Gruyter GmbH & Co. KG, Berlin, 2004.

[EMM98] A. Eskin, G. Margulis, S. Mozes. Upper bounds and asymptotics in a quantitative version of the Oppenheim conjecture. *Ann. of Math.* (2) **147** (1998), no. 1, 93–141.

[EMM05] A. Eskin, G. Margulis, S. Mozes. Quadratic forms of signature (2,2) and eigenvalue spacings on rectangular 2-tori. *Ann. of Math.* (2) **161** (2005), no. 2, 679–725.

[EMas01] A. Eskin, H. Masur. Asymptotic formulas on flat surfaces. *Ergodic Theory Dynam. Systems* **21** (2001), no. 2, 443–478.

[EMi11] A. Eskin, M. Mirzakhani. Counting closed geodesics in moduli space. *J. Mod. Dyn.* **5** (2011), no. 1, 71–105.

[EMi18] A. Eskin, M. Mirzakhani. Invariant and stationary measures for the SL(2,R) action on moduli space. *Publ. Math. Inst. Hautes Études Sci.* **127** (2018), 95–324.

[EMR12] A. Eskin, M. Mirzakhani, K. Rafi. Counting closed geodesics in strata, 2012, arXiv:1206.5574.

[EMM15] A. Eskin, M. Mirzakhani, A. Mohammadi. Isolation, equidistribution, and orbit closures for the SL(2,R) action on moduli space. *Ann. of Math.* (2) **182** (2015), no. 2, 673–721.

[EMS97] A. Eskin, S. Mozes, N. Shah. Non-divergence of translates of certain algebraic measures. *Geom. Funct. Anal.* **7** (1997), no. 1, 48–80.

[G04] F. Götze. Lattice point problems and values of quadratic forms. *Invent. Math.* **157** (2004), no. 1, 195–226.

[HLM17] J. Han, S. Lim, K. Mallahi-Karai. Asymptotic distribution of values of isotropic quadratic forms at S-integral points. *J. Mod. Dyn.* **11** (2017), 501–550.

[GM10] Friedrich Götze, Gregory Margulis. Distribution of values of quadratic forms at integral points, 2010, arXiv:1004.5123.

[KKLM17] S. Kadyrov, D. Y. Kleinbock, E. Lindenstrauss, G. A. Margulis. Entropy in the cusp and singular systems of linear forms, *J. Anal. Math.* **133** (2017), 253–277.

[Kh18] Osama Khalil. Bounded and divergent trajectories and expanding curves on homogeneous spaces, 2018, arXiv:1806.06832.

[KMar98] D. Y. Kleinbock, G. A. Margulis. Flows on homogeneous spaces and Diophantine approximation on manifolds. *Ann. of Math.* (2) **148** (1998), no. 1, 339–360.

[MV14] E. Malheiro, M. Viana. Lyapunov exponents of linear cocycles over Markov shifts, 2014, arXiv:1410.1411.

[Mar71] G. A. Margulis. *On the action of unipotent groups in the space of lattices.* In *Lie groups and their representations,* Proc. of Summer School in Group Representations, Bolyai Janos Math. Soc., Akademai Kiado, Budapest, 1971, pp. 365–370, Halsted, New York, 1975.

[Mar87] G. A. Margulis. Formes quadratiques indèfinies et flots unipotents sur les spaces homogènes, *C. R. Acad. Sci. Paris Ser. I* **304** (1987), 247–253.

[Mar89a] G. A. Margulis. Discrete subgroups and ergodic theory. In *Number theory, trace formulas and discrete subgroups,* a symposium in honor of A Selberg, pp. 377–398. Academic Press, Boston, MA, 1989.

[Mar89b] G. A. Margulis. *Indefinite quadratic forms and unipotent flows on homogeneous spaces.* In *Dynamical systems and ergodic theory,* Vol. 23, pp. 399–409, Banach Center Publ., PWN—Polish Scientific Publ., Warsaw, 1989.

[MM11] G. A. Margulis, A. Mohammadi. Quantitative version of the Oppenheim conjecture for inhomogeneous quadratic forms. *Duke Math. J.* **158** (2011), no. 1, 121–160.

[MT09] S. Meyn, R. Tweedie. Markov chains and stochastic stability. Second edition. With a prologue by Peter W. Glynn. Cambridge University Press, Cambridge, 2009.

[Mi96] Y. Minsky. *Extremal length estimates and product regions in Teichmüller space. Duke Math. J.* **83** (1996), no. 2, 249–286.

[MW02] Y. Minsky, B. Weiss. Nondivergence of horocyclic flows on moduli space. *J. Reine Angew. Math.* **552** (2002), 131–177.

[MW14] Y. Minsky, B. Weiss. Cohomology classes represented by measured foliations, and Mahler's question for interval exchanges. *Ann. Sci. Éc. Norm. Supér.* (4) **47** (2014), no. 2, 245–284.

[Ra91] M. Ratner. Raghunathan's topological conjecture and distributions of unipotent flows. *Duke Math. J.* **63** (1991), no. 1, 235–280.

[V14] M. Viana. *Lectures on Lyapunov exponents.* Cambridge University Press, Cambridge, 2014.

ELON LINDENSTRAUSS

RECENT PROGRESS ON RIGIDITY PROPERTIES OF HIGHER RANK DIAGONALIZABLE ACTIONS AND APPLICATIONS

Dedicated to G. A. Margulis

Abstract. The rigidity properties of higher rank diagonalizable actions is a major theme in homogeneous dynamics, with origins in work of Cassels and Swinnerton-Dyer in the 1950s and Furstenberg. We survey both results and conjectures regarding such actions, with emphasis on the applications of these results toward understanding the distribution of integer points on varieties, quantum unique ergodicity, and Diophantine approximations.

1 Introduction

The extensive theory of actions of unipotent groups on homogeneous spaces, to which G. A. Margulis made many pioneering contributions, gives very satisfactory qualitative (if not yet quantitative) understanding of these actions, with numerous and profound applications. The current state of the art regarding actions of diagonalizable groups is much less satisfactory, and indeed for most natural questions we only have partial results regarding the dynamics. Fortunately, these partial results already have fairly wide applicability. It is the purpose of the survey to present some of the rigidity results regarding such actions as well as their applications.

The motivation to study rigidity properties of higher rank diagonal actions comes from two different directions. One of these is from the geometry of numbers: the program, initiated by Minkowski, of using lattices in Euclidean spaces and their generalizations for understanding number theoretic questions. We shall make in the survey a distinction between *arithmetic questions*—that is to say, properties of integer points (or more generally rational or algebraic points), such as counting and distribution properties of integer

Elon Lindenstrauss. The Einstein Institute of Mathematics Edmond J. Safra Campus, Givat Ram, The Hebrew University of Jerusalem, Jerusalem, 91904, Israel
elon@math.huji.ac.il

Lindenstrauss acknowledges the support of ERC-2018-ADG project HomDyn and ISF grant 891/15.

points on varieties—and questions in *Diophantine approximations*, such as how well a point to a given variety can be approximated by integer or rational points. Both kinds of applications were already prominently present in the geometry of numbers since its inception by Minkowski.

A classical problem in the geometry of numbers is the study of the set of values attained at integer points by a homogeneous form F of degree d obtained by taking the products of d-linear forms in d-variables that is, one considers forms

$$F(x_1,\dots,x_d)=\prod_{i=1}^{d} l_i(x_1,\dots,x_d),$$

where $l_1,\dots,l_d$ are d linearly independent linear forms,[1] and investigate the values attained by F for $\mathbf{x}=(x_1,\dots,x_d)\in\mathbb{Z}^d$. For instance, one may study the quantity

$$\nu_F=\inf_{\mathbf{x}\in\mathbb{Z}^d\smallsetminus\{0\}}|F(\mathbf{x})|.$$

If we present the coefficients of the linear forms l_i in a $d\times d$-matrix g (one row for each linear form), then the map $F(g)$ assigning a product of d linear forms F to a $d\times d$-matrix g is left invariant under the action of the $(d-1)$–dimensional diagonal subgroup $A<\mathrm{SL}(d,\mathbb{R})$, whereas the map $F\mapsto\nu_F$ is invariant under composition of F by an element of $\mathrm{GL}(d,\mathbb{Z})$ (in the geometry of numbers literature, two forms that are the same up to the action of $\mathrm{GL}(d,\mathbb{Z})$ are said to be *equivalent*[2]). Thus we may view $g\mapsto\nu_{F(g)}$ as either a (left) A-invariant function on $\mathrm{GL}(n,\mathbb{R})/\mathrm{GL}(n,\mathbb{Z})$ or a (right) $\mathrm{GL}(n,\mathbb{Z})$-invariant function on $A\backslash\mathrm{GL}(n,\mathbb{R})$. It is convenient to use the normalized quantity $\bar{\nu}_{F(g)}=\nu_{F(g)}/\left|\det g\right|$, which is a well-defined function on $\mathrm{PGL}(d,\mathbb{R})$, left invariant under A and right invariant under $\mathrm{PGL}(d,\mathbb{Z})$.

Already the case $d=2$ is of some interest and was quite extensively studied [C2, section 2]. In this case the possible forms F considered are simply the set of nondegenerate indefinite quadratic forms in two variables. For any product of two linear forms in two variables F, the value of $\bar{\nu}_F$ is $\le\sqrt{5}$, with equality if and only if F is equivalent (up to a multiplicative scalar and the action of $\mathrm{GL}(2,\mathbb{Z})$) to $F(x,y)=x^2-xy-y^2$. Up to the same degrees of freedom, Markoff constructed a complete list of (countably many) binary form

[1] When considering a product of d linear forms in d variables, the forms will be implicitly assumed to be linearly independent even if this is not explicitly stated.

[2] Sometimes one makes a distinction between forms that are the same up to composition by an element of $\mathrm{SL}(d,\mathbb{Z})$, which are said to be properly equivalent, and forms that are the same under the action of the slightly bigger group $\mathrm{GL}(d,\mathbb{Z})$, which are said to be only improperly equivalent.

with $\bar{\nu}_F > \frac{1}{3}$ but there are uncountably many such indefinite binary forms with $\bar{\nu}_F = \frac{1}{3}$ and a set of full Hausdorff dimension of forms with $\bar{\nu}_F > 0$. Cassels and Swinnerton-Dyer investigated the possible values of $\bar{\nu}_F$ for forms that are a product of three linear forms in three variables [CSD]. They discovered that integral forms of this type satisfy a very strong isolation result, much stronger than the analogous isolation result of Remak and Rogers for a product of two linear forms in two variables. We emphasize that an integral form that is a product of d linear forms in d variables need not be presentable as a product of d integral linear forms in d variables. This led them to make the following remarkable conjecture (to be precise, Cassels and Swinnerton-Dyer state this conjecture in their paper for $d = 3$, but it is clear that they realized a similar phenomenon should hold for higher d; cf. also the much later remark in Swinnerton-Dyer's book [SD, p. 20]):

CONJECTURE 1a (Cassels and Swinnerton-Dyer [CSD]). *Let $d \geq 3$. Any form F that is a product of d linear forms in d variables that is not proportional to a form with integral coefficients has $\bar{\nu}_F = 0$.*

This farsighted paper [CSD], and in particular the above conjecture, was highlighted by Margulis in [M5]. Stated in terms of the homogeneous space $\mathrm{PGL}(d, \mathbb{R})/\mathrm{PGL}(d, \mathbb{Z})$, this conjecture is equivalent to the following:

CONJECTURE 1b ([CSD, M5]). *Let $d \geq 3$. Any orbit of the diagonal group A in $\mathrm{PGL}(d, \mathbb{R})/\mathrm{PGL}(d, \mathbb{Z})$ is either unbounded or periodic.*

Here and throughout we say that an orbit $L.x$ of a locally compact group L on a space X is periodic if the stabilizer of x is a lattice in L—that is, discrete and of finite covolume. Cassels and Swinnerton-Dyer show in [CSD] that Conjecture 1a implies a conjecture of Littlewood's from circa 1930:

CONJECTURE 2 (Littlewood). *For any $\alpha, \beta \in \mathbb{R}$, it holds that*

$$\inf_{n>0} n \,\|n\alpha\| \,\|n\beta\| = 0.$$

Here we use for $x \in \mathbb{R}$ the somewhat unfortunate but customary notation $\|x\| = \min_{n\in\mathbb{Z}} |x - n|$.

A second historical motivation comes from ergodic theory—namely, the work of Furstenberg on "transversality" of the $\times a$ and $\times b$ maps on $\mathbb{T} = \mathbb{R}/\mathbb{Z}$ for a and b multiplicatively independent. Recall that two integers a and b are said to be multiplicatively independent if they are not both powers of this same integer—that is, if $\log a/\log b \notin \mathbb{Q}$. In his landmark paper [F2], Furstenberg proved the following theorem:

THEOREM 1.1 (Furstenberg [F2]).
Let X be a closed subset of $\mathbb{T}$ invariant under the action of the multiplicative semigroup $S_{a,b}$, generated by two multiplicatively independent integers a and b—that is to say, $s.x \in X$ for any $s \in S$ and $x \in X$. Then X is either finite or $X = \mathbb{T}$.

In this paper Furstenberg introduced the notion of joinings and the related notion of disjointness of dynamical systems (which will be important for us later on in this survey) and deduced Theorem 1.1 from a particular disjointness principle, one of several enunciated in the paper.

He also presented the following highly influential conjecture that is still open, a natural analogue to Theorem 1.1 in the measure preserving category:

CONJECTURE 3 (Furstenberg, ca. 1967). *Let $S_{a,b} \subset \mathbb{N}$ be a semigroup generated by two multiplicatively independent integers as above, and let μ be an $S_{a,b}$-invariant and ergodic probability measure on $\mathbb{T}$. Then either μ is finitely supported or μ is the Lebesgue measure $m_{\mathbb{T}}$ on $\mathbb{T}$.*

In this survey we will use Greek letters to denote unknown probability measures and m (often decorated with subscripts) to denote a canonical probability measure such as Lebesgue measure or Haar measure. We stress that μ being $S_{a,b}$-ergodic does not imply it is ergodic under the action generated by multiplication by a or by b—only that any measurable subset $X \subset \mathbb{T}$ that is $S_{a,b}$-invariant has either $\mu(X) = 0$ or $\mu(\mathbb{T} \setminus X) = 0$.

Dealing with semigroup actions is somewhat awkward; this is easily remedied, though: it is easy to see that Conjecture 3 is equivalent to classifying the $\{a^k b^l : k, l \in \mathbb{Z}\}$-invariant and ergodic probability measures on $\prod_{p|ab \text{ prime}} \mathbb{Q}_p/\Lambda$ with $\Lambda = \mathbb{Z}[1/ab]$ diagonally embedded in $\prod_{p|ab} \mathbb{Q}_p$.

An important insight of Rudolph [R6], building on prior work of Lyons [L5], is that entropy plays an important role in understanding this measure classification question. Specifically, Rudolph proved for a, b relatively prime that Lebesgue measure is the only $S_{a,b}$-invariant and ergodic probability measure on $\mathbb{T}$ so that its entropy with respect to at least one element of $S_{a,b}$ is positive. This was extended to the more general multiplicative independent case by Johnson [J]. We now have quite a few other proofs of Rudolph's theorem (e.g., [F1], [H2], [H1] to name a few) that seem to me quite different, though all rely very heavily on the positive entropy assumption.

The key feature of the rigidity of higher rank abelian groups such as the action of $S_{a,b}$ on $\mathbb{T}$ is that the rigidity does not come from the action of an individual element. For any $s \in S_{a,b}$ there are uncountably many s-invariant and ergodic probability measures on $\mathbb{T}$ with any entropy in $[0, \log s]$ as well as uncountably many s-invariant closed subsets of $\mathbb{T}$ of any Hausdorff dimension in the range $[0, 1]$, though we do mention one important restriction: Lebesgue measure $m_{\mathbb{T}}$ is the unique s-invariant measure on $\mathbb{T}$ with entropy $\log s$, and $\mathbb{T}$ is only an s-invariant closed subset of $\mathbb{T}$ of Hausdorff dimension 1.

Furstenberg presented the $\times a$, $\times b$ problem as a special instance of a more general problem, and indeed the type of phenomena pointed out by Furstenberg exists also in the action of A on $\mathrm{PGL}(d, \mathbb{R})/\mathrm{PGL}(d, \mathbb{Z})$ and in many other high dimensional diagonal actions.

The key feature of rigidity of higher rank diagonalizable actions—rigidity of the action as a whole while no rigidity for the action of individual elements—is in contrast to the rigidity properties of unipotent groups and more generally actions of groups generated by unipotents.

DEFINITION 1.2. Let G be a locally compact group and $\Gamma < G$ a closed subgroup. A measure μ on G/Γ is said to be *homogeneous* if it is supported on a single orbit of its stabilizer $\mathrm{stab}_G\, \mu = \{g \in G : g.\mu = \mu\}$.

A landmark result of Ratner [R3, R2] gives that for groups generated by one-parameter unipotent subgroups, any invariant probability measure on a quotient space G/Γ has to be homogeneous. Here the rigidity is already exhibited in the action of individual one-parameter subgroups of the action (another proof of this measure classification result using entropy theory was given by Margulis and Tomanov in [MT1]). Ratner used her measure classification theorem to classify orbit closures under such actions [R4], which enabled her to prove Raghunathan's Conjecture (this conjecture, together with a related conjecture of Dani, appeared in [D1]). Several important nonhorospherical[3] cases of this conjecture were proved prior to [R4, R3] by Dani and Margulis [M3, DM1, DM2], including in particular Margulis's proof of the longstanding

[3]The horospherical case is more elementary and can be proved, e.g., using mixing of an appropriate one-parameter diagonalizable flow; this is not unrelated to the phenomenon of the uniqueness of measure of maximal entropy for a one-parameter diagonalizable flow we already encountered in the context of the $\times s$ map on $\mathbb{T}$. We note also that the horospherical case inspired Dani and Raghunathan to make their general conjectures on unipotent orbits—indeed, this is precisely what Dani's paper [D1] is about!

Oppenheim conjecture via the study of orbits of the group $SO(2,1) < SL(3,\mathbb{R})$ on $SL(3,\mathbb{R})/SL(3,\mathbb{Z})$.

There is another, less important wrinkle that requires some care in formulating general conjectures regarding rigidity of higher rank abelian groups, as the following simple example illustrates: Suppose one would like to classify invariant measures for the action of the complex diagonal matrices on $X_{\mathbb{C}} = SL(3,\mathbb{C})/SL(3,\mathcal{O})$ with $\mathcal{O}$ the ring of integers in an imaginary quadratic field; for instance, the Gaussian integers $\mathbb{Z}[i]$. Then $X_{\mathbb{R}} = SL(3,\mathbb{R})/SL(3,\mathbb{Z})$, considered as a homogeneous subspace of $X_{\mathbb{C}}$, is invariant under the real diagonal group. Let $m_{X_{\mathbb{R}}}$ denote the uniform measure on $X_{\mathbb{R}}$, and set

$$\mu = \fint_0^{2\pi} \fint_0^{2\pi} \begin{pmatrix} e^{i\theta_1} & & \\ & e^{i\theta_2} & \\ & & e^{-i(\theta_1+\theta_2)} \end{pmatrix} .m_{X_{\mathbb{R}}}\, d\theta_1 d\theta_2$$

(here we use the symbol $\fint$ to denote integration normalized by the measure of the set we integrate on that is, so that $\fint dx = 1$). The measure μ is invariant and ergodic under the action of the (complex) diagonal group in $SL(3,\mathbb{C})$, but it is not homogeneous.

DEFINITION 1.3. Let G be a locally compact group and $\Gamma < G$ a closed subgroup. A measure μ on G/Γ is said to be *almost homogeneous* if there is a homogeneous measure m_0 on G/Γ with stabilizer $H_0 = \mathrm{stab}_G\, m_0$ and a closed subgroup $L < G$ so that $L/(L \cap H_0)$ has finite L-invariant volume and

$$\mu = \fint_{L/(L\cap H_0)} \ell.m_0\, d\ell.$$

If the quotient $L/(L \cap H_0)$ is finite we say that μ is *virtually homogeneous.*

DEFINITION 1.4. *Let k be a local field (e.g., $\mathbb{R}$), and let $\mathbb{G}$ be an algebraic group defined over k. An element $g \in \mathbb{G}(k)$ said to be of class-$\mathcal{A}$ if it is diagonalizable over k, generates an unbounded subgroup of $\mathbb{G}(k)$, and moreover for any action of $\mathbb{G}(k)$ on a projective space $\mathbb{P}V(k)$ and $v \in \mathbb{P}V(k)$ any limit point of $\{g^n.v : n \in \mathbb{Z}\}$ is g-invariant. An element $g \in \prod_i \mathbb{G}_i(k_i)$ is of class-$\mathcal{A}$ if all of its components are of class-$\mathcal{A}$.*

For example, a $\mathbb{R}$-diagonalizable element of $\mathbb{G}(\mathbb{R})$ with positive eigenvalues is of class-$\mathcal{A}$. Another example that works in any local field k is taking an element $g \in \mathbb{G}(k)$, where all of its eigenvalues are integer powers of some fixed $\theta \in k$ with $|\theta| > 1$. This latter example has been called class-$\mathcal{A}$ by Margulis and Tomanov in [MT2], but it is precisely the invariance property of any limit point of elements for projective actions of the underlying group that was used there, and it seems convenient to enlarge this class using this property.

DEFINITION 1.5. We say that a topological group A is of *higher rank* if there is a homomorphism $\mathbb{Z}^2 \to A$ that is a proper map with respect to the discrete topology on $\mathbb{Z}^2$.

General conjectures regarding rigidity for invariant measures under higher rank abelian groups were made by Furstenberg (unpublished), Katok and Spatzier [KS2], and Margulis [M7]. The following is a variant of their conjectures:

CONJECTURE 4. *Let $\mathbb{G}$ be a linear algebraic group defined over $\mathbb{Q}$, and let S be a finite set of places for $\mathbb{Q}$ containing ∞. Let $\mathcal{O}_S = \mathbb{Z}[1/p : p \in S \setminus \infty]$ denote the ring of S-integers in $\mathbb{Q}$, $G = \prod_{v\in S} \mathbb{G}(\mathbb{Q}_v)$, and $\Gamma = \mathbb{G}(\mathcal{O}_S)$[4] diagonally embedded in G. Let $A < G$ be a closed subgroup consisting of elements of class-$\mathcal{A}$ of higher rank and let μ be an A-invariant and ergodic probability measure on G/Γ. Then either μ is virtually homogeneous or there is a $\mathbb{Q}$-subgroup $\mathbb{L} \leq \mathbb{G}$ and a proper normal $\mathbb{Q}$-subgroup $\mathbb{H} \lhd \mathbb{L}$ so that, if $H = \prod_{v\in S} \mathbb{H}(\mathbb{Q}_d)$ and $L = \prod_{v\in S} \mathbb{L}(\mathbb{Q}_d)$, then*

(1) $A \cap L$ has finite index in A,
(2) there is some $g \in G$ so that $\mu(g.[L]_\Gamma) > 0$ (with $[\bullet]_\Gamma$ denoting the image under the projection $G \to G/\Gamma$),[5] and
*(3) the image of $A \cap L$ in L/H is **not** of higher rank.*

Unlike the case of unipotent flows, where the classification of invariant measures and orbit closures go hand in hand and are very closely analogous, for diagonal flows the problem of classifying invariant measures seems, in general, better behaved than understanding orbit closures. This is somewhat surprising, as in the $\times a$, $\times b$ system considered by Furstenberg, with a and b multiplicatively independent integers, a complete orbit closure classification was obtained by Furstenberg already in 1967, whereas the measure classification question (without a positive entropy assumption) is Conjecture 3—a

---

[4]To be more precise: we fix a realization of $\mathbb{G}$ as a $\mathbb{Q}$-subgroup of $SL(d)$ for some d and set $\Gamma = \mathbb{G}(\mathbb{Q}) \cap SL(d, \mathcal{O}_S)$.

[5]We will also use the notation $[g]$ for $[g]_\Gamma$ when Γ is understood.

notoriously hard open problem. However, already in the slight generalization of considering $\times a$, $\times b$ for a and b multiplicatively independent (rational) integers on $\mathbb{C}/\mathbb{Z}[i]$, not much is known (see section 3.2 for more details).

While this is not immediately clear from the formulation, Conjecture 3 is essentially a special case of Conjecture 4. For simplicity, assume that a and b are distinct primes (the modification for general multiplicatively independent a and b is left to the imagination of the reader). Let $\mathbb{G} = \left\{ \begin{pmatrix} * & * \\ 0 & 1 \end{pmatrix} \right\}$—that is, the semidirect product of the multiplicative group $\mathbf{G}_m$ with the additive group $\mathbf{G}_a$—and take $S = \{\infty, a, b\}$. Let $G = \prod_{v \in S} \mathbb{G}(\mathbb{Q}_v)$, $\Gamma = \mathbb{G}(\mathbb{Z}[1/ab])$, and $A < \prod_{v \in S} \mathbf{G}_m(\mathbb{Q}_v) < G$ be the group $\{a^k b^l : k, l \in \mathbb{Z}\}$ diagonally embedded in $\prod_{v \in S} \mathbf{G}_m(\mathbb{Q}_v)$. For any $y \in \mathbb{R}^\times$ we have that

$$Y_y := \left\{ \left[\begin{pmatrix} y & x_\infty \\ 0 & 1 \end{pmatrix}, \begin{pmatrix} y_a & x_a \\ 0 & 1 \end{pmatrix}, \begin{pmatrix} y_b & x_b \\ 0 & 1 \end{pmatrix} \right]_\Gamma \;\middle|\; \begin{array}{l} y_v \in \mathbb{Z}_v^\times \text{ for } v = a, b \\ x_v \in \mathbb{Q}_v \text{ for } v = a, b, \infty \end{array} \right\}$$

is a compact A-invariant subset of G/Γ; hence any A-invariant and ergodic probability measures μ on G/Γ are supported on a single Y_y. Without loss of generality we can assume it is supported on Y_1. Let π be the projection of Y_1 to $X_{a,b} = \mathbb{R} \times \mathbb{Q}_a \times \mathbb{Q}_b/\mathbb{Z}[1/ab]$ given by

$$\left[\begin{pmatrix} 1 & x_\infty \\ 0 & 1 \end{pmatrix}, \begin{pmatrix} y_a & x_a \\ 0 & 1 \end{pmatrix}, \begin{pmatrix} y_b & x_b \\ 0 & 1 \end{pmatrix} \right]_\Gamma \mapsto [x_\infty, x_a, x_b]_{\mathbb{Z}[1/ab]}.$$

For any A-invariant and ergodic probability measure μ supported on Y_1, the measure $\pi_*\mu$ is a $\times a$, $\times b$-invariant and ergodic probability measure on $X_{a,b}$. Conversely, since the fibers of the map $\pi : Y_1 \to X_{a,b}$ are compact, any $\times a$, $\times b$-invariant and ergodic probability measures on $X_{a,b}$ can be lifted to an A-invariant and ergodic measure on Y_1.

In this survey we focus on *S-arithmetic quotients*: quotients of a finite index subgroup G of the $\mathbb{Q}_S$ points $\mathbb{G}(\mathbb{Q}_S)$ of a $\mathbb{Q}$-group $\mathbb{G}$ by a subgroup Γ commensurable to $\mathbb{G}(\mathcal{O}_S)$. By restriction of scalars, this implicitly also includes the case of algebraic groups defined over any number field, but because of issues related to those pointed out above for $\mathrm{SL}(3, \mathbb{C})/\mathrm{SL}(\mathbb{Z}[i])$, it is more convenient to work with the smaller field $\mathbb{Q}$.

DEFINITION 1.6. An S-arithmetic quotient G/Γ is *saturated by unipotents* if it has finite volume and the group generated by one-parameter unipotent subgroups of G acts ergodically on G/Γ (with respect to the Haar measure on G/Γ).

When working with real algebraic groups $\mathbb{G}(\mathbb{R})$, where $\mathbb{G}(\mathbb{C})$ is generated by unipotents (equivalently, the radical of $\mathbb{G}$ is equal to the unipotent radical of $\mathbb{G}$), a quotient G/Γ satisfies the saturated by unipotents property if and only if it is connected in the Hausdorff topology (cf. [M4, chapter 2]).

A very interesting and active direction we do not cover in this survey is actions on quotients of algebraic groups defined over global fields of positive characteristic. The key feature here is that there is no analogue to $\mathbb{Q}$: there is no minimal global field. This type of issue makes analyzing even the analogue of the $\times a$, $\times b$-system in positive characteristic quite intricate (cf. [KS4, construction 5.2] and [E]). For quotients of semisimple groups the situation is better, and a measure classification theorem for positive entropy measures analogous to what Einsiedler, Katok, and this author [EKL] proved for $\mathbb{Q}$ has been proved by Einsiedler, Mohammadi, and this author in [ELM]. However, even in this case, it is far from clear to which extent one should expect an analogue to Conjecture 4; in this context we mention the paper [ANL] by Adiceam, Nesharim, and Lunnon, where a very interesting example is constructed.

ACKNOWLEDGMENTS. This essay is dedicated with admiration to Gregory Margulis, whose deep and profound work has been, and continues to be, an inspiration to me ever since I started getting interested in homogeneous dynamics. Indeed, Margulis's deep work has been a big part of what drew me to the subject to begin with. On a more personal level, I would like to thank him for his kindness and generosity over the years.

Many of the results I describe in this work are joint with Manfred Einsiedler; it is a pleasure to express my gratitude to him for this collaboration; I would also like to thank him for comments on earlier versions of this survey. I also thank Ilya Khayutin for helpful comments and corrections. Finally, I would like to thank the editors of this volume for inviting me to contribute to it and for their patience.

2 Measure rigidity of higher rank diagonal actions

While Conjectures 3 and 4 are still wide open, significant progress was obtained regarding classifying invariant measures under a positive entropy condition. In subsections 2.4–2.6 we survey some applications of these results. Typically we are given a sequence of A-invariant probability measures μ_i on G/Γ and would like to understand what are the weak-* limit points of the sequence μ_i. Suppose μ is such a limit. A priori it seems very difficult to control any kind of ergodicity or mixing condition for the limiting measure. On the other hand, entropy is fairly well behaved with respect to weak-* limits. For example, for an A-invariant measure ν, let $h(\nu, a)$ denote the ergodic theoretic (also known as Komogorov-Sinai or metric) entropy of ν with respect to the action $[g]_\Gamma \mapsto a.[g]_\Gamma$. Then if G/Γ is compact, if $\mu_i \to \mu$ weak-*, then for any $a \in A$,

$$h(\mu, a) \geq \overline{\lim}_i\, h(\mu_i, a)$$

(cf., e.g., [EKL, section 9]). This actually also holds if G/Γ is not compact *assuming* μ is a probability measure.

Rudolph's theorem [R6], discussed above, regarding $S_{a,b} = \{a^k b^l : k, l \in \mathbb{N}\}$-invariant and ergodic measure μ on $\mathbb{T}$ has been a prototype for many subsequent theorems. We remark that a simple yet important lemma in Rudolph's proof implies that if $h(\mu, s) > 0$ for one $s \in S_{a,b}$, then $h(\mu, s) > 0$ for all $s \in S_{a,b}$. Katok and Spatzier ([KS2, KS3]; cf. also Kalinin and Katok [KK1]) pioneered the study following Rudolph of higher rank abelian actions by automorphisms on $\mathbb{T}^d$ and by translations on quotients G/Γ (similarly to what we have seen for the $\times a$, $\times b$ case, the former can be viewed as a special case of the latter, where the group $\mathbb{G}$ is a semidirect product of a torus and an abelian additive group). In some cases, Katok and Spatzier were able to obtain a full analogue of Rudolph's theorem, but in most cases (e.g., for $\mathbb{Z}^k$ actions on G/Γ with G semisimple) an additional ergodicity condition is needed, a condition that unfortunately is not stable under weak-* limits.

2.1 RIGIDITY OF JOININGS. Arguably the most complete result regarding the classification of higher rank abelian actions on arithmetic quotients does not explicitly mention entropy, though entropy plays an important role in the proof. In the same paper [F2] in which Furstenberg proved Theorem 1.1, thereby introducing higher rank rigidity from the dynamical perspective, Furstenberg also introduced joinings as a key tool in the study of measure preserving and topological dynamical systems. Suppose H is a topological group acting in a measure preserving way on two probability measures spaces (X, μ) and (X', μ'). Then H also acts on the product space $X \times X'$ by setting $h.(x, x') = (h.x, h.x')$. A *joining* of (X, μ, H) and (X', μ', H) is an H-invariant probability measure on $X \times X'$ that projects to the measure μ and μ' on X and X' respectively. There is always at least one joining between any two such actions—namely, the product measure $\mu \times \mu'$. Existence of other joinings can be interpreted as evidence of some communality between (X, μ, H) and (X', μ', H); an extreme form of this would be if these two measure preserving H-actions would be isomorphic (as H-actions; i.e., there is a measure preserving 1-1 and onto map ϕ between subsets of full measure of X and X' commuting with the H-action), in which case the pushforward under (id, ϕ) of μ would be a nontrivial joining supported on the graph of ϕ.

The following general joining classification theorem is the main result of [EL3] by Einsiedler and this author:

THEOREM 2.1 ([EL3]).
Let $r, d \geq 2$, and let $\mathbb{G}_1, \ldots, \mathbb{G}_r$ be semisimple algebraic groups defined over $\mathbb{Q}$ that are $\mathbb{Q}$ almost simple, $\mathbb{G} = \prod_{i=1}^r \mathbb{G}_i$, and S be a finite set of places of $\mathbb{Q}$. Let $X_i = \Gamma_i \backslash G_i$ be S-arithmetic quotients[6] saturated by unipotents for $G_i \leq \mathbb{G}_i(\mathbb{Q}_S)$, and let $X = \prod_i X_i$. Let $a_i : \mathbb{Z}^d \to G_i$ be proper homomorphisms so that $a = (a_1, \ldots, a_r) : \mathbb{Z}^d \to G = \prod_i G_i$ is of class-$\mathcal{A}$, and set $A = a(\mathbb{Z}^d)$. Suppose μ is an A-invariant and ergodic joining of the actions of $A_i = a_i(\mathbb{Z}^d)$ on X_i equipped with the Haar measure m_{X_i}. Then μ is homogeneous.

In fact, [EL3] gives slightly more precise information, in that μ is not just homogeneous but Haar measure on a finite index subgroup of the S-adic points of an algebraic group defined over $\mathbb{Q}$. Such a measure would be said to be *an algebraic measure defined over* $\mathbb{Q}$. This joining classification theorem can be extended to perfect groups. Recall that an algebraic group $\mathbf{G}$ is said to be perfect if $\mathbf{G} = [\mathbf{G}, \mathbf{G}]$.

THEOREM 2.2 ([EL3]).
Let $r, d \geq 2$, and let $\mathbb{G}_1, \ldots, \mathbb{G}_r$ be perfect algebraic groups defined over $\mathbb{Q}$, $\mathbb{G} = \prod_{i=1}^r \mathbb{G}_i$, and S be a finite set of places of $\mathbb{Q}$. Let $X_i = \Gamma_i \backslash G_i$ be S-arithmetic quotients for $G_i \leq \mathbb{G}_i(\mathbb{Q}_S)$ saturated by unipotents, and let $X = \prod_i X_i$. Let $a_i : \mathbb{Z}^d \to G_i$ be homomorphisms so that $a = (a_1, \ldots, a_r) : \mathbb{Z}^d \to G = \prod_i G_i$ is of class-$\mathcal{A}$, such that the projection of a_i to every $\mathbb{Q}$ almost simple factor of $\mathbb{G}_i(\mathbb{Q}_S)$ is proper. Suppose μ is an A-invariant and ergodic joining of the action of $A_i = a_i(\mathbb{Z}^d)$ on X_i equipped with the Haar measure m_{X_i}. Then μ is homogeneous, indeed an algebraic measure defined over $\mathbb{Q}$.

We remark that for the action of a one-parameter unipotent group on quotients of $SL(2, \mathbb{R})$ by lattices, Ratner established a joining classification theorem in [R1]. A general joining classification result for actions of unipotent

[6] In particular, by our definition of S-arithmetic quotients, G_i has finite index in $\mathbb{G}_i(\mathbb{Q}_S)$.

groups was given by Ratner in [R2] as a by-product of her techniques to classify all invariant measures.[7]

The restriction to perfect groups in Theorem 2.2 is important. If α is a (faithful) $\mathbb{Z}^k$-action on a torus $\mathbb{T}^d$ by automorphisms, or more generally a $\mathbb{Z}^k$-action on the solenoid $\mathbb{T}^d_S = \left(\prod_{v\in S}\mathbb{Q}_v\right)^d$ (a prime example of the latter being the action generated by the $\times a$ and $\times b$ maps on $\mathbb{T}_S$ with S containing ∞ as well as all prime factors of ab) for $k \geq 2$, then any hypothetical nonatomic $\alpha(\mathbb{Z}^d)$-invariant and ergodic invariant measure on $\mathbb{T}^d_S$ of zero entropy would give rise to a nontrivial, nonhomogeneous self-joining of $(\mathbb{T}^d_S, m_{\mathbb{T}^d_S}, \alpha(\mathbb{Z}^k))$ given by the pushforward of the measure $m_{\mathbb{T}^d_S} \times \mu$ using the map $(x, y) \mapsto (x, x + y)$ from $\mathbb{T}^{2d}_S \to \mathbb{T}^{2d}_S$. This simple example shows that classifying self-joinings of such $\mathbb{Z}^k$-actions is (at least) as hard as Conjecture 3. However, one can classify joinings between such $\mathbb{Z}^d$-actions up to zero entropy quotients [KK2, KS1, EL1].

2.2 SOME MEASURE CLASSIFICATION THEOREMS FOR S-ARITHMETIC QUOTIENTS. Joinings between higher rank abelian actions have positive entropy coming from the factors being homogeneous, but in fact being a joining imposes additional restrictions on leafwise measures that are very useful for the analysis. If one wants a measure classification of positive entropy measures, some additional conditions are needed.

One condition that gives rise to a clean statement is when the acting group is a maximal split torus or more generally satisfies the following condition:

DEFINITION 2.3. Let $\mathbb{G}$ be an algebraic group defined over $\mathbb{Q}$ and S a finite set of places. A subgroup $A < \mathbb{G}(\mathbb{Q}_S)$ will be said to be a *partially maximal* $\mathbb{Q}_S$*-split torus* if there is for each $s \in S$ a (possibly trivial) algebraic normal subgroup[8] $\mathbf{H}_s \lhd \mathbb{G}(\mathbb{Q}_s)$ so that $(A \cap \mathbf{H}_s)$ is a maximal $\mathbb{Q}_s$-split torus in $\mathbf{H}_s$ and $A = \prod_{s\in S'}(A \cap \mathbf{H}_s)$.

THEOREM 2.4 (Einsiedler and Lindenstrauss [EL2]).
Let $\mathbb{G}$ be a $\mathbb{Q}$ almost simple algebraic group, S a finite set of places, and G/Γ an S-arithmetic quotient for $\mathbb{G}$ saturated by unipotents in the sense of Definition 369.

---

[7] Indeed, the joining classification follows directly from the measure classification theorem of Ratner in [R3], but in [R2] (which is part of the sequence of papers establishing the results in [R3]), this result is already noted.

[8] To be precise, $\mathbf{H}_s$ is a group of $\mathbb{Q}_s$-points of a $\mathbb{Q}_s$ group, however (in contrast to the global field case), when considering an algebraic group over a local field will not make the distinction between the abstract algebraic group and the groups of points.

Let A be a higher rank, partially maximal $\mathbb{Q}_S$-split torus. Let μ be an A-invariant and ergodic measure on G/Γ, and assume that

(1) $\mu(g[\mathbb{L}(\mathbb{Q}_S)\cap G]_\Gamma)=0$ *for every proper reductive subgroup* $\mathbb{L}<\mathbb{G}$ *and* $g\in G$ *and*

(2) $h(\mu,a)>0$ *for some* $a\in A$.

Then μ is the uniform measure on G/Γ.

Using Theorem 2.4 a decomposition theorem can be proved for measures on *S*-arithmetic quotients corresponding to a semisimple $\mathbb{Q}$-group as a product of four pieces that may well be trivial:

THEOREM 2.5 (Einsiedler and Lindenstrauss [EL2]).
Let $\mathbb{G}$ be a semisimple algebraic group defined over $\mathbb{Q}$, S a finite set of places, G/Γ an S-arithmetic quotient for $\mathbb{G}$, and A a partially maximal $\mathbb{Q}_S$-split torus. Let μ be an A-invariant and ergodic measure on G/Γ. Then there is a finite index subgroup $A'<A$ and a probability measure μ' so that $\mu=\frac{1}{|A/A'|}\sum_{a\in A/A'}a\mu'$ and so that μ' can be decomposed as follows. For $i\in\{1,2,3\}$ there is a semisimple $\mathbb{Q}$-subgroup $\mathbb{L}_i\le\mathbb{G}$ and an anisotropic $\mathbb{Q}$-torus $\mathbb{L}_0<\mathbb{G}$ so that $\iota:(l_0,\dots,l_3)\mapsto l_0\cdot\ldots l_3$ gives a finite-to-one map $\prod_{i=0}^{3}\mathbb{L}_i(\mathbb{Q}_S)\to\mathbb{G}(\mathbb{Q}_S)$ so that $\mu'=\iota_*(\mu_0\times\cdots\times\mu_3)$ with each μ_i an $A'\cap\mathbb{L}_i(\mathbb{Q}_S)$-invariant and ergodic probability measure on $(\mathbb{L}_i(\mathbb{Q}_S)\cap G)/(\mathbb{L}_i(\mathbb{Q}_S)\cap\Gamma)$, $A'=\prod_{i=0}^{3}(A\cap\mathbb{L}_i(\mathbb{Q}_S))$ and

(1) μ_1 *is the uniform measure on* $L/(\mathbb{L}_1(\mathbb{Q}_S)\cap\Gamma)$ *with* $L\le\mathbb{L}_1(\mathbb{Q}_S)$ *a finite index subgroup,*

(2) μ_2 *satisfies that* $h(\mu_2,a)=0$ *for every* $a\in A\cap\mathbb{L}_2(\mathbb{Q}_S)$, *and*

(3) $\mathbb{L}_3(\mathbb{Q}_S)$ *is an almost direct product of* $\mathbb{Q}$ *almost simple groups* $\mathbb{L}_{3,i}(\mathbb{Q}_S)$ *so that for all i the group* $A\cap\mathbb{L}_{3,i}(\mathbb{Q}_S)$ *is not of higher rank.*

The special cases of Theorems 2.4 and 2.5 for G/Γ, a quotient of $\prod_{i=1}^{k}\mathrm{SL}(2,\mathbb{Q}_{v_i})$ by an irreducible lattice (with $v_i\in\{\text{primes or }\infty\}$) or $G/\Gamma=\mathrm{SL}(n,\mathbb{R})/\mathrm{SL}(n,\mathbb{Z})$, were proven earlier by this author [L2] and Einsiedler, Katok, and this author [EKL], respectively.

Ideally, one would like to obtain measure classification results for measures invariant under a higher rank diagonalizable group in the more general context of Conjecture 4. This is the subject of ongoing work; in particular, in joint work with Einsiedler we have the following:

THEOREM 2.6 (Einsiedler and Lindenstrauss [EL4]).
Let $\mathbb{G}$ be an algebraic group over $\mathbb{Q}$ that is $\mathbb{Q}$ almost simple and a form of SL_2^k or of PGL_2^k with $k \geq 1$, S a finite set of places, and G/Γ an S-arithmetic quotient for $\mathbb{G}$. Let $A < G$ be a closed abelian subgroup of class-$\mathcal{A}$ and of higher rank. Let μ be an A-invariant and ergodic probability measure on $X = \Gamma \backslash G$ such that $h_\mu(a) > 0$ for some $a \in A$. Then one of the following holds:

- **(Algebraic)** *The measure μ is homogeneous.*
- **(Solvable)** *The space X is noncompact. There exists a nontrivial unipotent subgroup L such that μ is invariant under L. The measure μ is supported on a compact A-invariant orbit $x_0 M \cong \Lambda_{M,x_0} \backslash M$, where $M < G$ is a solvable subgroup and $\Lambda_{M,x_0} = \{m \in M \mid m.x_0 = x_0\}$ is the stabilizer of x_0 in M. The lattice $\Lambda_{x_0,M}$ in M intersects the normal subgroup $L \lhd M$ in a uniform lattice, and if $\pi : M \to M/L$ denotes the natural projection map, then the image of μ under the induced map $\Lambda_{M,x_0} \backslash M \to \pi(\Lambda_{M,x_0}) \backslash (M/L)$ has zero entropy for the action of A.*

2.3 A RIGIDITY THEOREM FOR MEASURES INVARIANT UNDER A ONE-PARAMETER DIAGONAL GROUP WITH AN ADDITIONAL RECURRENCE ASSUMPTION. For the application of measure rigidity to quantum unique ergodicity, a variant of the above results was essential, where the assumption of invariance under a higher rank group was relaxed.

DEFINITION 2.7. Let H be a locally compact group acting on a standard Borel space $(X, \mathcal{B})$. We say that a measure μ on X is H-recurrent[9] if for every $B \subset X$ with $\mu(B) > 0$ and any compact subset $F \subset H$, then for μ-a.e. $x \in X$, there is an $h \in H \setminus F$ with $h.x \in B$.

We stress that no assumption is made regarding H-invariance of μ or even the measure class of μ.

THEOREM 2.8 (Lindenstrauss [L2]).
Let $G = \prod_{i=1}^r \mathrm{SL}(2, \mathbb{Q}_{v_i})$ with $v_i \in \{\infty, \text{primes}\}$ and $r \geq 2$, and let $\Gamma < G$ be an irreducible lattice. Let $A < \mathrm{SL}(2, \mathbb{Q}_{v_1})$ be a one-parameter diagonal group, with $a \in A$

[9] An alternative terminology often used in this context is H-conservative; we prefer H-recurrent as it seems to us more self-explanatory.

generating an unbounded subgroup of A, and $H = \prod_{i=2}^{r} \mathrm{SL}(2, \mathbb{Q}_{v_i})$. *Suppose that* μ *is A invariant and H-recurrent and that for a.e. A-ergodic component* μ_ξ *of* μ, *the entropy* $h(\mu_\xi, a) > 0$. *Then* μ *is the uniform measure on* G/Γ.

We note that using recurrence as a substitute for invariance under a higher rank group was motivated by Host's proof of Rudolph's theorem in [H2].

3 Orbit closures: Many questions, a few answers

3.1 PROLOGUE: ORBIT CLOSURES AND EQUIDISTRIBUTION FOR UNIPOTENT FLOWS. For unipotent flows, there is a very close relationship between behavior of individual orbits and the ergodic invariant measures. This correspondence was used by Ratner [R4] to prove the Raghunathan conjecture:

THEOREM 3.1 (Ratner [R4]).
Let U be a connected unipotent subgroup of real algebraic group G and $\Gamma < G$ *a lattice. Then for any* $x \in G/\Gamma$ *there is a closed subgroup* $U \leq L \leq G$ *so that* $\overline{U.x} = L.x$, *with L.x a periodic orbit (i.e.,* $\mathrm{stab}_L(x)$ *is a lattice in L). Moreover, U acts ergodically on* $L/\mathrm{stab}_L(x)$.

In particular, the orbit closure of every U-orbit $U.x$ is the support of a U-invariant and ergodic measure on G/Γ.

One key ingredient used to prove this surprisingly tight correspondence is a nondivergence estimate for unipotent flows developed by Dani and Margulis [M2, D2, DM3]. In addition to establishing nondivergence of the U-trajectory, needed in order to obtain from an orbit some limiting probability measure that can be analyzed, to deduce Raghunathan's Conjecture from the measure classification theorem, one needs to establish that a trajectory of a point $x \in G/\Gamma$ does not spend a lot of time close to a tube corresponding to shifts of a given periodic L orbit, unless x itself is in this tube. A flexible way to establish such estimates, known as the *Linearization Method*, was developed by Dani and Margulis [DM4]; while [DM4] uses Ratner's measure classification theorem, the technique itself was developed earlier by Dani and Margulis (with closely related works by Shah) in order to prove some cases of Raghunathan's conjecture by purely topological means (see, e.g., [DM2]); an alternative approach to linearization was used by Ratner in her proof of Raghunathan's Conjecture. We mention that a stronger (more) explicitly effective version of the Dani-Margulis Linearization Method was given recently by Margulis, Mohammadi, Shah, and this author [LMMS].

We also recall the following theorem of Mozes and Shah that relies on Ratner's measure classification theorem and the linearization method:

THEOREM 3.2 (Mozes and Shah [MS]).
Let G be a linear algebraic group over $\mathbb{R}$, $\Gamma < G$ a lattice, and let μ_i be a sequence of probability measures on G/Γ, and $u_t^{(i)}$ a sequence of one-parameter unipotent subgroups of G so that for every i the measure μ_i is $u_t^{(i)}$-invariant and ergodic. Suppose μ_i converges in the weak- topology to a probability measure μ. Then μ is homogeneous, and there are $g_i \to e$ so that $g_i \operatorname{supp}(\mu_i) \subset \operatorname{supp}(\mu)$ for i large enough.*

This theorem was extended to the *S*-arithmetic setting by Gorodnik and Oh [GO].

3.2 ORBIT CLOSURES FOR HIGHER RANK DIAGONALIZABLE GROUPS IN A TORUS.

Actions of one-parameter diagonal groups display no rigidity, and most questions about behavior of individual orbits for one-parameter diagonal groups seem to be hopelessly difficult. For instance, it is a well-known open problem whether $\sqrt[3]{2}$ (or indeed any other irrational algebraic number of degree ≥ 3) has a bounded continued fraction expansion, which is completely equivalent to the question whether the half-orbit

$$\left\{ \begin{pmatrix} e^t & \\ & e^{-t} \end{pmatrix} \left[\begin{pmatrix} 1 & \sqrt[3]{2} \\ 0 & 1 \end{pmatrix} \right] : t \geq 0 \right\}$$

is bounded in $G/\Gamma = \mathrm{SL}(2,\mathbb{R})/\mathrm{SL}(2,\mathbb{Z})$.

One could hope that the situation would be better for actions of higher rank diagonal groups, which do have some rigidity, and to a certain extent this is true. However, any hope of obtaining as good an understanding of orbits of higher rank diagonal groups as we have for unipotent flows is doomed to failure, in large part stemming from the fact that the connection between individual orbits and invariant measures for diagonal flows is much weaker.

To illustrate this point, consider first Furstenberg's Theorem 1.1. This theorem gives a complete classification of orbit closures for the action of $S_{a,b} = \{a^n b^m : n, m \in \mathbb{N}\}$ on $\mathbb{T} = \mathbb{R}/\mathbb{Z}$: either a finite orbit on which $S_{a,b}$ acts transitively, or $\mathbb{T}$. This was significantly extended by Berend, who showed the following:

THEOREM 3.3 (Berend [B1, B2]).
Let K be a number field, S a finite set of places including all infinite places, and $\mathcal{O}_S$ the ring of S integers—that is, the ring of $k \in K$ satisfying that $|k| \leq 1$ for any place $v \notin S$ of K. Let Σ a higher rank subgroup of $\mathcal{O}_S^*$. Assume that

(1) *no finite index subgroup of Σ is contained in a proper subfield of K and*
(2) *for every $v \in S$, there is some $a \in \Sigma$ with $|a|_v > 1$.*

Then any Σ-invariant closed subset of $X = \prod_{v \in S} K_v / \mathcal{O}_S$ is either finite or X itself.

For $K = \mathbb{Q}$ and $\Sigma \subset \mathbb{N}$ (not including 0!) this reduces easily to Furstenberg's theorem. The proofs of Furstenberg and Berend (as well as a simple proof of Furstenberg's theorem by Boshernitzan [B3]) are purely topological, but one can deduce Theorems 1.1 and 3.3 from Rudolph's theorem and its analogue to solenoids [EL1] by Einsiedler and this author respectively by establishing that any infinite closed Σ-invariant subset $Y \subset X = \prod_{v \in S} K_v / \mathcal{O}_S$ has to support a Σ-invariant measure of *positive entropy*. The reason this can be shown is that it is not hard to show if Y is such a closed, infinite, invariant subset $Y - Y = X$. This approach was used by Bourgain, Michel, Venkatesh, and this author to give a quantitative version of Furstenberg's theorem in [BLMV].

Both conditions in Theorem 3.3 are needed in order to ensure that any closed invariant subset is either finite or X. However, dropping assumption (2) does not dramatically change the situation: if there is some $v \in S$ so that for every $a \in \Sigma$ we have that $|a|_v = 1$, then there would certainly be other possible orbit closures—for example, a Σ-invariant subset of X supported on the K_v-orbit of the origin on which Σ acts by generalized rotations. However, with the minor necessary changes needed to accommodate such obvious examples of orbit closures, the above classification also holds without assumption (2). This was shown by Wang [W1] for the case of X being a torus (essentially, the same proof also works for the more general class of X considered in Theorem 3.3); we also mention that a very interesting combinatorial application for the case of $K = \mathbb{Q}(i)$ was given by Manners in [M1] (who gave an independent treatment of the relevant orbit closure classification theorem). We further note that the approach outlined in the previous paragraph to proving Theorem 3.3 using the measure classification result of Einsiedler and this author

[EL1] works just as well in the case where assumption (2) does not necessarily hold.[10]

If one instead weakens the conditions of Theorem 3.3 by eliminating the irreducibility assumption (1) (even keeping assumption (2)), we already enter the realm of conjectures, where surprisingly difficult questions loom. For instance, suppose that Σ is contained in a subfield $L < K$ with $[K:L] = 2$ but that no finite index subset of Σ is contained in a proper subfield of L. Suppose even that Σ is the full group of units of the S integers of L (or more precisely, the S_L units of L, with S_L the set of places of L corresponding to those in S) and that S consists only of all infinite places of K (so that X is a torus[11]). Then if the rank of Σ is ≥ 3, Wang and this author [LW1] proved that any orbit closure is (at most) a finite union of cosets of closed (additive) subgroups of X. Surprisingly, this statement is false for Σ of rank 2! We make, however, the following conjecture:

CONJECTURE 5. *Let K be a number field, S a finite set of places including all infinite places, and $\mathcal{O}_S$ the ring of S integers—that is, the ring of $k \in K$ satisfying that $|k| \leq 1$ for any place $v \notin S$ of K. Let $L < K$, S_L the set of valuations of L corresponding to places in S, and Σ a higher rank subgroup of the group of S_L units of L; assume moreover that no finite index subgroup of Σ is contained in a proper subfield of L. Let $X = \prod_{v \in S} K_v / \mathcal{O}_S$, and let $Y = \overline{\Sigma.x}$ for $x \in X$. Then either $Y = X$ or there exists a finite collection X_i of closed proper subgroups of X and torsion points $p_i \in X$ so that $Y \subset \Sigma.x \cup \bigcup_i (X_i + p_i)$.*

We remark that (at least when $[K:L] = 2$) one can give a complete classification of the *support* of Σ-ergodic and invariant measures, and (at least to us) it seems that the key difficulty in proving Conjecture 5 is the weak correspondence between invariant measures and individual orbits in the diagonalizable case, in sharp contrast to section 3.1.

Conjecture 5 is somewhat close in spirit to a recent result of Peterzil and Strachenko [PS] (which they extended later to nilmanifolds) that proves a similar structure for the image of a definable subset of $\mathbb{R}^d$ with respect to an o-minimal structure in $\mathbb{T}^d = \mathbb{R}^d / \mathbb{Z}^d$.

---

[10] The paper [EL1] gives a full treatment of a measure classification theorem assuming positive entropy for irreducible actions, which is the case relevant here, as well as announces results for more general cases with some hints regarding proofs.

[11] The assumption that Σ is the full group of units of the S integers of L is a significant assumption; the assumption that X is a torus—i.e., S consists only of all infinite places of K—can be removed.

3.3 ORBIT CLOSURES AND LIMITS OF PERIODIC MEASURES FOR ACTIONS OF HIGHER RANK DIAGONALIZABLE GROUPS ON QUOTIENTS OF SEMISIMPLE GROUPS. We already mentioned in the introduction the important conjecture of Cassels and Swinnerton-Dyer regarding orbit closures of the full diagonal group A in the homogeneous space $\mathrm{SL}(d,\mathbb{R})/\mathrm{SL}(d,\mathbb{Z})$.[12] For the convenience of the reader, we recall it here:

CONJECTURE 1b ([CSD, M5]). *Let $d \geq 3$. Any orbit of the diagonal group A in $\mathrm{PGL}(d,\mathbb{R})/\mathrm{PGL}(d,\mathbb{Z})$ is either unbounded or periodic.*

One would like to say at least conjecturally something stronger about the orbit closure of an orbit $A.x$ for $A.x$ nonperiodic. For instance, in the same paper Cassels and Swinnerton-Dyer give a conjecture that can be phrased as saying that any orbit of $\mathrm{SO}(2,1)$ on $\mathrm{SL}(3,\mathbb{R})/\mathrm{SL}(3,\mathbb{Z})$ is either periodic or unbounded, a conjecture that is a special case of Raghunathan's conjecture and was proved by Margulis in the mid-1980s [M6, M3]. As we saw in section 3.1 for $\mathrm{SO}(2,1)$, one actually has that any orbit is either closed or dense. But this is false for A-orbits. A trivial example is the orbit $A.[e]$ of the identity coset that is a divergent orbit. Slightly less trivial is the example of an A-orbit of a point

$$x \in \left[\begin{pmatrix} * & 0 & 0 \\ 0 & * & * \\ 0 & * & * \end{pmatrix}\right],$$

where essentially the action of A degenerates to a rank-1 action (one direction in A acts in a trivial way sending every point to the cusp). The following example, due to Shapira [S2], of elements in $\mathrm{SL}(3,\mathbb{R})/\mathrm{SL}(3,\mathbb{Z})$ shows even this is not the only obstacle to $A.x$ being dense: Consider for any $\alpha \in \mathbb{R}$ the point

$$p_\alpha = \left[\begin{pmatrix} 1 & 0 & \alpha \\ 0 & 1 & \alpha \\ 0 & 0 & 1 \end{pmatrix}\right].$$

The A-orbit of p_α is certainly not A-periodic, but

$$\overline{A.p_\alpha} \subset A.p_\alpha \cup \left[\begin{pmatrix} * & 0 & * \\ 0 & * & 0 \\ * & 0 & * \end{pmatrix}\right] \cup \left[\begin{pmatrix} * & 0 & 0 \\ 0 & * & * \\ 0 & * & * \end{pmatrix}\right];$$

12 We implicitly identify between $\mathrm{SL}(d,\mathbb{R})/\mathrm{SL}(d,\mathbb{Z})$ and $\mathrm{PGL}(d,\mathbb{R})/\mathrm{PGL}(d,\mathbb{Z})$; while the underlying algebraic groups are different, the quotients are isomorphic.

see [LS] for details. Note the analogy to the possible behavior allowed in Conjecture 5. Related examples of orbits of higher rank diagonal groups exhibiting this phenomena were given earlier by Macourant [M8], though not for a maximal diagonal group; another very interesting class of examples is investigated by Tomanov in [T3].

For $x = [g] \in \mathrm{SL}(d, \mathbb{R})/\mathrm{SL}(d, \mathbb{Z})$, set

$$\alpha_1(x) = \left(\inf_{\mathbf{n} \in \mathbb{Z}^d \smallsetminus 0} \|g\mathbf{n}\| \right)^{-1}.$$

The following conjecture seems to us plausible:

CONJECTURE 6. *Let $d \geq 3$, and let $x \in \mathrm{SL}(d, \mathbb{R})/\mathrm{SL}(d, \mathbb{Z})$ be such that*

$$\overline{\lim_{a \in A}} \frac{\log \alpha_1(a.x)}{\log \|a\|} = 0. \tag{3.1}$$

Then $\overline{A.x} = L.x$ for $A \leq L \leq \mathrm{SL}(d, \mathbb{Z})$, and moreover $L.x$ is a periodic L-orbit (i.e., has finite volume).

As explained to us by Breuillard and Nicolas de Saxce [BdS], the Strong Subspace Theorem of Schmidt [S1, section 6.3] implies that if $g \in \mathrm{SL}(d, \overline{\mathbb{Q}})$, then (3.1) holds for $[g]$ unless g is in a proper $\mathbb{Q}$-parabolic subgroup of $\mathrm{SL}(d, \mathbb{R})$. In particular we conjecture that if $g \in \mathrm{SL}(d, \overline{\mathbb{Q}})$, not contained in any proper $\mathbb{Q}$-parabolic subgroup of $\mathrm{SL}(d, \mathbb{R})$, then $\overline{A.[g]}$ is homogeneous.

Despite these difficulties, there are some positive results (not only conjectures) about orbit closures in this case. The first result in this direction is arguably Cassels and Swinnerton-Dyer's result from their farsighted paper [CSD] that we already mentioned. In this paper, Cassels and Swinnerton-Dyer prove that for the full diagonal group in $A < \mathrm{SL}(d, \mathbb{R})$, every A-orbit $A.x$ that is itself nonperiodic, but so that its closure $\overline{A.X}$ contains a periodic A-orbit, is unbounded. This allowed them to prove that Littlewood's conjecture (Conjecture 2) follows from Conjecture 1a.

Using Ratner's Orbit Closure Theorem, Barak Weiss and this author were able to strengthen this as follows:

THEOREM 3.4 (Weiss and Lindenstrauss [LW2]).
Let $A.x$ be an orbit of the full diagonal group A in $\mathrm{SL}(d, \mathbb{R})/\mathrm{SL}(d, \mathbb{Z})$, and suppose that $\overline{A.x} \supset A.x_0$ with $A.x_0$ periodic. Then $\overline{A.x}$ is a periodic orbit of some group L with $A \leq L \leq \mathrm{SL}(d, \mathbb{R})$.

An analogous result to Theorem 3.4 for $\mathrm{SL}(2,\mathbb{Q}_p)\times\mathrm{SL}(2,\mathbb{Q}_q)/\Gamma$ (for Γ, an irreducible lattice arising from a quaternion division algebra) was established earlier by Mozes [M9]; Mozes's work is completely self-contained. Theorem 3.4 was extended to inner forms of $\mathrm{SL}(d)$ (i.e., lattices arising from central simple algebras over $\mathbb{Q}$) by Tomanov in [T1].

We already noted that deciding, for example, if, for $\alpha=\sqrt[3]{2}$,

$$\left\{\begin{pmatrix} e^t & \\ & e^{-t}\end{pmatrix}\left[\begin{pmatrix} 1 & \alpha \\ 0 & 1\end{pmatrix}\right] : t\geq 0\right\}$$

is bounded in $G/\Gamma=\mathrm{SL}(2,\mathbb{R})/\mathrm{SL}(2,\mathbb{Z})$ is a notoriously difficult question. Indeed, despite the fact that it is conjectured that for any irrational algebraic number α of degree ≥ 3 the orbit above should be unbounded, not a single example of such an α is known. For higher rank (e.g., for $\mathrm{SL}(3,\mathbb{R})/\mathrm{SL}(3,\mathbb{Z})$), one can at least give examples of explicit A-orbits of algebraic points that are dense. Indeed, Shapira and this author show in [S2] and [LS] that if α,β are such that $1,\alpha,\beta$ span over $\mathbb{Q}$ a number field of degree 3 over $\mathbb{Q}$, then

$$\overline{A.\left[\begin{pmatrix} 1 & 0 & \alpha \\ 0 & 1 & \beta \\ 0 & 0 & 1\end{pmatrix}\right]}=\mathrm{SL}(3,\mathbb{R})/\mathrm{SL}(3,\mathbb{Z}).$$

This is related to another old result of Cassels and Swinnerton-Dyer, which showed in [CSD] that Littlewood's Conjecture (Conjecture 2) holds for such α,β.

The strongest result to date regarding Conjecture 1a for general points is due to Einsiedler, Katok, and this author [EKL], where using the classification of A-invariant measures of positive entropy on $\mathrm{SL}(d,\mathbb{R})/\mathrm{SL}(d,\mathbb{Z})$ that we obtained in that paper, it was shown that for $d\geq 3$,

$$\dim_H\left\{x\in\mathrm{SL}(d,\mathbb{R})/\mathrm{SL}(d,\mathbb{Z}) : A.x \text{ is bounded}\right\}=d-1,$$

which implies that transverse to the flow direction (i.e., to A), the set of x with a bounded A-orbit has zero Hausdorff dimension. Conjecturally, of course, this set is supposed to be a countable union of periodic A-orbits.

Finally, we mention that Tomanov and Weiss [TW] classified all closed A-orbits in $\mathrm{SL}(n,\mathbb{R})/\mathrm{SL}(n,\mathbb{Z})$ and more generally maximally split tori in arithmetic quotients, showing that an A-orbit $A.[g]$ is closed if and only if there is a

$\mathbb{R}$-split maximal $\mathbb{Q}$-torus $\mathbb{T} < \mathrm{SL}(n)$ so that $A.[g] = g[\mathbb{T}(\mathbb{R})]$. Their work builds on the result of Margulis classifying all divergent A-orbits in $\mathrm{SL}(n,\mathbb{R})/\mathrm{SL}(n,\mathbb{Z})$ (such orbits correspond to $\mathbb{Q}$-split maximal $\mathbb{Q}$-tori). This has the striking consequence that if F is a product of n linearly independent linear forms in n-variables, then $F(\mathbb{Z}^n)$ is discrete iff F is proportional to an integral form [T2].

We now turn our attention to the question whether an analogue to the theorem of Mozes and Shah (Theorem 3.2) holds for higher rank diagonal groups, where it seems the answer is mostly negative (but see section 4 for some significant positive results).

For example, there are explicit examples of sequences of A-periodic orbits $A.x_i$ in $\mathrm{SL}(d,\mathbb{R})/\mathrm{SL}(d,\mathbb{Z})$ for A, the $(d-1)$–dimensional diagonal group in $\mathrm{SL}(d,\mathbb{R})$, so that the corresponding measures $m_{A.x_i}$ on $\mathrm{SL}(d,\mathbb{R})/\mathrm{SL}(d,\mathbb{Z})$ have escape of mass: there is a $0 \le c < 1$ so that for any compact K for all $\epsilon > 0$ and all large enough i, we have that $\mu_{A.x_i}(K) < c + \epsilon$, and it is even possible to give such examples with $c = 0$. Examples of A-periodic trajectories with escape of mass were noted in [ELMV1] (following a suggestion by Sarnak), with more elaborate examples (in particular with $c = 0$) given by Shapira [S3] and David and Shapira [DS1]; implicitly, these examples feature already in old work of Cassels [C1]. Escape of mass can also occur for a sequence of periodic measures for unipotent groups or, more generally, a sequence of periodic measures that can arise as ergodic measures for unipotent groups, *but only if the support of these measures, in its entirety, escapes to infinity*—that is, if we denote the sequence of measures by μ_i, then for every compact set $K \subset G/\Gamma$ for every i large enough, $K \cap \operatorname{supp} \mu_i = \emptyset$. For the periodic A-orbits considered above, there is a fixed set $K \subset \mathrm{SL}(d,\mathbb{R})/\mathrm{SL}(d,\mathbb{Z})$ intersecting every one of them, indeed intersecting every A-orbit whether periodic or not.

Furthermore, assuming the equidistribution results of [ELMV2] hold in a quantitative way with polynomial error rates (which they surely should!), one can construct sequences of A-periodic orbits $A.x_i$ in $\mathrm{SL}(3,\mathbb{R})/\mathrm{SL}(3,\mathbb{Z})$ with volumes $\to \infty$, which converge weak-* to a probability measure that gives positive mass to periodic orbit $A.y$ distinct from all the $A.x_i$.

We end this section with a conjecture analogous to Conjecture 5.

CONJECTURE 7. *Let $\mathbb{G}$ be an algebraic group over $\mathbb{Q}$, S a finite set of places for $\mathbb{Q}$ containing ∞, and G/Γ a corresponding S-arithmetic quotient saturated by unipotents (see Definition 1.6). Let $A < G$ be a closed subgroup consisting of elements of class-$\mathcal{A}$ so that the projection of A to $\prod_{v\in S}(\mathbb{G}/\mathbb{H})(\mathbb{Q}_v)$ for any proper* *normal $\mathbb{Q}$-subgroup $\mathbb{H} \lhd \mathbb{G}$ is of higher rank. Then for any $x \in G/\Gamma$, either $A.x$ is dense in G/Γ or there are finitely many proper $\mathbb{Q}$-subgroups $\mathbb{L}_i < \mathbb{G}$ and $g_i \in G/\Gamma$ such that*

$$\overline{A.x} \subset A.x \cup \bigcup_i g_i[\mathbb{L}_i(\mathbb{R})].$$

4 Applications regarding integer points and $\mathbb{Q}$-tori

The study of integer points on varieties is arguably the most basic problem in number theory. It seems at first sight rather surprising that the rigidity results for diagonalizable groups listed above could be relevant for such a problem. Fortunately they are, and perhaps a good point to start the discussion of this topic is by going back to the remarkable work of Linnik on the distribution of integer solutions to ternary quadratic equations, work that is presented in his book with the apt title *Ergodic Properties of Number Fields* [L4], but in fact Linnik's farsighted work in this direction started even earlier, in the late 1930s.

4.1 LINNIK'S ERGODIC METHOD FOR STUDYING TERNARY QUADRATIC FORMS USING A ONE-PARAMETER DIAGONALIZABLE ACTION.

Linnik considered several related problems: local to global results regarding which integers can be represented by an integer quadratic form in three variables; the distribution of integer points on a two-dimensional sphere of radius $\sqrt{m}$ for $m \not\equiv 0, 4, 7 \bmod 8$; and the analogous problem regarding distribution of integer points on one- and two-sheeted hyperboloids in 3 space.

Consider in particular the distribution of integer points on the hyperboloid

$$V_d = \left\{(a, b, c) : b^2 - 4ac = d\right\},$$

where d is an integer. Let $V_d(\mathbb{Z})$ denote the integer points on V_d (these correspond to integral quadratic forms $ax^2 + bxy + cy^2$ of discriminant $d = b^2 - 4ac$), and let $V_d^*(\mathbb{Z}) \subseteq V_d(\mathbb{Z})$ be the set of primitive points (i.e., triplets (a, b, c) with no nontrivial common denominator). Note that for $V_d^*(\mathbb{Z})$ to be nonempty, d has to be $\equiv 0, 1 \pmod 4$. The discriminant d is said to be a fundamental discriminant if $V_d^*(\mathbb{Z}) = V_d(\mathbb{Z})$—that is, if either d is square-free and $\equiv 1 \pmod 4$ or $d = 4m$ with m square-free satisfying $m \equiv 2$ or $3 \pmod 4$. The action of GL(2) on binary quadratic forms gives us a natural action of $\mathrm{GL}(2, \mathbb{Z})$ on $V_d^*(\mathbb{Z})$ for every integer d. It is classical that $V_d^*(\mathbb{Z})$ consists of finitely many $\mathrm{GL}(2, \mathbb{Z})$ orbits; indeed it is one of Gauss's remarkable discoveries that for a given d one can define a natural commutative group law (which in this survey we denote by $\odot$) on $V_d^*(\mathbb{Z})/\operatorname{GL}(2,\mathbb{Z})$. One way to characterize this group law is that if $[q_1]\odot[q_2]=[q_3]$ with $q_i\in V_d^*(\mathbb{Z})$, there are bilinear integral forms α,β so that

$$q_1(n,m)q_2(l,s)=q_3(\alpha(n,m;l,s),\beta(n,m;l,s)) \tag{4.1}$$

(cf. [C3, section 1.3]). As the identity in this group we take the $\operatorname{GL}(2,\mathbb{Z})$ coset $[q_e]$ of $q_e=x^2-d'y^2$ if $d=4d'$ or $q_e=x^2+xy-(d-1)/4y^2$ if $d\equiv 1 \pmod 4$. For example, if $d=4d'$, then Equation (4.1) applied to the triple $[q_e]\odot[q_e]=[q_e]$ is given explicitly by

$$(n^2-d'm^2)(l^2-d's^2)=(nl+d'ms)^2-d'(ns+ml)^2.$$

The following natural problem is a special case of an important class of counting questions raised by Linnik:

QUESTION 4.1. Let $d_i\to+\infty$. Let $\tilde{V}_{d_i}$ be the sets $d_i^{-1/2}V_{d_i}^*(\mathbb{Z})\subset V_1$. How are the points in these sets distributed? Let m_1 denote the unique (up to scalar) $\operatorname{SL}(2,\mathbb{R})$-invariant measure on V_1. Do the points in $\tilde{V}_{d_i}$ equidistribute in the sense that for any nice subsets $E_1,E_2\subset V_1$ (e.g., bounded open sets with m_1-null boundary),

$$\frac{\#(\tilde{V}_{d_i}\cap E_1)}{\#(\tilde{V}_{d_i}\cap E_2)}\to\frac{m_1(E_1)}{m_1(E_2)}?$$

Similarly, let $d_i\to-\infty$, and let $\tilde{V}_{d_i}=|d_i|^{-1/2}V_{d_i}^*(\mathbb{Z})\subset V_{-1}$. Do the points in $\tilde{V}_{d_i}$ become equidistributed in V_{-1}?

The answer to the question is *yes* for any $d_i\to\infty$. This was proved by Duke [D3] (building on work of Iwaniec [I]), at least when d_i is a sequence of fundamental discriminants (which is, as implied by the name, the most fundamental [and hardest] case). Duke's proof is quantitative and relies on estimates of Fourier coefficients of half integral weight Maass forms. Under an additional congruence condition on the sequence d_i—namely, that there is some fixed prime p so that d_i are quadratic residues mod p for all i (i.e., $(\frac{d_i}{p})=1$)—this equidistribution result was proved much earlier by Linnik and Skubenko [L4, S4]. In fact if $d_i\to+\infty$, a variant of Linnik's argument can be used to establish equidistribution with no (additional) side condition, as was shown by Einsiedler, Michel, Venkatesh, and this author in [ELMV3].

Set $G = \mathrm{PGL}(2,\mathbb{R})$, $A < G$ the group of diagonal matrices, and $\Gamma = \mathrm{PGL}(2,\mathbb{Z})$. Note that A is the stabilizer[13] of the quadratic form $q = xy$, and using this we can view V_1 as $A\backslash \mathrm{PGL}(2,\mathbb{R})$. The above equidistribution question regarding $\tilde{V}_{d_i}$, $d_i > 0$, which we recall is a finite union of Γ-orbits in $V_1 \cong A\backslash G$, can be recast as a question regarding equidistribution of finite collections of closed A-orbits in G/Γ as follows. Consider a quadratic form $q(x,y) = ax^2 + bxy + cy^2 = a(x - \xi_1 y)(x - \xi_2 y)$ with $\xi_{1,2} = -b \pm \sqrt{d}/2a$. We associate to q the A-orbit $A.p_{a,b,c}$ in G, where

$$p_{a,b,c} := \begin{pmatrix} a & \frac{-b+\sqrt{d}}{2} \\ a & \frac{-b-\sqrt{d}}{2} \end{pmatrix}$$

and where as before A is the (one-parameter) diagonal subgroup of G. Clearly if $\gamma \in \Gamma$ the A-orbit corresponding to $q \circ \gamma$ will be $A.p_{a,b,c}\gamma$, and vice versa: if $A.p_{a,b,c} = A.p_{a',b',c'}\gamma$ for $(a,b,c), (a',b',c') \in V_d^*(\mathbb{Z})$ and $\gamma \in \Gamma$, then the corresponding quadratic forms q, q' satisfy that $q = q' \circ \gamma$.

It follows that to each $\mathrm{PGL}(2,\mathbb{Z})$-orbit in $V_d(\mathbb{Z})$ there corresponds an A-orbit in G/Γ. A standard duality argument can be used to show (at least for d_i that are not perfect squares, though the case of d_i perfect squares can also be handled this way) that equidistribution of the sequence of sets $\tilde{V}_{d_i}$ in $V_1 \cong A\backslash G$ in the sense of Question 4.1 is equivalent to the equidistribution of the sequence of collections of closed A-orbits

$$\mathcal{T}_{d_i} = \left\{ A.p_{a,b,c} : (a,b,c) \in V_{d_i}^*(\mathbb{Z}) \right\}$$

in G/Γ—that is, for every $f, g \in C_0(G/\Gamma)$ with $\int g \, dm \neq 0$,

$$\frac{\sum_{A.p \in \mathcal{T}_{d_i}} \int_{A.p} f}{\sum_{A.p \in \mathcal{T}_{d_i}} \int_{A.p} g} \to \frac{\int_{G/\Gamma} f \, dm}{\int_{G/\Gamma} g \, dm}. \tag{4.2}$$

Note that we present the equidistribution in the above form to allow for sequences d_i that contain perfect squares, as in that case the A-orbits $A.p_{a,b,c}$ are divergent. For d_i a sequence avoiding perfect squares, one can take $g \equiv 1$ instead.

[13] Technically we are being slightly imprecise here, as $\mathrm{PGL}(2,\mathbb{R})$ acts only on proportionality classes of quadratic forms.

We give two more ways to look at the points $V_d^*(\mathbb{Z})/\operatorname{GL}(2,\mathbb{Z})$, which will be important for us later.

I. The cosets $\{[p_{a,b,c}]_\Gamma : (a,b,c) \in V_d^*(\mathbb{Z})\}$ correspond in an obvious way to the $\mathbb{Z}$-modules $\bar{p}_{a,b,c}$ spanned by a and $-b+\sqrt{d}/2$, which (for d not a perfect square) are in fact ideals in the order $\mathcal{O}_d$ of discriminant d (for fundamental discriminants, $\mathcal{O}_d$ is the ring of integers in $\mathbb{Q}(\sqrt{d})$; if $d = f^2 d'$ with d' fundamental, $\mathcal{O}_d$ is a subring of $\mathcal{O}_{d'}$ containing the identity of index f in $\mathcal{O}_{d'}$). For $(a,b,c), (a',b',c') \in V_d^*(\mathbb{Z})$, we have that $[p_{a',b',c'}]_\Gamma \in A.[p_{a,b,c}]_\Gamma$ if and only if the ideals $\bar{p}_{a',b',c'}$ and $\bar{p}_{a,b,c}$ are in the same ideal class—that is, for some $k \in \mathbb{Q}(\sqrt{d})$, we have that $(k\mathcal{O}_d) \cdot \bar{p}_{a',b',c'} = \bar{p}_{a,b,c}$. An observation that can be attributed essentially to Dirichlet is that the Gauss composition law on $\operatorname{GL}(2,\mathbb{Z})$ cosets in $V_d^*(\mathbb{Z})$ is the same group law as the group law in the ideal class group $\operatorname{cl}(\mathcal{O}_d)$ of the order $\mathcal{O}_d$.

II. For any positive $d \in \mathbb{Z}$ and $(a,b,c) \in V_d^*(\mathbb{Z})$, the group $p_{a,b,c}^{-1} A p_{a,b,c}$ is the group of $\mathbb{R}$-points of a $\mathbb{Q}$-torus $\mathbb{T}_{a,b,c} < \operatorname{PGL}(2)$. Moreover, it is not hard to see that $\mathbb{T}_{a,b,c}$ is $\mathbb{Q}$-split iff d is a perfect square. It follows that the orbit $A.[p_{a,b,c}]$ is a closed A-orbit in G/Γ, and this A-orbit $A.[p_{a,b,c}]$ is a periodic A-orbit if d is not a perfect square and divergent otherwise. The tori $\mathbb{T}_{a,b,c}$ are conjugate to each other in $\operatorname{PGL}(2)$ over $\mathbb{Q}$ but not over $\mathbb{Z}$: for $(a,b,c), (a',b',c') \in V_d^*(\mathbb{Z})$, the tori $\mathbb{T}_{a,b,c}$ and $\mathbb{T}_{a',b',c'}$ are conjugate to each other over $\mathbb{Z}$ if and only if $p_{a',b',c'} \in p_{a,b,c}\Gamma$.

These collections $\mathcal{T}_d$ of A-orbits in G/Γ can be described very succinctly in the language of the adeles. Let $\mathbb{A}$ denote the adele ring of $\mathbb{Q}$, and let

$$\pi : \operatorname{PGL}(2,\mathbb{A})/\operatorname{PGL}(2,\mathbb{Q}) \to \operatorname{PGL}(2,\mathbb{R})/\operatorname{PGL}(2,\mathbb{Z}) = G/\Gamma$$

be the natural projection, which takes the coset $[(g, g_2, g_3, \dots)]_{\operatorname{PGL}(2,\mathbb{Q})}$ of an element $(g, g_2, g_3, \dots) \in \operatorname{PGL}(2,\mathbb{A})$ with $g \in \operatorname{PGL}(2,\mathbb{R})$ and $g_p \in \operatorname{PGL}(2,\mathbb{Z}_p)$ for every (finite) prime p to the coset $[g]_\Gamma$ in G/Γ. Then for any $(a,b,c) \in V_d^*(\mathbb{Z})$,

$$\mathcal{T}_d = p_{a,b,c}\, \pi\left([\mathbb{T}_{a,b,c}(\mathbb{A})]\right);$$

since this is valid for any choice of (a,b,c), we may as well take the explicit choice of $(a,b,c) = (1, 0, -d/4)$ (for $4|d$) or $(a,b,c) = (1, 1, -(d-1)/4)$ (for $d \equiv 1 \pmod 4$).

Up to minor changes—replacing A with the compact group $K = \left\{ \begin{pmatrix} \cos\theta & \sin\theta \\ -\sin\theta & \cos\theta \end{pmatrix} \right\}$ and taking $p_{a,b,c}$ to be $\begin{pmatrix} a & -b/2 \\ 0 & \sqrt{|d|}/2 \end{pmatrix}$ — the correspondences described in I and II also hold for d negative. Of course, A and K are quite different $\mathbb{R}$-groups, with K being $\mathbb{R}$-anisotropic and compact while A is $\mathbb{R}$-split.

For the remainder of this subsection we restrict our attention to $d_i > 0$ not perfect squares (i.e., $\mathbb{T}_{a,b,c}$ $\mathbb{Q}$-anisotropic); for a discussion of the isotropic case, see [OS], [DS2], and [SZ]. As explained in [ELMV3], in modern terminology Linnik's approach can be interpreted as the following three-step strategy:

A. Let μ_i be the probability measure given by

$$\mu_i(f) = \frac{\sum_{A.p \in \mathcal{T}_{d_i}} \int_{A.p} f}{\sum_{A.p \in \mathcal{T}_{d_i}} \int_{A.p} 1}.$$

One needs to establish that this sequence of measures is *tight*—that is, for every $\delta > 0$ there is a compact $X_\delta \subset G/\Gamma$ so that $\mu_i(X_\delta) > 1 - \delta$ for all i. Linnik establishes this via analytic number theory, in a way that is closely related to a key step in [ELMV2], which we discuss below, but this can also be established using purely ergodic theoretic means (cf. [ELMV3]).

B. Linnik proves an upper bound on the measure of small tubes transverse to the A action of radius $r \geq d_i^{1/4}$, on average

$$\int_{X_\delta} \mu_i(B(r,1,x))\, d\mu_i(x) \ll_{\delta,\epsilon} r^2 d_i^{\epsilon} \tag{4.3}$$

for $\epsilon, \delta > 0$, X_δ, as above, and

$$B(r,1,x') = \left\{ \begin{pmatrix} 1 & s \\ 0 & 1 \end{pmatrix} \begin{pmatrix} e^t & \\ & e^{-t} \end{pmatrix} \begin{pmatrix} 1 & 0 \\ s' & 1 \end{pmatrix} .x' : |s|, |s'| < r, |t| < 1 \right\}.$$

An important point here is that the d_i^{ϵ} term in the right-hand side implies that this estimate is meaningful only for $r > d_i^{-c}$; hence, for each i we obtain information regarding the distribution of μ_i at a different scale.

The estimate of Equation (4.3), which Linnik called the Basic Lemma, is key to the whole approach. Note that the exponent 2 in the right inside of (4.3) is *sharp*. It is a deep bound that relies on results of Siegel and Venkov on quadratic forms and is closely related to the Siegel Mass Formula (see [ELMV3, appendix A] for a self-contained treatment).

C. Now somehow one needs to upgrade the sharp nonconcentration (on average) estimate of (4.3) to an equidistribution statement: to both lower and upper bounds on the measure of fixed-sized subsets of G/Γ. One way to proceed, explained in [ELMV3], is as follows: By passing to a subsequence if necessary, and in view of the tightness of the sequence of measures μ_i discussed in step A, we can assume that μ_i converges to some A-invariant measure μ as $i \to \infty$ in the weak-* topology. The action of A on G/Γ is a prime example of a rank-1 diagonalizable group actions, one which does *not* satisfy the type of rigidity provided by Theorem 2.5 of the other measure classification theorems discussed in section 2.2. Because there are so many A-invariant measures in G/Γ, it is in general very hard to prove that a limiting measure obtained from a number theoretic construction will be the uniform measure (see the discussion at the beginning of section 3.2). However, the fact that the estimate of (4.3) is *sharp* rescues us, as (4.3) together with the subadditivity of entropy allows us to deduce that the entropy of μ with respect to the action of $a_t \in A$ is maximal and that on G/Γ there is a *unique measure of maximal entropy*.

Linnik and Skubenko did not quite follow the method outlined in step C. To begin with, they considered the dynamics not for the diagonal subgroup in $\mathrm{PGL}(2, \mathbb{R})$ but for a diagonal group over $\mathbb{Q}_p$: Linnik and Skubenko assumed that for some fixed prime p, the sequence d_i satisfied $\left(\frac{d_i}{p}\right) = 1$, which implies that the measures μ_i (as well as any limiting measure μ) can be lifted to probability measures on $\mathrm{PGL}(2, \mathbb{R}) \times \mathrm{PGL}(2, \mathbb{Q}_p)/\mathrm{PGL}(\mathbb{Z}[1/p])$ that are invariant under the diagonal subgroup of $\mathrm{PGL}(2, \mathbb{Q}_p)$. This p-adic dynamics is symbolic in nature, which facilitated the analysis. Moreover they did not first pass to the limit, which allowed them to give rates of equidistribution (even if logarithmic rather than polynomial, as in the work of Duke). A third alternative[14] to step C using property τ (i.e., using spectral gaps) was given by Ellenberg, Michel, and Venkatesh in [EMV2], which gives another readable, modern interpretation of the Linnik method and in addition raises an important joining question to which they are able to give a partial answer (see below).

---

[14]Some may say this is more of a development of Linnik's original method.

Linnik's method is not limited to the discriminant form $b^2 - 4ac$ but is applicable to any integral ternary quadratic form, in particular to the form $a^2 + b^2 + c^2$—that is, to the distribution of integer points on the sphere. Both $b^2 - 4ac$ and $a^2 + b^2 + c^2$ are quadratic forms with the nice property that any other integral form that is equivalent to them over $\mathbb{R}$ and $\mathbb{Z}_p$ for all p is in fact equivalent to them over $\mathbb{Z}$. The collection of all integral quadratic forms equivalent over $\mathbb{R}$ and $\mathbb{Z}_p$ for all p to a given quadratic form is called the *genus* of the quadratic form.[15] For general integral quadratic forms, one needs to study all forms in the genus in order to prove equidistribution of integer points on each of the corresponding quadratic surfaces. The form $a^2 + b^2 + c^2$ is treated explicitly by Linnik in his book [L4] and earlier works and is also the case explained in [EMV2]; the case of general quadratic forms is discussed, for example, in Linnik's paper [L3]. See [W2] for a nice modern exposition by Wieser.

4.2 GOING BEYOND LINNIK: JOINT EQUIDISTRIBUTION USING RIGIDITY OF JOININGS FOR HIGHER-RANK DIAGONALIZABLE ACTIONS. In the previous section we considered how points on the one- or two-sheeted hyperbolic

$$V_d(\mathbb{Z}) = \left\{(a,b,c) : b^2 - 4ac = d\right\}$$

project onto the unit one- or two-sheeted hyperboloid V_1 or V_{-1}, respectively depending on the sign of d; similarly, regarding projection of points on the sphere of radius $\sqrt{-d}$ (for notational convenience, we will use negative integers to parameterize spheres),

$$S_d(\mathbb{Z}) = \left\{(a,b,c :\in \mathbb{Z}^3 : a^2 + b^2 + c^2 = -d\right\}$$

projects to the unit sphere $\mathbb{S}$. As explained in I and II in section 4.1, for the special case of the one-sheeted hyperboloid (i.e., $V_d(\mathbb{Z})$ for $d > 0$; see also the paragraph immediately afterward regarding the modification for $d < 0$), these distribution problems regarding the integer points $V_d(\mathbb{Z})$ and $S_d(\mathbb{Z})$ can be interpreted in terms of the ideal class group of $\mathbb{Q}(\sqrt{d})$ or translated into questions regarding the distribution of suitable translates of the adelic points of $\mathbb{Q}$-tori $\mathbb{T}_d(\mathbb{A})$ in $\mathbb{G}(\mathbb{A})/\mathbb{G}(\mathbb{Q})$ for $\mathbb{G}$ being the $\mathbb{Q}$-group PGL(2) or SO(3) in the hyperboloid and sphere cases respectively.

---

[15] A somewhat anachronistic terminology, as this genus has nothing to do with the genus of any surface.

We will be mainly interested in the harder case of $\mathbb{T}_d$ $\mathbb{Q}$-anisotropic—that is, d not a perfect square—in which case $\mathbb{T}_d$ will split over the quadratic extension $\mathbb{Q}(\sqrt{d})$ of $\mathbb{Q}$. In particular, the $\mathbb{Q}$-torus $\mathbb{T}_d$ will be split over $\mathbb{R}$ for $d > 0$—that is, for one-sheeted hyperboloids—and over $\mathbb{Q}_p$ iff $\left(\frac{d}{p}\right) = 1$. If one wants to follow a scheme as in A–C to prove equidistribution using dynamical ideas—particularly if one follows C to construct a limiting measure μ out of a sequence $d_i \to \pm\infty$ and use dynamics to study this limiting measure—one needs to assume either that $d_i > 0$ for all i or that there is a fixed prime p so that $\left(\frac{d_i}{p}\right) = 1$ for all i.

We can strengthen our assumptions and require *two* places $v, w \in \{\infty, \text{primes}\}$ at which the tori $\mathbb{T}_{d_i}$ splits—that is, $\left(\frac{d_i}{v}\right) = \left(\frac{d_i}{w}\right) = 1$ for all i; to allow also the case of v or $w = \infty$, we define $\left(\frac{d}{\infty}\right) = \operatorname{sign}(d)$. If we make this assumption we will obtain a limiting measure μ on $\mathbb{G}(\mathbb{A})/\mathbb{G}(\mathbb{Q})$ (or, if we prefer, on an S-arithmetic quotient of $\mathbb{G}$ for any S containing v, w, ∞) that is invariant under a *higher rank* diagonal group, on which we can try to apply the measure rigidity theorems presented in section 2, and in particular the joining classification theorem, Theorem 2.1.

We now describe an arithmetic consequence of the rigidity of higher rank diagonal groups obtained in this way by Aka, Einsiedler, and Shapira [AES2]. Let d be a negative integer. By the Three Squares Theorem of Legendre and Gauss, $S_d(\mathbb{Z})$ is nonempty, iff $d \not\equiv 1 \bmod 8$ (recall that in our conventions, d is negative!). Consider for any integer vector $\mathbf{n} \in S_{d_i}(\mathbb{Z})$ the lattice in the plane orthogonal to $\mathbf{n}$ (with respect to the standard inner product on $\mathbb{R}^3$),

$$\Lambda_{\mathbf{n}} = \left\{ \mathbf{x} \in \mathbb{Z}^3 : \mathbf{x} \cdot \mathbf{n} = 0 \right\}.$$

Let $\mathbf{v}_1, \mathbf{v}_2$ be generators of $\Lambda_{\mathbf{n}}$ (considered as an additive group); then $\mathbf{v}_1, \mathbf{v}_2$ give rise to a positive definite binary quadratic form $q(x, y) = \left\| x\mathbf{v}_1 + y\mathbf{v}_2 \right\|^2$. The integer quadratic form q will have (negative) discriminant $4d$ (we leave this as an exercise for the reader), and given $\mathbf{n}$ the form q is well defined up to the action of $\mathrm{GL}(2, \mathbb{Z})$ on $V_{4d}(\mathbb{Z})$. To be slightly more precise, $\mathbf{n}$ gives an orientation on the plane $\mathbf{n}^\perp$, so if we chose $\mathbf{v}_1, \mathbf{v}_2$ to be a basis with positive orientation, the form q is well defined up to the action of $\mathrm{SL}(2, \mathbb{Z})$. Thus we get a map $\alpha : S_d(\mathbb{Z}) \to V_{4d}(\mathbb{Z})/\mathrm{SL}(2, \mathbb{Z})$. This map is neither injective nor onto, but it is close to being both: up to a bounded integer factor, both the kernel and co-kernel of this map is of size 2^{r-1}, where r is the number of distinct primes dividing d (this is less obvious but was understood already by Gauss). Let m_Y denote the $\mathrm{SL}(2, \mathbb{R})$-invariant measure on $Y = V_{-1}/\mathrm{SL}(2, \mathbb{Z})$ and $m_{\mathbb{S}}$ the uniform measure on the unit sphere $\mathbb{S}$ (both normalized to be probability measures). We remark that Y can naturally be identified with $\mathbb{H}/\mathrm{PSL}(2,\mathbb{Z})$, with $\mathbb{H}$ the hyperbolic plane. The map α is close enough to being 1-1 that, for example, if one assumes that $d_i \to -\infty$ with $d_i \not\equiv 1 \bmod 8$ square-free and $\left(\frac{d_i}{p}\right) = 1$ for some fixed prime p, then it follows—for example, using Linnik's methods—that the projections of $\{\alpha(\mathbf{n}) : \mathbf{n} \in S_{d_i}(\mathbb{Z})\}$ to Y become equidistributed with respect to m_Y. Recall also that Linnik showed under these conditions that the collection of points $S_{d_i}(\mathbb{Z})$ projected to the unit sphere $\mathbb{S}$ becomes equidistributed with respect to $m_{\mathbb{S}}$. Using Theorem 2.1, Aka, Einsiedler, and Shapira were able to upgrade these two statements to a joint equidistribution statement:

THEOREM 4.2 (Aka, Einsiedler, and Shapira [AES2]).
Let p, q be two distinct finite primes and $d_i \to -\infty$ a sequence of square-free negative integers $\not\equiv 1 \bmod 8$ so that $\left(\frac{d_i}{p}\right) = \left(\frac{d_i}{q}\right) = 1$ for all i. Then the projection of the sets

$$\{(\mathbf{n}, \alpha(\mathbf{n})) : \mathbf{n} \in S_{d_i}\} \subset S_{d_i}(\mathbb{Z}) \times (V_{4d_i}(\mathbb{Z})/\mathrm{SL}(2,\mathbb{Z}))$$

to $\mathbb{S} \times Y$ becomes equidistributed with respect to the measure $m_{\mathbb{S}} \times m_Y$ on this space as $i \to \infty$.

In other words, for any nice subsets $E \subset \mathbb{S}$ and $F \subset Y$

$$\frac{\#\left\{\mathbf{n} \in S_{d_i}(\mathbb{Z}) : |d_i|^{-1/2}\,\mathbf{n} \in E \text{ and } |4d_i|^{-1/2}\,\alpha(\mathbf{N}) \in F\right\}}{\#S_{d_i}(\mathbb{Z})} \to m_{\mathbb{S}}(E) \cdot m_Y(F).$$

Unlike the individual equidistribution on $\mathbb{S}$ and $V_{-1}/\mathrm{SL}(2,\mathbb{Z})$, which can also be proved using analytic number theoretic tools (indeed, the analytic tools give significantly sharper results), there does not seem to be a plausible approach using currently available technology to proving this joint equidistribution statement using the techniques of analytic number theory or automorphic forms. For more information in this direction we refer the reader to [AES1, Appendix by Ruixiang Zhang[16]].

Theorem 4.2 turns out to be closely related to the following equidistribution result stated in terms of the class group:

---

[16] [AES1] is the arXiv version of [AES2].

THEOREM 4.3.
Let $d_i \to -\infty$ be a sequence of negative integers $\equiv 0$ or $1 \bmod 4$ so that there are two primes p, q for which $\left(\frac{d_i}{p}\right) = \left(\frac{d_i}{q}\right) = 1$ for all i. Identifying as before elements of $V_{d_i}^*(\mathbb{Z})$ with primitive integral quadratic forms, we fix for every i an arbitrary integral form $q_i \in V_{d_i}^*(\mathbb{Z})$ and define a collection of points $\tilde{V}_{d_i}^{(2)} \subset (V_{d_i}(\mathbb{Z})/\operatorname{GL}(2,\mathbb{Z}))^2$ by

$$\tag{4.4} \tilde{V}_{d_i}^{(2),q_i} = \left\{([q], [q_i] \odot [q] \odot [q]) : q \in V_{d_i}^*(\mathbb{Z})\right\}.$$

Then the projection of these collections to Y^2 becomes equidistributed with respect to $m_Y \times m_Y$ as $i \to \infty$.

Note: We restrict ourselves to $d_i < 0$ for purely aesthetic reasons, as in this case the relation of the equidistribution statement to integer points is cleanest. In fact, taking $d_i \to +\infty$ is even better, as then only one additional split place is needed—that is, one need only assume the existence of one prime p for which $\left(\frac{d_i}{p}\right) = 1$ for all i.

Sketch of proof. For simplicity of notations, assume $4|d_i$ for all i (the modification to $d_i \equiv 1 \pmod 4$ poses no additional difficulties). Let $q_e = x^2 - \frac{d_i}{4}y^2$, and let $\mathbb{T}_i$ be the corresponding adelic torus as in paragraph II in section 4.1 (adapted for $d < 0$). Concretely, we can take $\mathbb{T}_i$ to be the stabilizer of the proportionality class of q_e in $\operatorname{PGL}(2,\mathbb{R})$. The $\mathbb{Q}$-torus $\mathbb{T}_i$ is anisotropic over $\mathbb{Q}$ and even over $\mathbb{R}$ but by our assumption on $\left(\frac{d_i}{p}\right)$ and $\left(\frac{d_i}{q}\right)$ will be split over $\mathbb{Q}_p$ and $\mathbb{Q}_q$. Let $G = \operatorname{PGL}(2,\mathbb{R})$, $\Gamma = \operatorname{PGL}(2,\mathbb{Z})$, K be the maximal compact subgroup of $\operatorname{PGL}(2,\mathbb{Z})$ as in paragraph II, $S = \{\infty, p, q\}$, $G_S = \prod_{v \in S} \operatorname{PGL}(\mathbb{Q}_v)$, and $\Gamma_S = \operatorname{PGL}(\mathbb{Z}[1/pq])$ diagonally embedded in G_S.

Let $\tilde{V}_{d_i} = |d_i|^{-1/2}\, V_{d_i}^*(\mathbb{Z}) \subset V_{-1}$. Consider the natural projections π_S, π_Y

$$\mathbb{G}(\mathbb{A})/(\mathbb{Q}) \xrightarrow{\pi_S} G_S/\Gamma_S \xrightarrow{\pi_Y} K\backslash G/\Gamma \cong Y,$$

and let $\pi = \pi_Y \circ \pi_S$. We denote by π'; π'_Y the unnormalized form of π and π_Y—that is, the corresponding maps to $V_d/\operatorname{GL}(2,\mathbb{Z})$. For suitable choice of $g_i \in \operatorname{PGL}(2,\mathbb{A})$, we have that

$$\pi(g_i[\mathbb{T}_i(\mathbb{A})]) = \tilde{V}_{d_i}/\operatorname{GL}(2,\mathbb{Z})$$

and that $\pi_S(g_i\mathbb{T}_i(\mathbb{A}))$ is invariant under the diagonal groups $A_1 < \operatorname{PGL}(2,\mathbb{Q}_p)$ and $A_2 < \operatorname{PGL}(2,\mathbb{Q}_q)$. In particular, there is a $t_i \in \mathbb{T}_i(\mathbb{A})$ so that $\pi(g_i[t_i])$ is the point in Y that corresponds to the $\operatorname{GL}(2,\mathbb{Z})$-coset of q_i.

Reconciling the two points of view on the set $V_d^*(\mathbb{Z})$ given in I and II in section 4.1, one verifies that

$$\tilde{V}_{d_i}^{(2),q_i} = \left\{([q], [q_i] \odot [q] \odot [q]) : q \in V_{d_i}^*(\mathbb{Z})\right\}$$

is equal to

$$\left\{(\pi'(g_i[t]), \pi'(g_i[t_i \cdot t \cdot t])) : t \in \mathbb{T}_i(\mathbb{A})\right\};$$

in particular $\tilde{V}_{d_i}^{(2),q_i}$ is the projection of the $\left\{(a_1 a_2, a_1^2 a_2^2) : a_1 \in A_1, a_2 \in A_2\right\}$-invariant subset

$$\left\{(\pi_S(g_i[t]), \pi_S(g_i[t_i t^2])) : t \in \mathbb{T}_i(\mathbb{A})\right\} \leq (G_S/\Gamma_S)^2$$

to $(V_d/\operatorname{GL}(2,\mathbb{Z}))^2$.

The projection of the first coordinate in $\tilde{V}_{d_i}^{(2),q_i}$ to Y equidistributes by the work of Linnik [L4] or Duke [D3] (since we already assumed two split places—$\left(\frac{d_i}{p}\right) = \left(\frac{d_i}{q}\right) = 1$ for all i—we may as well use Linnik, who needs only one). The second coordinate in $\tilde{V}_{d_i}^{(2),q_i}$ does not run over all of $V_{d_i}^*(\mathbb{Z})/\operatorname{GL}(2,\mathbb{Z})$ but rather over a subcollection—say, $V_{d_i}^{\text{second}}$ of index equal to the 2-torsion in the class group $\operatorname{cl}(\mathcal{O}_{d_i})$ of $\mathcal{O}_{d_i}$.

Fortunately, already Gauss understood the 2-torsion in the class group of quadratic fields (remarkably, even today we do not understand 2-torsion of the class group in fields of higher degree!), and its size is nicely controlled by the number of divisors of d_i; in particular it has size $\ll |d_i|^{\epsilon}$ for all $\epsilon > 0$ (by a theorem of Siegel, the size of $\operatorname{cl}(\mathcal{O}_{d_i})$ is (noneffectively) $\gg |d_i|^{1/2-\epsilon}$ for all ϵ).

To prove the equidistribution of $V_{d_i}^{\text{second}}$, one can either quote a result of Harcos and Michel [HM] that can be viewed as an extension of Duke's work, or one can use ergodic theory: $V_{d_i}^{\text{second}}$ is the projection of a $\left\{a_1^2 a_2^2 : a_1 \in A_1, a_2 \in A_2\right\}$-invariant subset of G_S/Γ_S that can be treated using Linnik's method as the analogue of Equation (4.3) will also hold for $V_{d_i}^{\text{second}}$. For more details,[17] we refer the reader to [AES2, section 4].

Once the equidistribution of each component of $\tilde{V}_{d_i}^{(2),q_i}$ separately has been proved, Theorem 2.1 takes care of the rest. A key point is that there is no nontrivial algebraic joining in $(G_S/\Gamma_S)^2$ invariant under $\left\{(a_1 a_2, a_1^2 a_2^2) : a_1 \in A_1, a_2 \in A_2\right\}$. □

---

[17] At least for the case of fundamental discriminants, though the general case is not more complicated.

It is a folklore conjecture that for any integer k, the k-torsion in $\mathrm{cl}(\mathcal{O}_d)$ is $\ll |d|^{\epsilon}$ as $|d| \to \infty$ (see, e.g., [EV]). *Assuming this conjecture*, the method of Theorem 4.3 would give that for any k, any sequence of negative integers $d_i \to -\infty$ with $d_i \equiv 0$ or $1 \bmod 4$ and $\left(\frac{d_i}{p}\right) = \left(\frac{d_i}{q}\right) = 1$ for two fixed primes p, q and any choice of $q_{i,1}, q_{i,2}, \ldots q_{i,k} \in V^*_{d_i}(\mathbb{Z})$, we have that the projection of

$$\left\{([q_{i,1}] \odot [q], [q_{i,2}] \odot [q]^{\odot 2}, \ldots, [q_{i,k}] \odot [q]^{\odot k}) : q \in V^*_{d_i}(\mathbb{Z})\right\}$$

to Y^k becomes equidistributed.

Consider now another collection of points in $(V_{d_i} / \mathrm{GL}(2, \mathbb{Z}))^2$,

$$\tilde{V}^{\mathrm{join}, q_i}_{d_i} = \left\{([q], [q_i] \odot [q] : q \in V^*_{d_i}(\mathbb{Z})\right\}, \tag{4.5}$$

depending on the discriminant d_i as well as $q_i \in V^*_{d_i}(\mathbb{Z})$.

It would seem to be a simpler collection to study than the non-linear collection $\tilde{V}^{(2), q_i}_{d_i}$ defined by Equation (4.4), if only because the equidistribution of each of the two projections to Y follows literally from Duke's theorem. However, this intuition turns out to be misguided. Studying the distribution of the collections, $\tilde{V}^{\mathrm{join}, q_i}_{d_i}$ turns out to be substantially subtler than the collection $\tilde{V}^{(2), q_i}_{d_i}$ for a simple reason: after passing to a limit, *there are nontrivial joinings* that need to be ruled out. And indeed, without further assumptions, the projections of the collection of points $\tilde{V}^{\mathrm{join}, q_i}_{d_i}$ to $Y \times Y$ need not equidistribute. For instance, if one takes q_i to be $x^2 - (d_i/4)y^2$ or $x^2 + xy - \frac{(d_i - 1)}{4} y^2$ (so that $[q_i]$ is the identity for Gauss's group law on $V^*_{d_i}(\mathbb{Z}) / \mathrm{GL}(2, \mathbb{Z})$) the projection of this collection of points to Y^2 is supported on the diagonal $\{([v], [v]) : v \in Y\}$ and hence certainly does not equidistribute!

A similar problem holds if $[q_i]$ has small size $\mathfrak{N}([q_i])$. To define $\mathfrak{N}(\bullet)$ we use the correspondence in paragraph I in section 4.1 between elements of $\tilde{V}^*_{d_i}(\mathbb{Z}) / \mathrm{GL}(2, \mathbb{Z})$ and elements of the ideal class group $\mathrm{cl}(\mathcal{O}_{d_i})$. If $I \lhd \mathcal{O}_{d_i}$ is the ideal corresponding to the quadratic form $q \in \tilde{V}^*_{d_i}(\mathbb{Z}) / \mathrm{GL}(2, \mathbb{Z})$, we define

$$\mathfrak{N}([q]) = \min \left\{\mathfrak{N}(J) : J \lhd \mathcal{O}_{d_i}, J \sim I\right\}.$$

If we consider a sequence $d_i \to -\infty$ and $q_i \in V^*_{d_i}(\mathbb{Z})$ with $\mathfrak{N}([q_i])$ bounded (say, less than N), for similar reasons the collections $\tilde{V}^{\mathrm{join}, q_i}_{d_i}$ will be restricted to a subset of $Y \times Y$ of dimension $\dim Y = 2$: the union of the graphs of the Hecke correspondences on $Y \times Y$ of order $\leq N$.

Michel and Venkatesh conjectured this is the only obstruction:

CONJECTURE 8 (Michel and Venkatesh [MV1, conjecture 2]). *Let $d_i \to -\infty$ along the sequence of fundamental discriminants. For each i, let $q_i \in V^*_{d_i}(\mathbb{Z})$, and assume that $\mathfrak{N}([q_i]) \to \infty$. Then the projection of the collection $\tilde{V}^{\mathrm{join},q_i}_{d_i}$ to $Y \times Y$ equidistributes as $i \to \infty$.*

We recall that for $d < 0$ (ineffectively) the size of $\mathrm{cl}(\mathcal{O}_d)$, which as we have seen can be identified with $V^*_d(\mathbb{Z})/\operatorname{GL}(2,\mathbb{Z})$, is $|d|^{1/2+o(1)}$. The number of ideal classes $[I]$ in $\mathrm{cl}(\mathcal{O}_d)$ with $\mathfrak{N}([I]) < N$ can be easily seen to be $\ll N^{1+o(1)}$. Moreover, by a simple application of Minkowski's theorem, one can see that for any ideal class $[I] \in \mathrm{cl}(\mathcal{O}_d)$, one has $\mathfrak{N}([I]) \ll |d|^{1/2}$ (except for the lower bound on the size of $\mathrm{cl}(\mathcal{O}_d)$, all of these bounds are elementary and effective).

In [EMV2], Ellenberg, Michel, and Venkatesh prove Conjecture 8 for $d_i \to -\infty$ and $q_i \in V^*_d(\mathbb{Z})/\operatorname{GL}(2,\mathbb{Z})$ with $\mathfrak{N}([q_i]) \to \infty$ assuming one split prime *as long as* $\mathfrak{N}([q_i]) < |d_i|^{1/2-\epsilon}$ *for some fixed* $\epsilon > 0$. Essentially, their proof employed a variant of Linnik's method with a rather quantitative variant of step C in section 4.1, which gave first an equidistribution statement for the projection of an appropriate shift of the adelic torus $g_i[\mathbb{T}_i(\mathbb{A})]$ (notations as in the proof of Theorem 4.3) to the quotient $\operatorname{PGL}(2,\mathbb{R})/\Gamma_i$ with $\Gamma_i < \operatorname{PGL}(2,\mathbb{Z})$ an appropriate congruence subgroup with $[\Gamma_i : \operatorname{PGL}(2,\mathbb{Z})] \to \infty$. This equidistribution of a single orbit in a homogeneous space of increasing volume can then be coupled with the equidistribution of the natural embedding

$$\operatorname{PGL}(2,\mathbb{R})/\Gamma_i \hookrightarrow (\operatorname{PGL}(2,\mathbb{R})/\operatorname{PGL}(2,\mathbb{Z}))^2$$

as the uniform measure on a closed orbit[18] of a diagonally embedded $\operatorname{PGL}(2,\mathbb{R}) \hookrightarrow \operatorname{PGL}(2,\mathbb{R})^2$ to the uniform measure on the product space.

Assuming two split primes, Khayutin has been able to (essentially) prove Conjecture 8 using a combination of ergodic and analytic tools, in particular Theorem 2.1:

THEOREM 4.4 (Khayutin [K2, theorem 1.3]).
Let $d_i \to -\infty$ be a sequence of fundamental discriminants so that there are two primes p, q for which $\left(\frac{d_i}{p}\right) = \left(\frac{d_i}{q}\right) = 1$ for all i, and let $q_i \in V^*_{d_i}(\mathbb{Z})$ satisfy $\mathfrak{N}([q_i]) \to \infty$. Assume furthermore that the Dedekind ζ-function of the fields $\mathbb{Q}(\sqrt{d_i})$ have

[18] In fact, the graph of a Hecke correspondence.

no exceptional Landau-Siegel zero. Then the projection of the collection $\tilde{V}_{d_i}^{\mathrm{join},q_i}$ *to* $Y \times Y$ *equidistributes as* $i \to \infty$.

It is very widely believed that Landau-Siegel zeros do not exist, as this is a very special, albeit important, case of the Riemann hypothesis for Dedekind ζ-functions. Moreover, even if these notorious Landau-Siegel zeros were to exist, they would have to be exceedingly rare: by a theorem of Landau (see [IK, theorem 5.28]), for any $A > 1$ and D large enough there will be at most one fundamental discriminant d between $-D$ and $-D^A$ for which the Dedekind ζ-function of $\mathbb{Q}(\sqrt{d})$ has such a zero.

Theorem 2.1 allows one to deduce from Theorem 4.4 the following seemingly much more general theorem:

THEOREM 4.5 (Khayutin [K2, theorem 3.9]).
Fix $k \in \mathbb{N}$. *Let* $d_i \to -\infty$ *be a sequence of fundamental discriminants so that there are two primes* p, q *for which* $\left(\frac{d_i}{p}\right) = \left(\frac{d_i}{q}\right) = 1$ *for all* i, *and let* $q_{i,1}, \ldots, q_{i,k} \in V_{d_i}^*(\mathbb{Z})$ *so that for any* $1 \leq j < \ell \leq k$ *we have that* $\mathfrak{N}([q_{i,j}] \odot [q_{i,\ell}]^{-1}) \to \infty$. *Assume furthermore that the Dedekind* ζ*-function of the fields* $\mathbb{Q}(\sqrt{d_i})$ *have no exceptional Landau-Siegel zero. Then the projection of the collections*

$$\left\{([q_{i,1}] \odot [q], \ldots, [q_{i,k}] \odot [q] : q \in V_{d_i}^*(\mathbb{Z})\right\}$$

to Y^k *equidistributes as* $i \to \infty$.

We end this rather long subsection with a striking equidistribution result by Aka, Einsiedler, and Wieser on the space $\mathrm{gr}_{2,4}(\mathbb{R}) \times Y^4$, where $\mathrm{gr}_{2,4}(\mathbb{R})$ is the projective variety (known as the *Grassmannian*) of two-dimensional subspaces of a four-dimensional space over $\mathbb{R}$, arising due to the "accidental" local isomorphism between $\mathrm{SO}(4)$ and $\mathrm{SU}(2) \times \mathrm{SU}(2)$ (or equivalently $\mathrm{SO}(3) \times \mathrm{SO}(3)$).

Consider now the quaternary[19] integer quadratic form $Q(x, y, z, w) = x^2 + y^2 + z^2 + w^2$, and consider all the binary integral quadratic forms q of discriminant d that can be represented by Q—that is, so that there is a 4×2 integer matrix M so that

$$q(x, y) \equiv Q\left(M\begin{pmatrix} x \\ y \end{pmatrix}\right).$$

[19] I.e., of four variables.

Necessarily these binary quadratic forms will be positive definite, and hence $d < 0$. It is a classical theorem that there are such binary quadratic forms iff $-d \not\equiv 0, 7, 12, 15 \bmod 16$. The image of M is a rational subspace $L < \mathbb{R}^4$ of dimension 2 and hence in particular gives us a point in $\mathrm{gr}_{2,4}$. The space perpendicular to L also intersects the lattice $\mathbb{Z}^4$ in a lattice; hence, after choosing (arbitrarily) a basis for $L^\perp \cap \mathbb{Z}^4$, we obtain another binary quadratic form that can be shown to also have discriminant d. Thus we obtain for any $d < 0$, $-d \not\equiv 0, 7, 12, 15 \bmod 16$ a collection of triplets of points in $\mathrm{gr}_{2,4} \times (V_d(\mathbb{Z}) / \mathrm{GL}(2, \mathbb{Z}))^2$, which we can project to $\mathrm{gr}_{2,4} \times Y^2$. For each choice of a binary form q represented by Q, Aka, Einsiedler, and Wieser magically pull two more rabbits (actually, only points of Y) out of the hat using the *Klein map*, which assigns to any $L \in \mathrm{gr}_{2,4}$ a point in $(\mathbb{S} \times \mathbb{S})/\{\pm 1\}$, with the ± 1 acting by scalar multiplication on both factors. Identifying $\mathbb{R}^4$ with the Hamilton quaternions $\{x + iy + jz + ijw : x, y, z, w \in \mathbb{R}\}$, where $i^2 = j^2 = -1$, $ij = -ji$, pick two linearly independent vectors $v, w \in L$, and define

$$a = v\bar{w} - \mathrm{trace}(v\bar{w}) \qquad a' = w\bar{v} - \mathrm{trace}(w\bar{v}).$$

a, a' are two traceless quaternions; they hence lie in a three-dimensional space and have the same norm. After rescaling, they define a point in $(\mathbb{S} \times \mathbb{S})/\{\pm 1\}$ that turns out to be independent of the choice of generators v, w. If we chose v, w to be generators of $L \cap \mathbb{Z}^4$, then a, a' will be integral, and $Q(a) = Q(a')$ will be equal to $|d|$. The vectors a and a' give a continuous parameterization of L, so considering the joint distribution in the limit of the rescaled collections of triplets consisting of

- a binary integral quadratic form q of discriminant d represented by Q,
- the corresponding $L \in \mathrm{gr}_{2,4}$, and
- the quadratic form induced on $L^\perp$

implicitly also describes the distribution of a, a'. However, these are integral vectors in a three-dimensional space, and the quadratic form induced by choosing generators of the lattices perpendicular to a and a' in this space gives the desired two additional points of Y.

THEOREM 4.6 (Aka, Einsiedler, Wieser [AEW]).
Let $d_i \to -\infty$ along the sequence $-d \not\equiv 0, 7, 12, 15 \bmod 16$ so that $\left(\frac{d_i}{p}\right) = \left(\frac{d_i}{q}\right) = 1$ for two odd primes p, q. Then the collections of 5-tuples in $\mathrm{gr}_{2,4} \times Y^4$ defined above become equidistributed as $i \to \infty$ with respect to the $\mathrm{SO}(4, \mathbb{R}) \times \mathrm{PGL}(2, \mathbb{R})^4$–invariant measure on this space.

In a similar way to how Theorem 4.3 relates to Theorem 4.2, Theorem 4.6 is related to the distribution of the 6-tuples

$$\Big\{([q],[q'],[q_{i,1}]\odot[q]\odot[q'],[q_{i,2}]\odot[q]\odot[q']^{-1},[q_{i,3}]\odot[q]^2,[q_{i,4}]\odot[q']^2):\\ q\in V^*_{d_i}(\mathbb{Z})\Big\},$$

with $q_{i,1},\ldots,q_{i,4}\in V^*_{d_i}(\mathbb{Z})$ arbitrary, in Y^6. We refer the reader to [AEW, section 7] for more details. Similarly to Theorems 4.3 and 4.2, the joining theorem, Theorem 2.1, is a key ingredient.

4.3 LINNIK'S PROBLEM IN PGL(3) AND BEYOND. Equidistribution of integer points discussed in sections 4.1 and 4.2 are at the core questions about the distribution of adelic points of $\mathbb{Q}$-tori on arithmetic quotients of forms of PGL(2) and their joinings. In this section we consider the more general question of density and equidistribution of *adelic torus subsets*—sets of the form $g[\mathbb{T}(\mathbb{A}_F)]$ in $\mathbb{G}(\mathbb{A}_F)/\mathbb{G}(F)$, where F is a number field, $\mathbb{A}_F$ is the Adele ring of F, $\mathbb{G}$ is a reductive group over F, and $\mathbb{T}$ is an anisotropic F-torus. When $F=\mathbb{Q}$ we write $\mathbb{A}$ for $\mathbb{A}_{\mathbb{Q}}$.

Let S be a set of places for F (equivalence classes of embeddings of F in local fields—if v is such an embedding, we denote by F_v the corresponding local field; we implicitly assume F is dense in F_v) containing at least one place v in which $\mathbb{T}$ splits and all infinite places (embeddings of F in $\mathbb{R}$ or $\mathbb{C}$ up to identifying conjugate embeddings in $\mathbb{C}$). Let $G=\prod_{v\in S}\mathbb{G}(F_v)$.

In section 4.2 essential (though mostly implicit) use was made of the fact that there is a natural projection from $\mathrm{PGL}(2,\mathbb{A})/\mathrm{PGL}(2,\mathbb{Q})$ to $\mathrm{PGL}(2,\mathbb{R})/\mathrm{PGL}(\mathbb{Z})$ with compact fibers. At the level of generality we are discussing here, the picture is slightly more complicated: if one would like to project $\mathbb{G}(\mathbb{A}_F)/\mathbb{G}(F)$ to some quotient of G, in general one needs to take a finite number of lattices $\Gamma_1,\ldots,\Gamma_k<G$ with Γ_i all conjugate over $\mathbb{G}(\mathbb{Q})$ and all commensurable to $\mathbb{G}(\mathcal{O}_{F,S})$ to obtain a natural projection

$$\pi_S:\mathbb{G}(\mathbb{A}_F)/\mathbb{G}(F)\to\bigsqcup_{k}^{i=1}G/\Gamma_i \tag{4.6}$$

analogous to the projection from $\mathrm{PGL}(2,\mathbb{A})/\mathrm{PGL}(2,\mathbb{Q})$ to $\mathrm{PGL}(2,\mathbb{R})/\mathrm{PGL}(\mathbb{Z})$ (see [PR, section 5]).[20] We will also use π_S to denote the natural projection $\mathbb{G}(\mathbb{A}_F)\to G$.

[20]We already saw this phenomenon implicitly when discussing orthogonal groups in section 4.2—this is precisely the reason why for general ternary definite quadratic forms we need to consider not one quadratic form individually but the whole genus of quadratic forms locally equivalent to it.

Since $\mathbb{T}$ was assumed to be F anisotropic, the orbit $[\mathbb{T}(\mathbb{A}_F)]$ in $\mathbb{G}(\mathbb{A}_F)/\mathbb{G}(F)$ supports a $\mathbb{T}(A_F)$-invariant probability measure $m_{\mathbb{T}(\mathbb{A}_F)}$. The projection of $g[\mathbb{T}(\mathbb{A}_f)]$ to $\bigsqcup_k^{i=1} G/\Gamma_i$ is a finite union of periodic $A_{g,S} = \pi_S(g)\prod_{v\in S}\mathbb{T}(\mathbb{Q}_S)\pi_S(g^{-1})$-orbits, and if $v \in S$ is a place where $\mathbb{T}$ is split, then the uniform measure on $\pi_S(g[\mathbb{T}(\mathbb{A}_F)])$, which is simply the average of the periodic measure on each of the $A_{g,S}$-periodic orbits comprising $\pi_S(g[\mathbb{T}(\mathbb{A}_f)])$, is invariant under a nontrivial $\mathbb{Q}_v$-diagonalizable group.

Moreover, if either the $\mathbb{Q}_v$-rank of $\mathbb{T}(\mathbb{Q}_v)$ is ≥ 2 or $\mathbb{T}$ splits over at least one other place $v' \in S$, $\pi_S(g[\mathbb{T}(\mathbb{A}_f)])$ is invariant under a higher rank action, and one can hope to use the tools of section 2 to study this set as well as the corresponding probability measure.

As an explicit example, we show how periodic A-orbits in $\mathrm{PGL}(n,\mathbb{R})/\mathrm{PGL}(n,\mathbb{Z})$ fit in this framework, where $A < \mathrm{PGL}(n,\mathbb{R})$ is the full diagonal group. This corresponds to the above for $S = \{\infty\}$ and $\mathbb{G} = \mathrm{PGL}(n)$ when the shifting element is chosen appropriately. Indeed, an A-periodic orbit $A.[g_\infty]$ in $\mathrm{PGL}(n,\mathbb{R})/\mathrm{PGL}(n,\mathbb{Z})$ defines a $\mathbb{R}$-split $\mathbb{Q}$-tori in $\mathbb{G}$ by

$$\mathbb{T} = C_{\mathbb{G}}\left(\mathrm{PGL}(n,\mathbb{Z}) \cap g_\infty^{-1} A g_\infty\right).$$

Then

$$A.[g_\infty] \subseteq \pi(g[\mathbb{T}(\mathbb{A})]) \qquad \left(g = (g_\infty, e, e, \dots) \in \mathrm{PGL}(n,\mathbb{A})\right). \tag{4.7}$$

Equality does not always occur in (4.7)—in general, the projection $\pi(g[\mathbb{T}(\mathbb{A})])$ of the above adelic toral subset consists of a *packet* of several periodic A-orbits $A.[g_\infty^{(i)}]$, all with the same shape—that is, with the same

$$\mathrm{stab}_A([g_\infty^{(i)}]) = \left\{a \in A : a.[g_\infty^{(i)}] = [g_\infty^{(i)}]\right\}.$$

In this context, the following seems to be a natural conjecture. Conjecture 8 can be viewed as a special case of this conjecture.

CONJECTURE 9. *Let $\mathbb{G}$ be a semisimple algebraic group over a number field F, let $\mathbb{T}_i < \mathbb{G}$ be anisotropic F-tori, and let $g_i \in \mathbb{G}(\mathbb{A}_F)$. Let $\tilde{\mathbb{G}}$ be the simply connected cover of $\mathbb{G}$ with $j : \tilde{\mathbb{G}} \to \mathbb{G}$ the corresponding isogeny. Then we have these alternatives:*

(1) *Any weak-* limit of the uniform measures on $g_i[\mathbb{T}(\mathbb{A}_F)]$ is invariant under $j(\tilde{\mathbb{G}}(\mathbb{A}_F))$.*

(2) There exist a bounded sequence $h_i \in \mathbb{G}(\mathbb{A}_F)$ and a proper $\mathbb{Q}$-subgroup $\mathbb{H} < \mathbb{G}$ so that for infinitely many i,

$$g_i[\mathbb{T}(\mathbb{A}_F)] \subset h_i[\mathbb{H}(\mathbb{A}_F)].$$

We remark that if $\mathbb{G}$ is simply connected, alternative (1) is equivalent to $g_i[\mathbb{T}(\mathbb{A}_F)]$ becoming equidistributed in $\mathbb{G}(\mathbb{A}_F)/\mathbb{G}(F)$ and in general implies that any weak-* limit of the corresponding measures is homogeneous. We also note that the assumption that h_i be bounded in alternative (2) of Conjecture 9 is equivalent to the following: there is a finite set of places S so that for any $v \in S$ the $\mathbb{G}(F_v)$ component of h_i remains bounded, and for any $v \notin S$ the $\mathbb{G}(F_v)$ component of h_i is in $\mathbb{G}(\mathcal{O}_v)$, where $\mathcal{O}_v$ is the maximal compact subring of F_v.

Of interest are results not just for the full group of adelic points $\mathbb{T}(\mathbb{A}_F)$ but also for large subgroups. These occur naturally, in particular in the context of the study of special points on Shimura varieties: the orbit under the absolute Galois group on a special point, considered as a $\overline{\mathbb{Q}}$-point in an arithmetic model of $K\backslash G/\Gamma$, turns out to be such a group, though it is rather difficult to put one's finger on how big this group is; see, for example, Tsimerman's paper [T4], which proves an important special case of the André-Oort conjecture (namely, when $\mathbb{G} = \mathrm{SP}(n)$) via such an analysis.

The analogue of Conjecture 9 when $\mathbb{T}_i$ is not a torus (e.g., when it is a semisimple or reductive group) is also highly interesting.[21] If one assumes (implicitly or explicitly) that there is a fixed place v for which $\mathbb{T}_i(F_v)$ contains a unipotent subgroup, one can bring to bear deep tools on unipotent flows, in particular Ratner's measure classification theorem [R3] and its S-arithmetic generalizations by Ratner [R5] and by Margulis-Tomanov [MT1], as was done by Eskin and Oh in [EO]. Indeed, under some additional assumptions, and for $\mathbb{T}_i$ semisimple, Einsiedler, Margulis, and Venkatesh [EMV1] and these three authors jointly with Mohammadi [EMMV] were able to give a *quantitative* equidistribution statement; an exciting feature of [EMMV] is that thanks to the quantitative nature of the proof, it is even able to handle sequences of $\mathbb{Q}$-groups $\mathbb{T}_i$ for which there is no place v at which all (or even infinitely many) of these groups split. The discussion of these interesting works is unfortunately beyond the scope of this survey.

[21] In general, in the context of $\mathbb{Q}$-groups or more generally groups defined over a number field F, we reserve $\mathbb{T}$ to denote an algebraic torus; we make an exception to this convention in this paragraph in order to abuse the notations of Conjecture 9 to cover a wider context.

Results toward Conjecture 9 for $\mathbb{G} = \mathrm{PGL}(n)$ or inner forms of $\mathrm{PGL}(n)$ (see below) were obtained by Einsiedler, Michel, Venkatesh, and this author in [ELMV1] and [ELMV2] and strengthened in certain respects by Khayutin in [K1].

When considering periodic orbits of a fixed group H on a space X, one can fix a Haar measure on H and, once this is done consistently, measure the volume of all H periodic orbits. When one allows the acting group to vary, one needs a slightly more sophisticated notion of volume:

DEFINITION 4.7 ([ELMV2, definition 4.3]). Let $\mathbb{G}$ be a fixed group defined over a number field F and $\Omega \subset \mathbb{G}(\mathbb{A}_F)$ a fixed neighborhood of the identity. Let $\mathbb{H} < \mathbb{G}$ be an F-subgroup. We define the *size*[22] of an adelic shifted orbit $g_i[\mathbb{H}(\mathbb{A}_F)]$ to be ∞ if $\mathbb{H}(\mathbb{A}_F)/\mathbb{H}(F)$ does not have finite $\mathbb{H}(\mathbb{A}_F)$-invariant measure and

$$\mathrm{size}(g_i[\mathbb{H}(\mathbb{A}_F)]) = \frac{m_{\mathbb{H}(\mathbb{A}_F)}(\mathbb{H}(\mathbb{A}_F)/\mathbb{H}(F))}{m_{\mathbb{H}(\mathbb{A}_F)}(\mathbb{H}(\mathbb{A}_F) \cap g_i^{-1}\Omega g_i)}.$$

Note that changing Ω changes the size only up to a constant factor.

THEOREM 4.8 (Einsiedler, Michel, Venkatesh, and Lindenstrauss [ELMV2]).
Let $\mathbb{G} = \mathrm{PGL}(3)$, F be a number field, $g_i \in \mathbb{G}(\mathbb{A}_F)$, and $\mathbb{T}_i < \mathbb{G}$ be a maximal F-torus. Assume the following:

(1) *There is a place v of F so that (a) $\mathbb{T}_i$ is split over F_v and (b) F_v has no proper closed subfield—that is, either $F_v \cong \mathbb{R}$ or $F_v \cong \mathbb{Q}_p$ for some prime p.*[23]

(2) $\mathrm{size}(g_i[\mathbb{T}(\mathbb{A}_F)]) \to \infty$.

Then any limiting measure of the uniform distribution on $g_i[\mathbb{T}(\mathbb{A}_F)]$ is invariant under the image of $\mathrm{SL}(3, \mathbb{A}_F)$ in $\mathbb{G}(\mathbb{A}_F)$. In particular, if the class number of the integer ring in F satisfies $\#\,\mathrm{cl}(\mathcal{O}_F) = 1$, then the adelic torus sets $g_i[\mathbb{T}(\mathbb{A}_F)]$ become equidistributed in $\mathbb{G}(\mathbb{A}_F)/\mathbb{G}(F)$.

[22] In [ELMV2] we used the term *volume* of a periodic orbit to denote what we call here *size*. We decided to use a different terminology in this survey so we can unambiguously use *volume* to denote the volume of a periodic orbit with respect to a fixed Haar measure on the acting group.

[23] Part (b) of this assumption was omitted in [ELMV2] but is implicitly used in the proof. It is of course automatically satisfied for $F = \mathbb{Q}$.

Assumption (2) in Theorem 4.8 turns out to be equivalent to assumption (2) in Conjecture 9, as there are no proper F-subgroups of PGL(3) containing a maximal F-torus other than the torus itself, and the assumption that the volume $\mathrm{size}(g_i[\mathbb{T}(\mathbb{A}_F)]) \to \infty$ rules out the adelic torus subsets $g_i[\mathbb{T}(\mathbb{A}_F)]$ being all in bounded translate of the image of a fixed F torus in $\mathbb{G}(\mathbb{A}_F)/\mathbb{G}(F)$.

The following is an easier-to-digest (weaker) form of Theorem 4.8 for $F=\mathbb{Q}$, where the adeles are not explicitly mentioned:

THEOREM 4.9 (Einsiedler, Michel, Venkatesh, and Lindenstrauss [ELMV2]).
Let A be the maximal diagonal group in $\mathrm{PGL}(3,\mathbb{R})$. *Let V_i be the sequence of all possible volumes of A-periodic orbits in* $\mathrm{PGL}(3,\mathbb{R})/\mathrm{PGL}(3,\mathbb{Z})$ *with respect to the Haar measure on A. For every i, let C_i be the collection of A-periodic orbits of A of volume exactly V_i. Then these collections become equidistributed in* $\mathrm{PGL}(3,\mathbb{R})/\mathrm{PGL}(3,\mathbb{Z})$ *as $i \to \infty$.*

The proof of Theorem 4.8 goes via a combination of analytic and ergodic tools:

(1) Using the analytic theory of automorphic forms—specifically the subconvexity estimates of Duke, Friedlander, and Iwaniec [DFI] (or when F is a general number field, an extension by Michel and Venkatesh of these subconvex bounds [MV2]), which incidentally are closely related to the works of Duke and Iwaniec mentioned in section 4.1—one shows that for some rather special functions $f \in L^2(\mathbb{G}(\mathbb{A}_F)/\mathbb{G}(F))$,

$$\fint_{g_i[\mathbb{T}_i(\mathbb{A}_F)]} f \to 0. \tag{4.8}$$

Indeed, here the estimates are even quantitative.

(2) Consider the F_v-split tori $g_{i,v}\mathbb{T}_i(\mathbb{Q}_v)g_{i,v}^{-1}$, with $g_{i,v}$ denoting the F_v component of G_i and v a place as in Theorem 4.8(1). Without loss of generality these would converge to some F_v-group A_v. If this group contains unipotent elements, we can use Ratner's measure classification theorem (or more precisely its S-arithmetic extensions [R5, MT1]). Otherwise one can use Equation (4.8), established for a certain collection of special f, to ensure that *every ergodic component of any weak-* limit* of the probability measures attached to $g_i[\mathbb{T}_i(\mathbb{A}_F)]$ has to have positive entropy with respect to the action of A_v, whence one can use the measure classification results of [EKL] and [EL2] (e.g., Theorem 2.4), to conclude the theorem.

We note that Theorem 4.8 can be combined with the joining classification theorem (Theorem 2.1) to obtain joint equidistribution statements; see [EL3, theorem 1.8] for a precise statement.

In [ELMV1], a purely ergodic theoretic approach was used. This approach is not powerful enough to give a full equidistribution result, but on the other hand it is significantly more flexible and in particular gives information about rather small subsets of an adelic torus subset. For simplicity (and to be more compatible with the terminology in [ELMV1]), we work over $\mathbb{Q}$ (instead of a general number field F) and use the more classical language of A-periodic orbits employed in Theorem 4.9. We also assume for notational simplicity that the place where the $\mathbb{Q}$-tori we will consider is split is ∞, though the discussion below with minimum modification also holds for tori split over $\mathbb{Q}_p$ instead of $\mathbb{R}$.

Recall the relationship given in Equation (4.7) between periodic A-orbits and the projection under π_S of appropriate shift of the adelic points of $\mathbb{Q}$-tori (with π_S as in Equation (4.6) and $S=\{\infty\}$): a periodic orbit $A.[g]$ in $\mathrm{PGL}(n,\mathbb{R})/\mathrm{PGL}(n,\mathbb{Z})$ defines a $\mathbb{Q}$-torus $\mathbb{T}$, and $g.\pi_S([\mathbb{T}(\mathbb{A})])$ is a packet of periodic A-orbits of the same volume. In addition to the volume of a periodic A-orbit $A.[g]$ and its shape $\mathrm{stab}_A([g])$, we can attach to this orbit an order in a totally real degree n-extension of $\mathbb{R}$ embedded in the subring of (not necessarily invertible) diagonal $n\times n$ matrices $D < M_{n\times n}(\mathbb{R})$ as follows:

$$\mathcal{O}_{[g]} = \left\{ x \in D : xg\mathbb{Z}^n \subseteq g\mathbb{Z}^n \right\}. \tag{4.9}$$

The order $\mathcal{O}_{[g]}$ in D is best thought of as an abstract order $\mathcal{O}$ in a totally real number field K with $[K:\mathbb{Q}]=n$, together with an embedding τ of this order in D (essentially this amounts to giving an ordering on the n embeddings $K\to\mathbb{R}$). The *discriminant* $\mathrm{disc}(A.[g])$ of the periodic orbit $A.[g]$ is by definition the discriminant of the order $\mathcal{O}_D$—that is (up to sign), the square of the covolume of $\mathcal{O}_{[g]}$ in D: it is an integer, since if $\alpha_1 = 1$, then $\alpha_2,\dots,\alpha_n$ are independent generators of $\mathcal{O}_{[g]}$:

$$\mathrm{disc}(\mathcal{O}_{[g]}) = \det(\mathrm{trace}(\alpha_i\alpha_j))_{i,j=1}^n.$$

The relation between the volume of a periodic orbit $A.[g]$ (which is called by number theorists the *regulator*) and the size of the adelic toral subset

$\tilde{g}[\mathbb{T}(\mathbb{A})]$ (with $\tilde{g}$ the image of g in $\mathrm{PGL}(n, \mathbb{A})$ under the obvious embedding $\mathrm{PGL}(n, \mathbb{R}) \hookrightarrow \mathrm{PGL}(n, \mathbb{A})$) is rather weak. Assuming the field generated by $\mathcal{O}_{[g]}$ does not contain any nontrivial subfields,[24] then

$$\log(\mathrm{disc}(A.[g]))^{n-1} \ll \mathrm{vol}(A.[g]) \ll \mathrm{size}\left(\tilde{g}[\mathbb{T}(\mathbb{A})]\right) = \mathrm{disc}(A.[g])^{1/2+o(1)}, \tag{4.10}$$

the last "equality" being ineffective (see [ELMV1] and [ELMV2] for details).

Using the measure classification result in $\mathrm{PGL}(n, \mathbb{R})/\mathrm{PGL}(n, \mathbb{Z})$ of [EKL] (a special case of Theorem 2.4) and a rather crude entropy estimate, the following was proved in [ELMV1]:

THEOREM 4.10 (Einsiedler, Michel, Venkatesh, and Lindenstrauss [ELMV1, theorem 1.4]).
Let Ω be a compact subset of $\mathrm{PGL}(n, \mathbb{R})/\mathrm{PGL}(n, \mathbb{Z})$ and A the maximal diagonal subgroup of $\mathrm{PGL}(n, \mathbb{R})$ for $n \geq 3$. Then

$$\#\left\{\textit{periodic A orbits } A.[g] \subset \Omega \textit{ with } \mathrm{disc}(A.[g]) \leq D\right\} \ll_{\epsilon,\Omega} D^{\epsilon}.$$

Since the number of A-periodic orbits $A.[g]$ with $\mathrm{disc}(A.[g])$ is easily seen to be $\gg D^c$ for appropriate $c > 0$, Theorem 4.10 can be viewed as evidence to the following conjecture, implied by Conjecture 1a:

CONJECTURE 10. *Let $n \geq 3$. Any compact $\Omega \subset \mathrm{PGL}(n, \mathbb{R})/\mathrm{PGL}(n, \mathbb{Z})$ contains only finitely many A-periodic orbits.*

Conjecture 10 follows from Conjecture 1a using the Cassels and Swinnerton-Dyer isolation result; see [M5]. Note that for $n = 2$ the analogue of Theorem 4.10 is false; indeed, for any ϵ there is a compact $\Omega \subset \mathrm{PGL}(2, \mathbb{R})/\mathrm{PGL}(2, \mathbb{Z})$ so that

$$\#\left\{\text{periodic } A \text{ orbits } A.[g] \subset \Omega \text{ with } \mathrm{disc}(A.[g]) \leq D\right\} \gg D^{1-\epsilon};$$

see [ELMV1, theorem 1.5].

The noncompactness of $\mathrm{PGL}(n, \mathbb{R})/\mathrm{PGL}(n, \mathbb{Z})$ makes it harder to deduce a density statement from these rigidity results; however, for cocompact inner forms of $\mathrm{PGL}(n)$—namely, $\mathrm{PGL}(1, \mathbb{M})$ with $\mathbb{M}$ a central division algebra of

[24] This assumption is needed only for the first inequality.

degree n over $\mathbb{Q}$—one can say more. Assume $\mathbb{M}$ splits at $\mathbb{R}$ (i.e., $\mathbb{M}\otimes\mathbb{R}\cong M_{n\times n}(\mathbb{R})$), and let $\mathcal{O}_{\mathbb{M}}$ be a maximal order in $\mathbb{M}(\mathbb{Q})$. Then $\mathrm{PGL}(1,\mathbb{M}\otimes\mathbb{R})\cong\mathrm{PGL}(n,\mathbb{R})$ and $\mathrm{PGL}(1,\mathcal{O}_{\mathbb{M}})$ (or any subgroup of $\mathrm{PGL}(1,\mathbb{M}\otimes\mathbb{R})$ commensurable to it) can be viewed as a cocompact lattice in $\mathrm{PGL}(n,\mathbb{R})$.

Let A be a maximal $\mathbb{R}$-split $\mathbb{R}$-torus in $\mathrm{PGL}(1,\mathbb{M}\otimes\mathbb{R})$, and let D be the abelian subalgebra of $\mathbb{M}\otimes\mathbb{R}$ commuting with A. As in Equation (4.9), we can define for any periodic A-orbit $A.[g]$ in $\mathrm{PGL}(1,\mathbb{M}\otimes\mathbb{R})/\mathrm{PGL}(1,\mathcal{O}_{\mathbb{M}})$ an order in a totally real number field and an embedding of this order to the algebra D by considering

$$\mathcal{O}_{A.[g]}=D\cap g\mathcal{O}_{\mathbf{M}}g^{-1}.$$

While this will not be of relevance to our purposes, not all orders in totally real fields of degree n can appear in this way; indeed, an abstract order $\mathcal{O}\cong\mathcal{O}_{A.[g]}$ attached to a periodic A-orbit in $\mathrm{PGL}(1,\mathbb{M}\otimes\mathbb{R})/\mathrm{PGL}(1,\mathcal{O}_{\mathbb{M}})$ has to satisfy the local compatibility condition that $\mathcal{O}\otimes\mathbb{Q}_p$ can be embedded in $\mathbf{M}\otimes\mathbb{Q}_p$ for all prime p. In this context, one has the following:[25]

THEOREM 4.11 (Einsiedler, Michel, Venkatesh, and Lindenstrauss [ELMV1, theorem 1.6]).
Let $\mathbb{M}$ be a division algebra over $\mathbb{Q}$ of degree n so that $\mathbb{M}\otimes\mathbb{R}\cong M_{n\times n}(\mathbb{R})$. Let $\mathcal{O}_{\mathbb{M}}$ be a maximal order in $\mathbb{M}\otimes\mathbb{Q}$, and let A be a maximal $\mathbb{R}$-split $\mathbb{R}$-torus in $\mathrm{PGL}(1,\mathbb{M}\otimes\mathbb{R})$. Let $\alpha>0$, and for any i, let C_i be a collection of (distinct) A-periodic orbits $\{A.[g_{i,1}],\dots,A.[g_{i,k_i}]\}$ so that

$$k_i\geq\left(\max_j \operatorname{disc}(A\cdot[g_{i,j}])\right)^{\alpha}$$

and $k_i\to\infty$. Assume that there is no subgroup $A\leq H<\mathrm{PGL}(1,\mathbb{M}\otimes\mathbb{R})$ so that infinitely many $g_{i,j}$ lie on a single H-periodic orbit in $\mathrm{PGL}(1,\mathbb{M}\otimes\mathbb{R})/\mathrm{PGL}(1,\mathcal{O}_{\mathbb{M}})$. Then the collections C_i become dense in $\mathrm{PGL}(1,\mathbb{M}\otimes\mathbb{R})/\mathrm{PGL}(1,\mathcal{O}_{\mathbb{M}})$—that is, for every open $U\subset\mathrm{PGL}(1,\mathbb{M}\otimes\mathbb{R})/\mathrm{PGL}(1,\mathcal{O}_{\mathbb{M}})$, we have that there is an i_0 so that for $i>i_0$ there is a $j\in\{1,\dots,k_i\}$ so that $A.[g_{i,j}]\cap U\neq\emptyset$.

To prove Theorem 4.11 one uses in addition to the ingredients used in Theorem 4.10 (namely, Theorem 2.4 and an appropriate entropy estimate) a variant of the orbit closure/isolation theorems of Weiss and this author [LW2] and Tomanov [T1] (cf. Theorem 3.4). Applying Theorem 4.11 to the

[25] The phrasing here is a bit stronger than that in [ELMV1]; the proof in [ELMV1] gives this slightly stronger version.

collections $\mathcal{C}_i = \{A.[g] : \mathcal{O}_{A.[g]} = \tau_i(\mathcal{O}_i)\}$ for $\mathcal{O}_i$, a sequence of maximal orders in totally real degree n number fields, one gets the following, which can also be interpreted as a theorem about the projection of adelic toral subsets in $\mathrm{PGL}(1, \mathbb{M} \otimes \mathbb{A}) / \mathrm{PGL}(1, \mathbb{M} \otimes \mathbb{Q})$ to $\mathrm{PGL}(1, \mathbb{M} \otimes \mathbb{R}) / \mathrm{PGL}(1, \mathcal{O}_{\mathbb{M}})$:

COROLLARY 4.12. *In the notations of Theorem 4.11, let $\mathcal{O}_i$ be the ring of integers in totally real fields K_i, τ_i the embeddings of $\mathcal{O}_i \hookrightarrow D$, and*

$$\mathcal{C}_i = \{A.[g] : \mathcal{O}_{A.[g]} = \tau_i(\mathcal{O}_i)\};$$

assume that $\mathcal{O}_i$ is chosen so that the collections $\mathcal{C}_i$ are nonempty. Assume that there is no fixed field L of degree $d|n$, which is a subfield of infinitely many K_i. Then the collections $\mathcal{C}_i$ become dense in $\mathrm{PGL}(1, \mathbb{M} \otimes \mathbb{R}) / \mathrm{PGL}(1, \mathcal{O}_{\mathbb{M}})$.

In [K1], a substantially more refined entropy estimate was given. This entropy estimate is quite interesting in its own sake, and in particular implies the following:

THEOREM 4.13 (Khayutin [K1]).
Let K_i be a sequence of totally real degree n number fields and let $\mathcal{O}_i$ be the ring of integers K_i. Let ζ be a generator for K_i over $\mathbb{Q}$. Assume that n is prime and that the Galois group of the Galois extension of K_i acts two-transitively on the Galois conjugates of ζ. Let τ_i be embeddings of $\mathcal{O}_i \hookrightarrow D$ and let

$$\mathcal{C}_i = \{A.[g] : \mathcal{O}_{A.[g]} = \tau_i(\mathcal{O}_i)\};$$

assume again that $\mathcal{O}_i$ is chosen so that the collections $\mathcal{C}_i$ are nonempty. Then for any bounded continuous f on $X =$ $\mathrm{PGL}(1, \mathbb{M} \otimes \mathbb{R}) / \mathrm{PGL}(1, \mathcal{O}_{\mathbb{M}})$,

$$\lim_{i \to \infty} \frac{\sum_{A.[g] \in \mathcal{C}_i} \int_{A.[g]} f}{\sum_{A.[g] \in \mathcal{C}_i} \int_{A.[g]} 1} \geq \frac{1}{2(n-1)} \int_X f. \tag{4.11}$$

The techniques of [ELMV1] also imply an estimate of the form (4.11) but with a much worse bound. An important technical point is that the entropy bounds in [K1] apply with regard to singular one-parameter diagonal subgroups of A and hence would also be useful in the context of analyzing periodic orbits of a $\mathbb{Q}$-torus that is only partially split at a given place.

We remark that if one fixes a $\mathbb{Q}$-torus $\mathbb{T}$ and shifts it either in the real place or in one (or several) p-adic places, one also obtains interesting equidistribution results, though they are now related less to diagonal flows and more to unipotent ones. See the work by Eskin, Mozes, and Shah [EMS1, EMS2] for the former (with a nice application regarding counting matrices with a given characteristic polynomial in large balls in $\mathrm{SL}(n,\mathbb{R})$) and by Benoist and Oh [BO] for the latter (using these results to study rational matrices with a given characteristic polynomial). Finally, we mention the work of Zamojski, giving counting (and equidistribution) results for rational matrices in a given characteristic polynomial in terms of the height of these matrices [Z1]. This leads to subtler issues, where unipotent flows or equidistribution of Hecke points do not apply. Instead, Zamojski uses measure rigidity of diagonal flows, building on [ELMV2]. Notably, by fixing a $\mathbb{Q}$-torus, Zamojski is able to handle $\mathbb{Q}$-tori in $\mathrm{SL}(n,\mathbb{R})$ for a general n; the fact that the $\mathbb{Q}$-torus is fixed allows one to avoid the need to use subconvexity results, and an additional averaging that is present in the problem studied by Zamojski allows handling intermediate subvarieties.

5 Applications regarding quantum ergodicity

In this section, we consider applications of homogeneous dynamics—namely, diagonal flows—to the study of Hecke-Maass cusp forms on $\mathbb{H}/\Gamma$ and their generalizations. We note that by the Selberg trace formula, Hecke-Maass forms can be considered as a dual object to the periodic A-trajectories considered in section 4.1, and though we are not aware of a dynamical result that makes use of this duality, the analogy is quite intriguing.

Consider first the case of $\mathbb{G}=\mathrm{PGL}(2)$. Then $K\backslash\mathbb{G}(\mathbb{R})/\mathbb{G}(\mathbb{Z})$ for $K=\mathrm{PSO}(2,\mathbb{R})$ can be identified with the modular surface[26] $\mathbb{H}/\mathrm{PSL}(2,\mathbb{Z})$.

To any primes p there is a correspondence—the *Hecke correspondence*, which we will denote by C_p^{Hecke}—assigning to every $x\in\mathbb{H}/\mathrm{PSL}(2,\mathbb{Z})$ a set of $p+1$-points in this space. This correspondence can be described explicitly as follows: If $x=[z]$ for $z\in\mathbb{H}$, then

$$C_p^{\mathrm{Hecke}}([z])=\Big\{[pz],[z/p],[(z+1)/p],\ldots,[(z+p-1)/p]\Big\}; \tag{5.1}$$

while each one of the points on the right-hand side depends on the choice of representative z of $[z]$, the collection of $p+1$-points is well-defined. Moreover this correspondence lifts to $\mathrm{PGL}(2,\mathbb{R})/\mathrm{PGL}(2,\mathbb{Z})$, giving to each $[x]\in\mathrm{PGL}(2,\mathbb{R})/\mathrm{PGL}(2,\mathbb{Z})$ a set (also denoted by $C_p^{\mathrm{Hecke}}([x])$) of $p+1$-points in $\mathrm{PGL}(2,\mathbb{R})/\mathrm{PGL}(2,\mathbb{Z})$ so that if $\pi_Y:\mathrm{PGL}(2,\mathbb{R})/\mathrm{PGL}(2,\mathbb{Z})\to\mathbb{H}/\mathrm{PSL}(2,\mathbb{Z})$ is

[26]To some, the modular curve.

the natural projection, then

$$\pi_Y\left(C_p^{\text{Hecke}}([x])\right) = C_p^{\text{Hecke}}(\pi_Y([x])).$$

An important property of the Hecke correspondence is its reflexivity:

$$[y] \in C_p^{\text{Hecke}}([x]) \text{ iff } [x] \in C_p^{\text{Hecke}}([y]). \tag{5.2}$$

Moreover, $C_p^{\text{Hecke}}([\bullet])$ is equivariant under left translations on $\text{PGL}(2,\mathbb{R})/\text{PGL}(2,\mathbb{Z})$—that is,

$$C_p^{\text{Hecke}}(h.[x]) = h.C_p^{\text{Hecke}}([x])$$

—which implies that on $\mathbb{H}/\text{PGL}(2,\mathbb{Z})$, each branch of $C_p^{\text{Hecke}}([\bullet])$ is a local isometry.

In terms of the projection (for $S = \{\infty\}$)

$$\pi_S : \text{PGL}(2,\mathbb{A})/\text{PGL}(2,\mathbb{Q}) \to \text{PGL}(2,\mathbb{R})/\text{PGL}(2,\mathbb{Z}),$$

if $a_p \in \text{PGL}(2,\mathbb{A})$ is the element equal to $\begin{pmatrix} p & \\ & 1 \end{pmatrix}$ in the $\mathbb{Q}_p$-component and the identity in every other component, then for any $[x] \in \text{PGL}(2,\mathbb{R})/\text{PGL}(2,\mathbb{Z})$,

$$C_p^{\text{Hecke}}([x]) = \pi_S(a_p.\pi_S^{-1}([x])). \tag{5.3}$$

Phrased slightly differently, if we consider an a_p orbit

$$\{[\tilde{x}], a_p.[\tilde{x}], \ldots, a_p^k.[\tilde{x}]\} \subset \text{PGL}(2,\mathbb{A})/\text{PGL}(2,\mathbb{Q})$$

and project it to $\text{PGL}(2,\mathbb{R})/\text{PGL}(2,\mathbb{Z})$, we will get a sequence of points $[x_0], \ldots, [x_k]$ with $[x_i] \in C_p^{\text{Hecke}}([x_{i-1}])$; moreover, it can be shown that this discrete trajectory is non-backtracking in the sense that $[x_i] \neq [x_{i+2}]$.

Using the Hecke correspondences $C_p^{\text{Hecke}}(\bullet)$ on $\mathbb{H}/\text{PSL}(2,\mathbb{Z})$, we define for any prime p a self-adjoint operator T_p, called *Hecke operators*, on $L^2(\mathbb{H}/\text{PSL}(2,\mathbb{Z}))$ by

$$(T_p f)(x) = p^{-1/2} \sum_{y \in C_p^{\text{Hecke}}(x)} f(y).$$

It follows from the relation between the Hecke correspondences and actions of diagonal elements in $\text{PGL}(2,\mathbb{A})/\text{PGL}(2,\mathbb{Q})$ (or directly from the definition of these correspondences in Equation (5.1)) that for every prime p, q, the operators T_p and T_q commute. Moreover, using the symmetry of the Hecke correspondences in (5.2) and the fact that each branch of $C_p^{\text{Hecke}}(\bullet)$ is a locally isometry, one sees that operators T_p are self-adjoint operators on $L^2(\mathbb{H}/\mathrm{PSL}(2,\mathbb{Z}))$ commuting with the Laplacian. Thus using the well-known spectral properties of the Laplacian, the discrete spectrum of Δ in $L^2(\mathbb{H}/\mathrm{PSL}(2,\mathbb{Z}))$ is spanned by joint eigenfunctions of Δ and all Hecke operators. Moreover, except for finitely many of them (in fact only the constant function), these eigenfunctions will be *cusp forms*—that is, eigenfunctions of Δ with the property that their integral over any periodic horocycle on $\mathbb{H}/\mathrm{PSL}(2,\mathbb{Z})$ is zero. These joint eigenfunctions are called *Hecke-Maass cusp forms.*[27]

A similar setup works in cocompact quotients. Let $\mathbb{M}$ be a quaternionic division algebra over $\mathbb{R}$ with $\mathbb{M}\otimes\mathbb{R}\cong M_{2\times 2}(\mathbb{R})$. Let $\mathbb{G}=\mathrm{PGL}(1,\mathbb{M})$. Then $\mathbb{G}(\mathbb{R})\cong\mathrm{PGL}(2,\mathbb{R})$, and if $\mathcal{O}_{\mathbb{M}}$ is a maximal order in $\mathbb{M}$, then $\Gamma=\mathcal{O}_{\mathbb{M}}^{\times}/\mathbb{Q}^{\times}$ is a cocompact lattice in $\mathbb{G}(\mathbb{R})$ commensurable to $\mathbb{G}(\mathbb{Z})$.[28] Taking as before $S=\{\infty\}$, then by Equation (4.6)

$$\pi_S:\mathbb{G}(\mathbb{A})/\mathbb{G}(\mathbb{Q})\to\bigsqcup_{i=1}^{k}\mathrm{PGL}(2,\mathbb{R})/\Gamma_i.$$

Indeed, maximal orders in $\mathbb{R}$-split quaternion algebras have class number 1, so in fact the image is a single quotient $\mathrm{PGL}(2,\mathbb{R})/\Gamma$; though if one takes a nonmaximal order $\mathcal{O}$, a disjoint union is needed (cf. [RS, section 2.2] and references therein). At any place p in which $\mathbb{M}\otimes\mathbb{Q}_p\cong M_{2\times 2}(\mathbb{Q}_p)$ (in particular, for all but finitely many places), we can choose (noncanonically) an element a_p as in the paragraph above (5.3), and this allows us using (5.3) to define Hecke correspondences $C_p^{\text{Hecke}}(\bullet)$ on $\mathrm{PGL}(2,\mathbb{R})/\Gamma$ and $\mathbb{H}/\Gamma$ as well as a family of self-adjoint operators T_p commuting with each other and with Δ on $L^2(\mathbb{H}/\Gamma)$. Then $L^2(\mathbb{H}/\Gamma)$ is spanned by Hecke-Maass forms—joint eigenfunctions of Δ and all T_p.[29]

Motivated in part by the study of Hecke-Maass forms, Rudnick and Sarnak made the following bold conjecture regarding any hyperbolic surface:

---

[27] For us cuspidality of the forms is not relevant, but only that these are eigenfunctions of Δ and all T_p in $L^2(\mathbb{H}/\mathrm{PSL}(2,\mathbb{Z}))$.

[28] To define $\mathbb{G}(\mathbb{Z})$ (at least in the way we do it in the survey) one needs to choose a $\mathbb{Q}$-embedding of $\mathbb{G}$ in some $\mathrm{SL}(N)$; reasonable people might do this in different ways, but they would all agree that the Γ we define is commensurable to $\mathbb{G}(\mathbb{Z})$.

[29] In this case, these joint eigenfunctions, even the constant function, satisfy the condition of cuspidality automatically (if somewhat vacuously) since there are no periodic horospheres!

CONJECTURE 11 (Quantum Unique Ergodicity[30] Conjecture [RS]). *Let M be a compact manifold of negative sectional curvature. Let $\{\phi_i\}$ be a complete orthonormal sequence of eigenfunctions of the Laplacian Δ on M ordered by eigenvalue. Then the probability measures $|\phi_i(x)|^2 \, d\,\mathrm{vol}_M(x)$ converge weak-* to the uniform measure on M.*

In their paper, Rudnick and Sarnak focus on the case of Hecke-Maass forms, showing that any weak-* limit of a subsequence $\left|\phi_{i_j}(x)\right|^2 \, d\,\mathrm{vol}_M(x)$ cannot be supported on finitely many closed geodesics. The multiplicities in the spectrum of the Laplacian on arithmetic surfaces $\mathbb{H}/\Gamma$ with Γ as above (or more generally the $\bigsqcup_{i=1}^{k} \mathbb{H}/\Gamma_i$ on which the Hecke correspondences are defined if we work with nonmaximal orders) are not well understood. Empirically these multiplicities seem to be bounded; indeed in favorable cases, the multiplicity of every eigenvalue of Δ seems to be one, so one does not seem to lose much by using a sequence of Hecke-Maass forms instead of an arbitrary sequence of eigenfunctions of Δ. Let M be an arithmetic surface (or a more general local symmetric manifold $K\backslash G/\Gamma$ with $K < G$ maximal compact, $G = \mathbb{G}(\mathbb{R})$ for $\mathbb{G}$ a semisimple $\mathbb{Q}$-group, and $\Gamma < G$ a congruence lattice [in particular commensurable to $\mathbb{G}(\mathbb{Z})$]). We shall call the closely related question to Conjecture 11—whether on such an M the probability distributions $|\phi_i(x)|^2 \, d\,\mathrm{vol}_M(x)$ corresponding to Hecke-Maass forms ϕ_i (i.e., joint eigenfunctions of Δ and all Hecke operators) converge weak-* to the uniform measure—the *Arithmetic Quantum Unique Ergodicity Problem.*

Conjecture 11 is to be compared to the following quantum ergodicity theorem of Schnirelman, Colin de Verdiere, and Zelditch:

THEOREM 5.1 (Schnirelman [Ś5], Colin de Verdiere [CdV], Zelditch [Z2]).
Let M be a compact manifold so that the geodesic flow on the unit tangent bundle of M is ergodic. Let $\{\phi_i\}$ be a complete orthonormal sequence of eigenfunctions of the Laplacian Δ on M ordered by eigenvalue. Then there is a subsequence i_j of density 1 so that restricted to this subsequence $\left|\phi_{i_j}(x)\right|^2 \, d\,\mathrm{vol}_M(x)$ converge weak- to the uniform measure on M.*

Using Theorem 2.8 as well as an entropy estimate by Bourgain and this author [BL1] we are able to prove the following, in particular establishing Arithmetic Quantum Unique Ergodicity for compact hyperbolic surfaces:

---

[30] Abbreviated QUE below.

THEOREM 5.2 ([L2]).
Let ϕ_i be an L^2-normalized sequence of Hecke-Maass forms on an arithmetic surface[31] *$M = \mathbb{H}/\Gamma$ with the lattice Γ either a congruence subgroup of $\mathrm{PGL}(2,\mathbb{Z})$ or arising from an order in a quaternion division algebra over $\mathbb{Q}$ as above. Suppose $|\phi_i(x)|^2\, d\mathrm{vol}_M(x)$ converges weak-* to a measure μ on $\mathbb{H}/\Gamma$. Then μ is, up to a multiplicative constant, the uniform measure on $\mathbb{H}/\Gamma$. In particular, arithmetic quantum unique ergodicity holds for compact arithmetic surfaces.*

What is not addressed in that theorem is the question whether there can be escape of mass for the sequence of measures $|\phi_i(x)|^2\, d\mathrm{vol}_M(x)$ for Γ, a congruence sublattice of $\mathrm{PGL}(2,\mathbb{Z})$. What is shown by Theorem 5.2 is that whatever remains converges to the uniform measure. This difficulty was resolved by Soundararajan using an elegant analytic argument:

THEOREM 5.3 (Soundararajan [S6]).
Let ϕ be a Hecke-Maass form on $\mathbb{H}/\mathrm{PSL}(2,\mathbb{Z})$, normalized to have L^2-norm 1. Then

$$\int_{\substack{|x|\le 1/2 \\ y>T}} \left|\phi(x+iy)\right|^2\, d\mathrm{vol}_{\mathbb{H}}(x+iy) \ll \frac{\log(eT)}{\sqrt{T}},$$

with an absolute implicit constant.

Theorem 5.3 implies in particular that any weak-* limit of a sequence of measures corresponding to Hecke-Maass forms $|\phi_i(x)|^2\, d\mathrm{vol}_M(x)$ is a probability measure; hence using Theorem 5.2 arithmetic QUE holds also in the noncompact case, where $\Gamma = \mathrm{PSL}(2,\mathbb{Z})$ (a similar argument works for congruence sublattices of $PSL(2,\mathbb{Z})$).

The entropy bound by Bourgain and this author in [BL1] gives a uniform upper bound on measures of small balls in $\mathrm{PGL}(2,\mathbb{R})/\Gamma$ of an appropriate lift of the measures $|\phi_i(x)|^2\, d\mathrm{vol}_M(x)$ to $\mathrm{PGL}(2,\mathbb{R})/\Gamma$. An alternative approach by Brooks and this author using only one Hecke operator gives a less quantitative entropy statement that is still sufficient to prove quantum unique ergodicity:

THEOREM 5.4 (Brooks and Lindenstrauss [BL2]).
Let ϕ_i be an L^2-normalized sequence of smooth functions on $\mathbb{H}/\Gamma$ with Γ an arithmetic cocompact lattice arising from an order in a quaternionic division algebra

[31] More generally, a finite union of surfaces $\bigsqcup_{i=1}^{k} \mathbb{H}/\Gamma_i$.

over $\mathbb{Q}$ as above. Assume that for some sequences $\lambda_i \to \infty$, $\lambda_{i,p} \in \mathbb{R}$, $\omega_i \to 0$,

$$\|\Delta\phi_i - \lambda_i\phi_i\|_2 \leq \lambda_i^{1/2}\omega_i \qquad \left\|T_p\phi_i - \lambda_{i,p}\phi_i\right\|_2 \leq \omega_i.$$

Then $|\phi_i(x)|^2\, d\operatorname{vol}(x)$ converges weak- to the uniform measure on $\mathbb{H}/\Gamma$.*

A surprising link between quantum unique ergodicity and the number of nodal domains for Hecke-Maass forms ϕ on $\mathbb{H}/\Gamma$ was discovered by Jang and Jung. If $\phi: M \to \mathbb{R}$ is a Δ eigenfunction (say, $\Delta\phi + \lambda\phi = 0$) on a compact surface M, Courant's Nodal Domain Theorem and the Weyl law imply that the number of nodal domain $\mathcal{N}(\phi)$ for ϕ satisfies $N(\phi) \ll \lambda$. However, it is well-known that in general $\mathcal{N}(\phi)$ could be much less: indeed in the 2 sphere $\mathbb{S}$ there is a sequence of Δ eigenfunctions with eigenvalues $\to \infty$ with $\mathcal{N}(\phi) \leq 3$, and in general it is very hard to bound the number of nodal domains from below; for more details, see, for example, [JJ] and the references given by that paper.

THEOREM 5.5 (Jang and Jung [JJ]).
Let ϕ_i be a sequence of Hecke-Maass forms on $\mathbb{H}/\Gamma$ for Γ, an arithmetic triangle group. Then $\lim_{i\to\infty} \mathcal{N}(\phi_i) = \infty$.

Triangle groups are discrete subgroups of the isometry group of $\mathbb{H}$ generated by reflections in three sides of a triangle with angles $\pi/a, \pi/b, \pi/c$. To such a group Γ' we can attach the orbifold $\mathbb{H}/\Gamma$, where $\Gamma < \Gamma'$ is the group of orientation preserving isometries. Γ' is generated by Γ and a reflection σ with $\sigma\Gamma\sigma = \Gamma$; hence σ induces an orientation reversing involution on $\mathbb{H}/\Gamma$ (for convenience we will also call Γ a triangle group). An arithmetic triangle group (there are only finitely many of these) is a triangle group that is commensurable to $\mathrm{PSL}(2,\mathbb{Z})$ or a lattice coming from a quaternionic order over $\mathbb{Q}$ as above. Examples are $\Gamma = \mathrm{PSL}(2,\mathbb{Z})$ itself (giving the triangle group $(\infty, 3, 3)$) and the compact triangle group $(2, 6, 6)$. Quantitative results giving $\mathcal{N}(\phi) \gg \lambda^{\frac{1}{27}-\epsilon}$ were given by Ghosh, Reznikov, and Sarnak, *assuming the Lindelöf hypothesis for L-functions of* GL(2) *forms* [GRS2, GRS1]; quantum unique ergodicity is used as a (partial) substitute for the Lindelöf hypothesis in [JJ].

In addition to considering a sequence of Hecke-Maass forms with eigenvalue $\lambda_i \to \infty$, one can consider on a given quotient $\mathbb{H}/\Gamma$ a sequence of holomorphic Hecke forms of weight $\to \infty$. These correspond to an irreducible $\mathrm{PGL}(2,\mathbb{R})$ representation in $L^2(\mathrm{PGL}(2,\mathbb{R})/\Gamma)$, which has no

$PO(2,\mathbb{R})$-invariant vectors. Using analytic techniques, and in particular making heavy use of the Fourier expansion of such forms in the cusp, Holowinsky and Soundararajan [HS] were able to prove an arithematic quantum unique ergodicity theorem for these automorphic forms *for* Γ*, a congruence subgroup of* $PGL(2,\mathbb{Z})$. Their techniques seem to be restricted to the noncompact case; the analogous question for compact quotients, and even on the sphere $\mathbb{S}$, remain important open questions.

A related question involves fixing the weight (or bounding the Laplacian eigenvalue) but considering a sequence of *newforms* on a tower $\mathbb{H}/\Gamma(N)$ of congruence subgroups. This question also makes sense in a purely discrete setting: instead of taking a quaternion division algebra $\mathbb{M}$ over $\mathbb{Q}$ that is split at infinity, one can take a definite quaternion algebra such as the Hamilton quaternions,

$$\mathbb{M}=\mathbb{Q}+i\mathbb{Q}+j\mathbb{Q}+ij\mathbb{Q} \qquad i^2=j^2=-1, ij=-ji.$$

For such $\mathbb{M}$, the group $PGL(1,\mathbb{M}\otimes\mathbb{R})$ is compact (in fact, isomorphic to $SO(3,\mathbb{R})$), so for $S=\{\infty,p\}$, the S-arithmetic projection

$$\pi_S: PGL(1,\mathbb{M}\otimes\mathbb{A})/PGL(1,\mathcal{O}_{\mathbb{M}})\to PGL(1,\mathbb{M}\otimes\mathbb{R}) \times PGL(1,\mathbb{M}\otimes\mathbb{Q}_p)/\Gamma \tag{5.4}$$

can be composed with a further projection by dividing the right-hand side of (5.4) from the left by the compact group $PGL(1,\mathbb{M}\otimes\mathbb{R})\times K(n)$, with $K(n)\le PGL(1,\mathbb{M}\otimes\mathbb{Z}_p)$ a compact open subgroup. This gives a map from $PGL(1,\mathbb{M}\otimes\mathbb{A})/PGL(1,\mathcal{O}_{\mathbb{M}})$ to a finite set. For every $q\neq p$, the q-Hecke correspondence gives this finite set the structure of a $q+1$–regular graph—the Lubotzky, Phillips, and Sarnak "Ramanujan graphs" [LPS]. Taking Hecke newforms on a sequence of these graphs with decreasing $K(n)$, Nelson [N] was able to use an adaptation of the method of proof of Theorem 5.2 to prove arithmetic QUE in the level aspect for newforms corresponding to principle series representations of $PGL(2,\mathbb{Q}_p)$. The restriction to principle series representations is the analogue in this context of the restriction in Theorem 5.2 to Maass forms (i.e., Laplacian eigunfunctions) as opposed to holomorphic forms.

The dynamical approach to arithmetic quantum unique ergodicity can be extended to other arithmetic quotients. Notable work in that direction was done by Silberman and Venkatesh:

THEOREM 5.6 (Silberman and Venkatesh [SV2, SV1]).
Let $\mathbb{G} = \mathrm{PGL}(1, \mathbb{M})$, where $\mathbb{M}$ is a degree n-division algebra over $\mathbb{Q}$, split over $\mathbb{R}$, for $n \geq 3$. Let $G = \mathbb{G}(\mathbb{R}) \cong \mathrm{PGL}(n, \mathbb{R})$, $K < G$ be maximal compact, and Γ be a lattice in G arising from a maximal order in $\mathbb{M}$. Let ϕ_j be a sequence of L^2-eigenfunctions of the ring of invariant differential operators on $K\backslash G$ as well as of all Hecke operators. Assume that the irreducible G-representation $H_j < L^2(G/\Gamma)$ of G spanned by left translations of ϕ_j is a principle series representation with parameter $\mathrm{t}_j \in i\mathfrak{a}_{\mathbb{R}}$ that stays away (uniformly in j) from the edges of the positive Weyl chamber in $i\mathfrak{a}_{\mathbb{R}}$. Then $|\phi_j(x)|^2 \, d\,\mathrm{vol}_{K\backslash G/\Gamma}$ converges to the uniform measure on $|\phi_j(x)|^2 \, d\,\mathrm{vol}_{K\backslash G/\Gamma}$.

The proof of Theorem 5.6 proceeds, similarly to that of Theorem 5.2, by lifting the probability measures $|\phi_j(x)|^2 \, d\,\mathrm{vol}_{K\backslash G/\Gamma}$ to an (approximately) A-invariant probability measure on G/Γ. This part of the argument, which is carried out in [SV2], uses only the information about the behavior of ϕ_j at the infinite place. Then using the information about other places—explicitly, the fact that ϕ_i is a sequence of eigenfunctions of all Hecke operators—positive entropy of any ergodic component of any weak-* limit as above is derived in [SV1]; this argument is related to the entropy estimate of Bourgain and this author in [BL1]. Once this entropy estimate is established, one can employ the measure classification results of [EKL] (special case of Theorem 2.4) to deduce the above arithmetic quantum unique ergodicity result.

6 Applications regarding Diophantine approximations

We started this survey with a historical introduction concerning some origins of the study of the rigidity properties of higher rank diagonal actions. One important such work was the paper of Cassels and Swinnerton-Dyer [CSD], relating Littlewood's Conjecture (Conjecture 2) to Conjecture 1a regarding bounded A-orbits in $\mathrm{PGL}(3, \mathbb{R})/\mathrm{PGL}(3, \mathbb{Z})$.

It is therefore not surprising that the significant progress obtained toward understanding these higher rank diagonal actions in the half century since Cassels and Swinnerton-Dyer's seminal paper has shed some light on Diophantine questions, though Littlewood's Conjecture itself remains at present quite open. Indeed, in terms of concrete (e.g., algebraic) numbers $\alpha, \beta \in \mathbb{R}$, for which Littlewood's conjecture can be verified—that is, so that

$$\inf_{n>0} n \, \|n\alpha\| \, \|n\beta\| = 0 \tag{6.1}$$

—I am not aware of any nontrivial examples beyond that given by Cassels and Swinnerton-Dyer in [CSD]: namely, those $\alpha, \beta \in \mathbb{R}$ so that the field $\mathbb{Q}(\alpha, \beta)$ they generate is of degree 3 over $\mathbb{Q}$.

The following was proved by Einsiedler, Katok, and this author in [EKL] using measure rigidity of the action of the diagonal group on $PGL(3, \mathbb{R})/PGL(3, \mathbb{Z})$:

THEOREM 6.1 (Einsiedler, Katok, Lindenstrauss). *For any $\epsilon > 0$, the set*

$$\left\{(\alpha, \beta \in [0, 1]^2 : \inf_{n>0} n \, \|n\alpha\| \, \|n\beta\| \geq \epsilon\right\} \tag{6.2}$$

has zero (upper) box dimension.

Zero upper box dimension simply means that the set in (6.2) can be covered by $\ll_{\delta,\epsilon} N^{\delta}$ squares of diameter N^{-1} for any $N > 0$. This of course implies that the set of exceptions to Littlewood's conjecture has Hausdorff dimension zero, and moreover for any $\alpha \in \mathbb{R}$ outside a set of Hausdorff dimension zero, (6.1) holds for every $\beta \in \mathbb{R}$.

This latter statement can actually be made more explicit: Let $\alpha \in [0, 1]$ be given. Expand α to a continued fraction

$$\alpha = \cfrac{1}{n_1 + \cfrac{1}{n_2 + \cfrac{1}{n_3 + \cfrac{1}{n_4 + \ldots}}}} \tag{6.3}$$

If the sequence n_d is unbounded, then already $\inf_{n>0} n \, \|n\alpha\| = 0$, and hence (6.1) holds for every β. For any k let $N_k(\alpha)$ denote the number of possible k-tuples of integers $i_1, \ldots, i_k$ appearing in the continued fraction expansion of α—that is, so that there is some $\ell \in \mathbb{N}$ so that

$$(i_1, \ldots, i_k) = (n_\ell, \ldots, n_{\ell+q-1}).$$

The following proposition follows readily from the techniques of [EKL]; we leave the details to the imagination of the interested reader, but the key point is that the condition given in the proposition on α can be used to verify the positive entropy condition.

PROPOSITION 6.2. *Let $\alpha \in [0,1]$ be such that the continued fraction expansion of α satisfies that*

$$\lim_{q\to\infty} \frac{\log(N_k(\alpha))}{k} > 0$$

(this limit exists by subadditivity). Then for any $\beta \in \mathbb{R}$, Equation (6.1) holds.

De Mathan and Teulié gave the following analogue to Conjecture 2:

CONJECTURE 12 (de Mathan and Teulié [dMT]). *For any prime p and any $\alpha \in \mathbb{R}$,*

$$\inf_{n>0} n\,|n|_p\,\|n\alpha\| = 0. \tag{6.4}$$

Recall that $|n|_p = p^{-k}$ if $p^k \mid n$ but $p^{k+1} \nmid n$; hence (6.4) is equivalent to

$$\inf_{n>0,k\geq 0} n\left\|np^k\alpha\right\| = 0.$$

Note that by Furstenberg's theorem (Theorem 1.1)[32] for any two distinct primes p, q,

$$\inf_{n>0} n\,|n|_p\,|n|_q\,\|n\alpha\| = 0$$

since either α is rational, in which case $\underline{\lim}\,\|n\alpha\| = 0$, or $\{p^k q^\ell \alpha \bmod 1\}$ is dense in $[0,1]$, in particular $\inf_{k,\ell\geq 0}\left\|p^k q^\ell \alpha\right\| = 0$.

By a variant of the argument of Cassels and Swinnerton-Dyer, de Mathan and Teulié show that (6.4) holds for quadratic irrational $\alpha \in \mathbb{R}$. Interestingly, Adiceam, Nesharim, and Lunnon give in [ANL] a completely explicit (and nonobvious) counterexample to the function field analogue of Conjecture 12, also stated in [dMT]. Using a similar argument to [EKL], but using the measure classification result of [L2] instead of that in [EKL], Einsiedler and Kleinbock prove in [EK] that for any $\epsilon > 0$, the set of $\alpha \in [0,1]$ for which

$$\inf_{n>0} n\,|n|_p\,\|n\alpha\| \geq \epsilon$$

has zero box dimension.

Theorem 6.1, unlike many of the other applications we gave for the rigidity of higher rank diagonal actions, only tells us that something is true outside an

[32] Another result we cited in the introduction that played a central role in the development of the subject!

unspecified, but small, set of exceptions. The following interesting application of measure rigidity by Einsiedler, Fishman, and Shapira gives an *everywhere* statement, in the spirit of Conjecture 12:

THEOREM 6.3 (Einsiedler, Fishman, and Shapira [EFS]).
For any $\alpha \in [0,1]$, let $n_1(\alpha), n_2(\alpha)\dots$ be the digits in the continued fraction expansion of α as in Equation (6.3). Denote by $c(\alpha) = \overline{\lim}_{i\to\infty} n_i(\alpha)$. Then for every irrational $\alpha \in [0,1]$,

$$\sup_n c(n\alpha \bmod 1) = \infty.$$

Somewhat unusually, the proof of this theorem actually involves adelic dynamics [L1], a result closely related to Theorem 2.8 but with no explicit entropy assumption (the necessary entropy assumption is derived in [L1] from the dynamical assumptions by a variation on the argument of [BL1]).

David Simmons observed that Theorem 6.3 implies in particular that for any $\psi : \mathbb{N} \to \mathbb{R}$ with $\psi(t) \to \infty$ as $t \to \infty$, for any $\alpha \in [0,1]$,

$$\underline{\lim}_{Q\to\infty} Q \min_{q\le Q, m\le\psi(q)} \|qm\alpha\| = 0, \tag{6.5}$$

answering a question of Bourgain related to the work of Blomer, Bourgain, Radziwill, and Rudnick [BBRR], where they show that if α is a quadratic irrational (with some additional restrictions, removed later by Dan Carmon), for every $\epsilon > 0$, one has that $\underline{\lim}_{Q\to\infty} Q^{2-\epsilon} \min_{q,m\le Q} \|qm\alpha\| = 0$; they also show this for a.e. α (but their techniques do not show (6.5) for every α, even when $\psi(q) = q$).

Write

$$A(Q,Q') = Q \min_{q\le Q, m\le Q'} \|qm\alpha\|.$$

By an (easy) result of Dirichlet, $Q \min_{q\le Q} \|q\alpha\| < 1$ for all α, Q; hence for any Q, Q' we have the trivial estimate $A(Q,Q') \le 1$. By considering an $\alpha \in [0,1]$ that has a sequence of extremely good approximations $\frac{p_i}{q_i}$ with q_i prime, it is easy to see that there are uncountably many α for which $\overline{\lim}_{Q\to\infty} A(Q,Q) = 1$. However, one can still give the following strengthening of (6.5), whose details will appear in the forthcoming [EL4]:

THEOREM 6.4 (Einsiedler and Lindenstrauss).
For any $\psi \to \infty$ and any $\alpha \in \mathbb{R}$, one has that $A(2^k, \psi(2^k)) \to 0$ outside possibly a subsequence of density zero.

This theorem also relies on a measure classification result for higher rank diagonal actions, in this case Theorem 2.6.

References

[ANL] F. Adiceam, E. Nesharim, and F. Lunnon, *On the t-adic Littlewood conjecture*, arXiv e-prints (2018Jun), arXiv:1806.04478.

[AES1] M. Aka, M. Einsiedler, and U. Shapira, *Integer points on spheres and their orthogonal lattices (with an appendix by ruixiang zhang)*, arXiv e-prints (2015Feb), arXiv:1502.04209v1.

[AES2] M. Aka, M. Einsiedler, and U. Shapira, *Integer points on spheres and their orthogonal lattices*, Invent. Math. **206** (2016), no. 2, 379–396. MR3570295

[AEW] M. Aka, M. Einsiedler, and A. Wieser, *Planes in four space and four associated CM points*, arXiv e-prints (2019Jan), arXiv:1901.05833.

[BO] Y. Benoist and H. Oh, *Equidistribution of rational matrices in their conjugacy classes*, Geom. Funct. Anal. **17** (2007), no. 1, 1–32. MR2306651

[B1] D. Berend, *Multi-invariant sets on tori*, Trans. Amer. Math. Soc. **280** (1983), no. 2, 509–532. MR85b:11064

[B2] D. Berend, *Multi-invariant sets on compact abelian groups*, Trans. Amer. Math. Soc. **286** (1984), no. 2, 505–535. MR86e:22009

[BBRR] V. Blomer, J. Bourgain, M. Radziwill, and Z. Rudnick, *Small gaps in the spectrum of the rectangular billiard*, Ann. Sci. Éc. Norm. Supér. (4) **50** (2017), no. 5, 1283–1300. MR3720030

[B3] M. D. Boshernitzan, *Elementary proof of Furstenberg's Diophantine result*, Proc. Amer. Math. Soc. **122** (1994), no. 1, 67–70. MR1195714 (94k:11085)

[BL1] J. Bourgain and E. Lindenstrauss, *Entropy of quantum limits*, Comm. Math. Phys. **233** (2003), no. 1, 153–171. MR1957 735

[BLMV] J. Bourgain, E. Lindenstrauss, P. Michel, and A. Venkatesh, *Some effective results for* $\times a \times b$, Ergodic Theory Dynam. Systems **29** (2009), no. 6, 1705–1722. MR2563089 (2011e:37022)

[BdS] E. Breuillard and N. de Saxce, private communications.

[BL2] S. Brooks and E. Lindenstrauss, *Joint quasimodes, positive entropy, and quantum unique ergodicity*, Invent. Math. **198** (2014), no. 1, 219–259. MR3260861

[C1] J. W. S. Cassels, *The product of n inhomogeneous linear forms in n variables*, J. London Math. Soc. **27** (1952), 485–492. MR0050632

[C2] J. W. S. Cassels, *An introduction to Diophantine approximation*, Cambridge Tracts in Mathematics and Mathematical Physics, No. 45, Cambridge University Press, New York, 1957. MR0087708

[CSD] J. W. S. Cassels and H. P. F. Swinnerton-Dyer, *On the product of three homogeneous linear forms and the indefinite ternary quadratic forms*, Philos. Trans. Roy. Soc. London. Ser. A. **248** (1955), 73–96. MR17,14f

[CdV] Y. Colin de Verdiére, *Ergodicité et fonctions propres du laplacien*, Comm. Math. Phys. **102** (1985), no. 3, 497–502. MR87d:58145

[C3] D. A. Cox, *Primes of the form* $x^{+}ny^2$*: fermat, class field theory, and complex multiplication*, Second edition, Pure and Applied Mathematics (Hoboken), John Wiley & Sons, Inc., Hoboken, NJ, 2013. MR3236783

[D1] S. G. Dani, *Invariant measures and minimal sets of horospherical flows*, Invent. Math. **64** (1981), no. 2, 357–385. MR629475

[D2] S. G. Dani, *On orbits of unipotent flows on homogeneous spaces II*, Ergodic Theory Dynam. Systems **6** (1986), no. 2, 167–182. MR857195 (88e:58052)

[DM1] S. G. Dani and G. A. Margulis, *Values of quadratic forms at primitive integral points*, Invent. Math. **98** (1989), no. 2, 405–424. MR1016271 (90k:22013b)

[DM2] S. G. Dani and G. A. Margulis, *Orbit closures of generic unipotent flows on homogeneous spaces of* SL(3,R), Math. Ann. **286** (1990), no. 1–3, 101–128. MR1032925 (91k:22026)

[DM3] S. G. Dani and G. A. Margulis, *Asymptotic behaviour of trajectories of unipotent flows on homogeneous spaces*, Proc. Indian Acad. Sci. Math. Sci. **101** (1991), no. 1, 1–17. MR1101994

[DM4] S. G. Dani and G. A. Margulis, *Limit distributions of orbits of unipotent flows and values of quadratic forms*, I. M. Gelfand seminar, pp. 91–137. Advances in Soviet Mathematics, 16, Part 1, American Mathematical Society, Providence, RI, 1993. MR1237827 (95b:22024)

[DS1] O. David and U. Shapira, *Dirichlet shapes of unit lattices and escape of mass*, Int. Math. Res. Not. IMRN **9** (2018), 2810–2843. MR3801497

[DS2] O. David and U. Shapira, *Equidistribution of divergent orbits and continued fraction expansion of rationals*, J. Lond. Math. Soc. (2) **98** (2018), no. 1, 149–176. MR3847236

[dMT] B. de Mathan and O. *Teulié, Problémes diophantiens simultanés*, Monatsh. Math. **143** (2004), no. 3, 229–245. MR2103807

[D3] W. Duke, *Hyperbolic distribution problems and half-integral weight Maass forms*, Invent. Math. **92** (1988), no. 1, 73–90. MR931205 (89d:11033)

[DFI] W. Duke, J. B. Friedlander, and H. Iwaniec, *The subconvexity problem for Artin L-functions*, Invent. Math. **149** (2002), no. 3, 489–577. MR1923476 (2004e: 11046)

[E] M. Einsiedler, *Invariant subsets and invariant measures for irreducible actions on zero-dimensional groups*, Bull. London Math. Soc. **36** (2004), no. 3, 321–331. MR2040136

[EFS] M. Einsiedler, L. Fishman, and U. Shapira, *Diophantine approximations on fractals*, Geom. Funct. Anal. **21** (2011), no. 1, 14–35. MR2773102

[EKL] M. Einsiedler, A. Katok, and E. Lindenstrauss, *Invariant measures and the set of exceptions to Littlewood's conjecture*, Ann. of Math. (2) **164** (2006), no. 2, 513–560. MR2247967

[EK] M. Einsiedler and D. Kleinbock, *Measure rigidity and p-adic Littlewood-type problems*, Compos. Math. **143** (2007), no. 3, 689–702. MR2330443

[EL1] M. Einsiedler and E. Lindenstrauss, *Rigidity properties of* Z^d*-actions on tori and solenoids*, Electron. Res. Announc. Amer. Math. Soc. **9** (2003), 99–110.

[EL2] M. Einsiedler and E. Lindenstrauss, *On measures invariant under tori on quotients of semisimple groups*, Ann. of Math. (2) **181** (2015), no. 3, 993–1031. MR3296819

[EL3] M. Einsiedler and E. Lindenstrauss, *Joinings of higher rank torus actions on homogeneous spaces*, 2016, to appear in Publications Mathématiques de l'IHÉS.

[EL4] M. Einsiedler and E. Lindenstrauss, *Rigidity of non-maximal torus actions and unipotent quantitative recurrence*, 2019, in preparation.

[ELMV1] M. Einsiedler, E. Lindenstrauss, P. Michel, and A. Venkatesh, The *distribution of periodic torus orbits on homogeneous spaces*, Duke Math. J. **148** (2009), no. 1, 119–174.

[ELMV2] M. Einsiedler, E. Lindenstrauss, P. Michel, and A. Venkatesh, *Distribution of periodic torus orbits and Duke's theorem for cubic fields*, Ann. of Math. (2) **173** (2011), no. 2, 815–885. See also erratum, in preparation, for ground fields other than $\mathbb{Q}$.

[ELMV3] M. Einsiedler, E. Lindenstrauss, P. Michel, and A. Venkatesh, *The distribution of closed geodesics on the modular surface, and Duke's theorem*, Enseign. Math. (2) **58** (2012), no. 3–4, 249–313. MR3058601

[ELM] M. Einsiedler, E. Lindenstrauss, and A. Mohammadi, *Diagonal actions in positive characteristic*, 2017, arXiv:1705.10418, to appear in Duke Math. Journal.

[EMMV] M. Einsiedler, G. Margulis, A. Mohammadi, and A. Venkatesh, *Effective equidistribution and property tau*, 2015, to appear in JAMS.

[EMV1] M. Einsiedler, G. Margulis, and A. Venkatesh, *Effective equidistribution for closed orbits of semisimple groups on homogeneous spaces*, Inventiones mathematicae **177** (2009), no. 1, 137–212.

[EMS1] A. Eskin, S. Mozes, and N. Shah, *Non-divergence of translates of certain algebraic measures*, Geom. Funct. Anal. **7** (1997), no. 1, 48–80. MR1437473

[EMS2] A. Eskin, S. Mozes, and N. Shah, *Unipotent flows and counting lattice points on homogeneous varieties*, Ann. of Math. (2) **143** (1996), no. 2, 253–299. MR1381987 (97d:22012)

[EMV2] J. S. Ellenberg, P. Michel, and A. Venkatesh, *Linnik's ergodic method and the distribution of integer points on spheres*, Automorphic representations and L-functions, pp. 119–185. Tata Institute of Fundamental Research Studies in Mathematics, 22, Tata Institute of Fundamental Research, Mumbai, 2013. MR3156852

[EV] J. S. Ellenberg and A. Venkatesh, *Reflection principles and bounds for class group torsion*, Int. Math. Res. Not. IMRN **1** (2007), 18. MR2331900

[EO] A. Eskin and H. Oh, *Representations of integers by an invariant polynomial and unipotent flows*, Duke Math. J. **135** (2006), no. 3, 481–506. MR2272974

[F1] J. Feldman, *A generalization of a result of R. Lyons about measures on* $[0, 1)$, Israel J. Math. **81** (1993), no. 3, 281–287. MR95f:28020

[F2] H. Furstenberg, *Disjointness in ergodic theory, minimal sets, and a problem in Diophantine approximation*, Math. Systems Theory **1** (1967), 1–49. MR35#4369

[GRS1] A. Ghosh, A. Reznikov, and P. Sarnak, *Nodal domains of Maass forms I*, Geom. Funct. Anal. **23** (2013), no. 5, 1515–1568. MR3102912

[GRS2] A. Ghosh, A. Reznikov, and P. Sarnak, *Nodal domains of Maass forms, II*, Amer. J. Math. **139** (2017), no. 5, 1395–1447. MR3702502

[GO] A. Gorodnik and H. Oh, *Rational points on homogeneous varieties and equidistribution of adelic periods*, Geom. Funct. Anal. **21** (2011), no. 2, 319–392. With an appendix by Mikhail Borovoi. MR2795511

[HM] G. Harcos and P. Michel, *The subconvexity problem for Rankin-Selberg L-functions and equidistribution of Heegner points II*, Invent. Math. **163** (2006), no. 3, 581–655. MR2207235 (2007j:11063)

[H1] M. Hochman, *Geometric rigidity of* $\times m$ *invariant measures*, J. Eur. Math. Soc. (JEMS) **14** (2012), no. 5, 1539–1563. MR2966659

[HS] R. Holowinsky and K. Soundararajan, *Mass equidistribution for Hecke eigenforms*, Ann. of Math. (2) **172** (2010), no. 2, 1517–1528. MR2680499

[H2] B. Host, *Nombres normaux, entropie, translations*, Israel J. Math. **91** (1995), no. 1–3, 419–428. MR96g:11092

[I] H. Iwaniec, *Fourier coefficients of modular forms of half-integral weight*, Invent. Math. **87** (1987), no. 2, 385–401. MR870736

[IK] H. Iwaniec and E. Kowalski, *Analytic number theory*, American Mathematical Society Colloquium Publications, vol. 53, American Mathematical Society, Providence, RI, 2004. MR2061214

[JJ] S. U. Jang and J. Jung, *Quantum unique ergodicity and the number of nodal domains of eigenfunctions*, J. Amer. Math. Soc. **31** (2018), no. 2, 303–318. MR3758146

[J] A. S. A. *Johnson, Measures on the circle invariant under multiplication by a nonlacunary subsemigroup of the integers*, Israel J. Math. **77** (1992), no. 1–2, 211–240. MR93m:28019

[KK1] B. Kalinin and A. Katok, *Invariant measures for actions of higher rank abelian groups*, Smooth ergodic theory and its applications (Seattle, WA, 1999), pp. 593–637. Proceedings of Symposia in Pure Mathematics, 69, American Mathematical Society, Providence, RI, 2001. MR2002i:37035

[KK2] B. Kalinin and A. Katok, *Measurable rigidity and disjointness for* $\mathbb{Z}^k$ *actions by toral automorphisms*, Ergodic Theory Dynam. Systems **22** (2002), no. 2, 507–523. MR1898 802

[KS1] B. Kalinin and R. Spatzier, *Rigidity of the measurable structure for algebraic actions of higher-rank abelian groups*, Ergodic Theory Dynam. Systems **25** (2005), no. 1, 175–200. MR2122918 (2005k:37008)

[KS2] A. Katok and R. J. Spatzier, *Invariant measures for higher-rank hyperbolic abelian actions*, Ergodic Theory Dynam. Systems 16 (1996), no. 4, 751–778. MR97d:58116

[KS3] A. Katok and R. J. Spatzier, *Corrections to: "Invariant measures for higher-rank hyperbolic abelian actions" [Ergodic Theory Dynam. Systems 16 (1996), no. 4, 751–778; MR 97d:58116]*, Ergodic Theory Dynam. Systems **18** (1998), no. 2, 503–507. MR99c:58093

[K1] I. Khayutin, *Arithmetic of double torus quotients and the distribution of periodic torus orbits* (2015), arXiv:1510.08481

[K2] I. Khayutin, *Joint equidistribution of CM points*, Ann. of Math. (2) **189** (2019), no. 1, 145–276. MR3898709

[KS4] B. Kitchens and K. Schmidt, *Markov subgroups of* $(\mathbb{Z}/2\mathbb{Z}^{\mathbb{Z}^2})$, Symbolic dynamics and its applications (New Haven, CT, 1991), pp. 265–283. Contemporary Mathematics, 135, American Mathematical Society, Providence, RI, 1992. MR93k:58136

[L1] E. Lindenstrauss, *Arithmetic quantum unique ergodicity and adelic dynamics*, Current developments in mathematics 2004, pp. 111–139. International Press of Boston, Somerville, MA, 2006.

[L2] E. Lindenstrauss, *Invariant measures and arithmetic quantum unique ergodicity*, Ann. of Math. (2) **163** (2006), no. 1, 165–219. MR2195133

[LMMS] E. Lindenstrauss, G. Margulis, A. Mohammadi, and N. Shah, *Quantitative behavior of unipotent flows and an effective avoidance principle*, arXiv e-prints (2019Mar), arXiv:1904.00290

[LS] E. Lindenstrauss and U. Shapira, *Homogeneous orbit closures and applications*, Ergodic Theory Dynam. Systems **32** (2012), no. 2, 785–807, special volume in memory of D. Rudolph. MR2901371

[LW1] E. Lindenstrauss and Z. Wang, *Topological self-joinings of Cartan actions by toral automorphisms*, Duke Math. J. **161** (2012), no. 7, 1305–1350. MR2922376

[LW2] E. Lindenstrauss and B. Weiss, *On sets invariant under the action of the diagonal group*, Ergodic Theory Dynam. Systems **21** (2001), no. 5, 1481–1500. MR2002j:22009

[L3] Yu. V. Linnik, *The asymptotic geometry of the Gaussian genera; an analogue of the ergodic theorem*, Dokl. Akad. Nauk SSSR (N.S.) **108** (1956), 1018–1021. MR0081919

[L4] Yu. V. Linnik, *Ergodic properties of algebraic fields*, translated from Russian by M. S. Keane, Ergebnisse der Mathematik und ihrer Grenzgebiete, Band 45, Springer-Verlag New York Inc., New York, 1968. MR0238801 (39 #165)

[LPS] A. Lubotzky, R. Phillips, and P. Sarnak, *Ramanujan graphs*, Combinatorica **8** (1988), no. 3, 261–277. MR963118

[L5] R. Lyons, *On measures simultaneously 2- and 3-invariant*, Israel J. Math. **61** (1988), no. 2, 219–224. MR89e:28031

[M1] F. Manners, *A solution to the pyjama problem*, Invent. Math. **202** (2015), no. 1, 239–270. MR3402799

[M2] G. A. Margulis, *The action of unipotent groups in a lattice space*, Mat. Sb. (N.S.) **86(128)** (1971), 552–556. MR0291352 (45 #445)

[M6] G. A. Margulis, *Formes quadratriques indéfinies et flots unipotents sur les espaces homogénes*, C. R. Acad. Sci. Paris Sér. I Math. *304* (1987), no. 10, 249–253. MR882782 (88f:11027)

[M3] G. A. Margulis, *Discrete subgroups and ergodic theory*, Number theory, trace formulas and discrete groups (Oslo, 1987), pp. 377–398. Academic Press, Boston, MA, 1989. MR90k:22013a

[M4] G. A. Margulis, *Discrete subgroups of semisimple Lie groups*, Ergebnisse der Mathematik und ihrer Grenzgebiete (3) [Results in Mathematics and Related Areas (3)], vol. 17, Springer-Verlag, Berlin, 1991. MR1090825 (92h:22021)

[M5] G. A. Margulis, *Oppenheim conjecture*, Fields medallists' lectures, pp. 272–327. World Scientific Series in 20th Century Mathematics, 5, World Scientific Publishing, River Edge, NJ, 1997. MR99e:11046

[M7] G. Margulis, *Problems and conjectures in rigidity theory*, Mathematics: frontiers and perspectives, pp. 161–174. American Mathematical Society, Providence, RI, 2000. MR2001d:22008

[MT1] G. A. Margulis and G. M. Tomanov, *Invariant measures for actions of unipotent groups over local fields on homogeneous spaces*, Invent. Math. **116** (1994), no. 1–3, 347–392. MR95k:22013

[MT2] G. A. Margulis and G. M. Tomanov, *Measure rigidity for almost linear groups and its applications*, J. Anal. Math. **69** (1996), 25–54. MR1428093 (98i:22016)

[M8] F. Maucourant, *A nonhomogeneous orbit closure of a diagonal subgroup*, Ann. of Math. (2) **171** (2010), no. 1, 557–570. MR2630049

[MV1] P. Michel and A. Venkatesh, *Equidistribution, L-functions and ergodic theory: on some problems of Yu. Linnik*, International Congress of Mathematicians, vol. 2, 2006, pp. 421–457. European Mathematical Society, Zürich, 2006. MR2275604

[MV2] P. Michel and A. Venkatesh, *The subconvexity problem for* GL_2, Publ. Math. Inst. Hautes Études Sci. **111** (2010), 171–271. MR2653249 (2012c:11111)

[M9] S. Mozes, *On closures of orbits and arithmetic of quaternions*, Israel J. Math. **86** (1994), no. 1–3, 195–209. MR95k:22014

[MS] S. Mozes and N. Shah, *On the space of ergodic invariant measures of unipotent flows*, Ergodic Theory Dynam. Systems **15** (1995), no. 1, 149–159. MR1314973 (95k:58096)

[N] P. D. Nelson, *Microlocal lifts and quantum unique ergodicity on* $GL_2(\mathbb{Q}_p)0$, Algebra Number Theory **12** (2018), no. 9, 2033–2064. MR3894428

[OS] H. Oh and N. A. Shah, *Limits of translates of divergent geodesics and integral points on one-sheeted hyperboloids,* Israel J. Math. **199** (2014), no. 2, 915–931. MR3219562

[PS] Y. Peterzil and S. Starchenko, *Algebraic and o-minimal flows on complex and real tori,* Adv. Math. **333** (2018), 539–569. MR3818086

[PR] V. Platonov and A. Rapinchuk, *Algebraic groups and number theory,* Pure and Applied Mathematics, vol. 139, Academic Press, Inc., Boston, MA, 1994. Translated from the 1991 Russian original by Rachel Rowen. MR1278263

[R1] M. Ratner, *Horocycle flows, joinings and rigidity of products,* Ann. of Math. (2) **118** (1983), no. 2, 277–313. MR85k:58063

[R2] M. Ratner, *On measure rigidity of unipotent subgroups of semisimple groups,* Acta Math. **165** (1990), no. 3–4, 229–309. MR91m:57031

[R3] M. Ratner, *On Raghunathan's measure conjecture,* Ann. of Math. (2) **134** (1991), no. 3, 545–607. MR93a:22009

[R4] M. Ratner, *Raghunathan's topological conjecture and distributions of unipotent flows,* Duke Math. J. **63** (1991), no. 1, 235–280. MR93f:22012

[R5] M. Ratner, *Raghunathan's conjectures for Cartesian products of real and p-adic Lie groups,* Duke Math. J. **77** (1995), no. 2, 275–382. MR96d:22015

[RS] Z. Rudnick and P. Sarnak, *The behaviour of eigenstates of arithmetic hyperbolic manifolds,* Comm. Math. Phys. **161** (1994), no. 1, 195–213. MR95m:11052

[R6] D. J. Rudolph, $\times 2$ *and* $\times 3$ *invariant measures and entropy,* Ergodic Theory Dynam. Systems **10** (1990), no. 2, 395–406. MR91g:28026

[S1] W. M. Schmidt, *Diophantine approximation,* Lecture Notes in Mathematics, vol. 785, Springer, Berlin, 1980. MR568710

[S2] U. Shapira, *A solution to a problem of Cassels and Diophantine properties of cubic numbers,* Ann. of Math. (2) **173** (2011), no. 1, 543–557. MR2753608 (2011k:11094)

[S3] U. Shapira, *Full escape of mass for the diagonal group,* Int. Math. Res. Not. IMRN **15** (2017), 4704–4731. MR3685113

[SZ] U. Shapira and C. Zheng, *Limiting distributions of translates of divergent diagonal orbits,* arXiv e-prints (2017Dec), arXiv:1712.00630

[SV1] L. Silberman and A. Venkatesh, *Entropy bounds for Hecke eigenfunctions on division algebras,* 2006, unpublished preprint.

[SV2] L. Silberman and A. Venkatesh, *On quantum unique ergodicity for locally symmetric spaces,* Geom. Funct. Anal. **17** (2007), no. 3, 960–998. MR2346281

[S4] B. F. Skubenko, *The asymptotic distribution of integers on a hyperboloid of one sheet and ergodic theorems,* Izv. Akad. Nauk SSSR Ser. Mat. **26** (1962), 721–752. MR0151436

[Ś5] A. I. Šnirelman, *Ergodic properties of eigenfunctions,* Uspehi Mat. Nauk **29** (1974), no. 6(180), 181–182. MR53#6648

[S6] K. Soundararajan, *Quantum unique ergodicity for* $\mathrm{SL}_2(\mathbb{Z})/\mathbb{H}$, Ann. of Math. (2) **172** (2010), no. 2, 1529–1538. MR2680500

[SD] H. P. F. Swinnerton-Dyer, *A brief guide to algebraic number theory,* London Mathematical Society Student Texts, vol. 50, Cambridge University Press, Cambridge, 2001. MR1826558

[T1] G. Tomanov, *Actions of maximal tori on homogeneous spaces,* Rigidity in dynamics and geometry (Cambridge, 2000), pp. 407–424. Springer, Berlin, 2002. MR1919414

[T2] G. Tomanov, *Values of decomposable forms at S-integral points and orbits of tori on homogeneous spaces,* Duke Math. J. **138** (2007), no. 3, 533–562. MR2322686

[T3] G. Tomanov, *Locally divergent orbits on Hilbert modular spaces*, 2010, preprint.

[TW] G. Tomanov and B. Weiss, *Closed orbits for actions of maximal tori on homogeneous spaces*, Duke Math. J. **119** (2003), no. 2, 367–392. MR1997950 (2004g:22006)

[T4] J. Tsimerman, *The André-Oort conjecture for* $\mathcal{A}_g$, Ann. of Math. (2) **187** (2018), no. 2, 379–390. MR3744855

[W1] Z. Wang, *Rigidity of commutative non-hyperbolic actions by toral automorphisms*, Ergodic Theory Dynam. Systems **32** (2012), no. 5, 1752–1782. MR2974218

[W2] A. Wieser, *Linnik's problems and maximal entropy methods*, Monatsh. Math. **190** (2019), no. 1, 153–208. MR3998337

[Z1] T. Zamojski, *Counting rational matrices of a fixed irreducible characteristic polynomial*, ProQuest LLC, Ann Arbor, MI, 2010, Thesis (Ph.D.), University of Chicago. MR2941378

[Z2] S. Zelditch, *Uniform distribution of eigenfunctions on compact hyperbolic surfaces*, Duke Math. J. **55** (1987), no. 4, 919–941. MR89d:58129

12

MANFRED EINSIEDLER AND AMIR MOHAMMADI

EFFECTIVE ARGUMENTS IN UNIPOTENT DYNAMICS

Dedicated to Gregory Margulis

Abstract. We survey effective arguments concerning unipotent flows on locally homogeneous spaces.

1 Introduction

In the mid-1980s Margulis resolved the long-standing Oppenheim conjecture by establishing a special case of Raghunathan's conjecture. Further works by Dani and Margulis in the context of the Oppenheim conjecture and Ratner's full resolution of Raghunathan's conjectures have become a cornerstone for many exciting applications in dynamics and number theory.

Let us briefly recall the setup. Let G be a connected Lie group and let $\Gamma \subset G$ be a lattice (i.e., a discrete subgroup with finite covolume) and $X = G/\Gamma$. Let $H \subset G$ be a closed subgroup of G. This algebraic setup gives hope for the following fundamental dynamical problem.

. Describe the behavior of the orbit Hx for *every* point $x \in X$.

However, without further restrictions on H this question cannot have any meaningful answer; for example, if G is semisimple and H is a one-parameter $\mathbb{R}$-diagonalizable subgroup of G, then the time one map is partially hyperbolic (and in fact has positive entropy and is a Bernoulli automorphism) and the

---

Manfred Einsiedler. Departement Mathematik. ETH Zürich, Rämistrasse 101, 8092, Zürich, Switzerland
manfred.einsiedler@math.ethz.ch

Amir Mohammadi. Department of Mathematics. University of California, San Diego, CA 92093, USA
ammohammadi@ucsd.edu

Einsiedler acknowledges support by the SNF (grant 178958).
Mohammadi acknowledges support by the NSF.

behavior of orbits can be rather wild, giving rise to fractal orbit closures (see, e.g., [46]). There is, however, a very satisfying answer when H is generated by unipotent subgroups—for example, when H is a unipotent or a connected semisimple subgroup; in these cases Ratner's theorems imply that closures of all orbits are properly embedded manifolds; see section 5.

These results, however, are not effective; for example, they do not provide a rate at which the orbit fills out its closure. As is already stated by Margulis in his ICM lecture [55], it is much anticipated and quite a challenging problem to give effective versions of these theorems. It is worth noting that except for uniquely ergodic systems, such a rate would generally depend on delicate properties of the point x and the acting group H. Already for an irrational rotation of a circle, the *Diophanine properties* of the rotation enter the picture. The purpose of this essay is to provide an overview of effective results in this context of unipotently generated subgroups.

Let us further mention that there have been fantastic developments both for other choices of H and beyond the homogeneous setting. In fact the papers [50], [24], and [25] give partial solutions to the conjectures by Margulis [56] concerning higher rank diagonalizable flows; the papers [10] and [5, 6, 7] (inspired by the methods of Eskin, Margulis, and Mozes [31]) concern the classification of stationary measures; and [34] and [35] concern the $SL_2(\mathbb{R})$ action on moduli spaces and also apply the method developed for stationary measures. These works, with the exception of [10], are all qualitative, and an effective account of these would be very intriguing. This essay, however, will focus on the case where H is generated by unipotent elements.

We note that good effective bounds for equidistribution of unipotent orbits can have far-reaching consequences. Indeed the Riemann hypothesis is equivalent to giving an error term of the form $O_\epsilon(y^{\frac{3}{4}+\epsilon})$ for equidistribution of periodic horocycles of period $1/y$ on the modular surface [81, 71].

Given the impact of Margulis's work for the above research directions, and especially the research concerning effective unipotent dynamics on homogeneous spaces portrayed here, but more importantly, given our personal interests, it is a great pleasure to dedicate this survey to Gregory Margulis.

1.1 PERIODIC ORBITS AND A NOTION OF VOLUME. Suppose $x \in X$ is so that Hx is dense in X. As will be evident in the following exposition, and was alluded to above, the orbit Hx may fill up the space very slowly; for example, x may be very *close* to an H-invariant manifold of lower dimension. To have any effective account, we first need a measure of *complexity* for these intermediate behaviors.

We will always denote the G-invariant probability measure on X by vol_X. Let $L \subset G$ be a closed subgroup. A point $x \in X$ will be called *L-periodic* if

$$\mathrm{stab}_L(x) = \{g \in L : gx = x\}$$

is a lattice in L. Similarly, a periodic L-orbit is a set Lx where x is an L-periodic point. We note that a periodic L-orbit is automatically closed in X. Given an L-periodic point x we let μ_{Lx} denote the probability L-invariant measure on Lx. By a *homogeneous* measure on X we always mean μ_{Lx} for some L and x. Sometimes we refer to the support of a homogeneous measure, which is an L-periodic set for some L, as a homogeneous set.

Fix some open neighborhood Ω of the identity in G with compact closure. For any L-periodic point $x \in X$, define

$$\mathrm{vol}(Lx) = \frac{m_L(Lx)}{m_L(\Omega)}, \tag{1.1}$$

where m_L is any Haar measure on L and $m_L(Lx)$ is the covolume of $\mathrm{stab}_L(x)$ in L with respect to m_L. We will use this notion of volume as a measure of the complexity of the periodic orbit.

Evidently this notion depends on Ω, but the notions arising from two different choices of Ω are comparable to each other, in the sense that their ratio is bounded above and below. Consequently, we drop the dependence on Ω in the notation. See [26, section 2.3] for a discussion of basic properties of the above definition.

The general theme of statements will be a dichotomy of the following nature. Unless there is an explicit obstruction with *low complexity*, the orbit Hx fills up the whole space—the statements also provide rates for this density or equidistribution whenever available.

We have tried to arrange the results roughly in their chronological order.

2 Horospherical subgroups

Let $\mathbf{G}$ be a semisimple $\mathbb{R}$-group and let G denote the connected component of the identity in the Lie group $\mathbf{G}(\mathbb{R})$.

A subgroup $U \subset G$ is called a horospherical subgroup if there exists an ($\mathbb{R}$-diagonalizable) element $a \in G$ so that

$$U = W^+(a) := \{g \in G : a^n g a^{-n} \to e \text{ as } n \to -\infty\};$$

put $W^-(a) = W^+(a^{-1})$.

Horospherical subgroups are always unipotent, but not necessarily vice versa; for example, for $d \geq 3$, a one-parameter unipotent subgroup in $\mathrm{SL}_d(\mathbb{R})$ is never horospherical. In a sense, horospherical subgroups are *large* unipotent subgroups. For example, if $U \subset G$ is a horospherical subgroup, then $G/\mathrm{N}_G(U)$ is compact, where $\mathrm{N}_G(U)$ denotes the normalizer of U in G.

Let $\Gamma \subset G$ be a lattice and $X = G/\Gamma$. We fix a horospherical subgroup $U = W^+(a)$ for the rest of the discussion. The action of U on X has been the subject of extensive investigations by several authors—this action induces the horocylce flow, when $G = \mathrm{SL}_2(\mathbb{R})$ or the horospherical flow in the general setting of rank 1 groups.

Various rigidity results in this context are known thanks to the works of Hedlund, Furstenberg, Margulis, Veech, Dani, Sarnak, Burger, and others [12, 14, 15, 18, 39, 43, 57, 71, 79]. Many of these results and subsequent works use techniques that in addition to proving strong rigidity results do so with a *polynomially* strong error term—for example, the methods in [71], [12], [73], and [77], relying on harmonic analysis, or the more dynamical arguments in [57], [46], and [80]. See Theorems 2.1 and 2.2 for some examples.

Let $U_0 \subset U$ be a fixed neighborhood of the identity in U with smooth boundary and compact closure; for example, one can take U_0 to be the image under the exponential map of a ball around 0 in $\mathrm{Lie}(U)$ with respect to some Euclidean norm on $\mathrm{Lie}(U)$. For every $k \in \mathbb{N}$, put $U_k = a^k U_0 a^{-k}$.

We normalize the Haar measure, σ, on U so that $\sigma(U_0) = 1$.

THEOREM 2.1.

Assume X is compact. There exists some $\delta > 0$, depending on Γ, so that the following holds. Let $f \in C^\infty(X)$; then for any $x \in X$ we have

$$\left| \frac{1}{\sigma(U_k)} \int_{U_k} f(ux)\, d\sigma(u) - \int_X f \, d\mathrm{vol}_X \right| \ll_X \mathcal{S}(f) e^{-\delta k},$$

where $\mathcal{S}(f)$ denotes a certain Sobolev norm.

The constant δ depends on the rate of decay for matrix coefficients corresponding to smooth vectors in $L^2(X, \mathrm{vol}_X)$—in other words, on the rate of mixing for the action of a on X. In particular, if G has property (T) or Γ is a congruence lattice, then δ can be taken to depend only on $\dim G$.

As mentioned above there are different approaches to prove Theorem 2.1. We highlight a dynamical approach that is based on the mixing property of the action of a on X via the so-called *thickening* or *banana* technique; this idea is already present in [57]; the exposition here is taken from [46].

Making a change of variable, and using $\sigma(U_0)=1$, one has

$$\frac{1}{\sigma(U_k)}\int_{U_k} f(ux)\,d\sigma(u)=\int_{U_0} f(a^k uy)\,d\sigma(u),$$

where $y=a_{-k}x$.

The key observation now is that the translation of U_0 by a^k is quite well approximated by the translation of a *thickening* of U by a^k. To be more precise, let B be an open neighborhood of the identity so that

$$B=(B\cap W^-(a))(B\cap Z_G(a))\,U_0.$$

Then since $a^k\big(B\cap W^-(a)\big)a^{-k}\to e$ in the Hausdorff topology, we see that $a^k U_0 y$ and $a^k By$ stay *near* each other. This, in view of the fact that y *stays in the compact set* X, reduces the problem to the study of the correlation

$$\int_X \mathbb{1}_B(z) f(a^k z)\,\mathrm{dvol}_X,$$

which can be controlled using the mixing rate for the action of a on X.

As the above sketch indicates, compactness of X is essential for this unique ergodicity result (with a uniform rate) to hold. If X is not compact, there are intermediate behaviors that make the analysis more involved—for instance, if x lies on a closed orbit of U, or is very *close* to such an orbit, Theorem 2.1 as stated cannot hold. We state a possible formulation in a concrete setting; see [73] and [77] for different formulations.

THEOREM 2.2.
Let $G=\mathrm{SL}_d(\mathbb{R})$ and $\Gamma=\mathrm{SL}_d(\mathbb{Z})$. There exists some $\delta>0$ so that the following holds. For any $x=g\Gamma\in X$ and $n,k\in\mathbb{N}$ with $k>n$, at least one of the following holds.

1. *For any $f\in C_c^\infty(X)$ we have*

$$\left|\frac{1}{\sigma(U_k)}\int_{U_k} f(ux)\,d\sigma(u)-\int_X f\,\mathrm{dvol}_X\right|\ll_d \mathcal{S}(f)e^{-\delta n},$$

   where $\mathcal{S}(f)$ denotes a certain Sobolev norm.
2. *There exists a rational subspace $W\subset\mathbb{R}^d$ of dimension*

$$m\in\{1,\ldots,d-1\}$$

so that

$$\|ug\mathbf{w}\| \ll_d e^n \quad \text{for all } u \in U_k,$$

where $\mathbf{w} = w_1 \wedge \ldots \wedge w_m$ *for a* $\mathbb{Z}$*-basis* $\{w_1, \ldots, w_m\}$ *of* $W \cap \mathbb{Z}^n$*, and* $\|\cdot\|$ *is a fixed norm on* $\bigwedge^m \mathbb{R}^d$.

Similar results hold for any semisimple group G. In the more general setting, Theorem 2.2(2) needs to be stated using conjugacy classes of a finite collection of parabolic subgroups of G that describe the non-compactness (roughly speaking, the cusp) of X.

The proof of Theorem 2.2 combines results on quantitative non-divergence of unipotent flows [53, 16, 17, 21, 47], together with the above sketch of the proof of Theorem 2.1; see [48] and [61].

Recall that a subgroup $H \subset G$ is called symmetric if H is the set of fixed points of an involution τ on G—for example, $H = \mathrm{SO}(p, n-p)$ in $G = \mathrm{SL}_n(\mathbb{R})$. Translations of closed orbits of symmetric subgroups present another (closely related) setting where effective equidistribution results, with polynomial error rates, are available. In this case as well, the so-called *wave front lemma* [33, theorem 3.1] asserts that translations of an H-orbit stay *near* translations of a thickening of it. Therefore, one may again utilize mixing; see, for example, [33] and [4]. Analytic methods also are applicable in this setting; see [22].

We end this section by discussing a case that is beyond the horospherical case but closely related. Let $\hat{G} = G \ltimes W$, where G is a semisimple group as above and W is the unipotent radical of $\hat{G}$. Let $\hat{\Gamma} \subset \hat{G}$ be a lattice and put $\hat{X} = \hat{G}/\hat{\Gamma}$. Let $\pi : \hat{G} \to G$ be the natural projection.

PROBLEM 2.3. *Let* $U \subset \hat{G}$ *be a unipotent subgroup so that* $\pi(U)$ *is a horospherical subgroup of* G*. Prove analogues of Theorem 2.2 for the action of* U *on* $\hat{X}$.

Strömbergsson [78] used analytic methods to settle a special case of this problem—that is, $G = \mathrm{SL}_2(\mathbb{R}) \ltimes \mathbb{R}^2$ with the standard action of $\mathrm{SL}_2(\mathbb{R})$ on $\mathbb{R}^2$, $\Gamma = \mathrm{SL}_2(\mathbb{Z}) \ltimes \mathbb{Z}^2$, and U the group of unipotent upper triangular matrices in $\mathrm{SL}_2(\mathbb{R})$; his method has also been used to tackle some other cases.

3 Effective equidistribution theorems for nilflows

In this section we assume G is a unipotent group. That is, we may assume G is a closed connected subgroup of the group of strictly upper triangular $d \times d$ matrices. Let $\Gamma \subset G$ be a lattice and $X = G/\Gamma$—that is, X is a nilmanifold.

Rigidity results in this setting have been known for quite some time thanks to works of Weyl, Kronecker, L. Green, and Parry [3, 65] and more recently Leibman [49].

Quantitative results, *with a polynomial error rate,* have also been established in this context and beyond the abelian case; see [38] and [42]. The complete solution was given by B. Green and Tao [42]; here we present a special case from that work describing the equidistribution properties of pieces of trajectories.

THEOREM 3.1 ([42]).
Let $X = G/\Gamma$ be a nilmanifold as above. There exists some $A \geq 1$ depending on $\dim G$ so that the following holds. Let $x \in X$, let $\{u(t) : t \in \mathbb{R}\}$ be a one-parameter subgroup of G, let $0 < \eta < 1/2$, and let $T > 0$. Then at least one of the following holds for the partial trajectory $\{u(t)x : t \in [0, T]\}$.

1. *For every $f \in C^\infty(X)$ we have*

$$\left| \frac{1}{T} \int_0^T f(u(t)x)\, \mathrm{d}t - \int_X f\, \mathrm{dvol}_X \right| \ll_{X,f} \eta,$$

where the dependence on f is given using a certain Lipschitz norm.

2. *For every $0 \leq t_0 \leq T$ there exists some $g \in G$ and some $H \subsetneq G$ so that $H\Gamma/\Gamma$ is closed with $\mathrm{vol}(gH\Gamma/\Gamma) \ll_X \eta^{-A}$ and for $t \in [0, T]$ we have*

$$|t - t_0| \leq \eta^A T \implies \mathrm{dist}(u(t)x, gH\Gamma/\Gamma) \ll_X \eta,$$

where dist *is a metric on X induced from a right-invariant Riemannian metric on G.*

This is a consequence of a special case of a more general effective equidistribution result for polynomial trajectories on nilmanifolds [42, theorem 2.9], as we now explicate. Since $T > 0$ is arbitrary, we may assume that $u(t) = \exp(t\mathsf{z})$ for some z in the Lie algebra of G of norm one. We note that for $T \leq \eta^{-O(1)}$ the above is trivial. In fact, as is visible in the maximal abelian torus quotient, every point belongs to an orbit $gH\Gamma/\Gamma$ of bounded volume of a proper subgroup $H \subsetneq G$, and now (2) follows by the continuity properties of the one-parameter subgroup if $T \leq \eta^{-O(1)}$. Hence we will assume in the following $T > \eta^{-O(1)}$ for a constant $O(1)$ that will be optimized.

For every $1/2 \leq \tau \leq 1$, put $B_\tau = \{u(n\tau) : n = 0, 1, \ldots, N_\tau\}$, where $N_\tau = \lfloor T/\tau \rfloor$. We now apply [42, theorem 2.9] for the sequence (discrete trajectory) B_τ. Assume first that $B_\tau x$ is η-equidistributed for some $\tau \in [\frac{1}{2}, 1]$. That is,

$$\left| \frac{1}{N_\tau} \sum_{n=0}^{N_\tau - 1} h(u(n\tau)x) - \int_X h \, \mathrm{dvol}_X \right| \ll_{X,h} \eta \tag{3.1}$$

for all $h \in C^\infty(X)$. In this case Theorem 3.1(1) holds. Indeed,

$$\frac{1}{N_\tau \tau} \int_0^{N_\tau \tau} f(u(t)x)\, dt = \frac{1}{\tau} \int_0^\tau \frac{1}{N_\tau} \sum_{n=0}^{N_\tau - 1} f(u(s)u(n\tau)x)\, ds,$$

so the claim in (1) follows from (3.1) applied with $h(\cdot) = f(u(s)\cdot)$ for all $0 \le s \le \tau$.

The alternative in [42, theorem 2.9] to η-equidistribution as above is an obstruction to equidistribution in the form of a slowly varying character of G/Γ. To make a precise statement we need some notation. Fix a rational basis for Lie(G). Using this basis we put coordinates (also known as coordinates of the second kind) on G; the standing assumption is that Γ corresponds to elements with integral coordinates; see [42, definitions 2.1 and 2.4]—the estimates[1] will depend on the *complexity* of the structural constants for group multiplication written in this basis (which we assume to be fixed). Following [42] we denote the coordinates of $g \in G$ by $\psi(g)$. In this notation, given a character $\chi : G \to \mathbb{R}/\mathbb{Z}$ with $\Gamma \subset \ker(\chi)$, there exists a unique $k_\chi \in \mathbb{Z}^{\dim G}$ so that $\chi(g) = k_\chi \cdot \psi(g) + \mathbb{Z}$; see [42, definition 2.6].

Assume (3.1) fails for all $\tau \in [1/2, 1]$. Then, by [42, theorem 2.9]; see also lemma 2.8, we have this: there are constants $A_0, A_1 > 1$, and for every τ there is a character $\chi_\tau : G \to \mathbb{R}/\mathbb{Z}$ with $\Gamma \subset \ker(\chi_\tau)$ so that the following two conditions hold.

(a) Let $k_\tau \in \mathbb{Z}^{\dim G}$ be so that $\chi_\tau(g) = k_\tau \cdot \psi(g) + \mathbb{Z}$; then we have the bound $\|k_\tau\| \ll_{G,\Gamma} \eta^{-A_0}$; and
(b) $\|\chi_\tau(u(\tau))\|_{\mathbb{R}/\mathbb{Z}} \ll_{G,\Gamma} \eta^{-A_1}/T$, where $\|x\|_{\mathbb{R}/\mathbb{Z}} = \mathrm{dist}(x, \mathbb{Z})$.

Let H_τ denote the connected component of the identity in $\ker(\chi_\tau)$. Informally, (a) tells us that χ_τ defines a closed orbit $H_\tau\Gamma/\Gamma$ of not too large volume—indeed the latter covolume is bounded by $\|k_\tau\|$. Moreover, $\|k_\tau\|$ controls the continuity properties of χ_τ. On the other hand, (b) tells us that the character changes its values very slowly along the discrete trajectory (since we are allowed to think of a large T). We wish to combine these for various $\tau \in [1/2, 1]$ to obtain (2).

[1] Indeed the estimates in Theorem 3.1(1) also depend on this basis.

To that end, note that the number of characters χ_τ so that (a) holds is $\leq \eta^{-O(1)}$ for some $O(1)$ depending on A_0. Moreover, (b) implies that there exist some $C = C(G, \Gamma)$, some A_2 depending on A_0 and A_1, and for every $1/2 \leq \tau \leq 1$ there is some rational vector v_τ with $\|v_\tau\| \ll 1$ and a denominator bounded by $O(\eta^{-O(1)})$ so that the distance of $\psi(u(\tau))$ to $v_\tau + \psi(H_\tau)$ is $< C\eta^{-A_2}/T$. For every $1/2 \leq \tau \leq 1$, let I_τ be the maximal (relatively open) interval so that for all $s \in I_\tau$ the distance of $\psi(u(s))$ to $v_\tau + \psi(H_\tau)$ is $< C\eta^{-A_2}/T$. This gives a covering of $[1/2, 1]$ with $\eta^{-O(1)}$ many intervals. Therefore, at least one of these intervals—say, $I_0 = (a_0, b_0)$ defined by τ_0—has length $b_0 - a_0 \gg \eta^{O(1)}$. Let $\chi = \chi_{\tau_0}$. Then for any $\tau \in I_0$ we have that the distance of $\psi(u(\tau))$ to $v_{\tau_0} + \psi(H_{\tau_0})$ is $< C\eta^{-A_2}/T$. Since $b_0 - a_0 \gg \eta^{O(1)}$, we get that the distance of $\psi(u(\tau))$ to $v_{\tau_0} + \psi(H_{\tau_0})$ is $\ll \eta^{-O(1)}/T$ for all $0 \leq \tau \leq 1$. Hence we obtain the character estimate $\|\chi(u(\tau))\|_{\mathbb{R}/\mathbb{Z}} \ll_X \eta^{-O(1)}/T$ for all $0 \leq \tau \leq 1$.

Let $g \in G$ and $\gamma \in \Gamma$ be so that $u(t_0) = g\gamma$ and $\|\psi(g)\| \ll_X 1$. Let $H = H_{\tau_0}$. Then since $\gamma H\Gamma = H\Gamma$, the claim in (2) holds with g and H if we choose A large enough.

We now highlight some elements involved in the proof of [42, theorem 2.9] for our simplified setting of a *linear* sequence—that is, a discrete trajectory (the reader may also refer to [42, section 5], where a concrete example is worked out).

Let $\tau = 1$—that is, consider $\{u(n) : n = 0, 1, \ldots, N-1\}$ on X. The goal is to show that either Equation (3.1) holds or (a) and (b) must hold for a character χ; note first that replacing $\{u(t)\}$ by a conjugate, we will assume x in (3.1) is the identity coset, Γ, for the rest of the discussion. The proof is based on an inductive argument[2] that aims at decreasing the nilpotency degree of G. For the base of the induction—that is, when G is abelian—one may use Fourier analysis to deduce the result (e.g., see [42, proposition 3.1]).

The following corollary (see [42, corollary 4.2]) of the van der Corput trick plays an important role in the argument. Let $\{a_n : n = 0, 1, \ldots, N-1\}$ be a sequence of complex numbers so that $\frac{1}{N}\sum_{n=0}^{N-1} a_n \geq \eta$. Then for at least $\eta^2 N/8$ values of $k \in \{0, 1, \ldots, N-1\}$, we have $\frac{1}{N}\sum_{n=0}^{N-1} a_{n+k}\bar{a}_n \geq \eta^2/8$, where we put $a_n := 0$ for $n \notin \{0, 1, \ldots, N-1\}$.

Let $a_n = h(u(n)\Gamma)$, and suppose that (3.1) fails. One may further restrict to the case where h is an eigenfunction for the action of the center of G corresponding to a character ξ whose *complexity* is controlled by $\eta^{-O(1)}$; see [42, lemma 3.7]. Note that if ξ is trivial, then h is $Z(G)$ invariant; thus we have

[2]The reader may see the argument in [65, section 3].

already reduced the problem to $G/Z(G)$—that is, a group with smaller nilpotency degree. Hence, assume that ξ is nontrivial; in consequence $\int h\,\mathrm{dvol}_X = 0$. Now by the aforementioned corollary of the van der Corput trick, there are at least $\eta^2 N/8$ many choices of $0 \le k \le N$ so that

$$\frac{1}{N}\sum_{n=0}^{N-1} h(u(n+k)\Gamma)\overline{h(u(n)\Gamma)} \gg \eta^2/8. \tag{3.2}$$

Fix one such k and write $u(k)\Gamma = v\Gamma$ for an element v in the fundamental domain of Γ; note that $\{(v^{-1}gv, g) : g \in G\} \subset \{(g_1, g_2) \in G \times G : g_1 g_2^{-1} \in [G, G]\} =: G'$. Similarly, define Γ'. Two observations are in order.

- Equation (3.2) implies that $\frac{1}{N}\sum_{n=0}^{N-1} \tilde{h}(w_n\Gamma') \gg \eta^2/8$, where $\tilde{h}$ is the restriction of $\hat{h}(y, z) = h(vy)\overline{h(z)}$ to G' and $w_n = (v^{-1}u(n)v, u(n))$.
- Since h is an eigenfunction for the center of G, we have $\tilde{h}$ is invariant under $\Delta(Z(G)) = \{(g, g) : g \in Z(G)\}$; moreover, $\tilde{h}$ has mean zero.

The above observations reduce (3.2) modulo $\Delta(Z(G))$—that is, to the group $G'/\Delta(Z(G))$, which has smaller nilpotency degree; see [42, proposition 7.2]. There is still work to be done; for example, one needs to combine the information obtained for different values of k to prove (a) and (b), but this reduction, in a sense, is the heart of the argument.

4 Periodic orbits of semisimple groups

Beyond the settings discussed in sections 2 and 3, little was known until roughly a decade ago. The situation drastically changed thanks to the work of Einsiedler, Margulis, and Venkatesh [27], where a *polynomially effective* equidistribution result was established for closed orbits of semisimple groups.

We need some notation in order to state the main result. Let $\mathbf{G}$ be a connected, semisimple algebraic $\mathbb{Q}$-group, and let G be the connected component of the identity in the Lie group $\mathbf{G}(\mathbb{R})$. Let $\Gamma \subset \mathbf{G}(\mathbb{Q})$ be a congruence lattice in G and put $X = G/\Gamma$. Suppose $H \subset G$ is a semisimple subgroup without any compact factors that has a finite centralizer in G.

The following is the main equidistribution theorem proved in [27].

THEOREM 4.1 ([27]).
There exists some $\delta = \delta(G, H)$ so that the following holds. Let Hx be a periodic H-orbit. For every $V > 1$ there exists a subgroup $H \subset S \subset G$ so that x is S-periodic, $\mathrm{vol}(Sx) \le V$, *and*

$$\left|\int_X f\,\mathrm{d}\mu_{Hx} - \int_X f\,\mathrm{d}\mu_{Sx}\right| \ll_{G,\Gamma,H} \mathcal{S}(f)V^{-\delta} \quad \textit{for all } f \in C_c^\infty(X),$$

where $\mathcal{S}(f)$ *denotes a certain Sobolev norm.*

Theorem 4.1 is an effective version (of a special case) of a theorem by Mozes and Shah [63]. The general strategy of the proof is based on *effectively* acquiring extra *almost invariance* properties for the measure $\mu = \mu_{Hx}$. This general strategy (in qualitative form) was used in the topological context by Margulis [54], by Dani and Margulis [19], and by Ratner in her measure classification theorem [67, 68].

The polynomial nature of the error term—that is, a (negative) power of V—in Theorem 4.1 is quite remarkable; effective dynamical arguments often yield worse rates (see section 5). A crucial input in the proof of Theorem 4.1, which is responsible for the quality of the error, is a uniform spectral gap for congruence quotients.

The proof of Ratner's Measure Classification Theorem for the action of a semisimple group H is substantially simpler. A simplified proof in this case was given by Einsiedler [23]; see also [27, section 2]. This is due to complete reducibility of the adjoint action of H on Lie(G) as we now explicate. Let $\{u(t) : t \in \mathbb{R}\}$ be a one-parameter unipotent subgroup in H, and let L be a subgroup that contains H. Then one can show that the orbits $u(t)y$ and $u(t)z$ of two *nearby points in general position* diverge in a direction transversal to L. This observation goes a long way in the proof. Indeed starting from an H-invariant ergodic measure μ, one may use arguments like this to show that unless there is an algebraic obstruction, one can increase the dimension of the group, which leaves μ invariant. In a sense, the argument in [27] is an effective version of this argument; a different argument that is directly based on the mixing property of an $\mathbb{R}$-diagonalizable subgroup in H was given by Margulis (see [62]).

Let us elaborate on a possible effectivization of the above idea. The divergence of two nearby points $u(t)y$ and $u(t)z$ is governed by a polynomial function. For example, in the setting at hand, write $y = \exp(v)z$ for some $v \in \mathrm{Lie}(G)$; then this divergence is controlled by $\mathrm{Ad}(u(t))v$. In consequence, one has a rather good quantitative control on this divergence.

However, the size of T so that the piece of orbit, $\{u(t)y : 0 \le t \le T\}$, approximates the measure μ depends on y. More precisely, suppose μ is $\{u(t)\}$-ergodic.[3] Then it follows from Birkhoff's ergodic theorem that for μ-a.e.

[3] By the generalized Mautner phenomenon, when H has no compact factors, one can always choose $\{u(t)\}$ so that μ is $\{u(t)\}$-ergodic.

y and all $f \in C_c^\infty(X)$, we have $\frac{1}{T}\int_0^T f(u(t)y)\,\mathrm{d}t \to \int_X f\,\mathrm{d}\mu_{Hx}$. However, for a given $\epsilon > 0$, the size of T so that

$$\left|\frac{1}{T}\int_0^T f(u_t y)\,\mathrm{d}t - \int_X f\,\mathrm{d}\mu_{Hx}\right| \ll_f \epsilon \quad \text{where } f \in C_c^\infty(X)$$

depends on delicate properties of the point y—for example y may be too close to a $\{u(t)\}$-invariant submanifold in support of μ.

One of the remarkable innovations in [27] is the use of *uniform spectral gap* in order to obtain an effective version of the pointwise ergodic theorem. The required uniform spectral gap has been obtained in a series of papers [45, 64, 74, 44, 11, 13, 40]. This is then used to define an effective notion of generic points where the parameters ϵ and T above are *polynomially* related to cach other.

If H is a maximal subgroup, one can use the above (combined with bounded generation of G by conjugates of H and spectral gap for vol_X) to finish the proof. However, Theorem 4.1 is more general and allows for (finitely many) intermediate subgroups. The main ingredient in [27] to deal with possible intermediate subgroups is an *effective closing lemma* that is proved in [27, section 13]. In addition to being crucial for the argument in [27], this result is of independent interest—it is worth mentioning that the proof of [27, proposition 13.1] also uses spectral gap.

Theorem 4.1 imposes some assumptions that are restrictive for some applications: H is not allowed to vary; moreover, H has a finite centralizer. The condition that *H is assumed not to have any compact factors* is a splitting condition at the infinite place; this too is restrictive in some applications.

4.2 ADELIC PERIODS. In a subsequent work by Einsiedler, Margulis, Mohammadi, and Venkatesh [26], the subgroup H in Theorem 4.1 is allowed to vary. Moreover, the need for a splitting condition is also eliminated. The main theorem in [26] is best stated using the language of adeles; the reader may also see [28] for a more concrete setting.

Let $\mathbf{G}$ be a connected, semisimple, algebraic $\mathbb{Q}$-group[4] and set $X = \mathbf{G}(\mathbb{A})/\mathbf{G}(\mathbb{Q})$, where $\mathbb{A}$ denotes the ring of adeles. Then X admits an action of the locally compact group $\mathbf{G}(\mathbb{A})$ preserving the probability measure vol_X.

Let $\mathbf{H}$ be a semisimple, simply connected algebraic $\mathbb{Q}$-group, and let $g \in \mathbf{G}(\mathbb{A})$. Fix also an algebraic homomorphism

[4]The paper [26] allows for any number field, F, but unless X is compact, δ in Theorem 4.3 will depend on $\dim \mathbf{G}$ and $[F:\mathbb{Q}]$.

$$\iota: \mathbf{H} \to \mathbf{G}$$

defined over $\mathbb{Q}$ with finite central kernel. For example, we could have $\mathbf{G} = \mathrm{SL}_d$ and $\mathbf{H} = \mathrm{Spin}(Q)$ for an integral quadratic form Q in d variables.

To this algebraic data, we associate a homogeneous set

$$Y := g\iota(\mathbf{H}(\mathbb{A})/\mathbf{H}(\mathbb{Q})) \subset X$$

and a homogeneous measure μ; recall that we always assume $\mu(Y) = 1$.

The following is a special case of the main theorem in [26].

THEOREM 4.3 ([26]).
Assume further that $\mathbf{G}$ *is simply connected. There exists some* $\delta > 0$*, depending only on* $\dim \mathbf{G}$*, so that the following holds. Let* Y *be a homogeneous set and assume that* $\iota(\mathbf{H}) \subset \mathbf{G}$ *is maximal. Then*

$$\left| \int_X f \, d\mu - \int_X f \, d\mathrm{vol}_X \right| \ll_{\mathbf{G}} \mathcal{S}(f) \mathrm{vol}(Y)^{-\delta} \quad \text{for all } f \in C_c^\infty(X),$$

where $\mathcal{S}(f)$ *is a certain adelic Sobolev norm.*

As we alluded to above, Theorem 4.3 allows H to vary, and it also assumes no splitting conditions[5] on $\mathbf{H}$; this feature is crucial for applications: for example, $\mathbf{H}(\mathbb{R})$ is compact in an application to quadratic forms, which will be discussed momentarily. These liberties are made possible thanks to Prasad's volume formula [66] and the seminal work of Borel and Prasad [8]. Roughly speaking, the argument in [26, section 5] uses [66] and [8] to show that if at a prime p the group $\mathbf{H}(\mathbb{Q}_p)$ is either compact or too *distorted*, then there is at least a factor p contribution to $\mathrm{vol}(Y)$. Thus one can find a *small* prime p (compared to $\mathrm{vol}(Y)$) where H is not distorted.

The dynamical argument uses unipotent flows as described above: the source of a polynomially effective rate is again the uniform spectral gap.

Let us highlight two corollaries from Theorem 4.3. The method in [26] relies on uniform spectral gap. However, it provides an independent proof of property (τ), except for groups of type A_1—that is, if we only suppose property (τ) for groups of type A_1, we can deduce property (τ) in all other cases as well as our theorem. In particular it gives an alternative proof of the main result of Clozel in [13] but with weaker exponents; see [26, section 4].

[5] Compare this to the assumption that H has no compact factors in Theorem 4.1.

Another application is an analogue of Duke's theorem for positive definite integral quadratic forms in $d \geq 3$ variables, as we now explicate. Even in a qualitative form this result is new in dimensions 3 and 4 since the splitting condition prevented applying unipotent dynamics before.

Let $\mathcal{Q}_d = \mathrm{PO}_d(\mathbb{R})\backslash\mathrm{PGL}_d(\mathbb{R})/\mathrm{PGL}_d(\mathbb{Z})$ be the space of positive definite quadratic forms on $\mathbb{R}^d$ up to the equivalence relation defined by scaling and equivalence over $\mathbb{Z}$. We equip $\mathcal{Q}_d$ with the pushforward of the normalized Haar measure on $\mathrm{PGL}_d(\mathbb{R})/\mathrm{PGL}_d(\mathbb{Z})$.

Let Q be a positive definite integral quadratic form on $\mathbb{Z}^d$, and let genus(Q) (respectively spin genus(Q)) be its *genus* (respectively spin genus).

THEOREM 4.4 ([26]).
Suppose $\{Q_n\}$ varies through any sequence of pairwise inequivalent, integral, positive definite quadratic forms. Then the genus (and also the spin genus) of Q_n, considered as a subset of $\mathcal{Q}_d$, equidistributes as $n \to \infty$ (with speed determined by a power of $|$ genus$(Q)|$*).*

In the statement of Theorem 4.3 we made a simplifying assumption that **G** is simply connected; PGL_d, however, is not simply connected. Indeed the proof of Theorem 4.4 utilizes the more general [26, theorem 1.5]. In addition one uses the fact that

$$\mathrm{PGL}_d(\mathbb{A}) = \mathrm{PGL}_d(\mathbb{R})K\mathrm{PGL}_d(\mathbb{Q}) \quad \text{where } K = \textstyle\prod_p \mathrm{PGL}_d(\mathbb{Z}_p)$$

to identify $L^2(\mathrm{PGL}_d(\mathbb{R})/\mathrm{PGL}_d(\mathbb{Z}))$ with the space of K-invariant functions in $L^2(\mathrm{PGL}_d(\mathbb{A})/\mathrm{PGL}_d(\mathbb{Q}))$.

Similar theorems have been proved elsewhere (see, e.g., [37] where the splitting condition is made at the Archimedean place). What is novel here, besides the speed of convergence, is the absence of any type of splitting condition on the $\{Q_n\}$—this is where the effective Theorem 4.3 becomes useful.

Theorem 4.3 assumed $\iota(\mathbf{H}) \subset \mathbf{G}$ is maximal. This is used in several places in the argument. This assumption is too restrictive for some applications; see, for example, [30] where $\iota(\mathbf{H}) \subset \mathbf{G}$ has an infinite centralizer. The following is a much desired generalization.

PROBLEM 4.5. *Prove an analogue of Theorem 4.3 allowing $\iota(\mathbf{H})$ to have an arbitrary centralizer in* **G**.

The uniform spectral gap, which is the source of a polynomially effective error term, is still available in this case. However, in the presence of an infinite centralizer, closed orbits come in families; moreover, there is an abundance of intermediate subgroups. These features introduce several technical difficulties.

See [29], [28], and [1] for some progress toward this problem.

We note that a positive solution to Problem 4.5 would lead to strengthening of the results in [30]. Indeed, it is expected that one would obtain the result in [30] when the gap between the number of variables is three or more without requiring any congruent conditions.

5 The action of unipotent subgroups

We now turn to the general case of unipotent trajectories. Let $\mathbf{G}$ be an $\mathbb{R}$-group and let G be the connected component of the identity in the Lie group $\mathbf{G}(\mathbb{R})$; U will denote a unipotent subgroup of G. Let $\Gamma \subset G$ be a lattice and $X = G/\Gamma$.

Let us recall the following theorems of Ratner, which resolved conjectures of Raghunathan and Dani.

THEOREM 5.1 ([67, 68, 69]).

1. *Every U-invariant and ergodic probability measure on X is homogeneous.*
2. *For every $x \in X$ the orbit closure $\overline{Ux}$ is a homogeneous set.*

The above actually holds for any group that is generated by unipotent elements [76]. In the case of a unipotent one-parameter subgroup U, more can be said. Suppose $\overline{Ux} = Lx$ as in Theorem 5.1(2). Ratner [69] actually proved that the orbit Ux is equidistributed with respect to the L-invariant measure on Lx.

Prior to Ratner's work, some important special cases were studied by Margulis [54] and Dani and Margulis [19, 20]. The setup they considered was motivated by Margulis's solution to the Oppenheim conjecture—unlike Ratner's work, their method is topological and does not utilize measures. Let $G = \mathrm{SL}_3(\mathbb{R})$, let $\Gamma = \mathrm{SL}_3(\mathbb{Z})$, and let U be a generic one-parameter unipotent[6] subgroup of G. In this context, the paper [20] proves that $\overline{Ux}$ is homogeneous for all $x \in X$.

Theorem 5.1 has also been generalized to the S-arithmetic context—that is, the product of real and p-adic groups—independently by Margulis and Tomanov [59] and Ratner [70].

[6] A one-parameter unipotent subgroup of $\mathrm{SL}_d(\mathbb{R})$ is called *generic* if it is contained in only one Borel subgroup of $\mathrm{SL}_d(\mathbb{R})$.

Let us recall the basic strategy in the proof of Raghunathan's conjectures; see also the discussion after Theorem 4.1. The starting point, á la Margulis and Ratner, is a set of "generic points" (a dynamical notion) for our unipotent group U. The heart of the matter, then, is to carefully investigate divergence of the U-orbits of two nearby generic points; the slow (polynomial-like) nature of this divergence implies that nearby points diverge in directions that are *stable* under the action of U. That is, the divergence is in the direction of the normalizer of U—this is in sharp contrast to hyperbolic dynamics where points typically diverge along the unstable directions for the flow. The goal is to conclude that unless some explicit algebraic obstructions exist, the closure of a U-orbit contains an orbit of a subgroup $V \supsetneq U$.

As mentioned, the slow divergence of unipotent orbits is a major player in the analysis. This actually is not the only place where polynomial-like behavior of unipotent actions is used in the proofs. Indeed in passing from measure classification to topological rigidity (and more generally equidistribution theorem) nondivergence of unipotent orbits [53, 16, 17] plays an essential role; see also section 5.4.

Generic sets play a crucial role in the above study, and the existing notions—for example, minimal sets in the topological approach or a generic set for Birkhoff's ergodic theorem in Ratner's argument are rather non-effective. Providing an effective notion of a generic set that is also compatible with the nice algebraic properties of unipotent flows is the first step toward an effective account of the above outline. With that in place, one then may try to carry out the analysis in an effective fashion. The caveat is that the estimates one gets from such arguments are usually rather poor—that is, rather than obtaining a negative power of *complexity*, one typically gets a negative power of an iterated logarithmic function of the complexity; see the discussions in section 4 for an instance when this argument is carried out successfully and actually with a polynomial rate.

5.2 EFFECTIVE VERSIONS OF THE OPPENHEIM CONJECTURE. The resolution of the Oppenheim conjecture by Margulis [54] has played a crucial role in the developments of the field.

Let us recall the setup. Let Q be a nondegenerate, indefinite quadratic form in $d \geq 3$ variables on $\mathbb{R}^d$. The Oppenheim and Davenport conjecture stated that $\overline{Q(\mathbb{Z}^d)} = \mathbb{R}$ if and only if Q is not a multiple of a form with integral coefficients. Quantitative (or equidistribution) versions were also obtained [31, 32, 60, 58]; see also [72], [9], and [2], where effective results for generic forms (in different parameter spaces) are obtained.

On an effective level one might ask the following question. Given $\epsilon > 0$, what is the smallest $0 \neq v \in \mathbb{Z}^d$ so that $|Q(v)| \leq \epsilon$? Analytic methods, which were used prior to Margulis's work to resolve special cases of the Oppenheim conjecture, yield such estimates. Margulis's proof, however, is based on dynamical ideas and does not provide information on the size of such solutions.

The paper [41] proves a polynomial estimate for $n \geq 5$ and under explicit Diophantine conditions on Q—it combines analytic methods with some ideas related to systems of inequalities that were developed in [31]. In [72] and [9] analytic methods are used to obtain polynomial estimates for almost every form in certain families of forms in dimensions 3 and 4.

In general, however, the best-known results in dimension 3 are due to Lindenstrauss and Margulis, as we now discuss.

THEOREM 5.3 ([51]).
There exist absolute constants $A \geq 1$ and $\alpha > 0$ so that the following holds.

Let Q be an indefinite, ternary quadratic form with $\det Q = 1$ *and $\epsilon > 0$. Then for any $T \geq T_0(\epsilon)\|Q\|^A$, at least one of the following holds.*

1. *For any $\xi \in [-(\log T)^\alpha, (\log T)^\alpha]$ there is a primitive integer vector $v \in \mathbb{Z}^3$ with $0 < \|v\| < T^A$ satisfying*

$$|Q(v) - \xi| \ll (\log T)^{-\alpha}.$$

2. *There is an integral quadratic form Q' with* $|\det Q'| < T^\epsilon$ *so that*

$$\|Q - \lambda Q'\| \ll \|Q\| T^{-1},$$

where $\lambda = |\det Q'|^{-1/3}$.

The implied multiplicative constants are absolute constants.

The above theorem provides a dichotomy: unless there is an explicit Diophantine (algebraic) obstruction, a density result holds; in this sense the result is similar to Theorems 2.2 and 3.1. In particular, if Q is a reduced, indefinite, ternary quadratic form that is not proportional to an integral form but has algebraic coefficients, then Theorem 5.3(1) holds true for Q; see [51, corollary 1.12].

Note, however, that the quality of the effective rate one obtains in Theorem 5.3 is $(\log T)^{-\alpha}$—ideally, one would like to have a result where $(\log T)^{-\alpha}$

is replaced by $T^{-\alpha}$; however, such an improvement seems to be out of reach of the current technology.

We now highlight some of the main features of the proof of Theorem 5.3. An important ingredient in the proof is an explicit Diophantine condition [51, section 4]; this is used in place of the notion of minimal sets that was used in [54], [19], and [20]. We will discuss a related Diophantine condition in the next section; one important feature of the notion used in [51] is that *it is inherited for most points along a unipotent orbit*; see also Theorem 5.8.

The proof in [51] then proceeds by making effective and improving on several techniques from [54], [19], and [20]. If one follows this scheme of the proof, the quality of the estimates in Theorem 5.3(1) would be $(\log\log T)^{-\alpha}$. Instead [51] uses a combinatorial lemma about rational functions to *increase the density of points* (see [51, section 9])—this lemma, which is of independent interest, is responsible for the better error rate in Theorem 5.3(1).

5.4 EFFECTIVE AVOIDANCE PRINCIPLES FOR UNIPOTENT ORBITS.

Let $\mathbf{G}$ be a connected $\mathbb{Q}$-group and put $G=\mathbf{G}(\mathbb{R})$. We assume Γ is an arithmetic lattice in G. More specifically, we assume fixed an embedding $\iota:\mathbf{G}\to\mathrm{SL}_d$, defined over $\mathbb{Q}$ so that $\iota(\Gamma)\subset\mathrm{SL}_d(\mathbb{Z})$. Using ι we identify $\mathbf{G}$ with $\iota(\mathbf{G})\subset\mathrm{SL}_d$ and hence $G\subset\mathrm{SL}_d(\mathbb{R})$. Let $U=\{u(t):t\in\mathbb{R}\}\subset G$ be a one-parameter unipotent subgroup of G and put $X=G/\Gamma$.

Define the family of subgroups

$$\mathcal{H}=\Big\{\mathbf{H}\subset\mathbf{G}:\mathbf{H}\text{ is a connected }\mathbb{Q}\text{-subgroup and }\mathrm{R}(\mathbf{H})=\mathrm{R}_u(\mathbf{H})\Big\},$$

where $\mathrm{R}(\mathbf{H})$ (respectively $\mathrm{R}_u(\mathbf{H})$) denotes the solvable (respectively unipotent) radical of $\mathbf{H}$. Alternatively, $\mathbf{H}\in\mathcal{H}$ if and only if $\mathbf{H}$ is a connected $\mathbb{Q}$-subgroup and $\mathbf{H}(\mathbb{C})$ is generated by unipotent subgroups. *We will always assume that* $\mathbf{G}\in\mathcal{H}$.

For any $\mathbf{H}\in\mathcal{H}$ we will write $H=\mathbf{H}(\mathbb{R})$; examples of such groups are $H=\mathrm{SL}_d(\mathbb{R})$, $\mathrm{SL}_d(\mathbb{R})\ltimes\mathbb{R}^d$ (with the standard action), and $\mathrm{SO}_d(\mathbb{R})$. By a theorem of Borel and Harish-Chandra, $H\cap\Gamma$ is a lattice in H for any $\mathbf{H}\in\mathcal{H}$.

Define $\mathrm{N}_G(U,H)=\{g\in G: Ug\subset gH\}$. Put

$$\mathcal{S}(U)=\Big(\bigcup_{\substack{H\in\mathcal{H}\\ H\neq G}}\mathrm{N}_G(U,H)\Big)/\Gamma\quad\text{and}\quad\mathcal{G}(U)=X\setminus\mathcal{S}(U)$$

Following Dani and Margulis [21], points in $\mathcal{S}(U)$ are called *singular* with respect to U, and points in $\mathcal{G}(U)$ are called *generic* with respect to U. This notion of a generic point is a priori different from measure theoretically generic points for the action of U on X with respect to vol_X; however, any measure theoretically generic point is generic in this new sense as well.

By Theorem 5.1(2), for every $x \in \mathcal{G}(U)$ we have $\overline{Ux} = X$.

In [21], Dani and Margulis established strong avoidance properties for unipotent orbits; see also [75]. These properties, which are often referred to in the literature as *linearization* of unipotent flows, go hand in hand with Ratner's theorems in many applications; see, for example, [63] and [36].

In this section we will state a polynomially effective version of the results and techniques in [21]. These effective results as well as their S-arithmetic generalizations are proved in [52].

The main effective theorem, Theorem 5.8, requires some further preparation. Let us first begin with the following theorems, which are corollaries of Theorem 5.8.

THEOREM 5.5 ([52]).
There exists a compact subset $\mathcal{K} \subset \mathcal{G}(U)$ with the following property. Let $x \in \mathcal{G}(U)$; then $\overline{Ux} \cap \mathcal{K} \neq \emptyset$.

Theorem 5.5 is a special case of the following.

THEOREM 5.6 ([52]).
There exists some $D > 1$ depending on d and some $E > 1$ depending on G, d, and Γ so that the following holds.

For every $0 < \eta < 1/2$ there is a compact subset $\mathcal{K}_\eta \subset \mathcal{G}(U)$ with the following property. Let $\{x_m\}$ be a bounded sequence of points in X, and let $T_m \to \infty$ be a sequence of real numbers. For each m let $I_m \subset [-T_m, T_m]$ be a measurable set whose measure is $> E\eta^{1/D}(2T_m)$. Then one of the following holds.

1. $\overline{\bigcup_m \{u(t)x_m : t \in I_m\}} \cap \mathcal{K}_\eta \neq \emptyset$.
2. *There exists a finite collection $H_1, \ldots, H_r \in \mathcal{H}$ and for each $1 \le i \le r$ there is a compact subset $C_i \subset N(U, H_i)$, so that all the limit points of $\{x_m\}$ lie in $\cup_{i=1}^r C_i\Gamma/\Gamma$.*

Theorem 5.6 is reminiscent of the sort of dichotomy that we have seen in previous sections: unless an explicit algebraic obstruction exists, the pieces of the U-orbits intersect the *generic* set; see also Theorem 5.8, where this dichotomy is more apparent.

The polynomial dependence $E\eta^{1/D}$ in Theorem 5.6 is a consequence of the fact that Theorem 5.8 is *polynomially* effective. We note, however, that even for Theorem 5.5 it is not clear how it would follow from the statements in [21].

We now fix the required notation to state Theorem 5.8. Let $\|\cdot\|$ denote the maximum norm on $\mathfrak{sl}_N(\mathbb{R})$ with respect to the standard basis; this induces a family of norms, $\|\cdot\|$ on $\bigwedge^m \mathfrak{sl}_N(\mathbb{R})$ for $m=1,2,\ldots$. Let furthermore $\mathfrak{g}=\mathrm{Lie}(G)$ and put $\mathfrak{g}(\mathbb{Z}):=\mathfrak{g}\cap\mathfrak{sl}_N(\mathbb{Z})$.

For any $\eta>0$, set

$$X_\eta=\left\{g\Gamma\in X:\min_{0\neq v\in\mathfrak{g}(\mathbb{Z})}\|\mathrm{Ad}(g)v\|\geq\eta\right\}.$$

By Mahler's compactness criterion, X_η is compact for any $\eta>0$.

Let $\mathbf{H}\in\mathcal{H}$ be a proper subgroup and put

$$\rho_H:=\bigwedge^{\dim\mathbf{H}}\mathrm{Ad}\quad\text{and}\quad V_H:=\bigwedge^{\dim\mathbf{H}}\mathfrak{g}.$$

The representation ρ_H is defined over $\mathbb{Q}$. Let $\mathbf{v_H}$ be a primitive integral vector in $\wedge^{\dim\mathbf{H}}\mathfrak{g}$ corresponding to the Lie algebra of $\mathbf{H}$—that is, we fix a $\mathbb{Z}$-basis for $\mathrm{Lie}(H)\cap\mathfrak{sl}_N(\mathbb{Z})$ and let $\mathbf{v_H}$ be the corresponding wedge product.

We define the height of $\mathbf{H}\in\mathcal{H}$ by

$$\mathrm{ht}(\mathbf{H}):=\|\mathbf{v_H}\|. \tag{5.1}$$

Given $\mathbf{H}\in\mathcal{H}$ and $n\in\mathbb{N}$, define

$$\mathrm{ht}(\mathbf{H},n):=e^n\mathrm{ht}(\mathbf{H}).$$

Given a finite collection $\mathcal{F}=\{(\mathbf{H},n)\}\subset\mathcal{H}\times\mathbb{N}$, define

$$\mathrm{ht}(\mathcal{F})=\max\{\mathrm{ht}(\mathbf{H},n):(\mathbf{H},n)\in\mathcal{F}\}.$$

Using the element $\mathbf{v_H}\in\bigwedge^{\dim\mathbf{H}}\mathfrak{g}$ we define the orbit map

$$\eta_H(g):=\rho_H(g)\mathbf{v}_H\ \text{ for every } g\in G.$$

Given a nonzero vector $w\in\bigwedge^r\mathfrak{g}$, for some $0<r\leq\dim\mathbf{G}$, we define $\overline{w}:=\frac{w}{\|w\|}$.

Let $z\in\mathfrak{g}$ with $\|z\|=1$ be so that $u(t)=\exp(tz)$. Let $\mathbf{H}\in\mathcal{H}$; the definition of $N_G(U,H)$ then implies that

$$N_G(U,H)=\{g\in G: z\wedge\eta_H(g)=0\}.$$

We will need the following definition of an effective notion of a generic point from [52].

DEFINITION 5.7. *Let $\varepsilon : \mathcal{H} \times \mathbb{N} \to (0,1)$ be a function so that $\varepsilon(\mathbf{H}, \cdot)$ is decreasing and $\varepsilon(\cdot, n)$ is decreasing in* $\mathrm{ht}(\mathbf{H})$.

A point $g\Gamma$ is said to be ε Diophantine for the action of U if the following holds. For every nontrivial $\mathbf{H} \in \mathcal{H}$, with $\mathbf{H} \neq \mathbf{G}$, and every $n \in \mathbb{N}$, we have

$$\begin{aligned} &\text{for every } \gamma \in \Gamma \text{ with } \|\eta_H(g\gamma)\| < e^n \\ &\qquad \text{that } \|\varkappa \wedge \overline{\eta_H(g\gamma)}\| \geq \varepsilon(\mathbf{H}, n). \end{aligned} \tag{5.2}$$

Given a finite collection $\mathcal{F} \subset \mathcal{H} \times \mathbb{N}$, we say that $g\Gamma$ is $(\varepsilon; \mathcal{F})$ Diophantine if (5.2) *holds for all $(\mathbf{H}, n) \in \mathcal{F}$.*

This is a condition on the pair $(U, g\Gamma)$. We note that the definition of singular points $\mathcal{S}(U)$ using the varieties $N(U,H) = \{g \in G : g^{-1}Ug \subset H\}$ for various subgroups H is defined using polynomial equations. As such, its behavior may change dramatically under small perturbations. Definition 5.7 behaves in that respect much better.

Moreover, one checks easily that $\#\{\mathbf{H} \in \mathcal{H} : \mathrm{ht}(\mathbf{H}) \leq T\} \ll T^{O(1)}$; now for a given pair $(\mathbf{H}, n)$, the condition in (5.2) is given using continuous functions. This implies that any $x \in \mathcal{G}(U)$ is ε Diophantine for some ε as above.

Normal subgroups of $\mathbf{G}$ are fixed points for the adjoint action of G, and hence for U. Thus, we need to control the *distance* from them separately. For any $T > 0$, define

$$\sigma(T) = \min\left\{ \|\varkappa \wedge \overline{\mathbf{v_H}}\| : \begin{array}{c} \mathbf{H} \in \mathcal{H}, \mathbf{H} \lhd \mathbf{G}, \\ \mathrm{ht}(\mathbf{H}) \leq T, \{1\} \neq \mathbf{H} \neq \mathbf{G} \end{array} \right\}.$$

For every $(\mathbf{H}, n) \in \mathcal{H} \times \mathbb{N}$ and any $C > 0$, set

$$\ell_C(\mathbf{H}, n) := \min\left\{ \mathrm{ht}(\mathbf{H}, n)^{-C}, \sigma\left(\mathrm{ht}(\mathbf{H}, n)^C\right)\right\}. \tag{5.3}$$

Let $\| \ \|$ be a norm on $\mathrm{SL}_d(\mathbb{R})$ fixed once and for all. For every $g \in \mathrm{SL}_d(\mathbb{R})$, and in particular, for any $g \in G$, let

$$|g| = \max\{\|g\|, \|g^{-1}\|\}.$$

As we discussed after Theorem 5.3, an important property one anticipates from generic points is that *genericity* is inherited by many points along the orbit. The following theorem guarantees this for the notion defined in Definition 5.7.

THEOREM 5.8 ([52]).
There are constants C and D, depending only on d, and a constant E depending on d, G, and Γ so that the following holds. Let $\mathcal{F} \subset \mathcal{H} \times \mathbb{N}$ *be a finite subset. For any* $g \in G$, $k \geq 1$, *and* $0 < \eta < 1/2$, *at least one of the following holds.*

1. $$\left|\left\{t \in [-1,1] : \begin{array}{c} u(e^k t)g\Gamma \notin X_\eta \text{ or} \\ u(e^k t)g \text{ is not } (\eta^D \ell_C; \mathcal{F}) \text{ Diophantine} \end{array}\right\}\right| < E\eta^{1/D}.$$

2. *There exist a nontrivial proper subgroup* $\mathbf{H}_0 \in \mathcal{H}$ *and some* $n_0 \in \mathbb{N}$ *with*

$$\mathrm{ht}(\mathbf{H}_0, n_0) \leq E \max\{\mathrm{ht}(\mathcal{F}), |g|\eta^{-1}\}^D,$$

so that the following hold.

(a) *For all* $t \in [-1,1]$ *we have*

$$\|\eta_{H_0}(u(e^k t)g)\| \leq E e^{n_0}.$$

(b) *For every* $t \in [-1,1]$ *we have*

$$\left\|\varkappa \wedge \overline{\eta_{H_0}(u(e^{k-1}t)g)}\right\| \leq E e^{-k/D} \max\{\mathrm{ht}(\mathcal{F}), |g|\eta^{-1}\}^D.$$

As was alluded to before, the effective notion of a generic point (Definition 5.7) is one of the main innovations in [52]. In addition to this, the proof of Theorem 5.8 takes advantage of the role played by the subgroup $\mathbf{L} = \{g \in \mathbf{G} : g\mathbf{v_H} = \mathbf{v_H}\}$ to control the *speed* of unipotent orbits: the *distance* between U and subgroup $\mathbf{L}(\mathbb{R})$ controls the speed of $t \mapsto u_t \mathbf{v_H}$. Note that $\mathbf{L}$ is a $\mathbb{Q}$-subgroup of $\mathbf{G}$ whose height is controlled by $\mathrm{ht}(\mathbf{H})^{O(1)}$—it is defined as the stabilizer of the vector $\mathbf{v_H}$. However, $\mathbf{L}$ may not belong to the class $\mathcal{H}$. Actually, it turns out that one may use the fact that U is a unipotent group to replace $\mathbf{L}$ by a subgroup $\mathbf{M} \subset \mathbf{L}$ in $\mathcal{H}$, which already controls the aforementioned speed.

The general strategy of the proof of Theorem 5.8, however, is again based on polynomial-like behavior of unipotent orbits, and it relies on effectivizing the approach in [21].

References

[1] M. Aka, M. Einsiedler, H. Li, and A. Mohammadi, *On effective equidistribution for quotients of* SL(d, $\mathbb{R}$), Israel J. Math. **236** (2020), no. 1, 365–391. MR 4093891

[2] J. Athreya and G. Margulis, *Values of random polynomials at integer points*, J. Mod. Dyn. **12** (2018), 9–16. MR 3808207

[3] L. Auslander, L. Green, and F. Hahn, *Flows on homogeneous spaces*, with the assistance of L. Markus and W. Massey and an appendix by L. Greenberg, Annals of Mathematics Studies, No. 53, Princeton University Press, Princeton, NJ, 1963. MR 0167569

[4] Yves Benoist and Hee Oh, *Effective equidistribution of S-integral points on symmetric varieties*, Ann. Inst. Fourier (Grenoble) **62** (2012), no. 5, 1889–1942. MR 3025156

[5] Yves Benoist and Jean-François Quint, *Mesures stationnaires et fermés invariants des espaces homogènes*, Ann. of Math. (2) **174** (2011), no. 2, 1111–1162. MR 2831114

[6] ———, *Stationary measures and invariant subsets of homogeneous spaces (II)*, J. Amer. Math. Soc. **26** (2013), no. 3, 659–734. MR 3037785

[7] ———, *Stationary measures and invariant subsets of homogeneous spaces (III)*, Ann. of Math. (2) **178** (2013), no. 3, 1017–1059. MR 3092475

[8] Armand Borel and Gopal Prasad, *Finiteness theorems for discrete subgroups of bounded covolume in semi-simple groups*, Inst. Hautes Études Sci. Publ. Math. (1989), no. 69, 119–171. MR 1019963

[9] Jean Bourgain, *A quantitative Oppenheim theorem for generic diagonal quadratic forms*, Israel J. Math. **215** (2016), no. 1, 503–512. MR 3551907

[10] Jean Bourgain, Alex Furman, Elon Lindenstrauss, and Shahar Mozes, *Stationary measures and equidistribution for orbits of nonabelian semigroups on the torus*, J. Amer. Math. Soc. **24** (2011), no. 1, 231–280. MR 2726604

[11] M. Burger and P. Sarnak, *Ramanujan duals II*, Invent. Math. **106** (1991), no. 1, 1–11. MR 1123369

[12] Marc Burger, *Horocycle flow on geometrically finite surfaces*, Duke Math. J. **61** (1990), no. 3, 779–803. MR 1084459

[13] Laurent Clozel, *Démonstration de la conjecture* τ, Invent. Math. **151** (2003), no. 2, 297–328. MR 1953260

[14] S. G. Dani, *Invariant measures of horospherical flows on noncompact homogeneous spaces*, Invent. Math. **47** (1978), no. 2, 101–138. MR 0578655

[15] ———, *Invariant measures and minimal sets of horospherical flows*, Invent. Math. **64** (1981), no. 2, 357–385. MR 629475

[16] ———, *On orbits of unipotent flows on homogeneous spaces*, Ergodic Theory Dynam. Systems **4** (1984), no. 1, 25–34. MR 758891

[17] ———, *On orbits of unipotent flows on homogeneous spaces II*, Ergodic Theory Dynam. Systems **6** (1986), no. 2, 167–182. MR 857195

[18] ———, *Orbits of horospherical flows*, Duke Math. J. **53** (1986), no. 1, 177–188. MR 835804

[19] S. G. Dani and G. A. Margulis, *Values of quadratic forms at primitive integral points*, Invent. Math. **98** (1989), no. 2, 405–424. MR 1016271

[20] ———, *Orbit closures of generic unipotent flows on homogeneous spaces of* $\mathrm{SL}(3,\mathbf{R})$, Math. Ann. **286** (1990), no. 1–3, 101–128. MR 1032925

[21] ———, *Asymptotic behaviour of trajectories of unipotent flows on homogeneous spaces*, Proc. Indian Acad. Sci. Math. Sci. **101** (1991), no. 1, 1–17. MR 1101994

[22] W. Duke, Z. Rudnick, and P. Sarnak, *Density of integer points on affine homogeneous varieties*, Duke Math. J. **71** (1993), no. 1, 143–179. MR 1230289

[23] Manfred Einsiedler, *Ratner's theorem on* $\mathrm{SL}(2,\mathbb{R})$*-invariant measures*, Jahresber. Deutsch. Math.-Verein. **108** (2006), no. 3, 143–164. MR 2265534

[24] Manfred Einsiedler, Anatole Katok, and Elon Lindenstrauss, *Invariant measures and the set of exceptions to Littlewood's conjecture*, Ann. of Math. (2) **164** (2006), no. 2, 513–560.

[25] Manfred Einsiedler and Elon Lindenstrauss, *On measures invariant under tori on quotients of semisimple groups*, Ann. of Math. (2) **181** (2015), no. 3, 993–1031. MR 3296819

[26] M. Einsiedler, G. Margulis, A. Mohammadi, and A. Venkatesh, *Effective equidistribution and property* (τ). J. Amer. Math. Soc. 33 (2020), no. 1, 223–289.

[27] M. Einsiedler, G. Margulis, and A. Venkatesh, *Effective equidistribution for closed orbits of semisimple groups on homogeneous spaces*, Invent. Math. **177** (2009), no. 1, 137–212. MR 2507639

[28] M. Einsiedler, R. Rühr, and Wirth P., *Distribution of shapes of orthogonal lattices*, Ergodic Theory Dynam. Systems **39** (2019), no. 6, 1531–1607. MR 3944098

[29] M. Einsiedler and P. Wirth, *Effective equidistribution of closed hyperbolic surfaces on congruence quotients of hyperbolic spaces*, chapter 13 in this volume.

[30] Jordan S. Ellenberg and Akshay Venkatesh, *Local-global principles for representations of quadratic forms*, Invent. Math. **171** (2008), no. 2, 257–279. MR 2367020

[31] Alex Eskin, Gregory Margulis, and Shahar Mozes, *On a quantitative version of the Oppenheim conjecture*, Electron. Res. Announc. Amer. Math. Soc. **1** (1995), no. 3, 124–130. MR 1369644

[32] ———, *Upper bounds and asymptotics in a quantitative version of the Oppenheim conjecture*, Ann. of Math. (2) **147** (1998), no. 1, 93–141. MR 1609447

[33] Alex Eskin and Curt McMullen, *Mixing, counting, and equidistribution in Lie groups*, Duke Math. J. **71** (1993), no. 1, 181–209. MR 1230290

[34] Alex Eskin and Maryam Mirzakhani, *Invariant and stationary measures for the* $\mathrm{SL}(2,\mathbb{R})$ *action on moduli space*, Publ. Math. Inst. Hautes Études Sci. **127** (2018), 95–324. MR 3814652

[35] Alex Eskin, Maryam Mirzakhani, and Amir Mohammadi, *Isolation, equidistribution, and orbit closures for the* $\mathrm{SL}(2,\mathbb{R})$ *action on moduli space*, Ann. of Math. (2) **182** (2015), no. 2, 673–721. MR 3418528

[36] Alex Eskin, Shahar Mozes, and Nimish Shah, *Unipotent flows and counting lattice points on homogeneous varieties*, Ann. of Math. (2) **143** (1996), no. 2, 253–299. MR 1381987

[37] Alex Eskin and Hee Oh, *Representations of integers by an invariant polynomial and unipotent flows*, Duke Math. J. **135** (2006), no. 3, 481–506. MR 2272974

[38] Livio Flaminio and Giovanni Forni, *Equidistribution of nilflows and applications to theta sums*, Ergodic Theory Dynam. Systems **26** (2006), no. 2, 409–433. MR 2218767

[39] Harry Furstenberg, *The unique ergodicity of the horocycle flow*, Lecture Notes in Math., vol. 318 95–115, Springer, Berlin, Heidelberg, 1973. MR 0393339

[40] Alex Gorodnik, François Maucourant, and Hee Oh, *Manin's and Peyre's conjectures on rational points and adelic mixing*, Ann. Sci. Éc. Norm. Supér. (4) **41** (2008), no. 3, 383–435. MR 2482443

[41] Paul Buterus, Friedrich Götze, Thomas Hille, and Gregory Margulis, *Distribution of values of quadratic forms at integral points*, Preprint, arXiv:1004.5123.

[42] Ben Green and Terence Tao, *The quantitative behaviour of polynomial orbits on nilmanifolds*, Ann. of Math. (2) **175** (2012), no. 2, 465–540. MR 2877065

[43] Gustav A. Hedlund, *Fuchsian groups and transitive horocycles*, Duke Math. J. **2** (1936), no. 3, 530–542. MR 1545946

[44] H. Jacquet and R. P. Langlands, *Automorphic forms on* $\mathrm{GL}(2)$, Lecture Notes in Mathematics, vol. 114, Springer-Verlag, Berlin-New York, 1970. MR 0401654

[45] D. A. Każdan, *On the connection of the dual space of a group with the structure of its closed subgroups*, Funkcional. Anal. i Priložen. **1** (1967), 71–74. MR 0209390

[46] D. Y. Kleinbock and G. A. Margulis, *Bounded orbits of nonquasiunipotent flows on homogeneous spaces*, Sinai's Moscow Seminar on Dynamical Systems, Amer. Math. Soc. Transl. Ser. 2, vol. 171, pp. 141–172, Amer. Math. Soc., Providence, RI, 1996.

[47] ______, *Flows on homogeneous spaces and Diophantine approximation on manifolds*, Ann. of Math. (2) **148** (1998), no. 1, 339–360. MR 1652916

[48] ______, *On effective equidistribution of expanding translates of certain orbits in the space of lattices*, Number theory, analysis and geometry, pp. 385–396, Springer, New York, 2012. MR 2867926

[49] A. Leibman, *Pointwise convergence of ergodic averages for polynomial sequences of translations on a nilmanifold*, Ergodic Theory Dynam. Systems **25** (2005), no. 1, 201–213. MR 2122919

[50] Elon Lindenstrauss, *Invariant measures and arithmetic quantum unique ergodicity*, Ann. of Math. (2) **163** (2006), no. 1, 165–219. MR 2195133

[51] Elon Lindenstrauss and Gregory Margulis, *Effective estimates on indefinite ternary forms*, Israel J. Math. **203** (2014), no. 1, 445–499. MR 3273448

[52] E. Lindenstrauss, G. Margulis, A. Mohammadi, and N. Shah, *Effective avoidance principle for unipotent orbits*, in preparation.

[53] G. A. Margulis, *The action of unipotent groups in a lattice space*, Mat. Sb. (N.S.) **86(128)** (1971), 552–556. MR 0291352

[54] G. A Margulis, *Indefinite quadratic forms and unipotent flows on homogeneous spaces*, Dynamical systems and ergodic theory (Warsaw, 1986) **23** (1989), 399–409.

[55] Grigorii Margulis, *Dynamical and ergodic properties of subgroup actions on homogeneous spaces with applications to number theory*, ICM-90, Mathematical Society of Japan, Tokyo; distributed outside Asia by the Amer. Math. Soc., Providence, RI, 1990. Plenary address presented at the International Congress of Mathematicians held in Kyoto, August 1990. MR 1127160

[56] Gregory Margulis, *Problems and conjectures in rigidity theory*, Mathematics: frontiers and perspectives, pp. 161–174, Amer. Math. Soc., Providence, RI, 2000.

[57] Grigoriy A. Margulis, *On some aspects of the theory of Anosov systems*, Springer Monographs in Mathematics, Springer-Verlag, Berlin, 2004. With a survey by Richard Sharp: Periodic orbits of hyperbolic flows, translated from Russian by Valentina Vladimirovna Szulikowska. MR 2035655

[58] Gregory Margulis and Amir Mohammadi, *Quantitative version of the Oppenheim conjecture for inhomogeneous quadratic forms*, Duke Math. J. **158** (2011), no. 1, 121–160. MR 2794370

[59] Gregori Aleksandrovitch Margulis and Georges Metodiev Tomanov, *Invariant measures for actions of unipotent groups over local fields on homogeneous spaces*, Invent. Math. **116** (1994), no. 1–3, 347–392.

[60] Jens Marklof, *Pair correlation densities of inhomogeneous quadratic forms*, Ann. of Math. (2) **158** (2003), no. 2, 419–471. MR 2018926

[61] Taylor McAdam, *Almost-primes in horospherical flows on the space of lattices*, J. Mod. Dyn. **15** (2019), 277–327. MR 4042163

[62] A. Mohammadi, *A special case of effective equidistribution with explicit constants*, Ergodic Theory Dynam. Systems **32** (2012), no. 1, 237–247. MR 2873169

[63] Shahar Mozes and Nimish Shah, *On the space of ergodic invariant measures of unipotent flows*, Ergodic Theory Dynam. Systems **15** (1995), no. 1, 149–159. MR 1314973

[64] Hee Oh, *Uniform pointwise bounds for matrix coefficients of unitary representations and applications to Kazhdan constants*, Duke Math. J. **113** (2002), no. 1, 133–192. MR 1905394

[65] William Parry, *Dynamical systems on nilmanifolds*, Bull. London Math. Soc. **2** (1970), 37–40. MR 0267558

[66] Gopal Prasad, *Volumes of S-arithmetic quotients of semi-simple groups*, Inst. Hautes Études Sci. Publ. Math. (1989), no. 69, 91–117, with an appendix by Moshe Jarden and the author. MR 1019962

[67] Marina Ratner, *On measure rigidity of unipotent subgroups of semisimple groups*, Acta Math. **165** (1990), no. 3–4, 229–309.

[68] ———, *On Raghunathan's measure conjecture*, Ann. of Math. (2) **134** (1991), no. 3, 545–607.

[69] ———, *Raghunathan's topological conjecture and distributions of unipotent flows*, Duke Math. J. **63** (1991), no. 1, 235–280.

[70] ———, *Raghunathan's conjectures for cartesian products of real and p-adic lie groups*, Duke Math. J. **77** (1995), no. 2, 275–382.

[71] Peter Sarnak, *Asymptotic behavior of periodic orbits of the horocycle flow and Eisenstein series*, Comm. Pure Appl. Math. **34** (1981), no. 6, 719–739. MR 634284

[72] ———, *Values at integers of binary quadratic forms*, Harmonic analysis and number theory (Montreal, PQ, 1996), CMS Conf. Proc., vol. 21, Amer. Math. Soc., Providence, RI, 1997, pp. 181–203. MR 1472786

[73] Peter Sarnak and Adrián Ubis, *The horocycle flow at prime times*, J. Math. Pures Appl. (9) **103** (2015), no. 2, 575–618. MR 3298371

[74] Atle Selberg, *On the estimation of Fourier coefficients of modular forms*, Proc. Sympos. Pure Math., Vol. VIII, pp. 1–15, Amer. Math. Soc., Providence, RI, 1965. MR 0182610

[75] Nimish A. Shah, *Uniformly distributed orbits of certain flows on homogeneous spaces*, Math. Ann. **289** (1991), no. 2, 315–334. MR 1092178

[76] ———, *Invariant measures and orbit closures on homogeneous spaces for actions of subgroups generated by unipotent elements*, Lie groups and ergodic theory (Mumbai, 1996), Tata Inst. Fund. Res. Stud. Math., vol. 14, pp. 229–271, Tata Inst. Fund. Res., Bombay, 1998. MR 1699367

[77] Andreas Strömbergsson, *On the uniform equidistribution of long closed horocycles*, Duke Math. J. **123** (2004), no. 3, 507–547. MR 2068968

[78] ———, *An effective Ratner equidistribution result for* $\mathrm{SL}(2,\mathbb{R}) \ltimes \mathbb{R}^2$, Duke Math. J. **164** (2015), no. 5, 843–902. MR 3332893

[79] William A. Veech, *Minimality of horospherical flows*, Israel J. Math. **21** (1975), no. 2–3, 233–239. Conference on Ergodic Theory and Topological Dynamics (Kibbutz Lavi, 1974). MR 0385009

[80] Akshay Venkatesh, *Sparse equidistribution problems, period bounds and subconvexity*, Ann. of Math. (2) **172** (2010), no. 2, 989–1094. MR 2680486

[81] D. Zagier, *Eisenstein series and the Riemann zeta function*, Automorphic forms, representation theory and arithmetic (Bombay, 1979), Tata Inst. Fund. Res. Studies in Math., vol. 10, pp. 275–301, Tata Inst. Fund. Res., Bombay, 1981. MR 633666

13

MANFRED EINSIEDLER AND PHILIPP WIRTH

EFFECTIVE EQUIDISTRIBUTION OF CLOSED HYPERBOLIC SUBSPACES IN CONGRUENCE QUOTIENTS OF HYPERBOLIC SPACES

Dedicated to Professor Gregory Margulis

Abstract. We prove an effective equidistribution theorem for closed hyperbolic subspaces in congruence quotients of hyperbolic spaces. The argument relies on uniform decay of matrix coefficients and effective versions of arguments in unipotent dynamics. Along the way an effective Borel density theorem is proved for the cases at hand.

1 Introduction

Let $\mathbb{G}$ be a semisimple $\mathbb{Q}$-group and set $G = \mathbb{G}(\mathbb{R})$. Let Γ be a lattice in G such that Γ is a congruence subgroup of $\mathbb{G}(\mathbb{Q})$ and write $X = \Gamma\backslash G$. Furthermore, let H be a connected, semisimple subgroup of G that has no compact factors and assume that the centralizer of $\mathfrak{h} = \mathrm{Lie}(H)$ in $\mathfrak{g} = \mathrm{Lie}(G)$ is trivial. Building on the work of Margulis [17], of Ratner [23], and of Mozes and Shah [19], Einsiedler, Margulis, and Venkatesh proved the following result in [9]:

THEOREM 1.1.
There exists $\delta > 0$, $d \in \mathbb{N}$ depending only on G, H, and $V_0 > 1$ depending only on Γ, G, H with the following properties. Let μ be the H-invariant probability measure on a closed H-orbit x_0H inside $X = \Gamma\backslash G$. For any $V \geq V_0$ there exists an intermediate subgroup S (satisfying $H \subseteq S \subseteq G$) such that x_0S is a closed S-orbit with volume $\leq V$ and

$$\left|\int f\,\mathrm{d}\mu - \int f\,\mathrm{d}\mu_{x_0S}\right| < V^{-\delta}\mathcal{S}_d(f)$$

for any $f \in C_c^\infty(X)$, where μ_{x_0S} is the normalized invariant measure on the orbit x_0S and $\mathcal{S}_d(f)$ denotes an L^2-Sobolev norm of degree d.

We note that the restriction to congruence quotients implies a uniform spectral gap property, which is crucial to the argument. For this we refer to the

Einsiedler acknowledges the support of the SNF Grants 200021-152819 and 200020-178958.

paper by Burger and Sarnak [2] (which is sufficient for the theorem below in the case of noncompact quotients) respective to the general case in the work of Clozel [3] (see also [21], [8, section 4], and references therein).

For the desired extension of Theorem 1.1 to hyperbolic spaces, we recall some standard notation. We write Q_0 for the indefinite quadratic form

$$Q_0(x_1,\dots,x_{n+1}) = -x_1^2 + x_2^2 + \cdots + x_{n+1}^2$$

in $n+1$ variables of signature $(n,1)$. We will assume implicitly that $n \geq 3$. The special orthogonal group of signature $(n,1)$ is then given by

$$\begin{aligned}\mathrm{SO}_{Q_0}(\mathbb{R}) = \mathrm{SO}(n,1) &= \{g \in \mathrm{SL}_{n+1}(\mathbb{R}) \mid Q_0 \circ g = Q_0\}\\ &= \{g \in \mathrm{SL}_{n+1}(\mathbb{R}) \mid g^T J_0 g = J_0\},\end{aligned}$$

where $J_0 = \mathrm{diag}(-1,1,\dots,1)$ is the symmetric matrix representing Q_0.

Let Q be another quadratic form defined over $\mathbb{Z}$ with signature $(n,1)$ and fix $g_Q \in \mathrm{SL}_{n+1}(\mathbb{R})$ so that $Q_0 \circ g_Q = \lambda Q$ for some $\lambda > 0$. In this essay, we are going to prove a result similar to Theorem 1.1 for the special case where $G = \mathrm{SO}^\circ(n,1)$, $\Gamma < G$ corresponds to $\mathrm{SO}_Q(\mathbb{Z})$ under conjugation by g_Q and H is a fixed subgroup of G isomorphic to $\mathrm{SO}^\circ(2,1)$ (corresponding to the orthogonal group on the subspace $\mathbb{R}^3 \times \{0\}^{n-2}$). Here, $\mathrm{SO}^\circ(n,1)$ denotes the connected component of the identity, or for $n \geq 2$ equivalently the index two subgroup of $\mathrm{SO}(n,1)$ generated by its unipotent elements. We note that the connected component of the centralizer of H equals $\mathrm{SO}(n-2)$ (embedded in the lower right corner block of G). Therefore, the methods from [9] do not apply directly once $n \geq 4$. We say that a closed connected subgroup L of G is intermediate if $H < L$; we will classify these subgroups in section 3. The purpose of this paper is to generalize the arguments from [9] to handle the following new cases of effective equidistribution.

THEOREM 1.2.

There exist $\delta > 0$, $d \in \mathbb{N}$ depending only on the dimension n, and $V_0 > 1$ depending on Γ with the following properties. Let μ be the H-invariant and ergodic probability measure on $X = \Gamma\backslash G$, which equals the normalized Haar measure on a closed orbit of some $x_0 \in X$ and some intermediate subgroup. For any $V \geq V_0$ there exists an intermediate subgroup L such that x_0L is a closed L-orbit with volume $\leq V$ and

$$\left|\int f\,\mathrm{d}\mu - \int f\,\mathrm{d}\mu_{x_0L}\right| < V^{-\delta} S_d(f)$$

for any $f \in C_c^\infty(X)$.

Let $K = \mathrm{SO}(n) < G$ be the maximal compact subgroup (defined as the orthogonal group on the subspace $\{0\} \times \mathbb{R}^n$). Restricting to torsion-free lattices $\Gamma < G$ and K-invariant functions on X we may phrase the above theorem in geometric terms by saying that closed hyperbolic subspaces of large volume within the hyperbolic manifold $\Gamma\backslash\mathbb{H}^n \cong \Gamma\backslash G/K$ effectively equidistribute within closed hyperbolic submanifolds of smaller volume.

We refer to the survey in this volume [10] to put Theorems 1.1–1.2 in a broader context of effective arguments in unipotent dynamics.

We thank Andreas Wieser and Amir Mohammadi for discussions and their comments on an earlier draft. M. E. also thanks Klaus Schmidt for mentioning the permanent at precisely the right time.

It is a great pleasure to dedicate this paper to Professor Gregory Margulis, whose influence on the arguments of this essay and the theory of homogeneous dynamics in general could not be more obvious.

2 Quantitative Nondivergence

In this section, we are going to recall a quantitative nondivergence result for unipotent orbits that became an indispensible tool in homogeneous dynamics and goes back to the work of Margulis [16].

Suppose that $Q_0 \circ g_Q = \lambda Q$ as in the introduction. We note that $Q_0(v) = v^T J_0 v$ for all $v \in \mathbb{R}^{n+1}$ and also

$$Q_0(g^{-1} g_Q v) = v^T g_Q^T (g^{-1})^T J_0 g^{-1} g_Q v = \lambda Q(v) \in \lambda\mathbb{Z} \tag{1}$$

for all $v \in \mathbb{Z}^{n+1}$ and $g \in \mathrm{SO}(n,1)$. By the same argument we also have that $g_Q^{-1}\,\mathrm{SO}(n,1) g_Q$ equals the orthogonal group SO_Q associated to the quadratic form Q. Moreover, we may also make the description of Γ in the introduction formal by setting $\Gamma = g_Q\,\mathrm{SO}_Q(\mathbb{Z}) g_Q^{-1} \cap G$.

Our first lemma shows that, in contrast to the case of general elements of $\mathrm{SL}_n(\mathbb{R})/\mathrm{SL}_n(\mathbb{Z})$ for $n \geq 3$, every lattice corresponding to elements of the space $X = \Gamma\backslash\mathrm{SO}^\circ(n,1)$ contains (up to scalar multiples) at most one nonzero short vector. For this we identify $x = \Gamma g \in \Gamma\backslash\mathrm{SO}^\circ(n,1)$ with

$$g^{-1} g_Q\,\mathrm{SL}_{n+1}(\mathbb{Z}) \in \mathrm{SL}_{n+1}(\mathbb{R})/\mathrm{SL}_{n+1}(\mathbb{Z})$$

and in addition with the unimodular lattice

$$\Lambda_x = g^{-1} g_Q \mathbb{Z}^{d+1} < \mathbb{R}^{n+1}.$$

Then Equation (1) shows that $Q_0(\Lambda_x) \subseteq \lambda\mathbb{Z}$ and similarly $\langle \Lambda_x, \Lambda_x \rangle_{Q_0} \subseteq \frac{1}{2}\lambda\mathbb{Z}$ for all $x \in X$, where $\langle \cdot, \cdot \rangle_{Q_0}$ denotes the indefinite bilinear form induced by Q_0.

LEMMA 2.1 (Short vectors). *There exists some $\rho_Q > 0$ with the following property. For any $x \in X$ there exists, up to scalar multiples, at most one vector $v \in \Lambda_x \setminus \{0\}$ with Euclidean norm $\|v\| < \rho_Q$.*

Proof. Using continuity of Q_0 there exists some $\rho_Q > 0$ such that $v, w \in \mathbb{R}^{n+1}$ with $\|v\|, \|w\| < \rho_Q$ implies that $|Q_0(v)|, |Q_0(w)|, |\langle v, w \rangle_{Q_0}| < \frac{1}{2}\lambda$.

If now $v, w \in \Lambda_x$ for some $x \in X$ satisfy $\|v\|, \|w\| < \rho_Q$, then by (1) we must have $Q_0(v) = Q_0(w) = \langle v, w \rangle_{Q_0} = 0$. However, given that Q_0 has signature $(n, 1)$, its maximal isotropic subspace is one-dimensional, and this implies that v and w are linearly dependent. □

Since there is (up to sign) at most one short primitive lattice element, the proof of the following quantitative nondivergence result becomes substantially easier than the corresponding statement for $\mathrm{SL}_n(\mathbb{R})/\mathrm{SL}_n(\mathbb{Z})$. In what follows, we think of $H = \mathrm{SO}^\circ(2, 1)$ as being embedded in the top left corner of $\mathrm{SO}^\circ(n, 1)$. Moreover, we define

$$(2) \qquad u_t = \begin{pmatrix} 1+\frac{t^2}{2} & \frac{t^2}{2} & t \\ -\frac{t^2}{2} & 1-\frac{t^2}{2} & -t \\ t & t & 1 \end{pmatrix} \text{ and } v_t = \begin{pmatrix} 1+\frac{t^2}{2} & -\frac{t^2}{2} & t \\ \frac{t^2}{2} & 1-\frac{t^2}{2} & t \\ t & -t & 1 \end{pmatrix}$$

for $t \in \mathbb{R}$ and note that these define unipotent one-parameter subgroups $\{u_t \mid t \in \mathbb{R}\}$ and $\{v_t \mid t \in \mathbb{R}\}$ of H that together generate H.

LEMMA 2.2 (Invariant vectors). *If $v \in \Lambda_x$ is a primitive lattice element with $\|v\| < \rho_Q$ that is invariant under $\mathrm{SO}^\circ(2, 1)$, then $v = 0$.*

Proof. A short calculation using the above matrices reveals that a vector $v \in \Lambda_x$ invariant under $\mathrm{SO}^\circ(2, 1)$ must be of the form $(0, 0, 0, v_4, \ldots v_{n+1})^T$. Since $\|v\| < \rho_Q$ and $Q_0(v) \in \lambda\mathbb{Z}$, the same argument as in the proof of Lemma 2.1 implies that $0 = Q_0(v) = \sum_{k=4}^{n+1} v_k^2$. Hence $v = 0$ as required. □

Note that the previous lemma implies in particular that for every $x \in X$ there exists a unipotent one-parameter subgroup (either $\{u_t \mid t \in \mathbb{R}\}$ or $\{v_t \mid t \in \mathbb{R}\}$) of H that does not stabilize the (up to sign) unique short primitive vector from Lemma 2.1, if such a vector exists. For a lattice $\Lambda \subseteq \mathbb{R}^{n+1}$, we denote by $\lambda_1(\Lambda)$

the length of the shortest vector in $\Lambda \setminus \{0\}$. Moreover, for every $\varepsilon > 0$ we define the compact set

$$X(\varepsilon) = \{x \in X \mid \lambda_1(\Lambda_x) \geq \varepsilon\}.$$

PROPOSITION 2.3 (Quantitative nondivergence). *Assume that the unipotent one-parameter subgroup* $\{u_t\} < H$ *defined in Equation (2) has no nontrivial invariant vectors* $v \in \Lambda_x$ *of norm less than* ρ_Q. *Then there exists some constant* $T_x > 0$ *such that for all* $\varepsilon > 0$ *and for all* $T \geq T_x$, *we have*

$$\frac{1}{T}\,|\{t \in [0, T] \mid xu_t \notin X(\varepsilon)\}| \ll_Q \varepsilon^{1/2}.$$

This is a special case of the results proven in [16] and [4]. Proposition 2.3 also holds for the second unipotent subgroup in (2) and implies in particular that there exists a fixed compact subset $X_{\mathrm{cpct}} \subseteq X$, such that each H-orbit is of the form xH for some $x \in X_{\mathrm{cpct}}$ (and the same also holds for each L-orbit, where L is an intermediate group). Moreover, Proposition 2.3 and the Mautner phenomenon for $\{u_t\} < H$ allow us to choose the compact subset so that $\mu(X \setminus X_{\mathrm{cpct}}) \leq \frac{1}{10^{11}}$ for any H-invariant and ergodic measure μ on X.

3 Intermediate subgroups

The aim of this section is to identify all intermediate Lie algebras $\mathfrak{so}(2,1) \subseteq \mathfrak{l} \subseteq \mathfrak{so}(n,1)$, which then also gives us a classification of the closed, connected, intermediate subgroups $\mathrm{SO}^\circ(2,1) \leq L \leq \mathrm{SO}^\circ(n,1)$.

PROPOSITION 3.1. *Every intermediate Lie algebra* $\mathfrak{so}(2,1) \subseteq \mathfrak{l} \subseteq \mathfrak{so}(n,1)$ *is* Ad-*conjugated to* $\mathfrak{so}(k,1) \oplus \mathfrak{k}$ *for some* $2 \leq k \leq n$ *and* $\mathfrak{k}$ *being contained in the centralizer of* $\mathfrak{so}(k,1)$ *in* $\mathfrak{so}(n,1)$. *In fact, the conjugating element can be choosen in* $\mathrm{SO}(n)$.

Proof. There is a surjective algebraic homomorphism

$$\mathrm{SL}_2(\mathbb{R}) \longrightarrow \mathrm{SO}^\circ(2,1),$$

which extends to a homomorphism into $\mathrm{SO}(n,1)$. Note that the Lie algebra of $\mathrm{SO}^\circ(2,1)$ (defined using the quadratic form $-x_1^2 + x_2^2 + x_3^2$) is given by

$$\mathfrak{so}(2,1) = \left\{X \in \mathrm{Mat}_3(\mathbb{R}) \mid X^T J + JX = 0\right\}, \text{ where } J = \begin{pmatrix} -1 & 0 & 0 \\ 0 & 1 & 0 \\ 0 & 0 & 1 \end{pmatrix}.$$

We explicitly define the above homomorphism by giving the following $\mathfrak{sl}_2$-triple in $\mathfrak{so}(2,1)$:

$$H = \begin{pmatrix} 0 & 2 & 0 \\ 2 & 0 & 0 \\ 0 & 0 & 0 \end{pmatrix}, \quad X = \begin{pmatrix} 0 & 0 & 1 \\ 0 & 0 & 1 \\ 1 & -1 & 0 \end{pmatrix}, \text{ and } Y = \begin{pmatrix} 0 & 0 & 1 \\ 0 & 0 & -1 \\ 1 & 1 & 0 \end{pmatrix}.$$

It is easily checked that this indeed is an $\mathfrak{sl}_2$-triple—that is, it satisfies the relations

$$[H,X] = 2X, \quad [H,Y] = -2Y, \quad [X,Y] = H.$$

Consider now the Lie algebra

$$\mathfrak{so}(n,1) = \left\{ X \in \mathrm{Mat}_{n+1}(\mathbb{R}) \mid X^T J_0 + J_0 X = 0 \right\},$$

where $J_0 = \mathrm{diag}(-1,1,\ldots,1)$, and embed $\mathfrak{so}(2,1)$ in the top left corner of $\mathfrak{so}(n,1)$. Note that the centralizer of $\mathfrak{so}(2,1)$ in $\mathfrak{so}(n,1)$ is isomorphic to $\mathfrak{so}(n-2)$, embedded in the bottom right corner of $\mathfrak{so}(n,1)$. We will denote the corresponding elements when embedded into $\mathfrak{so}(n,1)$ by H, X, and Y as well. Moreover, we introduce the following shorthand notation for $k \in \{4,\ldots,n+1\}$:

$$(k;a,b,c) := \left(\begin{array}{ccc|ccc} 0 & & & & a & \\ & 0 & & & b & \\ & & 0 & & c & \\ \hline & & & 0 & & \\ a & -b & -c & & \ddots & \\ & & & & & 0 \end{array}\right),$$

where the entries a,b,c appear in the kth row (resp. column). Notice that $(k;a,b,c) \in \mathfrak{so}(n,1)$ and that for any element of $\mathfrak{so}(n,1)$ the entries below the diagonal are uniquely determined by the entries above the diagonal. Moreover, one easily checks that the following identities hold for all $k \in \{4,\ldots,n+1\}$:

$$\begin{aligned} \mathrm{ad}_X(k;a,b,c) &= (k;c,c,a-b) \\ \mathrm{ad}_Y(k;a,b,c) &= (k;c,-c,a+b) \\ \mathrm{ad}_H(k;a,b,c) &= (k;2b,2a,0). \end{aligned}$$

In particular, this shows that the algebra $\mathfrak{so}(2,1)$ acts irreducibly on

$$V^k := \{(k;a,b,c) \mid a,b,c \in \mathbb{R}\}$$

for all $k \in \{4, \ldots, n+1\}$ and that these representation spaces are isomorphic to the adjoint representation. Note that if we define

$$R = \bigoplus_{k=4}^{n+1} V^k \subseteq \mathfrak{so}(n,1),$$

then every nontrivial, irreducible $\mathfrak{sl}_2$-representation in R can be written in the form

$$V_\lambda = \left\{ \sum_{k=4}^{n+1} \lambda_k \cdot (k; a, b, c) \mid a, b, c \in \mathbb{R} \right\}$$

for some $\lambda = (\lambda_4, \ldots, \lambda_{n+1}) \in \mathbb{R}^{n-2}$ with $\|\lambda\| = 1$. Moreover, we have

$$\mathfrak{so}(n,1) = \mathfrak{so}(2,1) \oplus R \oplus \mathfrak{so}(n-2),$$

where $\mathfrak{so}(n-2)$ is the centralizer of $\mathfrak{so}(2,1)$ inside $\mathfrak{so}(n,1)$. In terms of the adjoint action of $\mathfrak{so}(2,1)$ on $\mathfrak{so}(n,1)$ the decomposition into the summands $\mathfrak{so}(2,1) \oplus R$ and $\mathfrak{so}(n-2)$ is canonical in the following sense: the former summand is as an $\mathfrak{so}(2,1)$-space isomorphic to $n-1$ copies of the adjoint representation, and on the latter the representation is trivial.

Let $\mathfrak{so}(2,1) \subseteq \mathfrak{l} \subseteq \mathfrak{so}(n,1)$ be an intermediate Lie algebra. Note that if $\mathfrak{l} \cap R = \{0\}$, then $\mathfrak{l}$ is of the required form $\mathfrak{so}(2,1) \oplus \mathfrak{k}$, where $\mathfrak{k} = \mathfrak{l} \cap \mathfrak{so}(n-2)$ is contained in the centralizer $\mathfrak{so}(n-2)$ of $\mathfrak{so}(2,1)$ in $\mathfrak{so}(n,1)$. We therefore assume that $\mathfrak{l} \cap R \neq \{0\}$. This means that

$$\mathfrak{l} \cap R = V_{\lambda^{(1)}} \oplus \cdots \oplus V_{\lambda^{(m)}}$$

is a direct sum of irreducible $\mathfrak{sl}_2$-representations of the form as above for some $1 \leq m \leq n-2$, where $\lambda^{(i)} \in \mathbb{R}^{n-2}$ satisfies $\|\lambda^{(i)}\| = 1$ for all $i \in \{1, \ldots, m\}$ and the vectors $\lambda^{(1)}, \ldots, \lambda^{(m)}$ are linearly independent.

We claim that there exists an element $g \in \mathrm{SO}(n-2)$, so that

$$\mathrm{Ad}_g(\mathfrak{l}) \cap R = \bigoplus_{k=4}^{m+3} V^k.$$

Indeed, let M be the $m \times (n-2)$-matrix whose rows are given by $\lambda^{(1)}, \ldots, \lambda^{(m)}$. By the Gram-Schmidt algorithm, there exists an element $g \in \mathrm{SO}(n-2)$, so that Mg is of the form $(T, 0)$, where T is an invertible, lower triangular $m \times m$-matrix. This implies the claim.

Let $E_{k,\ell}$ denote the square matrix with a 1 in the kth row and the ℓth column and with 0 otherwise. Then

$$(k; 1, 0, 0) = E_{1,k} + E_{k,1}$$
$$(k; 0, 1, 0) = E_{2,k} - E_{k,2}$$
$$(k; 0, 0, 1) = E_{3,k} - E_{k,3}$$
$$[(k; 1, 0, 0), (\ell; 1, 0, 0)] = E_{k,\ell} - E_{\ell,k}.$$

all belong to $\mathrm{Ad}_g(\mathfrak{l})$ for all $k, \ell \in \{4, \ldots, m+3\}$. This implies that $\mathfrak{so}(m+2, 1) \subseteq \mathrm{Ad}_g(\mathfrak{l})$, where the Lie algebra $\mathfrak{so}(m+2, 1)$ is embedded in the top left corner of $\mathfrak{so}(n, 1)$. Since $\mathrm{Ad}_g(\mathfrak{l}) \cap \bigoplus_{k=m+4}^{n+1} V^k = \{0\}$, we have $\mathrm{Ad}_g(\mathfrak{l}) = \mathfrak{so}(m+2, 1) \oplus \mathfrak{k}$, where $\mathfrak{k}$ is contained in the centralizer of $\mathfrak{so}(m+2, 1)$ in $\mathfrak{so}(n, 1)$. Therefore, $\mathfrak{l}$ is Ad-conjugated to $\mathfrak{so}(m+2, 1) \oplus \mathfrak{k}$ as claimed in the proposition. □

COROLLARY 3.2. Every closed, connected, intermediate group

$$\mathrm{SO}^\circ(2, 1) \le L \le \mathrm{SO}^\circ(n, 1)$$

is conjugated by some element of $\mathrm{SO}(n)$ to $\mathrm{SO}^\circ(m, 1) \times K'$, where $2 \le m \le n$ and $K' \subseteq \mathrm{SO}(n-m)$ is a subgroup of the compact centralizer of $\mathrm{SO}(m, 1)$ in $\mathrm{SO}(n, 1)$.

Proof. By the previous proposition we have after conjugation by an element of $\mathrm{SO}(n)$ that $\mathrm{Lie}(L) \cong \mathfrak{so}(m, 1) \oplus \mathfrak{k}$ for some $2 \le m \le n$ and $\mathfrak{k}$ contained in the centralizer of $\mathfrak{so}(m, 1)$ within $\mathfrak{so}(n, 1)$. Since L is connected, this immediately implies the result. □

COROLLARY 3.3. Let $\mathbb{L} < \mathrm{SO}_Q$ be an algebraic subgroup defined over $\mathbb{Q}$ so that its group of $\mathbb{R}$-points contains a subgroup conjugated to $\mathrm{SO}(2, 1)$. Then $\mathbb{L}$ contains a unique algebraic subgroup $\widetilde{\mathbb{L}}$ defined over $\mathbb{Q}$ that is $\mathbb{Q}$-simple and satisfies that over $\mathbb{R}$ the group $\widetilde{\mathbb{L}}$ is isomorphic to $\mathrm{SO}(m, 1) \times \mathrm{SO}(m+1)^p$ for some $m \ge 2$ and $p \ge 0$. In fact, $g_Q \widetilde{\mathbb{L}} g_Q^{-1}$ is conjugated by some element of $\mathrm{SO}(n)$ to $\mathrm{SO}(m, 1) \times \mathrm{SO}(m+1)^p$ embedded as block matrices into $\mathrm{SO}(n, 1)$.

Proof. Without loss of generality we may assume $\mathbb{L}$ is Zariski connected. By Corollary 3.2 $\mathbb{L}(\mathbb{R})$ contains a normal semisimple subgroup F conjugated to $\mathrm{SO}(m, 1)$ for some $m \ge 2$. Using the Levi decomposition of $\mathbb{L}$ it follows that $\mathbb{L}$ must contain a $\mathbb{Q}$-simple normal subgroup $\widetilde{\mathbb{L}}$ whose group of $\mathbb{R}$-points contains F as a normal almost direct factor. Also recall that $\widetilde{\mathbb{L}}$ after conjugation by g_Q and an element of $\mathrm{SO}(n)$ becomes the product of $\mathrm{SO}(m, 1)$ embedded in the upper left block of size $(m+1)$ and a subgroup of $\mathrm{SO}(n-m)$ embedded in the lower right block of size $n-m$. In particular, F is indeed a direct (not just almost direct) factor of $\widetilde{\mathbb{L}}(\mathbb{R})$. Taking Galois conjugates of this normal subgroup we see that the other factors of $\widetilde{\mathbb{L}}$ over $\mathbb{R}$ must also be forms of $\mathrm{SO}(m+1)$ and must have similar block structure corresponding to a collection of mutually orthogonal subspaces. We also note that over $\mathbb{R}$ all other factors must be compact and so must indeed be isomorphic to $\mathrm{SO}(m+1)$ over $\mathbb{R}$. Hence $\widetilde{\mathbb{L}}(\mathbb{R})$ is a direct product as stated in the corollary. □

4 Existence of many small lattice elements

4.1 ORBIT CLOSURES. Recall that $G=\mathrm{SO}^\circ(n,1)$ and let $\Gamma<\mathrm{SO}^\circ(n,1)$ be as in Section 1. Let $S=\mathrm{SO}^\circ(m,1)$ for some $2\leq m\leq n$ (embedded in the upper left block in G) and let $x_0=\Gamma g_0\in X$ for some $g_0\in\mathrm{SO}^\circ(n,1)$. By Ratner's orbit closure theorem [24] we have that $\overline{x_0S}=x_0L$ is a closed orbit (automatically with finite volume) in $\Gamma\backslash G$ for a connected intermediate subgroup L. Applying Corollary 3.2 we find a maximal $m'\in\{m,m+1,\ldots,n\}$ so that a conjugate of $\mathrm{SO}(m',1)$ is a normal subgroup of L. Applying [1, proposition 1.1] and Corollary 3.3 we see that there exists an algebraic $\mathbb{Q}$-group $\mathbb{L}<\mathrm{SO}_Q$ defined over $\mathbb{Q}$ and $\mathbb{Q}$-simple so that L is the connected component of $g_0^{-1}g_Q\mathbb{L}(\mathbb{R})g_Q^{-1}g_0$ and in particular conjugated to $\mathrm{SO}^\circ(m',1)\times\mathrm{SO}(m'+1)^p$ for some $p\geq 0$. To simplify the notation we conjugate S and L by some element of $\mathrm{SO}(n)$ if necessary, move the point x_0 by the same element, and assume that $m'=m$ and that $L=S\times\mathrm{SO}(m+1)^p$ embedded in the upper left block of G.

4.2 VOLUMES AND LATTICE ELEMENTS. We note that the restriction of the Riemannian metric on G to H or more generally a connected intermediate subgroup L induces a Riemannian metric on L, which we will use to define the volume of a closed L-orbit. Indeed the volume $\mathrm{vol}(x_0L)$ of the orbit x_0L is defined as the Haar measure (induced by the restriction of the Riemannian metric) of a measurable fundamental domain $F\subset L$ for the quotient map $g\in L\mapsto x_0g$. Moreover, we define the Haar measure μ_{x_0L} on a closed orbit x_0L as the normalized measure $(\mathrm{vol}(x_0L))^{-1}(x_0m_L|_F)$, where m_L denotes the Haar measure on L, $m_L|_F$ denotes its restriction to the fundamental domain $F\subset L$, and $x_0m_L|_F$ denotes the pushforward under the quotient map.

Denote the connected component of the normalizer of S in G by

$$N=N_G(S)^\circ\cong S\times\mathrm{SO}(n-m). \tag{3}$$

As already pointed out, we may assume that $L < N$. We will also write $S^{g_0} = g_0 S g_0^{-1}$ and $N^{g_0} = g_0 N g_0^{-1}$ for the conjugated groups. Since N/L is compact, we have $\mathrm{vol}(x_0 N) \ll \mathrm{vol}(x_0 L)$. In the next section we are going to prove a version of the converse inequality—that is, $\mathrm{vol}(x_0 N) \gg \mathrm{vol}(x_0 L)^{\kappa_1}$ for some $\kappa_1 > 0$ that only depends on the dimension n. This section will provide a crucial step toward that result.

For $r > 0$ we define the ball in N of radius r by

$$B_r^{\mathrm{HS}} \cap N = \{g \in N \mid \|g\|_{HS} \leq r\},$$

where $\|\cdot\|_{HS}$ is the Hilbert-Schmidt norm with respect to the standard representation of N on $\mathbb{R}^{n+1}$:

$$\|g\|_{HS}^2 = \sum_{i=1}^{n+1} \|ge_i\|_2^2$$

PROPOSITION 4.1 (Existence of lattice elements). *Let $x_0 = \Gamma g_0 \in X$ be as above so that* $\mathrm{vol}(x_0 L) < \infty$. *There exist $d \in \mathbb{N}$, lattice elements*

$$\gamma_1, \ldots, \gamma_d \in N^{g_0} \cap \Gamma,$$

and a constant $\kappa_2 > 0$ depending only on n so that the Zariski closure of the subgroup generated by $\{\gamma_1, \ldots, \gamma_d\}$ contains S^{g_0} and

$$\|\gamma_i\|_{HS} \ll \mathrm{vol}(x_0 N)^{\kappa_2} \text{ for } i = 1, \ldots, d.$$

Recall that there exists a torsion-free congruence subgroup $\Gamma_1 < \Gamma$ that has finite index in Γ. We are going to prove Proposition 4.1 by finding lattice elements $\gamma_1, \ldots, \gamma_d$ in $g_0^{-1}\Gamma_1 g_0 \cap N$ with norms bounded by $\ll \mathrm{vol}(x_0 N)^{\kappa_2}$. Using Proposition 2.3 we may choose the element g_0 to belong to a fixed compact subset of X. Therefore, conjugating these elements by g_0 gives lattice elements in $N^{g_0} \cap \Gamma$ with the desired properties.

We note that $B_r^{\mathrm{HS}} \cap N$ is invariant under $\mathrm{SO}(n-m)$ so that the following counting results are easily reduced to a counting of lattice elements in S. In fact the action of the group S on $L^2(x_0 N / \mathrm{SO}(n-m))$ has a uniform spectral gap and there exists $\delta > 0$ independent of x_0 such that we have the following effective lattice point counting result (see, e.g., [12], [9, proposition 12.1], [6], [11], [13], and [18]):

$$(4)\qquad |g_0^{-1}\Gamma_1 g_0 \cap B_r^{\mathrm{HS}} \cap N| = \frac{m_N(B_r^{\mathrm{HS}} \cap N)}{\mathrm{vol}(x_0 N)} + \mathcal{O}(m_N(B_r^{\mathrm{HS}} \cap N)^{(1-\delta)})$$
$$= \frac{m_S(B_r^{\mathrm{HS}} \cap S) m_{C_G(S)}(C_G(S))}{\mathrm{vol}(x_0 N)} + \mathcal{O}(m_S(B_r^{\mathrm{HS}} \cap S)^{(1-\delta)})$$
$$\asymp \frac{m_S(B_r^{\mathrm{HS}} \cap S)}{\mathrm{vol}(x_0 N)} + \mathcal{O}(m_S(B_r^{\mathrm{HS}} \cap S)^{(1-\delta)}).$$

We set $d(r) = \left|g_0^{-1}\Gamma_1 g_0 \cap B_r^{\mathrm{HS}} \cap N\right|$ for a given radius $r > 0$. This means that there exist precisely $d(r)$ distinct elements $\gamma_1, \ldots, \gamma_{d(r)} \in g_0^{-1}\Gamma_1 g_0 \cap N$ of norm at most r. Let

$$L_r = \overline{\langle \gamma_1, \ldots, \gamma_{d(r)} \rangle}^Z \cap N$$

be (the group of $\mathbb{R}$-points of) the Zariski closure of the subgroup generated by the lattice elements $\gamma_1, \ldots, \gamma_{d(r)}$ intersected with the connected component N of the normalizer $N_G(S)$ of S in Equation (3). In order to prove Proposition 4.1, it remains to show that we can choose r in such a way that S is contained in L_r. In fact, we are going to show that by setting r to be a certain power of $\mathrm{vol}(x_0 N)$, we have too many lattice elements than can fit in $M \times \mathrm{SO}(n-m)$ for a proper subgroup $M \lneq S$. We denote the projection to the first factor by $\pi_S : N = S \times C_G(S) \to S$.

We claim that $S \not\subseteq L_r$ implies $\pi_S(L_r) \subsetneq S$. To see this we suppose indirectly that $\pi_S(L_r) = S$. In this case $L_r \cap S$ is a normal subgroup of $S \lhd N$, and as S is a simple Lie group we see that $L_r \cap S = S$ or $L_r \cap S$ is finite. The former contradicts our assumption $S \not\subseteq L_r$. The latter leads together with $\pi_S(L_r) = S$ to a contradiction, since in this case the subgroup L_r would define a surjective homomorphism from a subgroup of the compact group $\mathrm{SO}(n-m)$ to the simple noncompact Lie group $S/L_r \cap S$.

Therefore, L_r is contained in $M \times \mathrm{SO}(n-m)$, where $M = M_r$ is a proper Zariski closed subgroup of $\mathrm{SO}(m,1)$. Since $\mathrm{SO}(n-m)$ is compact and Γ_1 is torsion-free, L_r cannot be contained in $\mathrm{SO}(n-m)$ or the product of a finite group with $\mathrm{SO}(n-m)$ if r is large enough. For such radii M is at least one-dimensional.

LEMMA 4.2. *Let $M \leq \mathrm{SO}(m,1)$ be a proper Zariski closed subgroup of positive dimension. Then one of the following holds for some $g \in S$:*

1. *$M \cap C_S(A_M)$ has at most index two in M, where A_M is the unique split torus in M.*
2. *$M \leq g(\mathrm{O}(\ell,1) \times \mathrm{O}(m-\ell))g^{-1} \cap \mathrm{SO}(m,1)$ for some $\ell \in \{2, \ldots, m-1\}$.*

3. *$M \le gPg^{-1}$, where P is a fixed parabolic subgroup of S.*
4. *M is compact and contained in $g\,\mathrm{SO}(m)g^{-1}$.*

In the proof we will make use of the structure theory of algebraic groups and special properties of $\mathrm{SO}(m,1)$. In particular, we note that $\mathrm{SO}(m,1)$ contains the split rank one torus A containing the elements

$$a_t = \begin{pmatrix} \cosh t & \sinh t & 0 \\ \sinh t & \cosh t & 0 \\ 0 & 0 & I_{m-1} \end{pmatrix} \tag{5}$$

for $t \in \mathbb{R}$, where I_{m-1} stands for the identity matrix in $m-1$ dimensions. We also define $P < \mathrm{SO}(m,1)$ as the parabolic subgroup defined by $\{a_t\}$ so that $\{u_t\} < P$.

Proof of Lemma 4.2. Suppose that the Lie algebra $\mathfrak{m}$ of M contains $\mathrm{Ad}_g(\mathfrak{so}(\ell',1))$ for some $g \in S$ and $2 \le \ell' < m$. To simplify the notation we assume without loss of generality that $\mathfrak{so}(\ell',1) \subseteq \mathfrak{m}$. By Proposition 3.1 there exists some integer ℓ with $\ell' \le \ell < m$ so that after another conjugation $\mathfrak{m}$ contains $\mathfrak{so}(\ell,1)$ and is contained in $\mathfrak{so}(\ell,1) \times \mathfrak{so}(m-\ell)$. Let $h \in M \subseteq \mathrm{SO}(m,1)$. Then we have $\mathrm{Ad}_h(\mathfrak{m}) = \mathfrak{m}$, $\mathrm{Ad}_h(\mathfrak{so}(\ell,1)) = \mathfrak{so}(\ell,1)$, and hence $\mathrm{Ad}_h(\mathfrak{so}(m-\ell)) = \mathfrak{so}(m-\ell)$. We now define the subspaces of invariant vectors

$$V_1 = (\mathbb{R}^{m+1})^{\mathrm{SO}(m-\ell)} = \mathbb{R}^{\ell+1} \times \{0\}^{m-\ell}$$
$$V_2 = (\mathbb{R}^{m+1})^{\mathrm{SO}^\circ(\ell,1)} = \{0\}^{\ell+1} \times \mathbb{R}^{m-\ell}$$

for the two subgroups and obtain $\mathbb{R}^{m+1} = V_1 \oplus V_2$ and $h(V_i) = V_i$ for $i = 1,2$. This, however, implies that $h \in \mathrm{O}(\ell,1) \times \mathrm{O}(m-\ell)$ and therefore that M is contained in

$$(\mathrm{O}(\ell,1) \times \mathrm{O}(m-\ell)) \cap \mathrm{SO}(m,1)$$

(up to conjugation), as claimed in case (2) of the lemma. We note that the Jacobson-Morozov theorem, the fact that S has rank one, and Proposition 3.1 can be used to show that any noncompact simple subgroup of $\mathrm{SO}(m,1)$ is isomorphic to $\mathrm{SO}(\ell,1)$ for some k with $2 \le \ell \le m$. Hence we will assume in the following that M does not contain any noncompact simple subgroup.

If M is compact, it is contained in a maximal compact subgroup. As these are conjugated to $\mathrm{SO}(m)$ we obtain in this case (4) of the lemma.

If the unipotent radical of M is trivial, then M is reductive. As mentioned above we may assume M has a trivial or compact semisimple part. Assuming also that M is not compact and recalling that S has rank one, it follows that M contains a unique one-dimensional split torus A_M. This implies that M is contained in $N_S(A_M)$. Since the only nontrivial algebraic automorphism of A_M is $(a \in A_M \mapsto a^{-1})$, it follows that $M \cap C_S(A_M)$ has index at most two in M as claimed in case (1) of the lemma.

Assume now that the unipotent radical U_M of M is nontrivial. Then there exists some $g \in S$ with $U_M \subseteq gPg^{-1}$. We again assume for simplicity that $U_M \subseteq P$ and recall that P is the parabolic subgroup defined by a_1. Let $v_- \in \mathrm{Lie}(U_M)$ be nontrivial and let $h \in M$. Applying the Iwasawa decomposition we have $h = kan$ for $k \in \mathrm{SO}(m)$, $a \in A$, and $u \in U$ belonging to the unipotent radical U of P. Clearly $an \in P$ so we restrict our attention to k. A calculation reveals that $v_- \in \mathrm{Lie}(U)$ has the form

$$v_- = \begin{pmatrix} 0 & 0 & \xi^T \\ 0 & 0 & -\xi^T \\ \xi & \xi & 0 \end{pmatrix}$$

for some $\xi \in \mathbb{R}^{m-1}$ (see also Equation (2)). By assumption we have $\mathrm{Ad}_k v_- \in \mathrm{Lie}(U_M) \subseteq \mathrm{Lie}(U)$. We claim that $k \in \mathrm{SO}(m)$, and $\mathrm{Ad}_k v_- \in \mathrm{Lie}(U)$ implies $k \in C(A)$. This gives $h = kan \in P$ as required for case (3) of the lemma.

To prove the claim we suppose for $k \in \mathrm{SO}(m)$ that $\mathrm{Ad}_k v_-$ belongs to $\mathrm{Lie}(U)$ and calculate[1]

$$\begin{aligned} \mathrm{Ad}_k v_- &= \begin{pmatrix} 1 & 0 \\ 0 & k \end{pmatrix} \begin{pmatrix} 0 & 0 & \xi^T \\ 0 & 0 & -\xi^T \\ \xi & \xi & 0 \end{pmatrix} \begin{pmatrix} 1 & 0 \\ 0 & k^T \end{pmatrix} \\ &= \begin{pmatrix} 0 & (0, \xi^T) k^T \\ k \begin{pmatrix} 0 \\ \xi \end{pmatrix} & k \begin{pmatrix} 0 & -\xi^T \\ \xi & 0 \end{pmatrix} k^T \end{pmatrix}. \end{aligned}$$

Focusing on the second column of the product, we note that the top row entry is supposed to vanish and the remaining column is supposed to equal

$$k \begin{pmatrix} 0 \\ \xi \end{pmatrix}.$$

However, if c_{11} denotes the top left entry of k (and equivalently of k^T), then the remainder of the second column has the form

[1] The dimensions of this matrix calculation are confusing, but k has m rows and columns and ξ has $m-1$ coordinates.

$$c_{11}k\begin{pmatrix}0\\ \xi\end{pmatrix}+\ldots,$$

where the ellipsis indicates a vector linearly independent of

$$k\begin{pmatrix}0\\ \xi\end{pmatrix}.$$

Therefore, $c_{11}=1$ and since $k\in\mathrm{SO}(m)$ we obtain that k (when embedded as before into the lower right block) belongs to $C(A)$ as claimed. □

Recall that we wish to prove that for r large enough, L_r cannot be contained in $M\times\mathrm{SO}(n-m)$ in any of the cases appearing in Lemma 4.2.

If M is compact, then so is $M\times\mathrm{SO}(n-m)$. Since Γ_1 is torsion-free no nontrivial element of Γ_1 can be contained in $M\times\mathrm{SO}(n-m)$. Hence L_r is not contained in $M\times\mathrm{SO}(n-m)$ if r is chosen sufficiently large to ensure that $d(r)>1$.

In order to show that L_r cannot be contained in $M\times\mathrm{SO}(n-m)$ if M is one of the noncompact groups appearing in Lemma 4.2, we want to show that the lattice elements $\gamma_1,\ldots,\gamma_{d(r)}$ cannot all be contained in a ball of radius r inside $M\times\mathrm{SO}(n-m)$ if r is large enough. For this we need an estimate for the asymptotic volume growth of a ball in the relevant subgroups of S.

Recall the Cartan decomposition of S: every element $g\in S$ can be written in the form $g=k_1a_tk_2$ for some $k_1,k_2\in\mathrm{SO}(m)$ and $t\geq 0$.

LEMMA 4.3. *Let* $S=\mathrm{SO}^\circ(m,1)$ *for some* $m\geq 2$. *Then*

$$r^{m-1}\ll m_S(B_r^{\mathrm{HS}}\cap S)\ll r^{m-1}$$

for all sufficiently large $r\geq 1$.

Proof. Recall that $B_r^{\mathrm{HS}}\cap\mathrm{SO}^\circ(m,1)=\{g\in\mathrm{SO}^\circ(m,1)\mid \|g\|_{\mathrm{HS}}\leq r\}$. Writing $g=k_1a_tk_2\in\mathrm{SO}^\circ(m,1)$ and using the fact that k_1 and k_2 belong to the compact subgroup of $\mathrm{SO}(m)<\mathrm{SO}^\circ(m,1)$, we have

$$\|g\|_{\mathrm{HS}}^2=\|k_1a_tk_2\|_{\mathrm{HS}}^2=\|a_t\|_{\mathrm{HS}}^2=4(\cosh t)^2+m-3\geq e^{2t}. \tag{6}$$

This shows that $\|g\|_{\mathrm{HS}}\leq r$ implies $t\leq\log r$. From this we deduce (using, e.g., [15, chapter 1, section 5]) that

$$m_S(B_r^{\mathrm{HS}} \cap S) = \int_{B_r^{\mathrm{HS}}} 1 \,\mathrm{d}m_S \ll \int_0^{\log r} \sinh^{m-1} t \,\mathrm{d}t$$
$$\ll \int_0^{\log r} e^{(m-1)t} \,\mathrm{d}t \ll r^{m-1}.$$

The reverse estimate for all sufficiently large $r \geq 1$ follows similarly using instead of (6) the estimate $\|g\|_{\mathrm{HS}}^2 \leq e^{2t} + m$. □

It turns out that if L_r is contained in a conjugate of $P \times \mathrm{SO}(n-m)$, where P is the parabolic subgroup of $\mathrm{SO}(m,1)$ and L_r has a non-trivial unipotent radical, then the following estimate for the volume growth of a ball in the unipotent radical U of P will be helpful in the proof of Proposition 4.1.

LEMMA 4.4. *Let U be the unipotent radical of $P < \mathrm{SO}(m,1)$. Then*

$$m_U(B_r^{\mathrm{HS}} \cap U) \ll r^{\frac{1}{2}(m-1)}$$

for all $r \geq 1$.

Proof. The unipotent radical U of P is given by

$$U = \left\{ u_\xi = \begin{pmatrix} 1 + \frac{1}{2}\xi^T\xi & \frac{1}{2}\xi^T\xi & \xi^T \\ -\frac{1}{2}\xi^T\xi & 1 - \frac{1}{2}\xi^T\xi & -\xi^T \\ \xi & \xi & \mathrm{I}_{m-1} \end{pmatrix} \mid \xi \in \mathbb{R}^{m-1} \right\}.$$

Note that we have

$$\|u_\xi\|_{\mathrm{HS}}^2 = \sum_{i=1}^{m+1} \|u_\xi e_i\|_2^2 = (m+1) + 4\,\|\xi\|_2^2 + \tfrac{1}{2}\,\|\xi\|_2^4.$$

With this we see that $\|u_\xi\|_{\mathrm{HS}} \leq r$ implies $\|\xi\|_2 \leq \sqrt{2r}$. Therefore,

$$m_U\left(B_r^{\mathrm{HS}} \cap U\right) \leq \mathrm{vol}\left\{\xi \in \mathbb{R}^{m-1} \mid \|\xi\|_2^2 \leq 2r\right\} \ll r^{\frac{1}{2}(m-1)}$$

for all $r \geq 1$. □

We briefly sketch the argument for Proposition 4.1 in the case of the diagonal subgroup (case (1) in Lemma 4.2). For this recall that the diagonal subgroup has logarithmic volume growth (since we use balls with respect to the Hilbert-Schmidt norm), but we have in fact a polynomial growth rate of the lattice points in Equation (4). Since each lattice point contains a ball around it that can be chosen of uniform measure, this rules out the possibility that the lattice elements all belong to a diagonal subgroup. However, the precise formulation of this volume argument is more delicate; indeed the subgroup A_M is only conjugated to A by some $g \in S$, and the conjugation by g may drastically change the group and the normalization of the Haar measure on it. This in turn affects the desired volume estimate, as we now explain.

In order to understand the meaning of the conjugating element g, we consider its action on the hyperbolic space $\mathbb{H}^m = \mathrm{SO}^\circ(m,1)/K$—respectively its frame bundle $\mathrm{PSO}^\circ(m,1)$. See also Figure 1. In fact the set $gAK = \{ga_tK \mid t \in \mathbb{R}\}$ is a geodesic in $\mathbb{H}^m$ that contains the point gK. Consider the point $ga_{-t_0}K \in gAK$, which is closest to K. We claim that

$$g = kb_sk_ca_{t_0} = kb_sa_{t_0}k_c,$$

where $k \in K$, $k_c \in K \cap C_S(A)$, and

$$b_s = \begin{pmatrix} \cosh s & 0 & \sinh s \\ 0 & I_{m-1} & 0 \\ \sinh s & 0 & \cosh s \end{pmatrix} \tag{7}$$

for some $s \in \mathbb{R}$. Indeed let us first move the frame g along the geodesic from gK to $ga_{-t_0}K$, which results in the frame ga_{-t_0}. Since the point $ga_{-t_0}K$ is closest to K from all points on the geodesic gAK, the geodesic connecting $ga_{-t_0}K$ to K is orthogonal to the geodesic gAK. Allowing for some $k_c \in K \cap C_S(A)$ we can assume that the elements $k_c^{-1}b_sk_c$ with $s \in \mathbb{R}$ is moving the frame ga_{-t_0} back to K along this geodesic. In other words there exists some $s \in \mathbb{R}$ so that $ga_{-t_0}k_c^{-1}b_{-s}k_c \in K$, which gives the claim.

Therefore the Hilbert-Schmidt norm of the element ga_tg^{-1} in A_M equals the Hilbert-Schmidt norm of $b_sa_tb_{-s}$. Using our definitions in Equations (5) and (7) and subtracting the identity matrix from a_t, the conjugate $b_s(a_t - I)b_{-s}$ is given by

$$\begin{pmatrix} \cosh^2 s(\cosh t - 1) & \cosh s \sinh t & 0 & \cosh s \sinh s(1 - \cosh t) \\ \cosh s \sinh t & \cosh t - 1 & 0 & -\sinh s \sinh t \\ 0 & 0 & 0 & 0 \\ \cosh s \sinh s(\cosh t - 1) & \sinh s \sinh t & 0 & -\sinh^2 s(\cosh t - 1) \end{pmatrix}, \tag{8}$$

where the third row with only zeroes stands for $m-1$ rows, and similarly the third column stands for $m-1$ columns.

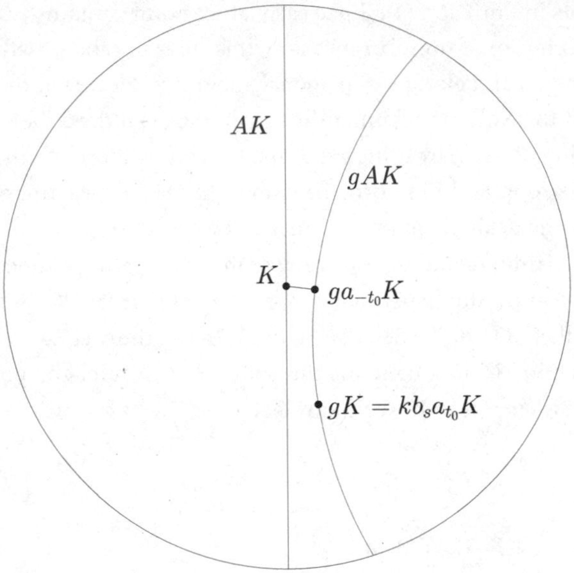

Figure 13.1. The image $gAK = \{ga_tK \mid t \in \mathbb{R}\}$ of the geodesic $AK = \{a_tK \mid t \in \mathbb{R}\}$ under application of the isometry g in $\mathbb{H}^m$

Here s is determined by g and we wish to estimate the volume of the Haar measure of all $ga_tg^{-1} \in A_M$ for which the Hilbert-Schmidt norm of (8) is bounded by r. In order to determine the normalization of the Haar measure on A_M, we may, for example, consider the intersection of A_M with the ball

$$\{ga_tg^{-1} \in A_M \mid \|b_sa_tb_{-s} - I\|_{\mathrm{HS}} \leq 1\}.$$

The Haar measure coming from the Riemannian metric is normalized so that this set has measure $\asymp 1$. One can then distinguish two cases. It could be that s is not that large (compared to r), in which case one obtains, roughly speaking, a logarithmic estimate. Or it could be that s is very large (compared to r) and the conjugated group $A_M = gAg^{-1}$ behaves on B_r^{HS} already almost like a unipotent subgroup. The latter leads to a polynomial estimate that is still sufficient. This outline has a gap, as we would also have to consider the conjugation of $K \cap C_S(A)$. We will give a more algebraic argument below for the diagonal case and give all details of the outlined volume argument in the case of noncompact semisimple group (case (2) of Lemma 4.2).

Proof of Proposition 4.1. Let $M \leq S$ be the intersection of one of the noncompact subgroups appearing in Lemma 4.2 with S, and assume that all the lattice elements in

$$g_0^{-1}\Gamma_1 g_0 \cap B_r^{\mathrm{HS}} \cap N$$

are contained in $M \times \mathrm{SO}(n-m)$. We are going to derive a contradiction to this assumption if $r = \mathrm{vol}(x_0 N)^{\kappa_2}$ and $\kappa_2 > 0$ is chosen correctly (only depending on n).

Let $v_1 = \mathrm{vol}(x_0 N)$ so that $r = v_1^{\kappa_2}$, and recall that by Equation (4) and Lemma 4.3 we have

$$d(r) = \left| g_0^{-1}\Gamma_1 g_0 \cap B_r^{\mathrm{HS}} \cap N \right| \gg v_1^{\kappa_2(m-1)-1} + O(v_1^{\kappa_2(m-1)(1-\delta)}).$$

Therefore there exist constants $c > 0$ and $\kappa_3 > 0$ so that

$$\left| g_0^{-1}\Gamma_1 g_0 \cap B_r^{\mathrm{HS}} \cap N \right| \geq c v_1^{\kappa_2(m-1)-1} \tag{9}$$

whenever $\kappa_2 \geq \kappa_3$ and $v_1 \geq 2$. As explained before, we may assume that M is noncompact.

Case (1) of Lemma 4.2. $C_S(A_M) \cap M$ has at most index two in M, where A_M is the unique split torus in M.

Let $g \in S$ be such that $A_M = gAg^{-1}$. We recall that $\Gamma < \mathrm{SO}(n, 1)$ has been obtained by intersection and conjugation from $\mathrm{SL}_{n+1}(\mathbb{Z})$. This implies that there exists some $c = c_n > 1$ such that for every $\gamma \in \Gamma$, either there is an eigenvalue of absolute value $\geq c$ or all eigenvalues of γ have absolute value one. Since $A_M < S$ is a split torus and $C(A_M)$ is an extension of A_M by a compact group and Γ_1 is torsion-free, the nontrivial lattice elements in $g_0^{-1}\Gamma_1 g_0 \cap C(A_M)$ have nontrivial components in A_M with absolute values of eigenvalues outside of (c^{-1}, c). This implies that the intersection must be cyclic. Moreover, note that all elements of B_r^{HS} have eigenvalues of absolute values $\ll r$—say, $\leq c' r$ for some $c' = c'_n > 1$. Using that, it follows that there are at most $\frac{\log(c' r)}{\log c}$ many elements in $g_0^{-1}\Gamma_1 g_0 \cap C(A_M) \cap B_r^{\mathrm{HS}}$.

It remains to handle the elements in $M \setminus C_S(A_M)$. If $g_0^{-1}\Gamma_1 g_0 \cap B_r^{\mathrm{HS}}$ intersects $M \setminus C(A_M)$, we suppose that γ_j belongs to the intersection and fix j. Multiplying now the elements in $g_0^{-1}\Gamma_1 g_0 \cap B_r^{\mathrm{HS}} \setminus C(A_M)$ with γ_j, we obtain a subset of $g_0^{-1}\Gamma_1 g_0 \cap B_{c'' r^2}^{\mathrm{HS}} \cap C(A_M)$ for some $c'' = c''_n > 1$. Applying the previous estimate we obtain therefore that $g_0^{-1}\Gamma_1 g_0 \cap B_r^{\mathrm{HS}} \setminus C(A_M)$ contains at most $\frac{\log(c' c'' r^2)}{\log c}$ many elements.

Together we arrive at the estimate

$$\left| g_0^{-1}\Gamma_1 g_0 \cap B_r^{\mathrm{HS}} \cap N \right| \ll \log r + C$$

for a constant $C = C_n$. This contradicts the lower bound in Equation (9) once ν_1 (and hence r) is sufficiently large.

Case (3) of Lemma 4.2. $M \le gPg^{-1}$ for some $g \in S$, where P is the parabolic subgroup within S.

If L_r has a nontrivial unipotent radical, then our assumption implies that the lattice elements $\gamma_1, \ldots, \gamma_{d(r)}$ are contained in $gUK_P g^{-1} \times \mathrm{SO}(n-m)$, where $K_P \cong \mathrm{SO}(m-1)$ is the maximal compact subgroup in P and, as before, U is the unipotent radical of P. Indeed, $L_r^{g_Q^{-1} g_0} = g_Q^{-1} g_0 L_r g_0^{-1} g_Q$ (with $L_r = \overline{\langle \gamma_1, \ldots, \gamma_{d(r)} \rangle}^Z$) is a $\mathbb{Q}$-group whose lattice points are Zariski dense, and so it admits no nontrivial $\mathbb{Q}$-character. However, if now L_r is not of the stated form, then the determinant of the adjoint representation of L_r on the Lie algebra of L_r would induce a nontrivial $\mathbb{Q}$-character on $L_r^{g_Q^{-1} g_0}$.

Next we apply the Iwasawa decomposition $g = kau$, where $k \in \mathrm{SO}(m)$, $a \in A$, and $u \in U$. We note that the UK_P is normalized by au. Therefore, the conjugation with g reduces to a conjugation with k belonging to a compact subgroup of S and is therefore irrelevant for the volume calculation of $B_r^{\mathrm{HS}} \cap (UK_P)^g$ below, where we again use the shorthand $(UK_P)^g = gUK_P g^{-1}$.

As mentioned before, the element g_0 belongs to a fixed compact set, and so there exists some uniform $\epsilon_0 > 0$ so that $g_0^{-1} \Gamma_1 g_0$ has no nontrivial elements in the Riemannian ball $B_{2\epsilon_0}(I)$ around the identity in S. Hence, the balls

$$B_{\epsilon_0}(\gamma) \text{ for } \gamma \in g_0^{-1} \Gamma_1 g_0 \cap B_r^{\mathrm{HS}} \cap N$$

are disjoint. We note that their union is contained in B_{2r}^{HS} (if $\epsilon_0 > 0$ is chosen sufficiently small depending only on n). Since $\mathrm{SO}(n-m)$ is compact, we may use the intersection of these balls with $(UK_P)^g \times \mathrm{SO}(n-m)$ to obtain from Lemma 4.4 that

$$\begin{aligned}\left| g_0^{-1} \Gamma_1 g_0 \cap B_r^{\mathrm{HS}} \cap N \right| &\le \frac{\mathrm{vol}(B_{2r}^{\mathrm{HS}} \cap ((UK_P)^g \times \mathrm{SO}(n-m)))}{\mathrm{vol}(B_{\epsilon_0}(I) \cap ((UK_P)^g \times \mathrm{SO}(n-m)))} \\ &\ll \mathrm{vol}(B_{2r}^{\mathrm{HS}} \cap (UK_P)) \ll r^{\frac{1}{2}(m-1)} = \nu_1^{\frac{1}{2}\kappa_2 (m-1)},\end{aligned}$$

with $r = \nu_1^{\kappa_2}$ and vol denoting the Haar measure on $(UK_P)^g \times \mathrm{SO}(n-m)$, respectively $(UK_P)^g$. It follows that there exists a $\kappa_4 > 0$ so that if $\kappa_2 \ge \kappa_4$ and $\nu_1 \ge 2$, then this upper bound contradicts the lower bound in Equation (9). In other words, we can ensure that there are too many lattice elements to fit in a conjugate of U or of P.

We note that in case L_r is contained in a conjugate M of P but L_r has a trivial unipotent radical, then L_r is contained in the centralizer of the split torus A_M of M and has already been treated in case (1) above.

Case (2) of Lemma 4.2. M is contained in $g(\mathrm{O}(\ell,1)\times \mathrm{O}(m-\ell))g^{-1}\cap S$ for some $g\in S$ and $2\leq \ell < m$.

We initially only consider the connected subgroup

$$\mathrm{SO}^\circ(\ell,1)\times \mathrm{SO}(m-\ell) < (\mathrm{O}(\ell,1)\times \mathrm{O}(m-\ell))\cap S \tag{10}$$

of index four. Similarly to the outline before the proof, we consider the subspace $g\,\mathrm{SO}^\circ(\ell,1)K$ and find the point $gh^{-1}K$ closest to K with $h\in \mathrm{SO}^\circ(\ell,1)$. The geodesic connecting $gh^{-1}K$ to K can again be parameterized by $gh^{-1}k_c^{-1}b_sk_cK$ for $s\in\mathbb{R}$ and $k_c\in K\cap C(\mathrm{SO}^\circ(\ell,1))$, which gives that

$$g = kb_sk_ch$$

for some $k\in K$, $s\in\mathbb{R}$, $k_c\in K\cap C(\mathrm{SO}^\circ(\ell,1))$, and $h\in \mathrm{SO}^\circ(\ell,1)$.

Since we are only interested in the effect that g has when the group in (10) is being conjugated by g and intersected with B_r^{HS}, we may ignore k, k_c, h and simply assume that $g=b_s$.

For an element $h\in \mathrm{SO}^\circ(\ell,1)$, we are going to consider the Cartan decomposition $h=k_1a_tk_2$ in $\mathrm{SO}^\circ(\ell,1)$, where k_1,k_2 are contained in the maximal compact subgroup of $\mathrm{SO}^\circ(\ell,1)$. Recall that $\mathrm{SO}^\circ(\ell,1)$ is embedded in the top left corner of $\mathrm{SO}^\circ(m,1)$, and hence b_s as in Equation (7) commutes with the maximal compact subgroup of $\mathrm{SO}^\circ(\ell,1)$ (which is isomorphic to $\mathrm{SO}(\ell)$ and is embedded in the coordinates $\{2,\ldots,\ell+1\}$). For an element $h=k_1a_tk_2\in \mathrm{SO}^\circ(\ell,1)$ with $k_1,k_2\in K\cap S$, we therefore have

$$\|ghg^{-1}\| = \|b_sk_1a_tk_2b_{-s}\| = \|b_sa_tb_{-s}\|.$$

Case (2), normalization of Haar measures. We continue with case (2) and estimate the normalization of the Haar measure.

In order to determine the normalization of the Haar measure on $g\,\mathrm{SO}^\circ(\ell,1)g^{-1}$, we first consider the set

$$\left\{h\in \mathrm{SO}^\circ(\ell,1) \mid h=k_1a_tk_2 \text{ and } \|b_sa_tb_{-s}-I_{\ell+1}\|_{\mathrm{HS}}\leq 1\right\} \tag{11}$$

Using Equation (8) we see that $\|b_s a_t b_{-s} - I_{\ell+1}\|_{\mathrm{HS}} \leq 1$ implies $|\cosh s \sinh t| \leq 1$ and hence $|t| \ll e^{-s}$. On the other hand, there exists an absolute constant c such that $|t| \leq ce^{-s}$ implies $\|b_s a_t b_{-s} - I_{\ell+1}\|_{\mathrm{HS}} \leq 1$. Using [15, chapter 1, section 5] again, it follows that the set in (11) has Haar measure

$$\asymp \int_0^{e^{-s}} \sinh^{\ell-1} t \, dt \asymp \int_0^{e^{-s}} t^{\ell-1} \, dt \asymp e^{-\ell s}.$$

After conjugating $\mathrm{SO}^\circ(\ell,1)$ to $g\,\mathrm{SO}^\circ(\ell,1)g^{-1}$ the set in (11) is the intersection of the group with a fixed neighborhood of the identity. Hence the Haar measure of this subset of $g\,\mathrm{SO}^\circ(\ell,1)g^{-1}$ is $\asymp 1$. In other words, the push-forward of the Haar measure of $\mathrm{SO}^\circ(\ell,1)$ needs to be multiplied with $\asymp e^{\ell s}$ to obtain the Haar measure on $g\,\mathrm{SO}^\circ(\ell,1)g^{-1}$.

Similarly, the compact group $\mathrm{SO}(m-\ell)$ is being embedded into S using the coordinates $\{\ell+2,\ldots,m+1\}$, and the subgroup $\mathrm{SO}(m-\ell-1)$ embedded into S using the coordinates $\{\ell+2,\ldots,m\}$ commutes with b_s. Hence the volume calculations reduce to a volume calculation on the quotient of these two subgroups—that is, to a volume calculation for a subset of $\mathbb{S}^{m-\ell-1}$. We will use matrices of the form

$$k_\phi = \begin{pmatrix} I_{m-1} & 0 & 0 \\ 0 & \cos\phi & -\sin\phi \\ 0 & \sin\phi & \cos\phi \end{pmatrix}$$

with $\phi \in \mathbb{R}$ as a replacement of our diagonal subgroup A. Conjugating $k_\phi - I$ by $g = b_s$ we obtain

$$\begin{pmatrix} \sinh^2 s(1-\cos\phi) & 0 & \sinh s \sin\phi & \cosh s \sinh s(\cos\phi - 1) \\ 0 & 0 & 0 & 0 \\ \sinh s \sin\phi & 0 & \cos\phi - 1 & -\cosh s \sin\phi \\ \cosh s \sinh s(1-\cos\phi) & 0 & \cosh s \sin\phi & \cosh^2 s(\cos\phi - 1) \end{pmatrix}. \tag{12}$$

Requiring this to have the Hilbert-Schmidt norm bounded by 1 (for a fixed s) amounts to an inequality of the form $|\phi| \ll e^{-s}$. Taking the dimension of $\mathbb{S}^{m-\ell-1}$ into account we see that the Haar measure of this set is of the order $\asymp e^{-(m-\ell-1)s}$.

Case (2), combining the subgroup. We continue with case (2) and explain how to put the groups $\mathrm{SO}(\ell,1)^\circ$ and $\mathrm{SO}(m-\ell)$ and their elements h, respectively k, together.

Since $(h-I)$ and $(k-I)$ are block matrices and in both cases three out of four blocks vanish, we see that $(h-I)(k-I)=0$ and the size of the conjugate of the product

$$\begin{aligned} b_s hkb_{-s} &= b_s\big((I+(h-I))(I+(k-I))\big)b_{-s} \\ &= I + b_s(h-I)b_{-s} + b_s(k-I)b_{-s} \end{aligned}$$

is basically the sum (instead of the product) of the sizes of the two conjugates. This shows that both our normalization of the Haar measure of the product $g(\mathrm{SO}(\ell,1)^\circ \times \mathrm{SO}(m-\ell))g^{-1}$ and the calculation of the Haar measure of B_r^{HS} below can be done in each group separately. Given a radius $r>0$, we want to estimate the volume of the ball $B_r^{\mathrm{HS}} \cap g\,\mathrm{SO}^\circ(\ell,1)g^{-1}$ and $B_r^{\mathrm{HS}} \cap g\,\mathrm{SO}(m-\ell)g^{-1}$ with the respective Haar measures (normalized as discussed above).

Case (2), volume estimate on $\mathrm{SO}^\circ(\ell,1)$. For $\mathrm{SO}^\circ(\ell,1)$ the requirement that the matrix in Equation (8) is of norm $\ll r$ is (apart from the precise choice of the implicit constants) equivalent to the two requirements $e^{2s}(\cosh t-1) \ll r$ and $e^s \sinh t \ll r$. If $e^s \le r^{1/2}$ we use [15, chapter 1, section 5] and the second inequality to obtain

$$\begin{aligned} m_{g\,\mathrm{SO}^\circ(\ell,1)g^{-1}}\big(B_r^{\mathrm{HS}} \cap g\,\mathrm{SO}^\circ(\ell,1)g^{-1}\big) &\ll e^{\ell s}\int_0^{\operatorname{arsinh}(ce^{-s}r)} \sinh^{\ell-1} t\,dt \\ &\ll e^{\ell s} e^{(\ell-1)\log(ce^{-s}r)} \\ &\ll e^s r^{\ell-1} \le r^{\ell-\frac{1}{2}} \end{aligned}$$

for some absolute constant $c>0$. If $e^s \ge r^{1/2}$ we use the first inequality $e^{2s}(\cosh t-1) \ll r$, which becomes $(\cosh t-1) \ll re^{-2s} \le 1$ and forces t to be in a bounded interval I (that can be chosen absolute). With $1+\frac{1}{2}t^2 \le \cosh t$ for all $t \in \mathbb{R}$ and $\sinh t \asymp t$ for all $t \in I$, we arrive at

$$\begin{aligned} m_{g\,\mathrm{SO}^\circ(\ell,1)g^{-1}}\big(B_r^{\mathrm{HS}} \cap g\,\mathrm{SO}^\circ(\ell,1)g^{-1}\big) &\ll e^{\ell s}\int_0^{cr^{\frac{1}{2}}e^{-s}} \sinh^{\ell-1} t\,dt \\ &\ll e^{\ell s}\int_0^{cr^{\frac{1}{2}}e^{-s}} t^{\ell-1}\,dt \\ &\ll e^{\ell s}(r^{\frac{1}{2}}e^{-s})^{\ell} = r^{\frac{1}{2}\ell} \le r^{\ell-\frac{1}{2}} \end{aligned}$$

for some absolute constant c.

Case (2), volume estimate on $\mathrm{SO}(m-\ell)$. For $\mathrm{SO}(m-\ell)$ we instead consider the matrix in Equation (12) and suppose that its norm is $\ll r$. If $e^s \leq r$, this inequality is essentially meaningless and we obtain the trivial estimate

$$m_{g\,\mathrm{SO}(m-\ell)g^{-1}}(B_r^{\mathrm{HS}} \cap g\,\mathrm{SO}(m-\ell)g^{-1}) \ll e^{(m-\ell-1)s} \leq r^{m-\ell-1}.$$

If $e^s \geq r$, then the required estimate for (12) implies $|\phi| \ll e^{-s}r$ and we obtain using the geometry of $\mathbb{S}^{m-\ell-1}$ that

$$m_{g\,\mathrm{SO}(m-\ell)g^{-1}}(B_r^{\mathrm{HS}} \cap g\,\mathrm{SO}(m-\ell)g^{-1}) \ll e^{(m-\ell-1)s}(e^{-s}r)^{m-\ell-1} = r^{m-\ell-1}.$$

Case (2), concluded. Putting these estimates together we obtain in any case that the Haar measure of B_r^{HS} for the conjugated group $g\,\mathrm{SO}^\circ(\ell,1) \times \mathrm{SO}(m-\ell)g^{-1}$ is bounded by $\ll r^{m-\frac{3}{2}}$, which beats the estimate for the group $\mathrm{SO}^\circ(m,1)$ in Lemma 4.3. Recalling that the subgroup in Equation (10) has index four, this volume estimate also generalizes to $(\mathrm{O}(\ell,1) \times \mathrm{O}(m-\ell)(\mathbb{R})) \cap S$ conjugated by g. Recalling moreover that $r = \nu_1^{\kappa_2}$ and applying (9) we obtain

$$\begin{aligned} \nu_1^{\kappa_2(m-1)-1} &\ll \left| g_0^{-1}\Gamma_1 g_0 \cap B_r^{\mathrm{HS}} \cap N \right| \\ &\ll \mathrm{vol}(B_{2r}^{\mathrm{HS}} \cap g((\mathrm{O}(\ell,1) \times \mathrm{O}(m-\ell)(\mathbb{R})))g^{-1} \cap S) \\ &\ll \nu_1^{\kappa_2(m-\frac{3}{2})}. \end{aligned}$$

Hence there exists a constant $\kappa_5 > 0$ so that for all $r = \nu_1^{\kappa_2}$ with $\kappa_2 \geq \kappa_5$, the above inequality cannot hold.

Conclusion of proof of Proposition 4.1. Choosing $r = \nu_1^{\kappa_2}$ with $\kappa_2 = \max\{\kappa_3, \kappa_4, \kappa_5\}$, we see that Equation (9) contradicts the assumption that the lattice elements belong to a proper subgroup of the form $M \times \mathrm{SO}(n-m)$. It follows that there are lattice elements $\{\gamma_1, \ldots, \gamma_d\} \subseteq g_0^{-1}\Gamma_1 g_0$ with $\|\gamma_i\|_{HS} \leq \nu_1^{\kappa_2}$ such that the Zariski closure of the subgroup generated by $\{\gamma_1, \ldots, \gamma_d\}$ contains S. Since g_0 belongs to a fixed compact subset, the lattice elements

$$\{g_0\gamma_1 g_0^{-1}, \ldots, g_0\gamma_d g_0^{-1}\} \subseteq \Gamma_1$$

satisfy $\left\| g_0\gamma_i g_0^{-1} \right\| \ll \nu_1^{\kappa_2}$, and the Zariski closure of their generated subgroup contains S^{g_0} as claimed. □

The following lemma strengthens the conclusion of Proposition 4.1 and shows that it suffices to consider a fixed number of lattice elements.

LEMMA 4.5. *There exists an absolute constant* $D \in \mathbb{N}$ *depending only on* n *with the following property. Suppose* $d \in \mathbb{N}$ *and* $g_1, \ldots, g_d \in \mathrm{SO}(n,1)$ *are nontorsion elements of a discrete subgroup that generate a subgroup with Zariski closure* $L \subseteq \mathrm{SO}(m,1) \times \mathrm{SO}(n-m)$ *containing* $\mathrm{SO}(m,1)$. *Then there exist*

$$j_1, \ldots, j_D \in \{1, \ldots, d\}$$

such that $g_{j_1}, \ldots, g_{j_D}$ *together generate a subgroup with Zariski closure containing* $\mathrm{SO}(m,1)$.

Proof. Denote by $\pi_S : L \to \mathrm{SO}(m,1)$ the projection to $\mathrm{SO}(m,1)$ and set $D_m = \dim(\mathrm{SO}(m,1))$. Since the element g_1 is nontorsion and no finite index subgroup of the subgroup generated by g_1 can belong to $\mathrm{SO}(n-m)$, the Lie algebra of the Zariski closure of the group generated by $\pi_S(g_1)$ is at least one-dimensional. We set $j_1 = 1$ and choose the indices $j_2, \ldots, j_{D_m}$ as follows: Assume we have already chosen $j_1, \ldots, j_\ell$ for some $1 \leq \ell < D_m$. If the elements $g_{j_1}, \ldots, g_{j_\ell}$ already generate a subgroup with Zariski closure containing $\mathrm{SO}(m,1)$, we are done (and, e.g., repeat the first element to obtain D_m elements). Otherwise, we look at the Lie algebra $\mathfrak{l} \subsetneq \mathfrak{so}(m,1)$ of the Zariski closure of the group generated by $\pi_S(g_{j_1}), \ldots, \pi_S(g_{j_\ell})$. Since $\mathfrak{l}$ is at least one-dimensional, it cannot be a Lie ideal and there exists some $g_{j_{\ell+1}}$ that does not normalize $\mathfrak{l}$. Adding this element to the list we see that the dimension of the Zariski closure of the group generated by the new list $g_{j_1}, \ldots, g_{j_{\ell+1}}$ increased at least by one. This gives the lemma for $D = \dim \mathrm{SO}(n,1)$ (by once more repeating elements to obtain a list of length D precisely). □

5 Bounding the volume in terms of the normalizer orbit

5.1 THE DISCRIMINANT OF A CLOSED S-ORBIT. Recall that $\Gamma = g_Q \mathrm{SO}_Q(\mathbb{Z}) g_Q^{-1}$ for some fixed $g_Q \in \mathrm{SL}_{n+1}(\mathbb{R})$ and that we are using the abbreviation $S = \mathrm{SO}^\circ(m,1)$ for some $2 \leq m \leq n$. We want to attach to each closed S-orbit an arithmetic invariant, called the *discriminant*. Let $V = \bigwedge^{m+1} \mathbb{R}^{n+1}$ and consider the integral lattice $V_\mathbb{Z} = \bigwedge^{m+1} \mathbb{Z}^{n+1} \subset V$. Note that the quadratic form Q on $\mathbb{R}^{m+1}$ induces a quadratic form $Q_\wedge$ on V so that

$$Q_\wedge(v_1 \wedge \ldots \wedge v_{m+1}) = \det\big(\langle v_i, v_j \rangle_Q\big).$$

Given a closed S-orbit $\Gamma g_0 S$ for some $g_0 \in \mathrm{SO}^\circ(n,1)$, we define

$$w_{g_0 S} = g_Q^{-1} g_0 (e_1 \wedge \ldots \wedge e_{m+1}).$$

and

$$\operatorname{disc}(\Gamma g_0 S) = \inf\{|Q_\wedge(dw_{g_0 S})| \,|\, d > 0 \text{ and } dw_{g_0 S} \in V_\mathbb{Z} \setminus \{0\}\}.$$

Note that $e_1 \wedge \ldots \wedge e_{m+1}$ is invariant under S, that $g_Q^{-1} g_0 \langle e_1, \ldots, e_{m+1} \rangle$ is a rational subspace by Borel density, and that $V_\mathbb{Z}$ is stable under $g_Q^{-1} \Gamma g_Q = \mathrm{SO}_Q(\mathbb{Z})$. Therefore, the discriminant of the orbit $\Gamma g_0 S$ is well-defined and finite. Moreover,

$$\begin{aligned} Q_\wedge\left(g_Q^{-1} g_0 (e_1 \wedge \ldots \wedge e_{m+1})\right) &= \lambda^{-(m+1)} Q_{0,\wedge}\left(g_0 (e_1 \wedge \ldots \wedge e_{m+1})\right) \\ &= -\lambda^{-(m+1)} \neq 0, \end{aligned}$$

which shows that $\operatorname{disc}(\Gamma g_0 S) > 0$.

We note that the ratio

$$\frac{\left|Q_\wedge\left(d g_Q^{-1} g (e_1 \wedge \ldots \wedge e_{m+1})\right)\right|}{\left\| d g_Q^{-1} g (e_1 \wedge \ldots \wedge e_{m+1}) \right\|^2}$$

is constant with respect to $d > 0$ and depends continuously on $g \in G$ so that it is bounded from above and below if g varies inside a compact subset of G. From this we see that

(13) $$\operatorname{disc}(\Gamma g_0 S) \asymp \inf\left\{ \|dw_{g_0 S}\|^2 \,\middle|\, d > 0 \text{ and } dw_{g_0 S} \in V_\mathbb{Z} \right\},$$

where we insist (as we may) on g_0 being chosen from a fixed compact subset of G that projects onto X_{cpct}.

5.2 THE DISCRIMINANT OF A CLOSED ***L***-ORBIT. More generally we already saw in section 4.1 that orbit closures of noncompact special orthogonal subgroups on X take (after applying on the right an element of $\mathrm{SO}(n)$) the form $x_0 L$ for some $x_0 = \Gamma g_0 \in X$ and $L = \mathrm{SO}^\circ(m,1) \times \mathrm{SO}(m+1)^p$ with $m \in \{2, \ldots, n\}$ and $p \geq 0$. We wish to define a discriminant of $x_0 L$ using as far as possible again the vector $\widetilde{w}_{g_0 L} = g_Q^{-1} g_0 (e_1 \wedge \ldots \wedge e_{m+1})$. Unlike the previous case this vector may not be a multiple of a rational vector; it is, however, an algebraic vector. In fact, the $\mathbb{R}$-simple factors of the $\mathbb{Q}$-group $\mathbb{L}$ define finitely many algebraic subspaces that are Galois conjugated to each other. The exterior tensor $g_Q^{-1} g_0 (e_1 \wedge \ldots \wedge e_{m+1})$ corresponds to one of them and by the above discussion is normalized to have $Q_\wedge$-value $-\lambda^{-(m+1)}$, which implies that the tensor product is algebraic and its Galois conjugates correspond to the algebraic subspaces on which the other $\mathbb{R}$-simple factors of $\mathbb{L}$ act nontrivially (possibly with repetitions).

We define $w_{g_0 L}$ as the symmetric tensor product of $g_Q^{-1} g_0(e_1 \wedge \ldots \wedge e_{m+1})$ and its p linearly independent Galois conjugates. More precisely, since we assume that L is embedded as block matrices in the first $(p+1)$ blocks of size $(m+1)$ along the diagonal, we can make the definition more concrete by setting

$$w_{g_0 L} = g_Q^{-1} g_0\big((e_1 \wedge \ldots \wedge e_{m+1}) \odot (e_{m+2} \wedge \ldots \wedge e_{2m+2}) \odot \cdots$$
$$\cdots \odot (e_{p(m+1)+1} \wedge \ldots \wedge e_{(p+1)(m+1)})\big) \in V_{\odot} = \bigodot^{p+1} V,$$

where we denote the symmetric tensor product by $\odot$. It follows that the line spanned by $w_{g_0 L}$ is invariant under the Galois action. Hence the line contains a rational and so also an integer vector in the canonical integer lattice $V_{\odot,\mathbb{Z}} = \bigodot^{p+1} V_{\mathbb{Z}} \subset V_{\odot}$.

We also note that for a given quadratic form q on $\mathbb{R}^N$ we can induce a quadratic form $q_{\odot}$ on $\bigodot^P \mathbb{R}^N$ by using the permanent of all possible inner products:

$$q_{\odot}(v_1 \odot \cdots \odot v_P) = \operatorname{perm}(\langle v_i, v_j \rangle) = \sum_{\sigma \in \mathcal{S}_P} \langle v_1, v_{\sigma(1)} \rangle \ldots \langle v_P, v_{\sigma(P)} \rangle.$$

We apply this to $q = Q_{\wedge}$ to define $Q_{\odot}$ on $\bigodot^{p+1} \bigwedge^{m+1} \mathbb{R}^{n+1}$. In our case this quadratic form is very easy to evaluate on $w_{g_0 L}$ since g_Q^{-1} again switches up to a power of λ to the quadratic form Q_0, g_0 preserves Q_0, and

$$Q_{0,\odot}\big((e_1 \wedge \ldots \wedge e_{m+1}) \odot \cdots \odot (e_{p(m+1)+1} \wedge \ldots \wedge e_{(p+1)(m+1)})\big) = -1.$$

We now define the discriminant of the closed orbit $\overline{\Gamma g_0 S} = \Gamma g_0 L$ by

$$\operatorname{disc}(\Gamma g_0 L) = \inf\big\{ |Q_{\odot}(d w_{g_0 L})| \mid d w_{g_0 L} \in V_{\odot,\mathbb{Z}} \setminus \{0\} \big\} > 0.$$

As in the previous case we have

$$\operatorname{disc}(\Gamma g_0 L) \asymp \inf\big\{ \|d w_{g_0 L})\|^2 \mid d w_{g_0 L} \in V_{\odot,\mathbb{Z}} \setminus \{0\} \big\} \tag{14}$$

if we insist that the representative g_0 is chosen from a fixed compact subset of G that projects onto X_{cpct}.

5.3 BOUNDING DISCRIMINANT IN TERMS OF VOLUME. The following represents the main result of the section and will relate the volume of the N-orbit and the discriminant of the closure of the associated S-orbit.

PROPOSITION 5.1. *There exists* $\kappa_6 > 0$ *such that* $S = \mathrm{SO}^\circ(m,1) \lhd L = S \times \mathrm{SO}(m+1)^p$ *and* $\overline{\Gamma g_0 S} = \Gamma g_0 L$ *implies*

$$\mathrm{disc}(\Gamma g_0 L) \ll \mathrm{vol}(\Gamma g_0 N)^{\kappa_6}.$$

Given lattice elements $\gamma_1, \ldots, \gamma_D \in N^{g_0} \cap \Gamma$ as in Proposition 4.1 and Lemma 4.5, we are going to study the set of $(m+1)$–dimensional subspaces of $\mathbb{R}^{n+1}$ that are invariant under all the elements $\eta_1 = g_Q^{-1}\gamma_1 g_Q, \ldots, \eta_D = g_Q^{-1}\gamma_D g_Q$. Notice that the subspace

$$V_0 = g_Q^{-1} g_0 \langle e_1, \ldots, e_{m+1} \rangle$$

satisfies this condition and that it is in some sense isolated from all other invariant subspaces (see Lemma 5.3). We will use this together with the bounds in Proposition 4.1 on the lattice elements to prove Proposition 5.1. In other words our goal in all of the following discussions is to bound the arithmetic complexity of the (potentially irrational but always algebraic) point

$$\mathbb{R}^\times g_Q^{-1} g_0 (e_1 \wedge \ldots \wedge e_{m+1})$$

in terms of the size of the parameters $\eta_1, \ldots, \eta_D$.

5.4 ALGEBRAIC SETUP FOR $L = S$. Let k be a field and $n \geq 1$, and denote by

$$\mathbb{P}_n(k) = \mathbb{P}_n\left(k^{n+1}\right) = \left(k^{n+1} \setminus \{0\}\right) / \sim$$

the projective space of dimension n over k, where the equivalence relation is defined by $(x_1, x_2, \ldots, x_{n+1}) \sim (y_1, y_2, \ldots, y_{n+1})$ if and only if there exists $\lambda \in k^\times$ with $x_i = \lambda y_i$ for all $i \in \{1, 2, \ldots, n+1\}$. We write $k^\times (x_1, x_2, \ldots, x_{n+1})$ for the corresponding equivalence class.

Recall that for $n \geq 2$ and $1 \leq r < n$, the Grassmannian $\mathrm{Gr}_{r,n}(k)$ is defined to be the set of all r-dimensional subspaces of k^n and has the structure of a projective variety; see, for example, [14, section 6] for a proof of the following.

PROPOSITION 5.2 (Plücker embedding). *For any field k, define the map*

$$\iota : \mathrm{Gr}_{m,n}(k) \longrightarrow \mathbb{P}\left(\bigwedge^{m} k^n\right)$$

as follows: For any m-dimensional subspace V of k^n, choose a basis $v_1, \ldots, v_m$ of V and define $\iota(V) := k^\times(v_1 \wedge \ldots \wedge v_m)$. Then the map ι is well-defined and injective. Moreover, the Grassmannian is a projective variety defined by quadratic equations. The map ι is called the Plücker Embedding.

We consider the set

$$(15)\quad \mathcal{Y} = \Big\{(V, \eta_1, \ldots, \eta_D) \in \mathrm{Gr}_{m+1,n+1}(\mathbb{R}) \times (\mathrm{Mat}_{n+1}(\mathbb{R}))^D \mid \eta_i V = V \text{ for } i = 1, \ldots, D\Big\},$$

which by Proposition 5.2 is a variety defined over $\mathbb{Q}$. We define

$$r = \dim\left(\bigwedge^{m+1} \mathbb{R}^{n+1}\right) = \binom{n+1}{m+1}.$$

We denote the variables corresponding to the standard basis of $\bigwedge^{m+1} \mathbb{R}^{n+1}$ obtained from the standard basis of $\mathbb{R}^{n+1}$ by $\mathbf{X} = (X_1, \ldots, X_r)$ and the variables corresponding to the entries of $\eta_1, \ldots, \eta_D$ by $\mathbf{T} = (T_1, \ldots, T_s)$, so $s = D(n+1)^2$. With that, we view $\mathcal{Y}$ as a subset of $\mathbb{P}(\mathbb{R}^r) \times \mathbb{R}^s$.

We will think of $\mathbf{T}$ as parameters while our main interest lies in the variables $\mathbf{X}$. In fact, notice that if we set $\eta_i = g_Q^{-1}\gamma_i g_Q$ in (15), where $\gamma_1, \ldots, \gamma_D$ are the lattice elements obtained from Proposition 4.1 and Lemma 4.5, then the fiber in $\mathcal{Y}$ above the point $(\eta_1, \ldots, \eta_D)$ corresponds to the set of all $(m+1)$–dimensional subspaces of $\mathbb{R}^{n+1}$ that are invariant under $g_Q^{-1}\gamma_1 g_Q, \ldots, g_Q^{-1}\gamma_D g_Q$. We denote this set, which is a projective variety, by

$$(16)\qquad \widetilde{\mathcal{Y}} = \Big\{V \in \mathrm{Gr}_{m+1,n+1}(\mathbb{R}) \mid (g_Q^{-1}\gamma_i g_Q)V = V \text{ for } i = 1, \ldots, D\Big\},$$

where we keep the dependence on $\gamma_1, \ldots, \gamma_D$ implicit.

LEMMA 5.3. *Let W be an $(m+1)$–dimensional, S-invariant subspace of $\mathbb{C}^{n+1}$. Then either $W = \mathbb{C}^{m+1} \times \{0\}^{n-m}$ or W is contained in $\{0\}^{m+1} \times \mathbb{C}^{n-m}$. In particular, $\mathbb{R}^\times w_{g_0 S}$ (as defined in section 5.1) constitutes a zero-dimensional irreducible component of $\widetilde{\mathcal{Y}}$.*

Proof. Assume indirectly that there is a vector $v=(v_1,v_2)\in W$, where $v_1\in \mathbb{C}^{m+1}\setminus\{0\}$ and $v_2\in(\{0\}^{m+1}\times\mathbb{C}^{n-m})\setminus\{0\}$. Let $s\in S$ be an element that acts nontrivially on v_1; then $(sv_1,v_2)\in W$ and $(\tilde{v}_1,0):=(v_1,v_2)-(sv_1,v_2)=(v_1-sv_1,0)\in W$, where $\tilde{v}_1\in\mathbb{C}^{m+1}\setminus\{0\}$. Since S acts irreducibly on $\mathbb{C}^{m+1}$, this implies that $\mathbb{C}^{m+1}\subseteq W$ and thus $\dim(W)\geq m+2$, which is a contradiction.

Recall that by our construction of $\gamma_1,\ldots,\gamma_D$ we have that S^{g_0} is contained in the Zariski closure of the group they generate. For the irreducible component $\mathcal{W}$ of $\widetilde{\mathcal{Y}}$ that contains our point, this forces $\dim\mathcal{W}=0$. □

5.5 ALGEBRAIC SETUP FOR GENERAL L. For orbit closures $\overline{\Gamma g_0 S}=\Gamma g_0 L$ with $S\lhd L=\mathrm{SO}^\circ(m,1)\times\mathrm{SO}(m+1)^p$ and $p>0$, we define $\mathcal{Y}$ as the collection of all tuples

$$(\mathbb{R}^\times v_1\odot v_2\odot\cdots\odot v_{p+1},\eta_1,\ldots,\eta_D),$$

where $v_1,\ldots,v_{p+1}\in\bigwedge^{m+1}\mathbb{R}^{n+1}$ are pure tensors, each representing a subspace that is fixed by the matrices $\eta_1,\ldots,\eta_D$. Once more this defines a variety defined over $\mathbb{Q}$.

Moreover, if we specify $\eta_i=g_Q^{-1}\gamma_i g_Q$ for $i=1,\ldots,D$ we obtain the variety $\widetilde{\mathcal{Y}}\subset\mathbb{P}(\bigodot^{p+1}\bigwedge^{m+1}\mathbb{R}^{n+1})$ corresponding to unordered $(p+1)$-tuples of $(m+1)$–dimensional subspaces of $\mathbb{R}^{n+1}$ that are fixed by the conjugates of our lattice elements.

LEMMA 5.4. *The point $\mathbb{R}^\times w_{g_0L}$ defined in section 5.2 constitutes a zero-dimensional irreducible component of $\widetilde{\mathcal{Y}}$.*

Proof. By Lemma 5.3 the subspace $g_Q^{-1}g_0\langle e_1,\ldots,e_{m+1}\rangle$ is an isolated point of the variety of all $S^{g_Q^{-1}g_0}$-invariant subspaces. We recall that the Zariski closure of $\eta_1,\ldots,\eta_D$ equals the $\mathbb{Q}$-group $\mathbb{L}$ so that $L^{g_Q^{-1}g_0}$ is the connected component of its group of $\mathbb{R}$-points. Hence the Galois conjugates of $g_Q^{-1}g_0\langle e_1,\ldots,e_{m+1}\rangle$ are also isolated points of the variety of subspaces that are invariant under $\eta_1,\ldots,\eta_D$. Since w_{g_0L} corresponds precisely to this unordered tuple, the lemma follows. □

5.6 THE TREE OF AFFINE VARIETIES. Let $\mathbf{X}=(X_1,\ldots,X_r)$ and $\mathbf{T}=(T_1,\ldots,T_s)$ be two lists of variables. Let $\mathcal{I}\subseteq\mathbb{Q}[\mathbf{X},\mathbf{T}]$ be an ideal. In this section, we are going to construct a finite tree of ideals in $\mathbb{Q}[\mathbf{X},\mathbf{T}]$, only depending on the ideal $\mathcal{I}$.

Let

$$\mathcal{I} = \mathfrak{q}_1 \cap \ldots \cap \mathfrak{q}_{n(\mathcal{I})}$$

be a minimal primary decomposition of $\mathcal{I}$ and denote by $\mathfrak{p}_1, \ldots, \mathfrak{p}_{n(\mathcal{I})}$ the corresponding prime ideals. For every $i \in \{1, \ldots, n(\mathcal{I})\}$ we define the prime ideal $\mathfrak{a}_i = \mathfrak{p}_i \cap \mathbb{Q}[\mathbf{T}]$, the ring $R_{\mathfrak{a}_i} = \mathbb{Q}[\mathbf{T}]/\mathfrak{a}_i$ of regular functions, and the field $k_{\mathfrak{a}_i}$ of rational functions on the variety $V(\mathfrak{a}_i)$. Identifying $\mathfrak{p}_i$ with its image in $R_{\mathfrak{a}_i}[\mathbf{X}]$ we also define the prime ideal $\mathfrak{p}_i' = k_{\mathfrak{a}_i}\mathfrak{p}_i \leq k_{\mathfrak{a}_i}[\mathbf{X}]$. We note that the dimension of the variety defined by $\mathfrak{p}_i$ (over $\mathbb{C}$) equals the dimension of the variety defined by $\mathfrak{a}_i$ (over $\mathbb{C}$) plus the dimension of the variety defined by $\mathfrak{p}_i'$ over $k_{\mathfrak{a}_i}$. This is a simple consequence of the definition of the dimension of a variety as the transcendence degree of its associated field of rational functions.

Denote by $P_{\mathcal{I}} \subseteq \mathbb{Q}[\mathbf{T}]$ the finite set of polynomials obtained by varying $i \in \{1, \ldots, n(\mathcal{I})\}$ and $j \in \{1, \ldots, r\}$ as follows. If $\mathfrak{p}_i'$ is zero-dimensional (referred to as a *zero-dimensional case* in the following discussions), then $X_j + \mathfrak{p}_i' \in k_{\mathfrak{a}_i}[\mathbf{X}]/\mathfrak{p}_i'$ is algebraic over $k_{\mathfrak{a}_i}$. Hence there exists a polynomial in $\mathfrak{p}_i$ in the variables X_j and $\mathbf{T}$ that is nonzero even when considered modulo $\mathfrak{a}_i$; we choose one such polynomial, assume that the leading coefficient (with respect to the variable X_j and belonging to $\mathbb{Q}[\mathbf{T}]$) does not belong to $\mathfrak{a}_i$, and add this leading coefficient to our set $P_{\mathcal{I}}$.

This defines the first level of our tree: the ideal $\mathcal{I}$ together with the set $P_{\mathcal{I}}$ and in any zero-dimensional case the polynomials in the variables X_j and $\mathbf{T}$ for $j = 1, \ldots, L$.

To find the next level of the graph, we let $p \in P_{\mathcal{I}}$ be one of the leading coefficients for a polynomial in X_j and $\mathbf{T}$ belonging to $\mathfrak{p}_i$ (in the zero-dimensional case). Now consider $\mathcal{I}_p = \mathfrak{p}_i + \mathbb{Q}[\mathbf{X}, \mathbf{T}]p$ (which is strictly bigger than $\mathfrak{p}_i$ and so also than $\mathcal{I}$). Whenever this ideal is proper we repeat the above discussion for it to find a new vertex of our tree.

Since the ring $\mathbb{Q}[\mathbf{X}, \mathbf{T}]$ is Noetherian, this process eventually terminates, leading to a finite tree of affine varieties. We also note that for the terminal leaves $\mathcal{I}$ we have that $p \in P_{\mathcal{I}}$ corresponding to $\mathfrak{p}_i$ implies that $\mathfrak{p}_i + \mathbb{Q}[\mathbf{X}, \mathbf{T}]p = \mathbb{Q}[\mathbf{X}, \mathbf{T}]$.

5.7 ESTIMATING THE HEIGHT. We return to our study of the Grassmannian $\mathrm{Gr}_{m+1,n+1}(\mathbb{R})$ and the varieties $\mathcal{Y}$ in Equation (15) and $\widetilde{\mathcal{Y}}$ in Equation (16), respectively its generalizations in section 5.5. For $i_0 \in \{1, 2, \ldots, r\}$, define the affine charts

$$\widehat{A}_{i_0} = \left\{\mathbb{R}^\times (x_1, x_2, \ldots, x_r) \in \mathbb{P}(\mathbb{R}^r) \mid x_{i_0} \neq 0\right\} \subseteq \mathbb{P}\left(\mathbb{R}^r\right)$$

and set $A_{i_0} := \mathcal{Y} \cap (\widehat{A}_i \times \mathbb{R}^s)$. Note that the sets A_{i_0} are affine varieties and that

$$\mathcal{Y} = \bigcup_{i_0=1}^{r} A_{i_0}. \tag{17}$$

We will always assume that i_0 is chosen so that A_{i_0} contains the point

$$\widetilde{X} := (\mathbb{R}^\times w_{g_0 L}, g_Q^{-1}\gamma_1 g_Q, \ldots, g_Q^{-1}\gamma_D g_Q) \in \mathcal{Y}, \tag{18}$$

where $\gamma_1, \ldots, \gamma_D$ are the lattice elements from Proposition 4.1 and Lemma 4.5.

DEFINITION 5.5 (Height). Given a point $\mathbb{R}^\times v \in \mathbb{P}(V)$ in the projective space associated to a vector space V with an integral structure $V_\mathbb{Z}$, we define its height $\operatorname{ht}(\mathbb{R}^\times v) = \operatorname{ht}(v)$ by

$$\operatorname{ht}(v) = \inf\{\|dv\| \mid dv \in V_\mathbb{Z} \text{ and } d \in \mathbb{R}^\times\}.$$

We note that for us this definition is useful since

$$\operatorname{ht}(\mathbb{R}^\times w_{g_0 L})^2 \asymp \operatorname{disc}(\Gamma g_0 L)$$

by Equations (13) and (14).

LEMMA 5.6 (Rational roots of rational polynomials). *For every $d \geq 1$ we have that any rational root $x = p/q \in \mathbb{Q}$ of a nonzero polynomial*

$$f(X) = a_d X^d + \ldots + a_1 X + a_0 \in \mathbb{Q}[X]$$

with $\gcd(p, q) = 1$ *satisfies*

$$\operatorname{ht}(x) := \max\{|p|, |q|\} \leq \operatorname{ht}(a_d) \ldots \operatorname{ht}(a_0).$$

Proof. We may suppose $a_0 a_d \neq 0$. For $i = 0, 1, \ldots, d$, write $a_i = \frac{\alpha_i}{\beta_i}$ with $\gcd(\alpha_i, \beta_i) = 1$. Multiplying f with $A = \beta_0\beta_1 \ldots \beta_d$ yields a polynomial $c_d X^d + \ldots + c_1 X + c_0 \in \mathbb{Z}[X]$ with coefficients $c_i = \beta_d \ldots \beta_{i+1}\alpha_i\beta_{i-1} \ldots \beta_0$ and therefore $\operatorname{ht}(c_i) \leq \operatorname{ht}(a_d) \ldots \operatorname{ht}(a_0)$. The claim now follows from the fact that given a rational root $x = p/q \in \mathbb{Q}$ of a polynomial $c_d X^d + \ldots + c_1 X + c_0 \in \mathbb{Z}[X]$ with $\gcd(p, q) = 1$, then q divides the leading coefficient c_d and p divides the constant term c_0. In particular, $\operatorname{ht}(x) \leq \max\{|c_d|, |c_0|\}$. □

In the proof of Proposition 5.1, we are going to use the following result; see, for example, [22, chapter 4, section 2].

PROPOSITION 5.7. *Let $\mathcal{V}$ be an irreducible affine variety of dimension n and let $f_1, \ldots, f_r$ be elements of the ring of regular function on $\mathcal{V}$. If $\mathcal{W}$ is an irreducible component of $\mathcal{V}(f_1, \ldots, f_r) = \{x \in \mathcal{V} \mid f_i(x) = 0$ for $1 \le i \le r\}$, then we have* $\dim \mathcal{W} \ge n - r$.

Proof of Proposition 5.1. Let $\mathcal{I} = \{f \in \mathbb{Q}[\mathbf{X}, \mathbf{T}] \mid f(y) = 0 \text{ for all } y \in A_{i_0}\}$ be the ideal of relations for all unordered $(p+1)$-tuples of $(m+1)$–dimensional subspaces within the affine chart $\widehat{A}_{i_0}$ and group elements that preserve (each of) the subspaces. Following the construction from section 5.6, we consider the finite tree corresponding to $\mathcal{I}$, and we want to choose one variety containing $\widetilde{X}$ in Equation (18) from this tree. It is important to note that the ideal $\mathcal{I}$ is independent of $\Gamma g_0 L$ except for the choice of $i_0 \in \{1, \ldots, r\}$. Hence all polynomials appearing below are from a finite list of polynomials constructed in section 5.6.

Let $\mathfrak{p}_i$ be one of the minimal prime ideals associated to $\mathcal{I}$ so that

$$\widetilde{X} \in V(\mathfrak{p}_i) = \{y \in A_{i_0} \mid f(y) = 0 \text{ for all } f \in \mathfrak{p}_i\},$$

and let $\mathbf{t} = (t_1, \ldots, t_M)$ be the entries of the lattice elements $\eta_1, \ldots, \eta_D$ conjugated to $\gamma_1, \ldots, \gamma_D$ from Lemma 4.5 by g_Q. Since g_Q is fixed, Proposition 4.1 gives that $\|\mathbf{t}\| \ll \mathrm{vol}(x_0 N)^{\kappa_2}$.

We now wish to introduce the additional relations $T_1 - t_1, \ldots, T_M - t_M$ to the ideal $\mathfrak{p}_i$, or equivalently specify the last M coordinates in the variety $\mathcal{V} = V(\mathfrak{p}_i)$. By definition of $\mathfrak{a}_i$ we have $\mathbf{t} \in V(\mathfrak{a}_i)$. Let $d_i = \dim V(\mathfrak{a}_i)$ so that precisely d_i elements of the list of variables $T_1, \ldots, T_M$ form a transcendence basis for the field of rational functions $k_{\mathfrak{a}_i}$ over $\mathbb{Q}$. For technical reasons we do not use these but instead apply the Noether normalization lemma (see [25, section 2.5.2, theorem 4]) for the variety $V(\mathfrak{a}_i)$ to define the regular functions $f_1, \ldots, f_{d_i} \in \mathbb{Q}[\mathbf{T}]$. We apply Proposition 5.7 to the variety $\mathcal{V} = V(\mathfrak{p}_i)$ and the regular functions $f_1(\mathbf{T}) - f_1(\mathbf{t}), \ldots, f_{d_i}(\mathbf{T}) - f_{d_i}(\mathbf{t})$. Hence the resulting subvariety of $\mathcal{V}$ defined by these equations has the property that all its irreducible components $\mathcal{W}$ have dimension at least $\dim V(\mathfrak{p}_i) - d_i$. Also note that for all of these varieties $\mathcal{W}$, the $\mathbf{T}$-coordinates are determined up to finitely many possibilities since the morphism

$$\varphi(\mathbf{T}) = (f_1(\mathbf{T}), \ldots, f_{d_i}(\mathbf{T}))$$

is finite on the variety $V(\mathfrak{a}_i)$ by construction of $f_1, \ldots, f_{d_i}$.

Suppose our point $\widetilde{X}$ belongs to the irreducible component $\mathcal{W}$ of $\mathcal{V}$. By construction, all elements of $\mathcal{W}$ consist of tuples of an (unordered $(p+1)$-tuple of) $(m+1)$–dimensional subspaces and group elements that preserve the subspaces. Since the group elements $\eta_1, \ldots, \eta_D$ are rational, the $\mathbf{T}$-coordinates in $\mathcal{W}$ are completely determined by these, and it follows that the subspaces appearing in $\mathcal{W}$ are contained in the variety $\widetilde{\mathcal{Y}}$ appearing in Equation (16) (respectively the variety discussed in section 5.2). By Lemmas 5.3 and 5.4 it follows that $\mathcal{W}$ is zero-dimensional. By the above dimension estimate we must therefore have $\dim V(\mathfrak{p}_i) = d_i$ or equivalently $\dim V(\mathfrak{p}_i') = 0$, which puts us into the zero-dimensional case of the construction of the tree.

We first suppose that $p(\mathbf{t}) \neq 0$ for all of the leading coefficients $p \in P_{\mathcal{I}}$ of the polynomials in X_j and $\mathbf{T}$ belonging to $\mathfrak{p}_i$. We apply this to the rational vector we obtain by considering $\mathbb{R}^\times w_{g_0 L}$ in the affine chart A_{i_0}. Hence each component of this rational vector satisfies a nontrivial polynomial relation whose degree is bounded and whose coefficients are rational with heights bounded by a multiple of a power of $\operatorname{vol}(x_0 N)$. Lemma 5.6 now shows that each component has height bounded by a power of $\operatorname{vol}(x_0 N)$. Clearing denominators we see that there exists an integral representative of $\mathbb{R}^\times w_{g_0 L}$ of size bounded by a power of $\operatorname{vol}(x_0 N)$.

In the second case we have that $p(\mathbf{t}) = 0$ for one of the leading coefficients $p \in P_{\mathcal{I}}$ of the polynomials in X_j and $\mathbf{T}$. However, this shows that $\widetilde{X}$ belongs to the variety defined by $\mathcal{I}_p = \mathfrak{p} + \mathbb{Q}[\mathbf{X}, \mathbf{T}]p$ and we may start the argument with this ideal again. Since the tree is finite we will eventually reach the first case, and Proposition 5.1 follows. □

5.8 AN UPPER BOUND FOR THE VOLUME OF A CLOSED S-ORBIT.

We start by recalling Siegel's lemma; see, for example, [20, lemma 6.1]:

LEMMA 5.8. *Let $M < N$ and let a_{ij} be integers with $|a_{ij}| \leq C$ for some $C \geq 0$ and all $1 \leq i \leq M$, $1 \leq j \leq N$. Then there is a $\kappa_7 > 0$ and a nonzero integral solution x to the homogeneous system*

$$\sum_{j=1}^{N} a_{ij} x_j = 0, \quad 1 \leq i \leq M$$

satisfying $|x_j| \ll_{M,N} C^{\kappa_7}$.

The next lemma specifies the principle that the orthogonal complement of a subspace of *low arithmetic complexity* is also of *low complexity*.

LEMMA 5.9. *Let $R > 0$ and let $v_1, \ldots, v_m \in \mathbb{Z}^n$ be vectors linearly independent over $\mathbb{R}$ with $\|v_i\| \le R$ for $i = 1, \ldots, m$. Then there exists a $\kappa_8 > 0$ depending only on n so that the orthogonal complement (for the bilinear form coming from Q) of $V = \mathrm{Span}(v_1, \ldots, v_m) \subseteq \mathbb{R}^n$ has a basis $v_{m+1}, \ldots, v_n \in \mathbb{Z}^n$ with $\|v_j\| \ll R^{\kappa_8}$ for $j = m+1, \ldots, n$.*

Proof. We apply Lemma 5.8 inductively in order to find a basis of $V^{\perp}$. Siegel's lemma immediately gives us a first vector $v_{m+1} \in V^{\perp} \cap \mathbb{Z}^n$ with $\|v_{m+1}\| \ll R^{\kappa_7}$ as a solution to the m linear conditions that v_{m+1} is orthogonal to $v_1, \ldots, v_m$. In order to complete the induction, assume that we have found linearly independent vectors

$$v_{m+1}, \ldots, v_{m+\ell} \in V^{\perp} \cap \mathbb{Z}^n$$

for some $1 \le \ell < n - m$ with $\|v_j\| \ll R^{\kappa_9}$ for some $\kappa_9 > 0$. Lemma 5.8 produces a vector $v_{m+\ell+1} \in V^{\perp} \cap \mathbb{Z}^n$, being linearly independent of $v_{m+1}, \ldots, v_{m+\ell}$ and satisfying $\|v_{m+\ell+1}\| \ll R^{\kappa_7 \kappa_9}$, as a solution to the $m + \ell$ linear conditions that $v_{m+\ell+1}$ is orthogonal to $v_1, \ldots, v_{m+\ell}$. □

We fix some Riemannian metric on G and write $d(g_1, g_2)$ for $g_1, g_2 \in G$, respectively $d(x_1, x_2)$ for $x_1, x_2 \in X$ for the resulting metrics.

PROPOSITION 5.10 (Transverse discreteness of orbits). *Suppose that g_1 and g_2 are two points defining the same S-orbit closure, chosen from a fixed compact subset that projects onto X_{cpct}. Moreover, assume that $\overline{\Gamma g_1 S} = \overline{\Gamma g_2 S} = \Gamma g_1 L$ (for an intermediate group L) has discriminant D and that $g_1 L \neq g_2 L$. Then there exists $\kappa_{10} > 0$ so that $d(g_1, g_2) \gg D^{-\kappa_{10}}$.*

Proof. We suppose first that $\Gamma g_1 S$ is a closed orbit as in section 5.1. Write $\gamma g_1 s = g_2$ for some $\gamma \in \Gamma$ and $s \in S$ and let $h = g_1^{-1} g_2$ so that

$$g_1 h = g_2 = \gamma g_1 s.$$

If h is not contained in $N_G(S)$, then h does not fix the subspace spanned by $e_1, \ldots, e_{m+1}$ and so $\mathbb{R}^{\times} w_{g_1 S} \neq \mathbb{R}^{\times} w_{g_2 S}$, where

$$w_{g_i S} = g_Q^{-1} g_i e_1 \wedge \ldots \wedge g_Q^{-1} g_i e_{m+1}$$

for $i=1,2$. By the definition of the discriminant, $\mathbb{R}w_{g_1S}$ and $\mathbb{R}w_{g_2S}$ contain two different lattice elements of $V_{\mathbb{Z}}$ of equal quadratic value and size approximately $D^{1/2}$. Therefore $\|w_{g_1S}-w_{g_2S}\| \gg D^{-1/2}$. Since the map $g\mapsto w_{gS}$ is smooth, since g_Q is fixed, and since g_1, g_2 belong to a fixed compact subset, this implies that $d(g_1, g_2) \gg D^{-1/2}$.

If on the other hand $h\in N_G(S)$, we consider the orthogonal complement (for the bilinear form coming from Q) of the subspace

$$W_1 = g_Q^{-1}g_1\langle e_1,\dots,e_{m+1}\rangle \subseteq \mathbb{R}^{n+1}.$$

By Minkowski's theorem on successive minimas and Equation (13) it follows that W_1 has a basis $w_1,\dots,w_{n-m}\in\mathbb{Z}^{n+1}$ with $\|w_i\|\ll D^{1/2}$. By Lemma 5.9, there exists $\kappa_{10}>0$ and a basis $w_1,\dots,w_{n-m}\in\mathbb{Z}^{n+1}$ of $W_1^{\perp}$ with $\|w_i\|\ll D^{\kappa_{10}}$ for $i=1,\dots,n-m$.

Note that $g_Q^{-1}\gamma g_Q w_i$ for $i=1,\dots,n-m$ are lattice elements as well. If $g_Q^{-1}\gamma g_Q w_i = w_i$ for $i=1,\dots,n-m$, then the conjugate $h^{g_Q^{-1}}$ of $h=g_1^{-1}\gamma g_1 s$ acts trivially on $\langle e_{m+2},\dots,e_{n+1}\rangle$. Therefore h belongs to S and $g_1S=g_2S$, which contradicts the assumption of the proposition. So there exists $i\in\{1,\dots,n-m\}$ with $g_Q^{-1}\gamma g_Q w_i\neq w_i$, which means that w_i and $g_Q^{-1}\gamma g_Q w_i$ are two different lattice elements. Note that given $v\in\mathbb{R}^{n+1}$ with $\|v\|=1$, then the map $g\mapsto gv$ is smooth and g_1, g_2 belong to a fixed compact subset. Therefore,

$$\begin{aligned}1\le Q\left(g_Q^{-1}\gamma g_Q w_i - w_i\right) &\ll Q_0\left(g_1 h s^{-1} g_1^{-1} g_Q w_i - g_Q w_i\right)\\ &= Q_0\left(g_1 h g_1^{-1} g_Q w_i - g_Q w_i\right) \ll D^{\kappa_{10}} d(g_2 g_1^{-1}, e)\end{aligned}$$

since s^{-1} fixes $g_1^{-1}g_Q w_i\in\langle e_{m+1},\dots,e_{n+1}\rangle$. It follows that

$$d(g_1, g_2) \gg D^{-\kappa_{10}}.$$

We now suppose that we are in the situation considered in section 5.2. We again write $\gamma g_1\ell = g_2$ for $\gamma\in\Gamma$, $\ell\in L$, and we set $h=g_1^{-1}g_2$. If h is not in $N_G(L)$, then $\mathbb{R}^\times w_{g_1L}\neq\mathbb{R}^\times w_{g_2L}$ and we can argue as in the first case considered above. So suppose $h\in N_G(L)$ or equivalently $w_{g_1L}=w_{g_2L}$. Here too the argument is very similar to the case considered above once we have shown that the orthogonal complement of $g_Q^{-1}g_1\langle e_1,\dots,e_{(p+1)(m+1)}\rangle$ has a basis consisting of integer vectors of size bounded by $\ll \mathrm{disc}(g_0L)^*$. To prove this we have to use that the symmetric tensor $dw_{g_1L}\in V_{\odot,\mathbb{Z}}$ is integral, is of size $\ll \mathrm{disc}(\Gamma g_1 L)^{1/2}$, and roughly speaking corresponds to $(p+1)$ linearly independent (algebraic) $(m+1)$–dimensional subspaces.

For this we start by noting that any tensor $w \in \bigodot^{p+1} \bigwedge^k \mathbb{R}^{n+1}$ can be used to induce a homogeneous polynomial map

$$w \wedge (\cdot) : v \in \mathbb{R}^{n+1} \mapsto w \wedge v \in \bigodot^{p+1} \bigwedge^{k+1} \mathbb{R}^{n+1}$$

of degree $p+1$. In fact, for any pure tensor $w = w_1 \odot w_2 \odot \cdots \odot w_{p+1}$ with $w_1, \ldots, w_{p+1} \in \bigwedge^k \mathbb{R}^{n+1}$, we define

$$w \wedge v = (w_1 \wedge v) \odot (w_2 \wedge v) \odot \cdots \odot (w_{p+1} \wedge v),$$

which extends by symmetry and multilinearity to a definition of $w \wedge v$ for all $w \in \bigodot^{p+1} \bigwedge^k \mathbb{R}^{n+1}$. Moreover, for integral w this polynomial has integer coefficients.

Since the form Q is nondegenerate it induces an isomorphism between $\mathbb{R}^{n+1}$ and its dual space. Moreover, identifying $\bigwedge^{n+1} \mathbb{R}^{n+1}$ with $\mathbb{R}$ gives an isomorphism between $V = \bigwedge^{m+1} \mathbb{R}^{n+1}$ with $V^* = \bigwedge^{n-m} \mathbb{R}^{n+1}$ (only depending on Q) so that, for example, $g_Q^{-1} g_1(e_1 \wedge \ldots \wedge e_{m+1})$ is identified with $g_Q^{-1} g_1(e_{m+2} \wedge \ldots \wedge e_{n+1})$ (apart from a sign and a power of λ that only depends on m and n). We let $w \in V_{\mathbb{Z}}^*$ be the image of $dw_{g_1 L}$ under this isomorphism, multiplied by a power of λ to ensure integrality of w. With this, we wish to consider the equation

$$\partial_\alpha (w \wedge v) = 0, \tag{19}$$

where $\alpha \in \mathbb{N}_0^{n+1}$ is assumed to satisfy $\sum_{j=1}^{n+1} \alpha_j = p$ and ∂_α stands for the iterated partial derivatives with respect to the coordinates of v. Since $w \wedge v$ is homogeneous of degree $p+1$, we see that (19) actually amounts to linear equations in the variable v. Moreover, by construction we have that the coefficients to these equations are integral and bounded by a power of $\operatorname{disc}(\Gamma g_1 L)$. We claim that these define precisely the subspace $g_Q^{-1} g_1 \langle e_{(p+1)(m+1)+1}, \ldots, e_{n+1} \rangle$. Using this and Lemma 5.8 it follows that this subspace has indeed a basis consisting of integer vectors of size that is bounded by a power of $\operatorname{disc}(\Gamma g_1 L)$, which in turn implies the remaining case of the proposition.

To see the claim, it is best to switch again from Q to Q_0 and using that g_1 belongs to the orthogonal group we may also drop g_1. After this, w becomes a scalar multiple of the symmetric tensor product of the alternating tensors

$$e_{m+2} \wedge \ldots \wedge e_{(p+1)(m+1)} \wedge e_{(p+1)(m+1)+1} \wedge \ldots \wedge e_{n+1},$$

$$e_1 \wedge \ldots \wedge e_{m+1} \wedge e_{2(m+1)+1} \wedge \ldots \wedge e_{(p+1)(m+1)} \wedge e_{(p+1)(m+1)+1} \wedge \ldots \wedge e_{n+1},$$

$$\vdots$$

$$e_1 \wedge \ldots \wedge e_{p(m+1)} \wedge e_{(p+1)(m+1)+1} \wedge \ldots \wedge e_{n+1},$$

where in each line a different block of $(m+1)$ consecutive basis vectors is missing. From this we see that $w \wedge v$ vanishes of order $p+1$ on the desired subspace $\langle e_{(p+1)(m+1)}, \ldots, e_{n+1} \rangle$ and hence all partial derivatives in (19) of order p vanish too. For $v = v_1 + \cdots + v_{p+1}$ with v_j belonging to the subspace spanned by the basis vectors with indices in $(j-1)(m+1)+1, \ldots, j(m+1)$, we have that $w \wedge v$ equals the symmetric tensor product of the first alternating tensor in the above list with v_1, the second with v_2, and so on. In particular, if we expand the product into linear combinations of the standard basis vectors, we obtain that each coefficient is itself a product of $p+1$ coordinates with one coordinate of each $v_1, \ldots, v_{p+1}$ (and not a sum of such products). Taking now the partial derivative ∂_α as above we can isolate all coordinates of v_j to obtain $v_j = 0$ for $j = 1, \ldots, p+1$. This implies the claim. □

The above transverse separation of an orbit to itself implies a bound for the volume of a closed S-orbit in terms of its discriminant.

PROPOSITION 5.11. *There exists a constant $\kappa_{11} > 0$ (depending only on n) so that $\overline{\Gamma g_0 S} = \Gamma g_0 L$ for some $g_0 \in G$ implies*

$$\mathrm{vol}(\Gamma g_0 L) \ll \mathrm{disc}(\Gamma g_0 L)^{\kappa_{11}}.$$

We only sketch the argument and refer to [7, proposition 2.8] for the details. If X is compact, we can cover it by finitely many balls with radius below the injectivity radius. In each one of these balls B the separation property in Proposition 5.10 shows that at most $\ll D^*$ many local pieces of the orbit $\Gamma g_0 L$ can go through B, where the exponent $*$ depends on n and κ_{10} only. Putting this together gives the result. If X is noncompact, one applies the same argument but within X_{cpct}, which suffices for the volume estimate.

Taking Propositions 5.1 and 5.11 together we obtain the desired partial converse to the trivial estimate $\mathrm{vol}(\Gamma g_0 N) \ll \mathrm{vol}(\Gamma g_0 S)$.

PROPOSITION 5.12 (Volume comparison). *There exists $\kappa_1 > 0$ such that*

$$\mathrm{vol}(\Gamma g_0 N) \ll \mathrm{vol}(\overline{\Gamma g_0 S}) \ll \mathrm{vol}(\Gamma g_0 N)^{\kappa_1}$$

for any $g_0 \in G$ for which $\overline{\Gamma g_0 S} = \Gamma g_0 L$ and $S \lhd L$.

5.9 NUMBER OF CLOSED N-ORBITS. Proposition 5.10 also applies to closed N-orbits. In fact we can define the discriminant of the N-orbit as the discriminant of any S-orbit closure it contains. Moreover, this shows that the number of closed N-orbits with $\mathrm{disc}(\Gamma g_0 N) \leq D$ is bounded by $\ll D^{\kappa_{12}}$ for some constant $\kappa_{12} > 0$ depending only on n. In fact, our discussion associates to each orbit $\overline{\Gamma g_0 S} = \Gamma g_0 L \subset \Gamma g_0 N$ (with the representative g_0 chosen from a fixed compact subset of G) the integer vector $w_{g_0 L}$ of size $\ll D^{1/2}$. Since this vector determines the orbit $\Gamma g_0 L$ up to the normalizer of L (which is contained in N), this gives the claim.

6 The dynamical argument

Throughout this section we let μ be an H-invariant and ergodic probability measure on X as in Theorem 1.2. In order to prove Theorem 1.2, we are going to iteratively increase the group H to a bigger group S under which μ is *almost invariant* until there exists a closed orbit $\overline{x_0 S}$ of *small volume*. We then conclude that μ is *close* to the Haar measure on $\overline{x_0 S}$.

While doing this iteration, we will always have that μ is almost invariant under a subgroup $S \leq G$ that is conjugated by an element of the maximal compact subgroup of G to $\mathrm{SO}^\circ(m,1)$ for some $2 \leq m \leq n$. If $\overline{x_0 S}$ does not have *small volume*, we are going to produce additional invariance *transversal* to the normalizer N of S, which leads to μ being almost invariant under $S' \cong \mathrm{SO}^\circ(m',1)$ for some $m < m' \leq n$ with $S \subsetneq S'$. The exact notions of *small* and *close* will be specified in the proofs by concrete exponents that only depend on m and will be denoted by κ with various indices.

The arguments in this section are very similar to the corresponding arguments in [9], except that we have to take care of the existence of the compact centralizer. Therefore, after some preparations of general nature, the material from sections 4 and 5 will become crucial for the arguments in section 6.5. Moreover, quite often in the argument it is useful to recall that, up to conjugation by elements of a *compact* subgroup, only finitely many subgroups $S < G$

need to be considered. In this sense the subgroups $S < N$ that we care about can be thought of as undistorted.

6.1 SOBOLEV NORMS. For an integer $d \geq 0$, we define the Sobolev norm of degree d by

$$\mathcal{S}_d(f)^2 = \sum_{\mathcal{D}} \left\| \mathrm{ht}(x)^d \mathcal{D} f \right\|^2_{L^2(X)},$$

where the sum runs over all monomials $\mathcal{D}$ in the universal enveloping algebra of $\mathfrak{g}$ of degree at most d, using a fixed basis of $\mathfrak{g}$, and the height $\mathrm{ht}(x)$ of a point $x \in X$ is defined by

$$\mathrm{ht}(x) = \sup \left\{ \left\| \mathrm{Ad}(g^{-1}) v \right\|^{-1} \mid \Gamma g = x, v \in \mathfrak{g}_{\mathbb{Z}} \setminus \{0\} \right\}.$$

For this we also let $\|\cdot\|$ be a fixed Euclidean norm on $\mathfrak{g}$ that is invariant under the maximal compact subgroup $\mathrm{SO}(n)$ of $G = \mathrm{SO}^\circ(n,1)$. Notice that there exists $R > 0$ so that

$$X_{\mathrm{cpct}} \subseteq \mathfrak{S}(R) := \left\{ x \in X \mid \mathrm{ht}(x) \leq R \right\},$$

where X_{cpct} is the compact subset of X chosen in section 2. Therefore, we may and will choose $X_{\mathrm{cpct}} = \mathfrak{S}(R_0)$ for some $R_0 \geq 0$ such that it still has the properties that each S-orbit is of the form xS for some $x \in X_{\mathrm{cpct}}$ and $\mu(X \setminus X_{\mathrm{cpct}}) \leq \frac{1}{10^{11}}$. We note that ht is $\mathrm{SO}(n)$-invariant, which implies the same for sets of the form $\mathfrak{S}(R)$ for $R > 0$.

We recall some basic properties of the Sobolev norm from [9, section 3.7]:

PROPOSITION 6.1. *The following properties hold:*

(S1) *There exists a constant $d_0 > 0$ such that for all $d \geq d_0$ and $f \in C_c^\infty(X)$,*

$$\|f\|_\infty \ll_d \mathcal{S}_d(f).$$

(S2) *For every $d \geq d_0$ there exist integers $d_2 > d_1 > d$ and an orthonormal basis $\{e_k\}$ of the completion of $C_c^\infty(X)$ with respect to $\mathcal{S}_{d_2}$, which is also orthogonal with respect to $\mathcal{S}_{d_1}$ so that*

$$\sum_k \mathcal{S}_{d_1}(e_k)^2 < \infty \text{ and } \sum_k \frac{\mathcal{S}_d(e_k)^2}{\mathcal{S}_{d_1}(e_k)^2} < \infty.$$

(S3) *There exists a constant* $\kappa_{13} > 0$ *so that for any* $g \in G$ *and* $d \geq 1$ *we have*

$$\mathcal{S}_d(g \bullet f) \ll_d \|g\|^{\kappa_{13} d}\, \mathcal{S}_d(f),$$

where $g \bullet f(x) = f(xg)$ *for all* $x \in X$.

(S4) *If* $d > d_0$, *then we have*

$$\|f - g \bullet f\|_\infty \ll_d d(I, g)\mathcal{S}_d(f),$$

where I *denotes the identity in* G.

(S5) *If* $f_1, f_2 \in C_c^\infty(X)$ *and* $d' > d + d_0$, *then*

$$\mathcal{S}_d(f_1 f_2) \ll_d \mathcal{S}_{d'}(f_1)\mathcal{S}_{d'}(f_2).$$

We refer to [9, section 3.7] for a proof of Proposition 6.1.

6.2 ALMOST INVARIANCE. For a measure μ on X and an element $g \in G$, we denote by μ^g the pushforward of μ under the map $x \mapsto xg$—that is,

$$\mu^g(f) = \mu(g \bullet f) = \int_X f(xg)\, d\mu \text{ for } f \in C_c(X)$$

—where we again write $g \bullet f$ for the function $x \in X \mapsto f(xg)$.

DEFINITION 6.2. Let μ be a measure on X, $d > 0$, and $\varepsilon > 0$. Then, μ is called ε-almost invariant under an element $g \in G$ with respect to $\mathcal{S}_d$ if

$$\left|\mu^g(f) - \mu(f)\right| \leq \varepsilon \mathcal{S}_d(f) \text{ for all } f \in C_c^\infty(X).$$

The measure μ is called ε-almost invariant under a connected subgroup $S \leq G$ with respect to $\mathcal{S}_d$ if it is ε-almost invariant with respect to $\mathcal{S}_d$ under every element $g \in S$ with $d(g, I) < 1$. We say that μ is ε-almost invariant under an element $v \in \mathfrak{g}$ if it is ε-almost invariant under $\exp(tv)$ for all $|t| \leq 1$.

An important tool concerning almost invariance is the following lemma; see also [9, lemma 8.2].

LEMMA 6.3. *There exists a constant* $\kappa_{14} > 0$ *so that the following holds: Let* S *be a simple connected intermediate subgroup—that is,* $H \leq S \leq G$ *with* $S \cong \mathrm{SO}^\circ(m, 1)$

for some $m \in \{2, \dots, n\}$. Assume that μ is ε-almost invariant under S with respect to $\mathcal{S}_d$. Then

$$\left|\mu^s(f) - \mu(f)\right| \ll \varepsilon \, \|s\|^{d\kappa_{14}} \, \mathcal{S}_d(f) \text{ for all } s \in S.$$

The proof of [9, lemma 8.2] uses the fact that there is no nontrivial centralizer and more precisely that there are only finitely many intermediate subgroups, which is not the case in our setting. Note, however, that by Corollary 3.2, there are, up to conjugation by an element of the maximal compact subgroup of G, only finitely many simple connected intermediate subgroups $H \leq S \leq G$. It therefore suffices to prove Lemma 6.3 individually for each possible conjugacy class.

Let $S = \mathrm{SO}^\circ(m, 1)$ for some $m \in \{2, \dots, n\}$. Using the Cartan decomposition of S, there are constants c_1, κ_{15} depending only on m, so that for every $r \geq 2$, the following holds: Every $s \in S$ with $\|s\| \leq r$ can be written as the product of $\leq c_1 + \kappa_{15} \log r$ elements $h \in S$ with $d(h, I) < 1$. This allows one to iterate the almost invariance while keeping track of the accumulated errors (using Proposition 6.1(S3)); see the proof of [9, lemma 8.2] for more details.

Finally, notice that this extends uniformly to all conjugates of $\mathrm{SO}^\circ(m, 1)$ by elements of $\mathrm{SO}(n)$. Indeed suppose S is conjugated to $\mathrm{SO}^\circ(m, 1)$ under some $k \in \mathrm{SO}(n)$. If now μ is almost invariant under S, then μ^k is almost invariant under $\mathrm{SO}^\circ(m, 1)$. Indeed the Sobolev norm of a smooth function $f \in C_c^\infty(X)$ changes by Proposition 6.1(S3) at most by a bounded multiplicative factor if we compose f with application of k. Applying the established result to μ^k then gives the same result for μ and S (possibly with a slightly worse but still uniform implicit constant).

6.3 GENERIC POINTS.

In what follows, we again use the unipotent one-parameter subgroup $\{u(t)\} \leq H$ defined in Equation (2).

DEFINITION 6.4 (Generic points). Fix some $M > 0$. For a function $f \in C^\infty(X)$, we define the discrepancy $D_\ell(f)$ for $\ell \geq 1$ by

$$D_\ell(f)(x) = \frac{1}{(\ell+1)^M - \ell^M} \int_{\ell^M}^{(\ell+1)^M} f(xu(t)) \, \mathrm{d}t - \int_X f \, \mathrm{d}\mu.$$

Let $T_0 < T_1$ be positive real numbers. A point $x \in X$ is said to be $[T_0, T_1]$-generic with respect to the Sobolev norm $\mathcal{S}_d$ if for all integers $\ell \in [T_0, T_1]$ and for all $f \in C^\infty(X)$, we have

$$\left|D_\ell(f)(x)\right| \leq \ell^{-1} \mathcal{S}_d(f). \tag{20}$$

A point $x \in X$ is called T_0-generic with respect to the Sobolev norm $\mathcal{S}_d$ if (20) holds for all integers $\ell \geq T_0$.

Since the action of $\mathrm{SO}^\circ(2,1)$ on $L^2_\mu(X)$ has a spectral gap independent of μ (see [2] and [3]), it turns out that most points are generic if only M is chosen sufficiently large. More precisely, we have the following proposition, which is proven in [9, section 9] as a consequence of the uniform spectral gap and Proposition 6.1(S2). We will use the shorthand $B_S(R) = S \cap B_R^{\mathrm{HS}}$ for any $R > 0$.

PROPOSITION 6.5 (Effective ergodic theorem). *Let $d \geq 1$. Then there exists $\beta \in (0, 1/2)$ and $d' > d$ so that the following holds. Let $S \cong \mathrm{SO}^\circ(m,1)$ for some $2 \leq m \leq n$, and assume that μ is ε-almost invariant under S with respect to $\mathcal{S}_d$ for some $\varepsilon > 0$. For $R, T_0 \in (0, \varepsilon^{-\beta})$, the fraction of points $(x, s) \in X \times B_S(R)$ with respect to $\mu \times m_S$ for which xs is not $[T_0, \varepsilon^{-\beta}]$-generic with respect to $\mathcal{S}_{d'}$ is $\ll_d T_0^{-1}$.*

6.4 TUPLES OF GENERIC POINTS LEAD TO ADDITIONAL INVARIANCE. We denote the Lie algebra of G by $\mathfrak{g} = \mathfrak{so}(n,1)$. Fixing some $m \in \{2, \ldots, n\}$ we let $S \cong \mathrm{SO}^\circ(m,1)$ be an intermediate subgroup with Lie algebra $\mathfrak{s} = \mathfrak{so}(m,1)$, and let $N = N_G(S)^\circ$ with Lie algebra $\mathfrak{n}$. Write $\mathfrak{g} = \mathfrak{n} \oplus \mathfrak{r}$, where $\mathfrak{r}$ is an $\mathrm{Ad}(H)$-invariant complement of $\mathfrak{n}$ in $\mathfrak{g}$. Moreover, let $\mathfrak{r}_0$ be the centralizer of $\{u_t\}$ in $\mathfrak{r}$ (or equivalently the highest weight subspace in $\mathfrak{r}$) and let $\mathfrak{r}_1$ be the orthogonal complement of $\mathfrak{r}_0$ in $\mathfrak{r}$ with respect to $\|\cdot\|$. We also write $r = (r_0, r_1)$ for $r = r_0 + r_1 \in \mathfrak{r}$ and assume $r_0 \in \mathfrak{r}_0$ and $r_1 \in \mathfrak{r}_1$. Finally we use the norm on $\mathfrak{g}$ and the interval $[0, 2] \subseteq \mathbb{R}$ to induce a supremums norm on the space of $\mathfrak{g}$-valued polynomials on $\mathbb{R}$.

PROPOSITION 6.6 (Additional invariance). *Let $d \geq d_0 + 1$. Then there exist constants $\kappa_{16} > 0$ and $\kappa_{17} > \kappa_{18} > 0$ with the following property:*

Suppose that $x_1, x_2 \in X$ satisfy $x_2 = x_1 g_C \exp(r)$ for some

$$r = (r_0, r_1) \in \mathfrak{r} \text{ with } r_1 \neq 0,$$
$$g_C \in C_G(S),$$

and that x_1, x_2 are $[\|r_1\|^{-\kappa_{18}}, \|r_1\|^{-\kappa_{17}}]$-generic with respect to μ and the Sobolev norm $\mathcal{S}_d$. Then there exists a polynomial $q : \mathbb{R} \longrightarrow \mathfrak{r}_0$ of degree ≤ 2 and norm 1 such that

$$\left| \mu^{g_C \exp q(s)}(f) - \mu(f) \right| \ll_d \|r_1\|^{\kappa_{16}} \mathcal{S}_d(f), \quad 1 \leq s \leq 2^{1/M}.$$

In fact μ is $\ll_d \|r\|^{\kappa_{16}/2}$-almost invariant under some $v \in \mathfrak{r}_0$ with $\|v\| = 1$.

Proof. The proof is very similar to the proof of [9, proposition 10.1], so we will be brief. Moreover, as we have seen in section 3 the invariant complement $\mathfrak{r}$ is isomorphic as a representation space over $H = \mathrm{SO}^\circ(2,1)(\mathbb{R})$ to several copies of the adjoint representation. This makes the following discussion easier than the general case considered in [9].

Let $r = (r_0, r_1)$ with $r_1 \neq 0$. Note that $\mathrm{Ad}(u(-t))r_0 = r_0$, whereas $\mathrm{Ad}(u(-t))r_1$ is a polynomial of degree at most 2 and with coefficients bounded up to a constant by $\|r_1\|$. Moreover, since $r_1 \neq 0$ the component of $\mathrm{Ad}(u(-t))r_1$ in the highest weight subspace $\mathfrak{r}_0$ is nontrivial with a trivial constant term. This implies that there exist a polynomial $q : \mathbb{R} \longrightarrow \mathfrak{r}_0$ and a constant $T > 0$ (used as a time-lapse parameter) with $\|r_1\|^{-1/2} \ll T \ll \|r_1\|^{-1}$ such that the following properties hold:

- The image of q is centralized by $u(\mathbb{R})$ (since $q(\mathbb{R}) \subseteq \mathfrak{r}_0$),
- $\mathrm{Ad}(u(-t))r = q(t/T) + \mathcal{O}(\|r\|^{1/2})$ for all $t \le 3T$, and
- $q(0) = 0$ and $\max_{s\in[0,2]} \left\|q(s)\right\| = 1$ (by definition of T).

We fix positive $\kappa_{18} < \frac{1}{2M}$ and $\kappa_{17} > \frac{1}{M}$ so that we have

$$\left[T^{1/M}, (2T)^{1/M}\right] \subseteq \left[\|r_1\|^{-\kappa_{18}}, \|r_1\|^{-\kappa_{17}}\right]$$

whenever $\|r_1\|$ is sufficiently small.

We now assume as in the proposition that $x_2 = x_1 g_C \exp(r)$ and x_1 are $\left[\|r_1\|^{-\kappa_{18}}, \|r_1\|^{-\kappa_{17}}\right]$-generic. This gives

$$\left| \int f \,\mathrm{d}\mu - \frac{1}{(\ell+1)^M - \ell^M} \int_{\ell^M}^{(\ell+1)^M} f(x_i u(t)) \,\mathrm{d}\mu \right| \le \ell^{-1} \mathcal{S}_d(f) \tag{21}$$

for any integer $\ell \in \left[T^{1/M}, (2T)^{1/M}\right]$ and for any $f \in C_c^\infty(X)$. Notice also that for $t, t_0 \in \left[(\ell-1)^M, (\ell+1)^M\right]$, we have

$$|t - t_0| \ll \ell^{M-1} \asymp T^{1-1/M}. \tag{22}$$

Therefore, using the properties of the polynomial q and (S4) of Proposition 6.1, we have

$$\begin{aligned} f(x_2 u(t)) &= f(x_1 u(t) g_C \exp(\mathrm{Ad}(u(-t))r)) \\ &= f(x_1 u(t) g_C \exp(q(t/T))) + \mathcal{O}(\|r_1\|^{1/2} \mathcal{S}_d(f)) \\ &= f(x_1 u(t) g_C \exp(q(t_0/T))) + \mathcal{O}(\|r_1\|^{1/2} \mathcal{S}_d(f)) + \mathcal{O}(T^{-1/M} \mathcal{S}_d(f)) \end{aligned}$$

for every $f\in C_c^\infty(X)$. Applying Equation (21) for x_2 with f and x_1 with $(g_C\exp(q(t_0/T))\bullet f$, we obtain

$$\mu(f)=\mu(g_C\exp(q(t_0/T))\bullet f)+\mathcal{O}((T^{-1/M}+\|r_1\|^{1/2})\mathcal{S}_d(f)). \tag{23}$$

Let us summarize the situation: We have already shown that there exists a linear or quadratic polynomial $q:\mathbb{R}\longrightarrow\mathfrak{r}_0$ of norm 1, such that for any integer

$$\ell\in[T^{1/M},(2T)^{1/M}]\subset[\|r_1\|^{-\kappa_{18}},\|r_1\|^{-\kappa_{17}}]$$

and for any $t_0\in[\ell^M,(\ell+1)^M]$, we have (23). In other words, there exists $\kappa_{16}>0$ so that μ is $\ll\|r_1\|^{\kappa_{16}}$-almost invariant under $g_C\exp(q(s))$ for all[2] $s\in[1,2^{1/M}]$. This proves the first assertion of the proposition.

If we denote by $\{a_t\}$ the diagonal one-parameter subgroup in H and use $\|r_1\|\le\|r\|$, then μ is also $\ll\|r\|^{\kappa_{16}}$-almost invariant under the conjugated element

$$a_{-\log 2}g_C\exp(q(s))a_{\log 2}=g_C\exp(2q(s))$$

and therefore, by Proposition 6.1(S3), under the element

$$\left(g_C\exp(q(s))\right)^{-1}g_C\exp(2q(s))=\exp(q(s))$$

for all $s\in[1,2^{1/M}]$. In fact, since q takes only values in the highest weight subspace $\mathfrak{r}_0$, we may use that $\mathfrak{r}_0$ is abelian (see also the discussions in section 3) in the above calculation.

Let $\delta=\|r\|^{\kappa_{16}/2}$ and notice that the conclusion of the proposition is trivial if δ is large enough—in particular, if $\delta>\frac{1}{2}(2^{1/M}-1)$. Using that the derivative q' is linear with $\max_{s\in[0,2]}|q'(s)|\gg 1$ and choosing either $s=1$ or $s=2^{1/M}-\delta$, we may assume that $|q'(s)|\gg 1$. Moreover, s and $s+\delta$ are both contained in $[1,2^{1/M}]$. Therefore, μ is $\ll\|r\|^{\kappa_{16}}$-almost invariant under both $\exp(q(s))$ and $\exp(q(s+\delta))$ and so also, by Proposition 6.1(S3), under the element

$$\exp(-q(s))\exp(q(s+\delta))=\exp(w^*)$$

for $w^*=q(s+\delta)-q(s)\in\mathfrak{r}_0$ satisfying $\|w^*\|\asymp\delta$.

Proposition 6.1(S3) then implies that μ is $\delta^2=\|r\|^{\kappa_{16}}$-almost invariant under $\exp(w^*)$. Iterating this statement ℓ times for all $\ell\ll\delta^{-1}$ and noting that

---

[2] For $t\in[T,\lceil T^{1/M}\rceil^M]$ we can use the given argument for $t_0=\lceil T^{1/M}\rceil^M$ together with Equation (22) to also obtain (23) in this case.

μ is $\ll \delta$-almost invariant under sw^* for all $s \in [-1, 1]$ by Proposition 6.1(S4), we see that μ is $\delta = \|r\|^{\kappa_{16}/2}$-almost invariant under $v = \frac{1}{\|w^*\|} w^*$, which proves the proposition. □

6.5 EXISTENCE OF TUPLES OF GENERIC POINTS. In this section we are going to prove that there exist two generic points $x_1, x_2 \in X$ that are close by in the direction of $\mathfrak{r}$. As in the previous section, they are allowed to differ by an element in the direction of the centralizer, whose size we do not control. This is the analogue to [9, section 14].

The existence of the nontrivial centralizer creates another technical wrinkle in the argument. In fact, the group S constructed inductively and almost preserving the measure is not well-defined within its conjugacy class; we may simply perturb it using a conjugation by a small element from $M = C_G(H)$. For this reason we also have to allow for this conjugation in the following discussion, which will be important for the construction of the tuple of generic points. As before we let $m \in \{2, \ldots, n\}$, $S \cong \mathrm{SO}^\circ(m, 1)$, and $N = N_G(S)^\circ$.

LEMMA 6.7. *There exist constants $V_0 > 0$ and $\kappa_{19} > 0$ with the following property: Let $V \geq V_0$ and suppose that $\mu(xNk) = 0$ for all closed N-orbits xN of volume $\leq V$ and all $k \in M$. Then*

$$\mu(\{x \in X \mid \textit{there exists } x' \in X \textit{ and } k \in M \textit{ with } d(x'k, x) \leq V^{-\kappa_{19}} \textit{ such that } x'Nk \textit{ is closed of volume } \leq V\}) \leq 1/2.$$

The argument behind this lemma is known as the linearization technique; see [9, section 11] and [5] and [19] for the general case. The proof of the above is significantly easier due to the concrete setup, and we only outline the proof but skip the details. In fact, due to the formulation of the lemma, it suffices to consider $S = \mathrm{SO}^\circ(m, 1)$. By the connection between discriminant and volume and the discussion in section 5.9, we have a concrete upper bound for the number of closed N-orbits of volume $\leq V$ in terms of a power of V. Hence it suffices to get a sufficiently good estimate for one closed N-orbit xN. Next we cover xNM by sufficiently thin tubular sets of the form $xN \exp(B_\delta^W) \exp(B_\delta^{W'})k$, where $k \in M$, W' is an invariant complement of $\mathfrak{n}$ within $\mathfrak{n} + \mathrm{Lie}(M)$, and W is an invariant complement of $\mathfrak{n} + W'$ within $\mathfrak{g}$. Here $\delta > 0$ is a negative power of V to ensure that these sets are canonically homoemorphic to the direct product of xN and the neighborhood $B_\delta^W \times B_\delta^{W'}$ of 0 in $W \times W'$. Combining the pointwise ergodic theorem for the characteristic function of these sets with polynomial divergence, we can bound the measure of the set of points for which the W_1-component is smaller than $V^{-\kappa_{19}}$. In fact, if κ_{19} is large enough the measure of this set is itself bounded by a negative power of V. Taking the union over $\ll \delta^{-\dim M}$ many $k \in M$ and all closed N-orbits of volume less than V and choosing κ_{19} large enough, this gives the lemma.

We let $\beta > 0$ be as in the effective ergodic theorem (Proposition 6.5). For $T > 0$ we define $B_N(T) = N \cap B_T^{\mathrm{HS}}$ and write m_N for the Haar measure on N.

PROPOSITION 6.8. *There exist constants $\kappa_{20} > 0$, $Q_0 \in (0, \beta)$, and $\delta > 0$ with the following properties for all $q \in (0, Q_0)$. Suppose μ is ε-almost invariant under $S = \mathrm{SO}^\circ(m, 1)$ for some $\varepsilon > 0$ and $m \in \{2, \dots, n\}$, set $T = \varepsilon^{-q}$, and suppose $x \in X_{\mathrm{cpct}}$ satisfies the following two conditions.*

1. *The set*
$$\mathcal{B}_x = \{h \in B_N(T) \mid xh \text{ has property } (\mathcal{G}')\}$$
has measure larger than $(1 - 10^{-6})v$, where $v = m_N(B_N(T))$ and $(\mathcal{G}')$ is defined below.
2. *There does not exist $x' \in X$ and $k \in M$ with $d(x'k, x) \le v^{-\kappa_{19}}$ such that $x'Nk$ is a closed orbit of volume $\le v$.*

Then there exist $b_1, b_2 \in \mathcal{B}_x$, $r' \in \mathfrak{r}$ so that $xb_1 = xb_2 \exp(r')$, and $T^{-\kappa_{20}} \ll \|r'\| \ll v^{-\frac{\delta}{2\dim(G)}}$. Moreover, the component $r_1' \in \mathfrak{r}_1$ of r' satisfies $\|r_1'\| \gg \|r'\|$.

In [9, proposition 14.1], Proposition 6.8 is formulated for S with the property

$$(\mathcal{G}) \qquad xh \text{ is } [T_0, \varepsilon^{-\beta}]\text{-generic},$$

whereas we are going to use it for N with the property

$$(\mathcal{G}') \qquad xhg_C \text{ is } [T_0, \varepsilon^{-\beta}]\text{-generic for some } g_C \in C_G(S).$$

We note that the proof of Proposition 6.8 in [9, section 14.2] remains essentially unchanged by this, so we will only outline the argument.

The proof of Proposition 6.8 relies on an effective closing lemma for actions of semisimple groups. The reader is referred to [9, section 13] for a statement and a proof of this closing lemma. Roughly speaking, the closing lemma says that if the orbit of a point returns very often very close to itself within a certain ball of the acting group, then the original point is in fact close to a closed orbit of small volume. The proof of the closing lemma relies on the arithmetic nature of the lattice to turn the close returns into actual returns for a nearby point and on techniques similar to the lattice point counting used in section 4 to estimate the volume of this orbit.

So how is this closing lemma relevant? Since we assume in Proposition 6.8 that our point is not close to a closed orbit of small volume, we know that the assumptions to the closing lemma cannot hold. This shows that the N-orbit cannot clump together too much. However, with this, a geometric version of a pigeonhole argument allows us to find the tuple of generic points as in the conclusion of Proposition 6.8. We refer to [9, section 14.2] for a more formal argument.

PROPOSITION 6.9 (Tuples of generic points). *Let $\zeta \in (0,1)$ and $d \geq 1$; then there exist ξ and d' (depending only on n, ζ, and d) and $\varepsilon_0 > 0$ with the following property: Suppose for some $\varepsilon \in (0, \varepsilon_0)$ that*

- *μ is ε-almost invariant under S with respect to $\mathcal{S}_d$, and*
- *$\mu(xNk) = 0$ for all closed N-orbits of volume $\leq \varepsilon^{-\zeta}$ and all $k \in M$.*

Then there exist $x_1, x_2 \in X$ so that $x_2 = x_1 g_C \exp(r)$, where $r \in \mathfrak{r}$, $g_C \in C_G(S)$, $\|r\| \leq \varepsilon^{\xi}$, and x_1, x_2 are both $[\|r_1\|^{-\kappa_{18}}, \|r_1\|^{-\kappa_{17}}]$-generic with respect to $\mathcal{S}_{d'}$.

Note that the constants κ_{18} and κ_{17} are coming from Proposition 6.6; below we will also use again $\kappa_{19} > 0$ from Lemma 6.7.

Proof. We again set $v = m_N(B_N(T))$ for $T = \varepsilon^{-q}$ and $q > 0$ to be chosen below. In order to apply Proposition 6.8, we start by establishing that there exists a point $x \in X_{\mathrm{cpct}}$ with the following two properties:

1. The set

$$\mathcal{B}_x = \{h \in B_N(T) \mid xhg_C \text{ is } [T_0, \varepsilon^{-\beta}]\text{-generic for some } g_C \in C_G(S)\}$$

   has measure larger than $(1 - 10^{-6})v$.
2. There does not exist $x' \in X$ and $k \in M$ with $d(x'k, x) \leq v^{-\kappa_{19}}$ such that $x'N$ is closed of volume $\leq v$.

Consider the set

$$E_1 = \{x \in X \mid \text{there exists } x' \in X \text{ and } k \in M \text{ with } d(x'k, x) \leq v^{-\kappa_{19}} \\ \text{such that } x'Nk \text{ is closed of volume } \leq v\},$$

and notice that by Lemma 6.7 we have $\mu(E_1) < 1/2$ whenever $\nu \in [V_0, \varepsilon^{-\zeta}]$ (which is the case if ε is small enough and we choose q small enough).

We claim that the fraction of points $(x, h) \in X \times B_N(T)$ (with respect to the measure $\mu \times m_N$) for which $xh \notin X_{\mathrm{cpct}}$ is $\leq \frac{1}{10^{10}}$ if ε is small enough. Indeed, recall that $X_{\mathrm{cpct}} = \mathfrak{S}(R_0)$ for some $R_0 \geq 0$ and that X_{cpct} is invariant under $\mathrm{SO}(n)$. Therefore, in order to prove the claim, it suffices to show that the fraction of points $(x, s) \in X \times B_S(T)$ (with respect to the measure $\mu \times m_S$) for which $xs \notin \mathfrak{S}(R_0)$ is $\leq \frac{1}{10^{10}}$ if T is big enough.

This argument is identical to the one in the proof of [9, proposition 14.1]. Consider a smooth function F on X so that

$$ {}_{X\setminus\mathfrak{S}(R_0)} \leq F \leq {}_{X\setminus\mathfrak{S}(R_0/2)}, $$

where we may assume that $\mathfrak{S}(R_0/2)$ satisfies similar estimates as $\mathfrak{S}(R_0)$ for all H-invariant and ergodic probability measures. Lemma 6.3 together with the fact that μ is ε-almost invariant under S with respect to $\mathcal{S}_d$ implies that

$$ \left| \int_X F(xs)\, \mathrm{d}\mu(x) - \int_X F\, \mathrm{d}\mu \right| \ll \varepsilon T^{d\kappa_{14}} \mathcal{S}_d(F) \text{ for all } s \in B_S(T) $$

and therefore

$$ \frac{1}{m_S(B_S(T))} \int_{s\in B_S(T), x\in X} F(xs)\, \mathrm{d}\mu(x)\, \mathrm{d}m_S(s) - \mu(X \setminus \mathfrak{S}(R_0/2)) $$
$$ \ll \varepsilon^{1-qd\kappa_{14}} \mathcal{S}_d(F). $$

Notice that F can be fixed depending on G and Γ only. Since we also have $1 - qd\kappa_{14} > 1/2$ if q is chosen sufficiently small, the claim follows for ε small enough.

For $x \in X$, consider the function

$$ f(x) = \frac{1}{\nu} m_N(\{h \in B_N(T) \mid xh \in X_{\mathrm{cpct}} \text{ and there exists } $$
$$ g_C \in C_G(S) \text{ so that } xhg_C \text{ is } [T_0, \varepsilon^{-\beta}]\text{-generic}\}). $$

Note that for every $h \in B_N(T)$ and $g_C \in C_G(S)$ we have $hg_C \in B_S(T)$. If $T_0 > 0$ is large enough, Proposition 6.5 together with Proposition 2.3 then implies that $\int (1 - f(x))d\mu \leq \frac{2}{10^{10}}$ and, therefore, the set $E_2 = \{x \in X \mid f(x) < 1 - 10^{-6}\}$ satisfies $\mu(E_2) < 1/10$. Define $X_{\mathrm{good}} = X \setminus (E_1 \cup E_2)$ (which is a set of positive measure) and note that any $x \in X_{\mathrm{good}}$ satisfies the properties at the beginning of the proof—that is, the set $\mathcal{B}_x$ has measure larger than $(1-10^{-6})v$ and there does not exist $x' \in X$ and $k \in M$ with $d(x, x'k) \le v^{-\kappa_{19}}$ such that $x'N$ is closed of volume $\le v$.

Proposition 6.8 now implies that there exist $b_1, b_2 \in B_N(T)$ and $c_1, c_2 \in C_G(S)$ so that xb_ic_i is $[T_0, \varepsilon^{-\beta}]$-generic for $i = 1, 2$ and we have

$$\begin{aligned} xb_1c_1 &= xb_2 \exp(r')c_1 \\ &= (xb_2c_2)c_2^{-1}c_1(c_1^{-1}\exp(r')c_1) \\ &= (xb_2c_2)g_C(\exp(r)), \end{aligned}$$

where $r \in \mathfrak{r}$ and $g_C \in C_G(S)$. Indeed, since $\mathfrak{r}$ is Ad_N-invariant and $c_1 \in N$, we have $c_1^{-1}\exp(r')c_1 = \exp(r)$ for some $r \in \mathfrak{r}$. Moreover, since $C_G(S)$ is compact, we still have the estimates

$$T^{-\kappa_{20}} \ll \|r\| \ll v^{-\frac{\delta}{2\dim(G)}} \text{ and } \|r_1\| \gg \|r\|.$$

If q is chosen sufficiently small (depending only on n) and $\varepsilon > 0$ is sufficiently small, we also have that

$$[\|r_1\|^{-\kappa_{18}}, \|r_1\|^{-\kappa_{17}}] \subseteq [T_0, \varepsilon^{-\beta}].$$

In fact for the inequality $\|r_1\|^{-\kappa_{17}} \le \varepsilon^{-\beta}$ we recall that by definition $T = \varepsilon^{-q}$, and so also, together with the above,

$$\|r_1\|^{-\kappa_{17}} \ll \|r\|^{-\kappa_{17}} \ll T^{\kappa_{20}\kappa_{17}} = \varepsilon^{-q\kappa_{20}\kappa_{17}}.$$

We now choose q so that $q\kappa_{20}\kappa_{17} < \beta$. For sufficiently small ε the implicit constants do not matter as we have a strict inequality in the exponents. □

6.6 MORE TOOLS FROM HOMOGENEOUS DYNAMICS. We cite two more results from [9] that we are going to use for the proof of Theorem 1.2. The first result can be found in [9, proposition 8.1]:

PROPOSITION 6.10. *Let $d > 0$ and suppose that μ is ε-almost invariant under a connected, intermediate subgroup $H \le S \le G$ and under $Z \in \mathfrak{r}$ with respect to $\mathcal{S}_d$, where $\|Z\| = 1$. Then there exists $\kappa_{21} > 0$ so that μ is also $\ll_d \varepsilon^{\kappa_{21}}$-almost invariant under some subgroup S' conjugated to $\mathrm{SO}(m', 1)$ with $m' > m$ and $H \subseteq S'$.*

Note that the proof as written in [9, section 8] establishes that under the assumption of Proposition 6.10, the measure μ is almost invariant under an intermediate Lie algebra $\mathfrak{s} \subset \mathfrak{s}' \subset \mathfrak{g}$ with $\dim(\mathfrak{s}') > \dim(\mathfrak{s})$ and containing $\mathfrak{h}$. Using the second assertion of [9, proposition 7.1], we can also deduce that $\mathfrak{s}$ and Z are *almost* contained in $\mathfrak{s}'$. By the discussion in section 3 of Lie subalgebras containing $\mathfrak{h}$, this implies that $\mathfrak{s}'$ contains a Lie subalgebra conjugated to $\mathfrak{so}(m', 1)$ for some $m' > m$.

In order to pass from the Lie algebra $\mathfrak{s}'$ to the Lie group S', we may apply the exponential map and argue as in Lemma 6.3, which completes the proof in our setting.

The next proposition tells us that a probability measure on an S-orbit $x_0 S$, which is *almost* invariant under S, is *close* to the invariant measure on $x_0 S$. This is [9, proposition 15.1].

PROPOSITION 6.11. *Let $\overline{x_0 S}$ be a closed orbit of volume V and suppose that μ is a probability measure on $\overline{x_0 S}$ that is ε-almost invariant under S with respect to a Sobolev norm $\mathcal{S}_d$. Let ν be the Haar probability measure on $\overline{x_0 S}$. Then there are $\kappa_{22}, \kappa_{23} > 0$ so that*

$$|\mu(f) - \nu(f)| \ll_d V^{\kappa_{22}} \varepsilon^{\kappa_{23}} \mathcal{S}_d(f) \text{ for all } f \in C_c^\infty(X).$$

In particular there exist $\kappa_{24}, \kappa_{25} > 0$ with the property that if $V \leq \varepsilon^{-\kappa_{24}}$, then

$$|\mu(f) - \nu(f)| \ll_d \varepsilon^{\kappa_{25}} \mathcal{S}_d(f) \text{ for all } f \in C_c^\infty(X).$$

Note that the proof of [9, proposition 15.1] is a simple application of spectral gap and does not make use of the triviality of the centralizer. It therefore remains unchanged in our setting.

6.7 PROOF OF THEOREM 1.2. We prove Theorem 1.2 by induction, using the following induction hypothesis: Let $S \cong \mathrm{SO}^\circ(m, 1)$ for some $2 \leq m \leq n$, $\varepsilon > 0$, and $d > 0$, and assume that μ is an H-invariant and H-ergodic measure on X which is ε-almost invariant under S with respect to the Sobolev norm $\mathcal{S}_d$. As before, let N denote the normalizer of S in G. The following lemma will use our results from sections 4 and 5.

LEMMA 6.12. *Let $d > 0$. Then there are constants $\kappa_{26}, \kappa_{27}, \kappa_{28} > 0$ and d' depending only on d so that for any sufficiently small $\varepsilon > 0$, the induction hypothesis implies that one of the following properties holds true:*

- $\overline{x_0 Sk}$ *is a closed orbit of volume* $\leq \varepsilon^{-2\kappa_1\kappa_{26}}$ *and*

$$\left|\mu(f) - \mu_{\overline{x_0 Sk}}(f)\right| \ll \varepsilon^{\kappa_{27}} \mathcal{S}_d(f) \text{ for all } f \in C_c^\infty(X),$$

or

- *the measure* μ *is* $\ll \varepsilon^{\kappa_{28}}$*-almost invariant under* $S' \leq G$ *with respect to* $\mathcal{S}_{d'}$*, where* $S' \cong \mathrm{SO}^\circ(m',1)$ *for some* $m < m' \leq n$ *and* $H \subseteq S'$.

Proof. Suppose first that $\mu(x_0 Nk) = 1$ for a closed orbit $x_0 Nk$ of volume $\leq \varepsilon^{-\kappa_{26}}$ and some $k \in M$, where the constant $\kappa_{26} > 0$ is chosen in a moment. Since S is conjugated to $\mathrm{SO}^\circ(m,1)$ by an element of the maximal compact subgroup of G by Corollary 3.2, we can apply Proposition 5.12 to see that the orbit $\overline{x_0 Sk}$ satisfies $\mathrm{vol}(\overline{x_0 S}) \ll \varepsilon^{-\kappa_1\kappa_{26}}$. Assuming that $\varepsilon > 0$ is small enough and worsening the exponent slightly, we can get rid of the implicit constant and may assume that the closed orbit $\overline{x_0 S}$ has $\mathrm{vol}(\overline{x_0 S}) \leq \varepsilon^{-2\kappa_1\kappa_{26}}$. Choosing κ_{26} small enough, we can apply Proposition 6.11 to obtain $\kappa_{27} > 0$ such that

$$\left|\mu(f) - \mu_{\overline{x_0 S}}(f)\right| \ll \varepsilon^{\kappa_{27}} \mathcal{S}_d(f) \text{ for all } f \in C_c^\infty(X).$$

Assume now that $\mu(xNk) = 0$ for all closed N-orbits of volume $\leq \varepsilon^{-\kappa_{26}}$ and $k \in M$. We apply Proposition 6.9 to produce $\xi > 0$ and $d' > d$, as well as $x_1, x_2 \in X$ with $x_2 = x_1 g_C \exp(r)$, where $r \in \mathfrak{r}$, $g_C \in C_S(G)$, $\|r\| \leq \varepsilon^{\xi}$, and x_1, x_2 are both $\left[\|r_1\|^{\kappa_{18}}, \|r_1\|^{\kappa_{17}}\right]$-generic with respect to $\mathcal{S}_{d'}$. Now apply Proposition 6.6 to see that μ is $\ll \varepsilon^{\min(1/2, \kappa_{16}\xi/2)}$-almost invariant with respect to $\mathcal{S}_{d'}$ under an element $Z \in \mathfrak{r}$ with $\|Z\| = 1$. Proposition 6.10 now implies that μ is $\ll \varepsilon^{\kappa_{21}\min(1/2, \kappa_{16}\xi/2)}$-almost invariant with respect to $\mathcal{S}_{d'}$ under some subgroup $S' \subseteq G$ conjugated to $\mathrm{SO}(m',1)$ with $m' > m$ and $H \subseteq S'$. Setting

$$\kappa_{28} = \kappa_{21} \min(1/2, \kappa_{16}\xi/2)$$

implies that μ is $\ll \varepsilon^{\kappa_{28}}$-almost invariant with respect to $\mathcal{S}_{d'}$ under $S' \cong \mathrm{SO}^\circ(m',1)$ with $m' > m$, and hence the result. □

Proof of Theorem 1.2. By assumption, μ is invariant under H. This means in particular that it is ε-almost invariant under H with respect to $\mathcal{S}_d$ for every $\varepsilon > 0$ and $d \geq d_0 + 1$. By Lemma 6.12, we get for ε small enough and with $\kappa_{29} = 2\kappa_1\kappa_{26}$ that either

- $\overline{x_0 H}$ is a closed orbit of volume $\leq \varepsilon^{-\kappa_{29}}$, or
- there exist constants $c_1 > 0$ and $\xi_1 > 0$ depending only on H and G so that μ is $c_1 \varepsilon^{\xi_1}$-almost invariant under a subgroup $S_1 \cong \mathrm{SO}^\circ(m, 1)$ with respect to $\mathcal{S}_{d_2}$, where $2 < m \leq n$.

In the second case, we have established the induction hypothesis with ε^{ξ_1} instead of ε and for ε small enough. Notice that we may iterate this process at most $n-2$ times until we arrive at a situation where

$$\left|\mu(f) - \mu_{\overline{x_0 S_j}}(f)\right| \ll \varepsilon^{\xi_j \kappa_{27}} \mathcal{S}_{d_j}(f) \text{ for all } f \in C_c^\infty(X),$$

where $S_j \cong \mathrm{SO}^\circ(m, 1)$ for some $2 \leq m \leq n$ and $\overline{x_0 S_j}$ is a closed orbit of volume $\ll \varepsilon^{-\xi_j \kappa_{29}}$. Here, the constant ξ_j and the Sobolev degree d_j only depend on H and G. Define

$$\Delta = \max_j \xi_j \kappa_{29}\,, \quad \delta = \min_j \xi_j \kappa_{27}\,, \quad d = \max_j d_j,$$

where the index j runs over all the stages occurring in the above process. Then, there exists $\varepsilon_0 > 0$ such that for all $\varepsilon < \varepsilon_0$ there exists an intermediate subgroup $S_j \cong \mathrm{SO}^\circ(m, 1)$ for some $2 \leq m \leq n$ so that $\overline{x_0 S_j}$ is a closed S_j-orbit of volume $\leq c_2 \varepsilon^{-\Delta}$ and

$$\left|\mu(f) - \mu_{\overline{x_0 S_j}}(f)\right| \leq c_3 \varepsilon^{\delta} \mathcal{S}_d(f) \text{ for all } f \in C_c^\infty(X)$$

for some constants $c_2, c_3 > 0$. Now choose ε so that $c_2 \varepsilon^{-\Delta} = V$. If V is large enough (i.e., $V \geq (c_3 \varepsilon^{\delta})^{-\Delta/\delta}$), then

$$\left|\mu(f) - \mu_{\overline{x_0 S}}(f)\right| \ll V^{-\delta/\Delta} \mathcal{S}_d(f) \text{ for all } f \in C_c^\infty(X),$$

where $S \cong \mathrm{SO}^\circ(m, 1)$ for some $2 \leq m \leq n$ and $\overline{x_0 S}$ is a closed orbit of volume $\leq V$. In order to get rid of the implicit constant, we worsen the exponent slightly and choose an appropriate $V_0 > 0$ so that for all $V \geq V_0$,

$$\left|\mu(f) - \mu_{\overline{x_0 S}}(f)\right| \leq V^{-2\delta/\Delta} \mathcal{S}_d(f) \text{ for all } f \in C_c^\infty(X),$$

which proves Theorem 1.2. □

References

[1] Armand Borel and Gopal Prasad. Values of isotropic quadratic forms at S-integral points. *Compositio Math.*, 83(3):347–372, 1992.

[2] M. Burger and P. Sarnak. Ramanujan duals. II. *Invent. Math.*, 106(1):1–11, 1991.

[3] Laurent Clozel. Démonstration de la conjecture τ. *Invent. Math.*, 151(2):297–328, 2003.

[4] S. G. Dani. On orbits of unipotent flows on homogeneous spaces. II. *Ergodic Theory Dynam. Systems*, 6(2):167–182, 1986.

[5] S. G. Dani and G. A. Margulis. Limit distributions of orbits of unipotent flows and values of quadratic forms. In *I. M. Gelfand Seminar*, volume 16 of *Advances in Soviet Mathematics*, pages 91–137. Amer. Math. Soc., Providence, RI, 1993.

[6] W. Duke, Z. Rudnick, and P. Sarnak. Density of integer points on affine homogeneous varieties. *Duke Math. J.*, 71(1):143–179, 1993.

[7] Manfred Einsiedler, Elon Lindenstrauss, Philippe Michel, and Akshay Venkatesh. Distribution of periodic torus orbits on homogeneous spaces. *Duke Math. J.*, 148(1):119–174, 2009.

[8] M. Einsiedler, G. Margulis, A. Mohammadi, and A. Venkatesh. Effective equidistribution and property (tau). *J. Amer. Math. Soc.*, 33(1):223–289, 2020.

[9] M. Einsiedler, G. Margulis, and A. Venkatesh. Effective equidistribution for closed orbits of semisimple groups on homogeneous spaces. *Inventiones mathematicae*, 177(1):137–212, 2009.

[10] M. Einsiedler and A. Mohammadi. Effective arguments in unipotent dynamics. Chapter 12 in this volume.

[11] Alex Eskin and Curt McMullen. Mixing, counting, and equidistribution in Lie groups. *Duke Math. J.*, 71(1):181–209, 1993.

[12] Alexander Gorodnik and Amos Nevo. Counting lattice points. *J. Reine Angew. Math.*, 663:127–176, 2012.

[13] Alexander Gorodnik, Hee Oh, and Nimish Shah. Integral points on symmetric varieties and Satake compactifications. *Amer. J. Math.*, 131(1):1–57, 2009.

[14] Joe Harris. *Algebraic geometry*, volume 133 of *Graduate Texts in Mathematics*. Springer-Verlag, New York, 1995. A first course, Corrected reprint of the 1992 original.

[15] Sigurdur Helgason. *Groups and geometric analysis*, volume 83 of *Mathematical Surveys and Monographs*. Amer. Math. Soc., Providence, RI, 2000. Integral geometry, invariant differential operators, and spherical functions, Corrected reprint of the 1984 original.

[16] G. A. Margulis. On the action of unipotent groups in the space of lattices. In *Lie groups and their representations (Proc. Summer School, Bolyai, János Math. Soc., Budapest, 1971)*, pages 365–370. Halsted, New York, 1975.

[17] G. A Margulis. Indefinite quadratic forms and unipotent flows on homogeneous spaces. *Dynamical systems and ergodic theory (Warsaw, 1986)*, 23:399–409, 1989.

[18] François Maucourant. Homogeneous asymptotic limits of Haar measures of semisimple linear groups and their lattices. *Duke Math. J.*, 136(2):357–399, 2007.

[19] S. Mozes and N. Shah. On the space of ergodic invariant measures of unipotent flows. *Ergodic Theory and Dynamical Systems*, 15:149–159, 1995.

[20] M. Ram Murty and Purusottam Rath. *Transcendental numbers*. Springer, New York, 2014.

[21] Hee Oh. Uniform pointwise bounds for matrix coefficients of unitary representations and applications to Kazhdan constants. *Duke Math. J.*, 113(1):133–192, 2002.

[22] Daniel Perrin. *Algebraic geometry*. Universitext. Springer-Verlag London, Ltd., London; EDP Sciences, Les Ulis, 2008. An introduction, Translated from the 1995 French original by Catriona Maclean.

[23] M. Ratner. On Raghunathan's measure conjecture. *Annals of Mathematics*, 134(3):545–607, 1991.

[24] Marina Ratner. Raghunathan's topological conjecture and distributions of unipotent flows. *Duke Math. J.*, 63(1):235–280, 1991.

[25] Igor R. Shafarevich. *Basic algebraic geometry. 1*. Springer-Verlag, Berlin, second edition, 1994. Varieties in projective space, Translated from the 1988 Russian edition and with notes by Miles Reid.

14

HEE OH

DYNAMICS FOR DISCRETE SUBGROUPS OF $SL_2(\mathbb{C})$

Dedicated to Gregory Margulis with affection and admiration

Abstract. Margulis wrote in the preface of his book *Discrete Subgroups of Semisimple Lie Groups* [29]: "A number of important topics have been omitted. The most significant of these is the theory of Kleinian groups and Thurston's theory of 3-dimensional manifolds: these two theories can be united under the common title *Theory of discrete subgroups of* $SL_2(\mathbb{C})$."

In this essay, we will discuss a few recent advances regarding this missing topic from his book, which were influenced by his earlier works.

1 Introduction

A discrete subgroup of $PSL_2(\mathbb{C})$ is called a Kleinian group. In this essay, we discuss dynamics of unipotent flows on the homogeneous space $\Gamma\backslash PSL_2(\mathbb{C})$ for a Kleinian group Γ that is not necessarily a lattice of $PSL_2(\mathbb{C})$. Unlike the lattice case, the geometry and topology of the associated hyperbolic 3-manifold $M=\Gamma\backslash\mathbb{H}^3$ influence both topological and measure theoretic rigidity properties of unipotent flows.

Around 1984–1986, Margulis settled the Oppenheim conjecture by proving that every bounded $SO(2,1)$-orbit in the space $SL_3(\mathbb{Z})\backslash SL_3(\mathbb{R})$ is compact [27, 28]. His proof was topological, using minimal sets and the polynomial divergence property of unipotent flows. With Dani [11, 12], he also gave a classification of orbit closures for a certain family of one-parameter unipotent subgroups of $SL_3(\mathbb{R})$. On the basis of Margulis's topological approach, Shah [49] obtained a classification of orbit closures for the action of any connected closed subgroup generated by unipotent elements in the space $\Gamma\backslash PSL_2(\mathbb{C})$ when Γ is a lattice. This result in a much greater generality, as conjectured by Raghunathan, was proved by Ratner using her measure rigidity theorem [43, 44].

Hee Oh. Department of Mathematics. Yale University, New Haven, CT 06511 and Korea Institute for Advanced Study, Seoul, Korea
hee.oh@yale.edu

Supported in part by NSF Grant #1900101.

The relation between invariant measures and orbit closures for unipotent flows is not as tight in the infinite volume case as it is in the finite volume case. Meanwhile, the topological approach in the orbit closure classification can be extended to the class of rigid acylindrical hyperbolic 3-manifolds, yielding the complete classification of orbit closures for the action of any connected closed subgroup generated by unipotent elements. This was done jointly with McMullen and Mohammadi [36, 37]. Much of this essay is devoted to explaining these results, although we present slightly different viewpoints in certain parts of the proof. Remarkably, this approach can handle the entire quasi-isometry class of rigid acylindrical hyperbolic 3-manifolds, as far as the action of the subgroup $PSL_2(\mathbb{R})$ is concerned [38]. An immediate geometric consequence is that for any convex cocompact acylindrical hyperbolic 3-manifold M, any geodesic plane is either closed or dense inside the interior of the convex core of M, thereby producing the first continuous family of locally symmetric manifolds for which such a strong rigidity theorem for geodesic planes holds. This result extends to geometrically finite acylindrical hyperbolic 3-manifolds as shown in joint work with Benoist [4]. We also present a continuous family of quasi-Fuchsian 3-manifolds containing geodesic planes with wild closures [38], which indicates the influence of the topology of the associated 3-manifold in the rigidity problem at hand.

We call a higher dimensional analogue of a rigid acylindrical hyperbolic 3-manifold a convex cocompact hyperbolic d-manifold with Fuchsian ends, following Kerckhoff and Storm [21]. For these manifolds $\Gamma\backslash\mathbb{H}^d$, in joint work with Lee [22], we have established a complete classification of orbit closures in $\Gamma\backslash SO^\circ(d,1)$ for the action of any connected closed subgroup of $SO^\circ(d,1)$ generated by unipotent elements. The possibility of accumulation on closed orbits of intermediate subgroups presents new challenges, and the avoidance theorem and the induction arguments involving equidistribution statements are major new ingredients in higher dimensional cases (Theorems 9.10 and 9.11). We note that these manifolds do not admit any nontrivial local deformations for $d \geq 4$ [21].

ACKNOWLEDGMENT. This survey is mostly based on the papers [37], [38], [36], [4], and [22]. I am grateful to my coauthors Curt McMullen, Amir Mohammadi, Yves Benoist, and Minju Lee. I would like to thank Yair Minsky and Amir Mohammadi for helpful comments on the preliminary version of this essay.

2 Kleinian groups

We give a brief introduction to Kleinian groups, including some basic notions and examples. General references for this section include [43], [26], [24], [33], [47], [17], and [8]. In particular, all theorems stated in this section with no references attached can be found in [26] and [33].

We will use the upper half-space model for hyperbolic 3 space:

$$\mathbb{H}^3 = \{(x_1, x_2, y) : y > 0\}, \quad ds = \frac{\sqrt{dx_1^2 + dx_2^2 + dy^2}}{y}.$$

In this model of $\mathbb{H}^3$, a geodesic is either a vertical line or a vertical semi-circle. The geometric boundary of $\mathbb{H}^3$ is given by the Riemann sphere $\mathbb{S}^2 = \hat{\mathbb{C}}$, when we identify the plane $(x_1, x_2, 0)$ with the complex plane $\mathbb{C}$.

The group $G := \mathrm{PSL}_2(\mathbb{C})$ acts on $\hat{\mathbb{C}}$ by Möbius transformations:

$$\begin{pmatrix} a & b \\ c & d \end{pmatrix} z = \frac{az+b}{cz+d} \quad \text{with } a, b, c, d \in \mathbb{C} \text{ such that } ad - bc = 1.$$

This action of G extends to an isometric action on $\mathbb{H}^3$ as follows: each $g \in G$ can be expressed as a composition $\mathrm{Inv}_{C_1} \circ \cdots \circ \mathrm{Inv}_{C_k}$, where Inv_C denotes the inversion with respect to a circle $C \subset \hat{\mathbb{C}}$.[1] If we set $\Phi(g) = \mathrm{Inv}_{\hat{C}_1} \circ \cdots \circ \mathrm{Inv}_{\hat{C}_k}$, where $\mathrm{Inv}_{\hat{C}}$ denotes the inversion with respect to the sphere $\hat{C}$ in $\mathbb{R}^3$, that is orthogonal to $\mathbb{C}$ and $\hat{C} \cap \mathbb{C} = C$, then $\Phi(g)$ preserves $(\mathbb{H}^3, ds)$. Moreover, the Poincaré extension theorem says that Φ is an isomorphism between the two real Lie groups

$$\mathrm{PSL}_2(\mathbb{C}) = \mathrm{Isom}^+(\mathbb{H}^3),$$

where $\mathrm{PSL}_2(\mathbb{C})$ is regarded as a six-dimensional real Lie group and $\mathrm{Isom}^+(\mathbb{H}^3)$ denotes the group of all orientation preserving isometries of $\mathbb{H}^3$.

DEFINITION 2.1. *A discrete subgroup Γ of G is called a Kleinian group.*

For a (respectively torsion-free) Kleinian group Γ, the quotient $\Gamma\backslash\mathbb{H}^3$ is a hyperbolic orbifold (respectively manifold). Conversely, any complete hyperbolic 3-manifold M can be presented as a quotient

$$M = \Gamma\backslash\mathbb{H}^3$$

[1] If $C = \{z : |z - z_0| = r\}$, then $\mathrm{Inv}_C(z)$ is the unique point on the ray $\{tz : t \geq 0\}$, satisfying the equation $|z - z_0| \cdot |\mathrm{Inv}_C(z) - z_0| = r^2$ for all $z \neq z_0$, and $\mathrm{Inv}_C(z_0) = \infty$.

for a torsion-free Kleinian group Γ. The study of hyperbolic manifolds is therefore directly related to the study of Kleinian groups.

Throughout the remainder of the essay, we assume that a Kleinian group Γ is nonelementary—that is, Γ does not contain an abelian subgroup of finite index. By Selberg's lemma, every Kleinian group has a torsion-free subgroup of finite index. We will henceforth treat the torsion-free condition loosely.

2.1 LATTICES. The most well-studied Kleinian groups are lattices of G: a Kleinian group $\Gamma < G$ is a lattice if $M = \Gamma\backslash\mathbb{H}^3$ has finite volume. When M is compact, Γ is called a uniform or cocompact lattice. If $d > 0$ is a square-free integer, then $\mathrm{PSL}_2(\mathbb{Z}[\sqrt{-d}])$ is a nonuniform lattice of G. More lattices, including uniform ones, can be constructed by number theoretic methods using the Lie group isomorphism $G \simeq \mathrm{SO}^\circ(3,1)$.

Let $Q(x_1, x_2, x_3, x_4)$ be a quadratic form with coefficients over a totally real number field k of degree n such that Q has signature $(3,1)$ and for any non-trivial embedding $\sigma : k \to \mathbb{R}$, Q^σ has signature $(4,0)$ or $(0,4)$; the orthogonal group $\mathrm{SO}(Q^\sigma)$ is thus compact.

Then for $G = \mathrm{SO}^\circ(Q)$ and for the ring $\mathfrak{o}$ of integers of k, the subgroup

$$\Gamma := G \cap \mathrm{SL}_4(\mathfrak{o}) \tag{2.1}$$

is a lattice in G by a theorem of Borel and Harish-Chandra [6]. Moreover, if Q does not represent 0 over k (which is always the case if the degree of k is bigger than 1), then Γ is a uniform lattice in G by the Godement's criterion. These examples contain all *arithmetic* lattices (up to a commensurability) that contain cocompact Fuchsian subgroups—that is, uniform lattices of $\mathrm{SO}^\circ(2,1) \simeq \mathrm{PSL}_2(\mathbb{R})$ [24].

Take two arithmetic noncommensurable hyperbolic 3-manifolds N_1 and N_2 that share a common properly imbedded closed geodesic surface S, up to an isometry. We cut each N_i along S, which results in one or two connected components. Let M_i be the metric completion of a component of $N_i - S$, which has geodesic boundary isometric to one or two copies of S. We now glue one or two copies of M_1 and M_2 together along their geodesic boundary and get a connected finite volume hyperbolic 3-manifold with no boundary. The resulting 3-manifold is a nonarithmetic hyperbolic 3-manifold, and its fundamental group is an example of the so-called hybrid lattices constructed by Gromov and Piatetski-Schapiro [16].

The Mostow rigidity theorem says that any two isomorphic lattices of G are conjugate to each other. Since a lattice is finitely presented, it follows that a

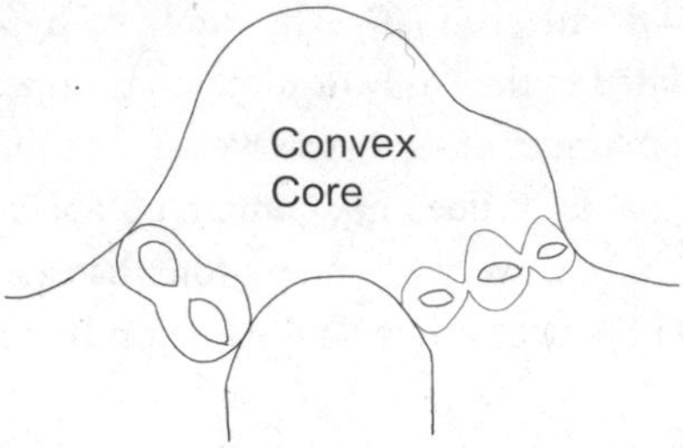

Figure 14.1. Convex core

conjugacy class of a lattice is determined by its presentation. Hence, despite the presence of nonarithmetic lattices in G, there are only countably many lattices of G up to conjugation, or equivalently, there are only countably many hyperbolic manifolds of finite volume up to isometry.

2.2 FINITELY GENERATED KLEINIAN GROUPS. We will mostly focus on *finitely generated* Kleinian groups. When studying a finitely generated Kleinian group Γ, the associated limit set and the convex core play fundamental roles.

Using the Möbius transformation action of Γ on $\mathbb{S}^2$, we define the following:

DEFINITION 2.2. *The limit set $\Lambda \subset \mathbb{S}^2$ of Γ is the set of all accumulation points of $\Gamma(z)$ for $z \in \mathbb{H}^3 \cup \mathbb{S}^2$.*

This definition is independent of the choice of $z \in \mathbb{H}^3 \cup \mathbb{S}^2$, and Λ is a minimal Γ-invariant closed subset of $\mathbb{S}^2$.

DEFINITION 2.3. *The convex core of M is the convex submanifold of M given by*

$$\text{core}\, M := \Gamma \backslash \text{hull}\, \Lambda \subset M = \Gamma \backslash \mathbb{H}^3$$

where $\text{hull}\, \Lambda \subset \mathbb{H}^3$ *is the smallest convex subset containing all geodesics connecting two points in Λ.*

If $\text{Vol}(M) < \infty$, then $\Lambda = \mathbb{S}^2$ and hence M is equal to its convex core.

DEFINITION 2.4.

(1) *A Kleinian group Γ is called geometrically finite if the unit neighborhood of* core M *has finite volume.*

(2) A Kleinian group Γ is called convex cocompact if core M *is compact or, equivalently, if Γ is geometrically finite without any parabolic elements.*

An element $g \in G$ is either hyperbolic (if it is conjugate to a diagonal element whose entries have modulus not equal to 1), elliptic (if it is conjugate to a diagonal element whose entries have modulus 1), or parabolic (if it is conjugate to a strictly upper triangular matrix). By discreteness, an element of a torsion-free Kleinian group is either hyperbolic or parabolic.

Geometrically finite (respectively convex cocompact) Kleinian groups are natural generalizations of (respectively cocompact) lattices of G. Moreover, the convex core of a geometrically finite hyperbolic manifold admits a thick-thin decomposition: there exists a constant $\varepsilon > 0$ such that core M is the union of a compact subset of injectivity radius at least $\varepsilon > 0$ and finitely many cusps. In the class of geometrically finite groups, lattices are characterized by the property that their limit sets are the whole of $\mathbb{S}^2$, and the limit sets of other geometrically finite groups have Hausdorff dimension strictly smaller than 2 [52, 53].

The group $G = PSL_2(\mathbb{C})$ can be considered as a real algebraic subgroup—more precisely, the group of real points of an algebraic group $\mathbf{G}$ defined over $\mathbb{R}$. A subset $S \subset G$ is called *Zariski dense* if S is not contained in any proper real algebraic subset of G. The Zariski density of a Kleinian group Γ in G is equivalent to the property that its limit set Λ is not contained in any circle of $\mathbb{S}^2$. When Λ is contained in a circle, Γ is conjugate to a discrete subgroup of $PSL_2(\mathbb{R})$; such Kleinian groups are referred to as *Fuchsian* groups. Geometrically finite Kleinian groups are always finitely generated, but the converse is not true in general; see section 2.6.

2.3 EXAMPLES OF GEOMETRICALLY FINITE GROUPS.

Below we give examples of three different kinds of geometrically finite groups that are relevant to subsequent discussion. Their limit sets are respectively totally disconnected, Jordan curves, and Sierpinski carpets. We note that a geometrically finite nonlattice Zariski dense Kleinian group Γ is determined by its limit set Λ up to commensurability; more precisely, Γ is a subgroup of finite index in the *discrete* subgroup $\mathrm{Stab}(\Lambda) = \{g \in G : g(\Lambda) = \Lambda\}$.

2.3.1 Schottky groups

The simplest examples of geometrically finite groups are Schottky groups. A subgroup $\Gamma < G$ is called (classical) Schottky if Γ is generated by hyperbolic elements $g_1, \cdots, g_k \in G$, $k \geq 2$, satisfying that there exist mutually disjoint closed round disks $B_1, \cdots, B_k$ and $B'_1, \cdots, B'_k$ in $\mathbb{S}^2$ such that each g_i maps the exterior of B_i onto the interior of B'_i.

If $g_1, \cdots, g_k$ are hyperbolic elements of G whose fixed points in $\mathbb{S}^2$ are mutually disjoint, then $g_1^N, \cdots, g_k^N$ generate a Schottky group for all N large enough. A Schottky group Γ is discrete and free; the common exterior of the hemispheres bounded by B_i, B'_i is a fundamental domain F of Γ. Since the limit set of Γ, which is totally disconnected, is contained in the union of interiors of B_i and B_i's, it is easy to see that the intersection of the hull of Λ and the fundamental domain F is a bounded subset of F. Hence Γ is a convex cocompact subgroup. Its convex core is the handlebody of genus k; in particular, the boundary of core M is a closed surface of genus k.

Any Kleinian group Γ contains a Schottky subgroup that has the same Zariski closure. If Γ is Zariski dense, take any two hyperbolic elements γ_1 and γ_2 of Γ with disjoint sets of fixed points. Suppose that all of four fixed points lie in a circle—say, $C \subset \mathbb{S}^2$; note that C is uniquely determined. Since the set of fixed points of hyperbolic elements of Γ forms a dense subset of Λ, there exists a hyperbolic element $\gamma_3 \in \Gamma$ whose fixed points are not contained in C. Now, for any $N \geq 1$, the subgroup generated by $\gamma_1^N, \gamma_2^N, \gamma_3^N$ is Zariski dense, as its limit set cannot be contained in a circle. By taking N large enough, we get a Zariski dense Schottky subgroup of Γ. This in particular implies that any Kleinian group contains a convex cocompact subgroup that is as large as itself in the algebraic sense.

2.3.2 Fuchsian groups and deformations: Quasi-Fuchsian groups

An orientation preserving homeomorphism $f : \mathbb{S}^2 \to \mathbb{S}^2$ is called κ-quasiconformal if for any $x \in \mathbb{S}^2$,

$$\limsup_{r \to 0} \frac{\sup\{|f(y) - f(x)| : |y - x| = r\}}{\inf\{|f(y) - f(x)| : |y - x| = r\}} \leq \kappa.$$

The 1-quasiconformal maps are precisely conformal maps [26, section 2]. The group $G = \mathrm{PSL}_2(\mathbb{C})$ is precisely the group of all conformal automorphisms of $\mathbb{S}^2$.

A Kleinian group Γ is called *quasi-Fuchsian* if it is a quasiconformal deformation of a (Fuchsian) lattice of $\mathrm{PSL}_2(\mathbb{R})$—that is, there exists a quasiconformal map f and a lattice $\Delta < \mathrm{PSL}_2(\mathbb{R})$ such that $\Gamma = \{f \circ \delta \circ f^{-1} : \delta \in \Delta\}$. Any quasiconformal deformation of a geometrically finite group is known to be geometrically finite; so a quasi-Fuchsian group is geometrically finite.

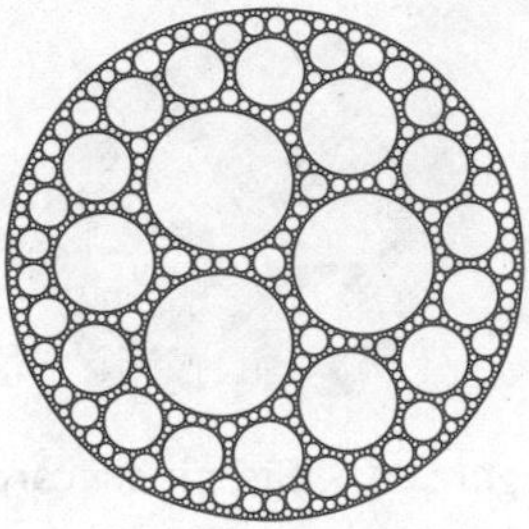

Figure 14.2. Limit set of a rigid acylindrical group (McMullen)

A quasi-Fuchsian group is also characterized as a finitely generated Kleinian group whose limit set Λ is a Jordan curve and which preserves each component of $\mathbb{S}^2 - \Lambda$. If $\Omega_\pm$ are components of $\mathbb{S}^2 - \Lambda$, then $S_\pm := \Gamma\backslash\Omega_\pm$ admits a hyperbolic structure by the uniformization theorem, and the product $\text{Teich}(S_+) \times \text{Teich}(S_-)$ of Teichmüller spaces gives a parameterization of all quasi-Fuchsian groups that are quasiconformal deformations of a fixed lattice of $\text{PSL}_2(\mathbb{R})$.

2.3.3 *Rigid acylindrical groups and their deformations*

A Kleinian group $\Gamma < G$ is called *rigid acylindrical* if the convex core of the associated hyperbolic manifold $M = \Gamma\backslash\mathbb{H}^3$ is a compact manifold with nonempty interior and with totally geodesic boundary. If core M has empty boundary, then M is compact and hence Γ is a uniform lattice. Rigid acylindrical nonlattice groups are characterized as convex cocompact Kleinian groups whose limit set satisfies that

$$\mathbb{S}^2 - \Lambda = \bigcup B_i,$$

where B_i's are round disks with mutually disjoint closures.

If M is a rigid acylindrical hyperbolic 3-manifold of infinite volume, then the double of core M is a closed hyperbolic 3-manifold; hence any rigid acylindrical group is a subgroup of a uniform lattice of G that contains a cocompact Fuchsian lattice $\pi_1(S)$ for a component S of ∂ core M. Conversely, if Γ_0 is a torsion-free uniform lattice of G such that $\Delta := \Gamma_0 \cap \text{PSL}_2(\mathbb{R})$ is a uniform lattice in $\text{PSL}_2(\mathbb{R})$, then $M_0 = \Gamma_0\backslash\mathbb{H}^3$ is a closed hyperbolic 3-manifold that contains a properly immersed totally geodesic surface $\Delta\backslash\mathbb{H}^2$. By passing to a finite cover of M_0, M_0 contains a properly embedded totally geodesic surface—say, S [24, theorem 5.3.4]. Now the metric completion of a component of $M_0 - S$ is a compact hyperbolic 3-manifold with totally geodesic boundary, and

Figure 14.3. Sierpinski carpet

its fundamental group, which injects to $\Gamma_0 = \pi_1(M_0)$, is a rigid acylindrical Kleinian group.

Rigid acylindrical Kleinian groups admit a huge deformation space comprised of convex cocompact acylindrical groups. We begin with the notion of acylindricality for a compact 3-manifold. Let D^2 denote a closed 2-disk and let $C^2 = S^1 \times [0,1]$ be a cylinder. A compact 3-manifold N is called *acylindrical*

(1) if ∂N is incompressible—that is, any continuous map $f:(D^2, \partial D^2) \to (N, \partial N)$ can be deformed into ∂N—or, equivalently, if the inclusion $\pi_1(S) \to \pi_1(N)$ is injective for any component S of ∂N; and

(2) if any essential cylinder of N is boundary parallel—that is, any continuous map $f:(C^2, \partial C^2) \to (N, \partial N)$, injective on π_1, can be deformed into ∂N.

A convex cocompact hyperbolic 3-manifold M is called *acylindrical* if its convex core is acylindrical. When M has infinite volume, it is also described by the property that its limit set is a Sierpinski carpet: $\mathbb{S}^2 - \Lambda = \bigcup B_i$ is a dense union of Jordan disks B_i's with mutually disjoint closures and with $\operatorname{diam}(B_i) \to 0$. By Whyburn [55], all Sierpinski carpets are known to be homeomorphic to each other. We refer to [57] for a beautiful picture of the limit set of a convex cocompact (nonrigid) acylindrical group.

Any convex cocompact acylindrical Kleinian group Γ is a quasiconformal deformation of a unique rigid acylindrical Kleinian group Γ_0, and its quasiconformal class is parameterized by the product $\prod_i \operatorname{Teich}(S_i)$ where S_i's are components of $\partial\operatorname{core}(\Gamma_0\backslash\mathbb{H}^3)$ [54, 35]. In terms of a manifold, any convex cocompact acylindrical hyperbolic 3-manifold is quasi-isometric to a unique *rigid* acylindrical hyperbolic 3-manifold M, and its quasi-isometry class is parameterized by $\prod_i \operatorname{Teich}(S_i)$.

The definition of acylindricality can be extended to geometrically finite groups with cusps using the notion of a compact core. If M is a hyperbolic 3-manifold with finitely generated $\pi_1(M)$, then there exists a compact connected submanifold $C \subset M$ (with boundary) such that the inclusion $C \subset M$ induces an isomorphism $\pi_1(C) \simeq \pi_1(M)$; such C exists uniquely, up to homeomorphism, and is called the compact core of M. Now a geometrically finite hyperbolic 3-manifold M is called acylindrical if its compact core is an acylindrical compact 3-manifold.

2.4 THURSTON'S GEOMETRIZATION THEOREM.

The complement $\Omega := \mathbb{S}^2 - \Lambda$ is called the set of discontinuity. Let Γ be a finitely generated Kleinian group. Ahlfors finiteness theorem says that $\Gamma\backslash\Omega$ is a union of finitely many closed Riemann surfaces with at most a finite number of punctures. The Kleinian manifold associated with Γ is defined by *adding* $\Gamma\backslash\Omega$ to $\Gamma\backslash\mathbb{H}^3$ on the conformal boundary at infinity:

$$\mathcal{M}(\Gamma) = \Gamma\backslash\mathbb{H}^3 \cup \Omega, \quad \partial\mathcal{M}(\Gamma) = \Gamma\backslash\Omega.$$

The convex cocompactness of Γ is equivalent to the compactness of $\mathcal{M}(\Gamma)$. If Γ is geometrically finite with cusps, then $\mathcal{M}(\Gamma)$ is compact except possibly for a finite number of rank-1 and rank-2 cusps. We denote by $\mathcal{M}_0(\Gamma)$ the compact submanifold of $\mathcal{M}(\Gamma)$ obtained by removing the interiors of solid pairing tubes corresponding to rank-1 cusps and solid cusp tori corresponding to rank-2 cusps (cf. [26]).

The following is a special case of Thurston's geometrization theorem under the extra nonempty boundary condition (cf. [20]):

THEOREM 2.5.
Let N be a compact irreducible[2] *orientable atoroidal*[3] *3-manifold with nonempty boundary. Then N is homeomorphic to $\mathcal{M}_0(\Gamma)$ for some geometrically finite Kleinian group Γ.*

We remark that if ∂N is incompressible and N does not have any essential cylinders, then Γ is a geometrically finite acylindrical group.

By applying Thurston's theorem to the compact core of $\Gamma\backslash\mathbb{H}^3$, we deduce that every finitely generated Kleinian group Γ is isomorphic to a geometrically finite group.

2.5 DENSITY OF GEOMETRICALLY FINITE GROUPS.

The density conjecture of Bers, Sullivan, and Thurston says that most Kleinian groups are geometrically finite. This is now a theorem whose proof combines the work

---

[2] Every 2 sphere bounds a ball.

[3] Any $\mathbb{Z}^2$ subgroup comes from boundary tori.

of many authors with the proof in full generality due to Namazi-Souto and Ohshika (we refer to [26, section 5.9] for more details and background).

THEOREM 2.6 (density theorem).
The class of geometrically finite Kleinian groups is open and dense in the space of all finitely generated Kleinian groups.

In order to explain the topology used in the above theorem, let Γ be a finitely generated Kleinian group. By Thurston's geometrization theorem, there exists a geometrically finite Kleinian group Γ_0 and an isomorphism $\rho : \Gamma_0 \to \Gamma$. In fact, a more refined version gives that ρ is type preserving—that is, ρ maps a parabolic element to a parabolic element. Fix a finite generating set $\gamma_1, \cdots, \gamma_k$ of Γ_0. The density theorem says there exists a sequence of geometrically finite groups $\Gamma_n < G$ and isomorphisms $\rho_n : \Gamma_0 \to \Gamma_n$ such that ρ_n converges to ρ as $n \to \infty$, in the sense that $\rho(\gamma_i) = \lim_n \rho_n(\gamma_i)$ for each $i = 1, \cdots, k$.

Here is an alternative way to describe the density theorem: fix a geometrically finite Kleinian group Γ with a fixed set of generators $\gamma_1, \cdots, \gamma_k$ and relations $\omega_1, \cdots, \omega_r$. Define

$$\mathfrak{R}(\Gamma) := \{\rho : \Gamma \to G \text{ homomorphism}\}/\sim$$

with the equivalence relation given by conjugation by elements of G. The set $\mathfrak{R}(\Gamma)$ can be identified with the algebraic variety $\{(g_1, \cdots, g_k) \in G \times \cdots \times G : \omega_i(g_1, \cdots, g_k) = e \text{ for } 1 \le i \le r\}/\sim$, where $\sim$ is given by conjugation by an element of G under the diagonal embedding. This defines a topology on $\mathfrak{R}(\Gamma)$ called the algebraic convergence topology.

The discrete locus is then defined by the subcollection of discrete and faithful representations:

$$AH(\Gamma) := \{\rho \in \mathfrak{R}(\Gamma) : \text{ type preserving isomorphism to a Kleinian group}\}.$$

Then $AH(\Gamma)$ is a closed subset that parameterizes hyperbolic structures on $\Gamma \backslash \mathbb{H}^3$. The interior of $AH(\Gamma)$ consists of geometrically finite Kleinian groups, and the density theorem says that

$$\overline{\mathrm{Int} AH(\Gamma)} = AH(\Gamma).$$

When Γ is a lattice in G, $AH(\Gamma)$ is a single point by the Mostow rigidity theorem. For all other geometrically finite Kleinian groups, $AH(\Gamma)$ is *huge*; the quasiconformal deformation space of Γ given by

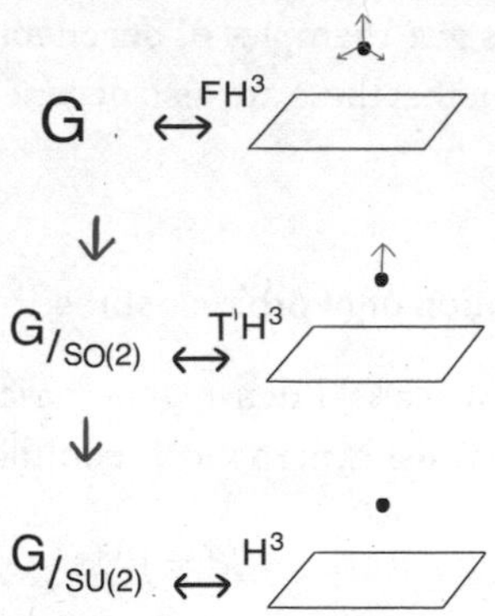

Figure 14.4. Frame bundle of $\mathbb{H}^3$

$$\mathfrak{T}(\Gamma) = \{\rho \in AH(\Gamma) : \rho \text{ is induced by a quasiconformal deformation of } \Gamma\}$$

is a connected component of the interior of $AH(\Gamma)$ and is a complex analytic manifold of dimension same as the dimension of Teich$(\Gamma\backslash\Omega)$—that is, $\sum_{i=1}^{m}(3g_i + n_i - 3)$, where g_i is the genus of the ith component of $\Gamma\backslash\Omega = \partial\mathcal{M}(\Gamma)$ and n_i is the number of its punctures [26, theorem 5.13]. Moreover, when Γ is rigid acylindrical, the interior of $AH(\Gamma)$, modulo the orientation (in other words, modulo the conjugation by elements of Isom$(\mathbb{H}^3)$ rather than by elements of $G = \text{Isom}^+(\mathbb{H}^3)$), is connected and hence equal to $\mathfrak{T}(\Gamma)$; this can be deduced from [9], as explained to us by Minsky. Therefore Int $AH(\Gamma)/\pm = \mathfrak{T}(\Gamma) = \text{Teich}(\Gamma\backslash\Omega)$.

2.6 EXAMPLES OF GEOMETRICALLY INFINITE GROUPS. Not every finitely generated Kleinian group is geometrically finite. An important class of finitely generated geometrically infinite Kleinian groups is given by the fundamental groups of $\mathbb{Z}$-covers of closed hyperbolic 3-manifolds. The virtual fibering theorem, proved by Agol, building on the previous work of Wise, says that every closed hyperbolic 3-manifold is a surface bundle over a circle, after passing to a finite cover [26, section 6.4]. This implies that, up to passing to a subgroup of finite index, any uniform lattice Γ_0 of G contains a normal subgroup Δ such that $\Gamma_0/\Delta \simeq \mathbb{Z}$ and Δ is a surface subgroup—that is, isomorphic to the fundamental group of a closed hyperbolic surface. Note that Δ is finitely generated (being a surface subgroup) but geometrically infinite as no normal subgroup of a geometrically finite group of infinite index is geometrically finite. In fact, any finitely generated geometrically infinite subgroup of a uniform lattice of G arises in this way, passing up to a subgroup of finite index (cf. [8]). These manifolds give examples of degenerate hyperbolic 3-manifolds with $\Lambda = \mathbb{S}^2$. We mention that there are also degenerate hyperbolic manifolds with $\Lambda \neq \mathbb{S}^2$.

3 Mixing and classification of *N*-orbit closures

Let $\Gamma < G = \mathrm{PSL}_2(\mathbb{C})$ be a Zariski dense geometrically finite Kleinian group and $M := \Gamma\backslash\mathbb{H}^3$ the associated hyperbolic 3-manifold. We denote by

$$\pi : \mathbb{H}^3 \to M = \Gamma\backslash\mathbb{H}^3$$

the quotient map.

We fix $o \in \mathbb{H}^3$ and a unit tangent vector $v_o \in \mathrm{T}_o(\mathbb{H}^3)$ so that $K = \mathrm{SU}(2)$ and $M_0 = \{\mathrm{diag}(e^{i\theta}, e^{-i\theta}) : \theta \in \mathbb{R}\}$ are respectively the stabilizer subgroups of o and v_o. The action of G on $\mathbb{H}^3$ induces identifications $G/K \simeq \mathbb{H}^3$, $G/M_0 \simeq \mathrm{T}^1(\mathbb{H}^3)$, and $G \simeq \mathrm{F}(\mathbb{H}^3)$, where $\mathrm{T}^1(\mathbb{H}^3)$ and $\mathrm{F}(\mathbb{H}^3)$ denote respectively the unit tangent bundle and the oriented frame bundle over $\mathbb{H}^3$.

Thus we may understand the oriented frame bundle $\mathrm{F}\, M$ as the homogeneous space $\Gamma\backslash G$. Denote by

$$p : \Gamma\backslash G \to M$$

the basepoint projection map.

Unless Γ is a lattice, the G-invariant measure on $\Gamma\backslash G$ is infinite and dissipative for natural geometric flows such as the geodesic flow and horospherical flow. Two locally finite measures on $\Gamma\backslash G$, called the Bowen-Margulis-Sullivan (BMS) measure and the Burger-Roblin (BR) measure, play important roles, and they are defined using the Patterson-Sullivan density on the limit set of Γ.

3.1 PATTERSON-SULLIVAN DENSITY. We denote by δ the critical exponent of Γ—that is, the infimum over all $s \geq 0$ such that the Poincaré series $\sum_{\gamma\in\Gamma} e^{-sd(o,\gamma(o))}$ converges. As Γ is geometrically finite, δ is equal to the Hausdorff dimension of Λ [52].

Bishop and Jones proved that δ is strictly bigger than 1, unless Λ is totally disconnected or contained in a circle [5]. As Γ is assumed to be Zariski dense, we have the following:

THEOREM 3.1.
If Λ is connected, then $\delta > 1$.

Recall that for $x, y \in \mathbb{H}^3$ and $\xi \in \mathbb{S}^2$, the Busemann function $\beta_\xi(x, y)$ is given by $\lim_{t\to\infty} d(x, \xi_t) - d(y, \xi_t)$ where ξ_t is a geodesic ray toward ξ.

DEFINITION 3.2. A Γ-invariant conformal density of dimension $s \geq 0$ is a family $\{\mu_x : x \in \mathbb{H}^3\}$ of finite measures on $\mathbb{S}^2$ satisfying

1. for any $\gamma \in \Gamma$ and $x \in \mathbb{H}^3$, $\gamma_* \mu_x = \mu_{\gamma(x)}$ and
2. for all $x, y \in \mathbb{H}^3$ and $\xi \in \mathbb{S}^2$, $\frac{d\mu_x}{d\mu_y}(\xi) = e^{s\beta_\xi(y,x)}$.

THEOREM 3.3 (Patterson-Sullivan).
There exists a Γ-invariant conformal density $\{\nu_x : x \in \mathbb{H}^3\}$ of dimension δ, unique up to a scalar multiple.

We call this Patterson-Sullivan density. Denoting by Δ the hyperbolic Laplacian on $\mathbb{H}^3$, the Patterson-Sullivan density is closely related to the bottom of the spectrum of Δ for its action on smooth functions on $\Gamma\backslash\mathbb{H}^3$. The function ϕ_0 defined by

$$\phi_0(x) := |\nu_x|$$

for each $x \in \mathbb{H}^3$ is Γ-invariant, and hence we may regard ϕ_0 as a function on the manifold $\Gamma\backslash\mathbb{H}^3$. It is the unique function (up to a constant multiple) satisfying $\Delta\phi_0 = \delta(2-\delta)\phi_0$; so we call ϕ_0 the base eigenfunction.

Set $\nu := \nu_o$ and call it the Patterson-Sullivan measure (viewed from o). When Γ is convex cocompact, the Patterson-Sullivan measure ν_o is simply proportional to the δ-dimensional Hausdorff measure on Λ in the spherical metric of $\mathbb{S}^2$.

3.2 MIXING OF THE BMS MEASURE. Consider the following one-parameter subgroup of G:

$$A := \left\{ a_t = \begin{pmatrix} e^{t/2} & 0 \\ 0 & e^{-t/2} \end{pmatrix} : t \in \mathbb{R} \right\}.$$

The right translation action of A on $F\mathbb{H}^3 = G$ induces the frame flow: if $g = (e_1, e_2, e_3)$, then ga_t for $t > 0$ is the frame given by translation in direction of e_1 by hyperbolic distance t. Let $v_o^{\pm} \in \mathbb{S}^2$ denote the forward and backward end points of the geodesic given by v_o respectively. In the upper half-space model of $\mathbb{H}^3$, choosing v_o to be the upward normal vector at $o = (0,0,1)$, we have $v_o^+ = \infty$ and $v_o^- = 0$.

For $g \in G$, we define

$$g^+ = g(v_o^+) \in \mathbb{S}^2 \quad \text{and} \quad g^- = g(v_o^-) \in \mathbb{S}^2.$$

The map $g \mapsto (g^+, g^-, s = \beta_{g^-}(o, g))$ induces a homeomorphism between $T^1(\mathbb{H}^3)$ and $(\mathbb{S}^2 \times \mathbb{S}^2 - \text{diagonal}) \times \mathbb{R}$ called the Hopf parameterization.

We define a locally finite measure $\tilde{m}^{\text{BMS}}$ on $\mathrm{T}^1(\mathbb{H}^3) = G/M_0$ as

$$d\tilde{m}^{\text{BMS}}(g) = e^{\delta\beta_{g^+}(o,g)}\, e^{\delta\beta_{g^-}(o,g)}\, d\nu(g^+)d\nu(g^-)ds,$$

where ds is the Lebesgue measure on $\mathbb{R}$.

Denote by m^{BMS} the unique M_0-invariant measure on $\Gamma\backslash G$ that is induced by $\tilde{m}^{\text{BMS}}$; we call this the Bowen-Margulis-Sullivan measure (BMS measure).

Sullivan showed that m^{BMS} is a *finite* A-invariant measure. The following is due to Babillot [2] for M_0-invariant functions and to Winter [56] for general functions:

THEOREM 3.4.
The frame flow on $(\Gamma\backslash G, m^{\text{BMS}})$ *is mixing—that is, for any* $\psi_1, \psi_2 \in L^2(\Gamma\backslash G, m^{\text{BMS}})$,

$$\lim_{t\to\infty} \int_{\Gamma\backslash G} \psi_1(ga_t)\psi_2(g)\, dm^{\text{BMS}}(g) = \frac{1}{|m^{\text{BMS}}|} m^{\text{BMS}}(\psi_1)\cdot m^{\text{BMS}}(\psi_2).$$

We define the renormalized frame bundle of M as

$$\text{RF}\, M = \{[g] \in \Gamma\backslash G : g^{\pm} \in \Lambda\}.$$

This is a closed A-invariant subset of $\Gamma\backslash G$, which is precisely the support of m^{BMS}, and an immediate consequence of Theorem 3.4 is the topological mixing of the A-action on RF M: for any two open subsets $\mathcal{O}_1, \mathcal{O}_2$ intersecting RF M, $\mathcal{O}_1 a_t \cap \mathcal{O}_2 \neq \emptyset$ for all sufficiently large $|t|$.

3.3 ESSENTIAL UNIQUE ERGODICITY OF THE BR MEASURE. We denote by $N := \{g \in G : a_{-t} g a_t \to e \text{ as } t \to +\infty\}$ the contracting horospherical subgroup for the action of A, which is explicitly given as

$$N = \left\{ u_t = \begin{pmatrix} 1 & t \\ 0 & 1 \end{pmatrix} : t \in \mathbb{C} \right\}.$$

The projection $\pi(gN)$ in $\mathbb{H}^3$ is a Euclidean sphere tangent to $\mathbb{S}^2$ at g^+, and gN consists of frames (e_1, e_2, e_3), whose last two vectors e_2, e_3 are tangent to $\pi(gN)$. That N is a contracting horospherical subgroup means geometrically that $\pi(gNa_t)$ for $t > 0$ is a Euclidean sphere based at g^+ but shrunk toward g^+ by the hyperbolic distance t.

We define $\tilde{m}^{\mathrm{BR}}$ on $G/M_0 = \mathrm{T}^1(\mathbb{H}^3)$ as

$$d\tilde{m}^{\mathrm{BR}}(g) = e^{\delta\beta_{g^+}(o,g)}\, e^{2\beta_{g^-}(o,g)}\, d\nu(g^+)dg^-ds,$$

where dg^- is the Lebesgue measure on $\mathbb{S}^2$. We denote by m^{BR} the unique M_0-invariant measure on $\Gamma\backslash G$ that is induced by $\tilde{m}^{\mathrm{BR}}$. We call this measure the Burger-Roblin measure (BR measure). If Γ is a lattice, m^{BR} is simply the G-invariant measure. Otherwise m^{BR} is an infinite, but locally finite, Borel N-invariant measure whose support is given by

$$\mathrm{RF}_+ M := \{[g] \in \Gamma\backslash G : g^+ \in \Lambda\} = \mathrm{RF}\, M \cdot N.$$

The projection of the BR measure to M is an absolutely continuous measure on M with a Radon-Nikodym derivative given by ϕ_o: if $f \in C_c(\Gamma\backslash G)$ is K-invariant, then

$$m^{\mathrm{BR}}(f) = \int_{\Gamma\backslash G} f(x)\phi_o(x)\, dx,$$

where dx is a G-invariant measure on $\Gamma\backslash G$. Using Theorem 3.4, Roblin and Winter showed the following measure classification of N-invariant locally finite measures, extending an earlier work of Burger [7]:

THEOREM 3.5. ([46], [56]).
Any locally finite N-ergodic invariant measure on $\Gamma\backslash G$ is either supported on a closed N-orbit or proportional to m^{BR}.

3.4 CLOSURES OF N-ORBITS. If $x \notin \mathrm{RF}_+ M$, then xN is a proper immersion of N to $\Gamma\backslash G$ via the map $n \mapsto xn$, and hence xN is closed. In understanding the topological behavior of xN for $x = [g] \in \mathrm{RF}_+ M$, the relative location of g^+ in the limit set becomes relevant. The hypothesis that Γ is geometrically finite implies that any $\xi \in \Lambda$ is either radial (any geodesic ray $\xi_t \in M$ converging to ξ accumulates on a compact subset) or parabolic (it is fixed by some parabolic element of Γ). Since this property is Γ-invariant, we will say that x^+ is radial (respectively parabolic) if g^+ is for $x = [g]$. When Γ is convex cocompact, Λ consists only of radial limit points.

The topological mixing of the A-action on $\mathrm{RF}\, M$ implies the following dichotomy for the closure of an N-orbit:

THEOREM 3.6. ([15], [56]).
For $x \in \mathrm{RF}_+ M$, xN is closed (if x^+ is parabolic) or dense in $\mathrm{RF}_+ M$ (if x^+ is radial).

4 Almost all results on orbit closures

Let $\Gamma < G = \mathrm{PSL}_2(\mathbb{C})$ be a Zariski dense geometrically finite Kleinian group and $M := \Gamma\backslash\mathbb{H}^3$ the associated hyperbolic 3-manifold.

We are mainly interested in the action of the following two subgroups on $\Gamma\backslash G$:

$$H := \mathrm{PSL}_2(\mathbb{R}) \text{ and} \tag{4.1}$$

$$U := \{u_t = \begin{pmatrix} 1 & t \\ 0 & 1 \end{pmatrix} : t \in \mathbb{R}\}.$$

Any one-parameter unipotent subgroup of G is conjugate to U, and any connected closed subgroup of G generated by unipotent one-parameter subgroups is conjugate to either N, H, or U. Note that the subgroups N, H, and U are normalized by the subgroup A, which is an important point for the following discussion as the measures m^{BMS} and m^{BR} are invariant and quasi-invariant under A respectively.

The first question is whether there exist almost all results for the closures of these orbits for appropriate measures.

We recall:

THEOREM 4.1 (Moore's ergodicity theorem).
Let $\Gamma < G$ be a lattice. For any unbounded subgroup W of G, xW is dense in $\Gamma\backslash G$ for almost all $x \in \Gamma\backslash G$.

When Γ is geometrically finite but not a lattice in G, no orbit of a proper connected subgroup W is dense in $\Gamma\backslash G$. Moreover, it is easy to verify that if $\partial(gW) \subset \mathbb{S}^2$ does not intersect Λ, then the map $W \to [g]W \subset \Gamma\backslash G$ given by $w \mapsto [g]w$ is a proper map, and hence $[g]W$ is closed.[4]

If W has the property that $\partial(gW) = (gW)^+$—for instance, if $W = H$ or U—then the nontrivial dynamics of the action of W on $\Gamma\backslash G$ exists only inside the closure of $\mathrm{RF}_+ M \cdot W$.

We will see that $\mathrm{RF}_+ M \cdot H$ is always closed; it is useful to understand the geometric description of $\mathrm{RF}_+ M \cdot H$ in order to understand its closedness.

[4] For a subset $S \subset G$, we use the notation ∂S to denote $\overline{\pi(S)} \cap \mathbb{S}^2$ under the projection $\pi : \mathrm{F}\,\mathbb{H}^3 \to \mathbb{H}^3 \cup \mathbb{S}^2$.

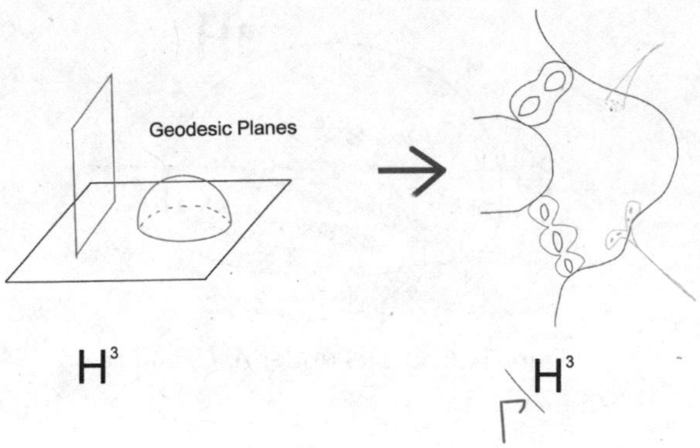

Figure 14.5. Geodesic planes in M

4.1 GEODESIC PLANES AND ALMOST ALL *H*-ORBITS. A geodesic plane in $\mathbb{H}^3$ is a totally geodesic imbedding of $\mathbb{H}^2$, which is simply either a vertical plane or a vertical hemisphere in the upper half-space model.

Let $\mathcal{P}$ denote the set of all oriented geodesic planes of $\mathbb{H}^3$ and $\mathcal{C}$ the set of all oriented circles in $\mathbb{S}^2$. The map $P \mapsto \partial P$ gives an isomorphism between $\mathcal{P}$ and $\mathcal{C}$.

On the other hand, the map

$$gH \mapsto P_g := \pi(gH)$$

gives an isomorphism between the quotient space G/H and the set $\mathcal{P}$, whose inverse can be described as follows: for $P \in \mathcal{P}$, the set of frames (e_1, e_2, e_3) based in P such that e_1 and e_2 are tangent to P and e_3 is given by the orientation of P is precisely a single H-orbit. Consequently, the map

$$gH \to C_g := \partial P_g$$

gives an isomorphism between G/H and $\mathcal{C}$.

DEFINITION 4.2. *An oriented geodesic plane $P \subset M$ is a totally geodesic immersion of an oriented hyperbolic plane $\mathbb{H}^2$ in M, or equivalently P is the image of an oriented geodesic plane of $\mathbb{H}^3$ under π.*

In this essay, geodesic planes and circles are always considered to be oriented. Note that any geodesic plane $P \subset M$ is of the form

$$P = p(gH) \quad \text{for some } g \in G.$$

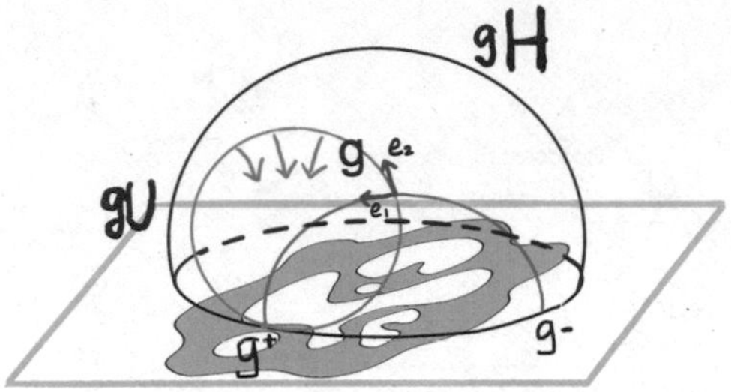

Figure 14.6. Orbits under A, U, and H

Therefore the study of H-orbits on $\Gamma\backslash G$ has a direct implication on the behavior of geodesic planes in the manifold $\Gamma\backslash\mathbb{H}^3$.

We set

$$F_\Lambda := \mathrm{RF}_+ M \cdot H \quad \text{and} \quad \mathcal{C}_\Lambda := \{C \in \mathcal{C} : C \cap \Lambda \neq \emptyset\}. \tag{4.2}$$

It follows from the compactness of Λ that $\mathcal{C}_\Lambda$ is a closed subset of $\mathcal{C} = G/H$. As

$$F_\Lambda / H = \Gamma\backslash\mathcal{C}_\Lambda,$$

we deduce the following:

LEMMA 4.3. *The set F_Λ is a closed H-invariant subset of $\Gamma\backslash G$.*

PROPOSITION 4.4. *For m^{BMS}-a.e. $x \in \mathrm{RF}\, M$,*

$$\overline{xH} = F_\Lambda;$$

in particular, the geodesic plane $p(xH)$ is dense in M.

Proof. We have $\overline{\mathrm{RF}\, M \cdot U} = \mathrm{RF}_+ M$ [34], and hence $\overline{\mathrm{RF}\, M \cdot H} = F_\Lambda$. Theorem 3.4 implies that m^{BMS} is ergodic; thus by the Birkhoff ergodic theorem, for almost all x, xA is dense in $\mathrm{RF}\, M$. Since $A \subset H$, we deduce

$$\overline{xH} \supset \overline{\mathrm{RF}\, M} \cdot H = F_\Lambda. \qquad \square$$

4.2 HOROCYCLES AND ALMOST ALL U-ORBITS. A horocycle in $\mathbb{H}^3$ is a Euclidean circle tangent to $\mathbb{S}^2$.

DEFINITION 4.5. *A horocycle χ in M is an isometrically immersed copy of $\mathbb{R}$ with zero torsion and geodesic curvature 1, or equivalently χ is the image of a horocycle of $\mathbb{H}^3$ under π.*

The right translation action of U on $\Gamma\backslash G$ is the horocyclic action: if $g = (e_1, e_2, e_3)$, then gu_t for $t > 0$ is the frame given by translation in the direction of e_2 by Euclidean distance t. In fact, any horocycle $\chi \subset M$ is of the form

$$\chi = p(gU) \quad \text{for some } g \in G.$$

Note that both gA and gU have their trajectories inside the plane $P_g = \pi(gH)$. In particular, $\pi(gU)$ is a Euclidean circle lying on P_g tangent to $\mathbb{S}^2$ at g^+.

We now discuss the almost all results for U-orbits in terms of the Burger-Roblin measure. It turns out that the size of the critical exponent δ matters in this question. The following was proved in joint work with Mohammadi for Γ convex cocompact [39] and by Maucourant and Schapira [34] for geometrically finite groups.

THEOREM 4.6.
If $\delta > 1$, m^{BR} is U-ergodic and conservative.

PROPOSITION 4.7. *Let $\delta > 1$ (e.g., Λ is connected). Then for m^{BR}-a.e. $x \in \mathrm{RF}_+ M$,*

$$\overline{xU} = \mathrm{RF}_+ M;$$

In particular, the horocycle $p(xU)$ is dense in M.

Proof. Since m^{BR} is an infinite measure, unless Γ is a lattice, the Birkhoff ergodic theorem does not apply. Instead we use the Hopf ratio theorem, which applies by Theorem 4.6, and hence the claim follows. □

In [34], it was proved that if $\delta < 1$, m^{BR} is totally U-dissipative, and hence almost all U-orbits are divergent (cf. [14]). Whether m^{BR} is ergodic or not at $\delta = 1$ remains an open question.

4.3 ORBIT CLOSURE THEOREM FOR LATTICES. The almost all results on orbit closures in Propositions 4.4 and 4.7 do not describe the topological behavior of a given individual orbit. In the lattice case, we have the following remarkable classification of all possible orbit closures, due to Ratner [45] and Shah [49] independently:

THEOREM 4.8.
Let $\Gamma < G$ be a lattice and $x \in \Gamma\backslash G$.

(1) *The closure $\overline{xH}$ is either xH or $\Gamma\backslash G$.*
(2) *The closure $\overline{xU}$ is either xU, $xv^{-1}Hv$ for some $v \in N$, or $\Gamma\backslash G$.*

This theorem immediately implies the first part of the following theorem; the rest follows from the results in the same papers by Ratner and Shah.

THEOREM 4.9.
If M has finite volume, the closures of a geodesic plane and a horocycle are properly immersed submanifolds of M. Moreover,

(1) *any properly immersed geodesic plane has finite area;*
(2) *there are at most countably many properly immersed geodesic planes in M; and*
(3) *any infinite sequence of properly immersed geodesic planes P_i becomes dense in M—that is, $\lim_{i\to\infty} P_i = M$.*[5]

The density statement (3), which is a topological version of the Mozes-Shah theorem [41], implies that every properly immersed geodesic plane P is topologically isolated, in the sense that there exists an open neighborhood of P that does not contain any other properly immersed geodesic plane in its entirety.

4.4 TOPOLOGICAL OBSTRUCTIONS TO ORBIT CLOSURE THEOREM. In this section, we describe a family of quasi-Fuchsian manifolds, some of whose geodesic planes have fractal closures; in particular, they have non-integral dimensions. These geodesic planes pass through the interior of the convex core of M but their boundaries meet the limit set Λ only at two points.

These examples can be seen easily for Fuchsian manifolds. By performing a small bending deformation along a simple closed geodesic far away from our fractal closures of a fixed plane, we will obtain quasi-Fuchsian manifolds keeping the fractal closure intact.

[5] For a sequence of closed subsets Y_i of a topological space X, we write $\lim_{i\to\infty} Y_i = Y$ if $\limsup_{i\to\infty} Y_i = \liminf_{i\to\infty} Y_i = Y$.

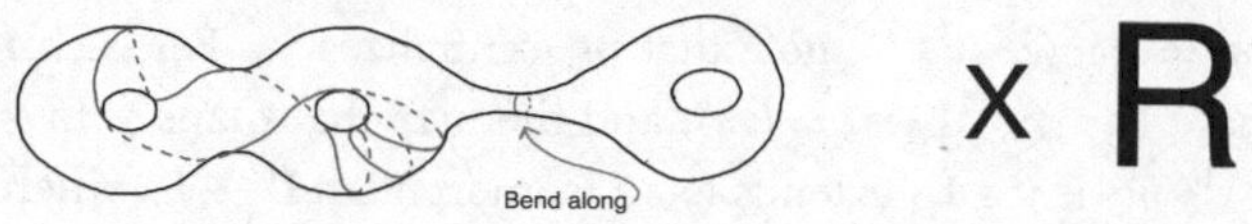

Figure 14.7. Bending deformation

4.4.1 *Fuchsian 3-manifolds*

Consider a Fuchsian 3-manifold M that can be expressed as

$$M = S \times \mathbb{R}$$

in cylindrical coordinates where S is a closed hyperbolic surface of genus at least 2. Or equivalently, take a torsion-free uniform lattice $\Gamma < \mathrm{PSL}_2(\mathbb{R})$, and consider Γ as a subgroup of G, so that $M = \Gamma \backslash \mathbb{H}^3 = (\Gamma \backslash \mathbb{H}^2) \times \mathbb{R}$. We have $\operatorname{core} M = S$.

It is well-known that geodesics on a closed hyperbolic surface S can behave as wildly as we wish for; in particular, for any $\beta \geq 1$, there exists a geodesic whose closure has Hausdorff dimension precisely β.

(1) The closure of a geodesic plane need not be a submanifold: if $\gamma \subset S$ is a geodesic and P is a geodesic plane orthogonal to S with $P \cap S = \gamma$, then

$$\overline{P} \simeq \overline{\gamma} \times \mathbb{R}.$$

Therefore, if we take a geodesic $\gamma \subset S$ whose closure $\overline{\gamma}$ is wild, then $\overline{P}$ is very far from being a submanifold.

(2) There are uncountably many properly immersed geodesic planes intersecting $\operatorname{core} M$; if $\gamma \subset S$ is a closed geodesic and P is a geodesic plane with $P \cap S = \gamma$, then P is a properly immersed geodesic plane. By varying angles between P and S, we obtain a continuous family of such P.

We can now use a small bending deformation of M to obtain quasi-Fuchsian manifolds in which the same phenomenon persists.

4.4.2 *Quasi-Fuchsian hyperbolic 3-manifolds*

Let $\gamma_0 \in \Gamma$ be a primitive hyperbolic element representing a separating simple closed geodesic β in S. Without loss of generality, we assume $\gamma_0 \in A$, up to conjugation. If S_1 and S_2 are components of $S - \beta$, then each $\Gamma_i := \pi_1(S_i)$ is a subgroup of Γ, and Γ can be presented as the amalgamated free product

$$\Gamma = \Gamma_1 *_{\langle \gamma_0 \rangle} \Gamma_2.$$

Setting $m_\theta = \mathrm{diag}(e^{i\theta}, e^{-i\theta})$, note that m_θ centralizes γ_0. For each non-trivial m_θ, we have $\Gamma_1 \cap m_\theta^{-1}\Gamma_2 m_\theta = \langle\gamma_0\rangle$, and the map that maps γ to γ if $\gamma \in \Gamma_1$ and to $m_\theta^{-1}\gamma m_\theta$ if $\gamma \in \Gamma_2$ extends to an isomorphism $\Gamma \to \Gamma_\theta$ where

$$\Gamma_\theta := \Gamma_1 *_{\langle\gamma_0\rangle} m_\theta^{-1}\Gamma_2 m_\theta.$$

If θ is sufficiently small, then

- Γ_θ is a discrete subgroup of G,
- $M_\theta := \Gamma_\theta \backslash \mathbb{H}^3$ is a quasi-Fuchsian manifold, and
- there is a path isometric embedding $j_\theta : S \to \partial \operatorname{core} M_\theta$ such that its image S_θ is bent with a dihedral angle of θ along the image of β and otherwise totally geodesic.

Fix $\varepsilon > 0$ sufficiently small that β has an embedded annular collar neighborhood in S of width 2ε. Let $\gamma \subset S_1$ be a geodesic whose closure $\overline{\gamma}$ is disjoint from a 2ε-neighborhood $\mathcal{O}(\beta, 2\varepsilon)$ of β. Now if we set $S_1(\varepsilon) := S_1 - \mathcal{O}(\beta, 2\varepsilon)$, then there is a unique orientation preserving isometric immersion

$$J_\theta : S_1(\varepsilon) \times \mathbb{R} \to M_\theta,$$

which extends $j_\theta|_{S_1(\varepsilon)}$ and sends geodesics normal to $S_1(\varepsilon)$ to geodesics normal to $j_\theta(S_1(\varepsilon))$. Now, if θ is small enough (relative to ε), then

$$J_\theta \text{ is a proper isometric embedding.}$$

This can be proved using the following observation. Let $\alpha = [a, b_1] \cup [b_1, b_2] \cup \cdots \cup [b_{n-1}, b_n] \cup [b_n, c]$ be a broken geodesic in $\mathbb{H}^3$, which is a union of geodesic segments and which bends by angle $0 \le \theta < \pi/2$ at each b_i. Suppose the first and the last segments have length at least $\varepsilon > 0$ and the rest have length at least 2ε. Let P_i denote the geodesic plane orthogonal to $[b_i, b_{i+1}]$ at b_i. If $\theta = 0$, then the distances among P_i's are at least ε. Now if θ is small enough so that $\sin(\theta/2) < \tanh \varepsilon$, then the planes P_i remain a positive distance apart, giving a nested sequence of half-planes in $\mathbb{H}^3$. This implies that J_θ is a proper imbedding.

It now follows that for the plane $P := \gamma \times \mathbb{R} \subset S_1(\varepsilon) \times \mathbb{R}$, its image $P_\theta := J_\theta(P) \subset M_\theta$ is an immersed geodesic plane whose closure $\overline{P}_\theta$ is isometric to $\overline{P} \simeq \overline{\gamma} \times \mathbb{R}$. Therefore by choosing γ, whose closure is wild, we can obtain a geodesic plane P_θ of M_θ with wild closure (see [37] for more details).

This example demonstrates that the presence of an essential cylinder in M gives an obstruction to the topological rigidity of geodesic planes. For the

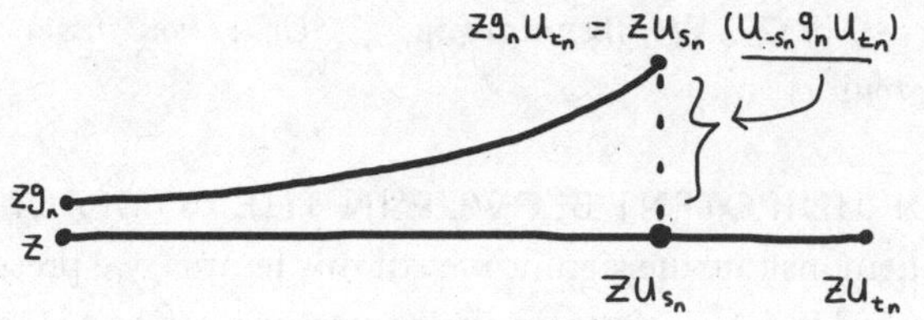

Figure 14.8. Divergence of U-orbits of two nearby points

behavior of an individual geodesic plane P, it also indicates that the *finite* intersection $\partial P \cap \Lambda$ can be an obstruction.

5 Unipotent blowup and renormalizations

The distinguished property of a unipotent flow on the homogeneous space $\Gamma\backslash G$ is the polynomial divergence of nearby points. Given a sequence $zg_n \in \Gamma\backslash G$, where $g_n \to e$ in G, the transversal divergence between two orbits $zg_n U$ and zU can be understood by studying the double coset $Ug_n U$ in view of the equality

$$zg_n u_t = zu_s(u_s^{-1} g_n u_t)$$

and the behavior of rational maps $t \mapsto u_{\alpha_n(t)} g_n u_t$ for certain reparameterizations $\alpha_n : \mathbb{R} \to \mathbb{R}$ so that $\limsup_{n\to\infty} \{u_{\alpha_n(t)} g_n u_t : t \in \mathbb{R}\}$ contains a non-trivial element of $G - U$.[6]

We denote by V the transversal subgroup

$$V = \{u_{it} : t \in \mathbb{R}\}$$

to U inside N, so that $N = UV$. Note that the normalizer $\mathrm{N}(U)$ of U is equal to AN, and the centralizer $\mathrm{C}(U)$ of U is equal to N.

The following unipotent blowup lemma (though stated in the setting of $\mathrm{SL}_3(\mathbb{R})$) was first observed by Margulis [28, lemma 5], in his proof of the Oppenheim conjecture.

LEMMA 5.1.

(1) *If $g_n \to e$ in $G - AN$, then $\limsup_{n\to\infty} Ug_n U$ contains a one-parameter semigroup of AV.*

[6]If Q_n is a sequence of subsets of G, $q \in \limsup_{n\to\infty} Q_n$ if and only if every neighborhood of q meets infinitely many Q_n, and $q \in \liminf_{n\to\infty} Q_n$ if and only if every neighborhood of q meets all but finitely many Q_n. If $\limsup_n Q_n = Q_\infty = \liminf Q_n$, then Q_n is said to be convergent and Q_∞ is the limit of Q_n [19].

(2) *If* $g_n \to e$ *in* $G - VH$*, then* $\limsup_{n\to\infty} Ug_nH$ *contains a one-parameter semigroup*[7] *of* V.

5.1 USE OF UNIPOTENT BLOWUP IN THE COMPACT $\Gamma\backslash G$ CASE.

In order to demonstrate the significance of this lemma, we present a proof of the following orbit closure theorem, which uses the notion of U-minimal subsets. A closed U-invariant subset $Y \subset \Gamma\backslash G$ is called U-minimal if every U orbit in Y is dense in Y. By Zorn's lemma, any compact U-invariant subset of $\Gamma\backslash G$ contains a U-minimal subset.

THEOREM 5.2. Let $\Gamma < G$ be a uniform lattice. For any $x \in \Gamma\backslash G$, xH is either closed or dense.

Proof. Set $X := \overline{xH}$. Suppose that $X \neq xH$. By the minimality of the N-action on $\Gamma\backslash G$ (Corollary 3.6), it suffices to show that X contains an orbit of V.

Step 1: For any U-minimal subset $Y \subset X$,

$$YL = Y \quad \text{for a one-parameter subgroup } L < AV.$$

It suffices to show that $Yq_n = Y$ for some sequence $q_n \to e$ in AV. Fix $y_0 \in Y$. As Y is U-minimal, there exists $t_n \to \infty$ such that $y_0u_{t_n} \to y_0$. Write $y_0u_{t_n} = y_0g_n$ for $g_n \in G$. Then $g_n \to e$ in $G - U$, because if g_n belonged to U, the orbit y_0U would be periodic, which is a contradiction to the assumption that Γ is a uniform lattice and hence contains no parabolic elements. If $g_n = a_nv_nu_n \in AN = AUV$, then we may take $q_n = a_nv_n$. If $g_n \notin AN$, then by Lemma 5.1, $\limsup_{n\to\infty} Ug_nU$ contains a one-parameter semigroup L of AV. Hence for any $q \in L$, there exist $t_n, s_n \in \mathbb{R}$ such that $q = \lim u_{t_n}g_nu_{s_n}$.

(5.1) Since Y is compact, $y_0u_{-t_n}$ converges to some $y_1 \in Y$

by passing to a subsequence. Therefore $y_0g_nu_{s_n} = y_0u_{-t_n}(u_{t_n}g_nu_{s_n})$ converges to $y_1q \in Y$. Since $q \in \mathrm{N}(U)$ and Y is U-minimal, we have

$$\overline{y_1qU} = \overline{y_1U}q = Yq = Y.$$

This proves the claim.

---

[7] A one-parameter semigroup of V is given by $\{\exp t\xi : t \geq 0\}$ for some nonzero $\xi \in \mathrm{Lie}(V)$.

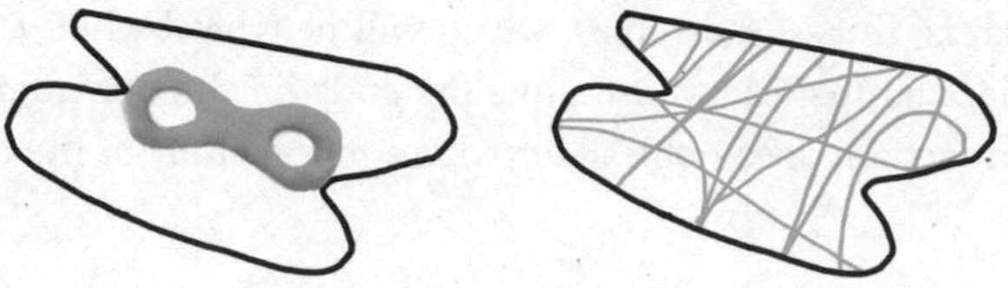

Figure 14.9. Closed or dense

Step 2: There exists a U-minimal subset $Y \subset X$ such that $X - y_0 H$ is not closed for some $y_0 \in Y$.

If xH is not locally closed—that is, $X - xH$ is not closed—then let Y be any U-minimal subset of X. If $Y \subset xH$, then for any $y_0 \in Y$, $X - y_0 H = X - xH$ is not closed. If $Y \not\subset xH$, then choose $y_0 \in Y - xH$. If xH is locally closed, then let Y be a U-minimal subset of $X - xH$. Then $X - y_0 H$ is not closed for any $y_0 \in Y$.

Step 3: For Y from step 2, we have

$$Yv \subset X \quad \text{for some nontrivial } v \in V.$$

By step 2, we have $y_0 g_n \in X$ for some $y_0 \in Y$ and a sequence $g_n \to e$ in $G - H$. If $g_n \in VH$ for some n, then the claim follows. If $g_n \notin VH$ for all n, then by Lemma 5.1(2), $\limsup_{n\to\infty} Ug_n H$ contains a nontrivial element $v \in V$. Since $v = \lim u_{t_n} g_n h_n$ for some $t_n \in \mathbb{R}$ and $h_n \in H$, we deduce $Yv \subset X$ as in step 1.

Step 4: X contains a V-orbit.

It suffices to show that X contains $x_0 V_+$ for a one-parameter semigroup V_+ of V, because if $v_n \to \infty$ in V_+ and $x_0 v_n \to x_1$, then

$$x_1 V = x_1 \cdot \limsup_n (v_n^{-1} V_+) \subset \overline{x_0 V_+} \subset X.$$

Let $Y \subset X$ be a U-minimal subset from step 2. By step 1, $YL \subset Y$ where L is either V or $v_0 A v_0^{-1}$ for some $v_0 \in V$. If $L = V$, this finishes the proof. If $L = A$, then by step 3, we get $Yv = Yv(v^{-1} A v) \subset X$. Hence we get $X \supset x_0 v^{-1} A v A$ for some $x_0 \in X$ and a nontrivial $v \in V$. Since $v^{-1} A v A$ contains a one-parameter semigroup of V, this finishes the proof. □

We highlight the importance of Equation (5.1) from the above proof: if q belongs to the set $\limsup_{n\to\infty} Ug_n U$ in Lemma 5.1—that is, $q = \lim_{n\to\infty} u_{t_n} g_n u_{s_n}$ for some $t_n, s_n \in \mathbb{R}$—then the size of t_n and s_n are essentially determined by the sequence $g_n \to e$, up to multiplicative constants. On the other hand, we need the convergence of the sequence $y_0 u_{-t_n}$ in order to derive

$Yq \subset Y$. That is, if $y_0 u_{-t_n}$ diverges, which will be typical when $\Gamma\backslash G$ has infinite volume, Lemma 5.1, whose proof depends on the polynomial property of unipotent action, does not lead anywhere in the study of the orbit closure problem.

5.2 UNIPOTENT BLOWUP AND RENORMALIZATIONS OF THE RETURN TIME. Loosely speaking, for a given $y_0 \in \Gamma\backslash G$, we now would like to understand the set

$$\limsup_{n\to\infty} \mathsf{T} g_n U,$$

where T is the recurrence time of $y_0 U$ into a fixed compact subset of $\Gamma\backslash G$. Most of the time, $\limsup \mathsf{T} g_n U$ may be empty. In order to make sure that this set is nontrivial enough for our purpose, we need a certain polynomial $\phi(t)$ (see proof of Lemma 5.5) not to vanish on the renormalized set $\limsup \lambda_n^{-1}\mathsf{T}$, where $\lambda_n > 0$ is a sequence whose size is dictated by the speed of convergence of the sequence $g_n \to e$. Since we do not have a control on g_n in general, the following condition on T, or more generally on a sequence T_n, is necessary for an arbitrary sequence $\lambda_n \to \infty$.

DEFINITION 5.3. *We say that a sequence $\mathsf{T}_n \subset \mathbb{R}$ has accumulating renormalizations if for any sequence $\lambda_n \to \infty$,*

$$\mathsf{T}_\infty := \limsup_{n\to\infty} \lambda_n^{-1}\mathsf{T}_n$$

accumulates at both 0 and ∞.

That is, T_∞ contains a sequence tending to 0 as well as a sequence tending to ∞. We allow a constant sequence T_n in this definition.

The following lemma is immediate:

LEMMA 5.4. *If there exists $\kappa > 1$ such that each T_n is κ-thick in the sense that for all $r > 0$, $\mathsf{T}_n \cap \pm[r, \kappa r] \neq \emptyset$, then the sequence T_n has accumulating renormalizations.*

We now present a refined version of Lemma 5.1, which will be a main tool in the study of U-orbits in the infinite volume homogeneous space: via the map $t \mapsto u_t$, we identify $\mathbb{R} \simeq U$.

We write $\mathfrak{g} = \mathfrak{h}^\perp \oplus \mathfrak{h}$ where $\mathfrak{h} = \mathfrak{sl}_2(\mathbb{R})$ is the Lie algebra of H and $\mathfrak{h}^\perp = i\mathfrak{sl}_2(\mathbb{R})$; note that $\mathfrak{h}^\perp$ is H-invariant under conjugation.

LEMMA 5.5 (unipotent blowup). Let $\mathsf{T}_n \subset U$ be a sequence with accumulating renormalizations.

(1) For any $g_n \to e$ in $G - AN$, the subset $AV \cap \left(\limsup_{n\to\infty} \mathsf{T}_n g_n U\right)$ accumulates at e and ∞.
(2) For any $g_n \to e$ in $G - VH$, the subset $V \cap \left(\limsup_{n\to\infty} \mathsf{T}_n g_n H\right)$ accumulates at e and ∞.
(3) For any $g_n \to e$ in $\exp \mathfrak{h}^{\perp} - V$, the subset $V \cap \left(\limsup_{n\to\infty} \{u_t g_n u_{-t} : t \in \mathsf{T}_n\}\right)$ accumulates at e and ∞.

Proof. For (1), we will find a sequence $\lambda_n \to \infty$ (depending on g_n) and a rational map $\psi : \mathbb{R} \to AV$ such that for $\mathsf{T}_\infty := \limsup_{n\to\infty} \lambda_n^{-1} \mathsf{T}_n$,

- $\psi(\mathsf{T}_\infty) \subset \limsup_{n\to\infty} \mathsf{T}_n g_n U$, and
- $\psi(\mathsf{T}_\infty)$ accumulates at e and ∞.

The construction of ψ follows the arguments of Margulis and Tomanov [32]. Since U is a real algebraic subgroup of G, by Chevalley's theorem, there exists an $\mathbb{R}$-regular representation $G \to \mathrm{GL}(W)$ with a distinguished point $p \in W$ such that $U = \mathrm{Stab}_G(p)$. Then pG is locally closed, and

$$\mathrm{N}(U) = \{g \in G : pgu = pg \text{ for all } u \in U\}. \tag{5.2}$$

Set $\mathcal{L} := VAMN^+$, where N^+ is the transpose of N. Then $U\mathcal{L}$ is a Zariski dense open subset of G, and $p\mathcal{L}$ is a Zariski open neighborhood of p in the Zariski closure of pG. We choose a norm on W so that $B(p,1) \cap \overline{pG} \subset p\mathcal{L}$, where $B(p,1) \subset W$ denotes the closed ball of radius 1 centered at p.

Without loss of generality, we may assume $g_n \in U\mathcal{L}$ for all n. For each n, define $\tilde{\phi}_n : \mathbb{R} \to W$ by

$$\tilde{\phi}_n(t) = pg_n u_t,$$

which is a polynomial of degree uniformly bounded for all n. Define $\lambda_n \geq 0$ by

$$\lambda_n := \sup\{\lambda \geq 0 : \tilde{\phi}_n[-\lambda, \lambda] \subset B(p,1)\}.$$

As $g_n \notin \mathrm{N}(U) = AN$, $\tilde{\phi}_n$ is nonconstant, and hence $\lambda_n < \infty$. As $\tilde{\phi}_n(0) = pg_n \to p$, we have $\lambda_n \to \infty$. We reparameterize $\tilde{\phi}_n$ using λ_n:

$$\phi_n(t) := \tilde{\phi}_n(\lambda_n t).$$

Then for all n,

$$\phi_n[-1,1] \subset B(p,1).$$

Therefore, the sequence ϕ_n forms an equicontinuous family of polynomials, and hence, after passing to a subsequence, ϕ_n converges to a polynomial

$$\phi : \mathbb{R} \to \overline{pG} \subset W$$

uniformly on every compact subset of $\mathbb{R}$. Note that ϕ is nonconstant, since $\phi(0) = p$ and $\max \|\phi(\pm 1)\| = 1$. As the map $\rho : \mathcal{L} \to p\mathcal{L}$ defined by $\ell \mapsto p\ell$ is a regular isomorphism and $p\mathcal{L}$ is a Zariski open neighborhood of p in the Zariski closure of pG, we now get a rational map $\psi : \mathbb{R} \to \mathcal{L}$ given by

$$\psi(t) = \rho^{-1}(\phi(t)).$$

If we define $\psi_n(t)$ as the unique $\mathcal{L}$-component of $g_n u_t$ in the $U\mathcal{L}$ decomposition—that is, $g_n u_t = u_{s_n} \psi_n(t)$ for some $s_n \in \mathbb{R}$—then

$$\psi(t) = \lim_{n \to \infty} \psi_n(\lambda_n t),$$

where the convergence is uniform on compact subsets of $\mathbb{R}$. It is easy to check that $\operatorname{Im} \psi \subset \mathrm{N}(U) \cap \mathcal{L} = AV$ using Equation (5.2).

Set

$$\mathsf{T}_\infty := \limsup_{n \to \infty} \lambda_n^{-1} \mathsf{T}_n.$$

By the hypothesis on T_n, T_∞ accumulates at 0 and ∞. Since $\psi : \mathbb{R} \to AV$ is a nonconstant rational map with $\psi(0) = e$, $\psi(\mathsf{T}_\infty)$ accumulates at e and ∞.

Letting $t \in \mathsf{T}_\infty$, choose a sequence $t_n \in \mathsf{T}_n$ such that $\lim_{n \to \infty} \lambda_n^{-1} t_n = t$. Since $\psi_n \circ \lambda_n \to \psi$ uniformly on compact subsets,

$$\psi(t) = \lim_{n \to \infty} (\psi_n \circ \lambda_n)\left(\lambda_n^{-1} t_n\right) = \lim_{n \to \infty} \psi_n(t_n) = \lim_{n \to \infty} u_{s_n} g_n u_{t_n}$$

for some sequence $s_n \in \mathbb{R}$. Hence,

$$\psi(\mathsf{T}_\infty) \subset \limsup_{n \to \infty} U g_n \mathsf{T}_n.$$

By applying this argument to g_n^{-1}, we may switch the position of U and T_n, and hence finish the proof of Lemma 5.5 (1).

To prove (2), by modifying g_n using an element of H, we may assume that $g_n = \exp q_n \in \exp \mathfrak{h}^{\perp} - V$. Hence, (2) follows from (3). We define a polynomial $\psi_n : \mathbb{R} \to \mathfrak{h}^{\perp}$ by

$$\psi_n(t) = u_t q_n u_{-t} \quad \text{for all } t \in \mathbb{R}.$$

Since $g_n \notin V$ and hence does not commute with U, ψ_n is a nonconstant polynomial. Define

$$\lambda_n := \sup\{\lambda \geq 0 : \psi_n([-\lambda, \lambda]) \subset B(0,1)\},$$

where $B(0,1)$ is the closed unit ball around 0 in $\mathfrak{h}^{\perp}$. Then $0 < \lambda_n < \infty$ and $\lambda_n \to \infty$.

Now the rescaled polynomials $\phi_n = \psi_n \circ \lambda_n : \mathbb{R} \to \mathfrak{h}^{\perp}$ form an equicontinuous family of polynomials of uniformly bounded degree and $\lim_{n\to\infty} \phi_n(0) = 0$. Therefore ϕ_n converges to a nonconstant polynomial

$$\phi : \mathbb{R} \to \mathfrak{h}^{\perp}$$

uniformly on compact subsets.

We claim that $\operatorname{Im}(\phi) \subset \operatorname{Lie}(V)$. For any fixed $s, t \in \mathbb{R}$,

$$\begin{aligned} u_s \phi(t) u_{-s} &= \lim_{n\to\infty} u_{\lambda_n t + s} q_n u_{-\lambda_n t - s} \\ &= \lim_{n\to\infty} u_{\lambda_n (t + \lambda_n^{-1} s)} q_n u_{-\lambda_n (t + \lambda_n^{-1} s)} \\ &= \lim_{n\to\infty} u_{\lambda_n t} q_n u_{-\lambda_n t} = \phi(t). \end{aligned}$$

Hence, $\phi(t)$ commutes with U. Since the centralizer of U in $\mathfrak{h}^{\perp}$ is equal to Lie V, the claim follows. Define $\psi : \mathbb{R} \to V$ by $\psi(t) = \exp(\phi(t))$, noting that $\exp : \operatorname{Lie} V \to V$ is an isomorphism. Setting

$$\mathsf{T}_{\infty} := \limsup_{n\to\infty} \lambda_n^{-1} \mathsf{T}_n,$$

we deduce that $\psi(\mathsf{T}_{\infty})$ accumulates at e and ∞. For any $t \in \mathsf{T}_{\infty}$, we choose $t_n \in \mathsf{T}_n$ so that $t = \lim \lambda_n^{-1} t_n$. Then

$$\psi(t) = \lim_{n\to\infty} u_{t_n} g_n u_{-t_n}$$

as $\phi_n(t) \to \phi(t)$ uniformly on compact subsets. Hence,

$$\psi(\mathsf{T}_{\infty}) \subset V \cap \left(\limsup_{n\to\infty} \{u_t g_n u_{-t} : t \in \mathsf{T}_n\} \right).$$

This completes the proof of Lemma 5.5 (3). ☐

5.3 RELATIVE MINIMAL SETS AND ADDITIONAL INVARIANCE. Let $\Gamma < G$ be a discrete subgroup. Let $X \subset \Gamma\backslash G$ be a closed H-invariant subset with no periodic U-orbits.[8] Let $W \subset \Gamma\backslash G$ be a compact subset such that $X \cap W \neq \emptyset$. We suppose that for any $y \in X \cap W$,

$$\mathsf{T}(y) := \{t \in \mathbb{R} : yu_t \in W\} \text{ has accumulating renormalizations.} \tag{5.3}$$

Under this hypothesis, we can obtain analogous steps to Theorem 5.2 steps 1 and 3 for relative U-minimal subsets of X. Since X is not compact in general, a U-minimal subset of X may not exist. Hence, we instead consider a relative U-minimal subset of X.

DEFINITION 5.6. *A closed U-invariant subset $Y \subset X$ is called U-minimal with respect to W if $Y \cap W \neq \emptyset$ and yU is dense in Y for every $y \in Y \cap W$.*

As W is compact, it follows from Zorn's lemma that X always contains a U-minimal subset with respect to W.

LEMMA 5.7 (translates of Y inside Y). *Let $Y \subset X$ be a U-minimal subset with respect to W. Then*

$$YL \subset Y$$

for some one-parameter semigroup $L < AV$.

Proof. It suffices to find a sequence $q_n \to e$ in AV such that $Yq_n \subset Y$.

Fix $y_0 \in Y \cap W$. We claim that there exists $g_n \to e$ in $G - U$ such that $y_0 g_n \in Y \cap W$. By the minimality assumption on Y, there exists $t_n \to \infty$ in $\mathsf{T}(y_0)$ so that $y_0 u_{t_n}$ converges to $y_0 \in Y \cap W$ (see [4, lemma 8.2]). Hence, there exists $g_n \to e$ such that

$$y_0 u_{t_n} = y_0 g_n.$$

Then $g_n \notin U$, because if g_n belonged to U, $y_0 U$ would be periodic, contradicting the assumption that X contains no periodic U-orbit.

Case 1: $g_n \in AN$. By modifying g_n with elements of U, we may assume that $g_n \in AV$. Since $g_n \in \mathrm{N}(U)$ and $y_0 \in Y \cap W$, we get $y_0 U g_n = y_0 g_n U \subset Y$, and hence $\overline{y_0 U g_n} = Y g_n \subset Y$.

[8]The case when X contains a periodic U-orbit turns out to be more manageable; see [4, proposition 4.2].

Case 2: $g_n \notin AN$. By Lemma 5.5, for any neighborhood $\mathcal{O}$ of e, there exist $t_n \in \mathsf{T}(y_0)$ and $s_n \in \mathbb{R}$ such that $u_{-t_n} g_n u_{s_n}$ converges to some $q \in (AV - \{e\}) \cap \mathcal{O}$. Since $y_0 u_{t_n} \in W$ and W is compact, $y_0 u_{t_n}$ converges to some $y_1 \in Y \cap W$ by passing to a subsequence. Therefore as $n \to \infty$,

$$y_0 g_n u_{-s_n} = (y_0 u_{t_n})(u_{-t_n} g_n u_{s_n}) \to y_1 q \in Y.$$

As $y_1 \in Y \cap W$ and $q \in \mathrm{N}(U)$, it follows that $Yq \subset Y$. Since such q can be found in any neighborhood of e, this finishes the proof. □

LEMMA 5.8 (one translate of Y inside X). *Let $Y \subset X$ be a U-minimal subset with respect to W such that $X - y_0 H$ is not closed for some $y_0 \in Y \cap W$. Then*

$$Yv \subset X \quad \textit{for some nontrivial } v \in V.$$

Proof. By the hypothesis, there exists $g_n \to e$ in $G - H$ such that $y_0 g_n \in X$.

If $g_n \in VH$ for some n, the claim is immediate as X is H-invariant. If $g_n \notin VH$ for all n, by Lemma 5.5, there exist $t_n \in \mathsf{T}(y_0)$ and $h_n \in H$ so that $u_{t_n}^{-1} g_n h_n$ converges to some nontrivial $v \in V$. Since $y_0 u_{t_n}$ belongs to the compact subset W, by passing to a subsequence, $y_0 u_{t_n}$ converges to some $y_1 \in Y \cap W$. Hence $y_0 g_n h_n = y_0 u_{t_n}(u_{t_n}^{-1} g_n h_n)$ converges to $y_1 v$. By the minimality of Y with respect to W, we get $Yv \subset Y$, as desired. □

For a subset $I \subset \mathbb{R}$, we write $V_I = \{u_{it} : t \in I\}$. When the conditions for Lemmas 5.7 and 5.8 are met, we can deduce that X contains some interval of a V-orbit:

LEMMA 5.9. *Let X be a closed H-invariant subset of $\Gamma \backslash G$ containing a compact A-invariant subset W. Let $Y \subset X$ be a U-minimal subset with respect to W. Suppose*

(1) *$YL \subset Y$ for some one-parameter semigroup $L < AV$ and*
(2) *$Yv \subset X$ for some nontrivial $v \in V$.*

Then X contains $x_0 V_I$ for some $x_0 \in W$ and an interval $0 \in I$.

Proof. Any one-parameter semigroup $L < AV$ is either a one-parameter semigroup $V_+ < V$ or $v_0 A_+ v_0^{-1}$ for some $v_0 \in V$ and a one-parameter semigroup $A_+ < A$.

Case 1: If $L = V_+$, we are done.

Case 2: If $L = v_0 A_+ v_0^{-1}$ for a nontrivial $v_0 \in V$, then

$$X \supset Y(v_0 A_+ v_0^{-1})A.$$

Since $v_0 A_+ v_0^{-1} A$ contains V_I for some interval $0 \in I$, the claim follows.

Case 3: If $L = A_+$, we first note that $YA \subset Y$; take any sequence $a_n \to \infty$ in A_+, and $y_0 \in Y \cap W$. Then $y_0 a_n \in Y \cap W$ converges to some $y_1 \in Y \cap W$. Now $\limsup_{n\to\infty} a_n^{-1} A_+ = A$. Therefore $Y \supset y_1 A$. Since $\overline{y_1 U} = Y$, we get $Y \supset YA$. Since AvA contains a semigroup V_+ of V, we deduce

$$X \supset YvA \supset YAvA \supset YV_+. \qquad \square$$

In the next section, we discuss the significance of the conclusion that X contains a segment $x_0 V_I$, depending on the relative location of x_0 to ∂ core M.

6 Interior frames and boundary frames

Let $\Gamma < G$ be a Zariski dense geometrically finite group, and let $M = \Gamma\backslash\mathbb{H}^3$. When M has infinite volume, its convex core has a nonempty boundary, which makes the dynamical behavior of a frame under geometric flows different depending on its relative position to ∂ core M.

Recall the notation F_Λ from Equation (4.2). We denote by F^* the interior of F_Λ and by ∂F the boundary of F_Λ. In order to show that a given closed H-invariant subset $X \subset F_\Lambda$ with no periodic U-orbits is equal to F_Λ, it suffices to show that X contains $x_0 V_I$ for some $x_0 \in F^* \cap \mathrm{RF}_+ M$ and an interval $0 \in I$ (Lemma 6.1). It is important to get $x_0 \in F^*$, as the similar statement is not true if $x_0 \in \partial F$. For example, in the rigid acylindrical case, if $x_0 \in \partial F$, then $x_0 H V_+ H$ is a closed H-invariant subset of ∂F for a certain semigroup $V_+ < V$ (cf. Theorem 7.1), and hence if V_I belongs to V_+, we cannot use $x_0 V_I$ to obtain useful information on $X \cap F^*$.

6.1 INTERIOR FRAMES.

In this section, we assume that Λ is connected. Under this hypothesis, the closed H-invariant set $F_\Lambda = \mathrm{RF}_+ M \cdot H$ has a nonempty interior that can be described as follows:

$$\begin{aligned} F^* &= \{[g] \in \Gamma\backslash G : \pi(P_g) \cap M^* \neq \emptyset\} \\ &= \bigcup \{xH \subset \Gamma\backslash G : p(xH) \cap M^* \neq \emptyset\}, \end{aligned}$$

where M^* denotes the interior of core M.

The condition $\pi(P_g) \cap M^* \neq \emptyset$ is equivalent to the condition that the circle $C_g = \partial P_g$ separates the limit set Λ—that is, both components of $\mathbb{S}^2 - C_g$ intersect Λ nontrivially. If we set

$$\mathcal{C}^* := \{C \in \mathcal{C} : C \text{ separates } \Lambda\},$$

we have

$$F^*/H = \Gamma \backslash \mathcal{C}^*.$$

We observe that the connectedness of Λ implies the following two equivalent statements:

(1) For any $C \in \mathcal{C}^*$, $\# C \cap \Lambda \geq 2$.
(2) $F^* \cap \mathrm{RF}_+ M \subset \mathrm{RF}\, M \cdot U$.

By the openness of F^* and (2), for any $x \in F^* \cap \mathrm{RF}_+ M$, there exists a neighborhood $\mathcal{O}$ of e in G such that

$$x\mathcal{O} \cap \mathrm{RF}_+ M \subset \mathrm{RF}\, M \cdot U. \tag{6.1}$$

Thanks to this stability, we have the following lemma:

LEMMA 6.1. *Let $X \subset F_\Lambda$ be a closed H-invariant subset intersecting* RF M *and with no periodic U-orbits. If X contains $x_0 V_I$ for some $x_0 \in F^* \cap \mathrm{RF}_+ M$ and an interval $0 \in I$, then*

$$X = F_\Lambda.$$

Proof. It suffices to find $z_0 V$ inside X for some $z_0 \in \mathrm{RF}\, M$ by Theorem 3.6. Without loss of generality, we may assume $I = [0, s]$ for some $s > 0$. We write $v_t := u_{it}$. Since $x_0 \in F^* \cap \mathrm{RF}_+ M$, there exists $0 < \varepsilon < s$ such that $x_0 v_\varepsilon \in X \cap \mathrm{RF}\, M \cdot U$ by Equation (6.1). Hence, there exists $x_1 \in x_0 v_\varepsilon U \cap \mathrm{RF}\, M \cap X$, so $x_1 v_\varepsilon^{-1} V_I = x_1 V_{[-\varepsilon, s-\varepsilon]} \subset X$. Since X has no periodic U-orbit, x_1^+ is a radial limit point of Λ, and hence there exists $t_n \to +\infty$ such that $x_1 a_{t_n}$ converges to some $z_0 \in \mathrm{RF}\, M$. Since

$$\limsup_{n\to\infty} a_{t_n}^{-1} V_{[-\varepsilon, s-\varepsilon]} a_{t_n} = V$$

and $x_1 V_{[-\varepsilon, s-\varepsilon]} a_{t_n} = x_1 a_{t_n} (a_{t_n}^{-1} V_{[-\varepsilon, s-\varepsilon]} a_{t_n}) \subset X$, we obtain $z_0 V \subset X$ as desired. □

6.2 BOUNDARY FRAMES. The geometric structure of the boundary $\partial F = F_\Lambda - F^*$ plays an important (rather decisive) role in the rigidity study. For instance, unless xH is bounded, xH is expected to accumulate on ∂F. In the most dramatic situation, all the accumulation of xH may fall into the boundary ∂F so that $\overline{xH} \subset xH \cup \partial F$. Unless we have some analysis on what possible closed H-invariant subsets of ∂F are, there isn't too much more we can say about such situation.

A geodesic plane $P \subset \mathbb{H}^3$ is called a supporting plane if it intersects hull(Λ) and one component of $\mathbb{H}^3 - P$ is disjoint from hull(Λ), or equivalently the circle $C = \partial P$ is a supporting circle in the sense that $\#C \cap \Lambda \geq 2$ and C does not separate Λ.

For $C \in \mathcal{C}$, we denote by Γ^C the stabilizer of C in Γ. The theory of bending laminations yields the following:

THEOREM 6.2 ([38, theorem 5.1]).
For any supporting circle $C \in \mathcal{C}$,

(1) Γ^C is a finitely generated Fuchsian group, and
(2) there exists a finite subset $\Lambda_0 \subset C \cap \Lambda$ such that

$$C \cap \Lambda = \Lambda(\Gamma^C) \cup \Gamma^C \Lambda_0,$$

where $\Lambda(\Gamma^C)$ denotes the limit set of Γ^C.

DEFINITION 6.3. We call $x \in \partial F$ a boundary frame, and we call $x = [g] \in \partial F$ a *thick* boundary frame if there exists a supporting circle C with a nonelementary stabilizer Γ^C such that $C_g = C$ or C_g is tangent to C at $g^+ \in \Lambda(\Gamma^C)$.

THEOREM 6.4.
If $x \in \partial F$ is a thick boundary frame such that xU is not closed, then $\overline{xU} \supset xvAv^{-1}$ for some $v \in V$. If $x \in \mathrm{RF}\,M$ in addition, then $\overline{xU} \supset xA$.

Proof. Choose $g \in G$ so that $[g] = x$. By the hypothesis on x, there exists a supporting circle C with Γ^C nonelementary and $g^+ \in \Lambda(\Gamma^C)$. The circle C_g is equal to C or tangent to C at g^+. It follows that there exists $v \in V$ such that $C = C_{gv}$. By Theorem 6.2, the stabilizer Γ^C is finitely generated and nonelementary. It now follows from a theorem by Dal'Bo [10] that xvU is either periodic (if $g^+ = (gv)^+$ is a parabolic fixed point of Γ^C) or $\overline{xvU}$ contains $xvH \cap \mathrm{RF}_+ M \supset xvA$. Since v commutes with U, the first claim follows. If $x \in \mathrm{RF}\,M$ in addition, then C_g must be equal to C, and hence $v = e$. ∎

LEMMA 6.5. *Let $X \subset F_\Lambda$ be a closed H-invariant subset intersecting $\mathrm{RF}\, M$ with no periodic U-orbits. If $X \cap F^*$ contains zv_0 for some thick boundary frame $z \in \partial F \cap \mathrm{RF}\, M$ and $v_0 \in V - \{e\}$, then*

$$X = F_\Lambda.$$

Proof. By Lemma 6.1, it suffices to find $x_0 V_I$ inside X for some $x_0 \in F^*$ and an interval $0 \in I$. By Theorem 6.4, we have $\overline{zU} \supset zA$. Therefore,

$$X \supset \overline{zv_0 UA} = \overline{zUv_0A} \supset \overline{zAv_0A} \supset zV_+,$$

where V_+ is the one-parameter semigroup contained in $V \cap Av_0A$. Since $zv_0 \subset zV_+ \cap F^* \neq \emptyset$ and F^* is open, $zv_0 V_I \subset zV_+ \cap F^*$ for some interval $0 \in I$, as desired. □

7 Rigid acylindrical groups and circular slices of Λ

Let $\Gamma < G$ be a rigid acylindrical Kleinian group and $M := \Gamma\backslash\mathbb{H}^3$ the associated hyperbolic 3-manifold. We assume $\mathrm{Vol}(M) = \infty$.

7.1 BOUNDARY FRAMES FOR RIGID ACYLINDRICAL GROUPS.

In this case, we have a complete understanding of the orbit closures in the boundary ∂F, which makes it possible to give a complete classification for *all* orbit closures in F.

When Γ is rigid acylindrical, *every* supporting circle C is contained in the limit set, so that $C \cap \Lambda = C$. It follows that Γ^C is a uniform lattice of G^C and the orbit $xH = p(gH)$ is compact whenever C_g is a supporting circle. This implies the following:

THEOREM 7.1 ([37]).
Let Γ be rigid acylindrical, and let $x \in \partial F$ be a boundary frame.

(1) If $x \in \mathrm{RF}_+ M$, then

$$\overline{xU} = xvHv^{-1} \quad \text{for some } v \in V.$$

(2) If $x \in \mathrm{RF}\, M$, then xH is compact.
(3) If $x \in \mathrm{RF}_+ M - \mathrm{RF}\, M$, there exist a one-parameter semigroup V_+ of V and a boundary frame $x_0 \in \partial F$ with $x_0 H$ compact such that

$$\overline{xH} = x_0 H V_+ H.$$

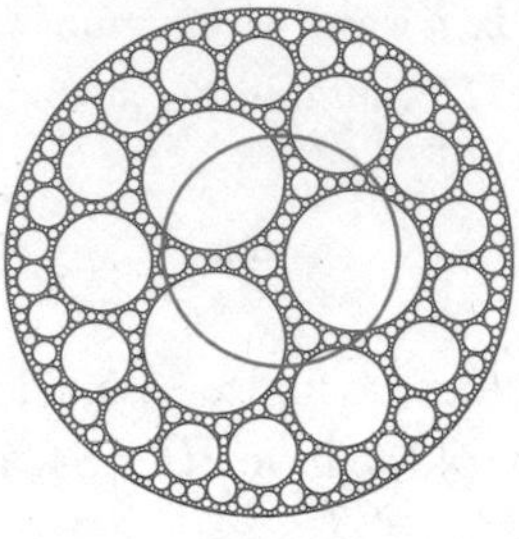

Figure 14.10. Circular slice of Λ

7.2 CIRCULAR SLICES OF Λ. Circular slices of the limit set Λ control the recurrence time of U-orbits into the compact subset RF M. For $x \in \mathrm{RF}\, M$, set

$$\mathsf{T}(x) := \{t \in \mathbb{R} : xu_t \in \mathrm{RF}\, M\}.$$

If $x = [g]$, then $(gu_t)^+ = g^+ \in C_g \cap \Lambda$, and hence

$$t \in \mathsf{T}(x) \text{ if and only if } (gu_t)^- \in C_g \cap \Lambda.$$

We will use the following geometric fact for a rigid acylindrical manifold M: if we write $S^2 - \Lambda = \bigcup B_i$ where B_i's are round open disks, then

$$\inf_{i \neq j} d(\mathrm{hull}(B_i), \mathrm{hull}(B_j)) \geq \varepsilon_0, \tag{7.1}$$

where $2\varepsilon_0$ is the systol of the double of the convex core of M. This follows because a geodesic in $\mathbb{H}^3$ that realizes the distance $d(\mathrm{hull}(B_i), \mathrm{hull}(B_j))$ is either a closed geodesic in M or the half of a closed geodesic in the double of core(M).

PROPOSITION 7.2. *Let Γ be rigid acylindrical. There exists $\kappa > 1$ such that for all $x \in \mathrm{RF}\, M$, $\mathsf{T}(x)$ is κ-thick. In particular, for any sequence $x_i \in \mathrm{RF}\, M$, $\mathsf{T}(x_i)$ has accumulating renormalizations.*

Proof. For $\varepsilon_0 > 0$ given by Equation (7.1), consider the upper half-plane model of $\mathbb{H}^2 = \{(x_1, 0, y) : x_1 \in \mathbb{R}, y > 0\}$. For $a < b$, $\mathrm{hull}_{\mathbb{H}^2}(a, b) \subset \mathbb{H}^2$ denotes the convex hull of the interval connecting $(a, 0, 0)$ and $(b, 0, 0)$. Define $\kappa > 1$ by the equation

$$d_{\mathbb{H}^2}(\mathrm{hull}(-\kappa, -1), \mathrm{hull}(1, \kappa)) = \varepsilon_0/2;$$

since $\lim_{s \to \infty} d_{\mathbb{H}^2}(\mathrm{hull}(-s, -1), \mathrm{hull}(1, s)) = 0$, such $\kappa > 1$ exists.

Since $z \mapsto tz$ is a hyperbolic isometry in $\mathbb{H}^2$ for any $t > 0$, we have

$$d_{\mathbb{H}^2}(\text{hull}(-\kappa t, -t), \text{hull}(t, \kappa t)) = \varepsilon_0/2.$$

We now show that $\mathsf{T}(x)$ is κ-thick for $x \in \text{RF}\,M$. It suffices to show the claim for $x = [g]$, where $g = (e_1, e_2, e_3)$ is based at $(0, 0, 1)$ with e_2 in the direction of the positive real axis and $g^+ = \infty, g^- = 0$. Note that $gu_t \in \text{RF}\,M$ if and only if $t = (gu_t)^- \in \Lambda$ and hence

$$\mathsf{T}(x) = \mathbb{R} \cap \Lambda.$$

Suppose that $\mathsf{T}(x)$ is not κ-thick. Then for some $t > 0$, $\mathsf{T}(x)$ does not intersect $[-\kappa t, -t] \cup [t, \kappa t]$—that is, $[-\kappa t, -t] \cup [t, \kappa t] \subset \bigcup B_i$. Since B_i's are convex and $0 \in \Lambda$, there exists $i \neq j$ such that $[-\kappa t, -t] \subset B_i$ and $[t, \kappa t] \subset B_j$. Hence,

$$d(\text{hull}(-\kappa t, -t), \text{hull}(t, \kappa t)) = \varepsilon_0/2 \geq d(\text{hull}(B_i), \text{hull}(B_j)) \geq \varepsilon_0,$$

yielding contradiction. The second claim follows from Lemma 5.4. □

7.3 CLOSED OR DENSE DICHOTOMY FOR H-ORBITS.

In Theorem 7.1, we have described all possible orbit closures for H and U-action inside ∂F. It remains to consider orbits of $x \in F^*$.

THEOREM 7.3 ([37]).
For any $x \in F^*$, xH is either closed or dense in F_Λ.

Proof. Set $X := \overline{xH}$, and assume that $X \neq xH$. We then need to show $X = F_\Lambda$. Since $F^* \cap \text{RF}_+ M \subset \text{RF}\,M \cdot U$ and $xH \subset F^* \cap \text{RF}_+ M \cdot H$, we may assume without loss of generality that $x = [g] \in \text{RF}\,M$.

Set $W := X \cap F^* \cap \text{RF}\,M$.

Case 1: W is not compact. In this case, there exists $x_n \in W$ converging to some $z \in \partial F \cap \text{RF}\,M$. Write $x_n = zg_n$ with $g_n \to e$ in $G - H$.

Suppose that $g_n = h_n v_n \in HV$ for some n. Since $zh_n \in zH \subset \partial F \cap \text{RF}\,M$ and $(zh_n)v_n \in F^* \cap \text{RF}\,M$, the claim follows from Lemma 6.5.

Now suppose that $g_n \notin HV$ for all n. By Lemma 5.5, there exist $t_n \in \mathsf{T}(x_n)$ and $h_n \in H$ such that $h_n g_n u_{t_n}$ converges to some $v \in V - \{e\}$. Since zH is compact, zh_n^{-1} converges to some $z_0 \in \partial F \cap \text{RF}\,M$ by passing to a subsequence. Hence, as $n \to \infty$,

$$x_n u_{t_n} = zh_n^{-1}(h_n g_n u_{t_n}) \to z_0 v.$$

Since $z_0 \in \partial F \cap \text{RF}\, M$ and $z_0 v \in \text{RF}\, M$, we get $z_0 v \in F^*$; hence the claim follows by Lemma 6.5.

Case 2: W is compact. It follows from the definition of W that for any $x \in W$,

$$\mathsf{T}(x) = \{t : xu_t \in W\}.$$

We claim that X contains a U-minimal subset Y with respect to W such that $X - y_0 H$ is not closed for some $y_0 \in Y \cap W$. We divide our proof into two further cases:

Case (a): Suppose that xH is not locally closed that is, $X - xH$ is not closed. In this case, any U-minimal subset $Y \subset X$ with respect to W works. First, if $Y \cap W \subset xH$, then choose any $y_0 \in Y \cap W$. Observe that $\overline{xH} - y_0 H = \overline{xH} - xH$ is not closed, which implies the claim. If $Y \cap W \not\subset xH$, choose $y_0 \in (Y \cap W) - xH$. Then $\overline{xH} - y_0 H$ contains xH and hence cannot be closed.

Case (b). Suppose that xH is locally closed and $X - xH$ intersects W non-trivially. Therefore $X - xH$ contains a U-minimal subset Y with respect to W. Then any $y_0 \in Y \cap W$ has the desired property; since $y_0 \in X - xH$, there exists $h_n \in H$ such that $xh_n \to y$. If we write $xh_n = yg_n$, then $g_n \to e$ in $G - H$, since $y \notin xH$.

By Lemmas 5.7, 5.8, and 5.9, X contains $x_0 V_I$ for some $x_0 \in W$ and for an interval $0 \in I$; since $x_0 \in F^*$, this finishes the proof by Lemma 6.1. □

7.4 TOPOLOGICAL RIGIDITY OF GEODESIC PLANES. In [36] and [37], the following theorem was obtained:

THEOREM 7.4.
Let M be a rigid acylindrical hyperbolic 3-manifold. Then

(1) *any geodesic plane P intersecting* core *M is either properly immersed or dense,*
(2) *the fundamental group of a properly immersed P intersecting* core *M is a nonelementary Fuchsian subgroup,*
(3) *there are at most countably many properly immersed geodesic planes in M intersecting* core *M, and*
(4) *any infinite sequence of geodesic planes P_i intersecting* core *M becomes dense in M—that is* $\lim P_i = M$.

REMARK 7.5.

(1) There exists a closed arithmetic hyperbolic 3-manifold $\Gamma\backslash\mathbb{H}^3$ without any properly immersed geodesic plane, as shown by Maclachlan-Reid [25].

(2) When M has finite volume and has at least one properly immersed geodesic plane, then M is arithmetic if and only if there are infinitely many properly immersed geodesic planes [31, 3].

(3) A natural question is whether a rigid acylindrical hyperbolic 3-manifold M necessarily covers an arithmetic hyperbolic 3-manifold if there exist infinitely many properly immersed (unbounded) geodesic planes intersecting its core. The reason for the word *unbounded* in parentheses is that in any geometrically finite hyperbolic 3-manifold of infinite volume, there can be only finitely many bounded geodesic planes [37, 4]. In view of the proofs given in [31, 3], the measure theoretic equidistribution of infinitely many closed H-orbits needs to be understood first.

7.5 CLASSIFICATION OF U-ORBIT CLOSURES. In the rigid acylindrical case, the complete classification of the U-orbit closures inside ∂F given in Theorem 7.1 can be extended to the whole space $\mathrm{RF}_+ M$:

THEOREM 7.6 ([36]).
For any $x \in \mathrm{RF}_+ M$,

$$\overline{xU} = xL \cap \mathrm{RF}_+ M,$$

where L *is either* $v^{-1}Hv$ *for some* $v \in N$, *or* G.

Our proof is based on two main features of a rigid acylindrical group. The first property is that

there exists a compact H-orbit in $\mathrm{RF}_+ M$

—namely, those $[g]H$ whose corresponding plane P_g is a supporting plane. This very important feature of M is a crucial ingredient of our proof. In particular, the following *singular* set is nonempty:

$$\mathscr{S}(U) = \bigcup zHV \cap \mathrm{RF}_+ M,$$

where the union is taken over all closed H orbits zH.

We set

$$\mathscr{G}(U) := \mathrm{RF}_+ M - \mathscr{S}(U)$$

and call it the generic set. Note that

$$\mathscr{G}(U) \subset F^*.$$

The second property is the following control on the prelimiting behavior of RF M-points, whose proof is based on the totally geodesic nature of ∂ core M.

LEMMA 7.7 ([36, lemma 4.2]). *If $x_n \in F^*$ converges to some $y \in \mathrm{RF}\,M$, then there exists a sequence $x_n' \in x_n U \cap \mathrm{RF}\,M$ converging to y or converging to some boundary frame $y' \in \partial F \cap \mathrm{RF}\,M$.*

For $x \in \mathscr{S}(U)$, Theorem 7.6 follows from a theorem of Hedlund [18] and Dal'Bo [10] on the minimality of horocyclic action on the Fuchsian case.

PROPOSITION 7.8 ([36]). *If $x \in \mathscr{G}(U)$, then*

$$\overline{xU} = \mathrm{RF}_+ M.$$

Proof. Setting $X := \overline{xU}$, we first claim that

$$X \cap \mathscr{S}(U) \neq \emptyset. \tag{7.2}$$

If $X \cap \partial F \neq \emptyset$, the claim follows from Theorem 7.1(1). Hence, we assume that $X \subset F^*$. Let Y be a U-minimal subset of X with respect to RF M. By Lemma 5.7, $YL \subset Y$, where $L < AV$ is a one-parameter semigroup. If L is a semigroup of V—say, V_+—then take a sequence $v_n \to \infty$ in V_+. Since $YV^+ \subset Y \subset F^* \subset \mathrm{RF}\,M \cdot U$, up to passing to a subsequence, there exists $y_n \in Y$ such that $y_n v_n$ converges to an RF M-point—say y_0. Then

$$y_0 V = \lim_{n\to\infty} (y_n v_n) \cdot \limsup_n (v_n^{-1} V_+) \subset \overline{YV_+} \subset Y.$$

Hence, $Y = X = \mathrm{RF}_+ M$, proving the claim. If $L = vA_+v^{-1}$ for some semigroup A_+ of A, since $\mathscr{S}(U)$ is V-invariant, we may assume that $L = A_+$. Take a sequence $a_n \to \infty$ in A_+. Then for any $y \in Y$, ya_n converges to an RF M-point—say, $y_0 \in Y$—by passing to a subsequence. So

$$y_0 A = \lim_{n\to\infty} (ya_n) \cdot \limsup_n (a_n^{-1} A_+) \subset \overline{YA_+} \subset Y.$$

On the other hand, either $y_0 \in \mathscr{S}(U)$ or $\overline{y_0 H} = F_\Lambda$ (Theorem 7.3). In the latter case, $\overline{y_0 H}$ contains a compact H-orbit zH. Since $\overline{y_0 AUM_0} = \overline{y_0 H}$, it follows that $\overline{y_0 AU} \cap zH \neq \emptyset$, proving the claim in Equation (7.2).

Therefore X contains $\overline{yU} = yvHv^{-1} \cap \mathrm{RF}_+ M$ for some $y \in \mathscr{S}(U)$. Without loss of generality, we may assume $X \supset yH \cap \mathrm{RF}_+ M$ by replacing x with xv. Set $Y := yH \cap \mathrm{RF}_+ M$, which is a U-minimal subset. There exists $s_n \in \mathbb{R}$ such that $y = \lim_{n\to\infty} xu_{s_n}$. In view of Lemma 7.7, we may assume that $xu_{s_n} \in \mathrm{RF}\,M$

for all n. Write $xu_{s_n} = yg_n$ for some sequence $g_n \to e$ in G. Since $y \in \mathscr{S}(U)$ and $x \in \mathscr{G}(U)$, it follows that $g_n \notin HV$ for all n. Hence by Lemma 5.5, there exist $t_n \in \mathsf{T}(yg_n)$ and $h_n \in H$ such that $h_n g_n u_{t_n}$ converges to some $v \in V$; moreover, v can be taken arbitrarily large. By passing to a subsequence, $yg_n u_{t_n}$ converges to some $y_0 \in \mathrm{RF}\, M$, and hence yh_n^{-1} converges to $y_1 := y_0 v^{-1} \in yH \cap \mathrm{RF}_+ M = Y$. Therefore $X \supset y_0 = y_1 v$, and hence $X \supset Yv$. As $y_1 v \in \mathrm{RF}\, M$, $Yv \cap \mathrm{RF}\, M \neq \emptyset$. As v can be taken arbitrarily large, there exists a sequence $v_n \to \infty$ in V such that X contains Yv_n. Choose $y_n \in Y$ so that $y_n v_n \in \mathrm{RF}\, M$ converges to some $x_0 \in \mathrm{RF}\, M$ by passing to a subsequence. Since Y is A-invariant and $\limsup_{n\to\infty} v_n^{-1} A v_n \supset V$, we deduce

$$X \supset \lim_{n\to\infty} (y_n v_n) \cdot \limsup_{n\to\infty} (v_n^{-1} A v_n) \supset x_0 V.$$

Therefore $X = \mathrm{RF}_+ M$. □

As an immediate corollary, we deduce the following:

COROLLARY 7.9 ([36]). *Let M be a rigid acylindrical hyperbolic 3-manifold. Then the closure of any horocycle is either a properly immersed surface, parallel to a geodesic plane, or equal to M.*

7.6 MEASURE RIGIDITY? If there exists a closed orbit xH for $x \in \mathrm{RF}_+ M$, then the stabilizer of x in H is a nonelementary convex cocompact Fuchsian subgroup and there exists a unique U-invariant ergodic measure supported on $xH \cap \mathrm{RF}_+ M$, called the Burger-Roblin measure m^{BR}_{xH} on xH.

QUESTION 7.10. Let M be a rigid acylindrical hyperbolic 3-manifold. Is any locally finite U-ergodic measure on $\mathrm{RF}_+ M$ either m^{BR} or m^{BR}_{xH} for some closed H orbit xH, up to a translation by the centralizer of U?

Theorem 7.6 implies the positive answer to this question at least in terms of the support of the measure: the support of any locally finite U-ergodic measure on $\mathrm{RF}_+ M$ is either $\mathrm{RF}_+ M$ or $\mathrm{RF}_+ M \cap xHv$ for some closed orbit xH and $v \in N$.

8 Geometrically finite acylindrical hyperbolic 3-manifolds

Let $\Gamma < G$ be a Zariski dense geometrically finite group, and let $M = \Gamma\backslash\mathbb{H}^3$. We assume $\mathrm{Vol}(M) = \infty$. In the rigid acylindrical case, we were able to give a complete classification of all possible closures of a geodesic plane in M; this is largely due to the *rigid* structure of the boundary of core M. In particular, the intersection of a geodesic plane and the convex core of M is either closed or dense in core M.

In general, the convex core of M is not such a natural ambient space to study the topological behavior of a geodesic plane because of its non-homogeneity property. Instead, its interior, which we denote by M^*, is a better space to work with: first of all, M^* is a hyperbolic 3-manifold with no boundary (although incomplete), which is diffeomorphic to M; second, a geodesic plane P that does not intersect M^* cannot come arbitrarily close to M^* as $\overline{P}$ must be contained in the ends $M - M^*$.

DEFINITION 8.1. A geodesic plane P^* in M^* is defined to be a nonempty intersection $P \cap M^*$ for a geodesic plane P of M.

Let $P = \pi(\tilde{P})$ for a geodesic plane $\tilde{P} \subset \mathbb{H}^3$, and set $S = \text{Stab}_\Gamma(\tilde{P}) \backslash \tilde{P}$. Then the natural map $f : S \to P \subset M$ is an immersion (which is generically injective), $S^* := f^{-1}(M^*)$ is a nonempty convex subsurface of S with $\pi_1(S) = \pi_1(S^*)$, and P^* is given as the image of the restriction of f to S^*. The group $\pi_1(S^*)$ will be referred to as the fundamental group of P^*. We note that a geodesic plane P^* is always connected as P^* is covered by the convex subset $\tilde{P} \cap$ Interior (hull Λ).

8.1 RIGIDITY OF GEODESIC PLANES IN M^*.

An analogous topological rigidity of planes to Theorem 7.4 continues to hold inside M^*, provided M is a geometrically finite *acylindrical* hyperbolic 3-manifold.

The following rigidity theorem was proved jointly with McMullen and Mohammadi for convex cocompact cases in [36] and extended to geometrically finite cases jointly with Benoist [4]:

THEOREM 8.2.
Let M be a geometrically finite acylindrical hyperbolic 3-manifold. Then geodesic planes in M^* are topologically rigid in the following senses:

(1) *Any geodesic plane P^* in M^* is either properly immersed or dense;*
(2) *the fundamental group of a properly immersed P^* is a nonelementary geometrically finite Fuchsian subgroup;*
(3) *there are at most countably many properly immersed geodesic planes in M^*; and*
(4) *any infinite sequence of geodesic planes P_i^* becomes dense in M^*—that is, $\lim P_i^* = M^*$.*

This theorem is deduced from the following results on H-orbits in F^*:

THEOREM 8.3. ([36], [4])
Let M be a geometrically finite acylindrical hyperbolic 3-manifold. Then

(1) *any H-orbit in F^* is either closed or dense;*
(2) *if xH is closed in F^*, then $\text{Stab}_H(x)$ is Zariski dense in H;*
(3) *there are at most countably many closed H-orbits in F^*; and*
(4) *any infinite sequence of closed H-orbits $x_i H$ becomes dense in F^*—that is, $\lim x_i H = F^*$.*

8.2 CLOSED OR DENSE DICHOTOMY FOR ACYLINDRICAL GROUPS.

In this section, we discuss the proof of the following closed or dense dichotomy:

THEOREM 8.4.
Let M be a geometrically finite acylindrical hyperbolic 3-manifold. Then any H-orbit in F^* is either closed or dense.

Indeed, the proof of Theorem 7.3 for the rigid acylindrical case can be modified to prove the following proposition.

PROPOSITION 8.5 (main proposition). *Let Γ be a Zariski dense convex cocompact subgroup of G with a connected limit set. Let $\mathcal{R}$ be a closed A-invariant subset of* RF M *satisfying that for any $x \in \mathcal{R}$, $\mathsf{T}(x) := \{t : xu_t \in \mathcal{R}\}$ has accumulating renormalizations. Then for any $x \in \mathcal{R} \cap F^*$, xH is either locally closed or dense in F. When xH is locally closed, it is closed in $\mathcal{R}H \cap F^*$.*

Proof. Let $X := \overline{xH}$ for $x \in \mathcal{R} \cap F^*$. Set $W := X \cap \mathcal{R} \cap F^*$. Suppose that either xH is not locally closed or $(X - xH) \cap W \neq \emptyset$. We claim that $X = F_\Lambda$.

Case 1: W is not compact. By repeating verbatim the proof of Theorem 7.3, we obtain $zv \in X \cap \mathcal{R}$ for some $z \in \partial F \cap \mathcal{R}$ and nontrivial $v \in V$. As $z = [g] \in \mathcal{R}$, Γ^{C_g} is nonelementary and hence z is a *thick* boundary frame. Since $zv \in F^*$, the claim follows from Lemma 6.5.

Case 2: W is compact. By repeating verbatim the proof of Theorem 7.3, we show that X contains a U-minimal subset Y with respect to W such that $X - y_0 H$ is not closed for some $y_0 \in Y \cap W$. Hence by applying Lemmas 5.7, 5.8, 5.9, and 6.1, we get $X = F_\Lambda$. □

When Γ is rigid acylindrical, note that $\mathcal{R} = \mathrm{RF}\, M$ satisfies the hypothesis of Proposition 8.5. In view of this proposition, Theorem 8.4 for a convex cocompact case now follows from the following theorem, and a general geometrically finite case can be proved by an appropriately modified version, taking account of closed horoballs, which is responsible for the noncompactness of RF M.

THEOREM 8.6.
Let M be a geometrically finite acylindrical hyperbolic 3-manifold. Then there exists a closed A-invariant subset $\mathcal{R} \subset \mathrm{RF}\, M$ such that for any $x \in \mathcal{R}$, $\mathsf{T}(x) := \{t : xu_t \in \mathcal{R}\}$ accumulating renormalizations. Moreover, $F^* \subset \mathcal{R}H$.

We use the notion of a conformal modulus in order to find a closed subset $\mathcal{R}$ satisfying the hypothesis of Theorem 8.6. An annulus $\mathcal{A} \subset \mathbb{S}^2$ is an open region whose complement consists of two components. If neither component is a single point, $\mathcal{A}$ is conformally equivalent to a unique round annulus of the form $\{z \in \mathbb{C} : 1 < |z| < R\}$. The modulus $\mathrm{mod}(\mathcal{A})$ is then defined to be $\log R$. If P is a compact set of a circle C such that its complement $C - P = \bigcup I_i$ is a union of at least two intervals with disjoint closures, we define the modulus of P as

$$\mathrm{mod}\, P := \inf_{i \neq j} \ \mathrm{mod}\, (I_i, I_j),$$

where $\mathrm{mod}\,(I_i, I_j) := \mathrm{mod}\left(\mathbb{S}^2 - (\bar{I}_i \cup \bar{I}_j)\right)$.

For $\varepsilon > 0$, define $\mathcal{R}_\varepsilon \subset \mathrm{RF}\, M$ as the following subset:

$$\mathcal{R}_\varepsilon := \{[g] : C_g \cap \Lambda \text{ contains a compact set of modulus} \geq \varepsilon \text{ containing } g^{\pm}\}.$$

LEMMA 8.7. *For $\varepsilon > 0$, the set $\mathcal{R}_\varepsilon$ is closed.*

Proof. Suppose that $g_n \in \mathcal{R}_\varepsilon$ converges to some $g \in \mathrm{RF}\, M$. We need to show $g \in \mathcal{R}_\varepsilon$. Let $P_n \subset C_{g_n} \cap \Lambda$ be a compact set of modulus $\geq \varepsilon$ containing $g_n^{\pm}$. Since the set of all closed subsets of $\mathbb{S}^2$ is a compact space in the Hausdorff topology on closed subsets, we may assume P_n converges to some P_∞ by passing to a subsequence. This means that $P_\infty = \limsup_n P_n = \liminf_n P_n$ [19].

Write $C_g - P_\infty = \bigcup_{i \in I} I_i$ as the disjoint union of connected components. As $g^{\pm} \in P_\infty$, $|I| \geq 2$. Let $i \neq j \in I$, and write $I_i = (a_i, b_i)$ and $I_j = (a_j, b_j)$. There exist $a_{i,n}, b_{i,n}, a_{j,n}, b_{j,n} \in P_n$ converging to a_i, b_i, a_j, and b_j, respectively. Set $I_{i,n}$ and $I_{j,n}$ to be the intervals $(a_{i,n}, b_{i,n})$ and $(a_{j,n}, b_{j,n})$ respectively. Since $I_{i,n} \to I_i$ and $I_{j,n} \to I_j$, $I_{i,n} \cup I_{j,n} \subset C_{g_n} - P_n$ for all large n. Since $a_{i,n}, b_{i,n} \in P_n$, $I_{i,n}$ is a

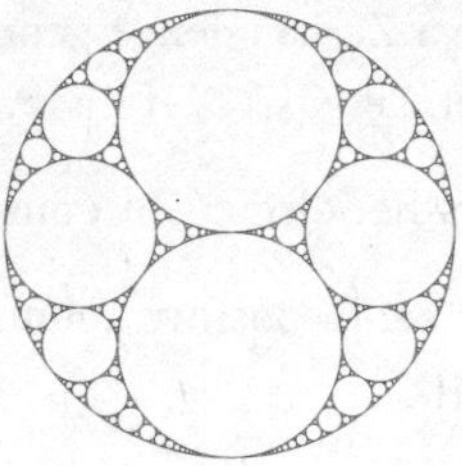

Figure 14.11. Apollonian gasket

connected component of $C_{g_n} - P_n$. Similarly, $I_{j,n}$ is a connected component of $C_{g_n} - P_n$. Since $\operatorname{mod}(I_{i,n}, I_{j,n}) \geq \varepsilon$ for all n, it follows that I_i and I_j have disjoint closures and $\operatorname{mod}(I_i, I_j) \geq \varepsilon$. This shows that P_∞ is a compact subset of $C_g \cap \Lambda$ of modulus at least ε containing $g^{\pm}$. Therefore, $g \in \mathcal{R}_\varepsilon$. □

There exists $\kappa = \kappa(\varepsilon) > 1$ such that for any $x \in \mathcal{R}_\varepsilon$, $\mathsf{T}(x) := \{t : xu_t \in \mathcal{R}_\varepsilon\}$ is κ-thick (see [36, proposition 4.3]); hence, $\mathcal{R}_\varepsilon$ satisfies the hypothesis of Theorem 8.6.

In general, $\mathcal{R}_\varepsilon$ may be empty! However, for geometrically finite acylindrical manifolds, there exists $\varepsilon > 0$ such that

$$F^* \subset \mathcal{R}_\varepsilon H \tag{8.1}$$

([36], [4]); hence, Theorem 8.4 follows. The inclusion (8.1) is proved using *bridge* arguments devised in [36] and the monotonicity of conformal moduli, based on the property that for a convex cocompact acylindrical manifold M, Λ is a Sierpinski carpet of positive modulus—that is,

$$\inf_{i \neq j} \operatorname{mod}(\mathbb{S}^2 - (\overline{B}_i \cup \overline{B}_j)) > 0,$$

where B_i's are components of $\mathbb{S}^2 - \Lambda$ (see [36] for details).

When M has cusps, the closures of some components of $\mathbb{S}^2 - \Lambda$ may meet each other, and hence Λ is not even a Sierpinski carpet in general. Nevertheless, under the assumption that M is a geometrically finite acylindrical manifold, Λ is still a *quotient of a Sierpinski carpet of positive modulus* in the sense that we can present $\mathbb{S}^2 - \Lambda$ as the disjoint union $\bigcup_\ell T_\ell$, where T_ℓ's are maximal *trees of components of* $\mathbb{S}^2 - \Lambda$ so that

$$\inf_{\ell \neq k} \operatorname{mod}(\mathbb{S}^2 - (\overline{T}_\ell \cup \overline{T}_k)) > 0.$$

QUESTION 8.8. Let Γ be a Zariski dense geometrically finite subgroup of G with a connected limit set. Let $C \in \mathcal{C}^*$. If $C \cap \Lambda$ contains a Cantor set,

$$\text{is } \Gamma C \text{ either discrete or dense in } \mathcal{C}^*?$$

If $C \cap \Lambda$ contains a Cantor set of positive modulus, this question has been answered affirmatively in [4].

One particular case of interest is when Λ is the Apollonian gasket. The correspdoning geometrically finite hyperbolic 3-manifold is not acylindrical in this case, because its compact core is a handlebody of genus 2, and hence it is not boundary incompressible; this can also be seen from the property that the Apollonian gasket contains a loop of three consecutively tangent disks.

QUESTION 8.9. Can we classify all possible closures of U-orbits in a geometrically finite acylindrical group? In order to answer this question, we first need to classify all possible H-orbit closures in ∂F, which is yet unsettled.

9 Unipotent flows in higher dimensional hyperbolic manifolds

Let $\mathbb{H}^d$ denote the d-dimensional hyperbolic space for $d \geq 2$ with $\partial(\mathbb{H}^d) = \mathbb{S}^{d-1}$, and let $G := \mathrm{SO}^\circ(d,1)$, which is the isometry group $\mathrm{Isom}^+(\mathbb{H}^d)$. Any complete hyperbolic d-manifold is given as the quotient $M = \Gamma\backslash\mathbb{H}^d$ for a torsion-free discrete subgroup $\Gamma < G$ (also called a Kleinian group). The limit set of Γ and the convex core of M are defined just like the dimension 3 case. As we have seen in the dimension 3 case, the geometry and topology of hyperbolic manifolds become relevant in the study of unipotent flows in hyperbolic manifolds of infinite volume, unlike in the finite volume case. Those hyperbolic 3-manifolds in which we have a complete understanding of the topological behavior of unipotent flows are rigid acylindrical hyperbolic 3-manifolds.

9.1 CONVEX COCOMPACT HYPERBOLIC MANIFOLDS WITH FUCHSIAN ENDS.

The higher dimensional analogues of rigid acylindrical hyperbolic 3-manifolds are as follows:

DEFINITION 9.1. *A convex cocompact hyperbolic d-manifold M is said to have Fuchsian ends if the convex core of M has nonempty interior and has totally geodesic boundary.*

The term *Fuchsian ends* reflects the fact that each component of the boundary of core M is a $(d-1)$–dimensional closed hyperbolic manifold, and each

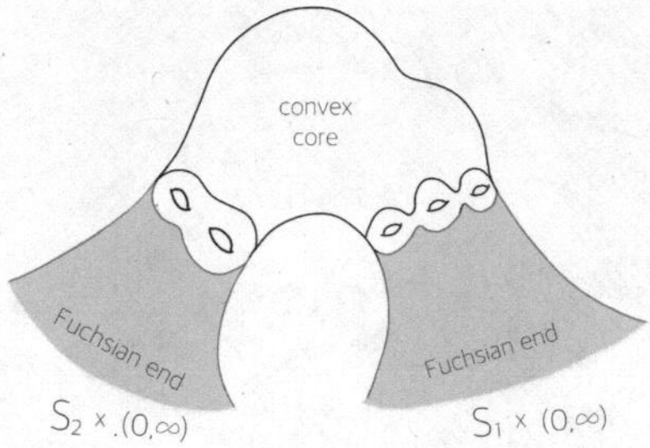

Figure 14.12. Convex cocompact manifolds with Fuchsian ends

component of the complement $M - \mathrm{core}(M)$ is diffeomorphic to the product $S \times (0, \infty)$ for some closed hyperbolic $(d-1)$-manifold S. For $d = 2$, any convex cocompact hyperbolic surface has Fuchsian ends. For $d = 3$, these are precisely rigid acylindrical hyperbolic 3-manifolds.

Convex cocompact hyperbolic manifolds with nonempty Fuchsian ends are constructed from closed hyperbolic manifolds as follows. Begin with a closed hyperbolic d-manifold $N_0 = \Gamma_0 \backslash \mathbb{H}^d$ with a fixed collection of finitely many, mutually disjoint, properly embedded totally geodesic hypersurfaces. Cut N_0 along those hypersurfaces and perform the metric completion to obtain a compact hyperbolic manifold W with totally geodesic boundary hypersurfaces. Then $\Gamma := \pi_1(W)$ injects to $\Gamma_0 = \pi_1(N_0)$, and $M := \Gamma \backslash \mathbb{H}^d$ is a convex cocompact hyperbolic manifold with Fuchsian ends.

Unlike the $d = 3$ case, Kerckhoff and Storm showed that if $d \geq 4$, a convex cocompact hyperbolic manifold $M = \Gamma \backslash \mathbb{H}^d$ with Fuchsian ends does not allow any nontrivial local deformation, in the sense that the representation of Γ into G is infinitesimally rigid [21].

9.2 ORBIT CLOSURE OF UNIPOTENT FLOWS ARE RELATIVELY HOMOGENEOUS. We let $A = \{a_t\}$ be the one-parameter subgroup of semisimple elements of G that give the frame flow, and let $N \simeq \mathbb{R}^{d-1}$ denote the contracting horospherical subgroup. We have a compact A-invariant subset $\mathrm{RF}\, M = \{x \in \Gamma \backslash G : xA \text{ is bounded}\}$.

The following presents a generalization of Theorems 7.3 and 7.6 to any dimension:

THEOREM 9.2 ([22]).
Let $d \geq 2$ and M be a convex cocompact hyperbolic d-manifold with Fuchsian ends. Let U be any connected closed subgroup of G generated by unipotent elements.

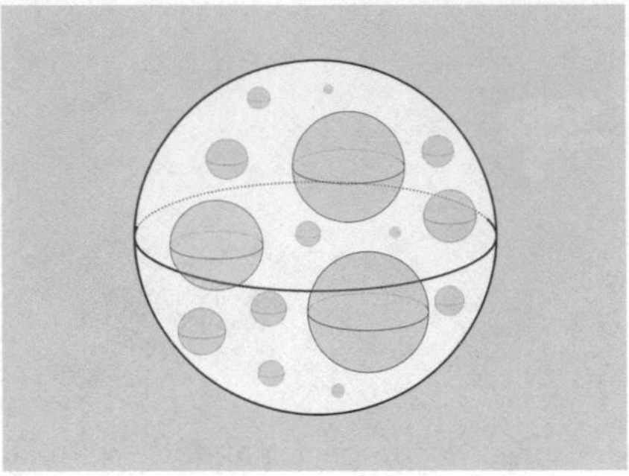

Figure 14.13. Limit set of a convex cocompact hyperbolic 4-manifold with Fuchsian boundary

Suppose that U is normalized by A. Then the closure of any U-orbit is relatively homogeneous in RF *M in the sense that for any* $x \in \mathrm{RF}\, M$,

$$\overline{xU} \cap \mathrm{RF}\, M = xL \cap \mathrm{RF}\, M$$

for a connected closed reductive subgroup $U < L < G$ *such that* xL *is closed.*

When M has finite volume, this is a special case of Ratner's orbit closure theorem [45]. This particular case was also proved by Shah by topological methods [48].

Theorem 9.2 and its refinements made in [22] yield the analogous topological rigidity of geodesic planes and horocycles. A geodesic k-plane of M is the image of a totally geodesic immersion $f : \mathbb{H}^k \to M$.

THEOREM 9.3 ([22]).

Let M be a convex cocompact hyperbolic d-manifold with Fuchsian ends. Then for any $2 \le k \le d-1$,

(1) *the closure of any geodesic k-plane intersecting* core *M is a properly immersed geodesic m-plane for some* $k \le m \le d$;

(2) *a properly immersed geodesic k-plane is a convex cocompact (immersed) hyperbolic k-manifold with Fuchsian ends;*

(3) *there are at most countably many maximal properly immersed geodesic planes intersecting* core *M; and*

(4) *any infinite sequence of maximal properly geodesic planes intersecting* core *M becomes dense in M.*

A k-horosphere in $\mathbb{H}^d$ is a Euclidean sphere of dimension k that is tangent to a point in $\mathbb{S}^{d-1}$. A k-horosphere in M is simply the image of a k-horosphere in $\mathbb{H}^d$ under the covering map $\mathbb{H}^d \to M = \Gamma\backslash\mathbb{H}^d$.

THEOREM 9.4 ([22]).
Let χ be a k-horosphere of M for $k \geq 1$. Then either

(1) χ is properly immersed or
(2) $\overline{\chi}$ is a properly immersed m-dimensional submanifold parallel to a convex cocompact geodesic m-plan of M with Fuchsian ends for some $m \geq k+1$.

9.3 AVOIDANCE OF SINGULAR SET. An important ingredient of the proof of Theorem 9.2 that appears newly for $d \geq 4$ is the avoidance of the singular set along the recurrence time of unipotent flows to RF M.

Let $U = \{u_t\}$ be a one-parameter unipotent subgroup of N. Extending the definition given by Dani-Margulis [13] to the infinite volume setting, we define the singular set $\mathscr{S}(U)$ as

$$\mathscr{S}(U) := \bigcup xL \cap \mathrm{RF}_+ M, \tag{9.1}$$

where $\mathrm{RF}_+ M = \mathrm{RF}\, M \cdot N$, and the union is taken over all closed orbits xL of proper connected closed subgroups L of G containing U. Its complement in $\mathrm{RF}_+ M$ is denoted by $\mathscr{G}(U)$ and called the set of *generic* elements of U.

The structure of $\mathscr{S}(U)$ as the countable union of *singular tubes* is an important property that plays crucial roles in both measure theoretic and topological aspects of the study of unipotent flows. Let $\mathscr{H}$ denote the collection of all proper connected closed subgroups H of G containing a unipotent element such that $\Gamma\backslash\Gamma H$ is closed and $H \cap \Gamma$ is Zariski dense in H. For each $H \in \mathscr{H}$, we define the singular tube:

$$X(H, U) := \{g \in G : gUg^{-1} \subset H\}.$$

We have the following:

(1) $\mathscr{H}$ is countable;
(2) $X(H_1, U) \cap gX(H_2, U) = X(H_1 \cap gH_2g^{-1}, U)$ for any $g \in G$; and
(3) if $H_1, H_2 \in \mathscr{H}$ with $X(H_1 \cap H_2, U) \neq \emptyset$, there exists a closed subgroup $H_0 \subset H_1 \cap H_2$ such that $H_0 \in \mathscr{H}$.

In particular $\mathscr{S}(U)$ can be expressed as the union of countable singular tubes:

$$\mathscr{S}(U) = \bigcup_{H \in \mathscr{H}} \Gamma\backslash\Gamma X(H, U) \cap \mathrm{RF}_+ M.$$

REMARK 9.5. If $\Gamma < G = \mathrm{PSL}_2(\mathbb{C})$ is a uniform lattice and U is the one-parameter subgroup as in Equation (4.2), then $H \in \mathscr{H}$ if and only if $H = g^{-1}\,\mathrm{PSL}_2(\mathbb{R})g$ for $g \in G$ such that Γ intersects $g^{-1}\,\mathrm{PSL}_2(\mathbb{R})g$ as a uniform lattice. It follows that if $H_1, H_2 \in \mathscr{H}$ and $X(H_1, U) \cap X(H_2, U) \neq \emptyset$, then $H_1 = H_2$.

We note that $\mathscr{H}$ and hence $\mathscr{S}(U)$ may be empty in general; see Remark 7.5(1).

When the singular set $\mathscr{S}(U)$ is nonempty, it is very far from being closed in $\mathrm{RF}_+ M$; in fact, it is dense, which is an a posteriori fact. Hence, presenting a compact subset of $\mathscr{S}(U)$ requires some care, and we will be using the following family of compact subsets $\mathscr{S}(U)$ in order to discuss the recurrence of U-flows relative to the singular set $\mathscr{S}(U)$. We define $\mathcal{E} = \mathcal{E}_U$ to be the collection of all subsets of $\mathscr{S}(U)$ that are of the form

$$\bigcup \Gamma\backslash\Gamma H_i D_i \cap \mathrm{RF}\, M,$$

where $H_i \in \mathscr{H}$ is a finite collection and D_i is a compact subset of $X(H_i, U)$.

The following theorem was obtained by Dani and Margulis [13] and independently by Shah [51] using the linearization method, which translates the study of unipotent flows on $\Gamma\backslash G$ to the study of vector-valued polynomial maps via linear representations.

THEOREM 9.6 (avoidance theorem for lattice case; [13]).
Let $\Gamma < G$ be a uniform lattice, and let $U < G$ be a one-parameter unipotent subgroup. Then for any $\varepsilon > 0$, there exists a sequence of compact subsets $E_1 \subset E_2 \subset \cdots$ in $\mathcal{E}$ such that $\mathscr{S}(U) = \bigcup_{n \geq 1} E_n$, which satisfies the following: Let x_j be a sequence converging to $x \in \mathscr{G}(U)$. For each $n \geq 1$, there exist a neighborhood $\mathcal{O}_n$ of E_n and $j_n \geq 1$ such that for all $j \geq j_n$ and for all $T > 0$,

$$\ell\{t \in [0, T] : x_j u_t \in \bigcup_{i \leq n} \mathcal{O}_i\} \leq \varepsilon T, \tag{9.2}$$

where ℓ denotes the Lebesgue measure.

If we set

$$\mathsf{T}_n := \{t \in \mathbb{R} : x_{j_n} u_t \notin \bigcup_{i \le n} \mathcal{O}_i\},$$

then for any sequence $\lambda_n \to \infty$, $\limsup \lambda_n^{-1} \mathsf{T}_n$ accumulates at 0 and ∞; hence, the sequence T_n has accumulating renormalizations.

When xL is a closed orbit of a connected closed subgroup of L containing U, the relative singular subset $\mathscr{S}(U, xL)$ *of* $xL \cap \mathrm{RF}_+ M$ is defined similarly by replacing $\mathscr{H}$ by its subcollection of proper connected closed subgroups of L, and $\mathscr{G}(U, xL)$ is defined as its complement inside $xL \cap \mathrm{RF}_+ M$. Theorem 9.6 applies in the same way to $\mathscr{G}(U, xL)$ with the ambient space $\Gamma\backslash G$ replaced by xL.

In order to explain some ideas of the proof of Theorem 9.6, we will discuss the following (somewhat deceptively) simple case when $G = \mathrm{PSL}_2(\mathbb{C})$ and Γ is a uniform lattice. Let $U = \{u_t\}$ be as in Equation (4.2).

PROPOSITION 9.7. *Let $E \in \mathcal{E}_U$. If $x \in \mathscr{G}(U)$, then xU spends most of its time outside a neighborhood of E; more precisely, for any $\varepsilon > 0$, we can find a neighborhood $E \subset \mathcal{O}$ such that for all $T > 0$,*

$$\ell\{t \in [0, T] : xu_t \in \mathcal{O}\} \le \varepsilon T. \tag{9.3}$$

Proof. Since xU is dense in $\Gamma\backslash G$ a posteriori, xu_t will go into any neighborhood of E for an infinite sequence of t's, but that the proportion of such t is very small is the content of Proposition 9.7. In view of Remark 9.5, we may assume that E is of the form $\Gamma\backslash\Gamma\,\mathrm{N}(H)D$, where $H = \mathrm{PSL}_2(\mathbb{R})$ and $D \subset V$ is a compact subset; note that $X(H, U) = \mathrm{N}(H)V$ and that $\mathrm{N}(H)$ is generated by H and $\mathrm{diag}(i, -i)$.

As remarked before, we prove this proposition using the linear representation and the polynomial-like behavior of unipotent action. As $\mathrm{N}(H)$ is the group of real points of a connected reductive algebraic subgroup, there exists an $\mathbb{R}$-regular representation $\rho : G \to \mathrm{GL}(W)$ with a distinguished point $p \in W$ such that $\mathrm{N}(H) = \mathrm{Stab}(p)$ and pG is Zariski closed. The set $pX(H, U) = pV$ is a real algebraic subvariety.[9]

Note that for $x = [g]$, the following are equivalent:

(1) $xu_t \in [e]\,\mathrm{N}(H)\mathcal{O}$.
(2) $p\gamma g u_t \in p\mathcal{O}$ for some $\gamma \in \Gamma$.

---

[9]We can explicitly take ρ and p as follows. Consider the adjoint representation of G on its Lie algebra $\mathfrak{g}$. We then let ρ be the induced representation on the wedge product space $\wedge^3 \mathfrak{g}$ and set $p = w_1 \wedge w_2 \wedge w_3$, where w_1, w_2, w_3 is a basis of $\mathfrak{h}$.

Therefore, we now try to find a neighborhood $p\mathcal{O}$ of pD so that the set

$$\{t \in [0, T] : xu_t \in [e]\,\mathrm{N}(H)\mathcal{O}\} \subset \bigcup_{q \in p\Gamma} \{t \in [0, T] : qgu_t \in p\mathcal{O}\}$$

is an ε-proportion of T. Each set $\{t \in [0, T] : qgu_t \in p\mathcal{O}\}$ can be controlled by the following lemma, which is proved using the property that the map $t \mapsto \|qgu_t\|^2$ is a polynomial of degree uniformly bounded for all $q \in p\Gamma$, and polynomial maps of bounded degree have uniformly *slow* divergence.

LEMMA 9.8 ([13, proposition 4.2]). *Let $\mathcal{A} \subset W$ be an algebraic variety. Then for any compact subset $C \subset \mathcal{A}$ and any $\varepsilon > 0$, there exists a compact subset $C' \subset \mathcal{A}$ such that the following holds: for any neighborhood Φ' of C' in W, there exists a neighbhorhood Φ of C of W such that for any $q \in W - \Phi'$ and any $T > 0$,*

$$\ell\{t \in [0, T] : qu_t \in \Phi\} \leq \varepsilon \cdot \ell\{t \in [0, T] : qu_t \in \Phi'\}.$$

Applying this lemma to $\mathcal{A} = pV$ and $C = pD$, we get a compact subset $C' = pD'$ for $D' \subset V$. Since $x \notin [e]\,\mathrm{N}(H)D'$, we can find a neighborhood $\mathcal{O}'$ so that $x \notin [e]\,\mathrm{N}(H)\mathcal{O}'$. Fix a neighborhood Φ' of C', so that $\Phi' \cap pG \subset p\mathcal{O}'$. We then get a neighborhood Φ of C such that if $\mathcal{O}$ is a neighborhood of D such that $p\mathcal{O} \subset \Phi$, then

$$\ell(J_q \cap [0, T]) \leq \varepsilon \cdot \ell(I_q \cap [0, T]), \tag{9.4}$$

where $J_q := \{t \in \mathbb{R} : qgu_t \in p\mathcal{O}\}$ and $I_q := \{t \in \mathbb{R} : qgu_t \in p\mathcal{O}'\}$.

We now claim that in the case at hand, we can find a neighborhood $\mathcal{O}'$ of D' so that all I_q's are mutually disjoint:

$$\text{If } q_1 \neq q_2 \text{ in } p\Gamma, \text{ then } I_{q_1} \cap I_{q_2} = \emptyset. \tag{9.5}$$

Using (9.4), this would finish the proof, since

$$\ell\{t \in [0, T] : xu_t \in [e]\,\mathrm{N}(H)\mathcal{O}\} \leq \sum_{q \in p\Gamma} \ell(J_q \cap [0, T]) \leq \varepsilon \cdot \sum_{q \in p\Gamma} \ell(I_q \cap [0, T]) \leq \varepsilon T.$$

To prove (9.5), we now observe the special feature of this example—namely, *no* singular tube $\Gamma\backslash\Gamma X(H, U)$ has *self-intersection*, meaning that

$$X(H, U) \cap \gamma X(H, U) = \emptyset \text{ if } \gamma \in \Gamma - \mathrm{N}(H). \tag{9.6}$$

If nonempty, by Remark 9.5, we must have $H \cap \gamma H \gamma^{-1} = H$, implying that $\gamma \in N(H)$. Now if $t \in I_{p\gamma_1} \cap I_{p\gamma_2}$, then $gu_t \in \gamma_1^{-1} HV \cap \gamma_2^{-1} HV$ and hence $\gamma_1 \gamma_2^{-1} \in N(H)$. So $p\gamma_1 = p\gamma_2$, proving (9.5). □

In the higher dimensional case, we cannot avoid self-intersections of $\Gamma X(H, U)$; so I_q's are not pairwise disjoint, which means a more careful study of the nature of the self-intersection is required. Thanks to the countability of $\mathscr{H}$, an inductive argument on the dimension of $H \in \mathscr{H}$ is used to take care of the issue, using the fact that the intersections among $\gamma X(H, U)$, $\gamma \in \Gamma$ are essentially of the form $X(H_0, U)$ for a proper connected closed subgroup H_0 of H contained in $\mathscr{H}$ (see [13] for details).

In order to illustrate the role of Theorem 9.6 in the study of orbit closures, we prove the following sample case: Let $G = SO^\circ(4,1)$, $H = SO^\circ(2,1)$, and $L = SO^\circ(3,1)$; the subgroups H and L are chosen so that $A < H < L$ and $H \cap N$ is a one-parameter unipotent subgroup. The centralizer $C(H)$ of H is $SO(2)$. We set $H' = H\,C(H)$.

PROPOSITION 9.9. *Let $\Gamma < G$ be a uniform lattice. Let $X = \overline{xH'}$ for some $x \in \Gamma\backslash G$. If X contains a closed orbit zL properly, then $X = \Gamma\backslash G$.*

A geometric consequence of this proposition is as follows: Let M be a closed hyperbolic 4-manifold, and let $P \subset M$ be a geodesic 2-plane. If $\overline{P}$ contains a properly immersed geodesic 3-plane P', then the closure $\overline{P}$ is either P' or M.

Proof. Let $U_1 = H \cap N$ and $U_2 = H \cap N^+$, where N^+ is the expanding horospherical subgroup of G. Then the subgroups U_1 and U_2 generate H, and the intersection of the normalizers of U_1 and U_2 is equal to $A\,C(H)$. Since zL is compact, each U_ℓ acts ergodically on zL by Moore's ergodicity theorem. Therefore, we may choose z so that zU_ℓ is dense in zL for each $\ell = 1, 2$.

It suffices to show X contains either N or N^+-orbit. Since zL is a proper subset of X, there exists $g_n \to e$ in $G - L\,C(H)$ such that $x_n = zg_n \in X$. As L is reductive, the Lie algebra of G decomposes into $\mathrm{Ad}(\mathfrak{l})$-invariant subspaces $\mathfrak{l} \oplus \mathfrak{l}^\perp$ with $\mathfrak{l}$ the Lie algebra of L. Hence we write $g_n = \ell_n r_n$ with $\ell_n \in L$ and $r_n \in \exp \mathfrak{l}^\perp - C(H)$. As $g_n \notin C(H)$, there exists $1 \le \ell \le 2$ such that no r_n belongs to the normalizer of U_ℓ by passing to a subsequence. We set $U = U_\ell$. Without loss of generality we assume $U = H \cap N$; otherwise, replace N by N^+ in the argument below.

Note that $\overline{zU} = zL$; in particular, z is a generic point: $z \in \mathcal{G}(U, zL) = zL - \mathscr{S}(U, zL)$. We replace the sequence $z\ell_n$ with $z\ell_{j_n}$, with j_n given by Theorem 9.6.

Define

$$\mathsf{T}_n := \{t \in \mathbb{R} : z\ell_n u_t \notin \bigcup_{i \le n} \mathcal{O}_i\}. \tag{9.7}$$

By Theorem 9.6 applied to $zL = z\,\mathrm{SO}^\circ(3,1)$, T_n has accumulating renormalizations.

Now by a similar argument as in the proof of Lemma 5.5(3), we can show that

$$\limsup\{u_t r_n u_{-t} : t \in \mathsf{T}_n\}$$

accumulates at 0 and ∞ in V, where V is the one-dimensional unipotent subgroup $(L \cap N)V = N$. In particular, there exists $v \in V$ of arbitrarily large size such that $v = \lim u_{-t_n} r_n u_{t_n}$ for some $t_n \in \mathsf{T}_n$.

Note that $z\ell_n u_{t_n}$ is contained in the compact subset $zL - \bigcup_{i \le n} \mathcal{O}_i$. Since $\bigcup_i \mathcal{O}_i$ is a neighborhood of $\mathscr{S}(U, zL)$, $z\ell_n u_{t_n}$ converges to some

$$z_0 \in \mathscr{G}(U, zL). \tag{9.8}$$

Therefore,

$$zg_n u_{t_n} = z\ell_n u_{t_n}(u_{-t_n} r_n u_{t_n}) \to z_0 v.$$

Since $z_0 \in \mathscr{G}(U, zL)$, by Proposition 7.8, we have

$$X \supset \overline{z_0 v U} = \overline{z_0 U} v = zLv.$$

As v can be taken arbitrarily large, we get a sequence $v_n \to \infty$ in V such that $X \supset zLv_n$. Using the A-invariance of X, we get $X \supset zL(Av_nA) \supset z(L \cap N)V_+$ for some one-parameter semigroup V_+ of V. Since $X \supset zv_n(L \cap N)v_n^{-1}V_+$ and $\limsup v_n^{-1}V_+ = V$, X contains an N orbit, finishing the proof. □

Roughly speaking, if H is a connected closed subgroup of G generated by unipotent elements, the proof of the theorem that $\overline{xH}$ is homogeneous uses an inductive argument on the codimension of $H \cap N$ in N and involves repeating the following two steps:

(1) Find a closed orbit zL inside $\overline{xH}$ for some connected reductive subgroup $L < G$.

(2) If $\overline{xH} \neq zL$, then enlarge zL—that is, find a closed orbit zL' inside $\overline{xH}$ with $\dim(L' \cap N) > \dim(L \cap N)$.

The proof of Proposition 9.9 is a special sample case of step (2), demonstrating the importance of getting accumulating renormalizations for the sequence of return time avoiding the exhausting sequence of compact subsets of the singular set.

The following version of the avoidance theorem in [22] is a key ingredient in the proof of Theorem 9.2:

THEOREM 9.10 (avoidance theorem).
Let $M=\Gamma\backslash\mathbb{H}^d$ *be a convex cocompact hyperbolic manifold with Fuchsian ends. Let* $U<N$ *be a one-parameter unipotent subgroup. There exists a sequence of compact subsets* $E_1\subset E_2\subset\cdots$ *in* $\mathcal{E}$ *such that* $\mathscr{S}(U)\cap \mathrm{RF}\,M=\bigcup_{n\geq 1}E_n$*, which satisfies the following: Let* $x_j\in \mathrm{RF}\,M$ *be a sequence converging to* $x\in\mathscr{G}(U)$*. For each* $n\geq 1$*, there exist a neighborhood* $\mathcal{O}_n$ *of* E_n *and* $j_n\geq 1$ *such that for all* $j\geq j_n$*,*

$$\mathsf{T}^{\diamond}(x_j):=\{t\in\mathbb{R}:x_ju_t\in \mathrm{RF}\,M-\mathcal{O}_n\} \tag{9.9}$$

has accumulating renormalizations.

Note that in the lattice case, one can use the Lebesgue measure ℓ to understand the return time away from the neighborhoods $\mathcal{O}_n$ to prove Theorem 9.6, as was done in [13] (see also the proof of Proposition 9.7). In the case at hand, the relevant return time is a subset of $\{t\in\mathbb{R}:x_nu_t\in \mathrm{RF}\,M\}$, on which it is not clear whether there exists any *friendly* measure. This makes the proof of Theorem 9.10 very delicate, as we have to examine each return time to RF M and handpick the time outside $\mathcal{O}_n$. First of all, we cannot reduce a general case to the case $E\subset\Gamma\backslash\Gamma X(H,U)$ for a single $H\in\mathscr{H}$. This means that not only do we need to understand the *self-intersections* of $\Gamma X(H,U)$, but we also have to control intersections among different $\Gamma X(H,U)$'s in $\mathscr{S}(U)$, $H\in\mathscr{H}$.

We also cannot use an inductive argument on the dimension of H. When $G=\mathrm{SO}(3,1)$, there are no intersections among closed orbits in $\mathscr{S}(U)$, and the proof is much simpler in this case. In general, our arguments are based on the k-thick recurrence time to RF M, a much more careful analysis on the graded intersections among $\Gamma X(H,U)$'s, $H\in\mathscr{H}$, and a combinatorial inductive search argument. We prove that there exists $\kappa>1$, depending only on Γ such that $\mathsf{T}^{\diamond}(x_n)$ is κ-thick in the sense that for any $r>0$,

$$\mathsf{T}^{\diamond}(x_n)\cap\pm[r,\kappa r]\neq\emptyset.$$

We remark that unlike the lattice case, we are not able to prove that $\{t\in\mathbb{R}:x_nu_t\in \mathrm{RF}\,M-\bigcup_{j\leq n}\mathcal{O}_j\}$ has accumulating renormalizations. This causes an issue in carrying out a similar proof as in Proposition 9.9, as we cannot conclude that the limit of $x_n u_{t_n}$ for $t_n \in \mathsf{T}^{\diamond}(x_n)$ belongs to a generic set as in (9.8).

Fortunately, to overcome this difficulty, we were able to devise an inductive argument (in the proof of Theorem 9.11) that involves an extra step of proving equidistribution of translates of maximal closed orbits.

9.4 INDUCTION. For a connected closed subgroup $U < N$, we denote by $H(U)$ the smallest closed simple Lie subgroup of G that contains both U and A. If $U \simeq \mathbb{R}^k$, then $H(U) \simeq \mathrm{SO}^{\circ}(k+1,1)$. A connected closed subgroup of G generated by one-parameter unipotent subgroups is, up to conjugation, of the form $U < N$ or $H(U)$ for some $U < N$.

We set $F_{H(U)} := \mathrm{RF}_+ M \cdot H(U)$, which is a closed subset. We define the following collection of closed connected subgroups of G:

$$\mathcal{L}_U := \left\{ L = H(\widehat{U})C : \begin{array}{c} \text{for some } z \in \mathrm{RF}_+ M,\ zL \text{ is closed in } \Gamma\backslash G \\ \text{and } \mathrm{Stab}_L(z) \text{ is Zariski dense in } L \end{array} \right\},$$

where $U < \widehat{U} < N$ and C is a closed subgroup of the centralizer of $H(\widehat{U})$. We also define

$$\mathcal{Q}_U := \{vLv^{-1} : L \in \mathcal{L}_U \text{ and } v \in N\}.$$

Theorem 9.2 follows from the following:

THEOREM 9.11 ([22]).
Let $M = \Gamma\backslash\mathbb{H}^d$ be a convex cocompact hyperbolic manifold with Fuchsian ends.

(1) *For any $x \in \mathrm{RF}\, M$,*

$$\overline{xH(U)} = xL \cap F_{H(U)},$$

where xL is a closed orbit of some $L \in \mathcal{L}_U$.

(2) *Let $x_0\widehat{L}$ be a closed orbit for some $\widehat{L} \in \mathcal{L}_U$ and $x_0 \in \mathrm{RF}\, M$.*

(a) *For any $x \in x_0\widehat{L} \cap \mathrm{RF}_+ M$,*

$$\overline{xU} = xL \cap \mathrm{RF}_+ M,$$

where xL is a closed orbit of some $L \in \mathcal{Q}_U$.

(b) *For any $x \in x_0\widehat{L} \cap \mathrm{RF}\, M$,*

$$\overline{xAU} = xL \cap \mathrm{RF}_+ M,$$

where xL is a closed orbit of some $L \in \mathcal{L}_U$.

(3) *Let $x_0\widehat{L}$ be a closed orbit for some $\widehat{L}\in\mathcal{L}_U$ and $x_0\in \mathrm{RF}\,M$. Let $x_iL_i\subset x_0\widehat{L}$ be a sequence of closed orbits intersecting* RF *M where $x_i\in \mathrm{RF}_+\,M$, $L_i\in\mathcal{Q}_U$. Assume that no infinite subsequence of x_iL_i is contained in a subset of the form y_0L_0D, where y_0L_0 is a closed orbit of $L_0\in\mathcal{L}_U$ with $\dim L_0<\dim\hat{L}$ and $D\subset \mathrm{N}(U)$ is a compact subset. Then*

$$\lim_{i\to\infty}(x_iL_i\cap \mathrm{RF}_+\,M)=x_0\widehat{L}\cap \mathrm{RF}_+\,M.$$

We prove (1) by induction on the codimension of U in N and (2) and (3) by induction on the codimension of U in $\widehat{L}\cap N$. Let us say $(1)_m$ holds if (1) is true for all U satisfying co-$\dim_N(U)\le m$. We will say $(2)_m$ holds if (2) is true for all U and $\widehat{L}$ satisfying co-$\dim_{\widehat{L}\cap N}(U)\le m$, and similarly for $(3)_m$.

We then deduce $(1)_{m+1}$ from $(2)_m$ and $(3)_m$; $(2)_{m+1}$ from $(1)_{m+1}$, $(2)_m$, and $(3)_m$; and $(3)_{m+1}$ from $(1)_{m+1}$, $(2)_{m+1}$, and $(3)_m$. In proving Theorem 9.2 for the lattice case, we do not need $(3)_m$ in the induction proof. In the case at hand, $(3)_m$ is needed since we could not obtain a stronger version of Theorem 9.10 with $\mathcal{O}_n$ replaced by $\cup_{j\le n}\mathcal{O}_j$.

We remark that in the step of proving $(2)_{m+1}$, the following geometric feature of convex cocompact hyperbolic manifolds M of Fuchsian ends is used to ensure that $\mathscr{S}(U,x_0\widehat{L})\neq\emptyset$.

PROPOSITION 9.12. *For any $2\le k\le d$, any properly immersed geodesic k-plane of M is either compact or contains a compact geodesic $(k-1)$-plane.*

This proposition follows from the hereditary property that any properly immersed geodesic k-plane P of M is a convex cocompact hyperbolic k-manifold of Fuchsian ends; hence, either P is compact (when P has empty ends) or the boundary of core P provides a codimension 1 compact geodesic plane.

References

[1] Jon Aaronson. *An introduction to infinite ergodic theory.* Mathematical surveys and monographs, Vol 50, AMS, Providence, RI, 1997.

[2] M. Babillot. *On the mixing property for hyperbolic systems.* Israel J. Math, 129, 61–76 (2002).

[3] U. Bader, D. Fisher, N. Miller and M. Stover. *Arithmeticity, superrigidity and totally geodesic submanifolds.* Preprint. arXiv:1903.08467.

[4] Y. Benoist and H. Oh. *Geodesic planes in geometrically finite acylindrical 3-manifolds.* Preprint. arXiv:1802.04423, to appear in the memorial issue for A. Katok in ETDS.

[5] C. Bishop and P. Jones. *Hausdorff dimension and Kleinian groups.* Acta Math. 179, no. 1, 1–39 (1997).

[6] A. Borel and Harish-Chandra. *Arithmetic subgroups of algebraic groups.* Annals of Math, 75, 485–535 (1962).

[7] M. Burger. *Horocycle flow on geometrically finite surfaces.* Duke Math. J., 61, 779–803 (1990).

[8] R. Canary. *Marden's tameness conjecture: history and applications.* Geometry, analysis and topology of discrete groups, 137–162, Adv. Lect. Math. (ALM), 6, Int. Press, Somerville, MA, 2008.

[9] R. Canary and D. McCullough. *Homotopy equivalences of 3-manifolds and deformation theory of Kleinian groups.* Mem. AMS, 812, AMS, Providence, RI, 2004.

[10] F. Dal'Bo. *Topologie du feuilletage fortement stable.* Ann. Inst. Fourier. 50, 981–993 (2000).

[11] S. G. Dani and G. A. Margulis. *Values of quadratic forms at primitive integral points.* Invent. Math. 98, no. 2, 405–424 (1989).

[12] S. G. Dani and G. A. Margulis. *Orbit closures of generic unipotent flows on homogeneous spaces of $SL(3, R)$.* Math. Ann. 286, no. 1–3, 101–128 (1990).

[13] S. G. Dani and G. A. Margulis. *Limit distributions of orbits of unipotent flows and values of quadratic forms.* Adv. soviet math., 16, Part 1, AMS, Providence, RI, 1993.

[14] L. Dufloux. *Projections of Patterson-Sullivan measures and the dichotomy of Mohammadi-Oh.* Israel J. Math., 223, no. 1, 399–421 (2018).

[15] D. Ferte. *Flot horosphérique des repères sur les variétés hyperboliques de dimension 3 et spectre des groupes kleiniens.* Bulletin Braz. Math. Soc., 33, 99–123 (2002).

[16] M. Gromov and I. Piatetski-Shapiro. *Non-arithmetic groups in Lobachevsky spaces.* Publications Mathématiques de l'IHÉS, Vol. 66, 93–103 Institut des Hautes 'Etudes Scientifiques, Bures-sur-Yvette, France, 1987.

[17] A. Hatcher. *Notes on basic 3-manifold topology.* http://www.math.cornell.edu/~ hatcher.

[18] G. A. Hedlund. *Fuchsian groups and transitive horocycles.* Duke Math. J, 530–542 (1936).

[19] J. Hocking and G. Young. *Topology.* Addison-Wesley Pub., 1961.

[20] M. Kapovich. *Hyperbolic manifolds and discrete groups.* Progress in Maths., 183, Birkhäuser 2001.

[21] S. P. Kerckhoff and P. A. Storm. *Local rigidity of hyperbolic manifolds with geodesic boundary.* J. Topol., 5, no. 4, 757–784 (2012).

[22] M. Lee and H. Oh. *Orbit closures of unipotent flows for hyperbolic manifolds with Fuchsian ends.* Preprint, arXiv:1902.06621

[23] M. Lee and H. Oh. *Topological proof of Benoist-Quint's orbit closure theorem for* $SO(d, 1)$. Journal of Modern Dynamics, 15, 263–276 (2019).

[24] C. Maclachlan and A. Reid. *The arithmetic of hyperbolic 3-manifolds.* Springer, New York, 2003.

[25] C. Maclachlan and A. W. Reid. *Commensurability classes of arithmetic Kleinian groups and their Fuchsian subgroups.* Math. Proc. Cambridge Philos. Soc., 102, no. 2, 251–257 (1987).

[26] A. Marden. *Hyperbolic manifolds: an introduction in 2 and 3 dimensions.* Cambridge. University Press, Cambridge, 2016.

[27] G. Margulis. *Formes quadratriques indéfinies et flots unipotents sur les espaces homogènes.* C. R. Acad. Sci. Paris Sér. I Math., 304, no. 10, 249–253 (1987).

[28] G. Margulis. *Indefinite quadratic forms and unipotent flows on homogeneous spaces.* In Dynamical systems and ergodic theory, Vol. 23 Banach Center Publ., 1989

[29] G. Margulis. *Discrete subgroups of semisimple lie groups.* Springer-Verlag, Berlin-Heidelberg-New York, 1991.

[30] G. Margulis. *On some aspects of the theory of Anosov systems.* Springer monographs in mathematics, Springer-Verlag, Berlin, 2004.

[31] G. Margulis and A. Mohammadi. *Arithmeticity of hyperbolic 3-manifolds containing infinitely many totally geodesic surfaces.* Preprint, arXiv:1902.07267.

[32] G. Margulis and G. Tomanov. *Invariant measures for actions of unipotent groups over local fields on homogeneous spaces.* Invent. Math., 116, no. 1–3, 347–392 (1994).

[33] K. Matsuzaki and M. Taniguchi. *Hyperbolic manifolds and Kleinian groups.* Oxford University Press, 1998.

[34] F. Maucourant and B. Schapira. *On topological and measurable dynamics of unipotent frame flows for hyperbolic manifolds.* Duke. Math. J., 168, 697–747 (2019).

[35] C. McMullen. *Iteration on Teichmuller space.* Inventiones Mathematicae, 99, 425–454 (1990).

[36] C. McMullen, A. Mohammadi and H. Oh. *Horocycles in hyperbolic 3-manifolds.* Geom. Funct. Anal., 26, no. 3, 961–973 (2016).

[37] C. McMullen, A. Mohammadi and H. Oh. *Geodesic planes in hyperbolic 3-manifolds.* Inventiones Mathematicae, 209, 425–461 (2017).

[38] C. McMullen, A. Mohammadi and H. Oh. *Geodesic planes in the convex core of an acylindrical 3-manifold.* Preprint. arXiv:1802.03853 2018. To appear in Duke Math. J.

[39] A. Mohammadi and H. Oh. *Ergodicity of unipotent flows and Kleinian groups.* Journal of the AMS, 28, 531–577 (2015).

[40] C. Moore. *Ergodicity of flows on homogeneous spaces.* American J. Math., 88, 154–178 (1966).

[41] S. Mozes and N. Shah. *On the space of ergodic invariant measures of unipotent flows.* Ergodic Theory Dynam. Systems 15, no. 1, 149–159 (1995).

[42] S. Patterson. *The limit set of a Fuchsian group.* Acta Mathematica, 136, 241–273 (1976).

[43] J. Ratcliffe. *Foundations of hyperbolic manifolds.* Springer-Varlag, 1994.

[44] M. Ratner. *On Raghunathan's measure conjecture.* Ann. of Math. (2), 134, no. 3, 545–607 (1991).

[45] M. Ratner. *Raghunathan's topological conjecture and distributions of unipotent flows.* Duke Math. J. 235–280 (1991).

[46] T. Roblin. *Ergodicité et équidistribution en courbure négative.* Mém. Soc. Math. Fr. (N.S.), 95, vi+96 (2003).

[47] C. Series. *A crash course on Kleinian groups.* Rend. Istit. Mat. Univ. Trieste, 37, 1–38 (2006).

[48] N. A. Shah, *Unpublished, 1992.*

[49] N. Shah. *Unipotent flows on homogeneous spaces.* Ph.D. Thesis, Tata Institute of Fundamental Research, Mumbai, 1994.

[50] N. Shah. *Closures of totally geodesic immersions in manifolds of constant negative curvature.* In Group theory from a geometrical viewpoint, 718–732, World Scientific, Singapore, 1991.

[51] N. A. Shah. *Uniformly distributed orbits of certain flows on homogeneous spaces*. Math. Ann., 289, no. 2, 315–334 (1991).

[52] D. Sullivan. *The density at infinity of a discrete group of hyperbolic motions*. Inst. Hautes Etudes Sci. Publ. Math., 50, 171–202 (1979).

[53] D. Sullivan. *Entropy, Hausdorff measures old and new, and limit sets of geometrically finite Kleinian groups*. Acta Math., 153, no. 3–4, 259–277 (1984).

[54] W. Thurston. *Hyperbolic structures on 3-manifolds. I. Deformation of acylindrical manifolds*. Ann. of Math., (2), 124, no. 2, 203–246 (1986).

[55] G. T. Whyburn. *Topological characterization of the Sierpinski curve*. Fund. Math., 45, 320–324 (1958).

[56] D. Winter. *Mixing of frame flow for rank one locally symmetric spaces and measure classification*. Israel J. Math., 210, 467–507 (2015).

[57] Y. Zhang. *Construction of acylindrical hyperbolic 3-manifolds with quasifuchsian boundary*. Preprint, 2019.